SECOND EDITION

Physiological Aspects of Sport Training and Performance

Jay Hoffman, PhD
University of Central Florida

Human Kinetics

Library of Congress Cataloging-in-Publication Data

Hoffman, Jay, 1961-
Physiological aspects of sport training and performance / Jay Hoffman. -- 2nd ed.
p. ; cm.
Includes bibliographical references and index.
I. Title.
[DNLM: 1. Exercise--physiology. 2. Sports--physiology. 3. Physical Education and Training. QT 260]
QP301
612'.044--dc23

2013015257

ISBN-10: 1-4504-4224-2 (print)
ISBN-13: 978-1-4504-4224-4 (print)

The web addresses cited in this text were current as of October 2013, unless otherwise noted.

Acquisitions Editor: Amy N. Tocco; **Developmental Editor:** Amanda S. Ewing; **Assistant Editor:** Casey A. Gentis; **Copyeditor:** Joyce Sexton; **Indexer:** Andrea J. Hepner; **Permissions Manager:** Dalene Reeder; **Graphic Designer:** Fred Starbird; **Graphic Artist:** Dawn Sills; **Cover Designer:** Keith Blomberg; **Photograph (cover):** © damiano fiorentini/age fotostock; **Photo Asset Manager:** Laura Fitch; **Visual Production Assistant:** Joyce Brumfield; **Photo Production Manager:** Jason Allen; **Art Manager:** Kelly Hendren; **Associate Art Manager:** Alan L. Wilborn; **Illustrations:** © Human Kinetics, unless otherwise noted; **Printer:** Sheridan Books

Printed in the United States of America 10 9 8 7 6 5 4 3 2 1

The paper in this book is certified under a sustainable forestry program.

Human Kinetics
Website: www.HumanKinetics.com

United States: Human Kinetics
P.O. Box 5076
Champaign, IL 61825-5076
800-747-4457
e-mail: humank@hkusa.com

Canada: Human Kinetics
475 Devonshire Road Unit 100
Windsor, ON N8Y 2L5
800-465-7301 (in Canada only)
e-mail: info@hkcanada.com

Europe: Human Kinetics
107 Bradford Road
Stanningley
Leeds LS28 6AT, United Kingdom
+44 (0) 113 255 5665
e-mail: hk@hkeurope.com

Australia: Human Kinetics
57A Price Avenue
Lower Mitcham, South Australia 5062
08 8372 0999
e-mail: info@hkaustralia.com

New Zealand: Human Kinetics
P.O. Box 80
Torrens Park, South Australia 5062
0800 222 062
e-mail: info@hknewzealand.com

E5840

Contents

Preface

The purpose of *Physiological Aspects of Sport Training and Performance, Second Edition* is to provide a focused and applied approach for those interested in exercise physiology and sport science. Students interested in the practical or scientific aspects of sport science will learn the importance of an evidence-based approach in developing exercise prescriptions. For sports medicine specialists and other health care providers who may have administrative responsibility for supervising strength and conditioning programs or consult with their patients on matters relating to sport performance, this book can be used as a primary reference on conditioning and performance of athletes. Each chapter is fortified with figures and tables that make it more attractive to the reader. Key terms are in bold print.

In comparison to other exercise physiology books on the market, *Physiological Aspects of Sport Training and Performance, Second Edition* provides an in-depth discussion of physiological adaptation from exercise. In addition, this textbook provides a thorough review of all components of an athlete's training program. A broad range of topic areas is covered, including environmental influences on performance, hydration status, sport nutrition, sport supplements, and performance-enhancing drugs. In contrast to other books, this one focuses primarily on training factors and how various environmental conditions can affect exercise performance. Furthermore, practical applications are provided so that exercise prescriptions can be developed for a number of different athletic populations.

Organization

Physiological Aspects of Sport Training and Performance, Second Edition is organized into five parts. Part I examines physiological adaptation and the effects of various modes of training (aerobic, anaerobic, and resistance) on biochemical, hormonal, muscular, cardiovascular, neural, and immunological adaptations. These adaptations are discussed as they relate to the training level of the athlete and their impact on sport performance.

Part II covers exercise training principles and prescription. Each chapter describes in detail the specific development of training programs for each mode of training (resistance, anaerobic, aerobic), including examples of exercises and training programs. The chapter on concurrent training explains how to incorporate several different modes of training into a periodized training program. In addition, some chapters include performance guidelines for various types of athletes to be used for comparison purposes. Expectations for performance improvements as athletes progress through their respective training programs are also included.

Part III discusses the relevance of nutrition, hydration status, sport supplementation, and performance-enhancing drugs to sport performance. Part IV focuses on environmental factors and their influence on sport performance. Specific areas of emphasis include exercise in the heat, in the cold, and at altitude, including the medical concerns related to exercise in these environments. The potential performance benefits of exercise at altitude are also considered.

Part V focuses on how certain medical and health conditions influence sport performance. Included in this section is a discussion of overtraining. Other chapters discuss conditions that are commonly encountered when one is working with athletes. Sudden death in sport is also covered.

Updates to the Second Edition

Many of the updates in this second edition focus on relevant issues that have come to the forefront in sport science and sports medicine over the past few years. This second edition presents a more in-depth discussion on sport supplementation and performance-enhancing drugs. New content has been added on power training, speed and agility development, and goals and program development. In addition, two new chapters have been added to this edition: a chapter on sudden death and a chapter providing a practical approach to developing the yearly training program.

This new edition also features two new pedagogical aids: chapter objectives, which help students and instructors tell at a glance what is covered in the chapter, and chapter review questions, which help students test their understanding of chapter content. A new sidebar feature—In Practice sidebars—brings practical aspects to life and helps students better understand the material. Each chapter features two In Practice sidebars.

Also new to the second edition is a new instructor resource—an image bank. The image bank provides all of the content photos, figures, and tables from the text, sorted by chapter. Instructors can use the image bank to enhance lecture notes, build a PowerPoint presentation, create student handouts, and so on.

Finally, the second edition also features a new web resource that provides 80 drills; these drills feature warm-up, flexibility, balance and stability, plyometric, speed, and agility exercises. Most drills are accompanied by at least one photo showing how to perform a key movement of the drill. Forty of the drills are accompanied by a video of the drill being performed in its entirety. Throughout chapters 7, 9, and 11, this icon alerts you to drill content that is available on the web resource:

The image bank and the web resource are available at www.HumanKinetics.com/PhysiologicalAspectsOfSportTrainingAndPerformance.

Acknowledgments

Behind the success of any great endeavor is the strong support system of the family. My wife, Yaffa, has been the root of my strength, and my children, Raquel, Mattan, and Ariel, have been the joy of my life. Their love, devotion, and unwavering support are cherished and, as always, help turn my dreams into reality. I only hope that the example that I have set as a husband and father has been able to guide them as they have turned into productive and successful adults.

Success does not just appear; it's the result of tremendous preparation and dedication. For that I wish to thank my parents who taught me that through hard work, intense desire, and persistence, great goals can be accomplished.

Finally, there are three things that I have found to be important for success: Be passionate in your work; commit yourself to excellence; and surround yourself with people who make you better.

Accessing and Using the Web Resource

The web resource for *Physiological Aspects of Sport Training and Performance, Second Edition,* features 80 drills in PDF format that show athletes how to perform warm-up, flexibility, balance and stability, plyometric, speed training, and agility exercises. Nearly all of the drills are accompanied by at least one photo showing a key movement of the drill. The drills can be printed out and distributed to athletes for their use. In addition, the web resource features high-definition online video of 40 of the drills in action plus a dynamic warm-up routine that shows how the warm-up drills can be combined.

To view the video clips, go to the web resource and select the appropriate player: Warm-Up, Flexibility, and Balance and Stability (chapter 7 drills); Plyometrics (chapter 9 drills); or Speed and Agility Training (chapter 11 drills). Within each player, clips are named by the drill titles used in the book. Use the scroll bar on the right of the player to find the video of the drill in action.

Here is a listing of the drills in the web resource, including which drills are demonstrated in video:

High-Knee Walk*
Stepping Trunk Twist
Trail Leg Walking*
Lunge Walk*
Lunge Walk With Reach
Leg Cradle*
Side Lunge*
Side Shuffle and Run
Carioca*
Backward Run*
Glute Kick*
Speed Skip*
Power Skip*
Lying Achilles Stretch
Partner-Assisted Calf Stretch
Semistraddle
Sitting Toe Touch
Straddle (Spread Eagle)
Butterfly Stretch
Side Quadriceps Stretch
Hurdler's Stretch
Forward Lunge
Single-Leg Lower Back Stretch
Double-Leg Lower Back Stretch
Spinal Twist
Pectoralis Stretch With Partner
Lateral Shoulder Stretch
Partner-Assisted Internal Rotator Stretch
External Rotator Stretch
Shoulder Flexor Stretch
Rotational Stretch of Neck
Neck Flexion and Extension
Behind-the-Neck Stretch
Balancing on a T-Board in the Frontal Plane
Single-Leg Stance on a Balance Board
Squatting on a Balance Board
Balance Pad Lunge
Walk or Run Over a Polygon of Balance Pads
Standing Long Jump**
Squat Jump**
Front Cone Hop**
Single-Leg Push-Off With Box**
Medicine Ball Throw**
Front Box Jump**
Depth Jump**
Tuck Jump With Knees Up**
Lateral Cone Hop**
Double-Leg or Single-Leg Zigzag Hop**
Standing Triple Jump**
Multiple-Box Jump**
Depth Jump to Prescribed Height**
Alternate-Leg Bounding**
Pike Jump**
Split Squat**
Single-Leg Hop**
Multiple-Box Squat Jump**

Single-Leg Depth Jump**
Single-Leg Bounding**
Arm Action (Seated)**
Arm Action (Standing Exchange)
Lean and Fall Run
Drop and Go**
Jump and Go**
Bound and Run**
Scramble Out**
Cone or Bag Jump and Sprint**
Resistive Runs**
Three-Person Tubing Acceleration Drill**
Double-Leg Lateral Hopping**
Pro-Agility**
Side Shuffle*, **
T-Drill**
Zigzag Drill or Z-Drill**
Four-Corner Drill**
Quick Feet**
L-Drill**
Bag or Cone Shuffle**
Ickey Shuffle**
In–Out Shuffle**
Snake Jump**

*Included in the dynamic warm-up video.
**Individual video clip.

The web resource can be accessed at www.HumanKinetics.com/PhysiologicalAspectsOfSportTrainingAndPerformance.

part I

PHYSIOLOGICAL ADAPTATIONS TO EXERCISE

This section focuses on the physiological adaptations from training. Discussion of neuromuscular, endocrine, metabolic, cardiovascular, and immune function is presented with special reference to how these systems are affected by acute and chronic training paradigms. The beginning of each chapter provides a brief review of the given physiological system. Training adaptations are discussed with specific reference to both resistance and endurance training. The influence of training status on physiological adaptations to exercise is also discussed when appropriate. This part of the book provides a basis for readers to understand physiological adaptations to exercise and to appreciate how changes in specific program variables can alter the training adaptation. This understanding will give the reader the tools necessary to develop training programs designed to elicit specific training adaptations.

chapter 1

Neuromuscular System and Exercise

CHAPTER OBJECTIVES

After reading this chapter you should be able to do the following:

- Define the structure of skeletal muscle.
- Understand how muscles contract.
- Describe how muscle responds to different exercise regimens.
- Distinguish how neuromuscular adaptations affect strength and power expression.

The ability of the muscle to provide force and maintain physical activity is the basis for athletic performance. The common goal for all athletic training programs is to improve the functional capability of the exercising muscles. The specific goals of the athlete, however, may be different depending on the needs of the particular sport. This chapter provides a brief review of the gross structure of skeletal muscle and the mechanism of muscle contraction. Further discussion focuses on both neurological and skeletal muscle adaptations seen subsequent to athletic conditioning programs.

Muscle Structure

A skeletal muscle consists of thousands of cylindrical muscle cells called **fibers**. Muscle fibers are long, thin, and multinucleated. They lie parallel to one another, and the force of a muscle contraction is along the long axis of the fibers. A layer of **connective tissue** called the **epimysium** surrounds the entire muscle. The muscle tapers at both its proximal and distal ends to form a **tendon**. A tendon is a strong, dense connective tissue that serves to connect muscle to bone. Tendons, although much smaller in magnitude than muscles, have a greater tensile strength, which allows them to withstand the forces generated by the relatively large muscles. Below the epimysium are bundles of up to 150 fibers bound together. Each bundle is termed a **fasciculus** and is surrounded by another layer of connective tissue called the **perimysium**. The **endomysium** is an additional layer of connective tissue that separates neighboring muscle fibers. A membrane called the **sarcolemma** surrounds the cellular contents that make up each muscle fiber. The number of fibers within each muscle varies considerably and depends on the size and function of the muscle. The fiber may extend the length of the muscle, or it may merge with another fiber. The structure of the muscle can be seen in figure 1.1.

A single muscle fiber is composed of many smaller units that lie parallel to the fiber. These units are called **myofibrils**. The myofibrils contain even smaller units, termed **myofilaments**, that consist mainly of two proteins, **actin** and **myosin**. The arrangement of these proteins gives the muscle fiber its striated appearance of light and dark bands (figure 1.2). The light band is referred to as the **I band**, and the dark band is called the **A band**. The **H zone** is in the middle of the A band and at rest contains only the myosin filaments. During contraction the actin filaments are pulled into this zone, causing this zone

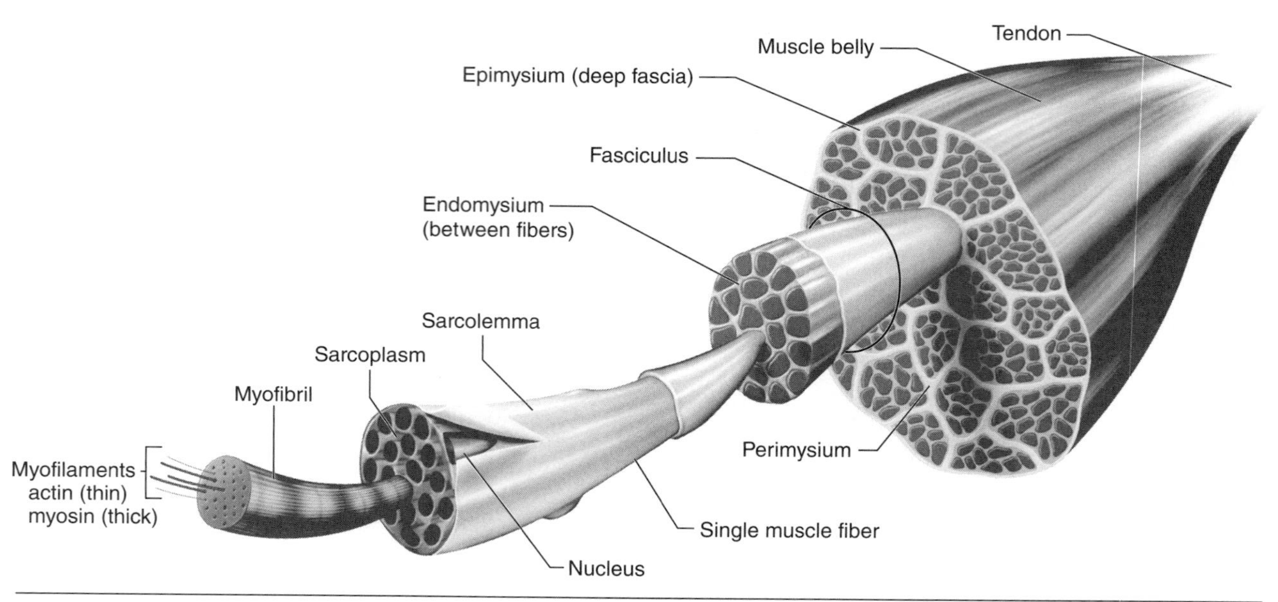

Figure 1.1 Structure of muscle.

to be similar in appearance to the A band. The **Z line** bisects the I band and is connected to the sarcolemma to give stability to the entire structure. The repeating unit between two Z lines is called the **sarcomere**, which is the functional unit of the muscle cell.

Muscle Contraction

The myosin filaments have projections, or **cross-bridges**, at the regions where the actin and myosin overlap. These projections, known as **myosin heads**, extend perpendicular from the thick myosin filaments to the thinner actin filaments. The interaction between actin and myosin can be seen in figure 1.3.

Actin filaments contain two additional proteins, **troponin** and **tropomyosin**, that regulate the contact between the myofilaments during contraction (see figure 1.2). Tropomyosin appears to inhibit contact between actin and myosin by covering the **active sites** located on the actin molecule. Troponin triggers the contraction of these filaments by uncovering the active sites when stimulated by calcium.

Calcium ions are stored in interconnecting tubules that lie within the muscle fibers. These tubules are arranged parallel to the myofibrils and are called the **sarcoplasmic reticulum**. The sarcoplasmic reticulum terminates in large vesicles known as the **terminal cisternae** (figure 1.4). The terminal cisternae abut an additional tubule system known as the **transverse tubules**. These tubules lie perpendicular to the myofibrils and are located in the lateral-most portion of the sarcoplasmic channels in the region of the Z line. The combination of the cisternae and the transverse tubule between them is referred to as the **triad**. The transverse tubules pass through the muscle fiber and form an opening to the inside of the muscle cell.

The transverse tubule system and triad function as the transportation network for the spread of the **action potential (depolarization)** from the outer membrane of the fiber inward. During depolarization, calcium ions are released from the sarcoplasmic reticulum and diffuse to the myofilaments. As mentioned previously, calcium ions bind to the protein troponin, causing a conformational change in the tropomyosin molecule. Its inhibiting action is impeded, allowing the active sites on the actin filament to be uncovered. This permits the myosin head to connect to the active site, allowing muscle contraction to proceed (see figure 1.4 for depiction of cross-bridge activity with binding of Ca^{2+} to the **troponin–tropomyosin complex**).

Muscle contraction is an active process that requires a constant influx of energy. For muscle contraction to continue, **ATP (adenosine triphosphate)** binds to its receptor on the myosin head. An enzyme (**myosin ATPase**) found on the myosin head splits ATP to yield **ADP (adenosine diphosphate)** and **Pi (inorganic phosphate)**. This process causes the myosin head to dissociate from its active site on the actin molecule and ready itself for another cycle.

The **sliding filament theory** proposes that a muscle shortens or lengthens because the actin and myosin filaments slide past each other without the filaments themselves changing length (Huxley 1969). The major structural change occurring within the sarcomere is the pulling together of the Z bands,

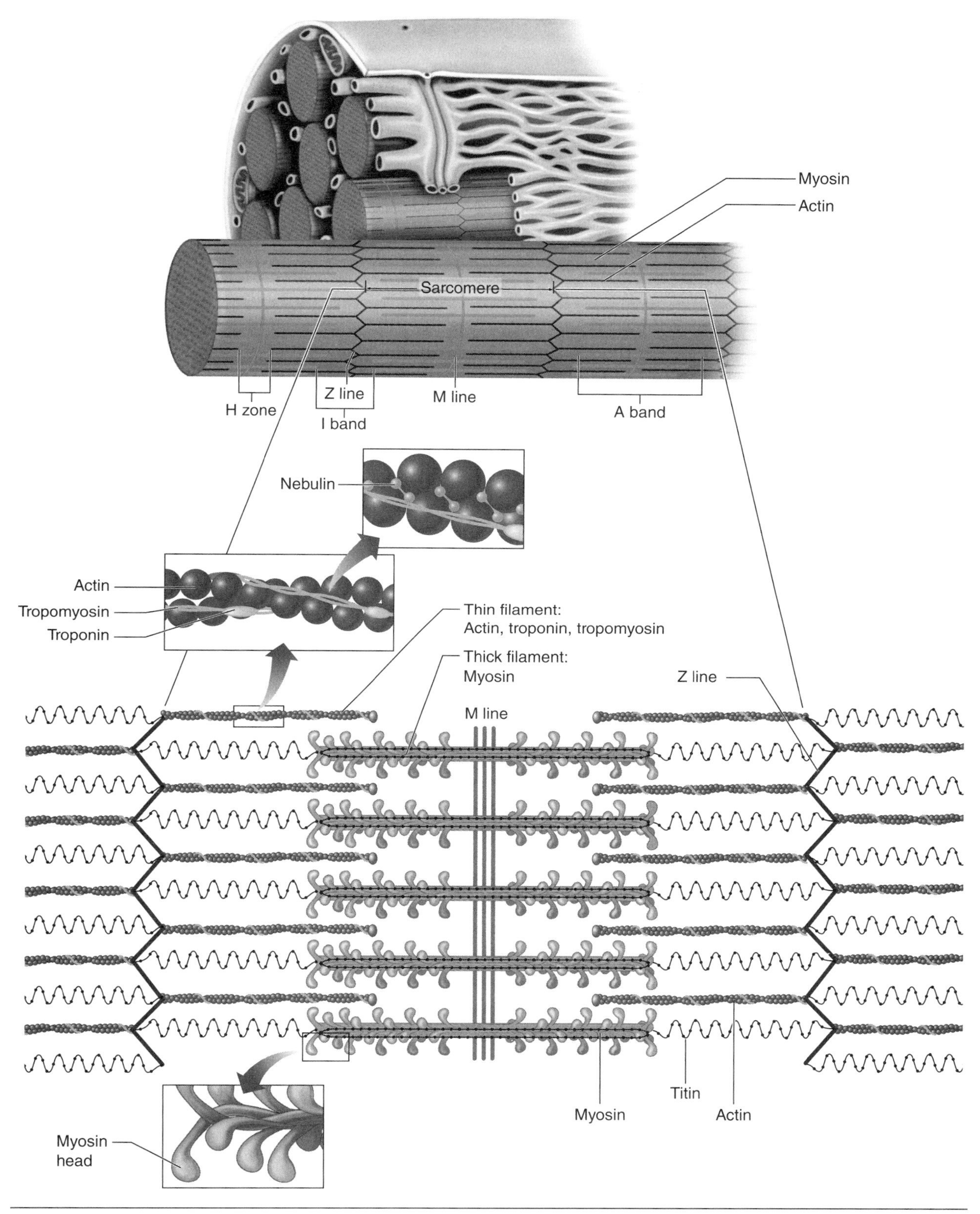

Figure 1.2 Sarcomere unit.

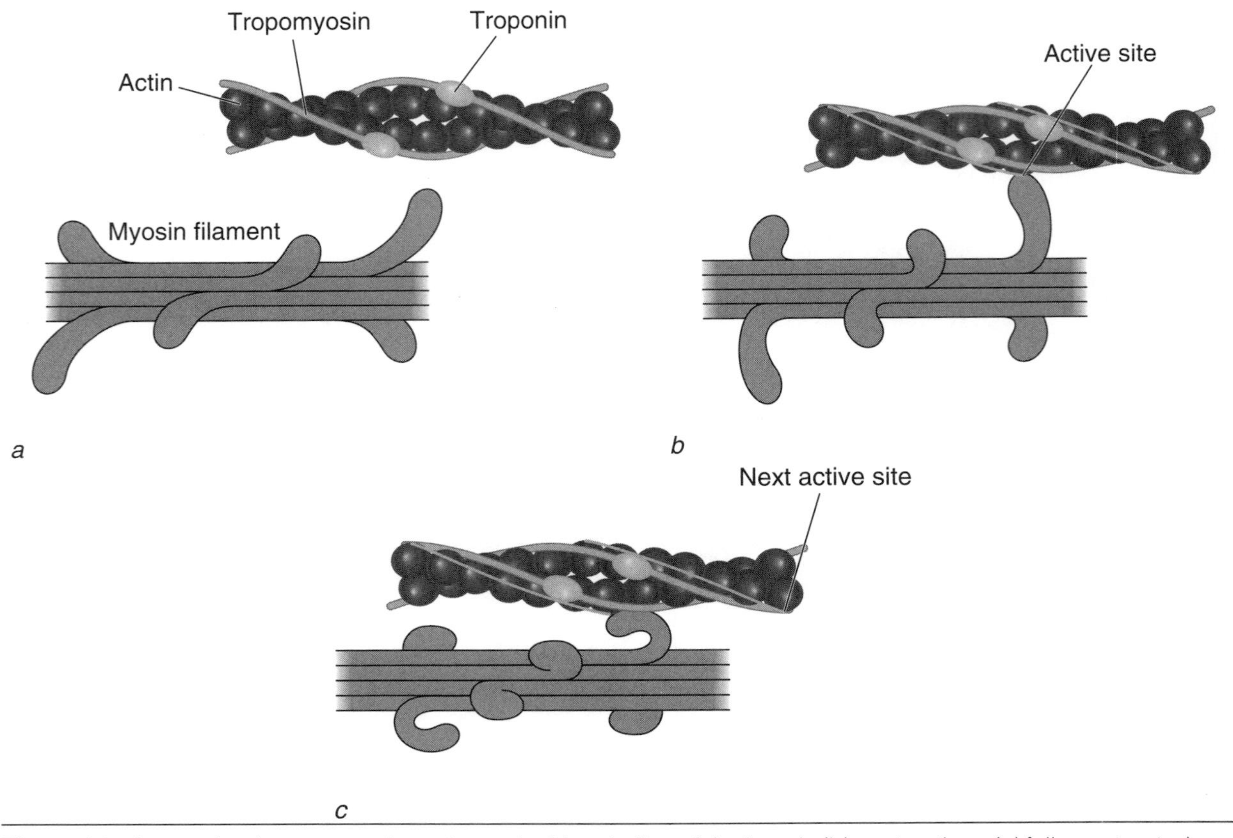

Figure 1.3 Interaction between actin and myosin. Muscle fiber *(a)* relaxed, *(b)* contracting, *(c)* fully contracted.

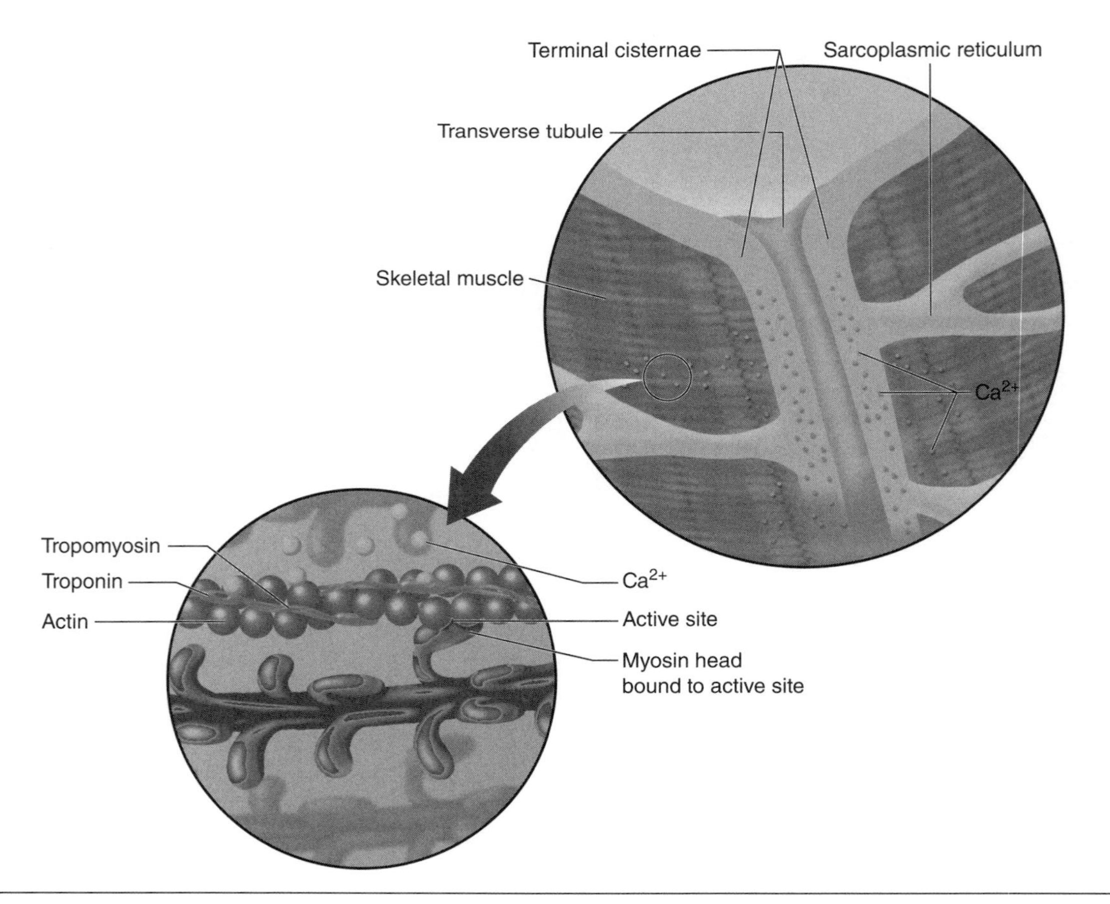

Figure 1.4 Sarcoplasmic reticulum transverse tubules and the binding of Ca^{2+} to the troponin–tropomyosin complex.

which decreases the region of the I band. During a muscle contraction, the cross-bridges do not move in a synchronous manner but may undergo repeated independent cycles of movement.

Neuromuscular System

Muscle contraction is the result of stimuli processed by the central nervous system. Stimulation of the central nervous system initiates an electrical impulse that is propagated along the length of a nerve cell, or **neuron**. The nerve cell consists of a cell body (**soma**), numerous short projections (**dendrites**), and a long projection (**axon**) that carry the electrical impulse from the soma toward the muscle. The connection between the nerve fiber and muscle is at the **neuromuscular junction**. The nerve fiber and the muscle that it innervates are known as the **motor unit**. Each motor unit is innervated by its own neuron, with its own specific contractile and metabolic characteristics. The individual fibers of a motor unit may be spread over a large portion of a muscle. Figure 1.5 is a schematic representation of the motor unit.

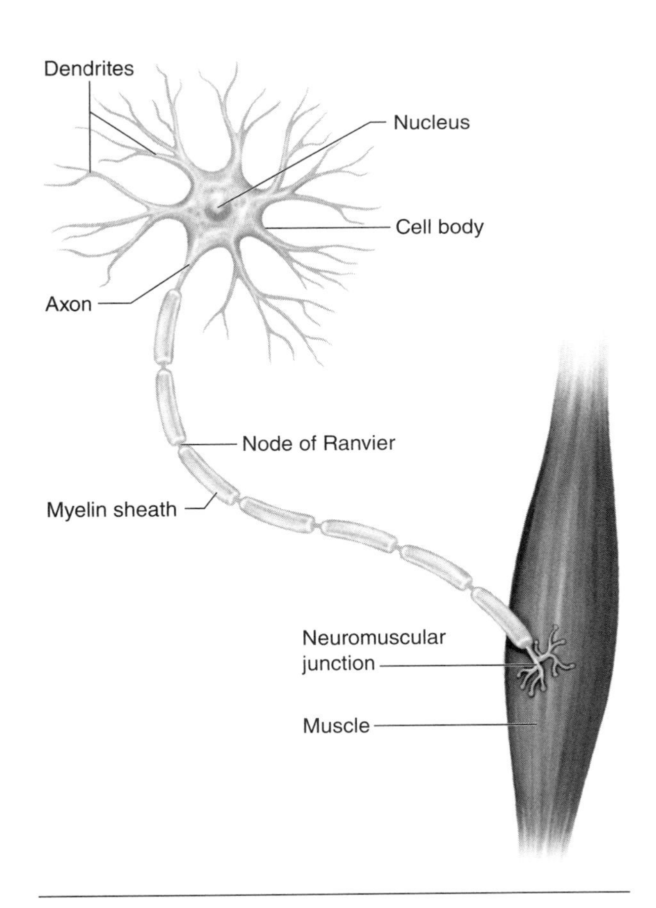

Figure 1.5 The motor unit.

A nerve is composed of many nerve fibers. Nerve fibers covered by a **myelin sheath** are called **medullary fibers**; they ensure that neural impulses meant for a specific muscle action do not activate other muscle groups. This myelin sheath does not run continuously along the length of the nerve fiber but is interrupted by small spaces known as the **nodes of Ranvier**. Nerve impulses jump from one node to another, allowing a fast action or impulse conduction called **saltatory conduction**. In contrast, a nerve fiber without the myelin sheath is referred to as **nonmedullary**, and the nerve impulse must travel the entire length of the nerve fiber.

Nerve impulses exist in the form of electrical energy. When there is no impulse, the inside of the nerve cell has a negative charge compared with the outside, basically because of a greater number of positive ions located outside of the cell (sodium) than inside the cell (potassium). This is termed the **resting membrane potential**. Excitation occurs when a nerve impulse causes the nerve cell membrane to become more permeable for both sodium and potassium. Sodium will move from an area of higher concentration to an area of lower concentration (from outside of the cell to inside), causing a positive charge on the inside of the cell in comparison with the outside. This is termed an **action potential** and lasts for a brief time. The action potential propagates along the length of the nerve, or motor neuron, until it reaches the muscle. The muscle fibers that the nerve innervates also have the capability of propagating an action potential and transmit the electrical impulse along its entire length.

The gap between the nerve fiber and muscle at the neuromuscular junction is called the **synapse**. The electrical impulse resulting from the action potential reaches the presynaptic side of the neuromuscular junction and causes the release of the neurotransmitter **acetylcholine**. Acetylcholine is stored in vesicles located at the presynaptic membrane and, on stimulation, diffuses across to the postsynaptic side of the neuromuscular junction. The interaction between acetylcholine and its receptor on the postsynaptic membrane causes an increase in the permeability of sodium and potassium ions. This propagation of an action potential causes a release of calcium from the sarcoplasmic reticulum, resulting in muscle contraction. Figure 1.6 depicts a flow chart from the moment of the action potential to muscle contraction.

The muscle contraction is limited to the muscle fibers of the motor unit. When the nerve of a particular motor unit is activated, contraction of all the muscle fibers that it innervates occurs. This is known as the **all-or-none law**. However, it is important to understand that not all the fibers within a muscle

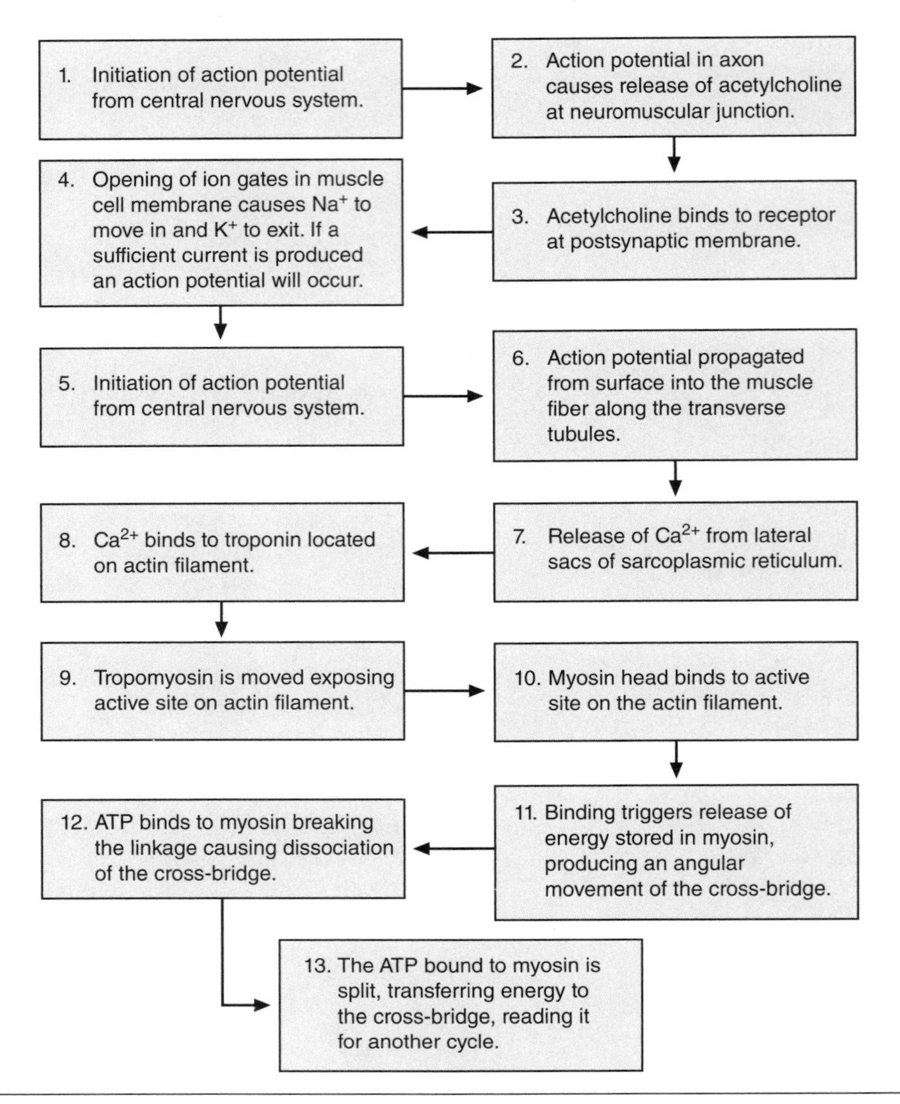

Figure 1.6 Sequence of events from action potential to skeletal muscle contraction.

are contracting. If this were true, there would be no ability to control the force output of the muscle. Gradations in force are accomplished according to the number of motor units recruited or stimulated. Maximal strength is accomplished when all the motor units of a muscle are activated.

Muscle Fiber Types

All motor units function in a similar manner although they may have different contractile and metabolic characteristics. Some motor units are more suited for **aerobic** metabolism, whereas others are more appropriate for **anaerobic** activity. Two distinct fiber types have been identified and classified by their contractile and metabolic characteristics. These have been termed **slow-twitch** and **fast-twitch** fibers, also referred to as type I and type II, respectively. The characteristics of these fiber types are shown in table 1.1. As can be seen in the table, these fibers possess certain distinguishable characteristics that make them suited for either prolonged, low- to moderate-intensity activity (slow-twitch fibers) or short-duration, high-intensity activity (fast-twitch fibers).

For many years, skeletal muscle fiber classification was limited to these two classifications (type I and type II), with type II fibers further subdivided into two distinct divisions: IIa and IIb. Type IIa fibers have a well-developed capacity for both aerobic and anaerobic metabolism and are commonly termed **fast oxidative glycolytic**. Type IIb fibers possess the greatest anaerobic capability and are termed **fast glycolytic**. With improvements in muscle-staining

Table 1.1 Characteristics of Type I and Type II Muscle Fibers

	Type I Slow twitch	Type IIa Fast twitch	Type IIb Fast twitch
Force production	Low	Intermediate	High
Contraction speed	Slow	Fast	Fast
Myofibrillar ATPase activity	Low	High	High
Fatigue resistance	High	Moderate	Low
Glycolytic capacity	Low	High	High
Oxidative capacity	High	Medium	Low
Capillary density	High	Intermediate	Low
Mitochondrial density	High	Intermediate	Low
Myoglobin content	High	Intermediate	Low
Endurance capability	High	Moderate	Low
Glycogen storage capability	No difference	No difference	No difference

techniques and improved electrophoretic procedures using immunohistochemical analyses via Western blotting, additional subtypes within each fiber type have been reported (Staron et al. 1991; Fry, Allemeier, and Staron 1994). Subtypes of type I oxidative fibers have been labeled type I and type Ic. Type Ic fibers are thought to have less oxidative capacity than type I fibers. Early studies suggested that there was a spectrum of different fiber subtypes identified as type II fibers. It was proposed that these fibers, labeled IIc, IIac, IIa, IIab, and IIb, represented a continuum of aerobic and anaerobic characteristics (Pette and Staron 1990). However, as research on human skeletal tissue has progressed, physiologists have generally reported on four major fiber types with distinct myosin heavy chain composition (Schiaffino and Reggiani 2011). Although the existence of additional fiber subtypes may be seen in various animal models, human skeletal muscle appears to have four distinct myosin heavy chain isoforms; type I, type IIa, type IIb, and type IIx (Baldwin and Haddad 2001). The type IIx fibers have twitch properties similar to those of types IIa and IIb and appear to have greater relevance in discussion of human muscle (Schiaffino and Reggiani 2011). Although the type IIb myosin heavy chain gene has been identified in the human genome (Weiss, Schiaffino, and Leinwand 1999), its expression and physiological role are debatable. In research on muscle fiber subtypes, it appears that older studies on fast-twitch muscle fibers focused on discussion of type IIa and IIb subtypes but that more recent investigations (within about the last 5 to 10 years) refer to the type IIa and type IIx subtypes.

It is believed that genetics largely determines muscle fiber-type distribution and that it is set at birth or early in life. The average individual (man or woman) has equal proportions of slow- and fast-twitch fibers. The composition of muscle fiber types (percentage of type I in relation to type II) is consistent among the major muscle groups in the body.

In elite athletes, the predominance of a particular fiber type appears to correspond to the metabolic requirements of the given sport. Endurance athletes have a large percentage of type I fibers, whereas highly anaerobic athletes (e.g., sprinters) have a predominance of type II fibers. Elite endurance athletes may have 90% of their skeletal muscle made up of type I fibers (Adams et al. 1993; Bergh et al. 1978), providing a large advantage for success in aerobic performance. Similarly, athletes with exceptional explosive power and speed would have a predominance of type II fibers. It is reported that in competitive weightlifters, approximately 60% of the skeletal muscle is composed of fast-twitch fibers (Fry et al. 2003; Tesch and Karlsson 1985). Interestingly, most of these are type IIa subtype, with less than 1.5% of the subtype reported as type IIb (Fry et al. 2003).

Muscle Recruitment

During muscle action, the muscle fibers are generally recruited according to the **size principle**, which suggests an orderly recruitment of the motor units. Motor unit recruitment patterns are related to the size of the neuron. Smaller motor units require a lower stimulus for activation. Thus, type I muscle fibers, which require a low stimulation, are recruited first during muscle activation. As greater force in the muscle action is required, higher-threshold units (type II fibers) are recruited. However, there may be an exception to the size principle. Type II fibers appear to be recruited first during powerful, high-velocity activities.

The gradations seen in recruitment patterns suggest that fatigue-resistant, slow-twitch fibers are recruited primarily during low-intensity, long-duration activities. As greater intensity is needed, the higher-threshold units (fast-twitch fibers) are stimulated. However, when muscle action requires immediate high-velocity, high-power movement (e.g., sprinting), the fast-twitch fibers may be recruited first. The benefit of the size principle is that it keeps the highly fatigable fast-twitch fibers in reserve until needed for a particular muscle action.

Muscle Proprioceptors

Muscle action is finely controlled by the interaction between motor and sensory activities. Sensory nerves, both **muscle spindles** and **Golgi tendon organs**, which are located in the muscles, tendons, and fibrous capsules of joints, provide sensory feedback to the brain at a conscious and subconscious level on movement and position of the muscles and joints.

Muscle spindles are arranged parallel to the skeletal muscle fibers and are composed of several small **intrafusal fibers**, whose nerve endings are attached to the sheaths of the surrounding skeletal muscle fibers. Muscle spindles are responsible for monitoring the stretch and length of the muscle and also initiate contraction to reduce the stretch in the muscle. During muscle contraction, force or tension develops as a result of direct stimulation of the motor unit and indirect stimulation through the muscle spindle.

Golgi tendon organs are sensory receptors located within muscle tendons. They are sensitive to tension in the **muscle–tendon complex** and form a protective mechanism to reduce the potential for injury. When tension in the activated muscle reaches levels that pose a potential risk of injury, the Golgi tendon organ is stimulated. It inhibits contraction of the contracting, or **agonist**, muscles and activates the **antagonist** muscle groups to reduce the force of the muscle contraction. Inhibition of the Golgi tendon organ during training is thought to result in greater strength expression by the exercising muscle.

Neuromuscular Adaptations to Exercise Training

There is a great deal of plasticity within the neuromuscular system. Participation in physical conditioning programs leads to improvements in muscle size or performance. These adaptations may occur fairly rapidly, depending on the training status of the individual. The type of training program also affects physiological adaptation. Strength training may have a particular effect on muscle adaptation whereas endurance training may cause a completely different type of adaptation. Likewise, if the training stimulus is removed, muscle tends to revert back to its pretraining state. An understanding of the type of alterations seen with a given training program helps the coach or athlete develop the most appropriate training program and set the most realistic training goals.

Neural Adaptation

Maximal strength expression is not determined solely by the quantity and quality of the muscle mass but also by the extent to which the muscle mass is activated (Sale 1988). Maximal strength is achieved when the primary muscle group is fully activated and the **synergists** and antagonists are appropriately activated (Sale 1988). A better coordination in the activation of these muscle groups allows for a better expression of strength. The neural adaptation to resistance training is thought to be the initial adaptation leading to increased strength (Moritani and deVries 1979; Komi 1986).

Electromyographic Changes

The most common method for assessing neural activity of the muscle is through electromyograph (EMG) recordings. The EMG measures the electrical activity within the muscle and nerves and indicates the neural drive to a muscle. An EMG recording is generally performed with surface electrodes on a prime mover. Recording the EMG activity of a muscle before and after a resistance training program can measure the neural adaptation resulting from the program. Once the motor unit activity is recorded, the signal can then be integrated in a number of different fashions, which is referred to as the integrated electromyogram (IEMG). Resistance training has been demonstrated to increase IEMG activity in both trained and untrained populations (Hakkinen and Komi 1983; Moritani and deVries 1979). Significant correlations between IEMG and increases in strength have been reported after strength training (Hakkinen and Komi 1986; Hakkinen, Komi, and Alen 1985). These findings support the idea that strength-trained subjects may be more capable of fully activating their primary muscles during maximal performance than untrained subjects. Further findings have shown that EMG activity significantly increases during the initial stages of a resistance training program but decreases, or increases at a diminished rate, during the later stages of training (figure 1.7). It is at this point in training that most strength increases are attributed to muscle **hypertrophy**, which is discussed later in the chapter.

Several potential neurological mechanisms are associated with enhanced muscle strength expression. These include changes in recruitment patterns, enhanced synchronization, and removal of inhibitory mechanisms, stimulating a more efficient activation of motor units (nerve and the muscle fibers that they innervate).

Recruitment Patterns

Neural adaptations may also be recognized by a decrease in the electrical activity in the muscle with a corresponding increase in force output (Sale 1988). The greater strength expression seen with lower EMG responses may reflect a more efficient recruitment pattern of the muscles responsible for the force production. It may also be an indicator of an increase in the size of the muscle fiber, resulting in a reduction of the motor unit pool needed for a given level of strength expression. Neural adaptation causing greater force production has also been attributed to an enhanced activation rate (Del Balso and Cafarelli 2007). An increase in motor unit activation rate may result in generating an increase in force development, which was found to be significantly correlated ($r = 0.95$) to force production (Del Balso and Cafarelli 2007).

Synchronization

Another neural adaptation that may result in greater force production is an increase in the **synchronization** of motor unit firing. A more synchronous pattern of motor unit firing simply means that the number of motor units active at any one time is increased, causing a greater number of fibers to contract at a given time and thus greater strength expression. Improved motor unit synchronization has been shown in a group of weightlifters after a short (6 weeks) training program (Cormie, McGuigan, and Newton 2011; Milner-Brown, Stein, and Yemm 1975). However, this effect may be depend on the loading used during training, as low-intensity resistance exercise may be unable to improve motor unit synchronization (Griffin et al. 2009).

Inhibitory Mechanisms

Inhibition of muscle contraction by the Golgi tendon organs is an inherent safety mechanism that limits force production to prevent injury at the muscle–tendon complex. This protective mechanism (activation of the Golgi tendon organ stimulates activation of antagonistic muscle groups) appears to be especially active when maximal contractions are performed at slow speeds of contraction (Caizzo, Perrine, and Edgerton 1981; Wickiewicz et al. 1984). Resistance training is thought to cause an inhibition of these protective mechanisms. Some athletes purposely contract the antagonistic muscle group immediately before a lift with the intent to partially inhibit the

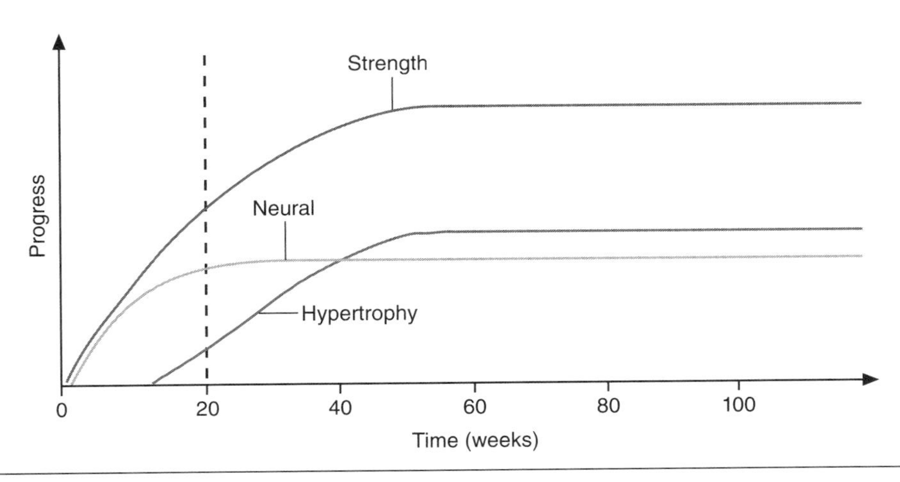

Figure 1.7 Relationship of neural and muscular adaptations to strength training.

Golgi tendon organ and allow for a more forceful contraction.

Time Course of Neural Activation

The rapid strength gains seen during the initial phases of a resistance training program are often accomplished without any noticeable increase in the cross-sectional area of the muscle. This suggests that the initial improvements in strength likely occur primarily from neurological adaptations. Moritani and deVries (1979) suggested that these neural adaptations are the primary mechanism underlying the increase in strength that occurs over the first 3 to 5 weeks of training. However, there is convincing evidence that increases in muscle protein synthesis and early stages of muscle hypertrophy do occur shortly following the onset of a resistance training program (Staron et al. 1991, 1994).

Neural Adaptation to Running

Maximal-effort sprints require a high level of neural activation (Ross, Leveritt, and Riek 2001). It is not well understood whether training has the ability to enhance neural activation and result in faster sprint speed. Improvements in running technique likely reflect a better-coordinated recruitment pattern, thus leading to a more explosive movement. Trained sprinters, though, do have greater nerve conduction velocity, greater EMG response, and more highly developed motor unit recruitment strategies than untrained runners and endurance athletes (Casabona, Polizzi, and Perciavalle 1990; Hakkinen et al. 1985; Ross, Leveritt, and Riek 2001), suggesting that a training response may occur.

Neural Adaptation to Concurrent Training

Changes in neural recruitment patterns occur with both resistance and endurance training. However, these responses appear to be in the opposite direction. Endurance training appears to delay fatigue (29.7 ± 13.4%) and is associated with a decrease in motor unit discharge rates, while resistance training is associated with an increase in force output (17.5 ± 7.5%) and an increase in motor discharge rates (33.3 ± 15.9%) (Vila-Chã, Falla, and Farina 2010). These adaptations appear to occur early on in the training program (~3 weeks); however, changes in motor unit conduction velocity appear to take longer (~6 weeks) and occur in both endurance and resistance training. If the two types of training programs are performed concurrently, the endurance training does not seem to impair the neural adaptations induced by resistance training in previously untrained individuals (McCarthy, Poszniak, and Agre 2002), but does appear to limit rapid voluntary neural activation in trained muscles (Hakkinen et al. 2003).

Skeletal Muscle Adaptation

Skeletal muscle is dynamic in its response to training and can adapt to a wide range of functional demands. When skeletal muscle is forced to work at intensities exceeding 60% to 70% of its maximal force-generating capacity, adaptations occur that may result in an increase in muscle size and strength (MacDougall 1992).

The initial increases seen in muscle strength after resistance training have been attributed primarily to the neurological adaptations discussed earlier in this chapter. Further increases in skeletal muscle strength appear to be the result of a growth in muscle size. An increase in the size of preexisting muscle fibers (hypertrophy) or an increase in the number of fibers within the muscle (**hyperplasia**) may result in skeletal muscle growth.

Changes in Muscle Cross-Sectional Area Following Resistance Exercise

Increases in muscle size are generally seen after 6 to 8 weeks of heavy resistance training. However, some evidence suggests that muscle growth may occur even earlier (Staron et al. 1991, 1994). Increases in muscle size have been attributed to increases in the cross-sectional area of existing muscle fibers (Alway, Grumbt, et al. 1989; MacDougall et al. 1984). This process of fiber growth appears to be related to the increased synthesis of contractile proteins (actin and myosin filaments) and to the increased formation of sarcomeres within the fiber (Goldspink et al. 1992; Alway, Grumbt, et al. 1989; MacDougall et al. 1979). The synthesis of these protein filaments, which constitute the contractile element of muscle fibers, may be related to the repeated trauma to the fibers from high-intensity resistance training. During recovery from the cellular damage caused by such training, an overcompensation of protein synthesis may occur, resulting in the noted anabolic effects (Antonio and Gonyea 1993; West et al. 2010).

Muscle hypertrophy resulting from a resistance training program occurs in both type I and type II muscle fibers (Campos et al. 2002). However, the type II fiber appears to undergo a greater relative hypertrophy (Staron et al. 1989). Since both type I and type II fibers are recruited during maximal contractions, the greater hypertrophy in the type II fiber may be related to an activation of high-threshold units greater than that normally seen during daily activity (MacDougall 1992). The magnitude of these increases varies considerably and depends on several factors, including the individual's responsiveness to training, the intensity of training, the duration of training, and the individual's training status. The hypertrophy response to resistance exercise appears to be dependent on the exercise intensity (percentage of the individual's maximal strength ability). Significant increases in the cross-sectional area of skeletal muscle were seen with moderate- to high-intensity resistance training (9-11 repetitions per set and 3-5 repetitions per set, respectively), but not with lower-intensity training (20-28 repetitions per set) (Campos et al. 2002).

Increases of 15.6%, 17.3%, and 28.1% in cross-sectional area have been reported in type I, type IIa, and type IIb muscle fibers, respectively, in novice female subjects after 6 weeks of high-intensity resistance training (Staron et al. 1991). This is similar to the 12.5%, 19.5%, and 26% increases seen in type I, type IIa, and type IIb fibers following 8 weeks of high- to moderate-intensity resistance training in trained men (Campos et al. 2002). As the length of the resistance training program increases (20 weeks), further increases in muscle size may be seen (15%, 45%, and 57% in type I, type IIa, and type IIb muscle fibers, respectively) (Staron and Johnson 1993). Similar increases in muscle hypertrophy have also been seen in untrained male subjects (Adams et al. 1993; Hather et al. 1991). Although sex differences in muscle growth exist, these differences become apparent only after longer periods of training.

As previously mentioned, the training status of the individual has an important effect on the morphological changes seen in muscle after resistance training. Experienced bodybuilders, both male and female, were examined during 24 weeks of training for competition. No significant improvements in muscle cross-sectional area were noted over the training period (Alway et al. 1992). This is consistent with what has been repeatedly reported in the literature concerning muscle growth in highly trained, experienced bodybuilders (Hakkinen, Komi, et al. 1987; Hakkinen et al. 1988). It should be understood, however, that there might be a large difference between statistical significance and practical significance with respect to understanding muscle growth. Generally, studies examining muscle morphological changes do not have a large sample number. Thus, a great difference is needed to achieve statistical significance. Alway and colleagues (1992) reported a 3.6% increase in the cross-sectional area of the biceps in five experienced male bodybuilders after 24 weeks of training. Although this was not statistically significant, it may represent an important component of success during competition. Muscle hypertrophy in experienced lifters may be attainable, but this requires a high-intensity training stimulus for a much longer training duration.

Changes in Muscle Cross-Sectional Area Following Endurance Exercise

Endurance training appears to contrast with resistance training with regard to skeletal muscle adaptation. Similar to what is generally seen during resistance training, changes are dependent on the type of running program and are specific to fiber type. High-intensity, short-duration running appears to be a stimulus for a reduction in the size of type IIa fibers, while low-intensity, long-duration runs result in a decrease in the size of both type I and type IIb fibers (Deschenes et al. 1995). Muscle fiber size is generally smaller in endurance athletes than in strength–power athletes (Tesch and Karlsson 1985), suggesting that mechanisms governing muscle size may be both intrinsic (genetic endowment) and extrinsic (training effect). The effect of training to reduce muscle fiber size may be related to duration of training. Seven weeks of endurance training (60 min per session, 5 days a week at 60% of $\dot{V}O_2max$) did not have any effect on the size of either type I or type II fibers in previously untrained men and women (Carter et al. 2001).

However, if the duration of training is prolonged, decreases in fiber size may become apparent. Previously recreationally trained male and female runners were examined during their preparation to compete in their first marathon. During the 16-week training program, which included progressive increases in training volume (24.3-58.3 km run per week), a significant decrease (~20%) was reported in the diameter of myosin heavy chain type I and type IIa muscle fibers (Trappe et al. 2006). The potential benefit of the smaller-diameter fibers would be to enhance oxygen diffusion from the capillaries to the cell. Even in trained runners, the muscle fiber atrophy seen consequent to a training season remains apparent. During the course of a 12-week competitive cross country season in male college runners, a significant decrease (~3%) in type I fibers was seen during the initial 8 weeks of the season (run volume: ~99 km/week), but no significant changes were noted in type IIa fibers (Harber et al. 2004). Following the 4-week taper (volume of runs was reduced by ~25%), type I fiber size continued to decrease (7% smaller than initial levels, $p < 0.05$). The ability of type IIa fibers to maintain their size may be related to the interval training performed during the initial 8 weeks of the season. The higher-intensity runs may have provided a sufficient training stimulus for the type IIa fibers to maintain their size. Decreases in type I fiber size were also accompanied by a significant decrease in shortening velocity (~23%) before the taper and (~17%) following the taper, while no change in shortening velocity was noted in the type IIa fibers. Similar changes were also noted in muscle fiber power performance.

Effect of Concurrent Training on Changes in Muscle Cross-Sectional Area

Although recreational exercisers may focus on a single mode of exercise, athletes often train multiple physiological systems to reach peak condition. This often complicates research designs, and as a result, fewer studies are published using commonly used training paradigms. Limited studies have examined combined strength–endurance or strength–sprint training programs. A classic paper in this area was published by Dr. William Kraemer and colleagues in 1995 (Kraemer, Patton, et al. 1995). They examined muscle fiber morphological changes in untrained subjects who exercised 4 days a week for 3 months in either a high-intensity resistance training program, an endurance training program, or a combined strength–endurance training program. All training programs were periodized (see chapter 14) to enhance recovery from exercise and prevent overtraining. Both strength and combined strength–endurance programs resulted in significant muscle fiber hypertrophy. However, the inclusion of endurance training stunted the growth of both type I and IIc fibers (figure 1.8). Endurance training alone caused a decrease in fiber size in the more oxidative fibers (type I and IIc). The attenuation of muscle hypertrophy with concurrent strength and endurance training had been previously noted (Dudley and Djamil 1985). Interestingly, Kraemer, Patton, and colleagues (1995) also included a study group that performed an upper body resistance training program in addition to the endurance training program. Those subjects were able to mitigate the decreases in fiber size of the legs through upper body training; this may have been attributable to isometric contractions of the leg musculature during upper body exercise.

Interestingly, the specific adaptation to muscle growth or atrophy may have important application to performance. Increases in fiber size do not appear to be accompanied by increases in mitochondrial number or in the capillary-to-fiber ratio. This lowering of the mitochondrial and capillary volume density in the fibers may not hinder strength or power performance, but it may have important implications for endurance capability in those muscles. This change might alter the oxygen kinetics within the muscle by delaying transport of oxygen from the vasculature to the exercising muscle. It is noteworthy that endurance training decreases fiber size (see previous paragraph) while causing increases to both mitochondrial and capillary density, thus potentially improving the aerobic capability of the muscle. In contrast, sarcoplasmic reticulum and transverse tubule volume density increase in proportion to the change in myofibrillar volume (Alway, MacDougall, and Sale 1989), thus maintaining or improving contraction capabilities of the muscle.

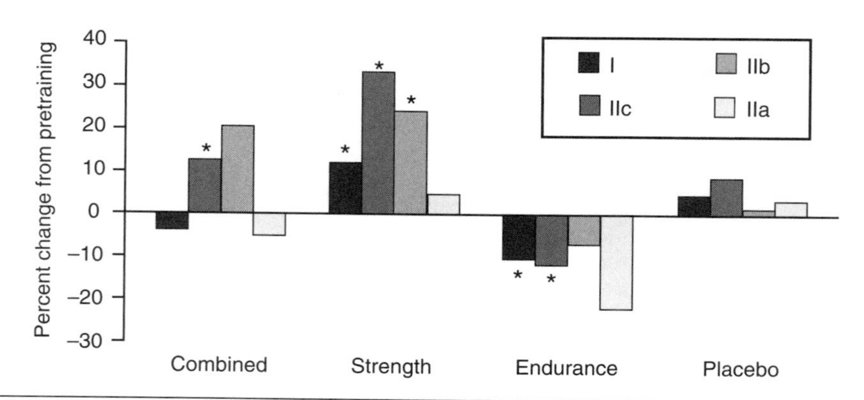

Figure 1.8 Muscle fiber changes as a consequence of strength, endurance, and combined strength–endurance training. * = Significant change.

Adapted from Kraemer et al. 1995.

In Practice

Concurrent Training and Muscle Remodeling: Effect on Protein Signaling

Muscle remodeling is a complicated process that involves activation of various intracellular pathways governing protein synthesis and modulating muscle growth. It has been demonstrated that resistance exercise enhances activation of the protein kinase B (Akt)/mammalian target of rapamycin (mTOR)/ribosomal protein S6 kinase ($p70^{s6k}$) pathway and stimulates protein synthesis and protein accretion (Bodine 2006; Dreyer et al. 2006), while endurance exercise activates signaling pathways governing mitochondrial biogenesis (Safdar et al. 2011). However, endurance exercise has also been reported to upregulate adenosine monophosphate (AMP)-activated protein kinase (AMPK) (Mounier et al. 2011), which has an antagonistic effect on the mTOR signaling pathway and may potentially impede muscle growth. These results have raised the question whether these two modes of exercise are physiologically compatible or incompatible. A recent study from Brazil examined the acute effects of strength, endurance, and combined strength and endurance training on the protein signaling responses in rat skeletal muscle (De Souza et al. 2013). Rats in the strength training group performed five sets of 10 repetitions with 75% of their maximal strength, while the rats in the endurance training group ran for 60 min at 60% of their maximum running velocity. The animals in the concurrent training group performed both training protocols but performed the endurance training protocol first.

Results of the study indicated that an acute concurrent exercise session did not activate opposing intracellular pathways. However, to interpret these results, one needs to examine them in the appropriate context. The animals in this study were not trained; thus the load used may not have been sufficiently intense to maximally stimulate the anabolic signaling pathways. In addition, the exercise chosen (leg extension) required the animals to exert effort only on the concentric movement of the exercise. The lack of any eccentric contraction eliminated the important mechanical stress associated with that phase of the lift. Without the eccentric load, it is possible that the mechanical stress associated with traditional resistance training was not sufficient to result in full activation of the mTOR signaling pathway. Under these specific conditions, concurrent training does not appear to cause any inhibition of the anabolic signaling processes. For the practitioner looking to interpret these results, the research model employed by the investigators was analogous to one in which untrained individuals use a relatively light to moderate training stimulus. Under these circumstances the incompatibility between endurance and resistance exercise may not be an issue. However, these results should not be extrapolated to experienced trained athletes.

If frequency of training is moderate (e.g., twice per week), the effects of concurrent training to attenuate muscle fiber hypertrophy in type II fibers may not be as drastic. In a study by Hakkinen and colleagues (2003) in middle-aged men (38 ± 5 years), training twice per week was able to stimulate increases in the fiber area of type I (46%), type IIa (26%), and type IIb (39%) fibers after 21 weeks. Increases in muscle size were also associated with a 22% increase in one-repetition maximum (1RM) strength of the leg extensors. In subjects who also added 2 days per week of progressive endurance training (from 30 min per session at the onset to 60 to 90 min per session by the end of the study), no difference in strength was observed (21% increase in 1RM leg extensor strength). Interestingly, increases in type II fibers remained similar (23% and 31% increase in type IIa and IIb, respectively), but type I fiber growth was attenuated (13% increase from baseline levels). Although strength performance and type II fiber size were not affected by concurrent training, the ability to increase the rate of force development was influenced by the added endurance exercise. Thus, endurance training even at moderate durations may negate the explosive expression of strength even when it does not impair strength development.

Despite the prevalence of combined resistance and sprint training among strength–power athletes such as American football, basketball, or hockey players, studies examining this type of concurrent

In Practice

Examining Muscle Architecture and Muscle Quality Noninvasively

A relatively new noninvasive method to investigate changes in muscle architecture is becoming popular. The use of ultrasonography to image muscle structure has proven to be a valid and reliable method of examining changes in muscle. The validity of ultrasound measures has been compared to that of both computed tomography (CT scans) and magnetic resonance imaging for analyzing large muscle groups (Reeves, Maganaris, and Narici 2004; Thomaes et al. 2012). The benefit of ultrasound compared to these other methods is related to the lower cost of operation and ease of use. With the other methods, physician prescription and the use of radiology technicians are often required.

Ultrasound measures are commonly used to assess muscle thickness, muscle cross-sectional area, fascicle length, and muscle fiber pennation angle. Improvements in our understanding and use of ultrasonography have resulted in a relatively new measure known as echo intensity. Echo intensity is the quantification of muscle quality using grayscale analysis from image analysis software (Cadore et al. 2012). The imaging software analyzes the pixel count that is obtained using the ultrasound image of skeletal muscle (generally the vastus lateralis, rectus femoris, or both) and provides a score that is based on an arbitrary units scale ranging from 0 to 256 (0: black; 256: white). Figure 1.9 provides an example of an echo intensity analysis. A lower number (lower echo intensity) is reflective of an increase in muscle quality. Echo intensity is indicative of the architectural characteristics of muscle, including the infiltration of adipose, connective, and other noncontractile tissue into the muscle. Skeletal muscles with greater lean tissue, less fat, and connective tissue have a lower echo intensity.

In a study of older (70.4 ± 5.5 years), healthy Japanese women, echo intensity values were found to significantly correlate with isometric knee extensor strength ($r = -0.40$, $p \leq 0.01$), independent of age or muscle thickness of the quadriceps femoris (Fukumoto et al. 2012). In addition to its use as a marker of muscle quality, echo intensity may provide a sensitive and reliable measure of muscle damage and recovery.

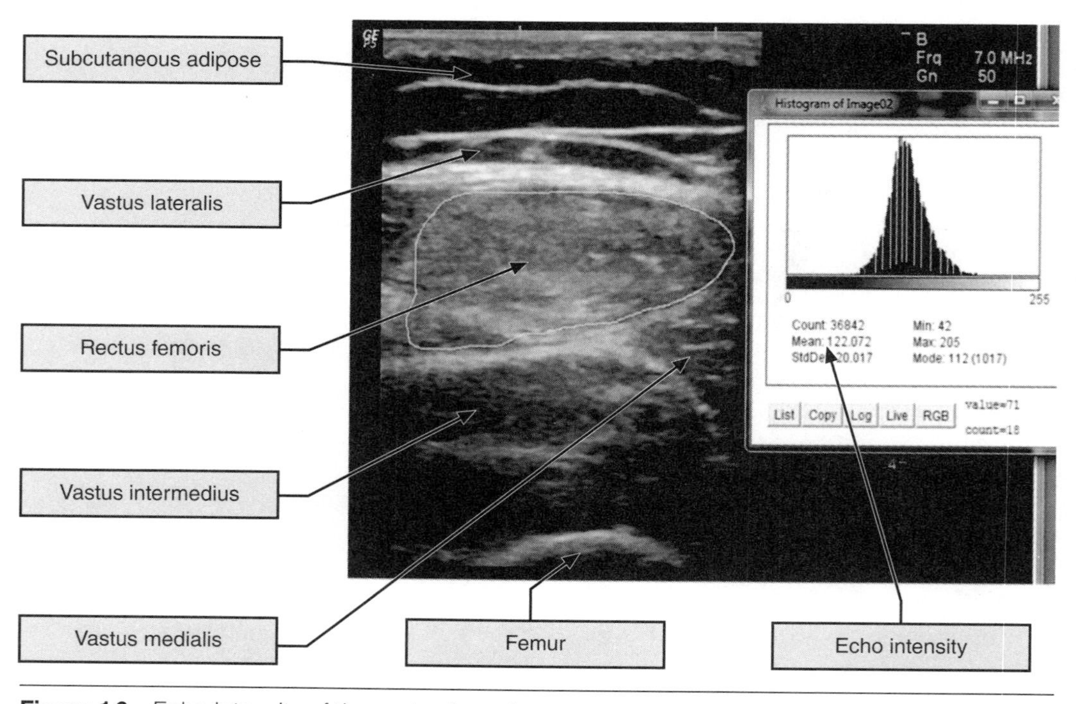

Figure 1.9 Echo intensity of the rectus femoris.

Radaelli and colleagues (2012) assessed echo intensity prior to the onset of exercise and following four sets of 10 repetitions at 80% of the subjects' maximal elbow flexor strength in the dominant arm at 24, 48, and 72 hours postexercise. Echo intensity values were significantly elevated from preexercise values at all three time points, whereas no change from baseline was seen in the nondominant (no-exercise) arm. These results suggested that echo intensity may be quite useful as a sensitive marker for recovery from resistance exercise. A number of studies on this unique measure of muscle quality will be published over the next few years.

training are more limited than those on concurrent resistance and endurance training. Interestingly, if the resistance training and sprint training are performed concurrently (~15 min of rest between exercise sessions), the anabolic response to resistance exercise (as determined by changes in the protein signaling response) may become mitigated (Coffey et al. 2009). In addition, the effect of concurrent sprint and resistance exercise appears to attenuate the anabolic hormone response to resistance exercise as reflected by a decrease in insulin-like growth factor 1 mRNA (messenger RNA) following 3-hour recovery. Although more research is needed to confirm these findings, existing evidence suggests that a longer recovery period may be needed between resistance and sprint exercises in training routines in order to maximize training adaptation.

Muscle Hyperplasia

It has been generally understood that muscle fiber number is fixed from birth and that skeletal muscle growth is a result of hypertrophy of existing muscle fibers. However, a number of studies have suggested that high-intensity resistance training may cause muscle hyperplasia (Gonyea 1980a, 1980b; Gonyea et al. 1986; Ho et al. 1980). Most of the earlier studies, which used animal models, met with criticism concerning the methods of data analysis (Gollnick et al. 1981, 1983). However, later studies that accounted for those concerns were still able to demonstrate hyperplasia of skeletal muscle subsequent to muscle overload (Alway, Winchester, et al. 1989; Gonyea et al. 1986).

The use of an animal model invoked further criticism directed at those who proposed skeletal muscle hyperplasia. The magnitude of hypertrophy seen in humans does not occur in many of the animal species (McArdle, Katch, and Katch 2010). Thus, for animals, muscle hyperplasia may be an important compensatory mechanism for combating muscle overload. Interestingly, MacDougall and coauthors (1982) and Tesch and Larson (1982) reported that elite bodybuilders had a greater number of muscle fibers than trained control subjects. These investigators suggested that the greater fiber number seen in the bodybuilders was attributable to years of high-intensity resistance training. However, these results were never duplicated in ensuing studies (MacDougall et al. 1984).

If muscle hyperplasia does occur, it is thought to be either through the development of new fibers from **satellite cells** (Antonio and Gonyea 1993; Appell, Forsberg, and Hollmann 1988) or through longitudinal splitting of existing muscle fibers (Antonio and Gonyea 1993; Ho et al. 1980; Gonyea et al. 1986). Satellite cells (located between the basement membrane and the plasma membrane) are thought to proliferate and grow to a **myoblast** and eventually **myotubes** that may develop into new muscle fibers. The myotube may also fuse with existing muscle fibers and remain incomplete along its length, leading to the incorrect impression of a split fiber (Appell, Forsberg, and Hollmann 1988). With longitudinal splitting, a hypertrophied muscle fiber that has reached some predetermined maximal ceiling of growth is thought to split into two or more smaller daughter cells through a process of lateral budding (Antonio and Gonyea 1994). There does not appear to be any convincing support for the occurrence of muscle hyperplasia in humans. However, conflicting results still make this issue controversial and its research potential appealing.

Fiber-Type Conversions

As mentioned earlier in the chapter, the proportion of type I to type II muscle fibers appears to be genetically determined and their expression set early in life. A number of studies have examined whether conditioning programs can alter the proportion of type I to type II muscle fibers. Some studies have suggested that aerobic training may be able to increase the percentage of type I fibers (Howald et al. 1985; Simoneau et al. 1985), while others have reported increases in type II fiber proportion after sprint training (Jansson, Sjodin, and Tesch 1978; Jansson et al. 1990). However, the overwhelming majority of investigations have not shown any alterations in fiber-type composition as a consequence of conditioning programs. It is generally believed that only fiber-type transformations within a fiber type can be accomplished through training.

High-intensity resistance training appears to be a potent stimulus for transformation of the type IIx or IIb subtype to the type IIa fiber subtype (Campos et al. 2002; Staron et al. 1989, 1991, 1994; Kraemer, Patton, et al. 1995). It has been reported that most of the type IIb fibers were converted to type IIa fibers after 20 weeks of resistance training (Staron et al. 1991). This is similar to the type II fiber conversions previously thought to be associated with aerobic exercise training (Staron and Hikida 1992). Kraemer, Patton, and colleagues (1995) have also demonstrated skeletal muscle fiber subtype transformations from IIb to IIa in subjects performing high-intensity resistance training and in those performing a combined high-intensity resistance training and endurance training program. Subjects who performed only endurance exercises also tended to increase the proportion of type IIa fibers but significantly elevated their type IIc fibers. This would be expected given that the type IIc fibers are the most oxidative of the type II subtypes.

Fiber subtype transformations appear to occur rapidly (within 2 weeks) during participation in physical conditioning programs. These adaptations, however, may be transient. During periods of inactivity or detraining, a transformation of fast-twitch fiber subtypes from type IIa back to type IIb is observed (Staron et al. 1991). A return to training will result in a fiber-type transformation back to its trained state in a relatively shorter period of time. These studies highlight the dynamic nature of skeletal fiber transformations.

Summary

This chapter presents a basic overview of muscle structure, motor units, muscle contraction, and muscle fiber types. Neuromuscular adaptations to training are specific to the type of training program (strength or endurance training). Initial improvements in strength are primarily associated with neurological adaptations, whereas further increases in strength are more dependent on increases in the cross-sectional area of the muscle. These increases are thought to occur either as a result of hypertrophy of existing muscle fibers or perhaps (although this is controversial) as a result of the splitting of muscle fibers (muscle hyperplasia). Finally, fiber-type transformations may be possible within a subtype, but conversion between fiber types (e.g., type I to type II) is not possible.

REVIEW QUESTIONS

1. Describe the role of the myofilaments during muscle contraction.
2. Explain the adaptive ability of muscle fiber types during intensive training sessions.
3. What types of neurological adaptations are typically observed during initial stages of resistance exercise?
4. What are the physiological consequences of performing both resistance and endurance training?
5. Explain the importance of exercise intensity in stimulating muscle growth.
6. What is the difference between muscle hypertrophy and muscle hyperplasia?

chapter 2

Endocrine System and Exercise

CHAPTER OBJECTIVES

After reading this chapter you should be able to do the following:

- Understand the difference between steroid and peptide hormones.
- Describe the regulation of hormone secretion.
- Explain the specific role of various hormones that are involved in muscle growth, metabolism, and fluid regulation.
- Understand the acute hormonal response to exercise stress for the various hormones discussed.
- Understand chronic changes in hormonal secretion patterns as they relate to different exercise training programs.

Hormones are chemical substances that circulate in the blood and interact with organs in the body to help combat various stresses. The primary role of hormones is to maintain internal equilibrium (**homeostasis**). Most hormones are synthesized in endocrine glands located throughout the body. Upon stimulation, the glands secrete their hormones into the surrounding extracellular space. The hormones then diffuse into the circulatory system and are transported to their respective target areas to perform their designated function. This chapter provides an overview of the endocrine system. It presents general information regarding the hormone–receptor interaction and how this interaction can be enhanced or reduced depending on various physiological changes within the body. Specific discussions focus on hormones related to growth, metabolism, and fluid regulation. In addition, the opioid hormones and their role in exercise are examined, as well as the acute and chronic response to exercise for each of the hormones reviewed.

Overview of the Endocrine System

Hormones influence the rate of specific cellular reactions by changing the rate of protein synthesis or **enzyme** activity and by inducing secretion of other hormones. In addition, hormones can facilitate or inhibit uptake of substances by cells. For example, **insulin** facilitates the uptake of **glucose** into the cell, and **epinephrine** inhibits glucose uptake to increase its concentration in the circulation.

Hormones can stimulate the production of enzymes or activate inactive enzymes. They can also combine with an enzyme to alter its shape (**allosteric modulation**), which will cause either an increase or a decrease in the effectiveness of the enzyme.

Regulation of Hormone Secretion

For hormones to function properly, their secretion rate must be precisely controlled. A signal needs to be received that triggers the necessary steps for hormone secretion. The initial step is the detection of an actual or threatened homeostatic imbalance. This imbalance must be able to activate a **secretory apparatus** (e.g., the **endocrine gland**), resulting in hormone secretion. The circulating hormone interacts with its target organ or tissue and exerts its effect. Once the hormonal effect has occurred, the hormonal signal has to be turned off and the hormone removed from the circulation. Finally, the secretory apparatus must replenish the hormone in its secretory cells. The regulation of hormone secretion is depicted in figure 2.1.

The secretion of most hormones is regulated by **negative feedback**, meaning that some consequence of the hormone secretion acts directly or indirectly on the secretory apparatus to inhibit further secretion. This type of secretory mechanism is self-limiting. **Positive feedback** mechanisms are rare in endocrine regulation. During this type of regulation, some consequence of the hormonal secretion causes an augmented secretory drive. Rather than being self-limiting, the stimulus for triggering hormonal secretion becomes stronger. An example of positive feedback is the release of **oxytocin** from the posterior pituitary gland caused by dilation of the uterine cervix during childbirth. The oxytocin causes a greater dilation that in turn creates a greater stimulus for further oxytocin release.

Changes in Circulating Hormonal Concentrations

Increases in the concentration of hormones can be attributed to a number of different physiological mechanisms. Exercise or other physical or psychological stresses appear to be potent stimulators in elevating the secretory patterns of hormones. Fluid volume shifts, changes in clearance rates, and venous pooling of blood are additional mechanisms that may increase circulating concentrations of hormones. Regardless of the mechanism, there is an increased potential for interaction with the receptor of the target tissue, leading to the desired cellular response.

Receptors are found in all types of cells within the body, and each hormone reacts with its specific receptor. The interaction of the hormone with its receptor can be explained by a theory called the **lock-and-key theory** (Kraemer 1994). The receptor is the lock, and the hormone is the key. There is some **cross-reactivity**, meaning that there may be more than one hormone that can bind with the receptor (this is depicted in figure 2.2, in which more than one hormone has an ability to cross-react with receptor C). When this occurs, the resulting biological actions are different from those induced by the primary hormone.

It is the hormone–receptor complex that results in delivery of a message to the cell nucleus for either inhibition or facilitation of **protein synthesis**. The number of receptors available for interactions with circulating hormones is considered another mechanism for initiating cellular action. Hormonal receptors are dynamic in that they also respond to physiological demand. They may increase in number to meet the demand of a rise in the circulating concentration of hormones. Such an increase in receptor number is termed **upregulation**. Similarly, the number of receptors can be decreased if adaptation is no longer possible or to prevent an overresponse by persistently increasing hormone levels. This adaptation is called **downregulation**. This type of control on the part of the receptor is as dramatic as the changes in hormonal secretory patterns. The endocrine glands and the hormones they secrete are shown in table 2.1. The normal blood concentrations of various hormones are listed in table 2.2.

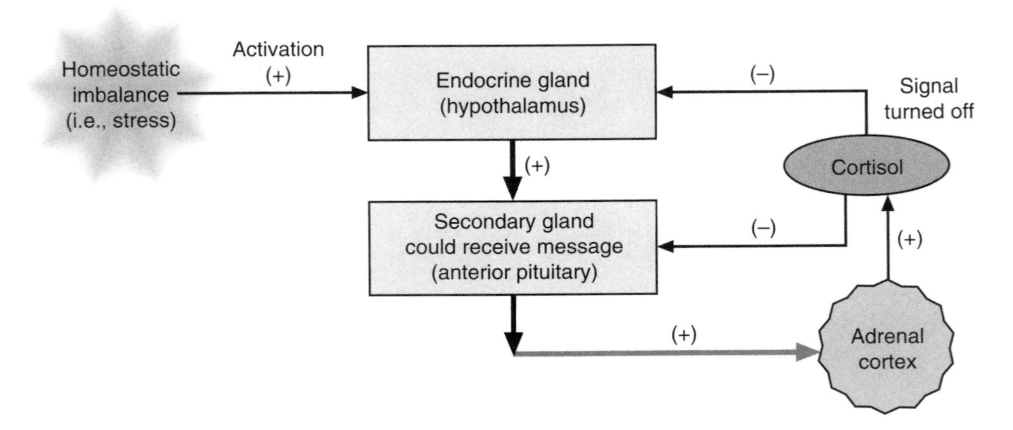

Figure 2.1 Regulation of hormone secretion.

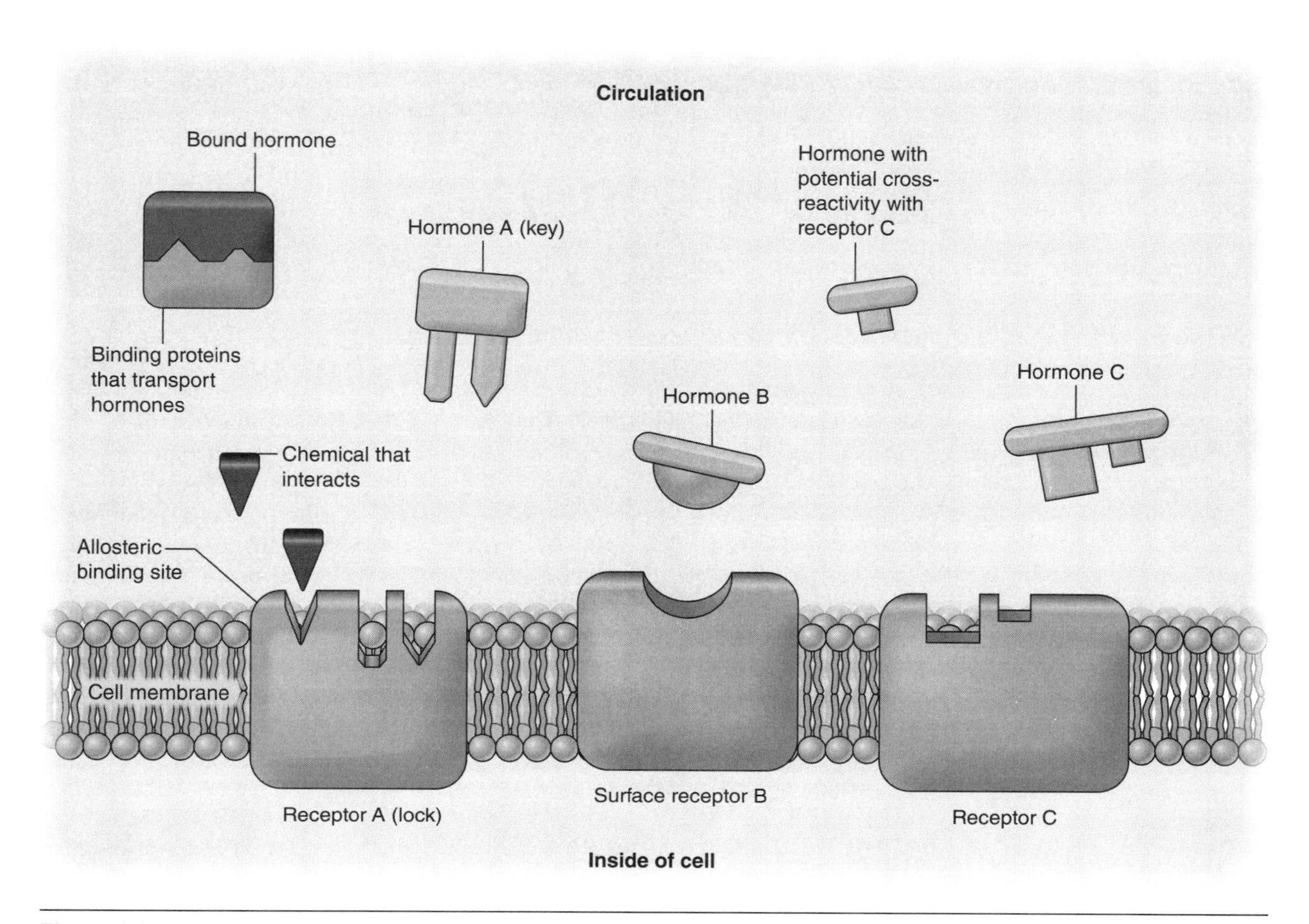

Figure 2.2 Hormone–receptor interaction.

Table 2.1 Endocrine Glands and Hormones

Gland	Hormone	Primary physiological action
Anterior pituitary	Growth hormone (GH)	Stimulates tissue growth, mobilizes fatty acids for energy, inhibits carbohydrate metabolism
	Proopiomelanocortin family: includes adrenocorticotropic hormone (ACTH)	Stimulates secretion of glucocorticoids and other adrenal hormones
	Endorphins, enkephalins, and dynorphins	Cause analgesia and postexercise euphoria
	Luteinizing hormone (LH) Follicle-stimulating hormone (FSH)	Act together to stimulate production of gonadal hormones (testosterone, estrogen, and progesterone)
	Thyroid-stimulating hormone (TSH)	Stimulates production and secretion of the thyroid hormones from thyroid gland
	Prolactin	Stimulates milk production in mammary glands
Posterior pituitary	Antidiuretic hormone (ADH), also called vasopressin	Stimulates reabsorption of water by kidneys
	Oxytocin	Stimulates uterine contraction and milk secretion in lactating breasts

(continued)

Table 2.1 *(continued)*

Gland	Hormone	Primary physiological action
Thyroid gland	Thyroxine and triiodothyronine	Stimulate metabolic rate and regulate cell growth and activity
Parathyroid gland	Parathyroid hormone	Increases blood calcium and lowers blood potassium
Pancreas	Insulin	Promotes glucose transport into cell, protein synthesis
	Glucagon	Promotes hepatic glucose release, increases lipid metabolism
Adrenal gland (adrenal cortex)	Glucocorticoids (cortisol)	Promote protein catabolism, stimulate conversion of protein into carbohydrates, promote lipid metabolism
	Mineralocorticoids (aldosterone)	Promote reabsorption of sodium and water by kidneys
Adrenal gland (adrenal medulla)	Catecholamines (epinephrine and norepinephrine)	Increase cardiac output, regulate blood vessels, increase glycogen catabolism and fatty acid metabolism
Liver	Insulin-like growth factors	Increase protein synthesis in cells
Ovaries	Estrogen Progesterone	Stimulates development of female sex characteristics
Testes	Testosterone	Stimulates growth and protein anabolism, development and maintenance of male sex characteristics

Table 2.2 Typical Adult Serum Concentrations in Both Système Internationale (SI) and Conventional Units

Variable	SI units	Conventional units
Testosterone Men Women	 10 to 35 nmol/L <3.5 nmol/L	 3 to 10 ng/ml <1 ng/ml
Cortisol	50 to 410 nmol/L	2 to 15 µg/dl
Growth hormone Men Women	 0 to 5 µg/L 0 to 10 µg/L	 0 to 5 ng/ml 0 to 10 ng/ml
Insulin-like growth factor 1 Men Women	 0.45 to 2.2 kU/L 0.34 to 1.9 kU/L	 0.45 to 2.2 U/ml 0.34 to 1.9 U/ml
Insulin (fasting)	35 to 145 pmol/L	5 to 20 µU/ml
Glucagon	50 to 100 ng/L	50 to 100 pg/ml
Epinephrine (resting, supine)	170 to 520 pmol/L	30 to 95 pg/ml
Norepinephrine (resting, supine)	0.3 to 2.8 nmol/L	15 to 475 pg/ml
Antidiuretic hormone	2.3 to 7.4 pmol/L	2.5 to 8.0 ng/L
Aldosterone	<220 pmol/L	<8 mg/dl
Thyroxine (T_4)	51 to 42 nmol/L	4 to 11 µg/dl
Triiodothyronine (T_3)	1.2 to 3.4 nmol/L	75 to 220 ng/dl
Calcitonin	<50 ng/L	<50 pg/ml
Parathyroid hormone	10 to 65 ng/L	10 to 65 pg/ml

From Wilson and Foster 1992; Young 1987.

Types of Hormones

There are three main categories of hormones: **steroids**, **peptides**, and **amines**. Steroid hormones (e.g., **testosterone**, **cortisol**) are synthesized from circulating cholesterol. They are also fat soluble and can diffuse across the cell membrane. Once across the membrane, the steroid hormone binds with its receptor, most probably within the cytoplasm of the cell, to form the **hormone–receptor complex**. The hormone–receptor complex then transports itself to the nucleus of the cell, where its message is delivered, transcribed, and translated into action.

Peptide hormones (e.g., **growth hormone**, insulin) are made up of **amino acids** and bind to their receptors located on the cell membrane. Because peptide hormones are unable to cross the cell membrane, they must rely on secondary messengers to send their message to the cell nucleus. When the hormone binds with its receptor within the membrane, it triggers the production of **cyclic AMP** (adenosine monophosphate) from ATP (adenosine triphosphate). This reaction is catalyzed by the enzyme **adenylate cyclase**. Cyclic AMP serves as the secondary messenger, activating a cascade of intracellular events and resulting in the cellular response.

Amine hormones are characterized by an amine ring. Since amine hormones are composed of amino acids, they are often classified as peptide hormones. These hormones are found as either hormones or neurohormones, meaning that they are secreted by an endocrine gland (e.g., adrenal gland) or nerve endings. Some of these compounds also function as a neurotransmitter. The amino acid tyrosine is generally the precursor for the synthesis of the amine hormones. However, phenylalanine can also be converted to tyrosine and subsequently synthesized into the amine hormones. The amine hormones are metabolized very quickly in circulation and thus exert their effect quite rapidly. Examples of amine hormones include epinephrine, norepinephrine, and dopamine. As with peptide hormones, they require a membrane receptor.

Hormonal Transport and Binding Proteins

When hormones are produced and secreted into the circulation, they can be released either free or bound to a **binding protein** (figure 2.3). In order to exert its biological effect, a hormone must be free. However, a problem can occur—the hormone can be metabolized or degraded as it travels through the circulatory system. Numerous factors can prevent hormone molecules from ever reaching their targets because of degradation. The time it takes for a hormone to be partially metabolized in the circulation, or for half of it to be degraded, is called its **half-life ($T_{1/2}$)**. Some hormones have a half-life measured in seconds, while the half-lives of others are measured in minutes or hours. In a process that enables a hormone to be preserved for longer periods, the hormone binds to a protein that provides protection from degradation and assists in its transport to the target cell (Mendel 1989). Steroid and thyroid hormones are bound to these transport proteins, while amines and protein hormones are not. Albumin, a binding protein made in the liver that helps to maintain blood volume in the arteries and veins, can bind numerous different hormones but does not exhibit a high affinity (attraction) to any of them. In any case, the hormone–binding protein complex can move through the circulation without the hormone's

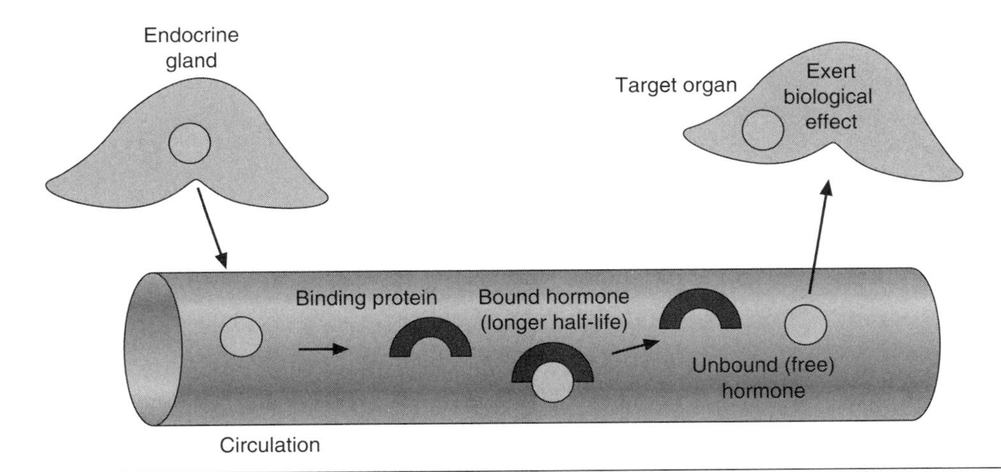

Figure 2.3 The role of binding proteins.

becoming degraded. The problem with this system is that the hormone is unable to bind to its target tissue and exert its physiological effect until it is released from the binding protein: Only when the hormone becomes free is it considered biologically active. For example, only approximately 1% to 2% of the total testosterone in the circulation is free testosterone, or not bound to its binding protein.

Hormones and Exercise

Exercise has been shown to be a potent stimulus to the endocrine system. The hormonal response to an acute exercise session suggests that hormones may be involved in the recovery and remodeling processes that occur after exercise. The exercise stimulus has an important role in the hormonal secretion pattern. Variables such as intensity of exercise, volume of exercise, rest intervals, choice of exercise, and recovery status of the muscle appear to influence the hormonal response.

The mechanisms of hormonal interaction with the remodeling of muscle tissue are based on several factors. The acute increase in hormonal concentration caused by the exercise stimulus allows for a greater interaction between the hormone and its receptors. Since the adaptations to exercise (particularly resistance exercise) are **anabolic** in nature, the recovery mechanisms involve tissue repair and remodeling (Kraemer 1992b). In instances in which training **intensity** or **volume** exceeds an individual's ability to recover, a situation of overtraining or overwork can occur, resulting in a greater **catabolic** effect. The hormonal response will either repair or remodel muscle tissue or perhaps impede this process.

Hormonal mechanisms may respond differently between trained and untrained individuals (Hakkinen et al. 1989). Furthermore, hormonal mechanisms may not be operational in both males and females (e.g., testosterone). In addition, program design, genetic predisposition, fitness level, training experience, and adaptational potential all seem to affect the endocrine mechanism for maintaining or improving muscle size and strength (Kraemer 1992a; Hakkinen et al. 1989).

This section discusses both the acute and chronic training responses of the hormones thought to be principally involved in the production of **muscle force** and muscle hypertrophy. It also addresses the role of the catabolic hormones in such mechanisms.

Testosterone

Testosterone is an androgen, a steroid hormone that has masculinizing effects. It is also anabolic because of its role in the maintenance and growth of muscle and bone tissue. Most of the circulating testosterone is produced in the testes, while small amounts are produced in the adrenal glands. Circulating testosterone binds to an androgen receptor located in the cytoplasm of skeletal muscle cells. This hormone–receptor interaction results in a migration of the hormone–receptor complex to the nucleus of the cell, leading to an increase in protein synthesis (see figure 2.2).

Testosterone is secreted primarily in the interstitial cells of Leydig in the testes. Although several other steroid hormones with anabolic–androgenic properties are produced in the testes (e.g., dihydrotestosterone and androstenedione), testosterone is produced in far greater quantities. In addition, testosterone and the other male sex hormones are secreted in significantly smaller amounts from the adrenal glands and ovaries. Circulating concentrations of testosterone range from 10.4 to 34.7 nmol/L in males and 0.69 to 2.6 nmol/L in females (Chattoraj and Watts 1987).

The biosynthesis of testosterone, and all other steroid hormones, begins with cholesterol and its conversion via a multistep process (figure 2.4). The primary area of testosterone synthesis is within the Leydig cells of the testes, but testosterone is also synthesized in small quantities within the adrenal cortex of the adrenal gland. The rate-limiting step for testosterone synthesis is the side-chain cleavage of cholesterol to form pregnenolone. The conversion of pregnenolone to testosterone occurs through one of two pathways. In the progesterone pathway (known as the Δ-4 pathway), pregnenolone is metabolized to progesterone by the enzyme 3-β-hydroxysteroid dehydrogenase and an isomerase (an enzyme that catalyzes the structural rearrangement of isomers). Progesterone is then converted to 17-α-hydroxyprogesterone by 17-α-hydroxylase and C17:C21-lyase to form androstenedione, which next is converted to testosterone via reduction of the 17-keto group by 17-β-hydroxysteroid dehydrogenase. The dehydroepiandrosterone (DHEA) pathway (known as the Δ-5 pathway) metabolizes pregnenolone to 17-alpha-hydroxypregnenolone, which is then converted to DHEA; DHEA is subsequently converted to 5-Δ-androstenediol by C17:C21-lyase. Understanding the pathway for the biosynthesis of testosterone is important in relation to many of the ergogenic aids and performance-enhancing drugs that are available to and used by athletes (see chapter 20). In the Δ-5 pathway, the double bond is between the fifth and sixth carbon atoms (i.e., C5 and C6), while in the Δ-4 pathway the double bond is between C4 and C5. In humans, the Δ-5 pathway appears to be the predominant pathway resulting in the synthesis of testosterone (Hedge, Colby, and Goodman 1987).

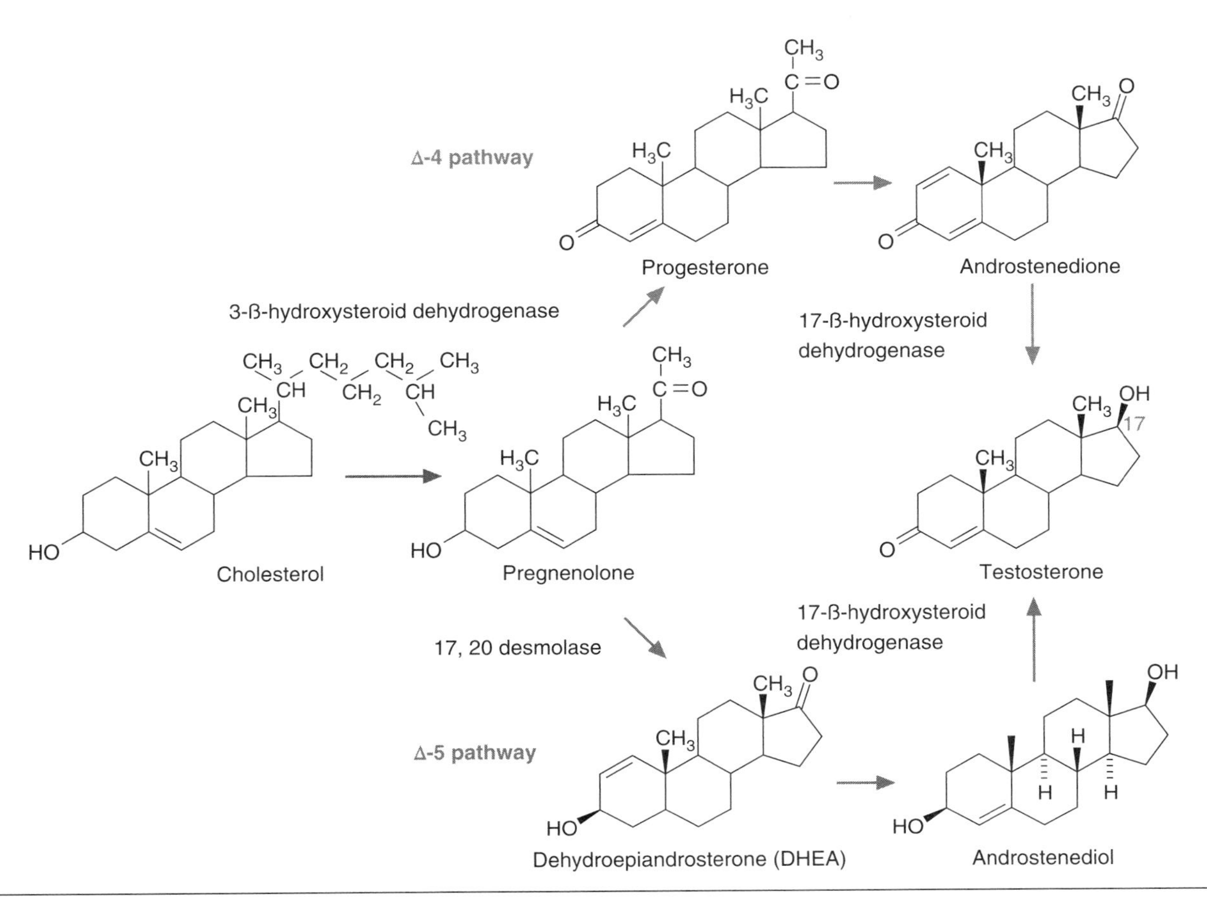

Figure 2.4 Synthesis of testosterone.

Little testosterone is produced during childhood until ages 10 to 13, when a rapid increase occurs under the stimulus of **luteinizing hormone** (a gonadotrophic hormone secreted from the anterior **pituitary gland**). This testosterone surge is responsible for the distinguishable characteristics and maturity of the male sex organs and the development of secondary sexual characteristics. The physiological roles of testosterone are as follows:

- Increase in protein synthesis, resulting in muscle growth
- Development and maturation of male sex organs
- Development of secondary sexual characteristics
  - Increase in body hair
  - Development of masculine voice
  - Development of male pattern baldness
  - Development of libido
  - Control of spermatogenesis
  - Aggressive behavior
- Interaction on epiphyseal growth plates, contributing to the longitudinal growth of long bones and subsequent cessation of growth caused by fusion of the epiphyseal plates
- Increase in secretions of sebaceous glands, contributing to acne
- Possible role in glycogen synthesis

Acute Exercise Response

A single training session of resistance exercise has been demonstrated to significantly increase the peripheral concentration of testosterone above resting levels in males (Hakkinen, Pakarinen et al. 1987; Hakkinen et al. 1988; Kraemer, Marchitelli, et al. 1990). However, no significant difference in the response of testosterone to a single exercise session was seen in a cross-sectional comparison between (a) males and females with little resistance training experience and (b) males and females who had trained with weights for at least 2 years (Fahey et al. 1976). Kraemer and colleagues (1992), on the other hand, reported that male weightlifters with more than 2 years of training experience had a significantly greater testosterone response to an exercise session than weightlifters with less than 2 years of lifting experience. Exercise response patterns of testosterone also appear to be related to the design of the exercise program. Single component variables (e.g., rest intervals, intensity) have been demonstrated to have a significant effect on the testosterone response to an acute bout of resistance exercise.

Significantly higher testosterone concentrations have been observed when rest periods between sets were reduced (3 min vs. 1 min between sets) or when the intensity of exercise was reduced (5RM [**repetition maximum**] vs. 10RM) (Kraemer, Marchitelli, et al. 1990). This may help explain the variability in muscle hypertrophy seen with different resistance training programs and the combination of high volume and short rest periods in the training programs of bodybuilders. Further discussion of repetition maximum appears in chapter 8.

Testosterone appears to show a biphasic response to an acute bout of aerobic exercise, which seems to be dependent on the duration of the exercise. In studies performed primarily on male subjects, exercise of relatively short duration (e.g., 10-20 min) did not appear to increase plasma testosterone concentrations (Bottecchia, Bordin, and Martino 1987; Galbo et al. 1977; Sutton et al. 1973). However, as exercise reaches 20 to 30 min in duration, significant elevations in testosterone are observed (Wilkerson, Horvath, and Gutin 1980; Hughes et al. 1996). As exercise continues, a biphasic response is seen (figure 2.5). Testosterone levels continue to increase as exercise is prolonged and then begin to decline toward baseline levels before exercise is completed. As exercise progresses past 3 h, significant declines (below resting levels) in testosterone have been reported (Dessypris, Kuoppasalmi, and Adlercreutz 1976; Guglielmini, Paolini, and Conconi 1984; Schurmeyer, Jung, and Nieschlag 1984; Urhausen and Kindermann 1987). These reduced levels may persist for 48 h after exercise (Urhausen and Kindermann 1987).

An increase in testosterone concentrations during an acute bout of anaerobic exercise appears also to depend on the duration. Increases in testosterone concentrations have been reported after both 90 s (Kindermann et al. 1982) and 2 min (Kuoppasalmi et al. 1980) of intermittent anaerobic exercise in male runners. However, exercise of shorter duration (a noncontinuous 15 s sprint) may not produce any change from resting levels (Kuoppasalmi et al. 1980). Interestingly, the immediate postexercise response of testosterone to anaerobic exercise may not be as important as its response during the recovery period. Decreases below resting levels have been seen after a 2 min bout of high-intensity anaerobic exercise (Kuoppasalmi et al. 1980). The physiological benefit or harm that may result from the depressed testosterone levels during recovery is not known.

Long-Term Response to Exercise

The relationships between strength, mass, and testosterone levels are not entirely understood. It is assumed that high resting concentrations of testos-

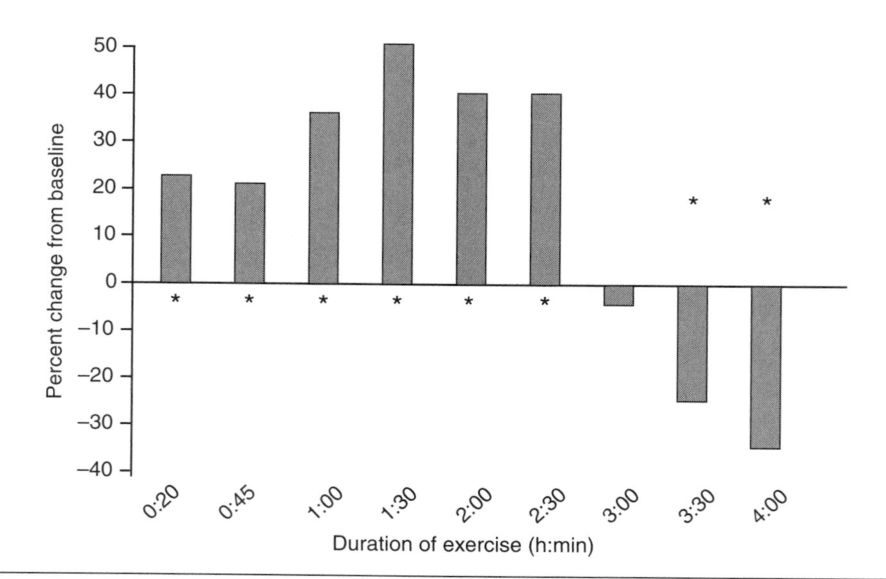

Figure 2.5 Response of testosterone to submaximal exercise of different durations. * = Significant differences from preexercise values, $p < 0.05$.

Data from Dessypris et al. 1976; Galbo et al. 1977; Guglielmini et al. 1984; Kuoppasalmi et al. 1980; Remes et al. 1980; Urhausen et al. 1987; Wilkerson et al.1980.

In Practice

Can Testosterone Influence Power Performance?

The anabolic effect of testosterone on muscle growth is well known, but its relationship to strength and power performance is less completely understood. Cardinale and Stone (2006) were interested in examining the relationship between testosterone concentrations and vertical jumping performance in elite male and female athletes. Considering that testosterone is produced in very small quantities in females, the investigators also examined whether these relationships varied by sex. Seventy competitive athletes (22 women and 48 men) participating in track and field (sprinters), handball, volleyball, and soccer at the national and international levels were examined. All athletes performed three countermovement vertical jumps on a force plate, and the highest vertical jump was recorded. Fasting testosterone concentrations were also measured. Resting testosterone concentrations in men (6.46 ± 0.37 ng/ml) were more than 10-fold greater than those seen in the women (0.62 ± 0.06 ng/ml). Vertical jump height of the female athletes was 86.3% of that in the male athletes. A significant correlation (r = 0.61) was reported between resting testosterone concentrations and vertical jump performance, indicating that 37% of the variance in testosterone concentrations was associated with vertical jump performance. When the relation between testosterone concentrations and vertical jump was compared within each sex separately, significant correlations were still seen. The correlation among men ($r = 0.62$, $p = 0.001$) was greater than among the female athletes (0.48, $p = 0.01$). The potential role that testosterone has in acute neuromuscular performance may be related to an enhancement of neurotransmission at the neuromuscular junction (Blanco, Popper, and Micevych 1997) or an increase in the number of acetylcholine receptors (Bleisch, Harrelson, and Luine 1982) or both. In addition, it has been suggested that testosterone changes myosin heavy chains toward more of the type II phenotype (Lyons, Kelly, and Rubenstein 1986). The difference between the sexes in the association between testosterone concentration and jump performance may be related to differences in muscle sensitivity to testosterone.

terone may enhance or facilitate the building of lean tissue (Hickson et al. 1994). This has been the primary reason for the widespread use of anabolic steroids by power athletes and bodybuilders. The potential of weightlifters to alter androgen levels through prolonged resistance training is unclear. Six months of resistance training in noncompetitive lifters did not alter resting testosterone levels (Hakkinen et al. 1985).

Even a full year of strength training was unable to alter resting testosterone concentrations in elite weightlifters, although strength increases were observed (Hakkinen, Pakarinen, et al. 1987). However, after 2 years of resistance training, elite weightlifters were able to significantly increase their resting testosterone concentrations while also improving their strength (Hakkinen et al. 1988). It is possible that changes in resting testosterone concentrations may be a reflection of an advanced adaptive strategy to increase force capability in subjects who have little potential for change in muscle hypertrophy (e.g., highly strength-trained athletes) (Kraemer 1992a).

Low resting levels of testosterone are frequently observed in endurance-trained athletes (Ayers et al. 1985; Hackney, Sinning, and Bruot 1988; Wheeler et al. 1984). Resting levels that are 31% (Hackney, Sinning, and Bruot 1988) and 40% (Ayers et al. 1985) below the level in sedentary controls have been reported in endurance athletes. The depressed levels of testosterone seen in these athletes may be insufficient to stimulate skeletal muscle growth and may also make it difficult to counteract the catabolic effects of **glucocorticoids** on skeletal muscle.

Growth Hormone

Growth hormone (GH) is a **polypeptide** hormone secreted from the anterior pituitary gland. It is also called somatotropin. Its secretion and release are controlled by neurotransmitters of the central nervous system. Physiological stimuli such as deep sleep, diet, hypoglycemia, nutritional intake, and stress (including exercise) can all stimulate a GH response. The primary physiological role of GH is its involvement in the growth processes of skeletal muscle and other tissues in the body. The actions of GH are mediated to a certain extent by secondary hormones known as insulin-like growth factors (IGF). The basic physiological actions of GH are as follows:

- Increase in protein synthesis
- Increase in amino acid transport across cell membrane
- Growth and development of bones
- Reduction of glucose utilization
- Decrease in glycogen synthesis
- Increase in utilization of fatty acids
- Increase in lipolysis
- Metabolic sparing of glucose and amino acids
- Collagen synthesis
- Stimulation of cartilage growth

The most commonly measured form of GH is the 191-amino acid (22 kD) isoform. However, there are other GH variants, including a 20 kD form that is produced from a gene deletion of 14 amino acids as well as other posttranslational isoforms of GH that have unknown physiological significance (Baumann, MacCart, and Amburn 1983). The release of GH is stimulated by growth hormone–releasing hormone and is inhibited by somatostatin, both hypothalamic hormones. However, many other factors affect GH regulation, most of which use these hypothalamic hormones as a common path. Inhibitory influences include obesity, ingestion of a carbohydrate-rich diet, and several pharmacological agents, such as β-2 adrenergic agonists. The release of GH from the anterior pituitary is pulsatile, meaning that it is not constant but occurs in bursts (Hymer et al. 2005). The largest peak GH secretion occurs about an hour after the onset of sleep, with subsequent smaller peaks occurring during the rest of the sleep period (Nindl, Hymer, et al. 2001).

The two primary physiological roles associated with GH secretion involve metabolism and bone and muscle growth. Its major metabolic effects include increasing fat mobilization through lipolysis (increased triglyceride hydrolysis to free fatty acids and glycerol, reduction in fatty acid re-esterification) and an augmentation of lipid oxidation. When GH secretions are deficient, an increase in body fat, adipocyte size, and lipid contents can be noted (Rogol 2011). Direct anabolic actions of GH include facilitating amino acid transport in muscle, leading to protein synthesis and an increase in nitrogen balance. However, it may also have an indirect effect on increasing the amino acid pool through its metabolic role by enhancing fatty acid oxidation. This may have a sparing effect on amino acid utilization (Norrelund et al. 2003). In addition, GH may influence muscle growth through stimulating protein synthesis via its activation of downstream effects from the mammalian target of rapamycin (mTOR) pathway; this stimulates the dephosphorylation and activation of specific enzymes involved in the translation, initiation, and elongation of the proteins synthesized (Hayashi and Proud 2007). A further role of GH in skeletal muscle growth is related to its ability to increase myonuclear number and to facilitate the fusion of myoblasts with myotubes (Sotiropoulos et al. 2006). GH exerts its growth-promoting effects in bone by its actions at the epiphysis and the differentiation of the osteoblasts (Reiter and Rosenfeld 2008). It shares some of these effects with IGF-1, and both are required for optimal linear growth (Rogol 2010).

Acute Response to Exercise

The acute GH response to a resistance exercise session is related to specific component variables of the training program. Both volume and intensity of training appear to be important factors in eliciting a GH response. At light exercise loads (28% of 7RM), no changes in GH concentrations are observed (Van Helder, Radomski, and Goode 1984). With a more moderate exercise intensity (10RM), a significant increase in GH is seen (Kraemer, Marchitelli, et al. 1990). This increase is significantly greater than that seen after a resistance training program of higher intensity (5RM). The most dramatic increases in GH were observed when exercise intensity was moderate (10RM) and the rest intervals were short (1 min) (Kraemer, Marchitelli, et al. 1990). The GH response, similar to that of testosterone as discussed previously, appears to be heightened when training programs of moderate intensity and short rest periods are used.

The volume of training also appears to be a potent stimulus in the GH response to exercise. Hakkinen and Pakarinen (1993) demonstrated a greater GH response (4.5-fold difference) in a high-volume, moderate-intensity training protocol (10 sets at 10RM) versus a low-volume, maximal-intensity training protocol (20 sets at 1RM). Another study demonstrated that a low-intensity, high-volume resistance training protocol (five sets of 15 repetitions at 60% of 1RM squat) increased the GH response over twofold more ($p < 0.05$) than a high-intensity, low-volume protocol (five sets of 4 repetitions at 90% of 1RM squat) (Hoffman, Im, et al. 2003). The greater acidosis (greater blood **lactate** concentrations) observed in these high-volume training programs most likely contributed to the elevated GH response. The importance of training volume in eliciting significantly greater GH responses has been confirmed in a number of other studies

(Kraemer, Marchitelli, et al. 1990; Craig and Kang 1994). Mulligan and colleagues (1996), controlling for the intensity of exercise, showed that a multiple-set training program (three sets at 10RM) elicited a greater GH response during an acute exercise session than a single-set design (one set at 10RM).

The GH response to resistance exercise in females also appears to be sensitive to changes in acute program variables (e.g., rest, intensity, volume of training) (Kraemer, Gordon, et al. 1991; Kraemer, Fleck, et al. 1993). However, these changes appear to be different from those observed in males and may be related to higher resting concentrations of GH in females during the early follicular stage of the menstrual cycle (Kraemer, Gordon, et al. 1991). Given the lack of any significant testosterone response in women, GH may have a more primary role in the anabolic adaptations in female skeletal muscle (Hakkinen, Pakarinen, et al. 1990; Kraemer, Gordon, et al. 1991; Kraemer, Fleck, et al. 1993).

Elevations in GH concentrations are typically reported during aerobic exercise, and these elevations are positively related to both the duration and the intensity of exercise (Bunt et al. 1986; Chang et al. 1986; Hartley et al. 1972; Karagiorgos, Garcia, and Brooks 1979; Sutton and Lazarus 1976; Van Helder et al. 1986). However, research on exercise intensity and increases in GH concentrations remains equivocal. Some researchers have suggested that a minimum exercise intensity may be needed to elicit increases in GH concentrations (Chang et al. 1986). Others have proposed that elevations in GH concentrations can occur without any change in blood lactate levels (Hansen 1973); and similar GH responses have been reported between continuous and intermittent exercise despite a greater lactate response during intermittent exercise (Lugar et al. 1992). It has been suggested that exercise above the lactate threshold needs to have a minimum duration (10 min), but the blood lactate levels cannot predict the amplitude and duration of the GH response (Felsing, Brasel, and Cooper 1992; Weltman et al. 1997). During repeat sprint activity (four sprints totaling 1000 m), GH concentrations were significantly elevated from resting concentrations during the activity and returned to baseline levels within an hour of recovery (Meckel et al. 2011). This was a quick elevation that was likely stimulated by a rapid rise in blood lactate concentrations.

Comparisons between the sexes have shown that the GH responses in male and female elite athletes are similar during a maximal exercise test (Ehrnborg et al. 2003). However, the maximal GH response occurred significantly earlier in the female athletes. Peak GH occurred immediately after exercise in the women but 15 min postexercise in the men. The GH response did not appear to be different in the female athletes during different stages of the menstrual cycle. GH concentrations at rest and postexercise are also reported to be higher in females than males (Ehrnborg et al 2003; Ubertini et al. 2008).

Long-Term Adaptations to Exercise

Significant increases in the GH response to exercise have been seen following resistance training programs (Hakkinen et al. 1985; Kraemer, Marchitelli, et al. 1990; Kraemer, Fleck, et al. 1993). However, resistance training does not appear to alter resting GH concentrations (Hakkinen et al. 1985; Kraemer et al. 1992). A year of exercise training at an exercise intensity above the lactate threshold has been shown to amplify the pulsatile release of GH at rest (Weltman et al. 1992), but the GH response to an acute exercise stimulus in trained subjects is ambiguous. Training has been reported to decrease (Hartley et al. 1972; Koivisto et al. 1982; Weltman et al. 1997), increase (Bunt et al. 1986), or not affect (Kjaer et al. 1988) the GH response to an exercise bout. A reduction in the GH response to exercise may occur within 3 weeks of training (Weltman et al. 1997). However, these changes may be related to a lower relative intensity used during the postexercise period. When trained subjects exercise at the same relative intensity, which accounts for improvements in performance, a greater GH response to the exercise stimulus is seen (Bunt et al. 1986). The pulsatile secretion of GH appears to be higher in elite athletes than nonelite or sedentary controls (Ubertini et al. 2008); this is seen in both male and female athletes.

Insulin-Like Growth Factors

Many of the effects of GH appear to be mediated through small polypeptide hormones called **insulin-like growth factors (IGF)** (Kraemer 1992b). The insulin-like growth factor system (also called **somatomedins**) is composed of the IGF-1 ligand (a 7.6 kD 70-amino acid polypeptide secreted from the liver), six binding proteins, an acid-labile subunit, and its two receptors. IGF is secreted in both hepatic and nonhepatic tissue. Originally, the liver was thought to be the only source of circulating IGF; however, it appears that IGF may also be secreted to a lesser extent in various tissues including skeletal muscle tissue (Deschenes et al. 1991). It has been reported that the synthesis of IGF-1 is regulated by GH release;

In Practice

Insulin-Like Growth Factor 1 as a Potential Metabolic Biomarker

Resting concentrations of IGF-1 have been shown to be a measure of physical fitness and health in young men (25 ± 5 years). Nindl and colleagues (2011) examined 846 healthy Finnish men on various fitness measures (peak $\dot{V}O_2$, leg and arm strength, and muscle endurance), health measures (blood lipids; blood pressure; waist circumference; drinking, smoking, and physical activity habits), and body composition. Subjects were divided into quintiles based on their IGF-1 concentrations. High concentrations of IGF-1 were reported to be associated with higher levels of fitness (both endurance and strength). In addition, IGF-1 concentrations were highest in young men who had the smallest waist circumference, lowest diastolic blood pressure, and lowest cholesterol and who did not smoke. No association was reported between IGF-1 concentrations and strength, triglyceride concentrations, or low-density lipoprotein (LDL) concentrations. The investigators concluded that IGF-1 was positively associated with aerobic fitness and muscular endurance, but not with measures of muscle strength. Nindl (2009) has also reported that the relationship between IGF-1 and body composition is a repeatable finding, as increases in fat-free mass are reflected by increased IGF-1 concentrations in female soldiers. Although the relationship between fat-free mass and IGF-1 concentrations may not be consistent among young males (Nindl et al. 2011), the IGF-1 system appears to be an important adjunct to the overall assessment of fitness and health and may offer prognostic information regarding body composition changes.

but exercise-related IGF-1 responses, especially at the local tissue level, appear to be independent of GH (Eliakim, Nemet, and Cooper 2005). Similar to GH, IGF-1 has multiple metabolic and hypertrophic roles, including insulin-like activity and direct stimulation of protein synthesis pathways. Somatomedins may also be involved in the growth processes of bone and connective tissue (Kraemer 1992b).

Exercise Response

The response of circulating IGF-1 to an acute resistance exercise bout is variable, with some investigations showing no change (Chandler et al. 1994; Consitt, Copeland, and Tremblay 2001; Kraemer, Aguilera, et al. 1995) and others reporting significant increases (Kraemer, Gordon et al. 1991; Kraemer, Marchitelli et al. 1990; Rubin et al. 2005). These differences may be due to the variability of study design and subjects' experience level, age, sex, and nutritional status. Kraemer, Gordon, and colleagues (1991) compared the responses of eight men and eight women to two different full-body acute resistance exercise protocols. One protocol used a 5RM load and 3 min rest periods, while the second used a 10RM load and 1 min rest periods plus higher total work than the first protocol. Both protocols were designed to control for load, rest period length, and total work (each protocol included a secondary workout surrounding the primary protocol to equalize total work). IGF-1 increased in response to both protocols in both sexes, but there were no significant differences between men and women. However, women had significantly higher resting values of IGF-1. This gives further support to the concept that women may rely on the IGF–GH axis to a greater extent than men as an anabolic signaller for strength gains whereas men probably rely more on testosterone.

Acute increases in IGF have been reported after high-intensity aerobic exercise (Cappon et al. 1994; De Palo et al. 2008; Ehrnborg et al. 2003). However, during short-duration, maximal-intensity exercise such as repeated sprints, both increases (Meckel et al. 2011) and no change from baseline (Stokes et al. 2010) have been reported. As previously noted, exercising IGF-1 levels are not under the control of GH, and increases may be due to the release of a muscle isoform of IGF-1. In addition, the long latency period (3-9 h) between GH-stimulated messenger RNA (mRNA) synthesis in hepatic tissue and the peak increase in IGF may be a factor in the lack of a consistent change with similar exercise stresses. Nindl and colleagues (2001) have also suggested that it is not the change in IGF-1 concentrations, but rather the manner in which IGF-1 is partitioned among its family of binding proteins, that is critical. The bioactive IGF-1 molecule is not bound to its binding protein. The greater the amount of IGF-1 that is free, the greater its bioactivity. However, some investigators have proposed that increases in IGF binding protein may serve to preserve IGF-1 bioavailability (Izquierdo et al. 2006). In addition, IGF-1 is highly dependent on nutritional status, which may account for some of the disparity between studies (Nemet et al. 2004), especially if diet is not controlled. Much research on the GH–IGF axis remains to be done.

Effect of Training

The effect of training on the IGF response remains equivocal. Kraemer, Aguilera, and colleagues (1995) reported a lack of change in the IGF response to an acute bout of resistance training in recreationally trained lifters, but others have suggested that trained lifters can increase IGF levels following resistance exercise (Rubin et al. 2005). This may reflect an ability of experienced lifters to tolerate higher-intensity workouts. In addition, experienced resistance-trained men are reported to have higher concentrations of IGF-1 at rest than untrained men (Rubin et al. 2005).

Chronic endurance training produces a biphasic response of IGF-1; decreases in the first few weeks of training are followed by an increase above pretraining values (Eliakim, Nemet, and Cooper 2005). In addition, cross-sectional studies have suggested that IGF-1 is highly correlated with $\dot{V}O_2max$ values, making it a possible biomarker for fitness status (Nindl and Pierce 2010; Nindl et al. 2011). Nindl and colleagues (2012) also reported that improvements in fitness were associated with body composition and fitness improvements.

Insulin

Insulin is a protein hormone secreted by the **β-cells** of the **islets of Langerhans** within the pancreas. The major function of insulin is to regulate glucose metabolism in all tissues except the brain. This is accomplished through facilitation of an increase in the rate of glucose uptake into both muscle tissue and fat cells. Glucose that is not used is converted into **glycogen**. If glycogen stores are full, excess **carbohydrates** are stored as **triglycerides** in adipose tissue. Insulin appears also to increase the rate of amino acid uptake by skeletal muscle and other tissue (Hedge, Colby, and Goodman 1987). This has important implications in regard to insulin's role in muscle remodeling, as it provides the needed nutrients to offset the rate of protein degradation within muscle tissue (Deschenes et al. 1991) and stimulate muscle growth (Florini 1985).

Exercise Response

Exercise appears to decrease the circulating concentrations of insulin. This is likely the result of the inhibitory effect of **catecholamines** on the β-cells of the pancreas. The reduction in insulin seems to be a function of the duration of exercise. As exercise duration lengthens, a greater decrease in insulin concentrations is seen (Galbo 1981; Koivisto et al. 1980). Insulin levels also appear to decrease during both mild and moderate exercise intensities. However, as the exercise stimulus approaches maximum intensity (e.g., 90% of $\dot{V}O_2max$), insulin concentrations may not decline (Galbo 1985). The decrease in insulin concentrations during exercise is likely controlled by other hormonal interactions (e.g., catecholamines). The action of insulin becomes more pronounced following exercise. A single bout of exercise enhances insulin sensitivity and skeletal muscle responsiveness to glucose uptake in exercised muscles (Richter et al. 1989). Thus, exercise-induced increased insulin sensitivity to glucose uptake serves to replenish depleted glycogen stores during the postexercise meal.

Effect of Training

Training appears to increase the sensitivity to insulin of both the skeletal muscle and the liver (Devlin, Calles-Escandon, and Horton 1986; Rodnick et al. 1987). Thus, less insulin is required to regulate blood glucose in trained individuals. Trained individuals also seem to show a less pronounced insulin reduction during exercise than untrained individuals (Bloom et al. 1976).

Glucagon

Glucagon is also produced in the pancreas, but in the α-cells. Similar to insulin, it is a polypeptide hormone, but it is a smaller molecule (29 amino acids long vs. 51 amino acids for insulin). Glucagon has a critical role in regulating blood glucose, but its effect is opposite to that of insulin. During periods of **hypoglycemia** (low blood glucose), glucagon secretion is stimulated to increase glucose concentrations in the blood; when glucose concentrations are elevated, glucagon secretion is inhibited. Inhibition of glucagon secretion becomes proportionally less as blood glucose concentrations decrease, and it disappears when blood glucose concentrations fall below 50 mg/dl (Goodman 1988). The α-cells are stimulated by several signals to increase glucagon production. Glucose concentrations in the blood appear to have a direct effect on α-cell stimulation (Marroqui et al.

2012), but secretion rates also increase with sympathetic stimulation. Sympathetic stimulation involves elevations of the catecholamines epinephrine and norepinephrine, which also respond to low blood glucose. A meal rich in protein stimulates glucagon secretion as well, likely to metabolize excess amino acids in the liver via gluconeogenesis (Goodman 1988).

Exercise Response

Data on the response of glucagon to various exercise stresses are limited. During endurance activity, the physiological role of glucagon is to maintain normal blood glucose concentrations. Although some scientists have suggested that circulating concentrations of glucagon are maintained during exercise (Hoene and Weigert 2010), this may be a function of exercise duration and possibly exercise intensity. Galbo (1985) suggested that glucagon concentrations appear to rise as duration of exercise increases. Glucagon concentrations were reported to remain unchanged following 60 min of cycle ergometer exercise performed at 45% of the subject's $\dot{V}O_2max$, but to increase 28% from resting levels when exercise intensity was elevated to 65% of $\dot{V}O_2max$ (Bergman et al. 1999). As exercise duration increases to 120 min (at 70% $\dot{V}O_2max$), plasma glucagon concentrations have been reported to increase to 53.9% from baseline (MacLaren et al. 1999). During high-intensity training such as sprint exercise on a cycle ergometer (130% $\dot{V}O_2peak$ until exhaustion), no changes in glucagon concentrations from baseline were observed (Harmer et al. 2006).

Effects of Training

Physiological adaptation involving glucagon secretion patterns in the context of chronic endurance or resistance training in healthy individuals has not been fully explored. Research to date suggests that a 7-week training program (training frequency of three times per week) does not alter the glucagon response to sprint exercise (Harmer et al. 2006). Considering the metabolic role of glucagon and the duration of high-intensity sprint exercise, these results would seem to be as expected. However, the glucagon response to endurance training (4 weeks) does suggest a training effect. Bergman and colleagues (1999) showed that postexercise glucagon concentrations were significantly lower after training at 65% $\dot{V}O_2max$.

Cortisol

Cortisol, a steroid hormone synthesized by and released from the **adrenal cortex** of the adrenal gland, is the primary glucocorticoid hormone found in humans. Its synthesis is stimulated by adrenocorticotropic hormone (ACTH), which is secreted by the anterior pituitary gland. The primary physiological function of cortisol is the stimulation of **gluconeogenesis** and the mobilization of fatty acids from body stores. Increases in cortisol appear to be stimulated by stress, diet, immobilization, inflammation, high-intensity exercise, and disease. The following are the physiological functions of cortisol.

- Conversion of amino acids to carbohydrates
- Increase in proteolytic enzymes
- Inhibition of protein synthesis
- Increase in protein degradation in muscle
- Stimulation of gluconeogenesis
- Increase in blood glucose concentrations
- Facilitation of lipolysis

Because cortisol is involved with protein degradation of skeletal muscle mass, it is considered a catabolic hormone. Much interest in comparing its response with that of testosterone has been generated. A ratio between the two hormones (testosterone/cortisol) has been used in an attempt to examine the anabolic/catabolic status of the body and to relate these measures to changes in performance (Kuoppasalmi and Adlercreutz 1985). Elevations of cortisol are considered a marker of catabolic activity in the muscle as well as a marker of physiological stress.

Exercise Response

The acute response of cortisol to a resistance training session appears to be related to the volume of training. In elite weightlifters performing 20 sets at 1RM, no increase in cortisol concentration was seen from its resting level (Hakkinen and Pakarinen 1993). However, when volume of training was increased (10 sets at 10RM), a significant increase was observed. Other studies have shown that postexercise cortisol concentrations are elevated in the initial stage of a resistance training program in novice weightlifters but remain at resting levels after several weeks of training (Hickson et al. 1994; Potteiger et al. 1995). The elevated cortisol levels seen in elite lifters may reflect an ability of these athletes to push themselves maximally during each training session.

Prolonged aerobic exercise appears to be a potent stimulator of the adrenocortical system. Increases in cortisol appear to be proportional to the intensity of exercise (Farrell, Garthwaite, and Gustafson 1983). However, cortisol levels may not change from baseline during exercise at mild to moderate intensities;

only when exercise is greater than 70% of $\dot{V}O_2max$ is a consistent increase in cortisol observed (Few 1974). Significant increases in cortisol concentrations also occur during a short bout (1 min) of exercise if the exercise is performed at maximal intensity (Buono, Yeager, and Hodgdon 1986).

Response to Training

Training appears to lower the cortisol response during prolonged endurance exercise (Tabata et al. 1990). These changes seem to reflect a better maintenance of blood glucose levels. In the strength–power athlete, resting cortisol concentrations may increase (Hakkinen and Pakarinen 1991), decrease (Hakkinen et al. 1985; Kraemer et al. 1998), or show no change (Hakkinen et al. 1990; Potteiger et al. 1995). This array of responses may reflect the various psychological and physiological stresses that can affect cortisol secretion patterns. Short-term elevations seen following prolonged training may reflect a training stress. This is discussed in greater detail in chapter 24. However, elevated cortisol concentrations may not only reflect a physiological stress but also be influenced by a psychological stress. An examination of college football players at the onset and end of preseason training camp demonstrated significantly lowered cortisol concentrations by the end of training camp (Hoffman, Cooper, Wendell, Im, et al. 2004). This reflected a decrease in training stress, which was a bit surprising considering that the physical demands of the training and the physical contact associated with football practices generally increase during training camp. However, there were several potential explanations. The athletes were in peak condition and were physically prepared to meet the rigors of training camp. Others have also reported no change in cortisol during periods of intense training in strength–power athletes (Fry, Kraemer, and Ramsey 1998). An additional explanation focused on a reduced anxiety level that could result from settling into a routine previously experienced by many of these players. Cortisol levels respond to changes in anxiety (Butki, Rudolph, and Jacobsen 2001), and reducing anxiety would potentially lower cortisol concentrations, especially in a veteran group of players.

Catecholamines

The catecholamines (epinephrine, **norepinephrine**, and **dopamine**) are secreted by the **adrenal medulla** of the adrenal gland and are controlled entirely by sympathetic nervous input. Epinephrine makes up 80% of the total catecholamine secretion from the adrenal medulla (Hedge, Colby, and Goodman 1987), while most of the circulating norepinephrine is derived from sympathetic neurons. Catecholamines are stimulated by hypoglycemia, physical or psychological trauma, circulatory failure, stress, exercise, illness, hypoxia, and cold exposure. The most potent of these stimuli is hypoglycemia. Decreases in blood glucose may elevate adrenal medulla secretion 10- to 50-fold (Hedge, Colby, and Goodman 1987). The direct and indirect actions of catecholamines (as suggested by Kraemer 1992b) on muscle function are as follows:

- Increase in force production
- Increase in contraction rate
- Increase in blood pressure
- Increase in energy availability
- Augmentation of secretion rates of other hormones

Catecholamines may be the most important hormones for the acute expression of strength (Kraemer 1992b). In addition, acute increases in catecholamines may be involved with the potentiation of other hormonal mechanisms (e.g., testosterone, IGF).

Exercise Response

Catecholamine concentrations appear to be elevated during both endurance and resistance exercises. These increases seem to reflect the acute demands and stresses of the training program (Kjaer 1989; Kraemer et al. 1987). The intensity rather than the duration of exercise appears to be the primary determinant of the catecholamine response (Brooks et al. 1990; Jezova et al. 1985). Even short-duration sprints (several seconds) of maximal intensity seem to be sufficient to elevate both epinephrine and norepinephrine concentrations (Brooks et al. 1988; Kraemer, Dziados, et al. 1990). Before high-intensity exercise, an anticipatory rise in plasma epinephrine has been reported and is thought to represent some preparatory mechanism in advance of the exercise (Kraemer, Gordon, et al. 1991).

During exercise at submaximal intensities, there may be a differential catecholamine response that appears to depend on the duration of exercise. During such exercise, increases in norepinephrine concentrations are seen within 15 min without any increase in epinephrine. As exercise duration increases, epinephrine concentrations may increase above resting levels (Pequignot et al. 1979). This most likely reflects a greater need for substrate mobilization (hepatic glucose production) during exercise of longer duration.

Response to Training

Training does not appear to alter resting catecholamine concentrations (Kjaer and Galbo 1988; Kraemer et al. 1985). However, trained individuals do appear to have a greater capacity to secrete epinephrine (Kjaer 1989; Kjaer and Galbo 1988) and perhaps norepinephrine as well (Kraemer et al. 1985).

Metabolic Hormones

Several hormones are involved with controlling metabolic function and to some degree may have an important role in body weight control. Although the discussion here focuses on the specifics of each hormone and how each responds to exercise and chronic training, the metabolic hormones generally function along with several other hormones to exert their effects. The past decade has seen increased focus on these hormones, partly as a result of new discoveries but also motivated by the obesity epidemic that is gripping many nations. Better understanding of mechanisms associated with body fat gain and loss provides for more productive interventions to reduce obesity and obesity-related disease.

Thyroid Hormones

The **thyroid gland** secretes three hormones. Calcitonin is secreted by parafollicular cells and is involved with the regulation of calcium balance. The thyroid gland also secretes **thyroxine (T_4)** and **triiodothyronine (T_3)**. These hormones are made up of both iodide and the amino acid tyrosine. Secretion of these hormones is stimulated by **thyroid-stimulating hormone**, released by the anterior pituitary gland. T_4 is secreted in greater quantity than T_3, but T_3 is the active and more potent form of the hormone. T_4 can be metabolized to T_3 in numerous tissues in the body, including skeletal muscle and the liver. In fact, 80% of the circulating T_3 is formed by extrathyroidal metabolism of T_4.

The primary function of the thyroid hormones is to increase the basal metabolic rate. Thyroid hormones may also potentiate the glucose uptake caused by insulin. However, an oversecretion of thyroid hormone, which is often seen in hyperthyroid patients, results in an increase in liver glycogen depletion. All aspects of lipid metabolism are stimulated by the thyroid hormones. Although these hormones are required for normal growth, high levels have a catabolic effect on skeletal muscle.

Exercise Response

Relatively little research is available concerning the response of thyroid hormones to both an acute exercise stress and prolonged training. The response of both T_3 and T_4 to acute exercise is inconclusive. These hormones have been reported to increase (Balsam and Leppo 1975) or not change from resting levels (Galbo et al. 1977) during acute exercise at varying intensities. However, the influence of acute exercise may not be detectable until several days after the exercise session (Galbo 1981). Following exhaustive endurance exercise, thyroid hormones have been reported to decrease for 24 h into recovery; this gives some indication of the importance of these hormones in energy balance (Hackney and Dobridge 2009).

Response to Training

Several investigations have examined the response of resting thyroid hormones to prolonged exercise training and as a possible hormonal indicator of overtraining (Alen, Pakarinen, and Hakkinen 1993; Hoffman, Epstein, Yarom, et al. 1999; Pakarinen et al. 1988). During prolonged periods of high-intensity training (specifically resistance training), decreases in resting levels of T_4 and T_3 may be observed (Alen, Pakarinen, and Hakkinen 1993; Pakarinen et al. 1988). Further research on the response of the thyroid hormones to prolonged training appears to be needed.

Leptin

Leptin is an adipocyte-derived hormone that has a controlling effect on satiety and hunger (McMurray and Hackney 2005). It plays a role in both energy intake and energy expenditure. When energy intake is restricted, leptin concentrations decrease, resulting in energy conservation and decreased thermogenesis (McMurray and Hackney 2005). As body fat begins to increase, leptin concentrations increase as well to reduce food intake and increase thermogenesis (Ahima 2000). Leptin's actions in relation to hunger and food intake appear to be regulated through its effect on the hypothalamic neuropeptide Y (Ahima 2000), but it has also been suggested to act through the sympathetic nervous system, possibly through galanin-like peptide (Hansen et al. 2003).

Obesity is associated with elevation in leptin production and may be related to the desensitization of adipose tissue to the sympathetic nervous system, making it more difficult for people to lose body fat (Florkowski et al. 1996; Monroe et al. 2000). A strong correlation ($r = 0.71$) has been shown between body fat percentage and plasma leptin concentrations (Ostlund et al. 1996). In addition, there appears to be a sex difference in circulating leptin concentrations, perhaps associated with the sex differences in gonadal steroids, that cannot be accounted for by differences in sex-specific fat (Rosenbaum et al. 2001). Ramis and colleagues (2005) reported a significant inverse correlation ($r = -0.79$) between plasma leptin concentrations and plasma testosterone concentrations after adjusting for sex, age, and body mass index in severely obese subjects. Thus, hypogonadal males (those with clinically low testosterone levels) have higher leptin concentrations and are at greater risk for obesity and obesity-related diseases. Elevations in leptin concentrations have been associated with a downregulation of receptors stimulating testicular steroidogenesis (Tena-Sempere et al. 2001), while androgen administration has been associated with decreasing leptin concentrations (Hislop et al. 1999). Interestingly, in the study by Hislop and colleagues (1999), when testosterone was suppressed, leptin concentrations became significantly elevated independent of changes in body fat.

Exercise Response

Exercise has a potent effect on inducing fat loss, even without energy restriction (Ballor and Keesey 1991; Tremblay, Simoneau, and Bouchard 1994). Endurance activity of moderate intensity (~60% $\dot{V}O_2max$) for an hour in obese individuals resulted in no change in circulating leptin concentrations immediately postexercise or up to 48 h postexercise (Kyriazis et al. 2007). Nindl and colleagues (2002) examined the effects of an acute resistance exercise session on leptin concentrations and found no difference for the first hour postexercise in comparison to values in a control group (except for a decrease at the 10 min mark postexercise). However, continued measures showed that at 9 h postworkout, leptin concentrations became significantly lower than in the control group and remained elevated for the next 4 h. The investigators suggested that the reduction in leptin concentrations was more a function of the greater energy expenditure resulting from the workout (excess postexercise oxygen consumption) than of the workout itself. Others who have examined leptin following several days of moderate-intensity exercise have been unable to show any change in fasting leptin concentrations (Hagobian, Sharoff, and Braun 2008).

Response to Training

With regard to chronic changes of leptin consequent to training programs, few data are available. One study demonstrated a dose–response relationship between leptin changes and the physical stress incurred during rowing practice (Jurimae, Maestu, and Jurimae 2003). Relatively low resting concentrations of leptin were observed to decrease even further during longer periods of higher training volume. These changes were found to be independent of changes in body composition, indicating that the physiological role of leptin may have to do with more than body weight regulation. The authors suggested that leptin could potentially be used to monitor training stress, but further research needs to be done in this area.

Ghrelin

Ghrelin is produced primarily in the stomach and is found in the stomach, the gastrointestinal tract, and blood. This hormone was discovered in 1999 (Kojima et al. 1999). It appears to have a strong effect on appetite by increasing the desire to eat—the opposite of what occurs with leptin. Ghrelin synthesis is stimulated by fasting, and the level increases before meals and decreases following food intake (Kraemer and Castracane 2007). Interestingly, ghrelin elevations appear to be more potent in stimulating GH release than growth hormone–releasing hormone is (Arvat et al. 2000). Thus, due to its effect on energy balance and GH release, ghrelin is a very interesting hormone in relation to sport and exercise science. During periods of weight gain, ghrelin concentrations tend to increase; following periods of weight loss, they decline (Kraemer and Castracane 2007).

Exercise Response

During treadmill running of various intensities, increases in GH were reported but without any change in ghrelin (Schmidt et al. 2004). Another study that provided a carbohydrate-rich meal before exercise showed a significant decrease in ghrelin compared to levels in subjects who performed the exercise protocol without having received a meal (Becker et al. 2012). In addition, hunger was suppressed for longer periods of time in the group that exercised and was inversely correlated with ghrelin concentrations. Clearly the effect of exercise was to suppress ghrelin concentrations, which reduced hunger sensations. Sprint exercise also appears to be a potent inhibitor of ghrelin release (Stokes et al. 2010). Thomas and colleagues (2012), examining the effect of acute resistance exercise on the ghrelin response, reported that obese individuals had higher ghrelin concentrations than lean individuals throughout the exercise period, but no changes from baseline levels were observed consequent to the resistance exercise session. Studies on competitive athletes have reported no change in ghrelin concentration from pre- to postexercise at intensities below and above the anaerobic threshold (Jurimae et al. 2007).

Response to Training

Only a limited number of studies have looked at the effect of training on resting concentrations of ghrelin or at the response of ghrelin to an exercise stress. Guelfi and colleagues (2012) used a 12-week (3 days per week) endurance and resistance training program in previously sedentary overweight or obese men. The endurance training program (stationary cycling or elliptical training) progressed from 40 to 60 min per workout during the fifth week. Exercise intensity progressed to 75% to 80% of the participant's maximal heart rate. The resistance training workout consisted of pulley weight machines and free weights and progressed from three sets of 10 repetitions at 75% of the participants' maximal strength (1RM) to four sets of eight repetitions at 85% of their 1RM. Both programs had limited effect on fasting concentrations of ghrelin. Interestingly, the endurance training program positively influenced feelings of hunger and satiation, while no change was noted in the resistance-trained group. A few studies on changes in ghrelin concentrations have also been performed with competitive athletes. Ramson and colleagues (2012), using a 4-week combined endurance and resistance training program in male rowers, showed a significant decrease (12.2%) in fasting ghrelin concentrations at week 3 (following a recovery week) compared to week 2 (following 2 weeks of high-volume training). Plinta and colleagues (2012) examined young female basketball players during a 3-month preseason training period, noting a significant decrease in plasma ghrelin during periods of moderate aerobic training. These changes occurred without any concomitant changes in body mass.

Fluid Regulatory Hormones

Maintenance of body fluid and **electrolyte balance** is regulated through the endocrine system and is very important during prolonged exercise or exercise in a hot environment, when individuals are at great risk of **dehydration** (loss of body water through excessive perspiration). In an attempt to conserve water, as well as electrolytes, the body stimulates hormonal action in both the renal and circulatory systems. The principal hormones involved in regulating blood volume and electrolyte balance are **antidiuretic hormone (ADH)**, **aldosterone**, and **angiotensin II**. Figure 2.6 depicts the fluid regulatory hormone response to dehydration.

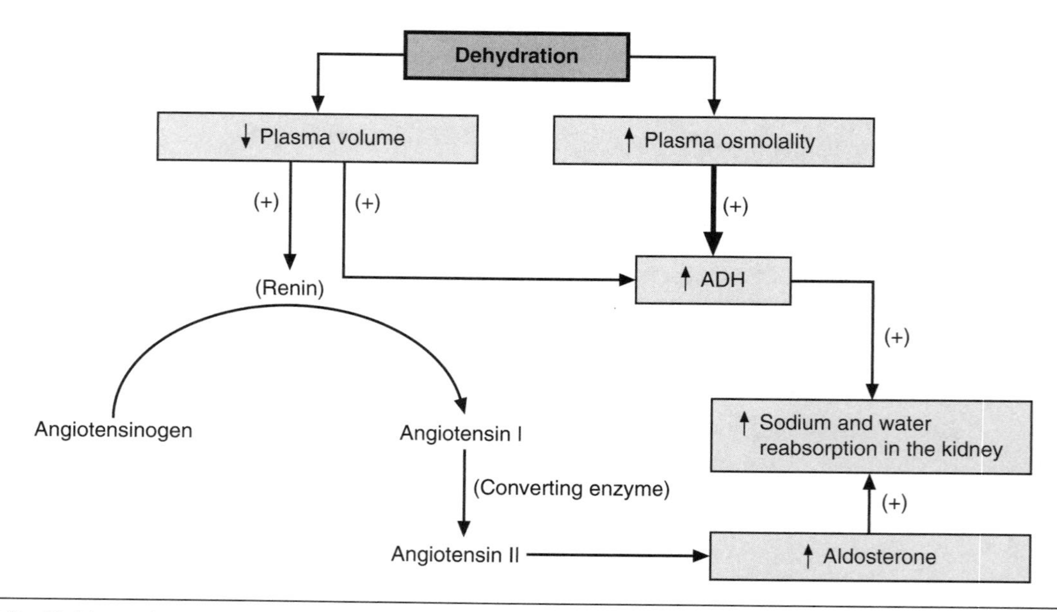

Figure 2.6 Fluid regulatory response to dehydration.

Antidiuretic Hormone

Antidiuretic hormone (ADH), or arginine vasopressin (AVP), is the hormone principally responsible for regulating fluid balance. ADH acts on the collecting ducts within the nephrons of the kidney, making them more permeable to water, and on arterioles in skeletal muscle and skin to produce its vasoconstrictor effects. It is synthesized and secreted in the posterior pituitary gland. The most important stimulus for ADH secretion is change in plasma **osmolality**. As long as plasma osmolality remains at or below 280 mOsm/L, little to no ADH is secreted. However, ADH is sensitive to changes in the plasma osmolality and responds proportionately.

ADH may be the most potent constrictor of vascular smooth muscle, perhaps 10 times more active than norepinephrine or angiotensin II for stimulating arteriole contractions (Goodman and Fray 1988). A change in blood volume is sensed by receptors in both the arterial and venous circulations. However, because of an extensive buffering capacity of the baroreceptors, changes in arterial pressure are seen only with large decreases in blood volume. At normal osmolality, ADH secretion is minimal as long as blood volume remains close to its physiological set point. A decrease in blood volume of 10% to 15% appears to be the minimum threshold to stimulate ADH secretion. ADH is considered an emergency responder rather than a fine-tuner of blood volume (Goodman and Fray 1988). Because ADH responds to two different inputs (osmolality and blood volume), its sensors need to integrate the signals and react appropriately. Volume depletion and increased osmolality, as might be seen during dehydration, result in a heightened sensitivity and the stimulation of ADH release.

Aldosterone

Aldosterone is a steroid hormone secreted by the adrenal cortex of the adrenal gland. It is primarily involved with salt and water balance and is stimulated mainly by angiotensin II as well as ACTH and high concentrations of potassium. Aldosterone is also stimulated by decreases in blood volume as detected by lowered blood pressure. Its response time is relatively slow (approximately 30 min lag time) compared with that of ADH (instantaneous). Its principal site of action is in the nephrons of the kidney, specifically on the cortical portion of the renal collecting ducts, to promote absorption of sodium and excretion of potassium. Under the influence of aldosterone, sodium is reabsorbed in exchange for potassium. This exchange of ions is not a one-to-one trade; a much greater concentration of sodium is reabsorbed compared with the amount of potassium lost. Aldosterone also affects the ratio of sodium to potassium in both sweat and saliva.

Angiotensin II

Angiotensin II is a peptide hormone whose primary role is maintaining salt and water balance. It accomplishes this through its role as the primary stimulator of aldosterone secretion. In addition, it is a powerful constrictor of vascular tissue and acts centrally to excite sympathetic vasomotor outflow, reinforcing its vasoconstrictor action. Angiotensin II causes elevations in blood pressure and is the most potent pressor agent known. The vasoconstricting action of angiotensin II is not uniform to all vascular beds, causing a redistribution of blood to the brain, heart, and skeletal muscle at the expense of visceral organs and the skin.

Angiotensin II is formed in the blood through a two-step process. Angiotensinogen is secreted by the liver and converted to angiotensin I by the enzyme **renin**. Renin is produced in the kidneys and catalyzes the proteolytic cleavage of angiotensinogen to angiotensin I. Angiotensin I is inactive and is converted to angiotensin II (active form) through additional proteolytic cleavage catalyzed by a converting enzyme. The rate-limiting step of this conversion process is the conversion of angiotensinogen to angiotensin I. Thus, the secretion of aldosterone is regulated by the secretion of renin by the kidneys (Goodman 1988).

Fluid Regulatory Hormones' Response to Exercise

Because the fluid regulatory hormones respond to similar stimuli and all act to defend fluid balance within the body, their responses to exercise are discussed together here. Increases in ADH, aldosterone, and plasma renin activity (PRA, indicator of angiotensin II activity) as a result of exercise are well documented (Convertino et al. 1981; Convertino, Keil, and Greenleaf 1983; Melin et al. 1980; Wade and Claybaugh 1980). As previously mentioned, these hormones conserve body fluid during periods of exercise when fluid and electrolyte loss occurs as a result of sweating. The responses of these hormones appear to be similar between males and females (Maresh, Wang, and Goetz 1985; DeSouza et al. 1989), but aldosterone concentrations may be significantly elevated in the midluteal phase of the menstrual cycle (DeSouza et al. 1989).

The fluid regulatory hormones respond in a graded fashion to the level of hypohydration (loss of body fluid), and the response is further magnified when exercise is performed in a hot environment (Francesconi et al. 1985; Montain et al. 1997). Exercise intensity is also a potent stimulator of AVP, aldosterone, and PRA secretion patterns. A significantly greater response of these hormones is observed in high-intensity exercise compared with low-intensity exercise (Freund et al. 1991; Montain et al. 1997). Rehydration with water during exercise reduces or abolishes the response of ADH and PRA, but aldosterone levels decline only when an isotonic solution is consumed (Brandenberger et al. 1986).

Fluid Regulatory Hormones' Response to Training

Training reduces the fluid regulatory hormone response when exercise is performed at a given pretraining exercise load (same absolute load). In trained versus untrained individuals performing exercise at the same relative intensity, a greater response of these hormones is seen in the trained group (Convertino, Keil, and Greenleaf 1983). It is well understood that prolonged training results in an increase in the fluid volume reserve, primarily because of plasma volume expansion. Thus, trained individuals are able to maintain a higher circulating volume even though they have a greater absolute loss in plasma volume because of an increased sweat rate.

Opioids and Exercise

Several endogenous peptides secreted by the anterior pituitary gland produce analgesia upon binding to their receptors in the brain. **Proopiomelanocortin (POMC)** is a large molecule from which other active molecules are split by enzymatic cleavage. POMC is the source of ACTH and endogenous **opioids**. These opioids can be subdivided into three groups (**endorphins**, **enkephalins**, and **dynorphins**) stemming from three major precursor molecules. Opioid receptors are seen in both the central and peripheral nervous systems (Akil et al. 1984). However, peripheral stimulation of these receptors is thought to be more prevalent when opioids are used in pharmacological dosages.

The primary function of the opioids is controlling pain. Stimulation of opioid receptors causes a reduction in the pain response to harmful stimuli, while pain relief is reversed when naloxone (an opioid antagonist) interacts with the opioid receptor. Opioids can also act as neurotransmitters by inhibiting the gonadotrophic hormones luteinizing hormone (LH) and follicle-stimulating hormone (FSH) (Kraemer et al. 1992) and stimulating GH and prolactin release (Molnar et al. 1990).

Increases in endogenous opioids, primarily **β-endorphins**, have been consistently reported during endurance exercise (Donevan and Andrew 1987; Farrell et al. 1987; Schwarz and Kindermann 1989). Elevations in β-endorphins have also been shown during both interval sprint (Fraioli et al. 1980) and resistance exercise (Kraemer, Dziados, et al. 1993). However, the β-endorphin response during these modes of exercise (sprint and resistance) depends on single component variables of the exercise program.

The magnitude of the β-endorphin response during and after exercise depends on the intensity of the exercise stimulus (Farrell et al. 1987; Goldfarb et al. 1990). During supramaximal exercise of very short duration, β-endorphins show no significant change from resting levels (Kraemer et al. 1989). A minimum duration of exercise appears to be necessary to stimulate β-endorphin release. An increase in β-endorphins is related to elevations in the stress hormones (e.g., ACTH and cortisol) and apparently requires that the individual reach a certain degree of fatigue. Similarly, during resistance exercise, elevations in endorphin levels are noted when the exercise paradigm calls for moderate intensity (10RM) of training with short rest periods (1 min) (Kraemer, Dziados, et al. 1993). Resistance training programs of greater intensity (5RM to 8RM) or of longer rest periods (3 min)—less fatiguing programs—do not appear to impose a stress sufficient to cause an endorphin response (Kraemer, Dziados, et al. 1993; Pierce et al. 1994).

The physiological significance of elevated levels of endorphin after exercise is unclear. However, it is associated with the "high" or euphoria felt after exercise. In addition, endorphins are thought to be involved in pain tolerance; improved appetite control; and reduction in anxiety, tension, anger, and confusion. All of these are considered psychological benefits of exercise (Morgan 1985; McArdle, Katch, and Katch 2010; O'Connor and Cook 1999).

Summary

The primary function of hormones is to influence the rate of cellular reactions. Both physical and psychological stresses are potent stimulators of elevation of the secretory patterns of hormones. The hormonal response to exercise is influenced by single component variables of the exercise program. These components (choice and intensity and volume of exercise, rest intervals, and recovery status) need to be well designed to optimize the desired endocrine response.

REVIEW QUESTIONS

1. What is meant by allosteric modulation?
2. Explain the negative feedback mechanism and provide examples.
3. Describe the lock-and-key theory as it relates to the hormone–receptor interaction.
4. Explain the differences between steroid and peptide hormones.
5. What is the difference in action between aldosterone and antidiuretic hormone during fluid regulation?
6. Describe the hormonal response to decreases in circulating concentrations of blood glucose.
7. What type of resistance training protocol would likely cause an increase in the anabolic hormone response?
8. What role does cortisol play in monitoring training stress?
9. Explain the metabolic role of growth hormone.

chapter 3

Metabolic System and Exercise

CHAPTER OBJECTIVES

After reading this chapter you should be able to do the following:

- Understand the different metabolic systems that fuel exercise.
- Differentiate between the specific metabolic adaptations that occur with endurance training compared to high-intensity anaerobic conditioning programs.
- Define appropriate rest intervals between sets of exercise to maximize recovery.
- Understand how different methods of exercise affect fat metabolism.
- Explain the lactate shuttle.

Many of the physiological adaptations during prolonged training relate to an improved ability to generate more **energy** and to use this energy more efficiently. The adaptations seen as a consequence of training are specific to the mode of training performed. **Endurance training** results in both **metabolic** and **morphological** changes that enhance the ability to bring nutrients to the muscle, allow the muscle to more efficiently use these nutrients, and increase the ability of the muscle to generate more energy. **Anaerobic training** causes metabolic adaptations that are much different from those commonly seen during **aerobic training** and that are more specific to the needs of the particular athlete. These adaptations are primarily aimed at enhancing the ability of the muscle to generate energy specific to the anaerobic energy system and at improving the athlete's tolerance to muscle acid–base imbalances through an improved buffering capacity.

This chapter focuses primarily on the metabolic adaptations subsequent to exercise training that enhance the athlete's ability to perform. The following brief review on **bioenergetics** will allow the reader to better appreciate the metabolic adaptations discussed in the remainder of the chapter. These adaptations are discussed in relation to the specific type of training program performed (i.e., aerobic or anaerobic).

Bioenergetics is primarily concerned with the source of energy for muscular contractions. There are three physiological systems in the body that yield energy. Two of these systems can function without the benefit of oxygen and are termed anaerobic. These energy systems are called the **phosphagen energy system (ATP-PC)** and the **glycolytic energy system**. The third system requires oxygen for its energy production and is termed aerobic, or the **oxidative energy system**.

The energy for all cellular functions is derived from the **metabolism** of various substances stored within the muscle (e.g., **glycogen** or **triglycerides**) or within storage sites in the body (e.g., **adipose** tissue). A set of metabolic reactions occurs within each cell to produce a potential store of chemical energy. This energy drives all cellular processes that do not proceed spontaneously. The principal energy component that governs all cellular functions is adenosine triphosphate (ATP). The ATP molecule has three inorganic phosphate (Pi) groups attached to an adenosine molecule. The enzyme ATPase causes the removal of a phosphate group from ATP to form adenosine diphosphate (ADP) and Pi plus the release of a large amount of energy. The process by which ATP can be formed is called **phosphorylation** and can proceed through the three different sources previously mentioned.

ATP-PC Energy Source

ATP-PC is stored within muscle and is available for immediate use. Like ATP, PC (**phosphocreatine**) has a phosphate group and a high-energy bond attached to a **creatine** molecule. Unlike the situation with ATP, in which the breakdown of ATP to ADP produces energy for direct use in cellular function, the Pi group removed from creatine, facilitated by the enzyme **creatine kinase**, can be used only to combine with an ADP molecule to re-form ATP.

The ATP-PC energy system is the simplest of the three energy systems. Oxygen is not required to release the energy from ATP-PC; thus this is considered an anaerobic energy source. However, only a limited amount of ATP and PC is available within the muscle; and during maximal exercise, the supply is exhausted within 30 s. Although the ATP-PC energy system is available for a relatively short period of time, there are several advantages to its use as an energy source. Basically, it's the energy source that is readily available for immediate use. It also has a large power capacity, providing the muscle with a large amount of energy within a short period of time. These characteristics make the ATP-PC energy source ideal for short-duration, high-intensity events (e.g., 100 m sprint, shot put, long jump). As such, one might expect that resting ATP concentrations in type II fibers would be greater than those in type I fibers. However, this does not seem to be the case. Greenhaff and colleagues (1994) demonstrated that resting ATP concentrations are similar between muscle fibers. However, resting PC concentrations in type II fibers are significantly greater than in type I fibers. During sprinting (30 s of maximal effort on a treadmill), the decline in ATP concentrations between fibers also appears to be similar; however, the decline in PC concentrations was shown by these authors to be 25% greater in type II fibers than in type I fibers. Even though resting PC concentrations were 10% higher in type II fibers, the depletion rate of PC in type II fibers exceeded that seen in type I. Table 3.1 compares pre- and postexercise ATP, PC, and glycogen concentrations.

An interesting and important question is whether the ATP content of the cell ever reaches a level at which the force-generating capacity of the muscle or the cycle rate of the actin–myosin cross-bridge is compromised. From a number of different studies, it appears that ATP-PC cellular concentrations do not reach such a critical level. Fatigue produced from other factors appears to reduce the ATP utilization rate before ATP concentrations become self-limiting (Bergstrom and Hultman 1988). In fact, the ATP concentration within skeletal muscle may not fall below 70% of its resting level even during extreme cases of fatigue (Fitts 1992).

Table 3.1 ATP, PC, and Glycogen Concentrations Immediately Before and Following a 30 s Maximal Sprint

	ATP (mmol/DM)	PC (mmol/DM)	Glycogen (mmol/DM)
Preexercise			
Type I muscle fiber	24.0 ± 0.8	71.3 ±3.0	375 ± 25
Type II muscle fiber	24.0 ± 0.4	79.3 ± 2.7**	472 ± 35**
Postexercise			
Type I muscle fiber	20.6 ± 0.5	12.2 ± 2.1	298 ± 30
Type II muscle fiber	19.0 ± 101	5.0 ± 1.3**	346 ± 27*

Data are reported as mean ± SEM, *$p < 0.05$, **$p < 0.01$. DM = dry matter.

Adapted from Greenhaff et al. 1994.

The decline in ATP utilization during maximal exercise is related to a large extent to the decrease in PC concentrations within the cell, as well as to an increase in cellular H^+ concentrations generated from the highly anaerobic event. Although ATP concentrations do not seem to be completely expended during maximal exercise, PC levels decline rapidly to the point of complete exhaustion as PC is used to replenish diminished ATP levels. This relationship between ATP and PC concentrations in skeletal muscle during maximal exercise is depicted in figure 3.1.

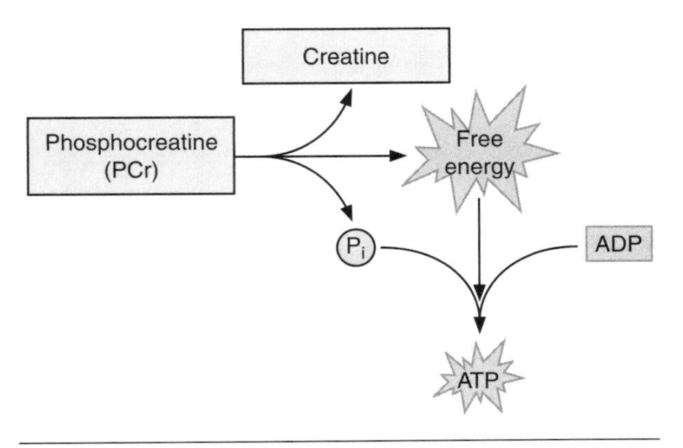

Figure 3.1 Relationship between ATP and PC.

The ability to perform repetitive, highly intensive exercise (e.g., football or basketball) requires a rapid regeneration of phosphagens. The resynthesis of PC is biphasic. Initially, there is a rapid recovery with a $t_{1/2}$ (half-time) of 20 to 30 s; this is followed by a slower phase of recovery that may require up to 20 min (Harris et al. 1976). However, most of the PC is regenerated by 3 min postexercise.

Glycolytic Energy Source

There is an additional energy source that produces ATP through the breakdown of a glucose molecule. This process of metabolizing glucose is called **glycolysis** and results in the net production of two or three molecules of ATP, depending on where the glucose molecule was derived from. Glycolysis results in the release of energy and the breakdown of a glucose molecule, through a chain of chemical reactions, into a compound called **pyruvic acid**. Glycolysis is depicted in figure 3.2. Because this energy system can produce ATP without the need for oxygen, it is also considered an anaerobic energy source.

The glucose metabolized during glycolysis comes from the blood either through the digestion of carbohydrates or from the breakdown of glycogen (storage form of glucose) from the liver. Glucose can also be metabolized from glycogen stored within the working muscle cells. The process of metabolizing glycogen into glucose is called **gluconeogenesis**. In all tissues, glycogen is metabolized into **glucose-1-phosphate** by the enzyme **phosphorylase** and is further broken down into **glucose-6-phosphate**. Once glucose-6-phosphate is formed, the process of glycolysis can begin. The importance of the phosphorylated glucose molecules should not be forgotten. The phosphate molecule attached to each glucose molecule prevents it from diffusing out of the cell. However, the liver possesses a specific enzyme called **phosphatase** that degrades glucose-6-phosphate into glucose and Pi. This allows the glucose molecule to diffuse into the circulation and reach tissues that require additional glucose. No other tissue has the ability to dephosphorylate glucose to allow it to be transported to tissues in need.

Metabolizing glucose into pyruvic acid is a 10-step process. If glycolysis begins with the breakdown of stored glycogen, a net of three ATP molecules is produced from its complete metabolism. However, if glycolysis begins from glucose, then a net total of only two ATP molecules is produced, because one ATP molecule is used for the conversion of glucose to glucose-6-phosphate (see figure 3.2). Because no oxygen is present during anaerobic glycolysis, the pyruvic acid is converted to **lactic acid**. Increases in lactic acid concentrations have traditionally been associated with an increase in acidosis (lowering of muscle pH) and increase in fatigue. However, there has been debate directed at this concept in the belief that it is not supported by fundamental biochemistry (Boning et al. 2005; Brooks 1986; Lindlinger, Kowalchuk, and Heigenhauser 2005; Robergs, Ghiasvand, and Parker 2004). The next section provides further discussion on lactic acid and lactate. In any case, the glycolytic energy system can produce a larger amount of energy than the ATP-PC system. However, it cannot supply as much energy per unit time and therefore is not as powerful as the ATP-PC energy source. Glycolysis is the primary energy source for high-intensity exercise lasting 1 to 3 min. Interestingly, similar to what is seen with PC concentrations and type II muscle fibers, glycogen concentrations appear to be significantly higher in type II than type I muscle fibers before and following high-intensity exercise (see table 3.1). Thus, there appears to be a strong relationship between anaerobic metabolism and muscle fibers primarily recruited for high-intensity activity.

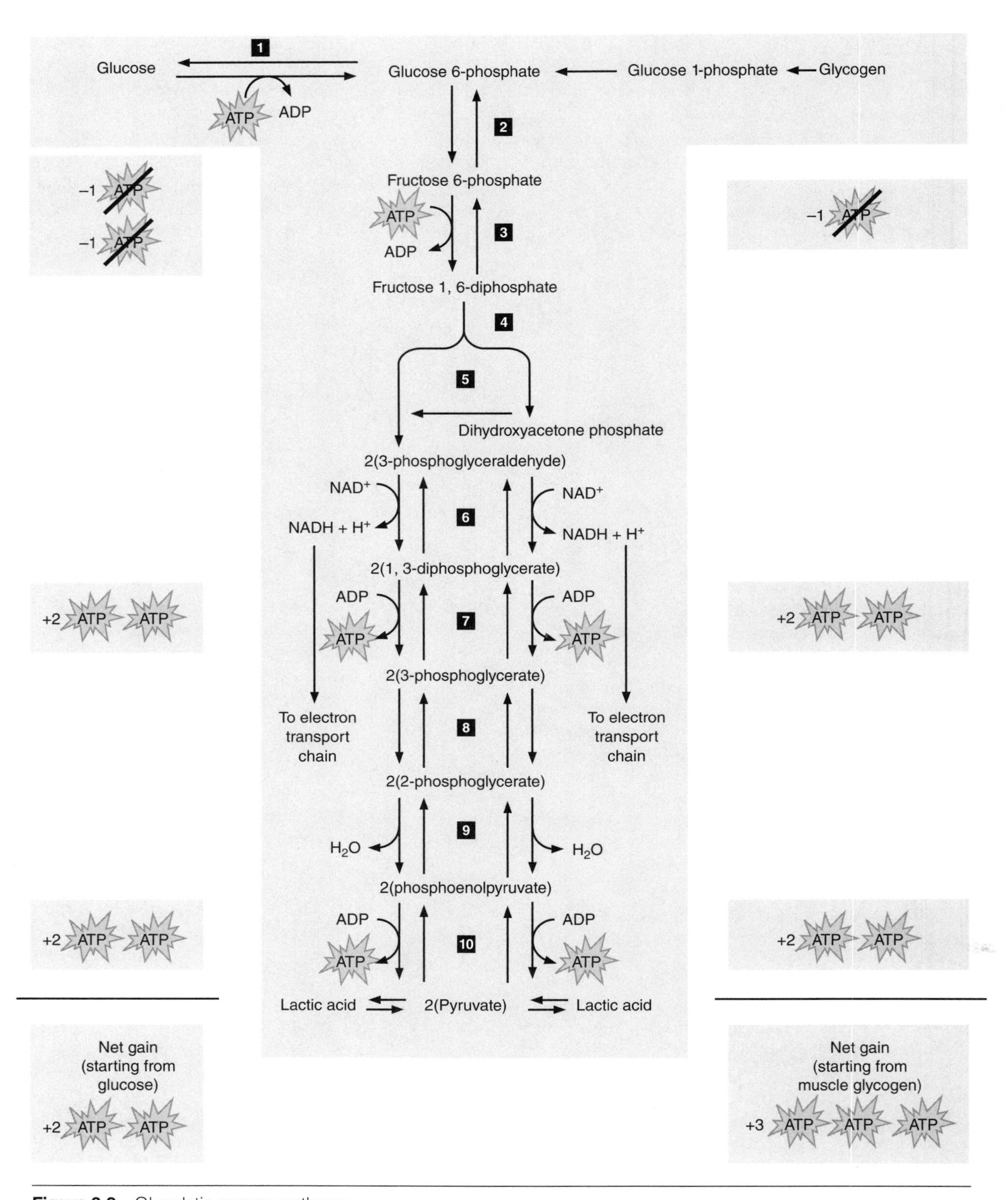

Figure 3.2 Glycolytic energy pathway.

Lactic Acid Controversy

For years, the accumulation of lactic acid within the muscle was believed to be primarily associated with fatigue and a decrease in muscle function. This was the well-accepted understanding of energy metabolism during high-intensity exercise (Robergs, Ghiasvand, and Parker 2004), providing a simple cause-and-effect explanation for muscle fatigue. According to this view, elevations of lactic acid were associated with a decrease in both blood and muscle pH, resulting in performance decrements. The increase in hydrogen ions (H^+) that results from lactic acid production was buffered by bicarbonate, producing carbon dioxide (CO_2) and increasing the rate of ventilation. The relationship between ventilatory and lactate thresholds and the inverse linear relationship observed between muscle lactate and pH provided indirect evidence to support this interpretation (Katz and Sahlin 1988; Sahlin et al. 1976). However, the biochemical reactions seen during glycolysis produce pyruvate, not pyruvic acid, which can then be converted to lactate (via the lactate dehydrogenase [LDH] reaction) (Lindlinger, Kowalchuk, and Heigenhauser 2005; Robergs, Ghiasvand, and Parker 2004). In fact, Robergs and colleagues (2004) indicated that none of the substrates formed during glycolysis are actual acids but are acid salts, meaning that they do not produce H^+. These authors suggested that the protons (H^+) produced from the hydrolysis of ATP during glycolysis result in metabolic acidosis, and that lactate actually facilitates proton removal from muscle. This may explain why blood lactate measurements provide a good measure of proton formation and are an indicator of decreased muscle and blood pH. Others have suggested that changes in the concentrations of strong ions (Na^+, K^+, Cl^-, Mg^{2+}, Ca^{2+}, lactate$^-$, pyruvate$^-$, PC^{2-}) and weak ions (HCO_3^-) produced during exercise have a greater influence on the acid–base balance of muscle (Lindlinger, Kowalchuk, and Heigenhauser 2005).

Lactate Shuttle

There is strong evidence that lactate may actually be an important fuel source (Brooks 1986, 1991, 2009). Lactate production is part of the process of glycolysis that occurs even during rest. During exercise, lactate production begins to rise to meet metabolic demands. The elevated lactate concentrations appear to have two important roles: to maintain blood gluconeogenesis and to shuttle an oxidizable substrate (lactate) from an area of production (through the process of glycogenolysis) to an area of cellular respiration (removal) (Brooks 1991). As lactate leaves the cell, it enters the circulation and enters the liver, where it is converted back to glucose and glycogen. The recycling of this metabolic substrate has been referred to as the **Cori cycle** (Wasserman et al. 1987), and it is an important source of fuel during exercise. During exercise, lactate concentration is significantly elevated, but only ~50% of lactate formed in tissue is released into the venous circulation (Brooks 1986). The other half of the lactate produced appears to diffuse from the active tissue to more oxidative tissue (i.e., from type IIx fibers to type I fibers) (Baldwin, Campbell, and Cooke 1977; Brooks 1986). The shuttling of lactate from active tissue to the circulation and more oxidative tissue has been referred to as the lactate shuttle (Brooks 1986). Interestingly, improved technology has permitted a greater understanding of the lactate shuttle, to include not only shuttling of lactate to the circulation and back, but also cell-to-cell shuttling and intracellular shuttling (Brooks 2009). The discovery of LDH within the mitochondria of heart and skeletal tissue provided a greater understanding of the ability of active cells that produce lactate to oxidize lactate (Brooks et al. 1999). Lactate produced within the cytosol (location of glycolysis) is shuttled to the mitochondria (where oxidation occurs) to be oxidized to pyruvate and converted to energy. The presence of cell-to-cell and intracellular lactate shuttles demonstrates the close link between the glycolytic and oxidative energy pathways.

Oxidative Energy Source

The oxidative energy system uses oxygen in the production of ATP and is therefore referred to as an aerobic energy source. The oxidative production of ATP occurs within the **mitochondria** of all cells. In skeletal muscle, the mitochondria are located adjacent to the myofibrils and throughout the sarcoplasm. The oxidative production of ATP cannot supply enough ATP per unit time to provide energy to sustain highly intense activity. However, because of the abundance of stored fat and carbohydrates in the body, the oxidative energy system can provide ample energy for prolonged periods of submaximal exercise. Thus, this is the primary energy system used for long-duration aerobic events.

Aerobic metabolism begins in the same way as glycolysis, with the breakdown of glycogen into glucose and the subsequent conversion of glucose to pyruvate. However, in the presence of oxygen, pyruvate is converted into **acetyl coenzyme A** (acetyl-CoA) and enters into a series of chemical reactions called the **Krebs cycle** and the **electron transport chain** (figures 3.3 and 3.4).

The Krebs cycle is a series of chemical reactions that produces **carbon dioxide** (expired through the lungs) and hydrogen. The hydrogen combines with the coenzymes nicotinamide adenine dinucleotide (NAD) and flavin adenine dinucleotide (FAD) and transports them from the cell cytoplasm to the mitochondria where they enter the electron transport chain. The hydrogen atoms involved in the electron transport chain are split into **protons** and **electrons**. The hydrogen protons combine with oxygen to form water, while the electrons pass through a series of reactions that phosphorylate ADP to form ATP. This process is also known as oxidative phosphorylation.

Oxidative metabolism uses primarily carbohydrates and fat. However, during periods of carbohydrate depletion, starvation, or prolonged exercise, significant amounts of protein can be metabolized for energy as well. While at rest, the body derives most of its energy from stored fat. However, the body begins to metabolize a greater percentage of stored carbohydrate during exercise. The oxidative metabolism of one molecule of glycogen produces a net gain of 39 ATP (table 3.2).

The use of stored fat as an energy source relies exclusively on the breakdown of triglycerides stored within both adipose sites and muscle. The process

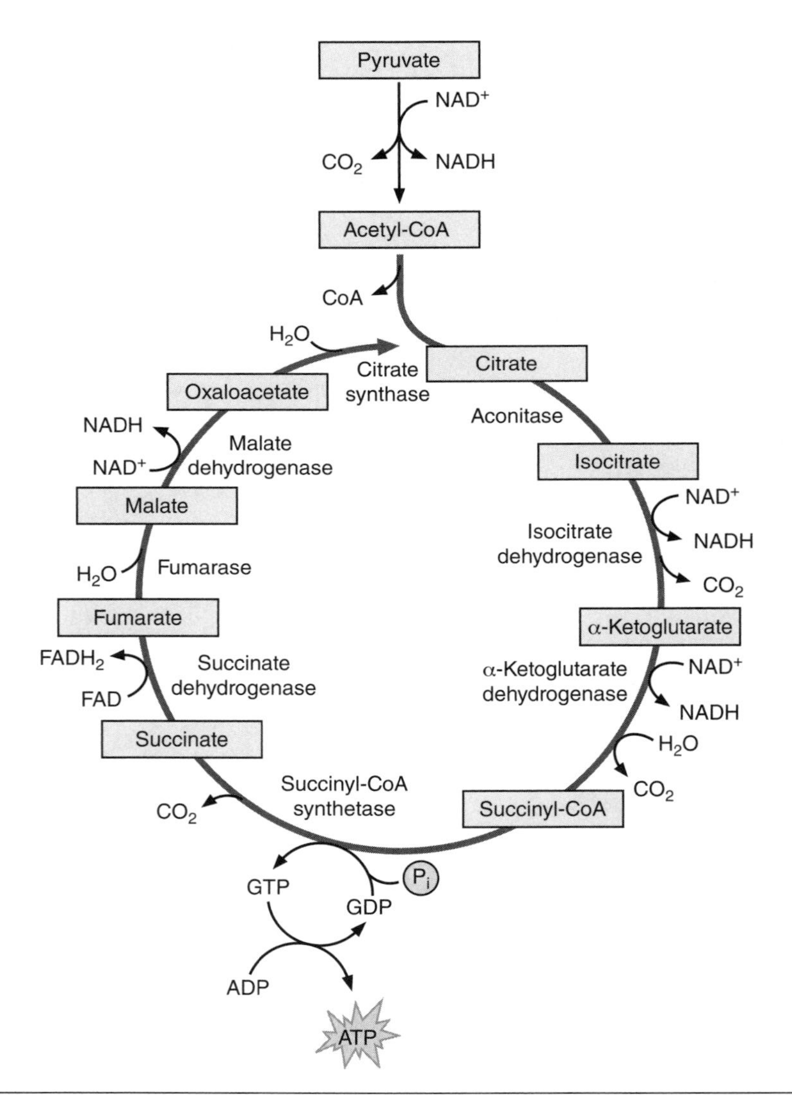

Figure 3.3 Oxidative metabolism: Krebs cycle.

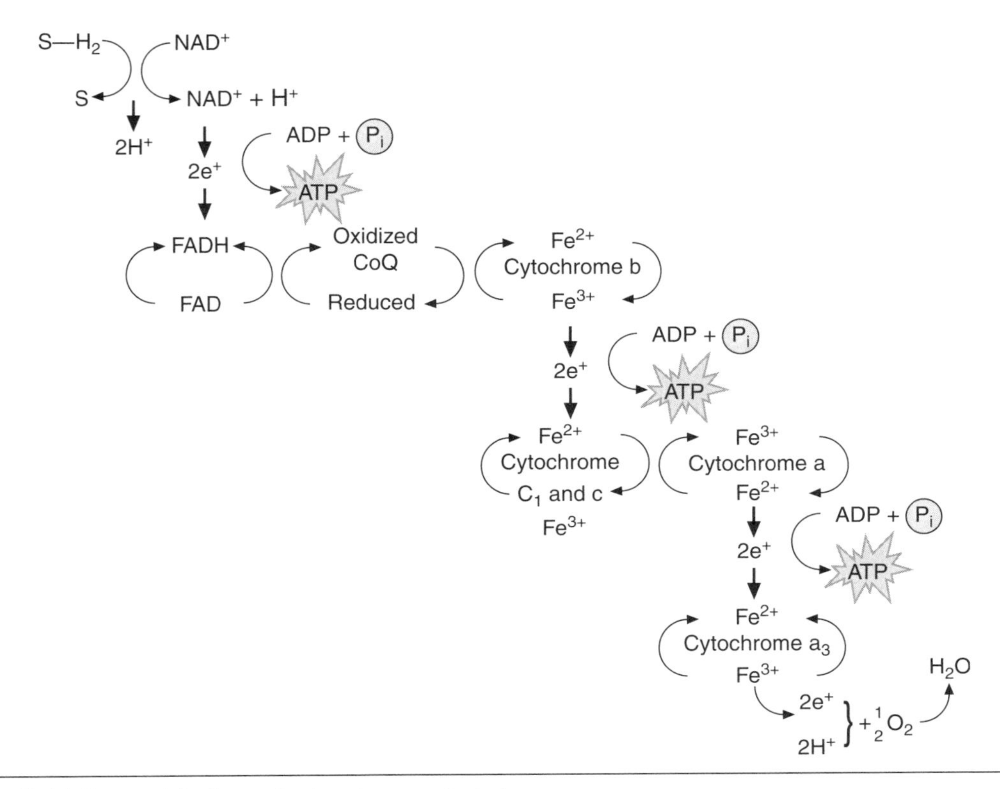

Figure 3.4 Oxidative metabolism: electron transport chain.

Table 3.2 Energy Production (ATP) From Glycolysis and Oxidative Metabolism

Energy source	Glycolysis	Oxidative phosphorylation (Krebs cycle and electron transport)
Glucose	2	38
Glycogen	3	39
Palmitic acid	–	129

of breaking down fat for energy is called **lipolysis** and results in the metabolism of triglycerides into a **glycerol** molecule and three **free fatty acids**. It is the free fatty acids that are used as the primary energy source. As the free fatty acids enter the mitochondria, they undergo a further catabolic process called **β-oxidation**. This involves the enzymatic cleavage of a free fatty acid molecule into acetyl-CoA. The acetyl group then enters the Krebs cycle through the **citrate synthase reaction** and is oxidized in the same way that carbohydrates are oxidized through aerobic glycolysis. The energy production from the oxidation of one molecule of a fatty acid such as **palmitic acid** yields 129 ATP (see table 3.2).

Interaction of the Energy Sources

Although one energy source may be the predominant system working at any given time, all three sources of energy supply a portion of the needed energy (ATP) for exercise at all times. Thus the ATP-PC source also provides energy at rest, and the oxidative energy source is also used during maximal exercise. The more intense the exercise, the greater the portion of ATP derived from anaerobic energy sources. As intensity of exercise decreases and exercise duration increases, energy production is primarily from aerobic metabolism. There is no exact point at which one energy source drops off and another energy source begins to provide more energy. Rather there is a gradual transition from one energy source to another.

Metabolic Adaptations to Endurance Training

Endurance training, such as prolonged running or cycling, results in profound physiological adaptations that cause significant improvements in exercise capacity. Improved endurance performance is the result of adaptations in a number of different physiological systems (e.g., cardiovascular, neuromuscular). This section examines the metabolic adaptations to endurance training.

Capillary Density

Endurance-trained men have been shown to have a 5% to 10% greater **capillary density** than sedentary control subjects (Hermansen and Wachtlova 1971; Ingjer 1979a). Other studies examining highly trained endurance athletes have reported an even larger disparity in capillary density (37-50% differences) in comparison with untrained individuals (Jansson, Sylven, and Sjodin 1983; Saltin and Rowell 1980). Although a greater capillary density in trained endurance athletes may be a function of genetics, a 28.8% improvement in capillary content of skeletal muscle has been observed following a 28-week endurance training program (Ingjer 1979b). These changes were more predominant within type I fibers and least in type IIb fibers. The greater capillary content has a direct effect on enhancing aerobic exercise performance by providing for a greater exchange of gases, heat, waste, and nutrients between the circulation and exercising muscle. Increases in capillary density appear to occur within 6 to 8 weeks of the onset of an endurance training program (Masuda et al. 2001; Shono et al. 2002).

Myoglobin Content

Myoglobin is the oxygen-transporting and storage protein of muscle. It shuttles the oxygen molecules from the capillaries to the mitochondria. Several animal studies have shown that the myoglobin content of skeletal muscle can be increased during prolonged endurance training (Hickson 1981; Froberg 1971). However, similar results have not been achieved in humans. Several studies have failed to show any changes in myoglobin concentration from baseline levels after an endurance training program (Coyle et al. 1985; Jansson, Sylven, and Sjodin 1983; Masuda et al. 2001). The specific role of myoglobin in improving aerobic capacity in humans remains unclear.

Mitochondrial Function and Content

Mitochondria are the cell's primary organelle for energy supply. Their functional role is to increase oxidative production of ATP. Endurance exercise provides a significant stimulus for enhancing mitochondrial function. Increases in the size and number of mitochondria have been reported after endurance exercise programs (Holloszy 1988; Holloszy and Coyle 1984; Ingjer 1979a, 1979b). In a study of rats, mitochondrial content increased by 15% while the size of the mitochondria increased by 35% during 27 weeks of endurance training (Holloszy et al. 1970). Exercise in untrained men has also been shown to be a potent stimulus for enhancing mitochondrial content (Ingjer 1979a, 1979b); type I fibers have the richest content of mitochondria and are most affected by endurance training (Ingjer 1979b).

Oxidative Enzymes

An important metabolic adaptation seen subsequent to endurance training is an increase in concentration of the enzymes involved in the Krebs cycle and electron transport chain and the enzymes responsible for the activation, transport, and β-oxidation of free fatty acids.

The increase in these enzymes allows for a more efficient metabolic system for oxidizing nutrients to form energy (ATP). In addition, the greater concentration of the oxidative enzymes is thought to spare muscle glycogen and reduce the production of lactate during exercise of a given intensity (Holloszy and Coyle 1984).

In untrained individuals, the concentration of mitochondrial enzymes appears to be twice as high in type I (slow-twitch) fibers as in type II (fast-twitch) fibers (Holloszy and Coyle 1984). Mitochondrial density may also be higher in type I versus type II fibers. During endurance training, the oxidative enzymes appear to increase at a greater rate in the type II oxidative fibers, making the difference between enzyme concentrations of type I and type II fibers negligible or even nonexistent in highly trained aerobic athletes (Holloszy 1988).

Succinate dehydrogenase (SDH) and citrate synthase are enzymes of the Krebs cycle that are often measured to provide a quantitative analysis of the improvement in oxidative potential of endurance-trained individuals. A moderate amount of daily exercise (20 min per day) appears to be an adequate stimulus to significantly increase oxidative enzymes (figure 3.5). Training for a longer duration (60-90 min per day), as performed by some endurance athletes, may cause an even greater increase in the oxidative enzymes. However, these changes appear to occur primarily during the initial stages (first few months) of training and may plateau even when training volume

In Practice

Can Resistance Exercise Enhance the Molecular Signaling of Mitochondrial Biogenesis?

A theme that is becoming very clear to the reader is that physiological adaptation is specific to the type of exercise that is being performed. However, a study by Wang and colleagues (2011) suggested that resistance exercise performed following an endurance training session may actually potentiate the signaling response for mitochondrial synthesis. The investigative team from Sweden examined 10 healthy men and women who were recreationally active but had not participated in any formal training program (either resistance or endurance) during the 6 months before the study. Subjects reported to the laboratory on two separate occasions. Upon reporting in the morning and following study preparation (muscle biopsy and blood sampling), they exercised on a cycle ergometer for 60 min (60-70 rpm at a load corresponding to 65% of their $\dot{V}O_2max$). A 3 min rest period occurred at the 30 min mark. Following the endurance training, subjects either performed a 25 min leg press training session or rested. The leg press exercise consisted of six sets using loads corresponding to 70%, 75%, 80%, 80%, 75%, and 70% of the subjects' maximal strength capability for that exercise. Subjects were required to perform as many repetitions as possible per set, up to 15 repetitions. A 3 min rest period was provided between each set. Blood samples and muscle biopsies were performed 1 h and 3 h after endurance exercise. Results indicated that the bout of resistance exercise actually enhanced the expression of genes involved in the signaling cascade for mitochondrial biogenesis and oxidative phosphorylation. This was the first study to demonstrate that resistance exercise may have a potentiating effect on aerobic adaptation. This study will provide much motivation and impetus to continue investigating potential benefits of combined endurance and resistance exercise for the endurance athlete.

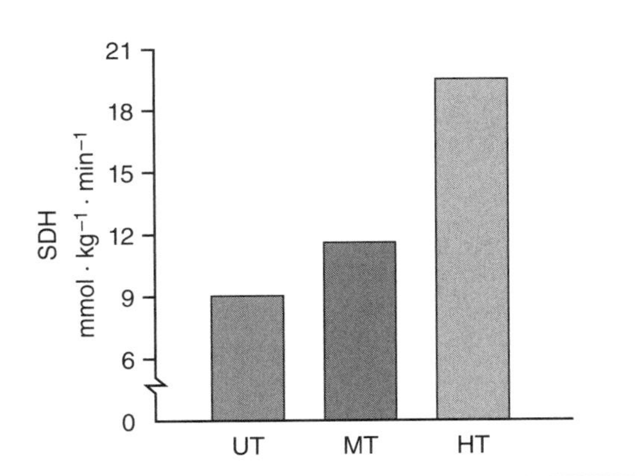

Figure 3.5 Succinate dehydrogenase activity of the gastrocnemius muscle in untrained (UT) and moderately trained (MT) joggers and highly trained (HT) marathon runners.

Reprinted, by permission, from W. L. Kenney, J.H. Wilmore, and D.L. Costill, 2012, *Physiology of sport and exercise*, 5th ed. (Champaign, IL: Human Kinetics), 262.

is further elevated (figure 3.6). Short-term endurance training has been shown to increase citrate synthase activity by 20% after 6 days (Chesley, Heigenhauser, and Spriet 1996). Others have suggested that it may take up to 4 to 6 weeks to see a significant increase in muscle oxidative enzymes (Phillips et al. 1995). Wibom and colleagues (1992) reported a 40% increase in citrate synthase activity following 6 weeks of endurance training in previously untrained men.

In comparisons among athletes, endurance athletes had a 30% to 60% higher concentration in citrate synthase activity in leg musculature than the other athletes (Boros-Hatfaludy, Fekete, and Apor 1986). A 10% to 15% difference in citrate synthase activity was reported between senior and junior Kenyan runners (Saltin et al. 1995), suggesting that even experienced endurance-trained athletes have the ability to enhance oxidative enzyme capacity.

Increases in oxidative enzymes do not correlate well with changes in maximal aerobic capacity ($\dot{V}O_2$ max), suggesting that other factors (e.g., circulation) may have a greater influence on improving aerobic capacity (Gollnick et al. 1972). Increases in the concentration of oxidative enzymes in endurance athletes may be more important for allowing exercise at a higher intensity (e.g., running at a faster pace) than for improving aerobic capacity.

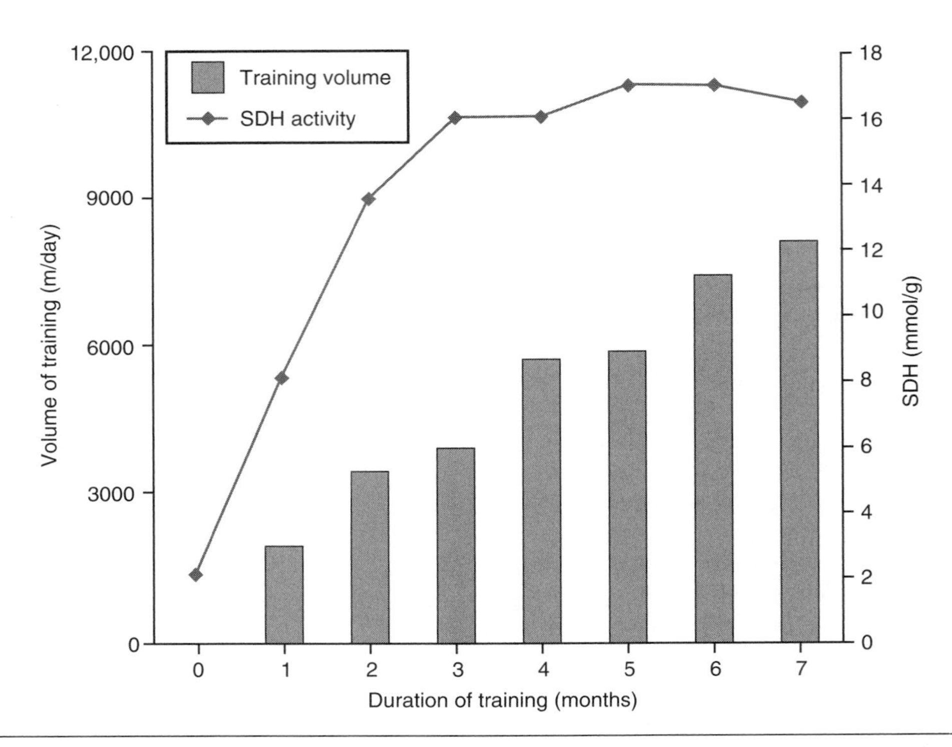

Figure 3.6 Changes in succinate dehydrogenase concentrations in the deltoid muscle during gradually increased swim training.

Adaptation of Glycolytic Enzymes to Endurance Exercise

Unlike the oxidative enzymes, the enzymes of glycolysis do not appear to be significantly affected by prolonged endurance exercise (Holloszy and Booth 1976). Endurance training places a greater demand on the oxidative energy system and therefore is more effective in increasing the mitochondrial enzymes.

Effect of Endurance Training on Carbohydrate Utilization

As previously mentioned, the benefit of an increase in oxidative enzymes and mitochondrial density appears to be in the sparing of muscle glycogen. This enhances the ability to transport the **pyruvate** produced during glycolysis into the larger mitochondrial volume and through the process of oxidative phosphorylation. The result is a reduced buildup of lactic acid and a lowering of the ATP-to-ADP ratio during submaximal exercise. Increases in citrate synthase activity have been reported to blunt muscle glycogenolysis after only 6 days of endurance training (Chesley, Heigenhauser, and Spriet 1996), supporting the muscle glycogen–sparing effect. Thus there will be a reduction in carbohydrate utilization (glycogen degradation) and an improvement in exercise tolerance at submaximal intensities.

Effect of Endurance Training on Fat Utilization

After endurance training programs one sees an increased reliance on stored fat as a source of energy during submaximal exercise (Gollnick and Saltin 1988), largely dependent on the mechanisms discussed in connection with the glycogen-sparing effect. In addition, the increase in capillary density after endurance training provides an enhanced opportunity for the exchange of free fatty acids from adipose tissue to the exercising muscle.

It has traditionally been accepted that fat diffuses across the plasma membrane of the cell. However, research has suggested that fat may also be transported across the plasma membrane through a carrier-mediated transport system (Turcotte 2000). This transport system, made up of proteins embedded in the plasma membrane, appears to respond to chronic endurance training (Turcotte et al. 1999). Thus, increases in fat utilization may also be related to changes in its uptake mechanism into the muscle cell.

Endurance training may also cause a greater increase in free fatty acid concentrations in plasma. This is thought to reflect a greater release from adipose storage sites, contributing to the sparing effect on muscle glycogen (Costill et al. 1977). However, elevated free fatty acid concentrations have not been consistently demonstrated in all studies (Gollnick and Saltin 1988); this may be related to the greater uptake and oxidation of fats seen in endurance-trained individuals compared to sedentary controls.

Metabolic Adaptations to Anaerobic Exercise

High-intensity training such as sprinting (running, swimming, or cycling) or competing in high-intensity sports such as basketball or hockey causes adaptations that are specific to the anaerobic energy system. As described earlier, the anaerobic energy system is made up of both the ATP-PC and glycolytic systems. This section discusses the metabolic adaptations seen in exercise programs that emphasize these energy systems.

ATP-PC System's Adaptations to High-Intensity Training

High-intensity training appears to cause little or no change in resting ATP or PC concentrations (Karlson, Dumont, and Saltin 1971; MacDougall et al. 1977; Troup, Metzger, and Fitts 1986). Whether the resting concentrations of the enzymes (creatine kinase and **myokinase**) that catalyze the phosphagen energy system can be positively altered by high-intensity training is unclear. Costill, Coyle, and colleagues (1979) were unable to demonstrate any change in these enzymes after 6 s of maximal exercise (knee extensions). In contrast, another study using a similar exercise protocol (5 s of knee extensions) showed significant elevations in these enzymes (Thorstensson 1975). Other research using a different mode of exercise (high-intensity cycle ergometer training) also yielded conflicting results. Parra and colleagues (2000) reported significant elevations in creatine kinase after 2 weeks of daily sprint training. However, when training was prolonged to 6 weeks with longer rest intervals between workouts, similar increases in resting enzyme concentrations were not seen. The findings were similar in other studies that examined high-intensity cycle ergometer exercise for 15 weeks and failed to show any changes in creatine kinase levels after training (Simoneau et al. 1987).

When exercise duration is increased, changes in the ATP-PC enzymes are still not consistent. When exercise consisted of 30 s of continuous knee extensions, significant elevations in both creatine kinase and myokinase were seen (Costill, Coyle, et al. 1979). In contrast, Jacobs and coauthors (1987) were unable to find any significant change in muscle creatine kinase levels after 6 weeks of high-intensity training (15 and 30 s maximal sprints on a cycle ergometer). The lack of any consistency in the ATP-PC enzyme response to exercise is difficult to explain. Minimal data are available regarding high-intensity, short-duration exercise programs and changes in these specific enzymes, and the available studies have used a variety of exercise protocols. It is possible that a longer duration of training may be needed to stimulate changes in resting creatine kinase levels or that there is an upper limit to creatine kinase concentrations within the muscle that cannot be altered with training. This latter hypothesis is explored further in the discussion of creatine kinase supplementation in chapter 19.

Adaptations of Glycolytic System to High-Intensity Training

As high-intensity exercise is prolonged, energy is derived primarily from the glycolytic energy system. Studies examining exercise training using bouts of exercise of 30 s or more have shown significant elevations in the glycolytic enzymes (**phosphofructokinase**, phosphorylase, **lactate dehydrogenase**) (Costill, Coyle, et al. 1979; Houston, Wilson, et al. 1981; Jacobs et al. 1987). The increase in the concentrations of these enzymes might enhance the glycolytic capacity, allowing the muscle to maintain a high intensity of exercise for a longer period of time. Increases between 10% and 31% in these glycolytic enzymes have been seen with high-intensity training programs varying in duration from 9 to 15 weeks (Linossier, Dormois, Perier, et al. 1997; Simoneau et al. 1987).

The increase in glycolytic enzymes appears to depend on the mode of exercise. In the studies just mentioned, the training involved either high-intensity running, cycling, or swimming. In contrast, studies examining the effect of resistance training on changes in glycolytic enzyme concentrations have been unable to show any significant alterations in these enzymes (Sale, MacDougall, and Garner 1990;

Sale et al. 1990; Tesch, Komi, and Hakkinen 1987). Apparently, resistance training alone is unable to stimulate any metabolic adaptation in the glycolytic enzymes. These studies suggest that athletes training for anaerobic sports with a large strength component (e.g., football) need to include both resistance training and sprint or interval exercises in their conditioning programs in order to maximize their physiological adaptation for the sport.

Adaptations of Oxidative Enzymes to High-Intensity Exercise

High-intensity exercise that stimulates increases in glycolytic enzymes also appears to significantly increase mitochondrial enzyme activity (oxidative enzymes) (Dudley, Abraham, and Terjung 1982; Troup, Metzger, and Fitts 1986). However, it appears that these increases are more prevalent when the duration of high-intensity exercise exceeds 3 min (Fitts 1992), but they are not consistently shown in all studies. Kohn and colleagues (2011), examining experienced endurance athletes, reported no change in oxidative enzymes following 6 weeks of high-intensity interval training (although significant increases in lactate dehydrogenase were noted). In addition, when increases in these enzymes resulting from high-intensity training are seen, they do not reach the magnitude typically observed after prolonged endurance training. An increase in oxidative enzymes with anaerobic training programs suggests that individuals who train anaerobically may still be able to generate some improvements in aerobic capacity.

High-Intensity Exercise and Buffering Capacity

High-intensity exercise (e.g., sprinting, cycling, swimming) results in an increase of acidity within the exercising muscle. The increase in H^+ causes a lowering of muscle pH and the onset of muscle fatigue. To offset the intramuscular acidity, there are several intrinsic buffering systems. Buffers such as **bicarbonate ($NaHCO_3$)** and muscle phosphates (Na_2HPO_4) combine with the H^+ released from high-intensity exercise to maintain **acid–base** balance within the exercising muscle. This prevents the onset

In Practice

Can High-Intensity Training Be Used to Enhance Fat Loss in Overweight Individuals?

The traditional exercise recommendation for individuals who desire to lose body fat is generally to have them train at a low intensity to maximize the duration of exercise. The greater duration of exercise, the greater the amount of fat oxidation. However, recent investigations have suggested that performing repeated high-intensity exercise may provide a significant benefit to fat oxidation. A group of investigators from the United Kingdom examined the effect of 2 weeks of repeated sprint training on fat oxidation in sedentary obese men (Whyte, Gill, and Cathcart 2010). During the 2-week training period, subjects performed six sessions of four to six repetitions of 30 s sprints on an electronically braked cycle ergometer, with a 4.5 min recovery between each repetition. Insulin sensitivity and resting fat oxidation were significantly elevated, while resting carbohydrate oxidation was significantly lower. The metabolic improvements provide evidence that high-intensity activity may be an alternative to prolonged endurance training for overweight individuals. This also supports the use of recreational sports such as soccer, basketball, or hockey as an effective means of improving body composition and health of sedentary individuals. A study from Copenhagen demonstrated that recreational soccer played for 1 h two or three times per week for 12 weeks was able to increase fat oxidation and lower body fat percent by 3% in previously untrained men (Krustrup et al. 2009). Thus, high-intensity competition may be a very effective method of eliciting body composition changes in overweight men. This may prove beneficial for people who either find low-intensity, long-duration exercise monotonous or simply respond better to competitive group activity or are more motivated to participate in this type of activity.

of fatigue and also allows the individual to exercise with a higher concentration of H^+ within the muscle. Training programs that stress the anaerobic energy system improve the buffering capacity within the muscle, changing the ability of the muscle to tolerate higher concentrations of acid buildup. Athletes with a high percentage of type II fibers can achieve a high buffering capacity and have a greater ability to sustain high-intensity exercise than athletes with predominantly type I fibers (Nakagawa and Hattori 2002).

Anaerobic exercise training (e.g., repeated sprints, high-intensity interval runs) appears to be a potent stimulator for enhancing buffering capacity. Increases of 12% to 50% in buffering capacity have been seen following 8 weeks of high-intensity exercise on a cycle ergometer (Sharp et al. 1986), while another study showed a 25% improvement following 5 weeks of interval training (6 to 10 × 2 min intervals on a cycle ergometer at ~130% of the subject's lactate threshold, 1 min rest between each sprint, 3 days per week) (Edge, Bishop, and Goodman 2006). A greater tolerance for high-intensity exercise is also reflected by an increase in blood lactate concentrations following maximal exercise. Jacobs and colleagues (1987) showed a 9.6% increase in blood lactate concentrations after 6 weeks of high-intensity cycling. An improved buffering capacity is also seen during exercise of submaximal intensity. Both blood and skeletal muscle lactate concentrations are lower in the trained state than in the untrained state (Holloszy and Booth 1976; Hurley et al. 1984). Comparisons of blood lactate concentrations in sedentary men before and after a 12-week endurance exercise training program are shown in figure 3.7. Common belief is that lower lactate production in the trained individual is related to a greater oxygenation of the exercising muscle, possibly from the increase in blood volume and capillary density seen after endurance training programs. However, some evidence suggests that during submaximal exercise, the blood flow per gram of muscle may actually be lower in the trained state compared with the untrained state (Holloszy 1988). This is thought to evolve from a reduced diversion of blood flow from the periphery (skin) or internal organs (liver) to the muscles during submaximal exercise. The exercising muscles in the trained state compensate for the lower blood volume by extracting more oxygen, resulting in a larger arteriovenous oxygen difference.

Lower lactate production in trained individuals during submaximal exercise may also be related to greater reliance on fat as the primary energy source for ATP generation. In addition, the greater mitochondrial content in the muscle after endurance training programs reduces the available pyruvate for conversion to lactate. A greater proportion of the pyruvate or lactate generated through glycolysis is channeled into the mitochondria for oxidative metabolism. Another mechanism that may contribute to reduced lactate concentrations in the blood (which reflect muscle lactate production) is an increase in the rate of lactate removal (Holloszy 1988) or possibly a greater conversion to energy substrate (Brooks 2009).

To maximize the physiological adaptations resulting from anaerobic exercise, the mode of training

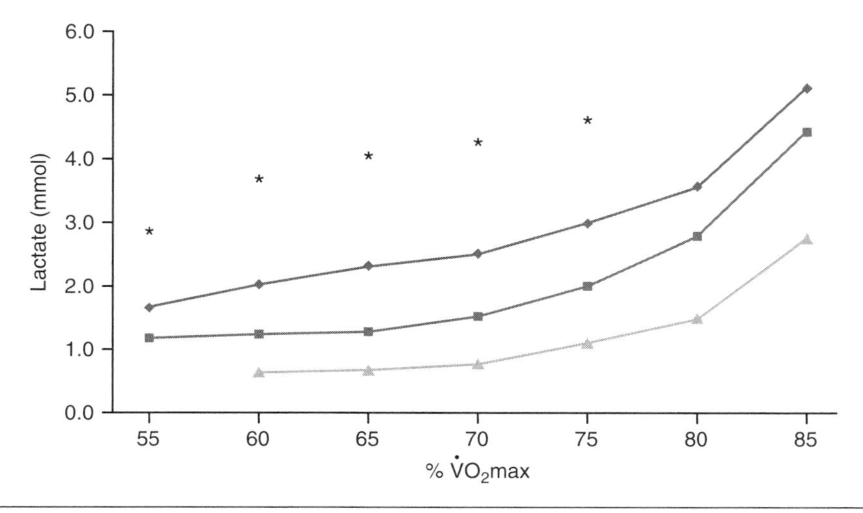

Figure 3.7 Blood lactate concentrations at the same relative exercise intensities. Sedentary men studied before (♦) and after (■) a 12-week endurance training program; competitive long-distance runners (▲) training regularly. * = Significant difference between pre and post 12-week training program.

Data from Hurley et al. 1984.

may have an important role. Many strength–power athletes rely primarily on resistance exercise to prepare for a season of competition. However, high-repetition, short-rest resistance training does not appear to enhance muscle buffering capacity, despite an improvement in repeated sprint ability (Edge, Bishop, and Goodman 2006). Although muscle buffering capacity was not improved, Edge, Hill-Haas, and colleagues (2006) reported a significant reduction in H^+ and lactate concentrations following 5 weeks of training, suggesting improved metabolic adaptations. These data emphasize that athletes preparing for a season of competition need to incorporate anaerobic conditioning programs (see chapter 10) into their training program to maximize physiological adaptation from training.

Summary

Metabolic adaptations are specific to the type of exercise training program employed. Endurance training results in improved capacity to generate energy through oxidative metabolism. This is reflected by increases in capillary density, mitochondrial size and content, and oxidative enzymes. Anaerobic training results in elevations in the glycolytic enzymes as well as enhanced buffering capacity within skeletal muscle. In addition, although high-intensity training does not rely on oxidative metabolism as its primary energy source, the increase in oxidative enzymes during high-intensity exercise programs suggests that slight improvements in aerobic capacity may also be seen with these programs.

REVIEW QUESTIONS

1. Provide justification for using 3 min of rest between sets during resistance exercise.
2. What are the metabolic adaptations that are common to prolonged endurance training?
3. What is the benefit of increasing capillary density for the endurance athlete?
4. Describe the role that mitochondria have in oxidative phosphorylation and how exercise affects mitochondrial function.
5. Describe the role that the bicarbonate buffering system has in enhancing repeat sprint performance.

chapter 4

Cardiovascular System and Exercise

CHAPTER OBJECTIVES

After reading this chapter you should be able to do the following:

- Understand the morphology of the heart and the cardiac cycle.
- Explain the importance of gas pressure differentials and how they control gas exchange in the lungs, heart, and tissue.
- Explain the acute cardiovascular response to exercise.
- Describe specific cardiovascular adaptations to chronic training and differentiate between adaptations resulting from endurance exercise and those resulting from resistance exercise.
- Describe the effect of physical exercise training on blood volume and red blood cell formation.

At rest, the heart provides approximately 5 L blood per minute to meet the energy demands of the average person. As the metabolic demands increase, as might be expected during exercise, the heart is able to compensate by increasing the volume of blood that it pumps into circulation. Cardiac output during exercise can increase more than fourfold in an average person; and in an elite endurance athlete, cardiac output may reach 40 L/min. Like any other muscle, the heart adapts to the increased demands placed on it during prolonged exercise training. These adaptations are specific to the type of exercise stimulus that is presented. This chapter reviews the cardiovascular adaptations seen during acute exercise and prolonged endurance and resistance training. The close relationship between the cardiovascular and respiratory systems and the effect that exercise has on improved cardiorespiratory function are also discussed.

Overview of Cardiovascular System

The cardiovascular system consists of an elaborate network of **vessels** (the circulatory system) and a powerful pump (**the heart**). It is responsible for delivering oxygen and nutrients to active organs and muscles and removing the waste products of metabolism. The heart is a four-chambered muscular organ located in the midcenter of the chest cavity. Its anterior border is the sternum, and its posterior border is the vertebral column. The diaphragm is inferior to the heart, and the lungs are situated on the lateral borders. Approximately two-thirds of the heart's mass lies to the left of the body's midline. The longitudinal axis of the heart from its base to its apex is directed anterior-inferior and 45° to the left of the midline.

Morphology of the Heart

The heart muscle, referred to as the **myocardium**, is similar in appearance to striated skeletal muscle. However, the fibers of the myocardium are multinucleated and interconnected end to end by **intercalated discs**. These discs contain **desmosomes**, which maintain the integrity of the cardiac fibers during contraction, and **gap junctions** that allow for a rapid transmission of the electrical impulse that signals for contraction. The structure of the myocardium can be thought of as consisting of three separate areas: **atrial**, **ventricular**, and **conductive**. The atrial and ventricular myocardium function similarly to skeletal muscle in that they contract in response to electrical stimuli. However, an electrical stimulus of only a single cell in either chamber results in the rapid spread of an **action potential** to the other cells of the atrial and ventricular myocardium, causing a coordinated contractile mechanism. In addition, the cardiac fibers in each of these areas can function separately. The conductive tissue found between these chambers provides a network for the rapid transmission of conductive impulses, allowing for coordinated action of the atrial and ventricular chambers.

The structural detail of the heart can be seen in figure 4.1. A striking difference in the anatomy and physiology of the right and left sides of the heart relates to their specific functions. The right side of the heart (right **atrium**) receives blood from all parts of the body, and the right **ventricle** pumps deoxygenated blood to the lungs through the **pulmonary circulation**. The left atrium receives oxygenated blood from the lungs and pumps this blood from the left ventricle into the **aorta** and through the entire **systemic circulation**. The left ventricle is an ellipsoidal chamber surrounded by thick musculature that provides the power to eject the blood through the entire body. The right ventricle, however, is crescent shaped with thin musculature, reflecting the reduced ejection pressures seen in this ventricle, 25 mmHg compared with approximately 125 mmHg in the left ventricle at rest. A thick solid muscular wall, or **interventricular septum**, separates the left and right ventricles.

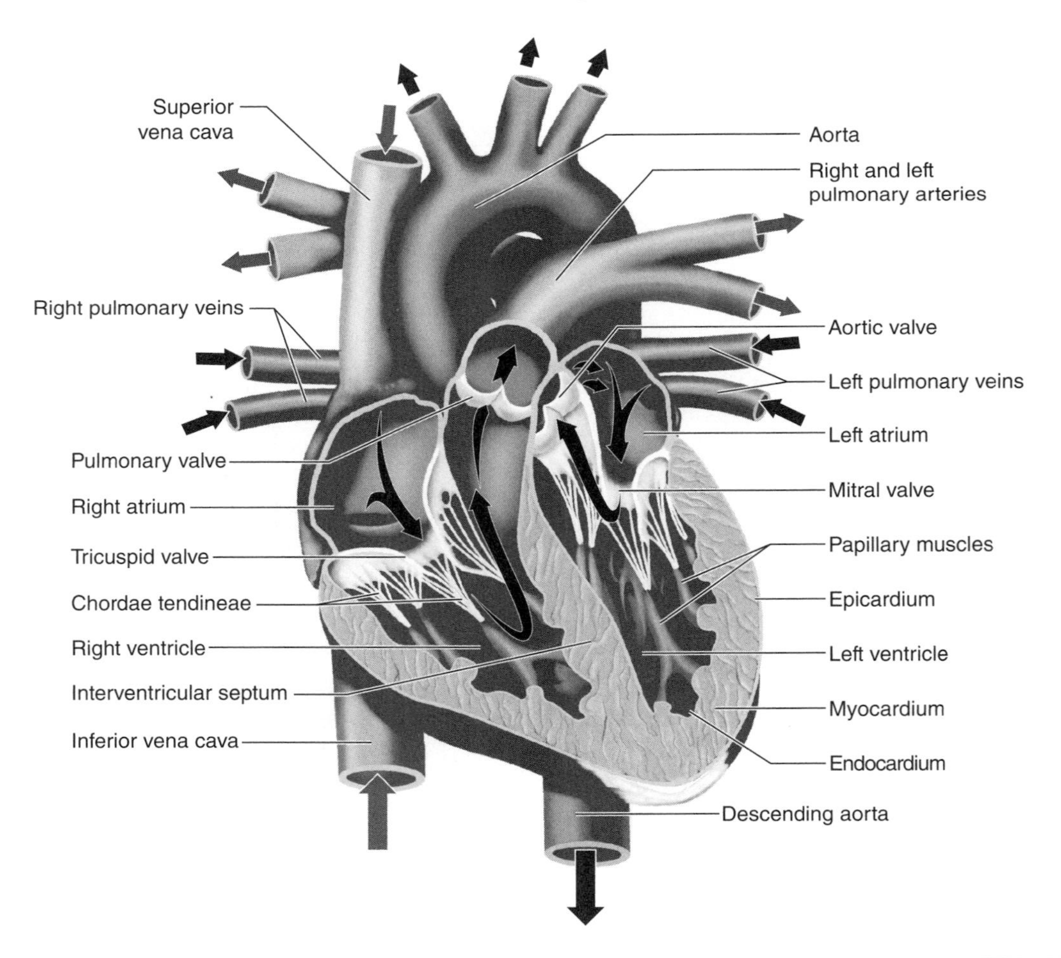

Figure 4.1 Structural detail of the heart.

Blood flow from the right atrium to the right ventricle goes through the **tricuspid valve** (consisting of three cusps or leaflets that allow only a unidirectional flow of blood). The **bicuspid** or **mitral valve** allows blood flow between the left atrium and left ventricle. The **semilunar valves**, located on the arterial walls of the outside of the ventricles, prevent blood from flowing back into the heart between contractions. During **systole**, the cusps lie against their arterial wall attachments; during **diastole** or during retrograde flow, the cusps fall passively inward, sealing the **lumen**.

Cardiac Cycle

The **contraction phase** in which the atria or ventricles expel the blood in their chambers is called systole. The **relaxation phase** in which these chambers refill with blood is referred to as diastole. The **cardiac cycle** is the total time spent in one complete revolution of systole and diastole. At rest, the heart spends most of its time (approximately 60%) filling with blood (diastole) and less time (approximately 40%) expelling the blood (systole). However, during exercise this situation is reversed, with most of the cardiac cycle spent in systole. During systole, the tricuspid and mitral valves are closed. However, blood flow from pulmonic and systemic circulation continues into the atria. As systole ends, the atrioventricular valves rapidly open and the blood that has accumulated in the atria flows quickly into the ventricles, accounting for 70% to 80% of the ventricular filling. This period of rapid filling accounts for one-third of diastole. The middle one-third of diastole is characterized by very little blood flow into the ventricle and is referred to as **diastasis**. During the last one-third of diastole, ventricle filling is completed, with an additional 20% to 30% of blood pumped into the ventricle as the result of atrial systole.

The volume of blood in the ventricle at the end of diastole is called the **end-diastolic volume (EDV)**. Two main phases occur during systole: **pre-ejection** and **ejection**. The pre-ejection phase includes an electromechanical lag, which is the time delay between the beginning of ventricular excitation (**depolarization**) and the onset of ventricular contraction and **isovolumic contraction**. Isovolumic contraction is the phase in which intraventricular pressure is raised before the onset of ejection. This part of the pre-ejection phase occurs between the closure of the mitral valve and the opening of the aortic semilunar valve. During the ejection phase, the blood within the ventricle is pumped into the systemic circulation through the opening of the semilunar valve. This phase ends with the closing of the semilunar valve. The blood remaining in the ventricle at the end of ejection is referred to as **end-systolic volume (ESV)**. The difference between EDV and ESV is called the **stroke volume (SV)**. The proportion of the blood pumped out of the left ventricle with each beat is called the **ejection fraction (EF)** and is determined by SV ÷ EDV. The ejection fraction averages about 60% at rest. This simply means that 60% of the blood in the left ventricle at the end of diastole will be ejected with the next contraction. Figure 4.2 depicts a complete cardiac cycle.

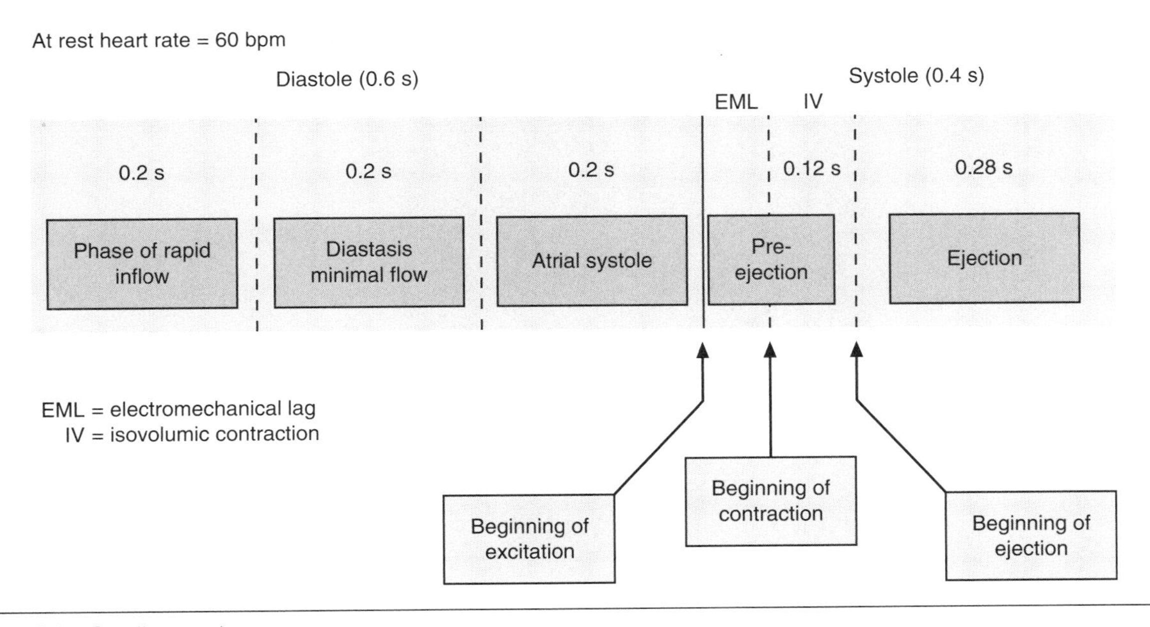

Figure 4.2 Cardiac cycle.

Heart Rate and Conduction

A unique feature of the heart is its ability to contract rhythmically without either neural or hormonal stimulation. This autorhythmicity is due to a specialized intrinsic conduction system that consists of the **sinoatrial node (SA node)**, **internodal pathways**, the **atrioventricular node (AV node)**, and **Purkinje fibers** (figure 4.3).

The SA node is located in the right atrium and is a collection of specialized cells that are capable of generating an electrical impulse. Because of this distinctive ability, the SA node is appropriately nicknamed the pacemaker of the heart. Once an impulse leaves the SA node, it propagates leftward and downward, spreading through the **atrial syncytium** of first the right and then the left atrium along internodal pathways to the AV node, which is located toward the center of the heart on the lower right atrial wall.

The **AV node**, or the AV junction (made up of the AV node and the **bundle of His**), delays transmission of the impulse for 0.1 s. This slight delay of ventricular excitation and contraction allows the atria to contract and also permits a limitation in the number of signals that are transmitted by the AV node. This appears to serve as a protective mechanism for the ventricles from atrial tachyarrhythmias. The bundle of His is found distally in the AV junction and divides into a right and a left segment (**bundle branches**) that transmit the electrical impulses to the right and left ventricles, respectively. The Purkinje fibers are found on the distal tips of the right and left bundle branches and extend into the walls of the ventricles, accelerating the conduction velocity of the impulse to the rest of the ventricle. The conduction velocity of the Purkinje fibers may increase fourfold compared with that of the bundle of His.

As mentioned earlier, the SA node, AV node, and Purkinje fibers have the inherent ability for spontaneous initiation of the electrical impulse. However, the autonomic nervous system can also influence the rate of impulse formation (**chronotropy**), the contractile state of the myocardium (**inotropy**), and the rate of spread of the excitation impulse. The **sympathetic** and **parasympathetic** nervous systems, as well as certain hormones, can influence cardiac contractility. The atria are well supplied with both sympathetic and parasympathetic neurons, whereas the ventricles are primarily innervated by sympathetic neurons. Sympathetic stimulation releases the catecholamines epinephrine and norepinephrine from sympathetic

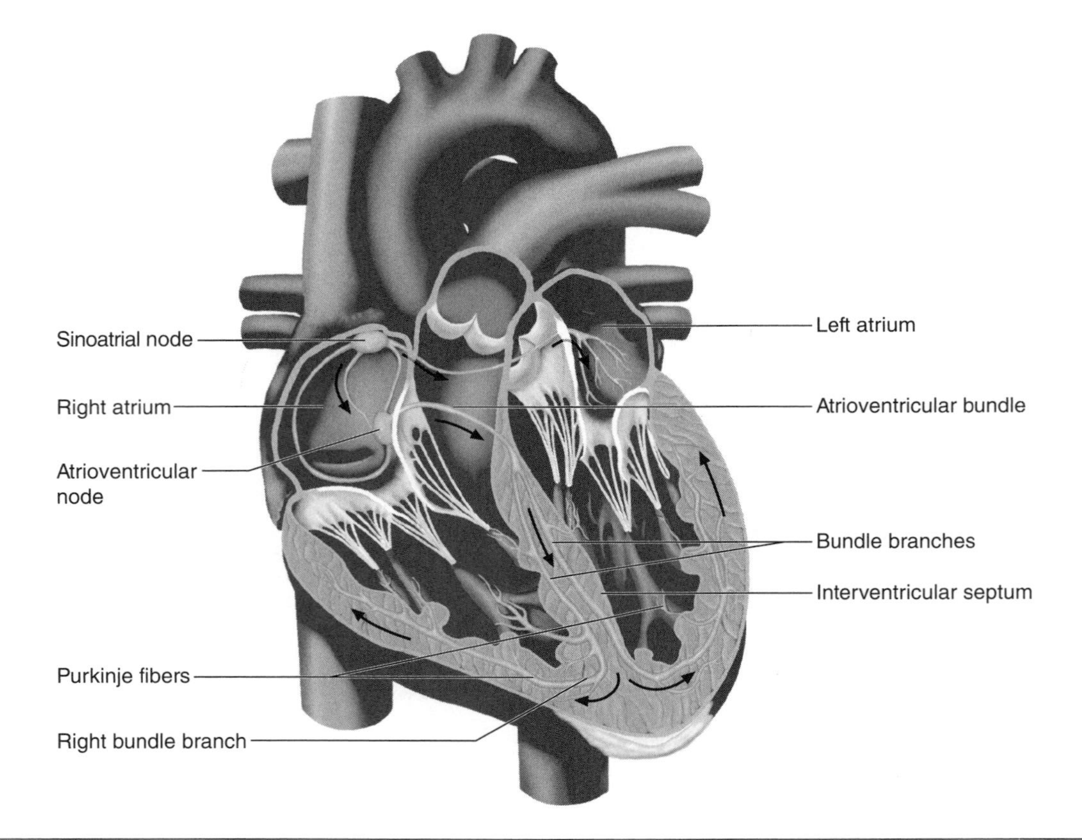

Figure 4.3 Conduction system of the heart.

neural fibers. These neural hormones accelerate heart rate by increasing SA node activity, and they increase both atrial and ventricular contractile force. Increases in heart rate are termed **tachycardia**.

Parasympathetic stimulation through the **vagus nerves** releases the neurohormone **acetylcholine**, which depresses SA node activity and decreases atrial contractile force. Decreases in heart rate are termed **bradycardia**. Sympathetic stimulation may increase heart rate by over 120 beats/min and strength of contraction by 100%, whereas maximal vagal stimulation may decrease heart rate by 20 to 30 beats/min and may lower strength of contraction by approximately 30% (Adamovich 1984).

Electrocardiogram

The electrocardiogram (ECG) is the electrical measure of heart activity from electrodes placed on the chest. The typical ECG uses 10 electrodes that are placed on specific locations on the chest, arms, and legs (figure 4.4). These leads or electrodes record different electrical signals that provide information on the heart's rhythm and conduction. Willem Einthoven, a Dutch physician and scientist, is credited with the invention of the ECG in the early part of the 20th century. He labeled the specific wave patterns P, Q, R, S, and T to note the deflections in the electrical activity of the heart. In contrast to the 12-lead ECG, a heart rate monitor typically used by a runner has a single lead.

Stimulation of cardiac tissue is a continuous sequence of a process referred to as depolarization and repolarization. The contraction of the atria and ventricles occurs when the cardiac cells are depolarized. As previously mentioned, this corresponds with the pumping of blood by the atria and ventricles. The relaxation phase, allowing the cardiac cells to return back to normal resting state, is the repolarization. In the ECG, this is represented by the depolarization of the atria (P wave), followed by the depolarization of the ventricles (QRS complex) and then a return to resting state or repolarization (S and T waves) (figure 4.5).

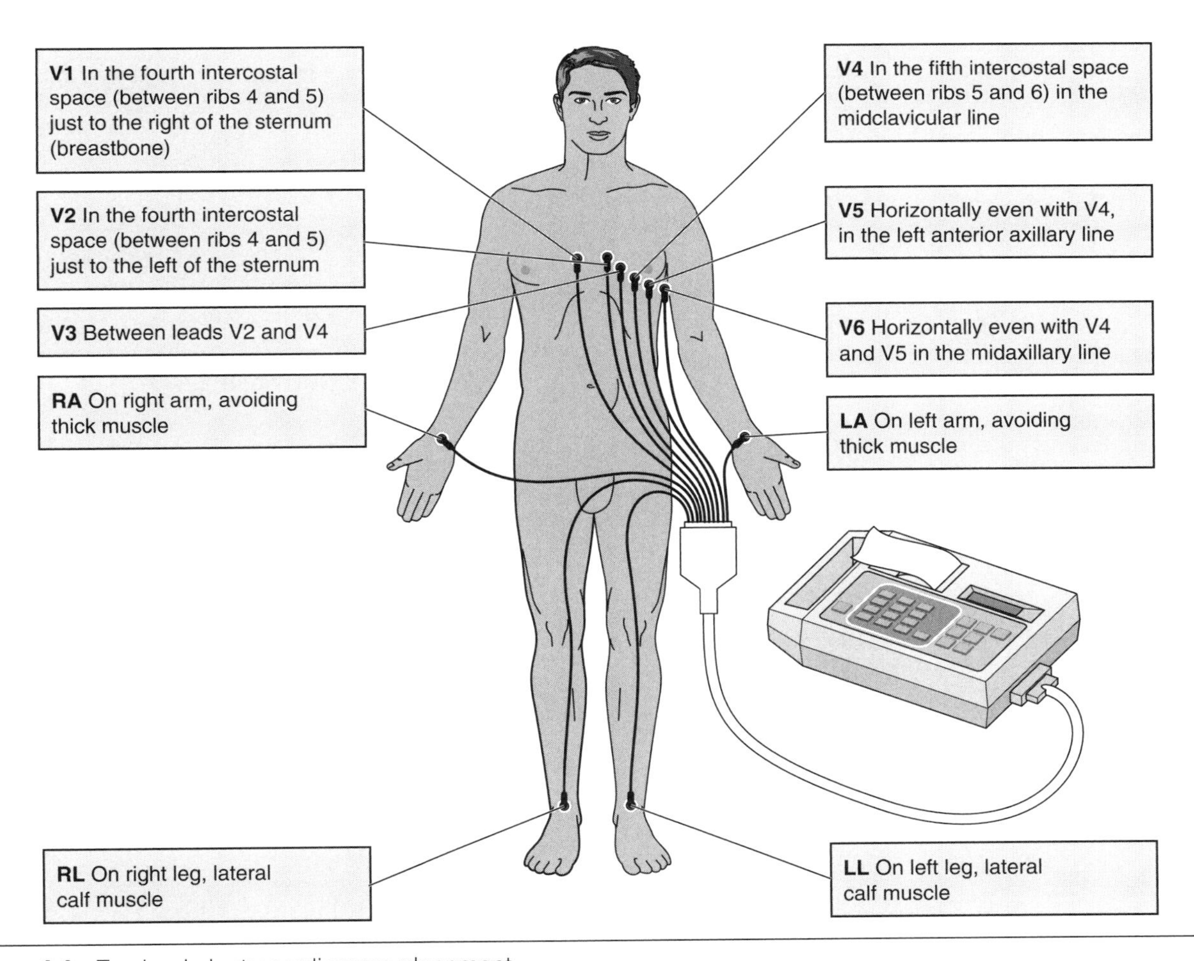

Figure 4.4 Ten-lead electrocardiogram placement.

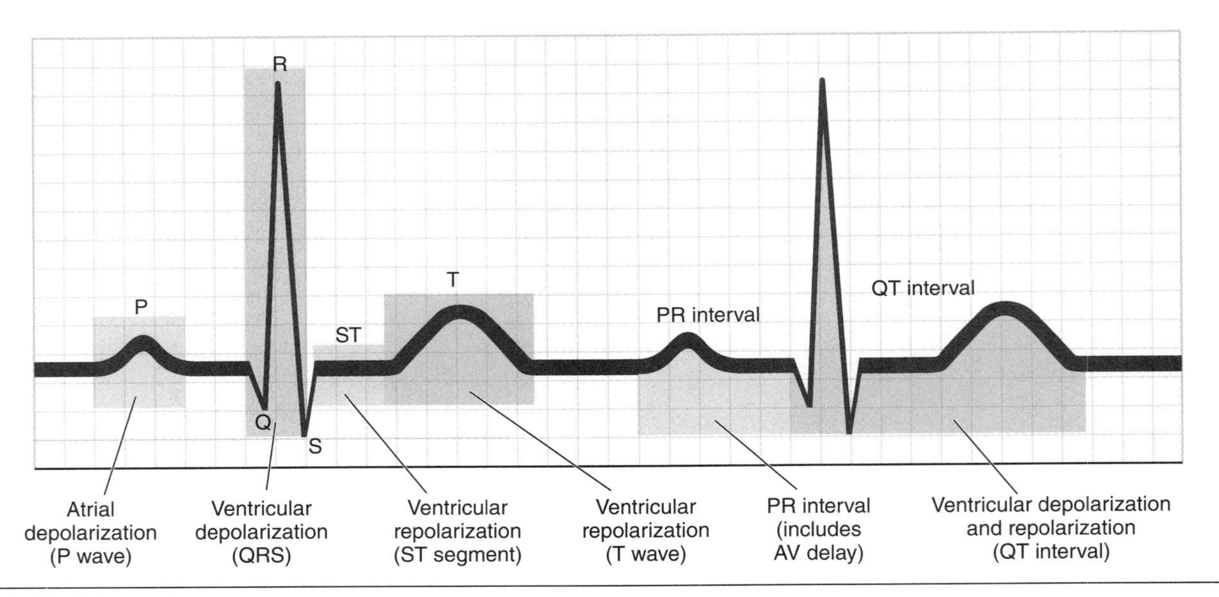

Figure 4.5 A graphic illustration of the various phases of the resting electrocardiogram.

- ***P wave.*** The P wave is the depolarization of the atria occurring at the SA node and spreading from the right atrium to the left atrium.
- ***QRS complex.*** The QRS complex represents the depolarization of the ventricles, where the blood is being ejected from the heart. The direction of the electrical impulse indicates whether the deflection will be positive or negative. The first negative deflection is the Q wave; the first positive deflection is the R wave; and the following negative deflection is the S wave.

Because of the fast response of the Purkinje fibers in the ventricles, the QRS complex depolarizes quickly. For this reason, we see in the QRS complex conduction delays resulting in an abnormal ECG. In the case of left and right ventricular hypertrophy, it is important to be aware that thin, young, and athletic individuals demonstrate tall voltage in the R waves of the chest leads. It is difficult sometimes to differentiate between someone who is a highly conditioned athlete and someone who has ventricular hypertrophy. It is usually recommended that the individual in question receive further testing in the form of an echocardiogram to make the differential diagnosis. Even then, diagnosing may still be difficult.

- ***T wave.*** Following the depolarization of ventricles and the resulting QRS complex, the T wave represents the beginning of ventricular repolarization. T waves are normally asymmetrical in shape, with the asymmetry occurring toward the end of the wave in the form of a peak. A symmetrical T wave may be indicative of a myocardial infarction. Frequently the S wave and the T wave are looked at collectively; the two together are referred to as the ST segment. The Q wave and T wave are also sometimes paired to provide the QT interval.

Cardiac Output

Cardiac output is the product of heart rate and stroke volume. The term generally refers to the amount of blood pumped by the heart in 1 min. Cardiac output responds to the energy demands of the body and varies considerably between people. On average, the total blood volume pumped out of the left ventricle is approximately 5 L/min for an adult male. This volume is similar for trained and sedentary males. In the untrained male, this 5 L blood is sustained with a heart rate of 70 beats/min. Thus, stroke volume would need to be approximately 71 ml/beat. In the endurance athlete, heart rate is generally much lower at rest because of greater vagal tone and reduced sympathetic drive. If the heart rate of the endurance athlete was 50 beats/min, the stroke volume would be 100 ml/beat. Table 4.1 compares the cardiac output in trained and sedentary males. The mechanism that drives this particular adaptation is not entirely clear but is likely related to the increased vagal tone seen after endurance training and to some of the morphological adaptations of the heart, which are discussed later.

Vasculature

The **vascular system** is composed of a series of vessels that carry oxygenated blood away from the heart to the tissues (**arterial system**) and return

In Practice

How Accurately Can We Predict Maximal Heart Rate?

Most students of exercise science, kinesiology, or related fields quickly learn that the formula to calculate maximal heart rate is 220 minus the person's age (220 – age). Most students, though, do not learn that there is absolutely no scientific evidence to support the formula. The formula appears to have originated from a superficial estimation of a linear best fit model based on raw and mean data compiled in 1971 (Robergs and Landwehr 2002). However, Robergs and Landwehr (2002), in a critical examination of heart rate prediction equations, clearly demonstrated that these equations could not be supported because of unacceptably high error. In reviewing the development of the heart rate prediction formulas, they linked the 220 – age formula with the concept of heart rate reserve developed by Karvonen and colleagues (1957). In a conversation with Robergs and Landwehr in 2000, Dr. Karvonen said that he had not developed the formula and suggested that they speak with Dr. Åstrand. In subsequent conversation with Dr. Åstrand, the authors soon realized that he had not published any data relating to this formula. They concluded in their study that the origin of the 220 – age heart rate prediction formula could be traced to Fox and colleagues (1971). Although the formula had been attributed to Robergs and Landwehr's study, these authors indicated that it was not based on their research. When reviewing the research on maximum heart rate, Robergs and Landwehr (2002) concluded that a large error occurs when age is the sole independent variable. The errors in heart rate prediction equations can be as high as 11 beats/min, while the error in determinations of maximal heart rate is reported to be 2 beats/min (Robergs and Landwehr 2002). In conclusion, Robergs and Landwehr (2002) stated that for the purposes of prescribing training heart rate ranges, an error of 8 beats/min may be considered acceptable, but for determination of maximal aerobic capacity ($\dot{V}O_2max$) they suggested that errors in predicting maximal heart rate should be <3 beats/min.

Table 4.1 Cardiac Output at Rest in Sedentary and Endurance-Trained Males

Cardiac output = heart rate × stroke volume	
Sedentary	4970 ml = 70 beats · min^{-1} · 71 ml
Trained	5000 ml = 50 beats · min^{-1} · 100 ml

deoxygenated blood from the tissues back to the heart (**venous system**). The heart has its own coronary vascular system responsible for supplying the myocardium with oxygen and nutrients. The arterial system receives the blood pumped from the left ventricle of the heart and distributes it throughout the body via a network of **arteries**, **arterioles** (small arterial branches), **metarterioles** (smaller branches), and **capillaries**. The left ventricle pumps the blood from the heart into a thick elastic vessel called the aorta. The blood is then circulated throughout the body via the previously mentioned arterial network. The walls of the arteries are strong and thick to withstand the rapid transport of blood under high pressure to the tissues. Their thickness prevents any exchange of gases between the arterial vessels and the surrounding tissues. In addition, these vessels are richly innervated by the sympathetic nervous system, which allows them to be effectively stimulated for regulating blood flow. As the blood reaches the tissues, it gets diverted to smaller branches of the arterial system. At the ends of the metarterioles (the smallest arterial vessels) are the capillaries, which consist of a single layer of **endothelial cells**. The capillaries are microscopic in size (approximately 0.01 mm in diameter) but may contain approximately 5% of the total blood volume at rest. Because of this small diameter, the rate of blood flow decreases as the blood circulates toward and into the capillaries. In addition, extensive branching of the capillary microcirculation allows it to reach a large surface area. The large surface area, the slow rate of blood flow, and the thin layer of endothelial cells make the capillaries an ideal place for gas exchange between the blood and the tissues.

As the blood leaves the capillaries, it enters the venous circulation. The venous system comprises vessels increasing in size as they get closer to the heart. Deoxygenated blood leaving the capillaries enters **venules** (small veins), and the rate of blood flow is increased (because of the smaller cross-sectional area of the venous system in comparison with the capillary system). The blood is transported back to the heart via the **superior vena cava** (venous blood returning from areas above the heart) and the **inferior vena cava** (venous blood returning from areas below the heart). The deoxygenated blood then enters the right atrium, goes through to the right ventricle, and is pumped to the lungs to be reoxygenated and subsequently transported back into the left side of the heart to be distributed through the arterial circulation.

During rest, blood flow is controlled by the **autonomic nervous system** and is primarily distributed to the liver, kidneys, and brain. However, during exercise, a redistribution of the blood flow to the exercising muscles occurs. The muscles may receive 75% or more of the available blood at the expense of the other organs. In combination with a greater cardiac output, the exercising muscles may receive up to a 25-fold increase in blood flow. The flow of blood to the muscles and organs at rest and during exercise is shown in figure 4.6.

Blood Pressure

In a resting state, as the blood is pumped into the aorta from the left ventricle during contraction (systole), the pressure within the aorta increases to approximately 120 mmHg. This measurement is referred to as the **systolic blood pressure** and represents the strain against the arterial walls during ventricular contraction. Since the pumping action or contraction of the left ventricle of the heart is pulsatile, the arterial pressure fluctuates between its high level during systole to a lower level during the relaxation phase of the heart (diastole). **Diastolic blood pressure** is approximately 80 mmHg at rest and provides an indication of the **peripheral resistance**, or ease with which blood flows into the capillaries. As blood flows through the systemic circulation, the pressure continues to progressively fall to approximately 0 mmHg as it reaches the right atrium. The decrease in arterial pressure during each segment of the systemic circulation is directly proportional to the vascular resistance in that segment. Changes in the resistance of the systemic circulation are important for the regulation of blood flow.

Overview of Respiratory System

The coordination between the cardiovascular and respiratory systems provides the body with an efficient means to transport oxygen to the tissues and remove carbon dioxide. During respiration, air is breathed in (**inspiration**) through the nasal cavity or mouth. From there the air travels through the **pharynx**, **larynx**, and **trachea** and into the lungs. Once in the lungs, the air flows through an elaborate system of branches termed **bronchi** and **bronchioles** that expand the surface area for gas exchange (figure 4.7). From the bronchioles the air reaches the **alveoli**, the smallest respiratory unit, where gas exchange with the pulmonary circulation occurs. The lungs are located in the chest cavity (**thorax**) but do not have any direct attachment to the ribs or any other bony structure. Instead, they are suspended by **pleural sacs** that connect to both the lungs and thoracic cavity. A fluid is present between the pleural sacs and lungs to prevent friction during respiration.

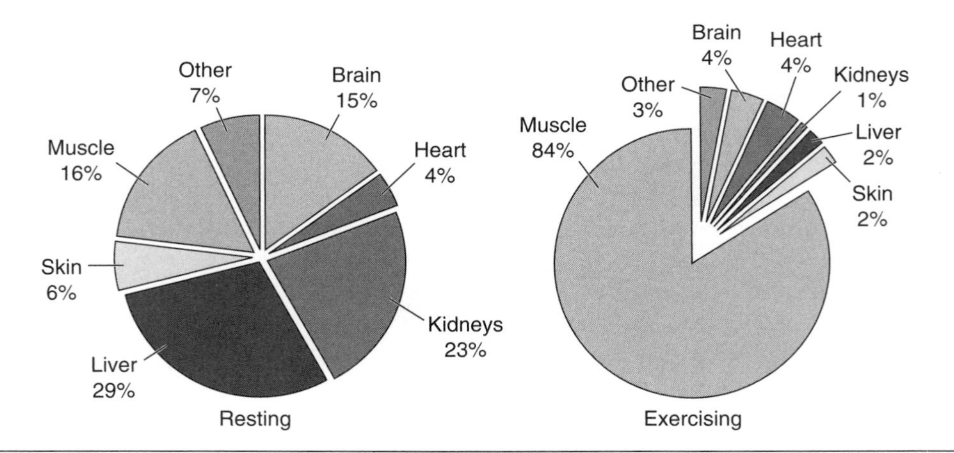

Figure 4.6 Distribution of cardiac output during resting and exercising.

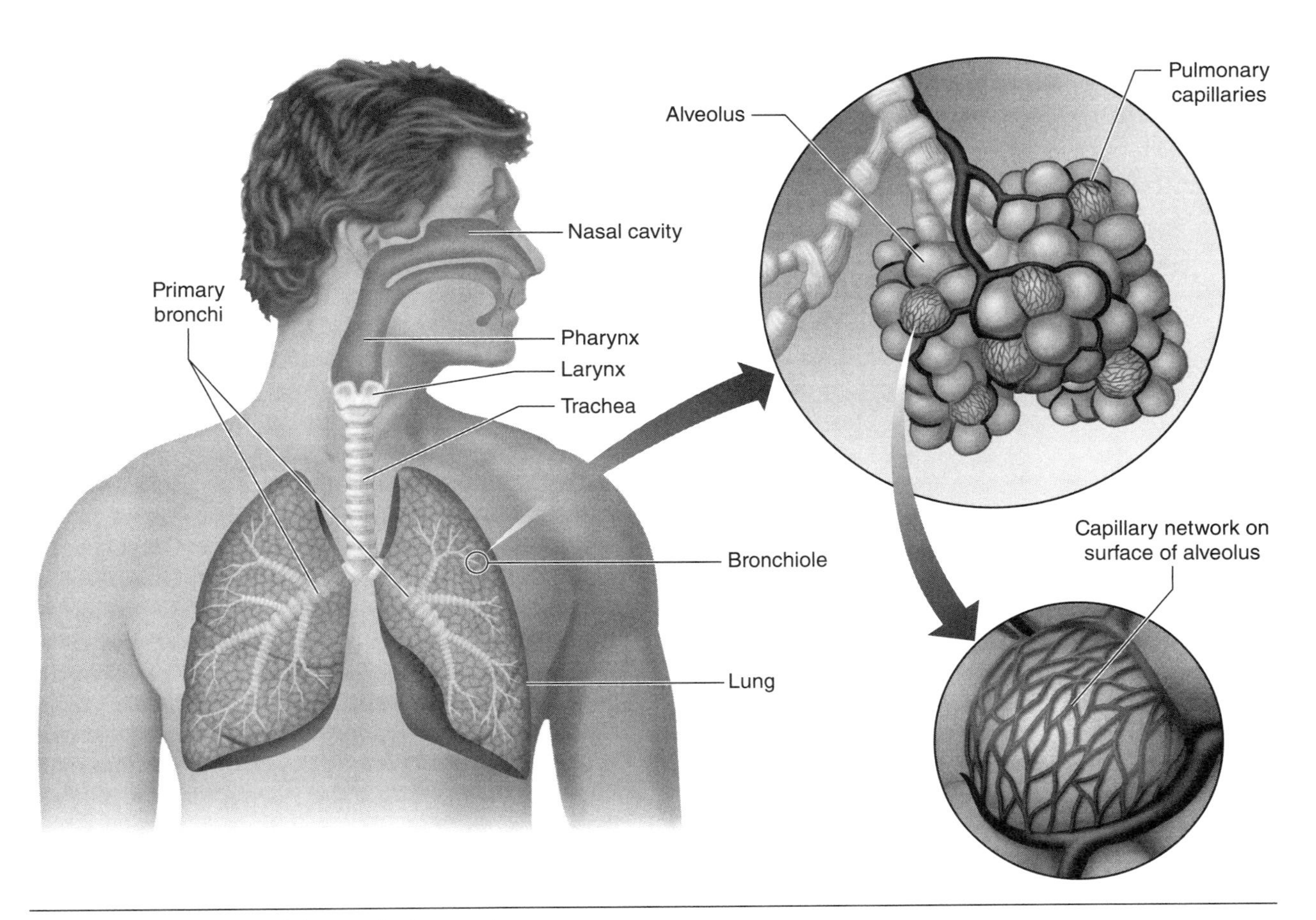

Figure 4.7 Anatomy of the respiratory system.

Inspiration and Expiration

During inspiration, the muscles of the thoracic cavity (**diaphragm** and **external intercostal**) contract, causing the thorax to expand and the lungs to stretch and fill with additional air. The lung expansion causes a reduced pressure gradient, and the pressure within the lungs is reduced to levels below that on the outside, causing air to rush in. During exercise, additional muscles (e.g., pectorals, sternocleidomastoid) may be recruited, causing a greater movement of the thorax and creating an even larger lung expansion.

When air is breathed out (**expiration**), the inspiratory muscles relax. In the case of forced expiration, contraction of the **internal intercostal** and abdominal muscles causes the thorax to return to its normal position. As a result, the pressure within the lungs expands to levels above that outside, and expiration occurs.

Change in pressure is the primary reason for air and gases to flow into and out of the lungs and through the entire respiratory and circulatory systems. For **ventilation** (the process of inspiration and expiration) to occur, only small changes in pressure between the lungs and the outside environment are required. For instance, standard atmospheric pressure is 760 mmHg, and only a slight change in **intrapulmonary pressure** (pressure within the lungs) causes air to be inhaled. This process is not as simple at altitude and is explained in much greater detail in chapter 23.

Pressure Differentials in Gases

In addition to changes in pressure that cause inspiration and expiration, pressure differentials in the air also result in both oxygen and carbon dioxide exchange. The air we breathe is a mixture of gases. Each gas exerts a pressure in proportion to its concentration in the gas mixture, known as its **partial pressure**. Air is made up of 79.04% nitrogen, 20.93% oxygen, and 0.03% carbon dioxide. Thus, at sea level where atmospheric pressure is 760 mmHg, the partial pressure of oxygen is 159.1 mmHg (20.93% of 760 mmHg) and carbon dioxide is 0.2 mmHg (0.03% of 760 mmHg).

As the air reaches the alveoli, the partial pressures of the gases in the alveoli and the partial pressures of the gases in the blood create a pressure gradient. This is the basis of gas exchange. If the partial pressures of the gases on either side of the membrane are equal, no gas exchange occurs. The greater the pressure gradient, the faster the gases will diffuse across the membrane. As the inspired air moves into the alveoli, the partial pressure of oxygen (PO_2) is between 100 and 105 mmHg (due to mixing of air within the alveoli). The pressure gradient between the capillaries and the alveoli is depicted in figure 4.8. At the pulmonary capillary, blood has been stripped of most of its oxygen by the tissues. Typically, the PO_2 at the pulmonary capillary is between 40 and 45 mmHg. As you can see, the pressure gradient favors oxygen going from the alveoli to the capillary. In addition, the pressure gradient of carbon dioxide favors exchange from the capillary to the alveoli, where it can be exhaled from the body during expiration. The pressure gradient for carbon dioxide is not as great at the capillary–alveoli membrane as it is for oxygen. Nevertheless, carbon dioxide diffuses easily across the membrane, despite the low pressure gradient, because of a greater membrane solubility compared to oxygen.

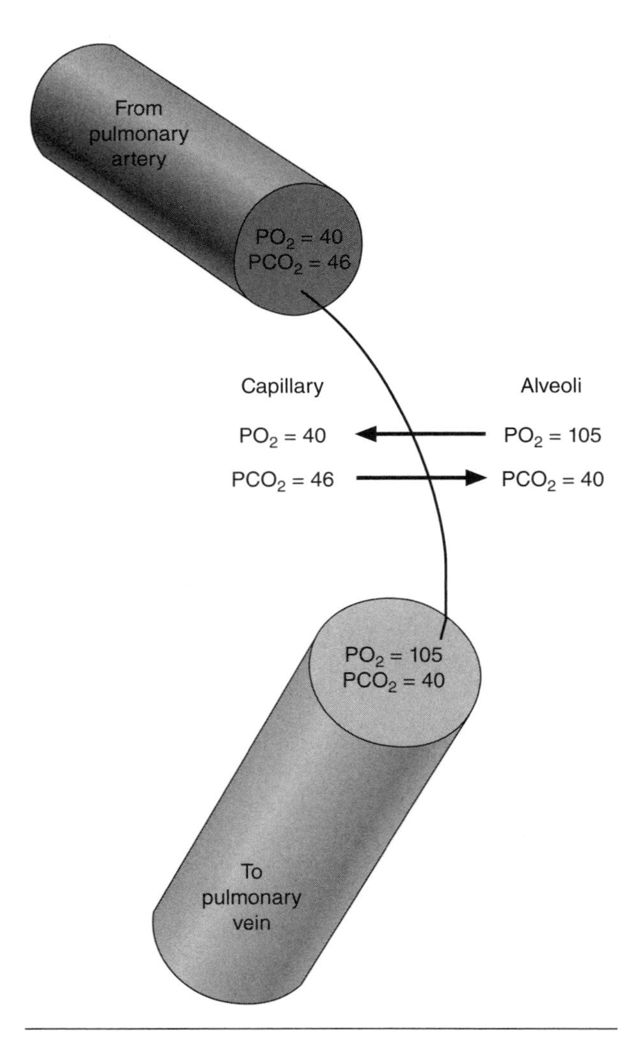

Figure 4.8 Pressure gradient between the capillary and alveoli within the lungs.

Oxygen and Carbon Dioxide Transport

Oxygen is transported in the blood either combined with **hemoglobin** (98%) or dissolved in blood plasma (2%). Each molecule of hemoglobin can carry four molecules of oxygen. The binding of oxygen to hemoglobin depends on the PO_2 in the blood and the affinity between oxygen and hemoglobin. The greater the PO_2, the more saturated the hemoglobin molecules are with oxygen. In addition, the temperature and pH of the blood (figure 4.9) also affect the affinity between oxygen and hemoglobin. As the pH of the blood decreases, the affinity that hemoglobin has to oxygen is decreased and oxygen is released. The rightward shift of the curve is known as the **Bohr effect** and is important during exercise when a greater amount of oxygen is needed in the exercising tissues. On the other hand, when the pH is high, as it would be in the lungs, there is a greater affinity between oxygen and hemoglobin. This is important in order to saturate the hemoglobin molecules with oxygen.

In the average male, there is approximately 14 to 18 g of hemoglobin in each 100 ml of blood. In the female, the concentration of hemoglobin ranges from 12 to 16 g per 100 ml of blood. Each gram of hemoglobin can bind 1.34 ml of oxygen. Thus, for males, the oxygen-carrying capacity of hemoglobin fully saturated with oxygen is approximately 18 to 24 ml per 100 ml of blood, while in females the range is approximately 16 to 22 ml per 100 ml of blood. At rest, normal oxygen saturation is approximately 95% to 98% (Pruden, Siggard-Anderson, and Tietz 1987).

Carbon dioxide transport in the blood occurs primarily in the form of **bicarbonate ions** (approximately 60-70%). Carbon dioxide is also transported dissolved in the **plasma** (7-10%) or bound to hemoglobin. When bound to hemoglobin it forms the molecule **carbaminohemoglobin**. However, it does not compete with oxygen since it has its own binding site on the **globin molecule**. In contrast, the binding site for oxygen is on the **heme molecule**. As carbon dioxide diffuses from the muscle to the blood, it combines with water to form **carbonic acid**. This very unstable acid quickly dissociates, releasing a hydrogen ion (H^+) and forming a bicarbonate ion (HCO_3^-). The H^+ binds to hemoglobin and causes the Bohr effect to occur, whereby hemoglobin loses its affinity for oxygen and increases the rate of diffusion of oxygen into the tissues. An example of this action is depicted in figure 4.10.

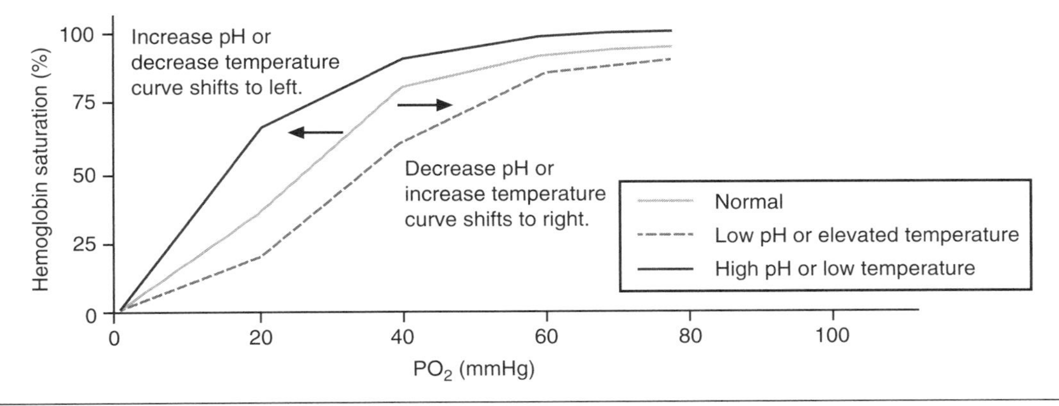

Figure 4.9 Oxygen saturation curve.

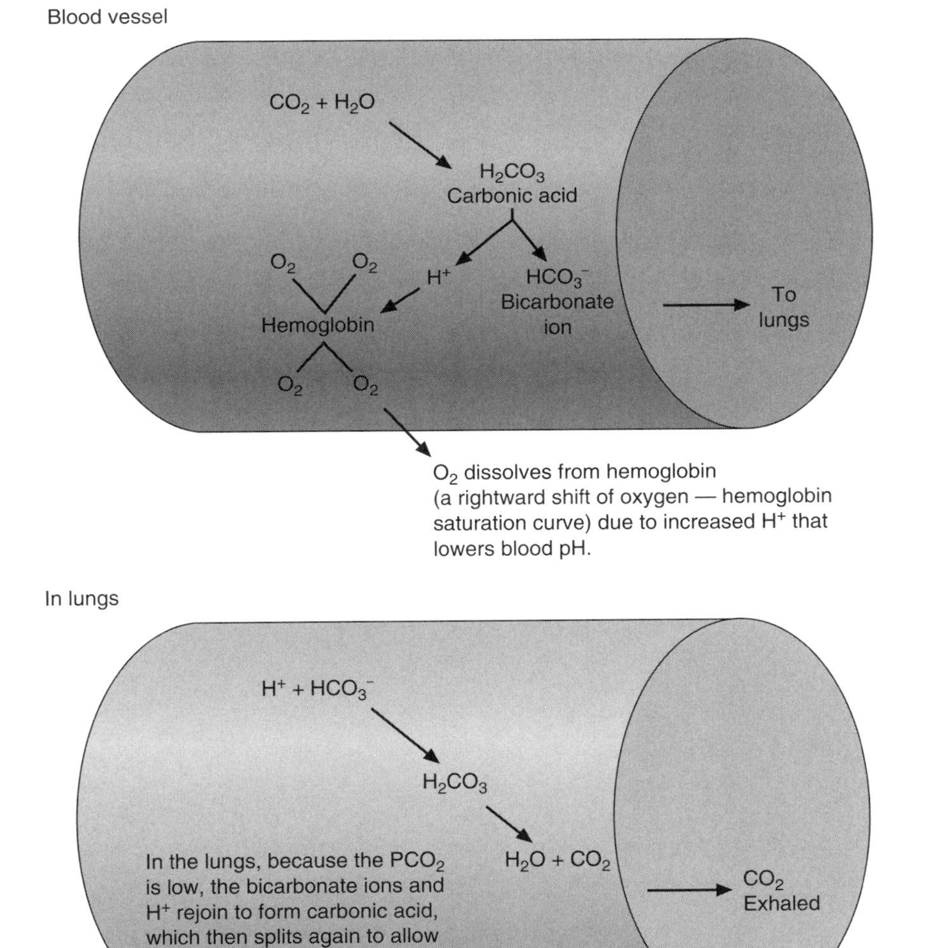

Figure 4.10 Effect of carbon dioxide release into the blood.

Cardiovascular Response to Acute Exercise

Oxygen consumption ($\dot{V}O_2$) is elevated during acute exercise to meet the higher energy needs of the exercising muscle. As exercise intensity increases, a greater demand for energy is met by an increase in the cardiac output or by a greater oxygen extraction from the vasculature [a greater $(a\text{-}\bar{v})O_2$ difference]. During the early stages of exercise, rapid increases in both heart rate and stroke volume bring about elevations in cardiac output. Figure 4.11 demonstrates the effects of varying intensities of exercise on heart rate, stroke volume, and cardiac output.

Cardiac Output During Acute Exercise

Cardiac output at rest is approximately 5 L. However, during maximal exercise, cardiac output may increase up to 20 L in young sedentary males; and in young endurance-trained male athletes, cardiac output may reach up to 40 L. In examining this considerable difference in cardiac output, we can see that the maximal heart rate for individuals from both these groups (assuming that both men are 20 years old) is approximately 200 beats/min (using the formula maximal heart rate = 220 – age). Thus, a difference in stroke volume must account for the large differences seen in cardiac output. In our example, the stroke volume of the sedentary male is approximately 100 ml/beat, whereas the stroke volume in the endurance-trained athlete may reach 200 ml/beat.

The importance of a large cardiac output for the endurance athlete is reflected by the linear relationship seen between cardiac output and oxygen consumption (Lewis et al. 1983). This relationship is seen not only in adults but also in children and adolescents (Cunningham et al. 1984) and between trained and untrained individuals (Saltin and Åstrand 1967).

Heart Rate During Acute Exercise

Heart rate elevation during exercise is primarily controlled by sympathetic stimulation from the higher somatomotor centers of the brain. The heart rate response is directly proportional and linear to the intensity of exercise. As intensity of exercise increases, the heart rate continues to increase until exercise reaches maximal intensity (figure 4.12). At maximal intensity, the heart rate plateaus, indicating that the individual is reaching his or her maximal level.

Initial increases in heart rate are also related to a withdrawal of parasympathetic input. This occurs during low-intensity exercise. As exercise continues in duration, or increases in intensity, a greater sympathetic stimulation becomes the driving force in elevating heart rate. Sympathetic activation occurs from feedback mechanisms in both peripheral mechanical and chemical receptors that monitor changes in pH, hypoxia, temperature, or other metabolic variables that can alter sympathetic drive.

During certain activities, an increase in heart rate can be seen before the onset of exercise. This anticipatory rise in heart rate appears to be primarily related to sprint or anaerobic-type events (McArdle, Katch, and Katch 2010). As the length of the exercise event increases (from a 60 yd [55 m] sprint to a 2.0 mi [3.2 km] run), the preexercise heart rate becomes lower. This pattern of an anticipatory heart rate response to high-intensity exercise may be a "feedforward" mechanism to provide for a rapid mobilization of bodily reserves, controlled by the central command

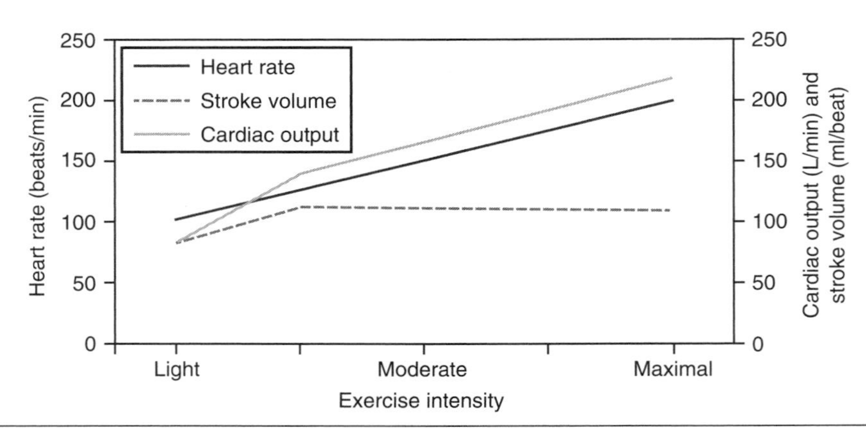

Figure 4.11 Effect of exercise intensity on heart rate, stroke volume, and cardiac output.

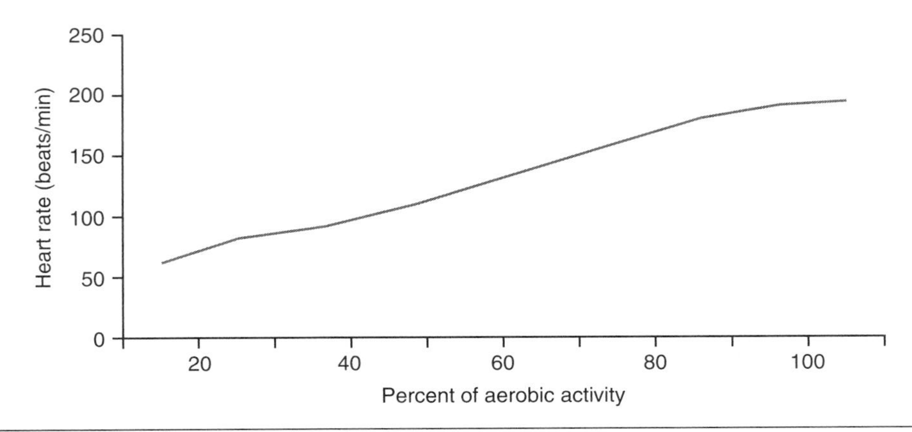

Figure 4.12 Relationship between heart rate and exercise intensity.

center in the medulla of the brain (McArdle, Katch, and Katch 2010). Such a mechanism does not appear warranted for longer-duration events.

The more times the heart beats per minute, the greater the volume of blood pumped into the circulation. However, there is a limit to this effect. As the heart rate rises above a certain level, the strength of each contraction may decrease because of metabolic overload. More important, the greater rate of contraction results in less time spent in diastole. The time between contractions becomes so reduced that there is not sufficient time for the blood to flow from the atria to the ventricles. Thus, the total volume of blood made available to the circulation is reduced. This is why during artificial electrical stimulation, the heart rate is elevated to only between 100 and 150 beats/min. However, sympathetic stimulation results in a stronger systolic contraction, decreasing the time during systole and thereby allowing a greater time for filling during diastole. Elevations in heart rate from sympathetic stimulation result in a heart rate between 170 and 250 beats/min.

Stroke Volume During Acute Exercise

Increases in stroke volume are accomplished early during exercise primarily through an increase in left ventricular EDV. This rapid augmentation of stroke volume is due to the **Frank-Starling mechanism**, which is related to the increased volume of blood that returns to the heart during exercise. With a greater volume of blood returning to the heart, the ventricles become stretched to a greater extent than normal and respond with a more forceful contraction. This stronger contraction results in a greater volume of blood entering the systemic circulation with each heartbeat. This mechanism appears to occur early during exercise and at a relatively low level of exercise intensity. The Frank-Starling mechanism may cause an approximate 30% to 50% increase in stroke volume (Bonow 1994). As exercise continues, increases in EDV reach a plateau while exercise intensity is still submaximal. Further increases in stroke volume are attributed to the enhanced left ventricular contractile function (controlled by enhanced sympathetic stimulation), resulting in a greater decrease in ventricular ESV.

Two mechanisms appear to be responsible for the increase in EDV during exercise. The initial mechanism uses the exercising muscles as a pump to increase the rate of return of blood to the heart. This would be expected to increase the pressures within the ventricular cavity during filling, thereby raising diastolic pressure. However, this does not occur in the healthy heart and, in contrast, the relaxation seen in the left ventricle reduces the ventricular pressure below that of the left atrium. This causes the mitral valve to open and the onset of ventricular filling. As mentioned earlier, the enhanced sympathetic response during exercise increases the relaxation time during diastole. During this time, the increase in size of the left ventricle causes a further reduction in pressure, creating a suctioning effect that draws additional blood into the chamber. This facilitation of the suctioning mechanism by sympathetic drive is the secondary mechanism that contributes to the increased stroke volume and is crucial in the recruitment of the Frank-Starling mechanism (Bonow 1994).

Cardiac Drift

As exercise duration is prolonged, or when exercise is performed in a hot environment, a gradual increase in heart rate and a decrease in stroke volume may occur even when exercise intensity is maintained. This has been referred to as **cardiac drift** and may also be accompanied by decreases in arterial and pulmonary pressures (Coyle and Gonzalez-Alonso 2001). The drift is thought to occur because a greater percentage of circulating blood is being diverted to the skin in an attempt to dissipate body heat caused by an increased core temperature. The greater concentration of blood in the periphery and a loss of some plasma volume to sweat result in a reduced blood return to the heart. This decrease in EDV results in a reduced stroke volume. Heart rate is elevated to compensate for the change in stroke volume and to maintain cardiac output.

Arteriovenous Oxygen Difference During Exercise

At rest, a person with normal hemoglobin concentration has approximately 200 ml oxygen in every 1 L blood. With a normal cardiac output of 5 L/min at rest, a potential of 1 L of oxygen is available to the body. However, only 250 ml, or 25%, of the available oxygen is extracted from arterial blood during rest, leaving the remaining 750 ml of oxygen available for reserve. This is called the $(a\text{-}\bar{v})O_2$ difference.

During exercise, the amount of oxygen extracted from the arterial blood is increased. Up to 75% of the available oxygen may be used by the exercising muscles. The increase in oxygen extraction appears to be related to the intensity of exercise and may be further enhanced after endurance training programs. The ability to extract oxygen from the blood and the total blood volume available to the muscles are critical for determining the aerobic capacity of the individual. This is reflected by the **Fick equation**:

$$\dot{V}O_2\text{max} = \text{Maximal cardiac output} \times \text{Maximal } (a\text{-}\bar{v})O_2 \text{ difference}$$

There may be very little difference between moderately trained individuals and endurance athletes in the ability to extract oxygen, despite large differences in $\dot{V}O_2$max. Therefore, the primary factor determining aerobic capacity appears to be cardiac output (McArdle, Katch, and Katch 2010).

Distribution of Cardiac Output During Exercise

As discussed earlier and depicted in figure 4.6, up to a 25-fold increase in blood flow occurs during exercise. However, most of the blood is diverted to the exercising muscles. The extent of this shunting depends on the environmental conditions and possibly other factors, including type of exercise and fatigue. The shunting of blood is generally accomplished through diversion of blood to the exercising muscles from organs or areas of the body that can tolerate a reduction in blood flow. However, certain organs such as the heart cannot function without a normal blood flow and do not compromise their blood supply during exercise.

Blood Pressure Response to Acute Exercise

Blood pressure typically increases during dynamic exercise such as walking, jogging, or running. In the healthy individual, this increase is seen only in the systolic response. Increases in systolic blood pressure are linear, and systolic pressure may exceed 200 mmHg during maximal exercise (either endurance or resistance). However, these increases do not parallel the four- to eightfold changes that may be seen in cardiac output. The systolic blood pressure appears to be buffered to a large extent by the decrease in peripheral resistance caused by vasodilation in the vasculature of the exercising muscles (MacDougall 1994). The decrease in peripheral resistance also appears to account for the minimal to no change observed in diastolic pressures. Diastolic pressure may also decrease during higher-intensity bouts of exercise. The blood pressure response to dynamic endurance exercise can be seen in figure 4.13.

During exercise that involves the upper body only, both systolic and diastolic blood pressures are higher than when exercise is performed with only the legs (Toner, Glickman, and McArdle 1990). This is thought to occur because of the relatively smaller muscle mass and vasculature of the arms. Even when these vessels are maximally dilated, they do not have the same effect on peripheral resistance as lower body exercise. Other possible explanations include greater involvement of a **Valsalva** (forced expiration against a closed or partially closed glottis) or partial Valsalva maneuver or the fact that a given absolute power output with arm exercise represents a greater relative exercise intensity than the same power output with lower body exercise (MacDougall 1994). Regardless of the mechanism, the higher pressor response seen with upper body exercise has important implications for determining the exercise prescription for individuals with coronary heart disease.

During resistance exercise, large increases in both systolic and diastolic blood pressures are evident (MacDougall et al. 1985, 1992; Sale et al. 1993). During maximal efforts that involve a large

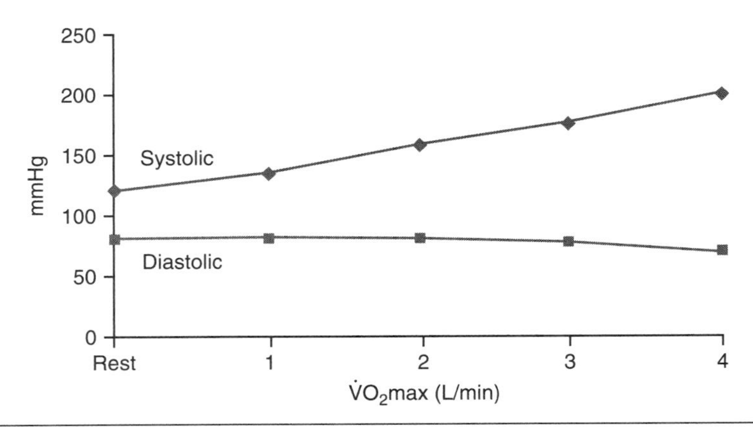

Figure 4.13 Blood pressure response to progressive endurance exercise.

muscle mass, intra-arterial blood pressures exceeding 350/250 mmHg in healthy young men have been reported (MacDougall et al. 1985). The large pressor response during resistance training is the result of a compression of the vasculature within the contracting muscles and the Valsalva maneuver. The magnitude of the pressor response is also related to the relative size of the muscle mass involved and the intensity of the effort. Blood pressure increases with each repetition in a set and then drops rapidly to below resting levels after the last repetition (MacDougall et al. 1985; Sale et al. 1993). This transient decrease is likely related to the large vasodilation of the vasculature that was occluded during muscle contraction and may contribute to the dizziness sometimes experienced after an intense exercise session.

A major portion of the large pressor response to resistance training is attributed to the Valsalva maneuver (MacDougall 1994). The Valsalva maneuver does not appear to take effect until exercise intensity reaches 80% to 85% of one-repetition maximum (1RM) or when fatigue sets in (MacDougall et al. 1985, 1992). During a Valsalva maneuver, a rapid increase in intrathoracic pressure has an immediate and direct effect on the arterial tree, resulting in an increase in both systolic and diastolic blood pressures (MacDougall et al. 1992). However, if the Valsalva maneuver is maintained, the systolic and diastolic pressures begin to drop within several seconds because of the reduced diastolic filling caused by impaired venous return. Although often contraindicated during resistance exercise, the Valsalva maneuver may in fact be beneficial and may have a protective effect in healthy resistance-trained individuals (MacDougall 1994; McCartney 1999). The increase in intrathoracic pressure during the Valsalva maneuver stabilizes the spinal column and has also been shown to decrease left ventricular transmural pressure (**afterload**) (Lentini et al. 1993), in contrast to the high afterload normally expected when systolic pressures are elevated. In addition, the increase in intrathoracic pressure is transmitted to the cerebral spinal fluid, which reduces the transmural pressures of the cerebral vessels and prevents vascular damage at the time of peak peripheral resistance (McCartney 1999).

Pulmonary Ventilation During Exercise

During exercise, an increase in ventilation results from several chemical and neural stimuli that appear to occur simultaneously (Eldridge 1994; Whipp 1994). Changes in ventilatory patterns appear to occur in three phases. In phase 1, there is a rapid increase in ventilation rate followed by a brief plateau. This initial increase in ventilation is thought to be the result of a central command drive (cerebral cortex in the brain) as well as feedback from the active muscles and the effects of increased K^+ concentrations (Eldridge 1994). Phase 2 occurs after approximately 20 s, when ventilation continues to increase as a result of these same stimuli; however, the neural component appears to increase as a result of an increasing drive from medullary short-term potentiation. As steady-state exercise is reached, all the mechanisms that controlled the increase in ventilatory pattern stabilize; this is phase 3. This phase can involve added input from peripheral sources (e.g., chemoreceptors and core temperature) that can fine tune the ventilatory response (McArdle, Katch, and Katch 2010). During recovery, the abrupt decrease in ventilation is the result of the removal of the central command drive and the afferent input from the active muscles. During the later stages of recovery, the slower return to resting ventilatory levels represents the gradual return to normal metabolic, thermal, and chemical levels.

During submaximal exercise, ventilation increases linearly with oxygen uptake. The increase in oxygen consumption is primarily the result of an increase in **tidal volume** (amount of air inspired or expired during a normal breathing cycle). As exercise intensity is elevated, the increase in oxygen consumption may rely more on increasing the breathing rate. During steady-state exercise, **minute ventilation** (liters of air breathed per minute) plateaus when the demand for oxygen is met by supply (Eldridge, 1994). The ratio of minute ventilation to oxygen consumption is termed the **ventilatory equivalent** and is referred to as $\dot{V}_E/\dot{V}O_2$. During submaximal exercise, the ventilatory equivalent in healthy individuals is approximately 25:1 (Wasserman, Whipp, and Davis 1981). That is, 25 L air is breathed in for every 1 L oxygen. This ratio may be slightly higher in children (Rowland and Green 1988) and also may be affected by the mode of exercise (swimming vs. running) (McArdle, Glaser, and Magel 1971). However, during maximal exercise, minute ventilation increases disproportionately in relation to oxygen uptake, and the ventilatory equivalent may get as high as 35 to 40 L air per liter of oxygen consumed in the healthy adult.

Cardiovascular Response to Training

Long-term physical conditioning results in a number of cardiovascular adaptations specific to the type of exercise program. Endurance training and resistance training are the exercise programs that are typically compared. These modes of training present distinctly different physiological demands on the cardiovascular system. Although many of the cardiovascular adaptations observed in these training programs are similar, others are very different. A summary of these adaptations is presented in this section (see table 4.2).

Table 4.2 Cardiovascular Adaptations to Prolonged Endurance and Resistance Training

	Endurance training	Resistance training
Resting adaptations		
Heart rate	↓	↓ or –
Stroke volume	↑↑	↑ or –
Cardiac output	–	–
Blood pressure		
Systolic	↓ or –	↓ or –
Diastolic	↓ or –	–
Exercise adaptations		
Heart rate	–	–
Stroke volume	↑↑	↑ or –
Cardiac output	↑↑	↑ or –
Blood pressure		
Systolic	↓ or –	↓ or –
Diastolic	↓ or –	↓ or –
Morphological adaptations		
Left ventricular mass	↑	↑
Left ventricular diameter	↑↑	↑ or –
Wall thickness		
Left ventricle	↑	↑↑
Septum	↑	↑↑

↑ = increase; ↓ = decrease; – = no change.

Effect of Training on Cardiac Output

Increases in $\dot{V}O_2$max are characteristic of endurance training programs. These increases are generally accompanied by increases in cardiac output and improved extraction capability of skeletal muscle [increase in (a-$\bar{v}$)O_2]. The improvement in oxygen extraction is related to the greater perfusion capabilities of the exercising muscle. The increase in cardiac output is primarily the result of improved stroke volume. Maximal heart rates are unaffected by training and do not differ between elite endurance athletes and age-matched sedentary individuals. Thus, improvements in cardiac output are directly related to the increase in stroke volume.

Effect of Training on Stroke Volume

Exercise has been consistently demonstrated to be a potent stimulus for increasing stroke volume at rest and during maximal exercise. It has recently been reported that Olympic-caliber athletes (endurance, power, and skill athletes) have a 43% greater stroke volume than age-matched control subjects (Caselli et al. 2011). However, endurance training may have a potent effect on stroke volume. Endurance-trained athletes have been shown to have a 60% greater stroke volume than sedentary control subjects, which is consistent with the relative difference in $\dot{V}O_2$max seen between these individuals (McArdle, Katch, and Katch 2010). The improved stroke volume is related to an enlarged ventricular chamber (referred to as **eccentric hypertrophy**) caused by chronic increased ventricular filling common to endurance exercise. This increased preload is thought to relate to the expanded plasma volume associated with such training (Carroll et al. 1995; Convertino 1991).

Resistance training results in little to no change in maximal aerobic capacity. Thus minimal changes would be expected in cardiac output. Significantly greater stroke volumes have been reported in elite-level weightlifters compared with recreational lifters (Pearson et al. 1986). However, when this study and others were examined in a meta-analysis (Fleck 1988), it appeared likely that the increase in stroke volume in these athletes was more an effect of larger body size than a training adaptation.

Effect of Training on Heart Rate

A decrease in resting heart rate and a relative decrease in heart rate at any given submaximal $\dot{V}O_2$ is a commonly found adaptation in endurance training programs (Blomqvist and Saltin 1983; Charlton and Crawford 1997). However, the magnitude of heart rate reduction during endurance training may be much smaller than that observed in some cross-sectional studies comparing elite endurance athletes with sedentary controls (Wilmore et al. 1996). Resistance training may or may not result in any significant change in resting heart rate. Several studies have reported significant decreases in resting heart rate after resistance training programs (Goldberg, Elliot, and Kuehl 1994; Kanakis and Hickson 1980; Stone, Nelson, et al. 1983), whereas others have failed to see any significant changes (Lusiani et al. 1986; Ricci et al. 1982; Stone, Wilson, et al. 1983). The mechanism regulating training-induced bradycardia is not thoroughly understood but is likely related to a change in the balance between sympathetic and parasympathetic activity. In addition, a decrease in the intrinsic rate of firing of the SA node after long-term training has been suggested to be a factor in the bradycardic response to long-term training (Schaefer et al. 1992).

Effect of Training on Blood Pressure

In individuals with normal resting systolic or diastolic blood pressure, exercise generally does not cause any changes. However, several epidemiological studies and clinical investigations have clearly indicated that exercise is a potent stimulus for reducing both systolic and diastolic blood pressure in hypertensive individuals (Cornelissen and Fagard 2005; Cornelissen et al. 2011; Pescatello et al. 2004). Endurance training appears to be most effective for reducing resting blood pressure at training frequencies between three and five sessions per week that are at least 30 min in duration and between 50% and 85% of $\dot{V}O_2$max (Fagard 2001; Pescatello et al. 2004). Although endurance exercise has been shown to decrease the blood pressure response for a given level of exercise intensity (MacDougall 1994), it is thought that training at lower intensities (40-70% of $\dot{V}O_2$max) may have a greater effect on the hypertensive response (Halbert et al. 1997; Pescatello et al. 2004).

Resistance training appears to result in no change or a slight decrease in resting blood pressure in the normotensive individual (Goldberg, Elliot, and Kuehl 1994) but a significant decrease in the blood pressure response during resistance exercise at a given absolute load (McCartney et al. 1993; Sale et al. 1994). Any decrease in resting blood pressure subsequent to resistance training is likely the result of a decrease in body fat and possible reduction in the sympathetic drive to the heart (similar to what may drive the reduction in blood pressure during endurance exercise) (Cornelissen et al. 2011; O'Sullivan and Bell 2000).

Effect of Training on Cardiac Morphology

The size of the heart in athletes is large in comparison with healthy individuals who participate only in recreational activities or are sedentary. The enlarged heart in athletes resembles what is seen in many pathological conditions of the heart, and some controversy existed in the past about whether it was a consequence of pathological versus physiological processes (Shapiro 1997). With the advent of M-mode echocardiographic studies in the 1970s, it became possible to achieve a better understanding of the physiological adaptations from prolonged training.

Ventricular Wall Thickness and Internal Diameter

During exercise, the hemodynamic demands on the heart are related to the dynamic changes in blood pressure and blood volume. During prolonged training, the heart adapts to match the workload placed on the left ventricle in order to maintain a constant relationship between systolic cavity pressure and the ratio of wall thickness to ventricular radius (Shapiro 1997). Adaptations to the morphology of the heart are governed by the law of Laplace, which states that wall tension is proportional to pressure and the radius of curvature (Ford 1976). During a pressure overload, common to resistance exercise programs, the septum and posterior wall of the left ventricle increase in size to normalize myocardial wall stress. During a volume overload, common to endurance training programs, the increase is predominantly in the internal diameter of the left ventricle (increasing the size of the cavity), with a proportional increase in both the septum and posterior wall of the ventricle. Endurance training and resistance training are at either end of the spectrum with regard to the volume and pressure stresses placed on the heart. However, most sports have a parallel impact on cavity dimension and wall thickness (Spirito et al. 1994). In these sports, athletes perform some combination of aerobic and anaerobic training, resulting in cardiovascular adaptations associated with both an enlarged diastolic cavity dimension and a larger wall thickness. In the sports that emphasize a single form of training, the morphological changes of the heart may be more extreme.

Endurance-trained athletes have been shown to have a greater than normal left ventricular internal diameter, with normal to slightly thicker walls (Maron 1986; Morganroth et al. 1975; Pelliccia et al. 1991; Spirito et al. 1994). These changes are in accordance with the law of Laplace in that a compensatory thickening of the walls of the ventricle occurs in response to the greater internal diameter. This type of left ventricular hypertrophy is termed eccentric hypertrophy and is considered a normal physiological response to a volume overload (greater EDVs) consistent with prolonged endurance training.

Resistance-trained athletes, on the other hand, have normal internal diameters but significantly thicker ventricular walls (Fleck, Henke, and Wilson 1989; Menapace et al. 1982; Morganroth et al. 1975; Pearson et al. 1986). This is referred to as **concentric hypertrophy** and at times may approach levels seen in hypertrophic cardiomyopathy (a disease of the myocardium associated with great thickening of the septum and posterior wall at the expense of cavity size, significantly impairing left ventricular function). The concentric hypertrophy in the resistance-trained athlete does not affect the internal diameter of the ventricle. In addition, the type of hypertrophy seen in cardiomyopathy is usually asymmetric, whereas in resistance-trained or power athletes, the change in wall size is generally symmetrical.

It appears that the type of resistance training program employed may determine the extent of cardiac morphological changes. Bodybuilders have been reported to have greater than normal cavity dimensions (Deligiannis, Zahopoulou, and Mandroukas 1988), but when these were examined relative to body surface area or lean body mass, no significant differences between bodybuilders or weightlifters were observed. However, right ventricular and left atrial volumes have also been shown to be greater in bodybuilders than in weightlifters. These differences were still evident when examined relative to body surface area or lean body mass (Deligiannis, Zahopoulou, and Mandroukas 1988; Fleck, Henke, and Wilson 1989). It appears that the high-volume resistance training programs common to bodybuilders may have the greatest potential to affect cardiac chamber size.

Left Ventricular Mass

Left ventricular mass is, on average, 45% greater in highly trained athletes than in age-matched control subjects (Maron 1986). This increase in mass is related to the increases in left ventricular internal diameter and ventricular wall thickness. When examined relative to changes in body mass or body surface area, the ventricular mass is still significantly greater. Some studies have suggested that differences are more prevalent in elite athletes than in athletes of lesser caliber (Fleck 1988).

In Practice

Cardiac Adaptations in Previously Untrained Individuals

The data on the specific morphological changes to left ventricular function as a result of either endurance or resistance training are very convincing. These studies are based primarily on cross-sectional designs that compare the morphological characteristics of various athletes and do not focus on a training response. What is less understood is what happens to the average individual who is performing either endurance or resistance exercise. Are these adaptations actually realized during a training program?

A study from Western Australia by Spence and colleagues (2011) examined the effect of 6 months (24 weeks) of either resistance or endurance training on previously untrained subjects who were in their mid-20s (27.4 ± 1.1 years). Left ventricular morphology was measured via cardiac magnetic resonance spectroscopy, and cardiac images were also obtained by echocardiography. Subjects in the endurance training group performed a periodized program that began with a mesocycle (weeks 1-12) during which they performed low- to moderate-intensity walking, jogging, and stretching to prepare for the more intense training in the later cycles. Training volume was gradually increased during this phase. During the second mesocycle (weeks 13-18), running intensity was increased and elements of hill running and short intervals were included. In the final training phase (weeks 19-24), exercise intensity was maintained but volume was somewhat reduced to prepare the subjects for a 12 km competition. Intensity of running for each training session was individualized through prescription of training paces based on $\dot{V}O_2$peak values and time trial performances. The subjects in the resistance training group also performed a periodized training routine. All cycles included Olympic weightlifting exercises. The "general preparatory phase" (weeks 1-12) focused on conditioning the body and developing correct lifting technique using low volume and load (e.g., two or three sets of 12 to 15 repetitions at 65% to 85% of maximal strength [1RM]). The next mesocycle (weeks 13-20) focused on improving technique in the Olympic lifts (clean and jerk and snatch) while increasing strength in assistance lifts (e.g., front squat, back squat, overheard squat, deadlift, and press). During the final mesocycle (weeks 21-24), volume of training was lowered and intensity was elevated.

Both training programs resulted in specific performance improvements (i.e., improved aerobic capacity in the endurance group and increased strength in the weightlifting group). However, results also indicated that the endurance-trained group experienced significant increases in left ventricular mass and wall thickness, while no changes in cardiac morphology were observed in the weightlifting group. This study provides support for the concept that cardiac morphology is very adaptable to endurance exercise, even in previously untrained individuals. However, it does not provide the same support for resistance training. In this regard it is possible that a longer duration of training is necessary, or that a difference in the exercise prescription may have resulted in a different outcome.

Respiratory Adaptations to Training

For the most part, the respiratory system is not a limiting factor in the provision of sufficient oxygen to the exercising muscles. However, similarly to most other physiological systems in the body, the respiratory system can adapt to physical exercise in order to maximize its efficiency. In general, lung volume and capacity change very little as a result of physical exercise. Maximal exercise may slightly increase **vital capacity**, but this may be related to greater strength of the inspiratory musculature (Coast et al. 1990).

Training Effects on Minute Ventilation and Ventilatory Equivalent

Endurance training appears to reduce the ventilatory equivalent (i.e., amount of air inspired at a particular rate of oxygen consumption) at a given intensity of exercise (Girandola and Katch 1976; Yerg et al. 1985). Consequently, the oxygen cost of exercise attributable to ventilation is reduced, indicating a greater economy of movement. This adaptation can reduce fatigue of the ventilatory musculature and provide for greater oxygen availability to the exercising muscles (Martin, Heintzelman, and Chen 1982). Findings on the effect of training on minute ventilation (V_E), the amount of gas inhaled and exhaled from the lungs in 1 min, appear to be inconclusive. Some have suggested no change in V_E during submaximal exercise (Itoh et al. 2002), while others have suggested that it can be lowered (McKenna et al. 1997). Hagberg and colleagues (1980) indicated that training may lower V_E during the recovery period, resulting in a smaller oxygen debt. Training does appear to result in an elevation in V_E during maximal exercise. An 11% increase in V_E was reported in recreationally trained men participating in a 7-week sprint training program (McKenna et al. 1997). Elevations in V_E during maximal exercise can assist in minimizing fatigue, which may be related to an enhanced neural or humoral respiratory control or both (Waldrop et al. 1996). Highly trained athletes appear to have a higher V_E than lesser-trained individuals, with the highest values seen in elite-level cyclists (Smith et. al. 1994). V_E can approach and exceed 200 L/min in the elite endurance athlete (Hagerman 1984; Smith et al. 1994).

Effect of Training on Blood Volume and Red Blood Cells

Endurance training appears to be a potent stimulus for causing **hypervolemia** (increases in blood volume). This has been demonstrated in both young and old populations (Carroll et al. 1995; Convertino 1991). During the initial 2 to 4 weeks of training, plasma volume expansion is thought to account for the hypervolemia (Convertino 1991). As training progresses, blood volume expansion appears to be the result of both continued plasma volume expansion and an increase in the number of red blood cells.

The increase in plasma volume is believed to be the result of increases in antidiuretic hormone and aldosterone (see chapter 2), which increase fluid retention by the kidney (Nagashima et al. 2001). In addition, exercise causes an increase in plasma proteins, primarily albumin (Yang et al. 1998). This increase in plasma proteins within the blood results in a greater osmotic pull, causing fluid retention in the blood.

Increases in blood volume appear to be the result of both plasma volume expansion and an increase in red blood cell number. However, plasma volume expansion seems to be a greater contributor to hypervolemia (Green et al. 1991). Figure 4.14 shows the effect of prolonged endurance training on blood volume, plasma volume, and red blood cell number. Although both plasma volume and red blood cell volume increase, they do not increase proportionally. Thus, **hematocrit** (% of red blood cells in relation to total blood volume) decreases as a response to training. A high hematocrit could be dangerous because of an increased blood viscosity. However, if hematocrit is reduced, the viscosity of the blood decreases, which may facilitate the blood flow through the circulation. Reductions in hematocrit do not appear to cause a concern about low hemoglobin concentrations. In fact, hemoglobin concentrations in endurance-trained athletes are typically above normal

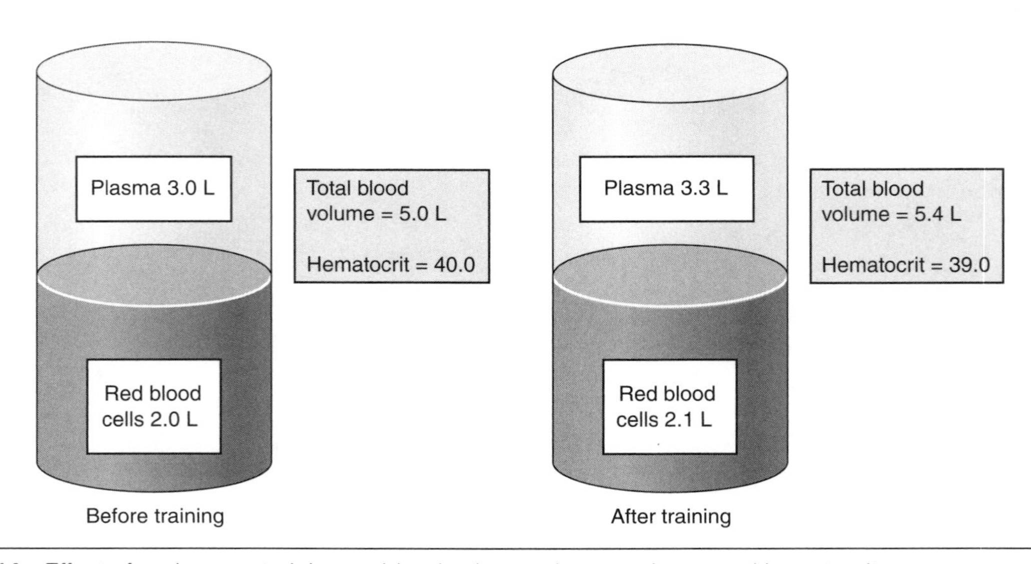

Figure 4.14 Effect of endurance training on blood volume, plasma volume, and hematocrit.

and provide an ample capacity of oxygen to meet the needs of the body during exercise.

Summary

This chapter demonstrates the effects of acute exercise on cardiac function and the ways in which the heart compensates for the increased energy demands of exercising muscles. This compensation is manifested by changes in cardiac output, which is regulated in part by enhanced sympathetic drive and increased venous return. In addition, blood flow is diverted to exercising muscles from nonexercising muscles and nonessential organs to provide for greater oxygen delivery. Differences in the acute cardiac response are evident between endurance and resistance training programs. The chapter also discusses the effects of prolonged training on cardiovascular adaptations and the dependence of these adaptations on the type of training program employed. Finally, the chapter reviews the coordinated relationship between the cardiovascular and respiratory systems and the effects of both acute and prolonged training on the respiratory system.

REVIEW QUESTIONS

1. Explain the difference between end-diastolic volume and end-systolic volume.
2. What does "partial pressure" of a gas refer to?
3. Describe the Bohr effect and its importance during exercise.
4. Contrast the response of stroke volume, heart rate, and cardiac output during endurance exercise to exhaustion.
5. What is cardiac drift?
6. Explain the differences between eccentric hypertrophy and concentric hypertrophy and whether any of these changes increase the risk for heart disease.
7. Describe changes typically observed in blood volume as the result of prolonged endurance training.

chapter 5

Immunological System and Exercise

CHAPTER OBJECTIVES

After reading this chapter you should be able to do the following:

- Understand the difference between innate immunity and acquired immunity.
- Differentiate among the various immune cells.
- Explain the acute response of various immune cells to exercise.
- Explain the immune response to prolonged exercise, and how the immune response in competitive athletes may differ from that in recreational participants in sport.
- Identify sex differences in the immune response to exercise.

The immune response to exercise has gained much interest and importance because of the implication that intense training increases the athlete's susceptibility to infection, especially upper respiratory tract infections. However, regular moderate exercise is also known to be beneficial to people's health and well-being and is thought to reduce the risk of many diseases and illnesses. Exercise appears to be beneficial to a certain point, beyond which an increase in either intensity or duration may impair the immune response. The focus of much of the research on exercise and the immune response has been directed at better understanding the intensity and volume of training that optimizes immune function.

This chapter briefly reviews the immune system, including the various types of cells and their functions. Further discussion focuses on the immune response to exercise and its implications for performance.

The immune system has two functional divisions: the **innate** immune system and the **adaptive** immune system. Innate immunity is the body's natural response and its first line of defense against infectious agents. This system combats all invading microorganisms seen for the first time. Natural immunity does not improve with additional exposures to these same microorganisms. The innate immune system includes the skin, which prevents infectious agents from penetrating, mucus membranes, pH of body fluids (e.g., stomach acid), and bodily secretions. The **complement** system, **lysozymes**, **phagocytes**, and **natural killer (NK) cells** are also part of the innate immune system response.

If the initial line of defense is not sufficient to kill invading pathogens, the adaptive immune system is activated. This immune system produces a specific reaction for each infectious agent. Normally, the adaptive system is successful in destroying the invading pathogen. The adaptive system is also capable of generating a memory of this exposure and producing **antibodies** or other immune cells to respond more effectively and quickly on subsequent exposures to the specific infectious agent.

Cells of the Immune System

Immune cells arise from **stem cells** through two lines of differentiation: the **lymphoid lineage**, which produces **lymphocytes**, and the **myeloid lineage**, which produces phagocytes and other cells. Two types of lymphocytes, **T cells** and **B cells**, have receptors for **antigens**. The T cells develop in the thymus, whereas the B cells develop in the bone marrow. These lymphocytes are produced at a very high daily rate (10^9). They migrate into the circulation and into secondary **lymphatic tissue**, such as the spleen, lymph nodes, tonsils, and unencapsulated lymphoid tissue. The lymphocytes represent approximately 20% of the total **leukocyte** (white blood cell) population in an adult.

Types of Leukocytes

There are three major types of leukocytes: **granulocytes**, **monocytes**, and lymphocytes (figure 5.1). Granulocytes (also referred to as polymorphonuclear leukocytes) make up 60% to 70% of the total leukocytes and include **neutrophils**, **eosinophils**, and **basophils**. The granulocytes, considered part of the initial response to foreign **pathogens**, are primarily involved in **phagocytosis** and destruction of infectious organisms. Neutrophils are the most common granulocyte found in the circulation. They are rapidly attracted by **chemotactic** factors to sites of infection or injury. Neutrophils are short-lived and act primarily by releasing **proteases** and **phospholipases** to eliminate the infectious organism. They are also involved in generating toxic molecules, such as **oxygen radicals**. Neutrophils are very active during tissue injury and inflammation and are believed to be involved in the degradation of damaged tissue. Eosinophils, basophils, and **mast cells** make up a very small percentage of the leukocytes. Eosinophils appear to be the most active in resistance to parasitic infection, whereas basophils and mast cells are primarily involved in allergic and inflammatory reactions.

Monocytes are also involved in phagocytosis during the early stages of the immune response. Once the monocytes enter the affected area, they differentiate further into **macrophages**. The monocytes-macrophages release proteases from **lysosomes** and generate oxygen radicals and **nitric oxide**, which are lethal to infectious agents. Monocytes also produce **cytokines**, which activate lymphocytes and stimulate the inflammatory process.

Phagocytosis

The phagocytic cells (e.g., neutrophils, monocytes, and macrophages) are brought to sites of infection or inflammation by chemotactic agents. The phagocytes have receptors on the surface that give them a nonspecific affinity to a variety of microorganisms. Attachment to these microorganisms may be enhanced if the microorganism has been opsonized by the **C3b** component of the complement system (Roitt, Brostoff, and Male 1993). Opsonization alters bacteria in a manner that allows them to be more readily and efficiently engulfed by phagocytes. The phagocytes have receptors that bind specifically to the C3b component, enhancing the phagocytic recognition of the infectious agent. After attachment, the phagocytes engulf and destroy the microorganism (figure 5.2).

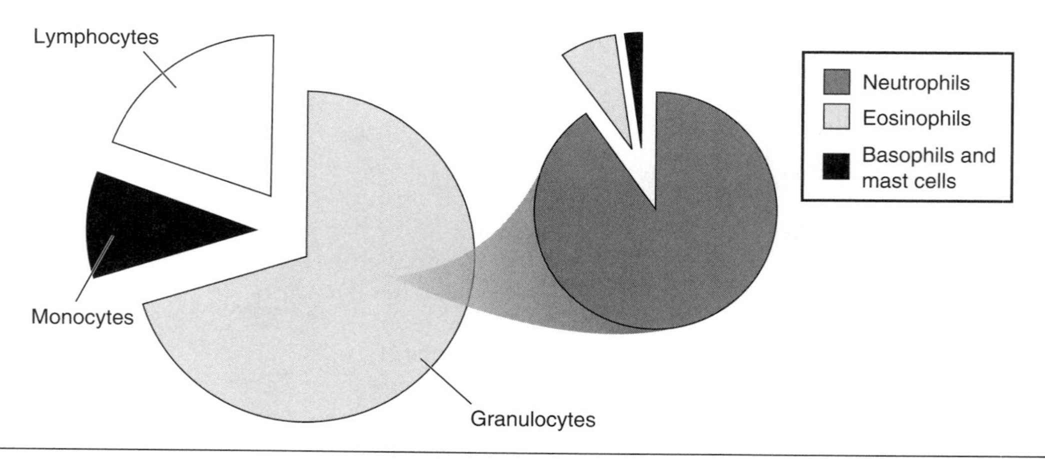

Figure 5.1 Leukocyte distribution.

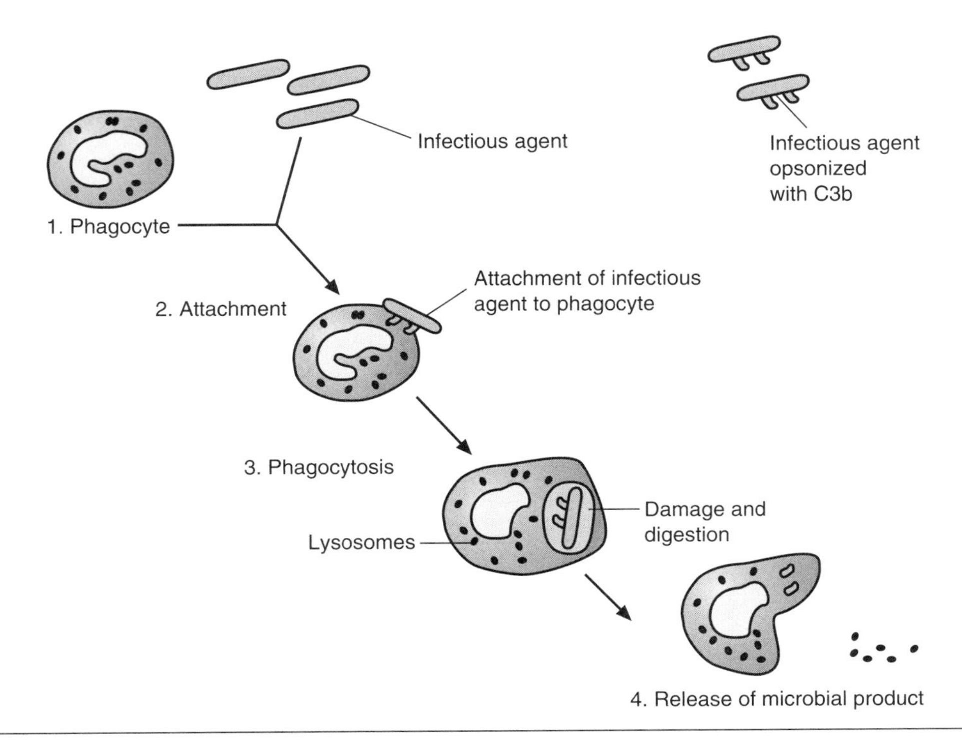

Figure 5.2 Phagocytic action.

Lymphocytes

The lymphocytes account for 20% to 25% of the total leukocyte population. They are made up of several subpopulations, each possessing a specific function. Lymphocytes are part of the initial immune response and are responsible for producing cytokines, antibodies, **cytotoxicity**, and memory of previous infection. There are three primary subpopulations of lymphocytes (T cells, B cells, and NK cells). They differ with regard to their size and morphological structure. The T cells and B cells are the major effectors of adaptive immunity, and natural killer cells have innate immunity capability.

T Cells

T lymphocytes are relatively small and agranular. The primary function of these cells is to initiate and regulate most of the immune response to infection and injury. This includes activation of B cells as well as use of their cytotoxic capability. All T cells are able to recognize their target by foreign antigen fragments attached to cell surface receptors on infected cells. These cells are called **major histocompatibility complex (MHC)** molecules (Janeway and Travers 1996). T cells have two distinct morphological patterns that differ in function and cell surface proteins.

T cells that have cytotoxic properties are referred to as cytotoxic T cells. These cells have a surface protein known as **CD8**. T cells with a **CD4** surface protein are further subdivided into T helper and T inflammatory cells. T helper cells activate B cells, and T inflammatory cells stimulate monocyte–macrophage phagocytic and antibacterial activities (Mackinnon 1999). CD4 cells also cause the production of several cytokines. The **clusters of differentiation (CD)** are briefly discussed in a later section.

B Cells

B cells are stimulated to action by T helper cells. On stimulation, they differentiate into plasma cells that produce and secrete large amounts of antibody. Each antibody recognizes a single antigen. The B cells are capable of memory, meaning that they will have a faster and more effective response to future exposures to the same antigen.

Natural Killer Cells

Natural killer cells are large granular lymphocytes with many cytoplasmic granules. They function as part of the initial immune reaction to defend against infectious agents. NK cells possess an ability to kill a wide variety of targets by releasing toxic substances similar to those released by cytotoxic T cells. They also have the ability to release some cytokines in response to stimulus by T cells.

Clusters of Differentiation

Leukocytes possess a number of different molecules on their surfaces. Some appear at particular stages of cell differentiation or are characteristic of different cell lineages (Roitt, Brostoff, and Male 1993). These molecules can be used as markers to identify particular cells through monoclonal antibodies binding to surface antigens of the cells. The numerical listing of the CD system has no specific meaning and is primarily related to the order in which its components were discovered and described.

Immunoglobulins

Immunoglobulins, or antibodies, are a group of **glycoproteins** found in all bodily fluids. They are produced in large amounts by plasma cells that have developed from precursor B cells. All antibodies are immunoglobulins, but not all immunoglobulins demonstrate antibody activities. Antibodies are critical for the recognition of previous exposures to an antigen. They combat infectious agents through both direct and indirect means. They act directly by binding to antigens on the microorganism to prevent it from entering host cells. Immunoglobulins may bind to both the antigen and the host tissue, including cells of the immune system. They act indirectly by stimulating recognition by other phagocytic cells that kill the invading organism. The latter method of action appears to be more prevalent.

There are five distinct classes of immunoglobulins (**IgG**, **IgA**, **IgM**, **IgD**, and **IgE**). Each differs from the others in size, amino acid components, and carbohydrate content. Their functions and properties can be seen in table 5.1.

Cytokines

The cytokines regulate growth factors and mediate the inflammatory response. They are involved in all aspects of the immune response and may have a role in nonimmune cells as well. There are several different types of cytokines. Although each cytokine acts on a number of target cells, the target cell has its own specific response. The cytokines are part of a group of soluble factors found in bodily fluids that mediate immune function. Other soluble factors include complement and acute-phase proteins, as well as the previously discussed immunoglobulins.

The cytokines consist of a group of different cells with various functions. Table 5.2 lists the primary producers and major functions of the cytokines frequently reported in the exercise science literature. The **interleukins (IL)** are a group of unrelated cytokines that are produced primarily by T cells but also by monocytes-macrophages and NK cells. The interleukins appear to have a wide array of roles, including inflammatory mediation, stimulation of further cytokine production, and enhancement of phagocytic function. The **interferons (IFN)** are released by both leukocytes and **fibroblasts** and are involved in preventing or limiting the spread of a virus between infected and noninfected cells. **Tumor necrosis factor (TNF)** is produced by both macrophages and T cells. TNF defends against both

Table 5.1 Immunoglobulins and Their Functions

Immunoglobulin	% of total immunoglobulin pool	Function
IgG	70-75%	Is distributed evenly between intravascular and extravascular spaces and is the major antibody of secondary immune response.
IgA	15-20%	Is seen predominantly in saliva, breast milk, and respiratory, genitourinary, and gastrointestinal secretions. It acts as a defense against infectious agents entering through mucosal secretions.
IgM	10%	Is primarily found in the mucosal secretions and is seen early in the immune response.
IgD	<1%	Is found on the membrane of circulating B cells and may be involved in antigen-triggered lymphocyte differentiation.
IgE	Trace	Is found on surface membrane of mast cells and basophils. Associated with immediate sensitivity to asthma and hay fever.

Table 5.2 Cytokines Frequently Reported in the Exercise Science Literature

Cytokine	Primary producer	Major function
IL-1α, IL-1β	Macrophages	These two have the exact same function. They are primarily involved as an inflammatory mediator (activating NK cells and TNF and inducing fever). They also potentiate the response of lymphocytes.
IL-2	T cells	Promotes T-cell division and B-cell growth. It is also involved with activating monocytes and NK cells.
IL-3	Activated T cells	Stimulates production of neutrophils and monocytes in bone marrow.
IL-4	Activated T cells	Activates T helper cells and B cells. Is also involved in IgE expression.
IL-5	Activated T cells	Eosinophil differentiation and maturation.
IL-6	Activated T cells, macrophages	T- and B-cell growth. Acts synergistically with IL-1 and TNF to stimulate acute-phase protein response. It is pyrogenic.
IFN-α IFN-β	Leukocytes Fibroblasts	These two have similar antiviral activity, activate NK cells, and enhance antigen recognition by increasing expression of MHC class I molecules.
IFN-γ	T cells, NK cells	Antiviral and antibacterial activity. Inhibits viral replication. Enhances antigen recognition by increasing expression of MHC I and II molecules. Stimulates phagocytic and cytotoxic activities.
TNF-α	Macrophages, NK cells	Stimulates cytotoxic activity of CD8 T cells, NK cells, and macrophages. Increases vascular permeability and migration of leukocytes to areas of inflammation and infection. Acts synergistically with IL-1 and IL-6 to stimulate acute-phase proteins.
TNF-β	CD4 T cells	Acts synergistically with IFN-α to increase direct killing of infectious agents. It also acts as a chemotactic agent, drawing macrophages to sites of infection and inflammation.

Data from Mackinnon 1999.

viral and bacterial infection by stimulating cytotoxic activity of leukocytes or acting synergistically with them to kill these microorganisms.

Complement System

The complement system is a group of proteins found in the blood whose primary function is to initiate and amplify the inflammatory response. The biological activity of the complement system includes recruitment of macrophages and neutrophils to sites of injury or infection, **lysis** of target cells (most likely bacteria), and **opsonization** of pathogens. Lysis refers to the rupture of the cell membrane and resultant loss of cytoplasm. Opsonization is a process that deposits an **opsonin** (e.g., antibody or C3b complement component) on bacteria, which enhances recognition by phagocytes. The complement system is made up of a number of components whose nomenclature is related to the order in which they were discovered.

The complement system is activated via one of two pathways, classical or alternative. Activation of these pathways occurs through the interaction of certain classes and subclasses of antibodies with antigens. The alternative pathway is thought to provide a nonspecific natural immunity, whereas the classical pathway is thought to represent an adaptive mechanism (Roitt, Brostoff, and Male 1993). Both of these pathways result in the production of C3b, the central component of the complement system. C3b in turn activates the membrane attack complex (C5-C9), which exerts a direct killing action on bacteria. C3a and C5a are **anaphylatoxins** that stimulate **chemotaxis** of leukocytes and **degranulation** of basophils and mast cells.

Acute-Phase Proteins

Acute-phase proteins, part of the innate immune response, are synthesized in the liver and circulate in the vascular fluid. These proteins increase rapidly (>100-fold) during both infection and inflammation. The **C-reactive protein** is the primary acute-phase protein. Its role during inflammation is thought to be opsonization of pathogens or damaged cells. **Ceruloplasmin** and **A_1-antitrypsin** are other acute-phase proteins thought to act away from the site of injury as a protective mechanism, neutralizing oxygen radicals and proteases (Evans and Cannon 1991).

Exercise and Immune Response

Exercise has a significant effect on immune function. The changes seen in the circulating concentration of the immune cells and their release pattern and distribution are governed to a large extent by changes in hormonal concentrations (e.g., catecholamines and cortisol) and cytokines (IL-1). This section examines the changes in immune cell function as they relate to acute exercise and to long-term training programs. Where possible, the response patterns are differentiated between different types of exercise programs (e.g., aerobic, anaerobic, or resistance exercise).

Acute Exercise and Leukocyte Response

Circulating leukocyte concentrations are consistently reported to increase after a wide range of exercise stresses (figure 5.3) varying in duration from several seconds to several hours (Gabriel, Urhausen, and Kindermann 1992; Gray et al. 1993; Nieman, Berk, et al. 1989; Nieman, Hensen, et al. 1995; Ndon et al. 1992; Shek et al. 1995). The magnitudes of these increases appear dependent on the intensity and duration of exercise. As the intensity or duration increases, a greater leukocyte concentration is observed (Mackinnon 1999; McCarthy and Dale 1988).

During short-duration, high-intensity exercise (e.g., sprints or resistance exercise), significant elevations in leukocyte concentrations (150-180% above baseline levels) are seen (Gabriel, Urhausen, and Kindermann 1992; Gray et al. 1993; Nieman, Hensen, et al. 1995). These increases persist during recovery, and the rate of decline appears to be related to the volume of the exercise protocol. Following brief maximal exercise, leukocyte concentrations begin to decline toward baseline levels after 30 to 60 min. As maximal exercise is maintained for a longer period (as might be seen with repeated interval training or multiple sets during resistance exercise), leukocyte concentrations may remain elevated for up to 2 h afterward (Gray et al. 1993; Nieman, Hensen, et al. 1995). As duration continues (up to 30 min at either moderate or high intensity), elevations in leukocyte concentrations may also remain for up to 2 h postexercise (Mackinnon 1999; Ndon et al. 1992).

During endurance exercise or prolonged exercise (up to 3 h), leukocytes reportedly increase rapidly and continue to rise for the duration of exercise (Mackinnon 1999). Total leukocyte concentrations may become 2.5- to 3-fold higher than resting levels

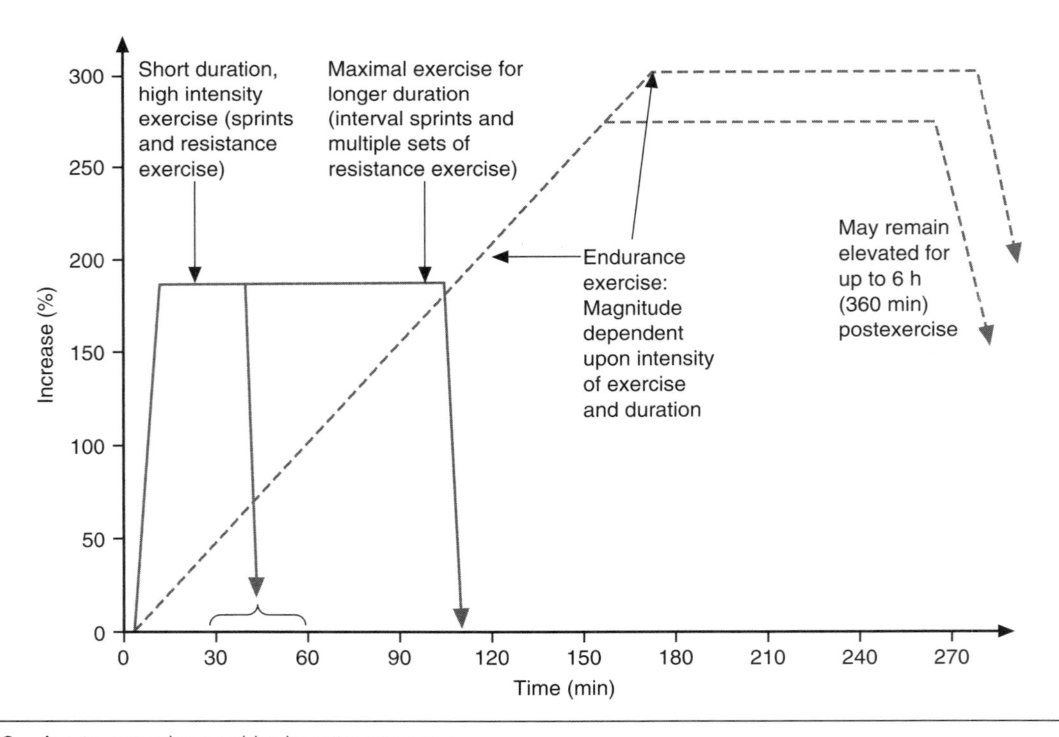

Figure 5.3 Acute exercise and leukocyte response.

(Nieman, Berk, et al. 1989; Shek et al. 1995) and remain elevated for up to 6 h postexercise (Nieman, Berk, et al. 1989). The magnitude of leukocyte elevation during endurance exercise may be a function of the intensity. Exercise performed at the anaerobic threshold produces a greater leukocyte response than exercise performed below the anaerobic threshold (Mackinnon 1999). Leukocyte numbers typically return to baseline levels between 3 and 24 h following endurance exercise (Fragala et al. 2011).

Given that neutrophils predominate in the leukocyte family, it is not surprising that their response to exercise parallels the changes generally seen in total leukocyte concentrations. During recovery, neutrophil content remains elevated and may possibly increase further for up to 6 h after high-intensity exercise (Nieman, Hensen, et al. 1995; Nieman, Simandle, et al. 1995; Shek et al. 1995). Other studies have shown that neutrophils may have a biphasic response. After exercise, neutrophil concentrations may return to resting levels within 30 min and increase again at 1 to 2 h postexercise (Hansen, Wilsgard, and Osterud 1991). The magnitude of the second response appears dependent on the duration of the exercise. A longer duration results in a greater secondary elevation in neutrophil response (Hansen, Wilsgard, and Osterud 1991).

Acute exercise of moderate intensity (60-85% of $\dot{V}O_2max$) does not appear to cause any change in monocyte concentrations from resting levels (Foster et al. 1986; Esperson et al. 1990). However, as exercise intensity increases to maximal effort, elevations in monocyte concentrations may be observed (Gray et al. 1993; Nieman, Hensen, et al. 1995). During prolonged exercise (>3 h), monocyte concentrations may increase 2.5-fold above resting levels (Gabriel et al. 1994) and may remain elevated between 2 and 6 h afterward (Gray et al. 1993; Nieman, Hensen, et al. 1995).

Increases in lymphocyte concentrations are consistently seen during and immediately after moderate- or high-intensity exercise of brief or prolonged duration. However, during prolonged exercise, the increase in lymphocyte concentrations is lower than that of neutrophils (Mackinnon 1999). In addition, lymphocyte concentrations return to baseline levels at a faster rate after exercise and may even decline below resting concentrations (Mackinnon 1999). This decline in lymphocyte concentrations below resting levels is reported after both brief (Gray et al. 1993; Nieman, Hensen, et al. 1995) and prolonged exercise (Gabriel et al. 1994; Nieman, Simandle, et al. 1995).

Lymphocyte subsets (T cells, B cells, and NK cells) appear to respond differently to acute exercise. Although all lymphocyte subsets increase during acute exercise, NK cells appear to show the greatest change (Gray et al. 1993; Mackinnon 1999). In addition, the ratio CD4:CD8 appears to be altered. A greater increase in concentrations of CD8 cells (cytotoxic T cells) is seen relative to changes in CD4 cells (helper and inflammatory T cells). Thus, the ratio CD4:CD8 declines during and immediately after both brief and prolonged exercise (Gabriel et al. 1992; Gray et al. 1993; Lewicki et al. 1988). Interestingly, the greater relative increase in the CD8 subset is also accompanied by a relatively greater decline during the recovery period. It appears that the CD8 cells are preferentially recruited into the circulation during exercise and removed after exercise (Mackinnon 1999).

Circulating B cells may increase or remain unchanged during or after exercise. Neither brief nor prolonged exercise, at either moderate or high intensity, appears to cause any elevations in B-cell numbers (Gabriel, Urhausen, and Kindermann 1992; Iverson, Arvesen, and Benestad 1994; Nielsen et al. 1996; Nieman, Hensen, et al. 1993). However, increases in B cells may occur after repeated maximal exercise (e.g., sprints or resistance training) (Gray et al. 1993; Nieman, Hensen, et al. 1995).

The largest increases in lymphocyte subsets are seen in the NK cells. Increases of up to 200% above resting levels have been reported during both brief (Nielsen et al. 1996) and prolonged (Mackinnon et al. 1988) exercise of high intensity. Increases are seen during and immediately after exercise, and the concentrations rapidly return to resting levels. After prolonged or more intense exercise, NK concentrations may decline below resting levels for several hours or even days (Berk et al. 1990; Mackinnon et al. 1988; Shek et al. 1995). Increases in NK cell concentrations appear related to the intensity of exercise. As intensity increases, the NK response is elevated (Nieman, Miller, et al. 1993; Tvede et al. 1993).

Long-Term Training and Leukocyte Response

Cross-sectional comparisons between athletes and nonathletes have generally not reported any significant differences in counts of resting total leukocytes or any of their subsets (neutrophils, monocytes, and lymphocytes) (Mackinnon 1999). Studies examining the effect of prolonged training on resting leukocyte concentrations have also been unable to demonstrate any significant alterations (Baum, Liesen, and Enneper 1994; Hooper et al. 1995; Gleeson

In Practice

What Happens to Neutrophil and Lymphocyte Functions During a Competition?

Japanese sport scientists examined the effect of a soccer game on changes in immune function in female university soccer players (Tsubakihara et al. 2011). Reactive oxygen species, production capability, phagocytic activity, and opsonic activity were measured to describe changes in neutrophil function. In addition, lymphocyte subpopulations were measured before and after the game. The stress of the match was assessed by changes in several muscle enzymes that are used to evaluate muscle damage (e.g., creatine kinase, lactate dehydrogenase, alanine aminotransferase, and aspartate aminotransferase). These markers were significantly elevated following the game, indicating that the level of intensity and stress during the contest was significant enough to cause muscle damage. Leukocytes and neutrophils were significantly elevated, while lymphocyte numbers were significantly depressed at the end of the game. Close examination of lymphocyte subpopulations revealed significant decreases in T cells (6.0%) and NK cells (22.1%), while B cells were significantly elevated by 45.3%. The immunoglobulins IgG, IgA, and IgM and complement were significantly elevated. Phagocytic activity was significantly depressed following the match, but no other changes were noted in neutrophil function. The decrease in neutrophil and lymphocyte function suggests that the stress and fatigue associated with a soccer game may result in a possible immunosuppression. Thus coaches and trainers should focus on ensuring adequate rest and recovery following competitions. This may have special relevance for intercollegiate soccer teams, who often play games separated by only 36 to 48 h. Considering the additional stresses of travel and academic class requirements, the total load may pose a significant challenge for athletes with regard to maintaining adequate immune function, and consequently health, throughout the competitive season.

et al. 1995; Tvede et al. 1989). Mackinnon (1999) has suggested that the magnitude of the leukocyte response to acute exercise remains unaffected by an individual's training status. However, several studies have reported that circulating leukocyte numbers may change during periods of training when the volume is elevated.

Lehmann and colleagues (1996) examined changes in resting leukocyte numbers in subjects performing prolonged training. When training volume was increased, a decrease in leukocyte count was seen. This decrease corresponded to an elevation in muscle stiffness and fatigue in these subjects. However, when training intensity was increased, no changes in leukocyte numbers were seen. It appears that the leukocyte response may be more sensitive to changes in training volume than to changes in training intensity. Other studies have also reported decreases in resting leukocyte concentrations after prolonged training (Ferry et al. 1990) or levels within the lower range of normal (Keen et al. 1995; Green et al. 1981). Although most studies have not shown any changes in total leukocyte count during prolonged training, those that have done so appear to have focused primarily on endurance athletes experiencing large increases in training volume. Long-term resistance exercise training has been reported to reduce resting cytokine concentrations and reduce low-grade inflammation (Calle and Fernandez 2010; Fragala et al. 2011; Mathur and Pedersen 2008).

Benoni and colleagues (1995), presenting findings that contrast with data on the response of endurance athletes, reported a significant increase in total leukocyte numbers over the duration of a competitive basketball season. In addition, they reported significant increases in neutrophils, monocytes, and lymphocytes. Neutrophil adhesion (an early event in the host defense mechanism) was also reduced. Thus, the leukocyte response may depend somewhat on the type of training stress (e.g., aerobic versus anaerobic).

In general, leukocyte subsets parallel the response seen in total leukocyte numbers during training. Most studies have not reported any training effects, either in resting counts or in the response to an acute bout of exercise, in neutrophils, monocytes-macrophages, or lymphocytes (including T cells, B cells, and NK cells). However, similar to findings on total leukocyte numbers during periods of prolonged high-volume training, a decrease in the resting concentrations of monocytes (Mackinnon et al. 1997; Ndon et al. 1992) or a reduction in the neutrophil response to exercise (Suzuki et al. 1996) has been reported. Other studies have reported elevations in leukocyte subsets over the duration of a competitive season (Baum, Liesen, and Enneper 1994; Benoni et al. 1995). The physiological significance of these responses is not clear, but as mentioned previously, may be related to the training mode.

Exercise Training and Phagocytic Cell Function

The number of circulating phagocytic cells provides information about the release of neutrophils and monocytes into the vasculature but does not provide any information on the functional capacity of these cells. These functions include migration of the phagocytes in response to chemotactic stimulation, **adherence**, and phagocytic activity.

Neutrophil function appears to be normal in athletes even during periods of intense training (Mackinnon 1999). Acute exercise is generally associated with enhanced neutrophil function (improved phagocytic activity, migration, and expression of complement receptors) (Mackinnon 1999). However, consistent with what we have seen previously, prolonged periods of endurance training may have a deleterious effect on neutrophil function. A 20% to 30% reduction in neutrophil migration has been reported (Esperson et al. 1991). In addition, several studies have shown a decrease in phagocytic activity in endurance athletes compared with control subjects (Blannin et al. 1996; Hack et al. 1994; Lewicki et al. 1987; Smith et al. 1990). The decreased functional capacity may be related to a decline in granule content, reducing the phagocytic ability of the neutrophils. Decreases are observed in

- ability to migrate in response to chemotactic stimulation,
- neutrophil adherence,
- granule content,
- phagocytic activity, and
- sensitivity to stimulation.

The reduction in neutrophil function may result in a greater susceptibility of endurance athletes to infection. However, Smith and colleagues (1990) have suggested that the lower sensitivity seen in neutrophil function may be a beneficial adaptation in these athletes that limits the inflammatory response to chronic tissue damage caused by continuous exercise training.

Much less research has focused on monocyte–macrophage function as it relates to exercise and training in humans. Lewicki and colleagues (1987) reported lower monocyte adherence at rest and after exercise in athletes compared with untrained individuals. Reduced monocyte activity in athletes at rest has also been reported (Osterud, Olsen, and Wilsgard 1989). However, in contrast to these findings, improvements between 20% and 60% in the metabolic and phagocytic activity of macrophages in trained athletes after exhaustive exercise have been reported (Fehr, Lotzerich, and Michna 1988, 1989). Whether these changes are related to the athletes' ability to adapt to an increased training stress (overtraining syndrome) is not clear. Further examination of monocyte–macrophage function as it relates to exercise training appears to be needed.

Complement, Acute-Phase Proteins, Cytokines, and Exercise

As previously mentioned, complement, acute-phase proteins, and cytokines are soluble mediators of innate immunity. They circulate within bodily fluids to inhibit or modulate the immune response to inflammation or infection.

Investigations on response of complement to an acute bout of aerobic exercise have yielded differing results. Several investigators have reported that complement levels are significantly elevated (Castell et al. 1997; Dufaux, Order, and Liesen 1991), whereas others have not seen any change from resting levels (Esperson et al. 1991). These differences may be related to the duration of exercise. Esperson and colleagues (1991) reported no change in complement levels after a 3 mi (5 km) run by elite athletes. However, in studies on exercise of longer duration, significant elevations in complement concentration (11-45% increase from resting levels) have been seen (Dufaux, Order, and Liesen 1991; Nieman, Berk, et al. 1989). Complement levels may remain elevated for several hours after exercise and may be responsible for cleaning **proteolytic fragments** released from damaged muscles (Mackinnon 1999).

Lower resting complement levels have been reported in endurance athletes compared with a nonathletic population (Nieman, Berk, et al. 1989). These lowered resting complement levels corresponded to a reduced complement response after a session of graded exercise in comparison with control subjects. The reduced resting response and the attenuated response to exercise seen in the athletic population may be considered a positive adaptation to training. The lower resting and postexercise complement concentrations are thought to reflect long-term adaptation to chronic inflammation from intense daily training (Mackinnon 1999).

The response of acute-phase proteins appears dependent on the duration of exercise. C-reactive protein (CRP), the predominant acute-phase protein, has been reported to increase when duration of exercise exceeds 2 to 3 h (Castell et al. 1997; Liesen, Dufaux, and Hollmann 1977; Strahan et al. 1984; Weight, Alexander, and Jacobs 1991). During exercise of shorter durations, CRP levels do not appear to be affected by the exercise stress (Hubinger et al. 1997; Nosaka and Clarkson 1996). Even when the exercise intensity is enough to cause significant muscle cell damage (e.g., maximal eccentric elbow flexion), CRP levels do not appear to become elevated (Nosaka and Clarkson 1996). The authors of this latter study suggested that the lack of acute-phase protein response may be related to the lack of a cytokine response seen during this particular exercise protocol. Prolonged endurance events (i.e., Ironman triathlon) have been reported to result in significant alterations to CRP for 5 days following the event (Neubauer, Konig, and Wagner 2008), indicating a prolonged inflammatory response associated with this endurance race.

The effects of training on resting acute-phase protein concentrations are not very clear. Mackinnon (1999) has reported that athletes may have CRP levels that are normal or higher or lower than those of nonathletes. However, the acute-phase protein response to endurance exercise may be reduced after a training program (Liesen, Dufaux, and Hollmann 1977). This may be an adaptation to control the inflammation that could result from daily training sessions and may also be related to the downregulation of neutrophil function after prolonged training programs.

The cytokine response to exercise is considered difficult to assess. Cytokines, especially the proinflammatory cytokines **IL-1**, **IL-6**, and **TNF-α**, appear to be released during and after exercise (Mackinnon 1999). However, because of the rapid clearance of cytokines from the blood and the local production of cytokines at sites of tissue damage, blood levels may be difficult to interpret and may not reflect the true response of the cytokines to exercise. Significant elevations of IL-1, IL-6, IFN, and TNF-α in the urine have been reported 3 to 24 h after endurance exercise (12.5 mi [20 km] run) despite limited changes in the plasma concentration of these cytokines (significant elevation seen in IL-6 only) (Sprenger et al. 1992). In addition, increases in IL-1 have been seen in skeletal muscle for up to 5 days after prolonged eccentric exercise (downhill running for 45 min) (Cannon et al. 1991; Fielding et al. 1993). IL-6 has been reported to remain significantly elevated for 5 days following an Ironman triathlon (Neubauer, Konig, and Wagner 2008). The elevated IL-1 levels in skeletal muscle after eccentric exercise or after prolonged endurance exercise suggest that the cytokines play some role in the inflammatory and repair process of tissue damage typically seen during this type of exercise.

It has also been shown that resistance exercise causes a significant inflammatory response mediated by cytokines (Pedersen et al. 2003). Elevations in cytokines, specifically IL-6, are thought to have a role in the muscle repair and remodeling process following an acute resistance exercise session (Pedersen et al. 2003; Steensberg et al. 2000). Izquierdo and colleagues (2009) reported that IL-6 concentrations were significantly elevated at 45 min postexercise, along with IL-1β (a proinflammatory cytokine). Following 7 weeks of training, IL-10 (an anti-inflammatory cytokine) was significantly elevated postexercise. The elevation in IL-10 subsequent to training indicates a positive adaptation that results in an attenuated inflammatory response and may explain the reduced soreness associated with consistent training programs.

The effect of training on resting cytokine response has not been well examined. Sprenger and colleagues (1992) showed that trained endurance athletes had higher resting concentrations of IL-1, IL-6, and TNF-α than control subjects. However, Smith and coauthors (1992), comparing endurance athletes to nonathletic controls, did not see any significant differences in resting cytokine levels. Although one can only speculate about the reason for these contrasting results, it may be that differences in the training program or training status of the athletes were a factor.

Exercise and Immunoglobulins

As previously discussed, immunoglobulins are important mediators of humoral (acquired) immunity. Immunoglobulins demonstrate antibody activity; bind to specific antigens; and stimulate phagocytosis, cytotoxicity, and complement binding. Immunoglobulins are found in both the blood and mucosal secretions, but each may respond differently to exercise.

Immunoglobulin concentrations in the blood appear to remain at resting levels or increase only slightly after exercise in athletes. No changes from resting levels were seen immediately or up to 24 h postexercise in male distance runners after runs of 8 mi (12.8 km) (Hansen and Flaherty 1981) and 13 mi (21 km) (Gmunder et al. 1990). Similar results were obtained in cyclists after a 2 h exercise bout (Mackinnon et al. 1989). Immunoglobulin levels also remained at baseline after 30 s of maximal exercise on a cycle ergometer (Wingate anaerobic test) when values were corrected for changes in plasma volume (Nieman et al. 1992). Slight increases in serum immunoglobulin levels were seen in overweight females after 45 min of walking. IgG levels remained elevated for 1.5 h postexercise, while IgA and IgM, after showing an increase immediately after exercise, declined to levels below baseline for 5 h postexercise.

Resting blood concentrations of immunoglobulins appear to be similar between athletic and sedentary populations (Nieman, Tan, et al. 1989; Nehlsen-Cannarella et al. 1991). However, several studies have shown reduced levels of immunoglobulins (low end of the clinically normal range) in various athletes (Garagioloa et al. 1995; Gleeson et al. 1995). The reduced levels of immunoglobulins seen in both endurance and nonendurance athletes is likely a function of intense exercise training. The lower immunoglobulin concentration may result in an increased susceptibility to infection. This is discussed in more detail later in the chapter.

Salivary IgA levels are generally used to measure mucosal immune states. Several studies have shown reduced resting salivary IgA levels in athletes involved in prolonged high-intensity training programs (Mackinnon and Hooper 1996; Tomasi et al. 1982). Tomasi and colleagues (1982) showed reduced IgA concentrations at rest (50% lower) in elite male and female Nordic skiers compared with age-matched controls. Mackinnon and Hooper (1996) reported lower resting salivary IgA levels in swimmers during a period of high-intensity training, which coincided with feelings of staleness. In that same study, it was also noted that no changes in salivary IgA levels were seen in athletes who apparently adapted well to the changes in training intensity. Tharp and Barnes (1990) and Gleeson and coauthors (1995) reported significant reductions in resting salivary IgA levels during the course of a swim season. These reductions were still observed even after the athletes began to reduce (taper) their training load. The lower resting IgA concentrations that were maintained during the taper suggest that the cumulative effect of high-intensity training over a prolonged period (6-7 months) necessitates a longer period of recovery for humoral immune function to return to its normal level. Other studies have also demonstrated the cumulative effect of training on reducing resting salivary IgA levels even during a relatively short duration of training (4-5 days) (Mackinnon and Hooper 1994; Mackinnon, Ginn, and Seymour 1991). It appears that intense exercise training, regardless of the duration, may suppress resting immunoglobulin concentrations. This immunosuppressive effect may result in a higher incidence of upper respiratory tract infections among athletes experiencing difficulty adapting to changes in the exercise stimulus.

The response of immunoglobulins to acute exercise appears dependent on the duration of exercise. Aerobic exercise exceeding 2 h in duration has been shown to cause reductions in salivary IgA concentrations between 40% and 60% immediately after exercise (Mackinnon et al. 1989; Tomasi et al. 1982). The lowered IgA levels may remain suppressed for up to 24 h. Salivary IgA levels may also be reduced after exercise of shorter duration and higher intensity (McDowell et al. 1992; Mackinnon, Ginn, and Seymour 1993; Tharp and Barnes, 1990), but the decrease is not as great as generally seen after prolonged endurance exercise. During exercise of moderate intensity and shorter duration (45-90 min of running), IgA concentrations may not be altered from their resting levels (McDowell et al. 1991; Mackinnon and Hooper 1994). In contrast to more intense and prolonged exercise, moderate exercise does not appear likely to stress the immune system.

Immune Response in Athletes

It has been demonstrated that exercise training in previously sedentary individuals has beneficial effects in terms of reducing the incidence of upper respiratory tract infections (URTI) (Nieman et al. 1990; Nieman, Hensen, et al. 1993). Nieman, Hensen, and colleagues (1993) reported a 50% reduction in URTI in women exercising 5 days per week compared with sedentary age-matched controls (figure 5.4). Exercise also enhances the perception of being healthy. A survey of 750 masters athletes (ranging in age from

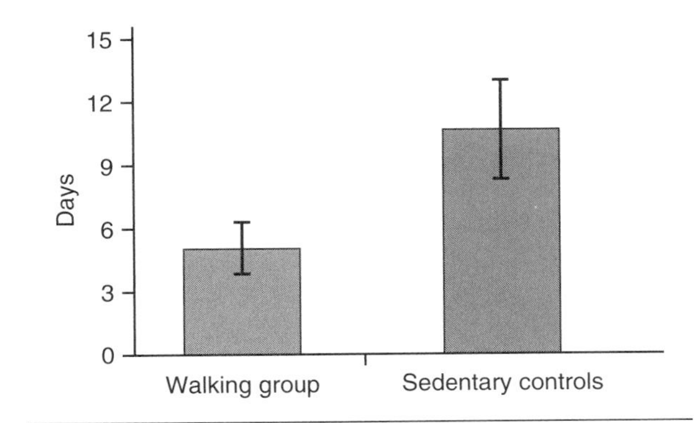

Figure 5.4 Effects of moderate exercise training on number of days with URTI in previously sedentary women.

Data from Nieman et al. 1990.

40 to 81 years) showed that these active individuals considered themselves less vulnerable to viral infection than their sedentary counterparts (Shephard et al. 1995). An additional survey reported that 90% of nonelite runners who had been running marathons for an average of 12 years agreed with the statement that they "rarely got sick" (Nieman 2000).

In contrast to the proposed health benefits of training, other reports have suggested that athletes may be at a greater risk for URTI than other populations. Nieman and coauthors (1990) reported that 12.9% of the runners in a marathon experienced URTI symptoms during the week after the race in comparison with 2.2% of the control runners (runners who did not run the marathon). Similarly, Peters (1990) reported a 28.7% incidence of URTI in runners after a 35 mi (56K) race compared with a 12.9% incidence rate in controls. The higher incidence of URTI appears to occur during the 2 weeks after the event. The elevated risk for URTI after acute athletic events may depend on the distance of the race. During the weeks after races of 3 mi (5K), 6 mi (10K), and 13 mi (21K), Nieman, Johanssen, and Lee (1989) were unable to see any increased prevalence in URTI compared with the week before the race. In that same study, the researchers also noted no differences in the incidence of URTI in runners who ran an average of 26 mi (42 km) per week in comparison with runners averaging 7.5 mi (12 km) per week.

The greater incidence of URTI appears to occur in highly stressful events such as marathons or ultramarathons. However, investigations have also suggested that intense prolonged training increases the risk of URTI. Linde (1987) reported that a greater incidence of URTI was seen in elite orienteers (2.5 episodes per subject) compared with age-matched controls (1.7 episodes per subject) during a year of training. Other epidemiological studies have also implied that intense, prolonged training is associated with an increased risk for URTI (Heath, Macera, and Nieman 1992; Nieman et al. 1994). The increased risk for URTI appears to be limited to endurance athletes. No known studies have reported an increase in URTIs in strength or power athletes. However, the amount

In Practice

Can Observation of Epstein-Barr Virus Antibodies Be Used to Assess Immunosuppression in Competitive Athletes?

The greater risk in competitive athletes for URTI or other illnesses has resulted in an interest on the part of scientists in exploring whether there is an underlying pathogen that increases the risk of these illnesses. Epstein-Barr virus antibody measurement is generally the first diagnostic test given by clinicians confronted with nonspecific symptoms of infectious mononucleosis such as pharyngitis, fever, lymphadenopathy, splenomegaly, or hepatomegaly and abnormal blood tests (Pottgiesser et al. 2012). It is performed to detect or rule out present or past Epstein-Barr virus. Pottgiesser and colleagues (2012) investigated changes in Epstein-Barr virus antibodies, specifically sequence of antibodies against early antigens, among competitive athletes during a competitive season to see if they could determine whether such measures could provide additional insight on risk or potentially identify susceptible athletes. Two groups of athletes were recruited; all were national- or international-level competitors, and some had competed at the Olympic Games. One group consisted of biathletes who trained approximately 25 h per week; the other group consisted of swimmers. University students who trained <8 h per week were recruited to serve as the control group. Epstein-Barr virus serostatus and antibody concentrations were collected between four and six time points over 13 to 15 months in the endurance athletes and for 6 months in the swimmers. The prevalence of Epstein-Barr virus seropositivity was 94% in the endurance athletes and 73% in the controls. The majority of the positive tests were attributed to past infections (67% in the athletes and 64% in the controls). Ninety-one percent of the swimmers tested positive for the virus, but only two cases (5%) were deemed to be recent infections. The results of the study demonstrated the relative stability of various antibody levels during the competitive season. Unfortunately, the investigators were unable to determine a sequence of antibody reactions that could identify Epstein-Barr virus reactivation or identify compromised immune function that may lead to infection among competitive athletes.

of research on this athletic population is considerably less than that on endurance athletes.

Evidence suggests that exercise training is beneficial for lowering the risk of infection. However, intense prolonged training in elite-level athletes performing long-distance endurance events (marathons or ultramarathons) may cause an elevated susceptibility to illness. Athletes in general do not appear to be immunosuppressed. The only illness to which they appear more susceptible is URTI (Mackinnon 1999). In the last few years, many studies cited in this chapter have indicated that several aspects of the immune system are affected by prolonged periods of intense training (reduced leukocyte counts, lower immunoglobulin concentrations, suppression of antimicrobial activity).

It has been proposed that athletes become more susceptible to infection in the days after intense exercise (Pedersen and Ullam 1994). If the athlete begins a training session before full recovery from the previous exercise session (during a period of potential immunosuppression), the risk of infection increases because the athlete is training at a lower baseline. If this pattern of beginning exercise sessions before complete recovery continues, a cumulative suppression of some aspects of immune function may be seen, causing a further elevation in the risk for infection.

Cox and colleagues compared the immune and inflammatory response in healthy and illness-prone runners (Cox et al. 2007, 2009). Male runners who had experienced no more than two episodes of URTI per year were classified as healthy, and runners who had experienced four or more episodes per year were classified as illness prone. Each subject was assessed by incremental endurance tests in the laboratory. In one test the runners performed a 30 min run at a speed corresponding to 65% of their $\dot{V}O_2max$; in another they performed a 60 min run at the same intensity; and in a third, they performed six 3 min interval runs at a speed corresponding to 90% of their $\dot{V}O_2max$. Resting and postexercise CRP concentrations were within normal ranges, and no differences were observed between the groups in any of the exercise protocols (Cox et al. 2009). However, the cytokine responses to the running protocols differed between the healthy and illness-prone athletes. Resting IL-8, IL-10, and IL-1ra concentrations were 19% to 38% lower in illness-prone subjects; postexercise IL-10 concentrations were 13% to 20% lower, and IL-1ra concentrations were 10% to 20% lower. In contrast, IL-6 elevations were 84% to 185% higher in illness-prone subjects. The illness-prone distance runners showed evidence suggestive of impaired inflammatory regulation in the hours after exercise that may account for the greater frequency of upper respiratory symptoms experienced.

Sex Differences in Immune Responses

There appears to be a difference in the response pattern between men and women to immune challenges (Fragala et al. 2011). Women seem to be able to develop a stronger immune response after infection (Verthelyi 2001). Males appear to be more susceptible to respiratory diseases and viral and some bacterial infections, whereas women may be more susceptible to many autoimmune diseases (e.g., multiple sclerosis, myasthenia gravis, thyroid disease) and have a greater immunoreactivity to specific pathogens (Fragala et al. 2011; Whitacre 2001). These differences may be explained in part by how the sex steroids act on the immune system. Differences in the concentration of testosterone and estrogen modulate the hypothalamic–pituitary–adrenal axis, which modulates the stress response. Androgens appear to suppress the production of autoantibodies, whereas estrogen appears to support their production (Fragala et al. 2011; Verthelyi 2001). Men also have a greater cytokine production than females, possibly related to the potential effect that testosterone has on cytokine production (Posma et al. 2004).

A lower inflammatory response to muscle-damaging exercise is seen in women compared to men, even when the muscle damage is similar (Stupka et al. 2000). This may be more relevant in trained women, since sex does not appear to be a factor in lymphocyte apoptosis in response to endurance exercise in untrained men and women (Miles et al. 2002). However, investigations on the immune response to resistance exercise in trained and untrained females is inconclusive, with some studies showing differences and others suggesting that these differences may be transient (Fragala et al. 2011).

Summary

Exercise may be beneficial for reducing the incidence of infection in previously untrained individuals. However, if exercise intensity remains high during a prolonged period of training or after a bout of prolonged exercise, the athlete may be at a greater risk for URTI. The increased risk for infection may be related to reduced mucosal IgA concentrations, leukocyte counts, and NK cell numbers. The higher incidence of URTI appears to be confined to endurance athletes. Further research on exercise and infection in strength and power athletes appears warranted.

REVIEW QUESTIONS

1. Discuss the difference between T cells and B cells.
2. What is the difference in immune function between a fitness enthusiast and a sedentary adult?
3. Describe what may occur with regard to neutrophil function in an overtrained athlete.
4. What is the role of the acute-phase proteins on muscle remodeling?
5. Discuss the benefit and risk associated with exercise and infection among competitive athletes.
6. What is the difference in the immune response to exercise between men and women?

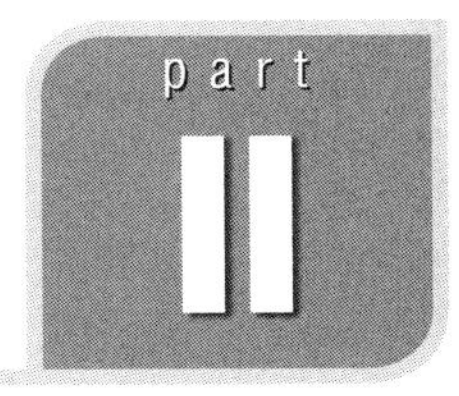

EXERCISE TRAINING PRINCIPLES AND PRESCRIPTIONS

This part of the book provides readers with the tools necessary to develop and design training programs for athletes in all sports. Specific emphasis is on training principles, warm-up and flexibility, resistance training, power training, anaerobic conditioning, speed and agility development, and endurance training. Most athletes tend to perform several different modes of training at the same time; the chapter on concurrent training discusses training modes that are compatible and those that are not. Finally, this section will give the reader the tools needed to develop a yearly training program. The chapter on periodization provides the theoretical basis of variation of training intensity and training volume to assist athletes in reaching peak performance at the appropriate time and to reduce the risk of overtraining. The chapter on program development and implementation discusses the practical aspects of program development, and the chapter on athletic performance testing and normative data gives an in-depth overview of both laboratory and field assessments. This chapter also provides normative data to use for comparison purposes.

chapter 6

Principles of Training

CHAPTER OBJECTIVES

After reading this chapter you should be able to do the following:

- Understand the basic principles of training.
- Describe the effects of detraining on both endurance and resistance exercise performance.
- Explain the effects of reduced training on bone mass and muscle cross-sectional area.
- Understand the effects of a taper on performance.

For an exercise training program to be properly designed, it is important that the goals and objectives of the program be clear first. Because the success of the training program is evaluated by its ability to accomplish these goals, it is imperative that the goals be both reasonable and attainable. Goals that are common to training programs include increasing muscle strength or size, improving sport performance, improving **aerobic capacity**, and improving **body composition**. Through a properly designed exercise program, the individual's training goals can be achieved by both acute and chronic alterations to the structure of the training program. The structure of the program includes specific component variables that can be manipulated from workout to workout, commonly known as **acute program variables**. These variables include the choice of exercise, order of exercise, **intensity** of exercise, **volume** of exercise, **frequency** of training, and length of rest period (between sets or between workouts). Each of these variables may have an important influence on the physiological adaptation to the training stimulus (see chapters 1-5). These acute program variables and their effects on specific training programs are reviewed in the next several chapters.

Unrealistic expectations of a training program can lead to a tremendous amount of frustration on the part of both the athlete and coach. To optimize the exercise program **prescription** and to set realistic training goals, the basic principles of exercise need to be understood. These principles are the basic tenets of exercise science and are valid whether one is designing a resistance training program or a running program. One of the principles discussed in this chapter involves the removal of the training stimulus. When this occurs, either through a designed reduction such as a taper or a forced cessation due to injury, a physiological response is seen. Depending on the type of training reduction and its duration, the positive adaptations that were accomplished will be lost, with values reverting back to their initial levels. This chapter also focuses on issues relating to detraining and taper and the use of the latter in program development.

Specificity Principle

According to the specificity principle, adaptations are specific to the muscles trained, the intensity of the exercise performed, the metabolic demands of the exercise, and the **joint angle** trained. For instance, if the goals of the training program were to maximize strength gains, then performing low-intensity, high-volume exercise would not be specific to the objectives of that particular program. Likewise, one would not prepare for a marathon by concentrating solely on running short sprints. **Resistance training** is often part of an athletic conditioning program with the primary objective to improve sport performance. For strength increases to positively affect sport performance, the training program must have a high **carryover** to the sport. Except for actual practice of the sport, no conditioning program has 100% carryover. To optimize the transfer of strength from the weight room to the field of play, it is important to select exercises that train the specific muscles recruited during performance. In addition, the exercises selected need to place a demand on the neuromuscular coordination of movement similar to that imposed during performance (i.e., one should choose exercises that best simulate the movement performed on the field of play). For example, a multijoint structural exercise like the push press requires a coordinated movement of both the upper and lower body musculature to press a barbell from shoulder height to a position above the head. This exercise is similar to specific movements seen in basketball, such as jumping for a rebound or attempting to score from close to the basket with defenders nearby.

Overload Principle

The basis of the **overload principle** is the idea that for training adaptations to occur, the muscle or physiological component being trained must be exercised at a level that it is not normally accustomed to. For instance, to maximize muscular strength gains, the muscle needs to be stimulated with a resistance of relatively high intensity (this is discussed in further detail in chapter 8). If the exercise prescription calls for an individual to perform a five-**repetition maximum (RM)** and the person uses a resistance that can be lifted for more than five repetitions, the individual may not be overloading the muscle. Consequently, strength gains may not be maximized. Another example is the endurance athlete who trains for a marathon. If the training goal is to maximize aerobic capacity in order to run a faster time, then training intensity must be near or at the individual's **anaerobic threshold** (this is explained in further detail in chapter 12). This can be expressed as a percentage of the individual's maximal heart rate. If the training intensity is not high enough (e.g., heart rate does not reach the required range), then the physiological adaptations that can result in an improved aerobic capacity will not occur.

Progression Principle

During the course of a training program, adaptations occur that change the relative intensity or volume of training. In order to maintain the same **absolute training stimulus** (i.e., intensity or volume of training), the resistance needs to be continually modified. As an example, if an exercise prescription requires a person to perform four sets of 8 to 10 repetitions of the squat exercise, the objective is for the individual to exercise with a resistance that can be lifted at least 8 times but not more than 10. At the beginning of the training program, the person may be able to squat 135 lb (61 kg) for 10 repetitions in the first set, 9 repetitions in the second set, and 8 repetitions in both the third and fourth sets. After several weeks, this individual is performing 10 repetitions of 135 lb (61 kg) for all four sets and feels able to do more than 10 repetitions per set. Obviously, the person has become stronger. To maximize further strength gains, the resistance needs to be increased for the next exercise session (perhaps to 145 lb [66 kg]). This process of applying **progressive overload** occurs continually throughout the training program. Figure 6.1 illustrates the importance of this principle during endurance training.

Individuality Principle

The **individuality principle** refers to the concept that people respond differently to a given training stimulus. The variability of the training response may be influenced by such factors as **pretraining status**, genetic predisposition, and sex. Many elite bodybuilders publish their training programs, and aspiring bodybuilders attempt to perform these regimens with hopes of duplicating the results. Unfortunately, more often than not, their results fall far short of the desired outcome. Although there may be many factors that relate to their disappointment, the primary factor is most likely the large variability between people in response to similar training stimuli.

Principle of Diminishing Returns

The **principle of diminishing returns** states that performance gains are related to the level of training

In Practice

The Overload Principle Is Critical for Maximizing Performance Gains, but Can It Be Abused?

There is strong scientific evidence demonstrating that to stimulate physiological adaptations, the system being trained must overcome a stimulus that it is not accustomed to. However, a few well-documented incidents suggest that strength and conditioning professionals may misinterpret the meaning of the principle or apply it incorrectly. In January 2011, thirteen student-athletes (football players) from the University of Iowa were hospitalized for exertional rhabdomyolysis (Drake et al. 2011). This incident occurred following the team's first off-season workout. This workout took place approximately 3 weeks after the team's bowl game (last game of the season). Players had been given individual workout cards for the 3-week winter break, but only 70% of the players had done 50% or more of the recommended workout. During the workout in question, players were to perform 100 squats at 50% of their previously measured one-repetition maximum (1RM). However, this measurement had been made 6 months earlier. In the interim period the players had completed a football season and had just returned from the 3-week winter break. Thus, it would be reasonable to assume that the load required for the workout would no longer represent 50% of the athlete's 1RM. In addition to the squat workout, the athletes were to perform barbell snatches, pull-ups, dumbbell rows, and a weighted sled-pushing exercise in the indoor training facility. Although players were allowed to take as much time as they needed to complete the 100 squats, many believed it was a competition.

Following the squat workout, many players had difficulty getting to the indoor training facility to complete the rest of the session. The primary question in review of this incident was whether the overload was appropriate for the athletes at that time of the training year. Interestingly, the incident report mentioned that the same or a similar workout had been performed by the coaching staff, albeit during the later parts of the training year. However, considering the time of the year (i.e., first workout), the overload used for the training session was excessive. More importantly, though, based on the workout, what was the goal of the training program? The exercise prescription needs to be based on a specific training goal. Once the goal is identified, the training program is developed to help accomplish that goal. The overload is specific to the goal. The primary problem identified regarding the incident at the University of Iowa was the lack of a training goal that could be associated with the specific training stress. This type of volume (very high) and intensity (low) overload would have absolutely no influence on strength or power development. In the development of the American football player, it is difficult to find support for this specific training program.

One of the motivating factors that became apparent from this specific workout was the desire of the coaching staff to "set the tone" for the off-season conditioning program. Unfortunately, it resulted in the hospitalization of 13 athletes. The use of extreme overload, especially relating to training volume, to develop the "mental toughness" of athletes should not be condoned or accepted. There is no evidence to suggest that extreme workouts can build mental toughness; they may reveal it, but they will not build it. There is no place for these types of training protocols in an evidence-based training program aimed at maximizing performance and minimizing risk for injury.

experience of the individual. Novice weightlifters experience large strength gains after a relatively short period of time. In contrast, athletes who have strength trained for several years make small strength gains over a long period of time. This point is illustrated in a theoretical training curve depicted in figure 6.2. At the onset of a training program, rapid strength gains are made. As training duration continues, the rate of strength improvement begins to slow down. As training continues further, changes in strength and performance are difficult to achieve, and it appears that a plateau has been reached. This plateau may be considered a **genetic ceiling**. It is at this point that many athletes become frustrated by the lack of

Figure 6.1 Importance of the progression principle during endurance training.

performance improvement and may experiment with anabolic steroids or some other **ergogenic aid** in hopes of pushing past the plateau.

The importance of pretraining status and its effect on strength improvement is reflected in a study performed on elite collegiate basketball players (Hoffman, Maresh, et al. 1991). The purpose was to examine the effectiveness of an in-season strength training program. The players were placed into two groups for analytical purposes. One group was made up of athletes who had previous strength training experience, and the other group consisted of athletes who had no previous strength training experience and were considered novice lifters. During the course of a season, no strength improvements were observed in the trained group, whereas a significant 4% increase in upper body strength (1RM bench press) was noted in the untrained group (see table 6.1). Although the two groups participated in identical strength training programs, the results clearly differed. The difference between the groups in strength improvement appeared to be related to the initial strength training experience of the athletes.

The principle of diminishing returns highlights the importance of being able to interpret performance results of the athlete who is training. Hak-

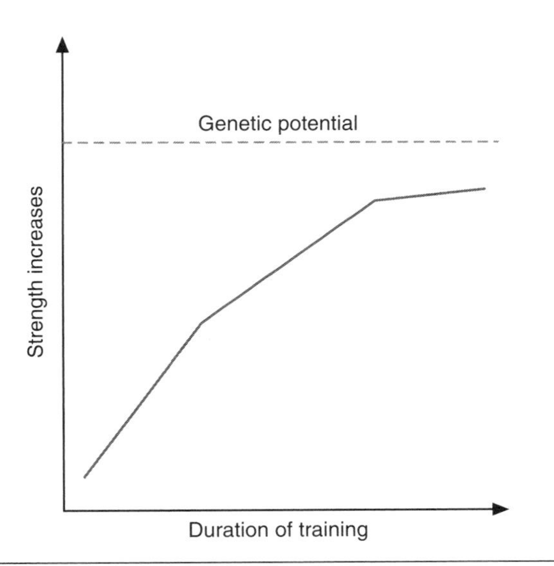

Figure 6.2 Theoretical training curve.

Table 6.1 Effect of an In-Season Resistance Training Program

Variable	Trained	Untrained
	Bench press	
Pre	101.6 ± 9.6	92.4 ± 26.0
Post	102.9 ± 11.0	96.2 ± 24.2*
	Squat	
Pre	161.4 ± 16.4	131.1 ± 23.3
Post	150.8 ± 13.1	133.7 ± 20.9

*Significantly different from Pre levels, $p < 0.05$.

kinen, Komi, and colleagues (1987) examined elite weightlifters for a 1-year period. Small increases in strength were observed, but none of these increases reached statistical significance. Although statistically speaking no changes were seen, practically speaking, the athletes and coaches could rate the training program a success. In a group of elite athletes, training improvements are so difficult to achieve that even small improvements can mean the difference between winning or not. In these situations, practical significance takes precedence over statistical significance.

Principle of Reversibility

When the training stimulus is removed or reduced, the ability of the athlete to maintain performance at a particular level is also reduced, and eventually the gains that were made from the training program will revert back to their original level. This is the basis for the **reversibility principle**. A common example of the effects of detraining is seen when an individual is forced to immobilize an injured body part in a cast. When the cast is removed, the muscles of the injured limb have undergone a significant reduction in size (referred to as **atrophy**) and strength. This drastic reduction in the size and strength of the muscle occurs because of the lack of activity while the limb is in the cast. Similar reductions in strength and size, as well as other performance measures, can be seen if the training stimulus that was used to produce the gains is removed or even just reduced.

When the training stimulus is removed or reduced for an extended duration, the athlete is said to be **detraining**. The removal or reduction of the training stimulus may lead to performance decrements. The extent of performance decrements relates to the length of the detraining or reduced training period and the type of activity. In addition, the pattern or time line of performance decrements appears to differ between endurance and strength–power activities.

Effect of Detraining on Endurance Performance

Performance decrements occur quite rapidly during periods of detraining in endurance athletes. Decreases in aerobic capacity (4-6% reduction in $\dot{V}O_2max$) have been noted after only 2 weeks of inactivity (Coyle, Hemmert, and Coggan 1986; Houston, Bentzen, and Larsen 1979), and longer periods of inactivity further increase aerobic capacity loss (Simoneau et al. 1987; Drinkwater and Horvath 1972). However, performance decreases resulting from inactivity or reduced training appear to affect highly trained endurance athletes to a greater degree than moderately trained athletes. Martin and colleagues (1986) reported a 20% reduction in maximal aerobic capacity following 3 to 8 weeks of deconditioning. The decrement in aerobic capacity following periods of training cessation appears to be related to the duration of the detraining period and the training level of the athlete. Figure 6.3 depicts the magnitude of performance decrements resulting from prolonged detraining. Despite a 16% loss in aerobic capacity following an 8-week detraining period, the aerobic capacity levels in these athletes were still 17% greater than those in sedentary controls (Coyle et al. 1984).

It appears that aerobic capacity declines between 4% and 20% during the initial 8 weeks of a detraining period. However, there does not appear to be

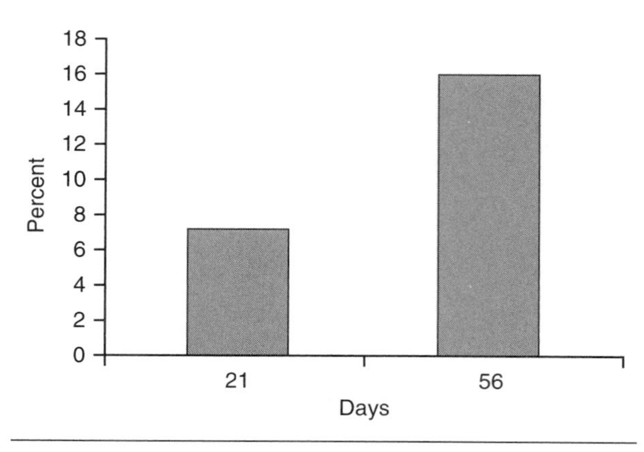

Figure 6.3 Loss of aerobic capacity dependent on duration of detraining period.

Adapted from Coyle et al.1984.

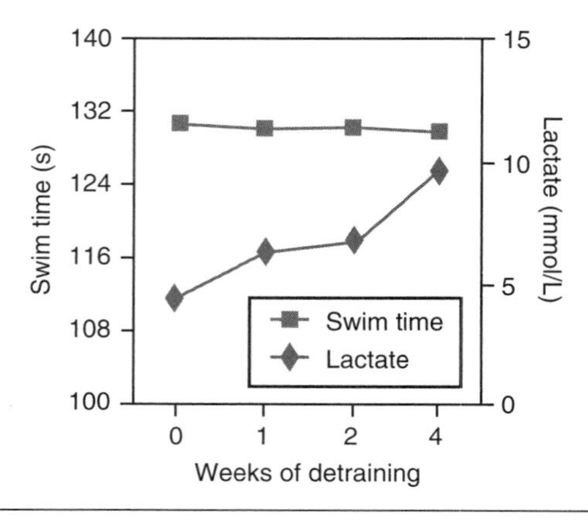

Figure 6.4 Changes in blood lactate and swim performance after 4 weeks of detraining.

Data from Neufer et al. 1987.

any further decrement after this time point (Mujika and Padilla 2001). Further, the aerobic capacity of detrained athletes still remains greater than that of their sedentary counterparts.

Decrements in aerobic capacity as the result of periods of detraining are thought to be partly related to changes in enzymatic activity and stroke volume (Coyle et al. 1984; Coyle, Hemmert, and Coggan 1986). A 12% decrease in stroke volume is evident after 2 to 4 weeks of detraining (Coyle, Hemmert, and Coggan 1986); and during a similar detraining period, decreases in oxidative enzymes (succinate dehydrogenase and cytochrome oxidase) are also seen. Comparing changes in glycolytic and oxidative enzymes after 3 months of detraining, Coyle and colleagues (1984) reported nearly a 60% decrease in the activities of various oxidative enzymes but no change in the activity of the glycolytic enzymes. Although anaerobic exercise performance may be maintained over longer periods of inactivity, metabolic efficiency appears to decrease.

Neufer and colleagues (1987), examining college swimmers, reported that a 4-week detraining period following the competitive season resulted in a 5.5 mmol/L increase in blood lactate during a standardized 183 m swim (see figure 6.4). Interesting, swim time performance was not significantly altered. Coyle and colleagues reported a similar response in endurance athletes following 12 weeks of training cessation: An increase in blood lactate from 1.9 mmol/L to 3.2 mmol/L was seen with the same workload. In these same athletes, lactate threshold was reported to decline from 79.3% of $\dot{V}O_2max$ during the competitive season to 74.7% after the 12-week detraining period. It is important to note that anaerobic threshold was still greater than that seen in sedentary controls (62.2% of $\dot{V}O_2max$). The magnitude of metabolic change appears to be dependent on the duration of training before the detraining period. No changes in blood lactate were seen following a 3-week period of inactivity in recreationally trained individuals who had recently completed a 6-week endurance training program (Wibom et al. 1992).

Periods of inactivity also appear to result in changes in the metabolic response to exercise. As discussed in chapter 3, prolonged exercise programs cause an increased reliance on fat oxidation and increased insulin sensitivity. During short durations of detraining (~5 weeks), a greater reliance on carbohydrate metabolism and less breakdown of stored fats are observed (Petibois and Deleris 2003). As the period of inactivity becomes elongated (5-12 months), the effects on both carbohydrate and fat metabolism are exacerbated. Petibois and Deleris (2003) demonstrated that long-term detraining impairs fatty acid delivery to muscle during subsequent endurance exercise, which forces an even greater reliance on glycolysis to fuel exercise. Based on these results, it appears that endurance-trained athletes should avoid prolonged periods of inactivity.

Effect of Detraining on Strength and Power Performance

Strength and power performances also decline during periods of detraining. The magnitude of the decline may depend on the training background, length of the training period before detraining, and the specific muscle group. Hortobagyi and colleagues (1993), examining power athletes (American football players or powerlifters), reported a nonsignificant decline in squat (0.9%) and bench press (1.7%) strength following a 2-week detraining period. Similar changes were noted in power performance. Hakkinen and coauthors (1989) showed that after a 2-week nontraining period, male strength athletes suffered a 3% decline in maximal isometric knee extension strength. Izquierdo and colleagues (2007) reported a 6% decrease in squat strength and a 9% decrease in bench press strength following a longer period of training cessation (4 weeks) in racket sport athletes, which was subsequent to a 16-week resistance training program. Upper and lower body power loss (17% and 14%, respectively) were also reported during the 4-week detraining period. Although the short-term detraining period caused significant declines in strength and power performance, it did not result in a return to baseline strength levels. This is consistent with other investigations as well. A 24-week strength program increased maximal isometric force strength by 27% in experienced strength-trained individuals (Hakkinen, Komi, and Alen 1985). After a 12-week detraining period, maximal isometric force was reduced, but it was still 12% higher than pretraining strength levels. Thus, even after 3 months of detraining, strength levels were higher than those seen before the 24-week training program. It appears that in short to moderate periods of training cessation, strength gains begin to decline, but not to the baseline levels. Interestingly, during short-duration detraining periods (~2 weeks), physically active men (not strength–power athletes) have been reported to increase their strength by 2% (Hakkinen et al. 1989). It is likely that in active individuals, a short term detraining or reduced training period may actually enhance recovery and result in greater performance. This is discussed in further detail in chapters 14 and 15.

Short-term periods of inactivity have been shown to affect type II fibers to a greater extent than type I fibers (Hortobagyi et al. 1993). A 6.4% decline in type II muscle fiber area suggests a change in activation patterns between the trained and detrained athlete. It has been suggested that the hypertrophied type II muscle fiber in the strength–power athlete is sensitive to the removal of the training stimulus and tends to atrophy within a relatively short period of time. Interestingly, a 2-week immobilization of the lower limb reportedly causes a similar change in muscle cross-sectional area (figure 6.5). Changes in strength and muscle activation have also been reported by Deschenes and colleagues (2002) in untrained individuals. Although these men and women were not previously trained, immobilization resulted in a 17.2% decrease in peak torque of the knee extensors and a 10% loss in the knee flexors. In addition, a 16% decline was observed in muscle electrical activity (as measured by integrated electromyography, iEMG). The investigators concluded that the loss of strength and atrophy of the muscle are the result of a decrease in neural activation, which likely plays a critical role in maintaining both muscle mass and strength output.

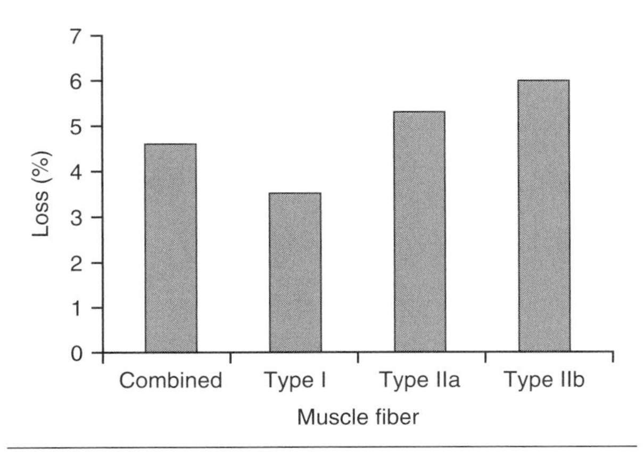

Figure 6.5 Muscle cross-sectional area changes following a 2-week immobilization period in untrained individuals.

Data from Deschenes et al. 2002.

Effect of Detraining on Bone Loss

An area of concern for athletes who retire or who experience a prolonged detraining period is the potential effect on bone loss. Within 4 years of competing at the collegiate level, female gymnasts have been reported to experience no loss in total bone mineral density, but significant declines in bone mineral density (ranging between 0.72% and 1.9% per year) were seen at the lumbar spine, femoral neck, Ward's triangle, and greater trochanter (Kudlac et al. 2004). A longitudinal study on adolescent male hockey players showed significant increases in bone density (in dominant and nondominant limbs) after 2.5 years of playing that were greater than those in age-matched controls (Nordstrom, Olsson, and Nordstrom 2005). A second follow-up nearly 6 years later compared those hockey players who had continued to play to those who had not. Players who maintained their hockey careers continued to increase bone mineral density, while those who retired or quit playing began to lose bone density. However, former hockey

In Practice

Do Long Athletic Careers Preserve Physiological Adaptations Into Retirement?

The athlete's competitive career is typically quite short. For American athletes, the majority of competitive athletics ends at the end of their college career. Few have the opportunity to extend their careers as professional athletes, and of those only a handful can make their careers last for a prolonged period of time. Similarly, athletes competing around the world generally compete for either professional or club teams; but like their American counterparts, few have the luxury of playing for an extended period of time. For athletes who have made playing sport their primary lifetime vocation, does the high level of training and competition have any effect on preserving the physiological changes that occur during their athletic careers? Very few studies have attempted to examine this question. A study published in Italy looked at whether cardiac adaptations to exercise are reversible upon retirement and whether a previous competitive athletic career influences the cardiac aging process or modifies ventricular function (Macchi 1987). Twenty-three former professional athletes (soccer players and boxers between the ages of 40 and 60) who had been active in their sport for at least 16 years, and who had not participated in any training for at least 10 years, were examined and compared to age- and weight-matched controls. Cardiac mass was significantly higher in the former athletes than the controls, with specific reference to concentric heart hypertrophy (wall thickening without any change in the diameter of the ventricle).

From this study it was apparent that cardiac adaptations are not completely reversible upon retirement. In a subsequent investigation in American football players who had remained physically active in retirement, the former football players did not differ from age-matched controls in aerobic capacity, but they had significantly greater lean body mass (13%) and bone mineral content (20%) and significantly lower fat mass (26%) (Lynch et al. 2007). In addition, the former football players had more favorable risk factors for heart disease and osteoporosis than the control group. It appears that prolonged experience in professional sport does preserve some of the adaptations seen during one's playing career. In addition, athletes who continue to be physically active during retirement appear to have more favorable body composition and health measures compared to age- and body mass index (BMI)-matched controls.

players' bone mineral density was still greater than that of the age-matched controls. Athletic participation results in significant increases in bone mineral density. When the training stimulus is removed, bone density decreases but still remains greater than in age-matched controls, suggesting that the benefits of exercise on bone density are maintained for several years following an athletic career.

Effect of a Taper on Athletic Performance

A taper is a planned reduction in the training load toward the end of a competitive season, with the aim of improving performance (Bosquet et al. 2007; Pyne, Mujika, and Reilly 2009). The taper strategy can be used in all sports. It is common in endurance sports such as swimming, running, and cycling but is often used to some extent in all sports. For instance, many endurance sports reduce the volume of training up to 85% (Houmard and Jones 1994). This is often accomplished through reducing the duration of each training session or reducing the frequency of training. The duration of the taper varies from 1 to 4 weeks, although the vast majority of tapers appear to be 2 weeks in length (Bosquet et al. 2007).

A meta-analysis examining the effect of a taper in competitive athletes showed that maximal gains are achieved from a taper of 2 weeks when the training volume is reduced by 41% to 60%, without changes in the frequency or intensity (Bosquet et al. 2007). Figure 6.6 depicts the effect sizes of different tapering strategies with regard to training volume (figure 6.6*a*) and duration of taper (figure 6.6*b*). Importantly,

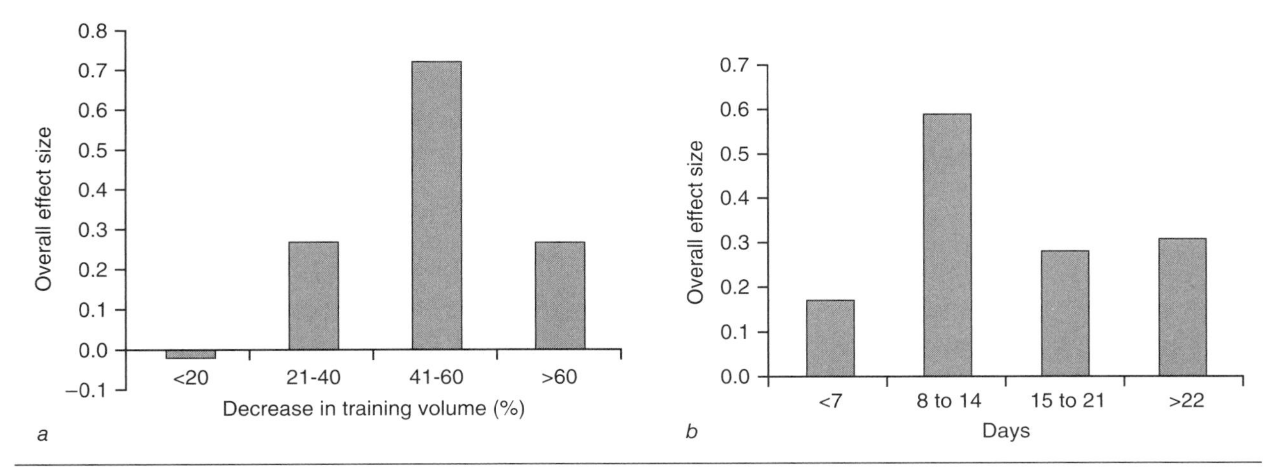

Figure 6.6 *(a)* Effects of decreases in training volume and *(b)* duration of taper on overall effect size for taper-induced changes in performance.

Adapted from Bosquet et al. 2007.

the results of the meta-analysis also suggested that manipulating training intensity does not have any effect on performance changes. Thus, the taper is very similar to, if not the same as, a peaking phase in the yearly training program for an athlete (see chapter 14). The taper, though, often occurs during the competitive season, whereas the peaking phase is the final mesocycle of the off-season conditioning program. Nevertheless, the concepts are very similar.

It has been suggested that the taper improves performance by approximately 2% to 3% (ranging from no improvement up to a 9% improvement) (Bosquet et al. 2007; Neary, McKenzie, and Bhambhani 2005; Pyne, Mujika, and Reilly 2009). Although the performance improvements may be the driving factor behind the taper, several other benefits are realized that may have both direct and indirect effects on performance. These benefits may be psychological or psychophysiological, in that the reduction in training volume gives the athlete an opportunity to recover without having to cease practicing (Pyne, Mujika, and Reilly 2009). This may provide the psychological recovery that may be necessary before the most important competitions of the year.

A taper or reduced training period also has important implications for strength and power performance. Izquierdo and colleagues (2007) demonstrated that short-term (4 weeks) tapering could increase strength in the upper and lower body by 2% but did not have any effect on power output. This has implications regarding situations following a season of competition. Many athletes take some time off for recovery from the end of a competitive season to the beginning of the off-season conditioning program for the next season. A reduced training schedule (similar to an active recovery) would provide an ability to minimize strength loss due to a detraining period between competitive seasons. A study on elite kayakers examined the difference between a 4-week taper (one resistance training session and two endurance training sessions per week) and total training cessation for 5 weeks (Garcia-Pallares et al. 2010). During the taper no changes were noted in strength, while significant declines were noted in strength (~9%) following the no-training period. In addition, a reduced training period also preserved aerobic capacity to a greater extent than training cessation. The benefits of tapering are apparent; and although performance improvements are small, in elite athletic performance they may mean the difference between winning a medal or not.

Summary

When designing training programs, it is important to have a clear understanding of the principles of training. This chapter discusses these principles (specificity, overload, progression, individuality, diminishing returns, and reversibility) to help the athlete and coach set realistic goals and develop training programs that will provide the greatest opportunity to achieve performance gains. Detraining is also covered as it relates to the principle of reversibility. Differences in how detraining affects endurance and strength performance are discussed, as are the potential training benefits of the taper (reduced training).

REVIEW QUESTIONS

1. Describe and provide an example of a progressive overload for a weightlifter.
2. Explain the theoretical training curve.
3. Would a taper programmed into the yearly training program for a collegiate swimmer competing in the 400 m medley be a benefit or detriment to maximizing performance? Explain your reasoning.
4. Removing the training stimulus from a 60-year-old male who began lifting weights a year ago would be expected to have what type of effect on muscle cross-sectional area?
5. Describe specificity of training.

chapter 7

Warm-Up, Flexibility, and Balance Training

CHAPTER OBJECTIVES

After reading this chapter you should be able to do the following:

- Understand the importance of using a warm-up before exercise.
- Explain how muscles stretch.
- Define the different flexibility techniques and how they should be incorporated into prepractice and postpractice or competition routines.
- Discuss the importance of balance and stability training and how to incorporate these training methods into a training protocol.
- Explain the potential consequences that flexibility exercises may have on power performance.

Before exercise or competition, the athlete typically performs a **warm-up** that may include stretching exercises. However, it is important to understand that the warm-up and flexibility training are separate. The warm-up prepares the athlete for more powerful and dynamic movements that will occur during exercise or competition and possibly minimizes the risk of musculoskeletal injuries. This chapter discusses the benefits of the warm-up and how it should be incorporated into the preexercise routine. Attention is also directed at **flexibility** training, including factors that affect flexibility, types of flexibility exercises, and the effect that flexibility training has on performance. Additionally, the chapter discusses the benefits and limitations associated with stability or balance training for competitive athletes and how such training can be incorporated into the athlete's warm-up routine.

Warm-Up

The goal of the warm-up period is to prepare the athlete both mentally and physically for exercise or competition. It generally consists of 5 to 10 min of low- to moderate-intensity jogging or cycling. The warm-up may be general and may consist of exercises that are not related to the specific activity that the athlete is preparing for. Besides jogging or cycling, jumping rope, brisk walks, or light calisthenics may also be used. The warm-up can also be more specific to the exercise or sport, and it may include movements that closely simulate actions used during the activity but are performed at a reduced intensity.

The benefits of performing a warm-up include increased blood flow to the exercising muscles, which increases muscle temperature and core body temperature. The increase in muscle and core body temperature has a significant positive effect on muscle **strength** and **power** (Bergh and Ekblom 1979) and also improves **reaction time** and the rate of **force** development (Asmussen, Bonde-Peterson, and Jorgensen 1976; Sargeant 1987). An increase in muscle temperature during the warm-up session may also increase muscle flexibility by 20% (Wright and Johns 1960). In addition, the warm-up increases heart rate, primes the nervous system by increasing the rate and effectiveness of **contraction** and **relaxation** of both **agonist** and **antagonist** muscle groups (**reciprocal inhibition**), and increases the **elasticity** and mobility of **connective tissue** and joints.

The warm-up is an integral part of the prepractice or pregame routine and does not require a large allocation of time. However, the warm-up should be of sufficient duration to allow the athlete to begin to perspire (an indication of an increase in body temperature). It is important that the warm-up not fatigue the athlete for the approaching performance. For years it was common practice for athletes to perform stretching exercises before the training or practice session. However, many coaches are now focusing more on a dynamic preparatory routine than on a static flexibility routine. This has been referred to as a **dynamic warm-up** and appears to allow more sport-specific preparation.

Dynamic warm-ups focus on movement patterns specific to the sport. In general, a continuous movement is required that activates the muscle through its full range of motion. Dynamic warm-up protocols typically use an exercise progression of low to moderate to high intensity, in which the high-intensity exercises provide a progression leading to the actual practice or competition. For example, volleyball athletes can begin their dynamic warm-up with a lunge walk and then progress to power skips and jump squats. Proper technique must be emphasized throughout the warm-up. It has been suggested that if a dynamic warm-up is performed correctly and is specific to the sport, performance improves between 2% and 10% (Tillin and Bishop 2009). Note that for the warm-up to be effective, it should occur within 18 min before exercise or competition (Faigenbaum 2012; Faigenbaum et al. 2010). Another benefit associated with a dynamic warm-up appears to be psychological preparation for the subsequent practice session or competition. A warm-up that is dynamic, engaging, and specific to performance may not only provide a physiological benefit but also enhance psychological readiness (i.e., focus, attention, and improved performance on task-relevant cues) (Faigenbaum 2012).

Coaches or conditioning professionals should use exercises that activate large muscle groups and use movement patterns similar to the ones that will be used by the athlete during practice or competition. In general, 10 to 15 exercises that progress from a simple, low-intensity movement (e.g., jog) to a more complex, higher-intensity movement (e.g., power skip) should be performed over a distance of 15 to 30 m. The number of repetitions per exercise can vary depending on the drill's complexity, but typically varies from two to four repetitions. The dynamic warm-up should not be more than 10 to 15 min in duration and should not be fatiguing. If the warm-up becomes too intense, it may negatively affect subsequent practice or performance. In certain sports, such as basketball or hockey, the warm-up may progress in intensity and include the use of implements of the sport (e.g., basketball, hockey stick and puck). Coaches have been successful in progressing from basic low-intensity warm-up drills to more sport-specific drills that complete the warm-up at an intensity that readies players for the remainder of practice. Table 7.1 provides an example of a dynamic warm-up for basketball players that progresses from low- to higher-intensity basketball-specific drills.

Table 7.1 Example of Basketball-Specific Dynamic Warm-Up

Full descriptions of how to perform select drills are available in the web resource at www.HumanKinetics.com/PhysiologicalAspectsofSportTrainingAndPerformance. Video of a dynamic warm-up routine featuring many of these drills also is available in the web resource.

Drill	Distance	Repetitions
Jog	Length of court (30 m)	4
High-knee walk	Half court (15 m)	2
Stepping trunk twist	Performed in place	2 with each side
Trail leg walking	Half court (15 m)	2
Lunge walk with or without reach	Half court (15 m)	2
Leg cradle	Half court (15 m)	2
Side lunge	Half court (15 m)	2
Side shuffle and run	Side shuffle 15 m, run 15 m	2
Carioca	Far foul line (~25 m)	2
Backward run	Backward run 15 m, run 15 m	2
Glute kick	Half court (15 m)	2
Speed skip	Far foul line (~25 m)	2
Power skip	Far foul line (~25 m)	2
Three-man weave	Full court twice (60 m)	4
3 on 2 to 2 on 1	Full court twice (60 m)	4

In Practice

Effect of Warm-Up on the Prevention of Lower Limb Injuries

A group of sports medicine professionals at the London School of Medicine examined 766 articles related to warm-up, stretching, and injury (Herman et al. 2012). After application of specific inclusion and exclusion criteria, which in part involved ensuring that the study addressed the effect of various warm-up strategies on injury rates in participants who had no previous history of injury, a total of nine studies qualified for inclusion. The number of participants per study averaged 1500 (range 1020-2020), and participant age ranged from 13 to 26 years. Study durations ranged from 12 weeks to 2 years. Subjects were either amateur athletes or army recruits. The injuries examined were those of the lower extremity, including foot, ankle, leg, knee, thigh, groin, and hip. The results indicated that various warm-up strategies, using or not using equipment (balance boards, sticks, medicine balls), are very effective in reducing injury risk. Warm-up strategies reduced overall lower limb injury risk (risk ratio of 0.67, with 95% confidence intervals ranging from 0.54 to 0.84) and knee injuries (risk ratio of 0.48, with 95% confidence intervals between 0.32 and 0.72) in female athletes. Similar results were seen for military recruits. This meta-analysis clearly demonstrated the benefits of warm-up strategies that use stretching, balance exercises, and sport-specific agility drills and landing techniques. The use of these drills for periods of at least 3 months appears to optimize the effects.

Web Resource Preview

Here is a preview of one of the web resource drills. Full drill content can be found at www.HumanKinetics.com/PhysiologicalAspectsofSportTrainingAndPerformance.

Trail Leg Walking

Purpose: Use for preexercise preparation

Type: Dynamic warm-up

Equipment: None

STEPS

1. Athlete stands tall with hands at the sides.
2. Athlete then walks forward, abducting one leg (rotation to the side) and up to waist level *(a)*.
3. The athlete then adducts the leg back to the midline of the body and lowers it to the ground *(b)*.
4. The athlete repeats with the opposite leg while moving forward with each step.
5. Athlete continues for 15 meters and repeats.

TIPS

- Focus on technique.
- Do not be concerned with speed of movement.
- Should be performed at a steady pace.

a

b

Flexibility

Flexibility is the ability to move a muscle or a group of muscles through the complete **range of motion**. All stretching exercises should be preceded by a warm-up. The elevated muscle temperature and increased mobility of connective tissue and joints generated by the warm-up allow for a greater range of motion during each stretching exercise. Flexibility exercises are generally performed before exercise or competition but may also be performed afterward during the cool-down period. Stretching during the postexercise cool-down period should be performed shortly after the conclusion of practice or competition (5-10 min) to take advantage of the elevated muscle temperatures. It is possible that postpractice stretching also decreases muscle soreness (Prentice 1983); however, there is little experimental evidence to support this contention.

How Do Muscles Stretch?

The muscle **proprioceptors**, **Golgi tendon organs** and **muscle spindles** (figure 7.1), are sensory neural fibers that relay information about the muscle stretch to the upper neural pathways. Their primary responsibility is to protect the muscle from injury. The Golgi tendon organs are located in the **tendons** of the muscle fiber; the muscle spindles are considered **intrafusal fibers** and are situated parallel to the muscle fiber. Golgi tendon organs are sensitive to **tension** development in the **muscle–tendon complex**. They inhibit contraction of agonist muscles and activate antagonist muscles when tension within the muscle–tendon complex is increased to a level that poses a risk of injury to the muscle. The muscle spindles are responsible for monitoring the stretch and length of the muscle and initiating contraction within the muscle to reduce the stretch if needed.

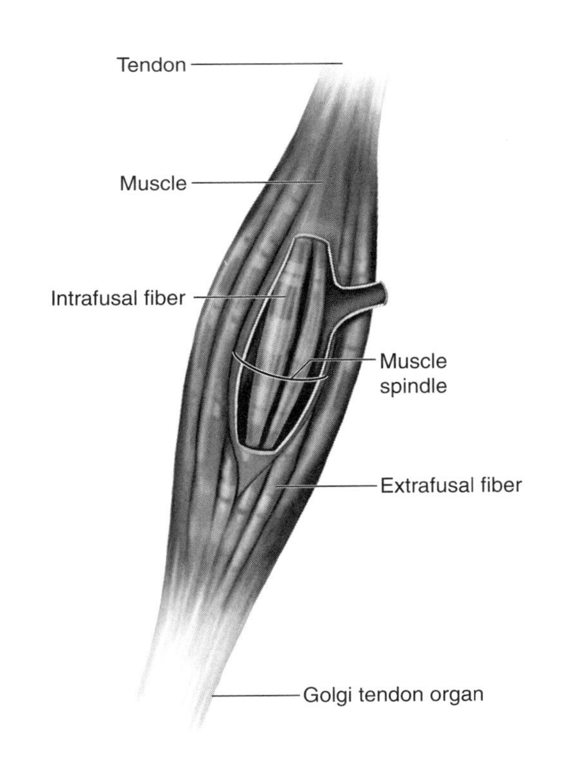

Figure 7.1 Muscle proprioceptors.

As the muscle lengthens during a stretching exercise, the muscle spindles become activated, causing a contraction of the muscle being stretched. During a rapid stretch that might be seen in a **ballistic** or bouncing type of movement, both the tension in the tendon and the stretch of the muscle are rapidly increased. In response, both the Golgi tendon organ and muscle spindles are activated, causing a rapid contraction of the muscle. (This **stretch reflex** is easily demonstrated by a light tap to the patellar tendon and the consequent contraction of the quadriceps muscles to ease the tension on the muscle spindles and Golgi tendon organs.) It is for this reason that **slow static** stretching, which results in a more relaxed and effective stretch, is recommended (Alter 1996).

Types of Stretching Techniques

There are three types of stretching exercises: slow static, ballistic, and **proprioceptive neuromuscular facilitation (PNF)**. All three types of stretching techniques improve range of motion within a muscle (Holt, Travis, and Okita 1970; Worrell, Smith, and Winegardner 1994). However, PNF stretches may result in the greatest improvements (Wallin et al. 1985). The benefits and risks of each of these stretching techniques are outlined in table 7.2.

Slow Static Stretching

Slow static stretching requires the individual to slowly stretch a muscle to its furthest point and hold that position for 10 to 30 s. Static stretching is the safest of all the stretching techniques and requires very little energy expenditure. In addition, because of its slow movement and long duration, the likelihood of pain and injury is minimized. The static stretch is common to most flexibility programs and is easy to learn.

Table 7.2 Benefits and Risks of Different Stretching Techniques

Factor	Slow static	Ballistic	PNF
Risk of injury	Low	High	Medium
Degree of pain	Low	Medium	High
Resistance to stretch	Low	High	Medium
Ease of performance	Excellent	Good	Poor
Effectiveness	Good	Good	Good

PNF, proprioceptive neuromuscular facilitation.

Adapted from Heyward 1997.

Ballistic Stretching

Ballistic stretching involves a bouncing, bobbing, or rhythmic movement that does not have an end point. This form of stretching produces the greatest amount of pain and risk of injury. Its dynamic movement initiates the stretch reflex, thereby activating muscle proprioceptors and increasing muscle contractility. Because of the short duration, there is insufficient time to cause either tissue or neural adaptation during the stretch (Alter 1996). Ballistic stretching can be performed using the same exercises as static stretching, the difference being the rate of the stretch. Although ballistic stretching may improve range of motion as effectively as static stretching (Holt, Travis, and Okita 1970; Worrell, Smith, and Winegardner 1994), the ballistic nature of the movement forces the muscle to exceed its range of motion, which increases the risk of injury. Nevertheless, ballistic stretching may have its place in the flexibility routines of athletes engaged in activities requiring dynamic movements, such as the martial arts (Alter 1996).

Proprioceptive Neuromuscular Facilitation

Proprioceptive neuromuscular facilitation is a mode of stretching that incorporates two separate techniques and is generally performed with the assistance of a partner. The first technique requires the individual to contract and then relax the stretched muscle, and the second technique requires the muscle to contract, relax, and contract again. Both techniques are effective in enhancing the range of motion in a muscle and are based on the concept of reciprocal inhibition. The contract–relax technique requires the individual to stretch the desired muscle gently. When the muscle is stretched to the point of slight discomfort, the individual isometrically contracts the muscle for 5 to 15 s against the partner's resistance. This is followed by a brief period of relaxation before the partner slowly moves the muscle through an extended range of motion. This enhanced range of motion is thought to occur because the isometric contraction causes a reflex facilitation and contraction of the agonist muscles (muscles that are not being stretched) (Heyward 1997). This action suppresses the contraction of the antagonist muscles (muscles being stretched) during the final phase of the stretch, allowing for a greater range of motion. Figure 7.2 provides an example of PNF contract–relax technique using the partner-assisted hamstrings muscle stretch.

The contract–relax–contract PNF technique begins similarly to the contract–relax technique. However, after the relaxation phase, the individual contracts the agonist muscle group (in the case of the hamstrings stretch, the quadriceps would be contracted). The partner can also assist with this movement. The theory behind the agonist contraction is that performing a submaximal contraction of the opposite (agonist) muscle group induces additional inhibitory input to the hamstrings through reciprocal inhibition and results in a greater range of motion (Moore and Hutton 1980). However, the exact mechanisms that underlie muscle stretch are still not completely clear.

Factors Affecting Flexibility

Flexibility is very individualized, and large variances may be seen within a relatively homogenous population such as an American football or a basketball team. Since most athletes on the team are typically in similar physical condition and of similar age, the wide range of flexibility among team members suggests that other factors cause differences in muscle flexibility. A host of factors may interact to determine the potential range of motion of a muscle. These factors can be classified as either kinesiologic or physiological. Kinesiologic factors are primarily associated with **joint structure**, muscle **origin** and **insertion**, muscle **cross-sectional area**, and connective tissue elasticity. Physiological factors may include age, sex, and physical activity level.

Kinesiologic Factors

Kinesiology is the study of the mechanics of human motion. It is generally understood that the structure of a joint and the origin and insertion of the muscle have a tremendous influence on range of motion. The range of motion about a joint is highly specific to the type of joint, as well as the muscles, tendons, and **ligaments** that cross that particular joint. Joints that have movement in all three **planes** (frontal, sagittal, and transverse), such as a ball and socket joint, generally have the greatest range of motion. The

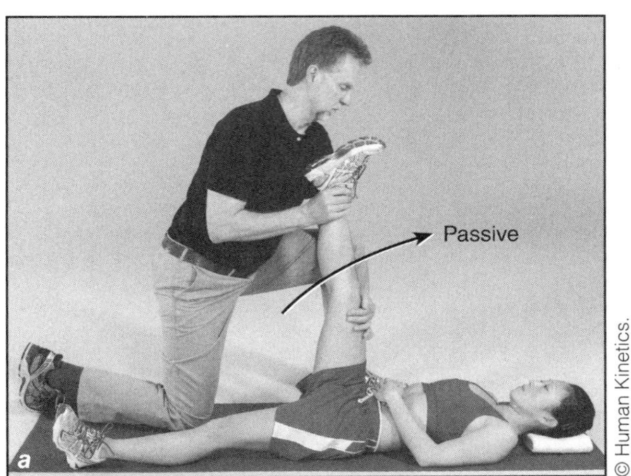

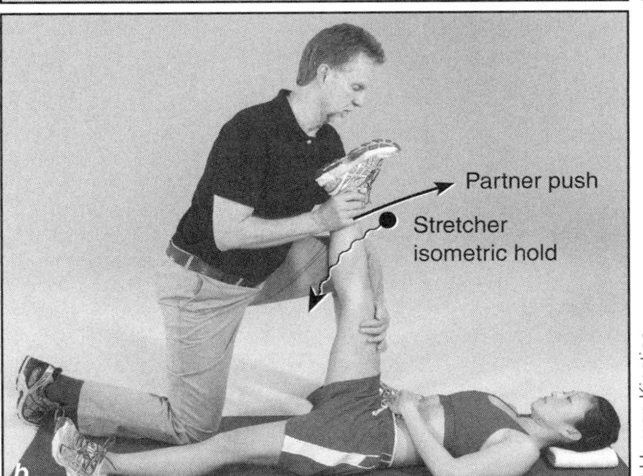

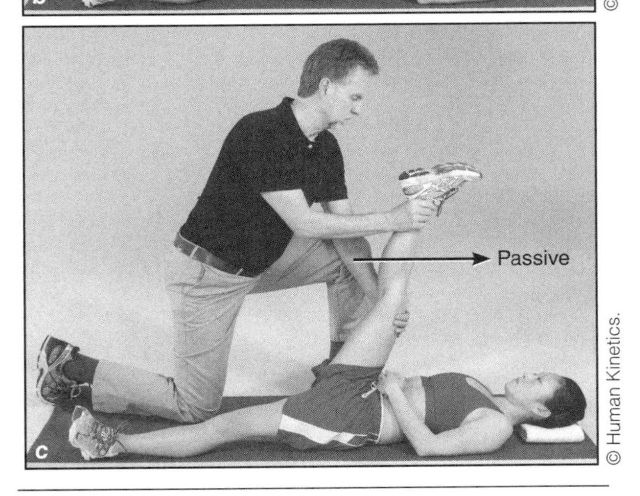

Figure 7.2 Proprioceptive neuromuscular facilitation: hamstrings stretch. *(a)* First phase: Subject is lying on the back with one leg raised 50° to 60° with knee locked and ankle at 90°. The partner then straddles the subject and, with the partner assisting the subject, the hamstrings approach the end of their range of motion. *(b)* Second phase: Subject isometrically contracts the muscle for 5 to 15 s against the manual resistance of the partner. *(c)* Third phase: Subject briefly relaxes muscle, and the partner slowly moves the muscle through an extended range of motion.

shoulder and hip are ball and socket joints and have a relatively large range of motion in comparison with hinge joints such as the elbow or knee (table 7.3).

The origin and insertion of the muscle may also affect range of motion. Although anatomical textbooks are precise in describing origin and insertion, the actual location for each individual may be very different. Depending on the extent of the difference and the relationship of the muscle to both bone and joint structure, this may have a large positive or negative effect on the range of motion about the joint.

Connective tissue also plays a large role in determining the range of motion about a joint. Connective tissue may be made up of either **collagen** or **elastic tissue**. When connective tissue is composed primarily of collagen, it is limited in its ability to stretch. In contrast, when it is composed primarily of elastic tissue, the range of motion has the highest potential. Ligaments account for 47% of the stiffness about a joint, followed by the **fascia** (41%), tendons (10%), and skin (2%) (Johns and Wright 1962). The fascia, which includes the membranes enveloping the individual muscle, muscle fibers, and fasciculi, has the greatest potential for stretch as it is composed primarily of elastic tissue. Tendons and ligaments are less elastic and have a greater amount of collagen. Thus, their ability to adapt is reduced. However, increasing the range of motion of both tendons and ligaments may not be highly desirable. Overstretching these connective tissues may result in joint laxity and increase the risk of musculoskeletal injuries.

A muscle with a large cross-sectional area may not be as flexible as one with a smaller cross-sectional area. Although resistance training is not detrimental to muscle flexibility (see chapter 8), individuals with significant bulk (whether through resistance training or obesity) may have a reduced range of motion about a joint. This may affect the ability of the individual to perform certain exercises or movements. For example, someone with large biceps and deltoids may experience difficulty racking a power clean or performing a front squat exercise.

Considering some of the structural anomalies that naturally occur, a large genetic component may limit the degree of flexibility that a person can achieve. However, even an individual who has poor flexibility that may or may not be attributable to any structural anomaly can improve flexibility through flexibility training.

Physiological Factors

As we age, the elasticity of the muscle is reduced, resulting in a decrease in range of motion. Reduced elasticity is caused by increased **fibrous cartilage** that replaces degenerated muscle fibers, increased

Table 7.3 Average Range of Motion (ROM) Values for Healthy Adults

Joint	ROM (degrees)	Joint	ROM (degrees)
Shoulder		**Thoracic-lumbar spine**	
Flexion	150-180	Flexion	60-80
Extension	50-60	Extension	20-30
Abduction	180	Lateral flexion	25-35
Medial rotation	70-90	Rotation	30-45
Lateral rotation	90		
Elbow		**Hip**	
Flexion	140-150	Flexion	100-120
Extension	0	Extension	30
		Abduction	40-45
		Adduction	20-30
		Medial rotation	40-45
		Lateral rotation	45-50
Radioulnar		**Knee**	
Pronation	80	Flexion	135-150
Supination	80	Extension	0-10
Wrist		**Ankle**	
Flexion	60-80	Dorsiflexion	20
Extension	60-70	Plantar flexion	40-50
Radial deviation	20		
Ulnar deviation	30		
Cervical spine		**Subtalar**	
Flexion	45-60	Inversion	30-35
Extension	45-75	Eversion	15-20
Lateral flexion	45		
Rotation	60-80		

Data from Heyward 1997.

adhesions and cross-links within the muscle, and increased calcium deposits (Alter 1996). However, flexibility training can still be beneficial in an older population, as demonstrated by the improved range of motion (ROM) in elderly subjects after 10 weeks of stretching exercises performed 3 days per week (Girouard and Hurley 1995).

Sex also appears to affect muscle and joint flexibility. Females tend to be more flexible than males at all ages. This is primarily attributable to sex differences in pelvic structure and hormonal concentrations that may affect the laxity of connective tissue (Alter 1996). However, the advantage that females have over males in flexibility may be joint and motion specific, as males have been reported to have a greater ROM than females in hip extension and spinal flexion-extension in the thoracolumbar region (Norkin and White 1995).

Physical activity is an important determinant of flexibility because active people tend to be more flexible than sedentary individuals (Kirby et al. 1981; McCue 1953). Inactivity causes a tightening or contraction of the inactive muscles. This is easy to understand considering the stiffness one feels after sitting for a prolonged period. During long durations of inactivity as a result of deconditioning or immobilization, the connective tissue of the muscles becomes shortened, reducing its ROM about a joint.

Flexibility Assessment

A number of tests can be used to measure the flexibility of a joint or group of muscles. Flexibility can be assessed either directly through measurement of the range of joint rotation in degrees or indirectly through measurement of static flexibility in linear units (Heyward 1997). The purpose of such testing is to identify joints or muscles that have poor ROM and are at a greater risk of injury.

Direct measurements of flexibility include the use of a **goniometer**, **flexometer**, or **inclinometer**, all of which measure the ROM about a joint in degrees. The goniometer is a protractor-like device that measures the angle of the joint at both extremes of the ROM. The center of the goniometer is placed at the axis of rotation (joint), and the arms of the goniometer are aligned with the longitudinal axis of each moving segment. First, the initial joint angle is recorded. The proximal limb is then moved through its complete ROM to its other extreme, and the angle is again recorded. The difference between the angles indicates the ROM of that joint. Table 7.3 presents the average ROM values for healthy adults.

The flexometer and inclinometer also provide direct ROM measurements of a joint or body segment. Both of these devices can be placed on the subject (either strapped or handheld), and the ROM for a particular joint or body segment is easily recorded. Although the validity and reliability of these tests have been established, it appears that reliability may be dependent on the skill of the technician and the joint being measured. Measurements of the upper extremities appear to have a greater reliability than measurements of the lower extremities (Norkin and White 1995).

Indirect measurements of a joint's ROM are easy to perform. The primary differences, reported as inches or centimeters rather than degrees of ROM, are also easily assessed. A commonly used indirect measurement of flexibility is the sit-and-reach test (figure 7.3). Although this test is used to evaluate lower back and hip flexibility, it appears to have greater validity for assessing hamstring flexibility than lower back flexibility (Minkler and Patterson 1994). This test can be performed with or without a sit-and-reach box. When a sit-and-reach box is used, the subject sits on the floor with the soles of the feet placed against the edge of the box.

With knees extended but not locked, the subject reaches as far forward as possible, with one hand positioned on top of the other, while keeping the back straight. A yardstick or other measuring device is used to mark the zero point. This procedure neutralizes the effect of differences in leg-to-trunk ratio (Hoeger, Hopkins, Button, et al. 1990). Finally, the subject reaches forward as far as possible, sliding the fingers along the measuring device. The most distant point reached is recorded (in either inches or centimeters). The percentile ranks for the sit-and-reach test for both men and women can be seen in table 7.4. This table does not indicate the flexibility requirements for athletes or predict performance capabilities. It is primarily used to evaluate normal adults. Scores for specific athletic populations are not presented.

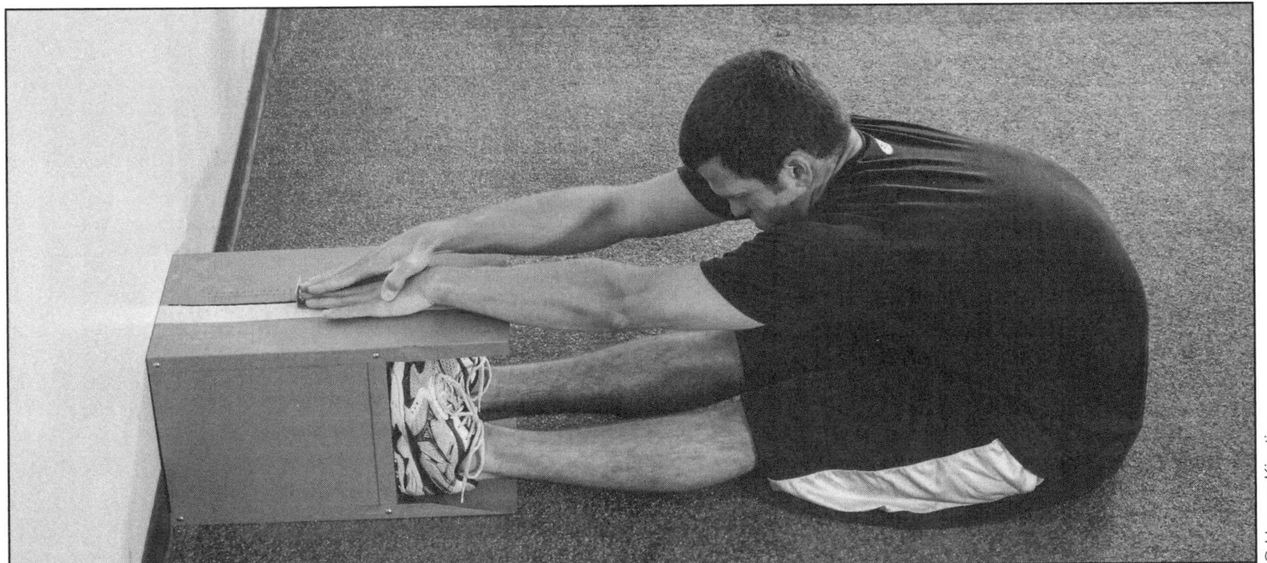

Figure 7.3 Sit-and-reach test.

Table 7.4 Percentile Ranks for the Sit-and-Reach Test in Inches

	Females				Males			
Percentile rank	18	19-35 years	36-49 years	50 years	18 years	19-35 years	36-49 years	50 years
99	22.6	21.0	19.8	17.2	20.1	24.7	18.9	16.2
95	19.5	19.3	19.2	15.7	19.6	18.9	18.2	15.8
90	18.7	17.9	17.4	15.0	18.2	17.2	16.1	15.0
80	17.8	16.7	16.2	14.2	17.8	17.0	14.6	13.3
70	16.5	16.2	15.2	13.6	16.0	15.8	13.9	12.3
60	16.0	15.8	14.5	12.3	15.2	15.0	13.4	11.5
50	15.2	14.8	13.5	11.1	14.5	14.4	12.6	10.2
40	14.5	14.5	12.8	10.1	14.0	13.5	11.6	9.7
30	13.7	13.7	12.2	9.2	13.4	13.0	10.8	9.3
20	12.6	12.6	11.0	8.3	11.8	11.6	9.9	8.8
10	11.4	10.1	9.7	7.5	9.5	9.2	8.3	7.8

Adapted from Hoeger and Hoeger 2000.

In Practice

Comparison Between Static and Dynamic Stretching on Jump Height, Reaction Time, and Flexibility

Sport scientists from Oregon State University examined the effect of static and dynamic stretching on jump power, reaction time, and flexibility (Perrier, Pavol, and Hoffman 2011). Twenty-one recreationally trained male students (24.4 ± 4.5 years) were recruited for this study. Participants reported to the lab on three different occasions and were randomly assigned to perform a no stretch, static stretch, or a dynamic stretch session. Participants performed the same general warm-up before each testing session (5 min treadmill jog at a self-selected pace). During the no-stretch session, participants sat quietly for 15 min. During the static stretching session, participants performed seven lower extremity stretching exercises for two repetitions of 30 s. The total time of the static stretching session was 14.8 ± 0.4 min. During the dynamic stretch session, participants performed 11 exercises of increasing intensities. The total time of the dynamic stretching was 13.8 ± 1.7 min. Following each session, participants were tested on low back–hamstring flexibility via a sit-and-reach box. Jump power was assessed by a maximal height countermovement vertical jump on a force plate, and reaction time was assessed from time to onset of motion following a visual stimulus. Results revealed that countermovement jump height was significantly highest following the dynamic stretching, with no differences between no stretching and static stretching. Both static and dynamic stretching resulted in greater flexibility than no stretching, but no difference was noted between the two stretching techniques. No differences were noted between any treatment effects with regard to reaction time. These results provide further evidence of the benefits of dynamic movement preceding exercise or competition.

Flexibility and Strength–Power Production

A number of investigations have indicated that stretching before activity may decrease both strength and power performance (Behm et al. 2004; Cornwell et al. 2001; Fletcher and Jones 2004; Kokkonen and Nelson 1996; Nelson and Heise 1996; Nelson et al. 1998; Winchester, Nelson, et al. 2008). Winchester, Nelson and colleagues (2008) reported that static stretching before sprint performance in college track and field athletes resulted in a 3% decrement in 40 m sprint performance. It should be noted, though, that the effect of static stretching on performance may be individual. Evidence has been reported of large individual variability in jump performance following various static, dynamic, and no-stretching routines in collegiate female volleyball athletes (Dalrymple et al. 2010). Although stretching makes the muscle more compliant and potentially reduces the risk of injury, it causes an increase in the muscle–tendon length (Magnusson et al. 1996). These changes in muscle–tendon length are quickly reversed, but they have been shown to decrease force output and rate of force development (Rosenbaum and Henning 1997). Athletes preparing for activities involving maximal strength and power performance should consider refraining from static stretching exercises until after the performance or should complete flexibility training at least 20 min before exercise or performance. In any situation, it is always vital to maintain the dynamic warm-up as a significant part of the prepractice or precompetition routine. The importance of the dynamic warm-up, as described at the beginning of this chapter, is not diminished by the results of these investigations. In addition, one should not conclude that flexibility training does not have a role for the strength–power athlete; only the timing or placement of flexibility training has been questioned. Given these studies, it may be that performing flexibility training during a postexercise cool-down period has greater merit.

Flexibility Exercises

The sidebar that follows lists examples of stretching exercises for specific muscle groups that can be used as part of a flexibility routine. This is a relatively small sample of the possible exercises available to the athlete and coach. The list does not represent any particular order of importance or effectiveness.

Selected Flexibility Exercises by Muscle Group

Full descriptions of how to perform all of these stretches are available in the web resource at www.HumanKinetics.com/PhysiologicalAspectsofSportTrainingAndPerformance.

Achilles Tendon and Posterior Lower Leg

Lying Achilles stretch

Partner-assisted calf stretch

Hamstrings

Semistraddle (hamstrings and gastrocnemius)

Sitting toe touch

Straddle (spread eagle) (hamstrings and adductors)

Adductors

Butterfly stretch

Quadriceps

Side quadriceps stretch (quadriceps and iliopsoas)

Hurdler's stretch

Hip Flexors

Forward lunge

Buttocks, Lower Back, and Hips

Single-leg lower back stretch

Double-leg lower back stretch

Spinal twist

Chest

Pectoralis stretch with partner

Shoulders

Lateral shoulder stretch

Partner-assisted internal rotator stretch

External rotator stretch

Shoulder flexor stretch

Neck

Rotational stretch of neck

Neck flexion and extension

Triceps

Behind-the-neck stretch

Stability and Balance Training

Over the last decade, balance and stability training has become more popular. In the past, it was primarily used as a part of rehabilitation programs, but more strength and conditioning professionals are beginning to incorporate these exercises as part of a preventive exercise routine that is referred to as **prehabilitation**.

The merits of balance and stability training have been established with regard to improving joint stability, static and dynamic balance, and movement awareness (Sarabon 2012; Taube, Gruber, and Gollhofer 2008). Considering the potential benefits related to reducing injury risk, many strength and conditioning professionals have begun to incorporate these drills into the conditioning programs of their athletes. It is important to acknowledge here and for conditioning professionals to be aware that balance and stability exercises have no beneficial effect on jump power, strength, or strength-endurance in competitive athletes (Kohler, Flanagan, and Whiting 2010). Thus these drills are often included in the healthy athlete's warm-up routine (Herman et al. 2012), although they can also be part of other aspects of the athlete's training program.

Adaptations to training for joint stabilization and balance appear to occur quickly (Sarabon 2012). Similar to what occurs with other training modes, the rate of improvement is reduced as training progresses. To maintain performance gains, as with any other physiological adaptation, it is recommended that the athlete continue to perform balance and stability exercises. The volume for maintenance of gains may be as low as one set of four repetitions (Sarabon 2012), but scientific examination of the ideal training volume has not been conducted.

When incorporating balance and stability exercises into the warm-up or training program, it is important for the coach or conditioning professional to avoid potentially dangerous movements that can either cause injury or exacerbate an existing injury. If one is training a joint that is injured or is susceptible to injury, the progression of movement (e.g., increasing ROM) should be slowed. Similarly, progression should be slowed for healthy athletes who are not familiar with stability training. Increasing the degree of instability to each exercise should be gradual.

Ways to provide instability include the use of monoaxial or multiaxial balance boards, sponges, medicine balls, and other unstable surfaces. The degree of instability should be gradated. For instance, instability can be achieved first with a monoaxial board and then with a multiaxial board. This will provide a progressive stimulus to the neuromuscular system to enhance balance and joint stability. Progression can also involve moving from training on a large surface to training on smaller support surfaces and incorporating longer work intervals. As the athlete masters each movement, a greater degree of instability can be added. An example of lower extremity balance and stability exercises is presented in table 7.5.

Table 7.5 Example of Lower Extremity Balance and Stability Exercises

Full descriptions of how to perform select drills are available in the web resource at www.HumanKinetics.com/PhysiologicalAspectsofSportTrainingAndPerformance.

Lower intensity	Higher intensity
Balancing on a T-board in the frontal plane	Double-leg landing on a balance pad following a forward jump
Single-leg stance on a balance board	Single-leg landing on a balance pad following a forward jump
Squatting on a balance board	Walk or run over a polygon of balance pads
Balance pad lunge	

Summary

This chapter discusses the importance of the warm-up and flexibility training. The dynamic warm-up is clearly a critical part of the preexercise or precompetition routine. Further discussion centers on the types of flexibility training and factors that influence flexibility. Flexibility is individualized and is dependent on factors that are often out of the athlete's control. Finally, new evidence is emerging that performing stretching exercises before strength–power performance may be detrimental. Given this, it may become more common for athletes to perform flexibility training during the postexercise cool-down period. Additional discussion deals with stability training. Although these exercises are often used as part of a routine for recovering from injury, evidence suggests that their use in a prehabilitation program may help to decrease the risk for sport-related injuries.

REVIEW QUESTIONS

1. Describe the differences between static and ballistic stretches.
2. How do kinesiologic and physiological factors affect flexibility?
3. Explain the difference between muscle spindles and the Golgi tendon organ and their roles during muscle stretch.
4. What is the physiological advantage of a dynamic warm-up compared to a static stretching program?
5. What are the benefits and limitations associated with balance and stability training?

chapter 8

Resistance Training

CHAPTER OBJECTIVES

After reading this chapter you should be able to do the following:

- Understand the development of a needs analysis.
- Explain how acute program variables can be manipulated to accomplish training goals.
- Understand the various modes of resistance training.
- Understand how improvements in strength can enhance other components of fitness.
- Explain the force–velocity curve and compare it to the power–velocity curve.

The idea of using resistance training to improve sport performance has gained acceptance and popularity only in the last 35 to 40 years. In the past, resistance exercises were discouraged as a form of training, even for football players, because it was thought that such exercises would limit range of mobility and decrease athletic performance. It was not until the mid to late 1970s and the 1980s that teams began hiring strength coaches to help train their athletes. During this same time period, research began to show the importance of a well-designed strength and conditioning program for helping both competitive and recreational athletes reach their desired goals. This chapter discusses the development of a resistance training program and how the manipulation of specific program variables can result in different physiological changes. Discussion also focuses on different modes of resistance training, including the benefits and limitations of each. Other topics to be explored are the changes in strength that occur during an athlete's career and the effects that resistance training has on the components of fitness (e.g., speed, vertical jump height).

Resistance Training Program Development

In the development of a resistance training program, several major program design components must be considered, including a **needs analysis**, acute program variables, chronic program manipulation, and administrative concerns. Each component includes several different variables that can be manipulated in order to achieve the desired goals. This section discusses the design components in detail with the exception of chronic program manipulation. This component, also known as periodization, is discussed in greater detail in chapter 14.

Needs Analysis

The needs analysis is the starting point of any training program. It consists of determining the basic needs of the individual in accordance with the target activity. The analysis generally focuses on three primary areas: physiological, biomechanical, and medical.

The **physiological analysis** focuses on determining the primary energy source used during the activity. The **duration** and **intensity** of the activity are the primary determinants of the energy contribution (see chapter 3). Although a continuum of the three energy sources exists, each source differs in its magnitude of contribution. As intensity of exercise increases and duration of activity is reduced, a greater reliance on anaerobic metabolism is seen. When intensity of exercise is reduced and duration of activity is elevated, aerobic metabolism appears to be the primary energy source. One can further analyze needs of the activity or sport by examining the emphasis placed on other major fitness components (**strength**, **power**, **speed**, **agility**, and **flexibility**). The physiological analysis provides information for acute program variables of resistance training, such as intensity (load) and rest period length.

The **biomechanical analysis** requires examination of the specific muscles and joint angles used during the activity or sport. A correct analysis will help determine the most appropriate exercise. The following factors are important to consider in developing a biomechanically specific training program (Kraemer et al. 2012):

- The joint around which movement occurs
- The **range of motion** around the joint
- The pattern of **velocity** throughout the range of motion
- The pattern of resistance throughout the range of motion
- The type of muscle action (**concentric**, **eccentric**, or **isometric**)

Training adaptations should be specific to the types of exercises chosen to increase the percentage of carryover to the specific activity. Strength training for any activity or sport should be performed throughout the full range of motion of the joint. This not only prevents loss of flexibility but may also improve both joint and muscle flexibility. However, exercises designed for sport-specific training (which may not involve exercising through the full range of motion) should be included to maximize the strength carryover to the respective sport.

One of the advantages of resistance training is its role in the prevention of injury (Fleck and Falkel 1986; Hoffman and Klafeld 1998). Therefore, the **medical analysis** is primarily concerned with locating previous sites of injury in the athlete or understanding the common sites of injury for a particular sport. Specific exercises could be selected that would strengthen particular joints or muscles with the purpose of preventing injuries or at least reducing their severity. This may improve the quality of performance by keeping the better players on the field or court longer.

Acute Program Variables

The acute program variables make up the exercise stimulus, and an infinite number of possible combinations can be used within an exercise session. A proper needs analysis and knowledge of the specific effects of variable manipulations on training adaptations are crucial to an intelligent exercise prescription.

Exercise Selection

The choice of exercise is related to the specific muscle actions that need to be trained. This decision is based on the biomechanical analysis and injury-site profile (medical analysis) performed during the needs analysis.

Exercises can be classified as structural or body part. Structural exercises require a coordinated action of many muscle groups. Examples of structural exercises include power cleans, deadlifts, and squats. Exercises that isolate a single joint or muscle group are known as body-part exercises. Examples of such exercises include biceps curl, knee extension, and triceps push-down. Because of the large muscle mass recruited, structural exercises may simulate sport action better and may result in a greater transfer of strength to the actual sport activity. Often, these types of exercises are the core of an athlete's training program. Structural exercises are multijoint exercises that require complex motor control; they should be performed by individuals ready for advanced lifting techniques and should be taught by experienced coaches or instructors. These exercises have not been shown to confer any greater risk for injury than isolated body-part exercises if properly taught. The isolated body-part exercises are often used as assistance exercises and are effective in training specific joints or muscles that are susceptible to injury. In addition, body-part exercises, because of their ability to isolate a specific joint or muscle, are commonly used in rehabilitation from injury.

Order of Exercise

The order of exercise has generally proceeded from large to small muscle groups. Because the large muscle group exercises are usually the structural exercises, it stands to reason that training in this direction (large to small) avoids fatigue before performance of the core lifts. Muscles that are preexhausted may not achieve the maximum possible intensity and may cause problems with exercises that require complex motor coordination (e.g., structural lifts). **Preexhaustion training**, however, may be used by athletes who do not feel that structural exercises are completely exhausting or that they are fully developing the targeted muscles.

As an example, if an individual is training legs, back, and biceps, the order generally proceeds from large to small among the muscle groups. In other words, the individual would begin by training the legs, followed by the back, and would finish the workout by exercising the biceps. Exercises for a specific muscle group are also performed from large to small. Thus, the order of exercise may be squats, leg extensions, leg curls, calf raises, lat pull-downs, seated rows, and dumbbell curls. Notice that the lat pull-downs are performed after the assistance exercises for the legs (i.e., leg extensions, leg curls, and calf raises), even though the lat pull-downs recruit a larger muscle mass. Because the muscles recruited by the leg exercises differ from those recruited by the back exercises, it is doubtful that performing leg assistance exercises would have any detrimental effects on large muscle mass exercises for the back.

Exercise Intensity

The intensity of exercise, synonymous with training load (the amount of weight lifted per repetition), is probably the most important variable in resistance training. Loading is most likely the major stimulus related to changes in muscle strength or muscle remodeling.

Exercise intensity is generally represented as a percentage of an individual's repetition maximum (RM) for an exercise. The RM refers to the maximum number of repetitions that can be performed with a given load. For example, if a person can perform 10 (but not 11) repetitions with 100 lb (45 kg), that individual is said to have a 10RM of 100 lb. Intensity can also be expressed relative to an individual's 1RM. A 1RM is the maximum amount of weight that a person can lift for a particular exercise. The exercise prescription can be written either as a percentage of an individual's 1RM (e.g., 80% of 1RM) or as a range of RM (e.g., 8-10RM). In the former scenario, if the 1RM for the bench press was 200 lb (90 kg), then the lifter would perform the required number of sets with 160 lb (73 kg). In the latter scenario, the lifter would select a weight that he or she can lift at least 8 times but not more than 10. If the individual could lift the weight more than 10 times, this weight would no longer be considered a 10RM. The weight to be used is often selected through trial and error.

The major stimulus for either strength or muscle endurance appears to be related to the number of repetitions performed (Anderson and Kearney 1982). There is an inverse relationship between the load lifted and the number of repetitions. This is referred to as the **repetition-maximum continuum** (see figure 8.1). This continuum relates RM loads to the training effects derived from their use. It appears that RM loads of six or fewer repetitions have the

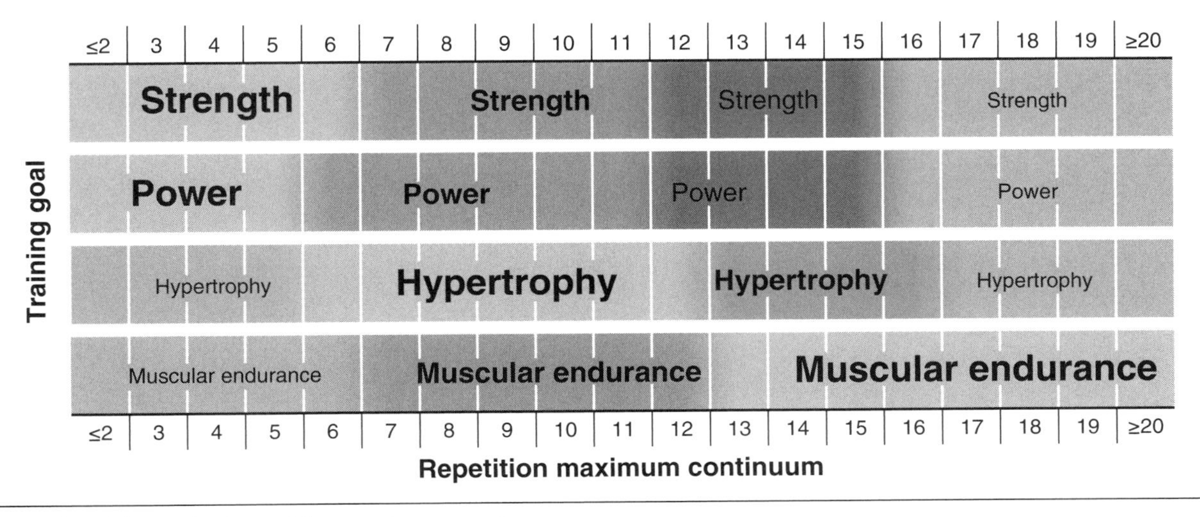

Figure 8.1 Theoretical repetition-maximum continuum.

greatest effect on maximal strength or power output. Loads exceeding 20 repetitions have the greatest effect on muscle endurance. This response may also be affected by **pretraining status**. An untrained individual may make significant gains in strength at a 15- to 20RM loading, which are likely related to neurological adaptations.

Determination of the appropriate resistance is crucial for optimal performance gains and may be achieved in several ways. The most effective method may be maximal strength (1RM) testing (Hoffman, Maresh, and Armstrong 1992). As already mentioned, once a 1RM is established for an exercise, the training load may be determined using a percentage of the 1RM. The protocol for determining a 1RM is depicted in figure 8.2. However, novice lifters may not have developed the skill, balance, and other neurological attributes that would allow for safe and effective 1RM testing (Baechle, Earle, and Wathen 2008). It may then be more appropriate to estimate the 1RM by using a submaximal weight and asking the individual to perform as many repetitions as possible with that resistance. Note, however, that there could be a large error between the predicted and actual maximal strength level (Hoeger et al. 1987; Hoeger, Hopkins, Barette, et al. 1990). It has been suggested that prediction models are able to maintain their validity of predicting maximal strength when the number of repetitions performed do not exceed 10 (Mayhew et al., 1999). This is discussed in greater detail in chapter 16. Table 8.1 provides an estimate of the number of repetitions performed at a percentage of the 1RM.

- Warm-up set of 5 to 10 repetitions with a resistance equaling approximately 40% to 60% of the estimated 1RM.
- Rest 1 to 3 min.
- A second warm-up set of 3 to 5 repetitions with a resistance equaling approximately 60% to 80% of the estimated 1RM.
- Rest 3 to 5 min.
- Attempt a 1RM lift.
- Rest 3 to 5 min.
- If the previous attempt was successful, add resistance and attempt another 1RM lift; continue until unable to achieve a complete repetition.
- Record the last successful lift as the 1RM.

Figure 8.2 Protocol for determining 1RM.

Table 8.1 Estimate of Number of Repetitions Performed at a Percentage of 1RM

% of 1RM	Repetitions
100	1
95	2
90	4
85	6
80	8
75	10

Length of Rest Period

The length of rest between sets is an important variable in the exercise design because it determines the amount of recovery of the anaerobic energy sources before the next set. Because the phosphagen (i.e., adenosine triphosphate-phosphocreatine [ATP-PC]) energy source is the most powerful, it is needed for maximal or near-maximal strength performance (1-4RM). It takes 2.5 to 3.0 min to replenish the phosphagen stores needed for the next set (Tesch and Larson 1982). Short rest periods (<1 min) appear to be related to the development of high-intensity muscular endurance and **hypertrophy**, and are commonly used during resting durations in training programs for bodybuilders who are interested in maximizing muscle hypertrophy. The short rest period, combined with a high volume of training (sets × repetitions performed), appears to be a primary stimulus for eliciting the most dramatic anabolic hormone response (see chapter 2). In contrast, longer periods of rest (2-3 min between sets), typical of lifters maximizing their strength potential (e.g., powerlifters or Olympic lifters), result in a lower anabolic hormone response (Kraemer, Marchitelli, et al. 1990).

The use of minimal rest between sets in the training programs of bodybuilders has led to various types of exercise routines that include **supersetting** and **compound setting**. Supersetting involves using agonist and antagonist muscle groups in an alternating fashion with little or no rest in between (e.g., biceps curls and triceps push-downs), while compound setting involves performing different exercises for the same muscle group in an alternating fashion with little or no rest in between (e.g., incline bench presses and incline dumbbell flys). The goal of these types of training routines is to enhance a greater anabolic response.

Rest intervals also have an important effect on the metabolic response to resistance training. The volume of exercise (number of repetitions performed) is proportionally reduced as the rest period length between sets is shortened (Ratamess, Falvo, et al. 2007; Ratamess et al. 2012). Figure 8.3 depicts lifting performance using a 10-repetition load over five sets with rest intervals of 30 s and 1, 2, 3, and 5 min. Similar responses were seen with a 5-repetition load. Fatigue rates are significantly correlated to the metabolic response to acute resistance exercise, indicating an inverse relationship between the length of the rest interval between sets and the mean acute metabolic response (Ratamess, Falvo, et al. 2007). The metabolic responses to the training protocol are greater with shorter rest intervals per set. In addition, metabolic responses are significantly higher during the 10-repetition protocol compared with the 5-repetition protocol. Interestingly, the fatigue rates may depend on strength levels. In addition to a greater metabolic response, a greater anabolic hormone response has been reported with short rest intervals (Buresh, Berg, and French 2009). However, these differences are not maintained. As individuals continue to train with short rest intervals, the difference in the anabolic hormone response to the training session becomes attenuated; and following 10 weeks of training, no significant differences in the hormonal response are noted between short (1 min) and long (2.5 min) rest intervals. Ratamess and colleagues (2012), studying sex differences and rest interval lengths, reported that the weaker groups (both women and men with low bench press strength) performed a greater number of repetitions than stronger groups (men). Thus, stronger individuals may need more rest between sets to provide the appropriate stimulus for maintaining a high-quality workout.

Training Volume

Volume of training can be characterized as the total amount of weight lifted or repetitions performed in a workout session. Heavy loads used during strength and power training cannot be lifted for many repetitions. Therefore, strength and power programs are generally of low volume. In contrast, training programs that aim to maximize muscle growth by using low resistance (for example, 10-15 repetitions per

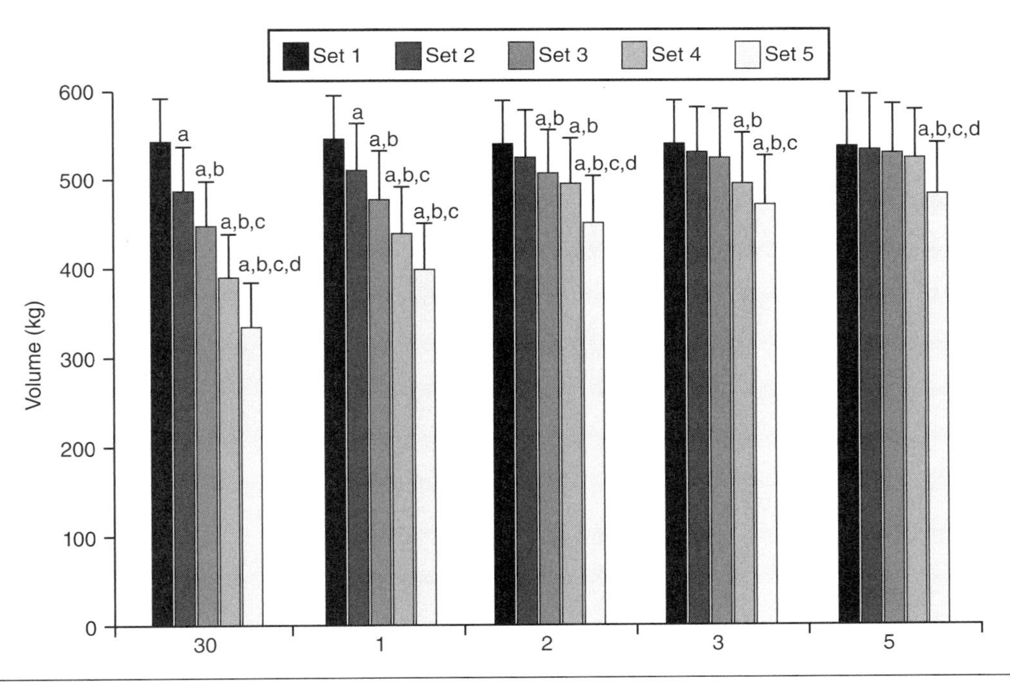

Figure 8.3 Lifting performance during 10-repetition sets of the bench press with 30 s and 1, 2, 3, and 5 min rest intervals. (a) Significantly less ($p < 0.05$) than set 1, (b) less than set 2, (c) less than set 3, and (d) less than set 4.

set) typically use a high volume of training. Volume is one of the controlling variables used in periodized training programs, which are discussed in more detail in chapter 14.

The volume–load method of calculating training volume is very appropriate for use in a weight room. Since the coach or athlete knows the load used and the number of repetitions performed, it is easy to determine training volume. But how can training volume be calculated when no external load is used (e.g., jump squats)? McBride and colleagues (2009) compared four different methods of calculating training volume: volume-load (as already explained), maximum dynamic strength volume-load, time under tension, and total work. The maximum dynamic strength volume-load uses the equation

$$\text{Number of repetitions} \times (\text{Body mass} - \text{Shank mass}) + \text{External load}$$

The authors defined the time under tension as the amount of time (in milliseconds) spent in the eccentric and concentric phases for every repetition. To calculate total work, they added the eccentric and concentric work for each repetition by integrating the area under curve of the force–displacement graphs of each contraction phase. Data for total work and time under tension involved measures using a force plate and position transducers. The investigators determined that calculation of total work may be the most appropriate method of determining resistance exercise volume, but the need for a force plate may preclude its use in most strength and conditioning facilities.

Number of Sets

The optimal number of sets per exercise depends on the specific goal of the program and the training level of the individual. Typically, the number of sets ranges from three to eight per exercise. However, novice lifters may experience strength gains from performing only one or two sets per exercise. This low number of sets may not be a stimulus sufficient to cause further adaptations in more experienced lifters (Kraemer 1997). Much information concerning the optimal number of sets is derived from empirical evidence. Typical training programs use between four and six sets for core lifts and slightly fewer (three or four sets) for assistance exercises.

Repetition Velocity

In general, the velocity of a repetition is dependent on the load. Use of heavier loads (greater percentage of one's maximal strength) results in a slower velocity of movement, and lighter loads (lower percentage of one's maximal strength) result in a faster velocity. However, some training protocols attempt to control velocity of movement by intentionally slowing down or accelerating the repetition. Working at a higher intensity of one's maximal strength capability or intentionally slowing down the repetition gives one the ability to increase the **time under tension**. It has been suggested that an increase in time under tension influences the mechanisms relating to strength and muscle adaptation. When comparing training protocols of equal volume, Tran, Docherty, and Behm (2006) demonstrated that greater time under tension influenced markers of fatigue. Increasing the duration of the concentric phase of a lift resulted in no differences in strength gain compared to a shorter duration of the concentric phase (but with a greater increase in the eccentric phase) (Gillies, Putman, and Bell 2006). However, a greater time under tension in the concentric phase resulted in significantly greater increases in both type I and IIa muscle cross-sectional area, while a greater time under tension in the eccentric phase resulted in hypertrophy of type I fibers only. A potential explanation for the difference in the hypertrophy response may be the lower muscle activation seen during the eccentric compared to the concentric phase of a lift with the same submaximal load (Nakazawa et al.1993). A greater time under tension has been shown to increase muscle protein synthesis rates and enhance mitochondrial and sarcoplasmic protein synthesis (Burd et al. 2012). Thus evidence does suggest a benefit of increasing time under tension, which is common to training programs using high-intensity loads.

Frequency of Training

Frequency of training generally refers to the number of training sessions per week. This number is determined by the training goals, time availability, type of training program, and experience level of the individual. The frequency of training appears to influence the extent of strength improvement. Surprisingly, very few studies have addressed the effect that frequency of training has on strength performance. A study examining the effects of 1 to 5 days per week of resistance training on nonresistance-trained male high school students showed that the highest frequency (5 days a week) produced the greatest strength gains (Gillam 1981). This study used the same training program (18 sets of 1RM of the bench press exercise for 9 consecutive weeks) for each session. Although this program style is not common in resistance training, the research did lay the groundwork for future work in this area. This study, as well as others (Hoffman, Maresh, et al. 1991; Hunter 1985), demonstrated that significant strength improvements can also be made with fewer training days (2-4) per week.

The optimal training frequency appears to be especially important for athletes who already possess a high level of strength. A study of strength-trained athletes (Hoffman et al. 1990) demonstrated that strength training three and six times per week did not result in any 1RM strength improvement in either the bench press or squat exercises. However, significant strength improvements were seen in athletes who trained 4 and 5 days per week. It should be noted that subjects exercising 3 days per week trained their entire body each training session. Subjects training 5 days per week trained their chest and legs three times a week and the rest of the body twice per week. Subjects exercising 4 and 6 days per week used a typical split routine and trained each body part twice during the week. It was speculated that training 3 days per week did not provide a sufficient training stimulus (because of a lack of assistance exercises) to improve an already high level of strength, and training 6 days per week may have caused a possible overtraining syndrome that blunted any strength improvement.

Administrative Concerns

The administration of a strength and conditioning program has a potentially major impact on the design of the training program. Be aware of the ease or difficulty of administration of any training. At times, compromises must be made to accommodate administrative limitations. A training facility may not be able to afford all types of equipment, often because of budgetary constraints. In addition, the training facility may be designed for a population with various training priorities. For example, an athlete may be a member of a health club that is designed primarily for a nonathletic population and lacks much of the equipment typically found in an athletic weight room (e.g., squat racks or lifting platforms). In this situation, the individual may need to make an exercise substitution in the training program.

An additional administrative concern involves time constraints, which may present a problem for completing a workout. This is typically seen in the training of student-athletes but may also be an issue for older individuals who are limited by work or family obligations. In such instances, individuals may need to prioritize the exercises in their training programs. Naturally, core exercises take priority over assistance exercises. Other possible solutions may involve designing the training program to include more multijoint, large muscle mass lifts or shortening the rest periods between sets. However, depending on the training goals, shortening the rest periods may not be an ideal solution. If this is the only choice, it may be best to use shortened rest periods for the assistance exercises and maintain the normal rest interval for the core lifts.

Various Modes of Resistance Training

Numerous exercises are available for a resistance training program. However, there are only a few modes of training to consider when one is developing the program. These modes of training—isometric, **dynamic constant resistance**, **variable resistance**, eccentric, and **isokinetic**—all have certain benefits and limitations. The selection of the mode of training to use is based on the performance needs and goals of the individual and the availability or cost of the equipment. This section addresses the benefits and limitations of each of these modes of exercise.

Isometric Training

Isometrics is also known as **static resistance training**. Isometric refers to a muscle contraction in which no change in the length of the muscle takes place. This type of training can be performed against an immovable object (e.g., a wall) or against a resistance that is greater than the concentric strength of the individual (e.g., a weight-loaded barbell or weight machine). In addition, several commercially produced machines or apparatuses (e.g., dynamometers) are designed for performance of isometric contractions.

Increases in static strength have been demonstrated with both submaximal and maximal contractions (Davies and Young 1983b; Fleck and Schutt 1985). However, maximal contractions are superior to submaximal contractions for producing strength gains. In addition, significant increases in both body weight and muscle hypertrophy have been reported after isometric training (Kanehisa and Miyashita 1983; Meyer 1967).

Benefits

Isometric strength increases are joint-angle specific. That is, if isometric contractions are performed at specific joints, strength increases will be seen at the specific joint angle trained. This strength increase will have a carryover of ±20° from the joint angle trained (Knapik, Mawdsley, and Ramos 1983). The benefit of such training may be related to the specific strength improvement observed at the **sticking point** during dynamic exercise. The sticking point is the joint-angle position of the contraction in which the muscle is at its weakest. If one examines a maximum repetition during an arm curl exercise, the muscle appears to be capable of moving a resistance throughout its full range of motion except at a particular angle in which the strength of the muscle is not enough to move the resistance. At this point the muscle will be unable to complete the repetition. Using isometric contractions at this specific sticking point increases the strength of the muscle at that angle, eventually allowing the individual to complete the lift with that resistance.

Limitations

Joint-angle specificity, which can be a great benefit of isometric training, is also its biggest limitation. Because strength is developed only at a specific angle and strength carryover is minimal, dynamic strength and power increases are not observed. For such increases to occur (strength increases throughout the range of motion of a joint), isometric training must be performed at several joint angles. This is not an efficient method of training.

An additional limitation to isometric training is its inability to increase motor performance ability (Fleck and Schutt 1985). Many individuals exercise to improve sport performance (e.g., to increase vertical jump height or sprint speed). Isometric strength training lacks the sport specificity to positively affect motor performance, thus negating its use as a primary training modality for athletes.

Dynamic Constant-Resistance Training

The term dynamic constant resistance (also known as free weights) pertains to a muscular contraction in which the muscle exerts a constant tension against a set resistance. This type of contraction has been generally referred to as an **isotonic** contraction. However, several problems arise with the use of this term in the context of training with free weights or weight machines. Although a muscle contraction is considered isotonic during such exercises, in reality the tension is not constant but varies with the mechanical advantage of the joint involved in the movement. A more appropriate definition of an isotonic exercise would be an exercise in which external resistance does not vary. Because of the confusion related to terminology, it has become more common to refer to these exercises as dynamic constant-resistance training.

Benefits

The primary benefit of dynamic constant-resistance exercise is its ability to simulate sport movement and recruit a large muscle mass. This provides an advantage over other training methods by offering a greater possibility of strength carryover to actual sport performance. The effect of strength training on motor performance improvements (e.g., vertical jump height, speed, and agility) is discussed later in the chapter.

Limitations

The primary limitation of training with dynamic constant-resistance exercises is the necessity of proper supervision. Sound safety protocols prohibit individuals from training alone (especially with free weights). Furthermore, adequate facility space is a must to ensure a safe lifting environment.

Variable-Resistance Training

Variable-resistance training equipment (primarily selectorized weight machines) operates through a lever arm, cam, or pulley arrangement. Its purpose is to alter the resistance throughout the range of motion of an exercise in an attempt to match the increases and decreases in strength throughout the entire range of motion. The goal of such a design is to force the muscle to contract maximally throughout its range of motion. However, because of differences in limb length, muscle–tendon attachment to the bone, and body size, it is very unlikely that a particular machine could match the mechanical arrangement of all individuals.

Benefits

Similar to dynamic resistance exercise, exercising on variable-resistance machines can lead to significant strength gains. Safety issues (e.g., need for spotters or risk of weights falling) do not present themselves as a major concern because of the controlled nature of the training.

Limitations

The controlled training environment, which offers a safety benefit, may create a hindrance to motor performance improvement by limiting the number of multijoint exercises that recruit a large muscle mass and simulate actual sport movements. Furthermore, the cost of variable-resistance machines may strain many budgets.

Eccentric Training

Eccentric training, which involves contractions in which the muscle lengthens, is frequently referred to in gyms as "negatives." This type of contraction is commonly seen in downhill running or in the lowering phase into a semi-squat position before jumping. The use of dynamic constant-resistance exercises requires an eccentric component. Whenever a resistance is lowered, the muscle contracts eccentrically (lengthens); as the resistance is raised to complete the repetition, the muscle contracts concentrically (shortens).

The **force capability** during an eccentric contraction is much greater than the force capability during either concentric or isometric contractions (see figure 8.4). For this reason, it has been suggested that eccentric training produces greater increases in strength than the other modes of training (Atha 1981). The optimal resistance for eccentric training appears to be 120% of the 1RM for a given exercise (Johnson et al. 1976). This resistance allows the lifter to lower the resistance slowly and to stop at will.

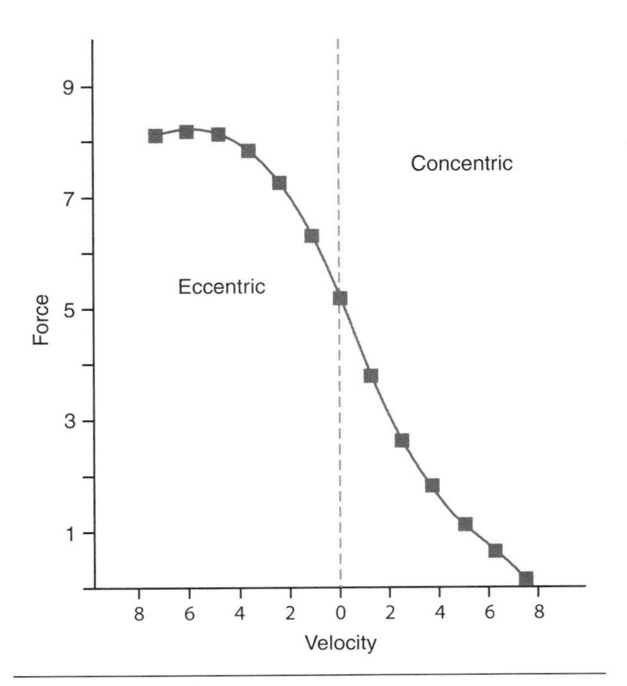

Figure 8.4 Force–velocity curve.
Data from Knuttgen and Kramer 1987.

Benefits

The primary benefit of eccentric training is its strength development capability. Specifically, this type of training may be appropriate for individuals interested in improving bench press and squat strength (e.g., powerlifters).

Limitations

Eccentric contractions are the primary stimulus for postexercise muscle soreness (Clarkson 1997). Soreness appears to peak 48 h after a training session. The effect that postexercise soreness has on performance is not completely understood. However, it stands to reason that the muscle damage and decreased flexibility associated with postexercise muscle soreness may negatively affect performance. Therefore, it would be prudent for individuals to avoid heavy eccentric exercise before important competitions. Needless to say, any sudden change in training intensity, even with dynamic constant-resistance training, may cause elevations in muscle soreness. Thus, careful consideration is warranted to ensure that all intensity changes occur at appropriate time points in the training program. An additional limitation posed by eccentric training is the requirement of a partner to prevent injuries from the supramaximal weight lifted.

Isokinetic Training

Isokinetics involves muscular contractions performed at a constant angular limb speed. In contrast to the situation with other types of strength training, there is no set resistance to overcome. Maximal force is applied to a controlled velocity throughout the full range of motion of the muscle.

Benefits

Proponents of isokinetic exercise hold that such training allows for maximal strength development throughout the full range of motion because of the constant maximal force applied by the muscle during a repetition. In addition, muscle contractions can be performed at velocities encountered during actual sport performance. The isokinetic device trains the subject in a controlled environment with isolation of a particular joint and, if needed, possible limitations on joint range of motion. Furthermore, because of the concentric–concentric nature of each repetition (most companies now offer an eccentric option), muscle and joint soreness is minimal. These benefits make isokinetic training an ideal tool for rehabilitation.

Limitations

An isokinetic machine is the most expensive of the equipment for any of the modes of training discussed here. It is not suitable for large groups of individuals training at the same time.

Alternative Methods of Resistance Training

Traditional methods of resistance training generally use free weights or selectorized weight machines; however, in the past 10 years, other implements for providing resistance have been incorporated into the training program of many athletes. Exercises that were generally associated with **strongman competitions** such as log (or stone or barrel) lifting, tire lifting, and farmer's walk are becoming a part of the training program of some athletes (Ratamess 2012). The benefit of these exercises is their ability to recruit large multijoint muscle groups, providing a whole-body stimulus for strength gains. Due to the large muscle mass recruitment, they provide a potential metabolic challenge. One study examined heart rate and blood lactate concentrations following two to four repetitions of the tire flip exercise, in sets of one or two repetitions, in 25-year-old resistance-trained men (Keogh et al. 2010). The tire weighed 232 kg (510 lb), and each repetition was completed in 2.9 ± 1.1 s. Heart rate and blood lactate increased during the two sets of tire flips. Mean heart rate reached approximately 180 beats/min, and mean lactate concentrations reached approximately 10.4 mmol/L. The lactate and heart rate responses were similar to those seen following high-intensity squat exercise (four sets of four repetitions), but less than those seen at low (four sets of 15 repetitions) or moderate (four sets of 10 repetitions) exercise intensity (Kang et al. 2005).

Many traditional exercises can also be performed with additional or alternative equipment such as kettle bells, sandbags, kegs, bands, chains, and thick bars. The use of rubber bands has become popular as it is often part of the exercise regimen for rehabilitation from injury. By varying the resting length and width of the band, the exerciser can vary the resistance. The use of **steel chains** attached to barbells is also becoming popular. Both of these devices are designed to provide a variable resistance to the traditional exercise. Regarding the efficacy of these alternative methods of training, McCurdy and colleagues (2009) demonstrated that chain-loaded bench press training did not provide any more benefit for strength improvement than traditional plate-loaded training. However, Swinton and colleagues (2011) demonstrated that the inclusion of chains equal to 20% to 40% of the subject's maximal deadlift strength resulted in maintenance of greater force to the end of the concentric phase of the lift and significantly increased peak force and impulse. On the other hand, these investigators also found that the inclusion of the chains decreased velocity, power, and the rate of force development. Rhea, Kenn, and Dermody (2009) suggested selection of a load that would allow the athlete to accelerate the resistance through the entire range of motion. If the load is too heavy, the athlete may not be able to maximize peak power benefits. Rhea and colleagues demonstrated that a fast movement with accommodated resistance (rubber bands) was able to enhance strength to the same magnitude as slow-velocity training (9.4% vs. 9.6%, respectively), but the fast movements with accommodated resistance resulted in a 17.8% improvement compared to a 4.8% improvement in power in the slow-velocity group. Thus, coaches should be aware of both the positive and potential negative effects of the use of implement training.

Another training issue that has been examined is strength attainment with the use of different bar thicknesses. Grip strength has been shown to have predictive ability with regard to baseball performance among professional ball players (Hoffman, Vazquez, et al. 2009). Thus, exercises that can enhance grip strength may provide an important practical benefit for athletes. Ratamess, Faigenbaum, and colleagues (2007) examined the effect of bars of different thicknesses (standard Olympic bar and 2 and 3 in. bars) on various strength measures. Significant performance reductions were observed in strength (ranging from 9% to 55%) with use of the 2 and 3 in. bars compared to a standard Olympic bar. The important information from this study was the magnitude of weight reduction needed with use of the thick bars. Whether the use of thick bars can enhance grip strength and provide any transfer benefits to sport performance is not known.

Vibration training is another alternate mode of training that is becoming more popular. One of the first studies in the English-language literature on the potential role of vibratory training of athletes was published by Issurin, Liebermann, and Tenenbaum (1994). They reported significantly greater improvements in strength in male athletes using the vibration training 3 days per week for three weeks rather than strength training only or the control group. A 44 Hz vibration was applied at an amplitude of 3 mm in these athletes, who had minimal resistance training experience. McBride and colleagues (2004) suggested that vibration applied to a traditional resistance exercise (e.g., vibratory dumbbell) may provide more efficient and effective recruitment of high-threshold motor units during fatiguing contractions. Sport scientists investigating whole-body vibration have presented inconclusive results on the efficacy of this method of training.

Several studies have suggested that whole-body vibration, involving unloaded static and dynamic leg exercises on a vibration platform (25-40 Hz, 1.7-2.5 mm, Power Plate) for 5 to 9 weeks, provided no benefit to speed–strength performance beyond that with traditional resistance training (Delecluse et al. 2005; Jones, Parker, and Cortes 2011; Preatoni et al. 2012). However, acute vibration training (used as part of a warm-up routine) may potentiate the muscle for a subsequent workout. Rhea and Kenn (2009) reported an approximate 5% improvement in squat power following 3 min of vibration compared to a rest period preceding squats performed at 75% of the subject's 1RM. Others have also confirmed an acute potentiating effect from vibration training on subsequent jump power performance (Turner, Sanderson, and Attwood 2011), but not on short sprint (40 m) performance (Guggenheimer et al. 2009). Still others have suggested that vibration training can accelerate recovery when applied postexercise (Lau and Nosaka 2011). When vibration is applied for 30 min per day for 4 days, an attenuation in muscle soreness and a faster return to full range of motion have been reported (Lau and Nosaka 2011). Thus, there appears to be some potential benefits to the use of vibration exercises with regard to potentiating subsequent workouts or during the recovery period. However, little evidence has been presented to suggest that it is an effective stimulus in improving strength and power gains when added to traditional training methods in competitive athletes.

Resistance Training Effects on the Components of Fitness

Besides increasing strength and improving muscle hypertrophy and body composition, resistance training programs can improve various components of fitness. The extent of many of the motor performance improvements is related to the design of the training program. Similar to what was discussed earlier regarding training experience and strength and size improvements, the ability of a resistance training program to affect performance variables, such as jump height or sprint speed, is related to the pretraining status of the individual. High-level athletes who begin a training program have a limited capacity for further improvements (remember, the closer the individual is to his or her genetic potential or performance ceiling, the more difficult it is to achieve performance improvements). Nevertheless, the importance of resistance training for these individuals is not diminished by their pretraining status. It is necessary only to adjust the pretraining expectations and to set realistic training goals. This section deals with the effect that resistance training has on strength improvement and its influence on the components of fitness.

Strength

Strength is the maximal force that a muscle or muscle group can generate at a specified velocity (Knuttgen and Kraemer 1987). If strength improvement is the measuring stick for program success, then several factors that can dramatically affect strength improvement need to be acknowledged (e.g., pretraining status, program duration, and training frequency). In addition, errors in the exercise prescription—either failure to provide a sufficient stimulus or the provision of too great a training volume—will influence strength improvements. As mentioned earlier, novice lifters experience relative strength gains much greater than those seen in strength-trained individuals. In a 12.5-year study of police recruits, Boyce and colleagues (2009) reported that the stronger recruits had the smallest improvements in strength (4.3% increase in 1RM bench press) and the weaker recruits experienced the greatest gains (21-34%). Short-term improvement goals in strength depend on the time of year. For instance, the goal of an in-season resistance training program (see chapter 14) is to maintain the strength gains made during the off-season. However, freshman football players working out twice per week on core lifts only (bench press, squat, power clean, and push press) may experience a 9.8% improvement in the 1RM squat (Hoffman, Wendell, et al. 2003); freshman performing the same workout as upperclassmen (sophomores through seniors) may experience only a 5.4% improvement in squat strength (Hoffman and Kang 2003). Interestingly,

In Practice

How Important Is Strength in Relation to Health?

Professionals in health and exercise science often focus on risk factors associated with disease and potential recommendations to reduce one's risk for disease. Traditional factors associated with disease risk include hypertension, cardiovascular fitness, obesity, blood lipids, and age. With limited exceptions, levels of strength have rarely been considered a potential risk factor except with respect to osteoporosis (Papaioannou et al. 2009). However, this appears to be changing dramatically. A cohort study on more than 1 million Swedish men was conducted over a 24-year period (Ortega et al. 2012). Subjects were Swedish military conscripts born between 1951 and 1976. Any subject who presented with an underlying disease was excluded, and subjects were required to be 19 years old or younger at the beginning of the study. The investigators reported on knee extension, handgrip and elbow flexion strength, blood pressure, and body mass index. The main study outcome was mortality; premature death was defined as death occurring before the age of 55. Subjects with less than 1 year of follow-up were excluded. During the course of the 24-year study, more than 26,000 participants died. Coronary heart disease accounted for 5.5% of the deaths, stroke 2.3%, cancer nearly 15%, suicide 22.3%, and nonintentional accidents 25.9%. The remaining deaths were attributed to other causes. Higher levels of strength were associated with lower risk of all-cause mortality. Although muscular strength was not associated with cardiovascular disease mortality or cancer, it was associated with a 20% to 30% lower risk of death from suicide. This study provided compelling evidence that muscle strength in adolescents is related to all-cause mortality, and that low levels of strength in adolescents should be treated as a potential risk factor.

in these same studies, freshman players were able to increase 1RM bench press by 2.5% during the season, but when they trained with upperclassmen they experienced a 0.6% decrease in strength. Clearly their window of opportunity for improvement in the squat exercise was greater than in upper body strength exercises.

In the off-season (generally 15 weeks of a spring semester), strength improvements in college football players have been reported to range from 2% to 8.7% in the bench press and between 5% and 20.7% in squat exercises (Hoffman, Kraemer, et al. 1990; Hoffman, Ratamess, Klatt, et al. 2009). The range in strength improvements may be attributable to differences between Division I and Division III off-season football programs. The lower improvements are generally seen at the higher level of football (Division I) and the greater improvements at the Division III level. Two potential factors are likely involved: first, the Division III athlete's greater window of opportunity for improvement, and second, the requirement of 4 weeks of spring football for the Division I athlete. Considering that strength levels are relatively similar between Division I and Division III football (Hoffman, Ratamess, and Kang 2011; Miller et al. 2002), it is likely spring football that prevents maximal performance gains during off-season conditioning. In female athletes or novices to training, the magnitude of strength improvements appears to be greater than for experienced strength–power athletes. Fry and colleagues (1991) reported a 10% improvement in bench press strength during the off-season conditioning of National Collegiate Athletic Association (NCAA) Division I female volleyball players, while Hoffman and Klafeld (1998) reported 23% and 27% improvements in a 10-week training program in the bench press and squat exercise, respectively, in female soldiers with no prior resistance training experience.

Studies on strength changes during an athlete's career are limited. Hunter, Hilyer, and Forster (1993) reported 24% and 32% increases in bench press and squat strength, respectively, in men over a 4-year intercollegiate basketball career. Most of these improvements appeared to occur during the athlete's freshman year of training (8% and 15% improvement in the bench press and squat strength, respectively). A study of female intercollegiate basketball players reported 20% to 25% improvements in strength measures in these athletes from their freshman year to senior year (Petko and Hunter 1997). Studies of strength changes in the career of NCAA Division III college football players reported improvements in the bench press and squat exercises ranging from 31% to 36% (Hoffman, Ratamess, and Kang 2011). These gains were similar to the improvements reported in NCAA Division I and Division II football players (Garstecki, Latin, and Cuppett 2004; Miller et al. 2002). The greatest gains in strength appear to occur between the first and second years of competition (7.9% and 9.1% strength increase in the bench press and squat exercises, respectively) and the second and third years (6.7% and 8.8% strength increase in the bench press and squat exercises, respectively). While the rate of strength improvements was reduced between the third and fourth years (3.1% in the 1RM bench press and 3.2% in the 1RM squat), for athletes who red-shirted and played a fifth year, the percent strength gains between the fourth and fifth years were the highest observed (13.3% strength gain in the 1RM bench press and 14.8% gain in 1RM squat).

This response pattern was different from that observed in other longitudinal investigations on

In Practice

Did the Strength Profile of Elite American College Football Players Change in the 2000 to 2010 Decade?

Robbins and colleagues (2013) compared the athletic profile of elite American college football players who participated in the National Football League (NFL) combine for the 3-year period from 1999 to 2001 and the 3-year period from 2008 to 2010. A total of 1712 players who were invited to the combine and drafted were included in the study. The analysis was performed on 15 offensive and defensive positions, but strength data were collected on only 13 positions (wide receivers and quarterbacks were excluded from the bench press exercise [102.3 kg or 225 lb] test for repetitions). The number of repetitions performed increased in nine of the positions, while body mass increased in only four positions. Body mass decreased in three positions over the 10-year period (centers, offensive tackles, and tight ends). Of the position players who decreased their body mass, two (centers and tight ends) were positions that failed to increase their strength. When subjects were grouped by linemen (offensive linemen, tight ends, defensive linemen, and linebackers) and skill position (wide receivers, defensive backs, running backs, and quarterbacks), increases in both body mass and repetitions were seen (see figure 8.5, *a* & *b,* respectively) but appeared to be moderate. Over the 10-year period, players entering the NFL appeared to have gotten heavier and performed more repetitions in the bench press test. Whether the greater number of repetitions reflects stronger athletes is debatable. The validity of a submaximal test to predict maximal strength with repetitions greater than 10 is questionable (see chapter 16).

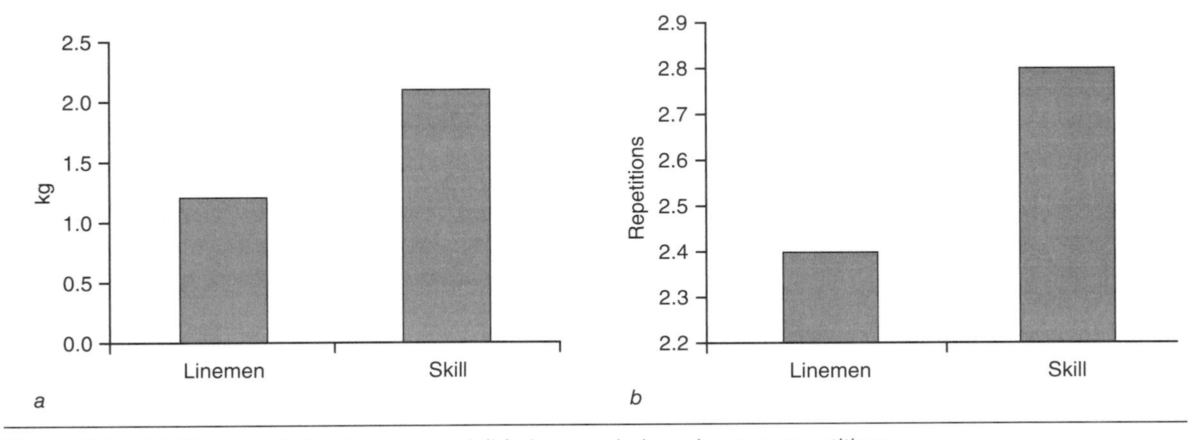

Figure 8.5 *(a)* Changes in body mass and *(b)* changes in bench press repetitions.

Data from Robbins et al. 2013.

collegiate football and basketball players. Miller and colleagues (2002) reported that the majority of the strength gains in their investigation of Division I football players occurred during the first 2 years of competition. This was similar to the responses observed in women's college basketball (Petko and Hunter 1997). In regard to the observed second peak in strength gains during the athletes' last year of competition (between the fourth and fifth years), the authors speculated on the use of performance-enhancing drugs. This was based on anecdotal reports emanating from team members that indicated that a number of fifth-year seniors experimented with such drugs before the start of their last competitive season. However, this was said about several fourth-year seniors as well.

Longitudinal strength data on professional athletes are sparse. Appleby, Newton, and Cormie (2012) reported on 2-years of strength changes in professional rugby union players. All players assessed had at least 2 years of experience as a professional athlete. During the first year of study, these strength–power athletes increased their 1RM bench press strength by 6.9% and their 1RM squat by 8.5%. In the second year of the study, 1RM bench press and 1RM squat were 10.8% and 8.8% greater than initial levels. Thus, the majority of strength gains were achieved in the athletes' first year. The greatest gains were observed in those athletes whose strength levels were the lowest. In addition, strength gains were associated with gains in lean body mass. The study clearly highlighted the limitations on strength improvements in experienced strength-trained athletes.

Anaerobic Power

Improvement in power may be the most important element with regard to enhancing athletic performance. Power (P) can be expressed as

$$P = \text{Force (strength)} \times \text{Velocity}.$$

Explosive muscle power is needed for throwing, jumping, and striking. In addition, power is needed to rapidly change direction or to accelerate from a run to a maximal sprint.

Maximal strength is attained at a very low shortening velocity. Although this strength expression is important in several sports (e.g., powerlifting and certain positions in football), most activities require strength at faster velocities. The strength capability of the muscle differs across its velocity of contraction. As seen in figure 8.4, the highest force outputs occur at the slowest velocities of concentric muscle contraction, while the lowest force outputs are produced at the fastest velocities of concentric muscle contraction. The maximal power output appears to occur at approximately 30% of the maximal shortening velocity (Knuttgen and Kraemer 1987). However, this may change with the type of exercise and equipment used, as discussed in further detail in chapter 9. Heavy resistance training programs have the greatest effect on strength improvement at the slow-velocity and high-force portion of the power–velocity curve shown in figure 8.6. However, peak power outputs occur at a relatively low force. During initial elevations in force, power is elevated, but soon the relationship between force and power becomes inverted. As force continues to increase, power decreases. This is related to the importance of velocity to power output. This principle can also explain why increasing only force outputs in the novice lifter causes significant improvements in power output but why athletes, once they have achieved a level of strength, need to include exercises designed to specifically enhance power performance (i.e., Olympic weightlifting exercises, plyometrics, ballistic exercises).

Improvement in 1RM strength, which is coveted by both athletes and coaches, is not without some merit for power development. It is true that slow-velocity strength capability has less impact on the ability of the muscle to produce force at rapid shortening velocities (Kanehisa and Miyashita 1983; Kaneko et al. 1983). However, all explosive movements begin at zero or slow velocities, and it is during these phases of the movement that slow-velocity strength may contribute to power development (Newton and Kraemer 1994).

Resistance Training and Jump Performance

Strength in both isokinetic and dynamic constant-resistance exercise has been shown to have a significant, positive correlation to vertical jump height (Bosco, Mognoni, and Luhtanen 1983; Podolsky et al. 1990). The relationship between jump height and isokinetic strength is seen when velocities of muscle contraction exceed 180°/s. In addition, the power clean, an exercise with a high power output and fast velocity of movement, has been shown to be a significant factor in predicting jumping ability (Mayhew et al. 1987). This relationship between jumping ability and strength in exercises with high speeds of joint movement is consistent with the angular velocity of the knee joint during the vertical jump (Eckert 1968).

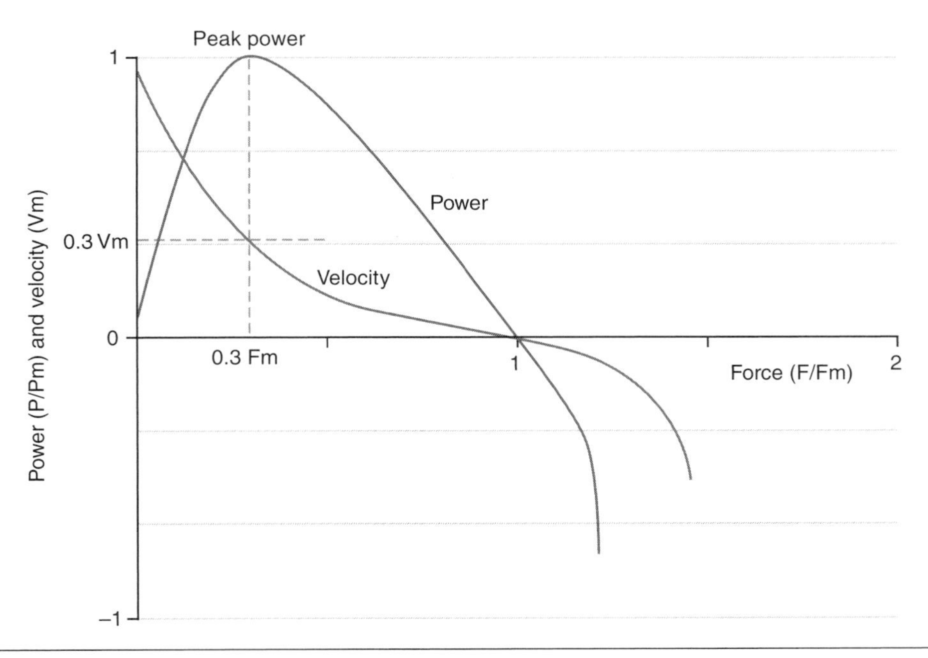

Figure 8.6 Relationship among force, velocity, and power for skeletal muscle. Vm, Pm, and Fm are maximum movement velocity, maximal power output, and maximum isometric force output, respectively.

Reprinted, by permission, from R.U. Newton, P. Cormie, and W.J. Kraemer, 2012, Power training, National Strength and Conditioning Association, In *NSCA's guide to program design*, edited by J. Hoffman (Champaign, IL: Human Kinetics), 100; Adapted, by permission, from J.A. Faulkner, D.R. Claflin, and K. K. McCully, 1986, Power output of fast and slow fibers from human skeletal muscles. In *Human muscle power*, edited by N.L. Jones, N. McCartney, and A.J. McComas (Champaign, IL: Human Kinetics), 81-94.

Squats, cleans, snatches, and push presses are the primary exercises that have been suggested to improve jumping ability (Garhammer and Gregor 1992; Young 1993). These exercises (e.g., as used in Olympic resistance training) appear to provide a greater advantage for vertical jump improvement than the traditional resistance training exercises (Channell and Barfield 2008: Hoffman, Cooper, Wendell, and Kang, 2004). The overall explosive nature of these exercises, and their ability to integrate strength, explosive power, and neuromuscular coordination among several muscle groups, suggest that they may be effective for improving jump performance. Increases in vertical jump height are generally seen when improvements in lower body strength are observed in resistance training programs. However, the relative improvement in vertical jump performance is much lower in magnitude than that in leg strength.

Acute program training variables (e.g., frequency, intensity, and volume of training) appear to be important factors influencing vertical jump performance. Resistance training programs of 5 or 6 days per week have been demonstrated to elicit greater vertical jump improvements (2.3-4.3%) than programs of 3 or 4 days per week (0.0-1.2%) in resistance-trained Division I college football players (Hoffman et al. 1990). Although these findings were not statistically different, they may have more practical significance for coaches of elite athletes. The potential benefit of a higher frequency of training may be the inclusion of a greater number of assistance exercises and their effect on sport performance changes (Hoffman et al. 1990).

The effect of training intensity (relative load of weight lifted) on vertical jump improvements is unclear. Several studies (Hakkinen and Komi 1985b; Wilson et al. 1993) have shown significant improvements in jump height using light loads (<60% of 1RM), which supports the theory of high-velocity training and consequent improvements in rate of force development. However, other reports suggest that increases in vertical jump height can also be achieved using higher intensities (>80% of 1RM) of training (Young 1993).

Resistance training programs using multiple sets per exercise have been shown to be superior for improving vertical jump performance than single-set training programs (23.1% vs. 6.9%, respectively) after 24 weeks of training (Kraemer 1997). In 12- and

15-week studies examining a periodized resistance training program (both linear and nonlinear) versus a nonperiodized resistance training program, no significant benefits for improving vertical jump performance were observed (Baker, Wilson, and Carlyon 1994; Hoffman, Ratamess, Klatt, et al. 2009). In addition, Hoffman, Ratamess, Klatt and colleagues (2009) reported no significant differences in 1RM squat strength between the training programs. Studies comparing periodized versus nonperiodized training programs of longer duration have not been conducted.

Many athletes are required to perform their off-season resistance training programs concurrently with a sport-specific conditioning program. These programs generally consist of agility, endurance, flexibility, speed, and **plyometric training**. The inclusion of these training variables in the athletes' regimen may have an effect on sport performance gains. The effect may be more related to the specific training program design. For instance, the addition of endurance training to a resistance training program may (Hennessy and Watson 1994; Hunter, Demment, and Miller 1987) or may not (McCarthy et al. 1995) reduce the relative improvement in vertical jump performance. However, the inclusion of plyometric training may provide a greater stimulus for improving vertical jump performance than resistance training alone (Komi et al. 1982; Newton and McEvoy 1994). This may be related to the increased neural stimulation (disinhibition of the Golgi tendon organ reflex) and a more efficient recruitment of agonist and synergistic muscle groups after plyometric training. Plyometric training is discussed further in chapter 9.

Resistance Training and Sprint Performance

Strength is also related to sprint performance (Alexander 1989; Anderson et al. 1991) and appears to be a better indicator of speed when strength testing is performed at velocities greater than 180°/s (Perrine and Edgerton 1978). Substantial strength of the hamstring and gluteal muscles is necessary to pull the body over the supporting foot during the ground contact phase of the sprint, while strength in the hip flexors (quadriceps and iliopsoas) allows for rapid acceleration of the thighs. Stepwise multiple regression analysis of isokinetic data has demonstrated that concentric knee extension strength at 230°/s (correlation coefficient [r] = 0.74) and eccentric hip flexor strength at 180°/s (r = 0.82) are the best predictors of speed performance in male sprinters (Alexander 1989). Less impressive but significant relationships (r = 0.57) have also been demonstrated between a slower speed of contraction (60°/s of concentric knee flexion strength) and sprint performance (Anderson et al. 1991). Absolute maximal strength in multijoint structural exercises (e.g., squats and power cleans) does not appear to be significantly related to sprint speed (Baker and Nance 1999). However, relative strength values in these exercises have been shown to have significant negative correlations (r = −0.66 and −0.72 in the squat and power clean, respectively) to sprint speed (Baker and Nance 1999).

The benefit of traditional high-intensity (70-100% of 1RM) resistance training programs on speed improvement is not well understood. Statistically insignificant improvements (<1% decrease in sprint time) for 30 yd (27 m) or 40 yd (37 m) sprints have been observed in collegiate athletes after off-season periodized resistance training programs (Fry et al. 1991; Hoffman et al. 1990). These very modest improvements in sprint speed may be more reflective of the pretraining speed ability of the athletes than any failure of the exercise prescription. It has been suggested that the inclusion of high-velocity movements (stretch–shortening exercises) in the athletes' resistance training program is important for achieving improvements in sprint speed (Delecluse et al. 1995). However, only slight ($p > 0.05$) changes in sprint speed have been shown when a high-velocity stretch–shortening program was added to a periodized resistance training program (Delecluse et al. 1995). The only notable training effect reported with the inclusion of high-velocity training was a significant improvement in initial sprint acceleration. Interestingly, speed skaters appear to have a higher peak torque output at 30°/s than land-based speed athletes (Smith and Roberts 1991). This may be the result of speed skating technique, which requires isometric contractions of the knee extensors during the glide phase. Thus, the use of traditional resistance training programs for athletes involved in ice sports (e.g., hockey and speed skating) may be potentially more beneficial to speed improvement than for land-based athletes.

Resistance Training and Agility Performance

Strength is an important factor in an athlete's ability to stop and change direction rapidly (Anderson et al. 1991; Hoffman, Maresh, and Armstrong 1992). The quick acceleration and deceleration of the body during these movements suggest that eccentric actions contribute to the ability to rapidly change direction. A significant relationship (r = 0.58) has been reported between peak eccentric hamstring force at 90°/s and agility run time and thus eccentric strength capability may be an important indicator of successful agility run time (Anderson et al. 1991).

Very few studies have reported on the effects of resistance training programs on agility measures. In studies on changes in agility performance after off-season resistance training programs, either no change (Hoffman, Maresh, et al. 1991) or an increase in time (Fry et al. 1991) in the T-drill was observed. This may be partially related to a greater emphasis on agility training during the preseason training period in contrast to the off-season training program.

Resistance Training and Swimming, Kicking, and Throwing

The importance of strength in swimming, kicking, and throwing has implications for the resistance training programs developed for swimmers, soccer players, and baseball players. A study on the importance of strength to swim performance showed that as the distance of the swim increased, the importance of strength decreased (Sharp, Troup, and Costill 1982). Thus, a greater emphasis on strength training appears to be warranted for swimmers competing in races of short duration.

For soccer players, the importance of strength in the kicking limb is reflected by a significant correlation (r = 0.82) between soccer ball velocity and isokinetic knee extensor strength (Poulmedis et al. 1988). The strength of the dominant limb in these athletes may contribute to the bilateral deficits reported in soccer players (Mangine et al. 1990). Therefore, resistance exercises that emphasize bilateral movements should be incorporated into their training programs.

The relationship between strength and throwing speed is complex because of the interaction of various muscle groups involved in the throwing action. However, significant correlations have been reported between strength of the wrist extensors (r = 0.71) and elbow extensors (r = 0.52) and throwing speed (Pedegna et al. 1982). These results clearly imply the importance of strength to throwing speed. Studies have demonstrated significant increases (2.0-4.1%) in throwing velocities of baseball players after 8 to 10 weeks of both traditional (Lachowetz, Evon, and Pastiglione 1998; Newton and McEvoy 1994) and ballistic (McEvoy and Newton 1998) resistance training programs. Whether these results can be reproduced in an elite athletic population needs to be examined.

Resistance Training and Cardiovascular Fitness

Resistance training is not considered a primary means of improving aerobic capacity as measured by $\dot{V}O_2max$. However, some forms of resistance training, specifically circuit training, have been shown to improve $\dot{V}O_2max$ (Gettman and Pollock 1981). Participation in a circuit resistance training program for up to 20 weeks may result in 5% to 8% improvements in aerobic capacity, although the improvements are moderate when compared with those seen after endurance-based exercise programs.

In a high-volume Olympic weightlifting program (i.e., exercises for the clean and jerk and the snatch), an 8% increase in aerobic capacity was observed after 8 weeks of training (Stone, Wilson, et al. 1983). Although resistance training is highly anaerobic, it appears that a greater aerobic effect occurs with a high volume of work. In addition, the use of large muscle mass exercises (e.g., power cleans, squats, and high pulls) and shorter rest periods between sets may contribute to improving aerobic capacity after resistance training programs.

Aerobic capacity improvements in weightlifters may be related to oxygen consumption during resistance training. Depending on the type of program, oxygen consumption may range from 38% to 60% of peak $\dot{V}O_2$ (Stone et al. 1991). The upper range is within the training threshold reported for aerobic training of untrained individuals. Additional mechanisms may involve the interaction of central (heart rate and stroke volume) and peripheral (arteriovenous oxygen difference) factors. Because resistance training causes cellular adaptations that are in direct contrast to those thought to promote aerobic adaptations (e.g., increase in mitochondrial and capillary densities), peripheral factors are less likely to be involved in improvements in aerobic capacity after resistance training. It is more probable that such improvements are the result of a central mechanism.

Resistance Training and Body Composition

Positive alterations in body composition have been demonstrated after short-term resistance training. Decreases in the percentage of body fat are achieved through increases in lean body mass as well as decreases in fat content (Kraemer, Deschenes, and Fleck 1988; Stone et al. 1991). Increases in lean body mass may result from increases in muscle tissue and possibly bone density. In addition, positive changes in body composition may be associated with increases in body weight by increasing the amount of lean body mass. The volume of training may be the key factor in these body composition changes.

Resistance Training and Flexibility

Flexibility is defined as the range of motion around a joint. It is related to both injury prevention and performance, and it has a large genetic component. Athletes with greater flexibility appear to be more resistant to muscle strains and pulls than athletes with less flexibility. A concern for many individuals is the loss of flexibility as a result of a heavy resistance training program. However, Olympic weightlifters are second only to gymnasts in flexibility tests (Fisher and Jensen 1990). If resistance training exercises are used in a way that permits full range of motion, then flexibility is greatly enhanced. In addition, to achieve maximal joint flexibility, exercises for both the agonist and antagonist muscle groups should be performed.

Women and Resistance Training

Large differences between the sexes are seen in examinations of strength performance. Women's mean total body strength is reported to be 63.5% that of men (Laubach 1976). However, there is a great deal of variation between body parts. Large absolute strength differences between the sexes are seen in the upper body (with women 55.8% as strong as men), while the difference is much less in the lower body (with women 71.9% as strong as men). When males and females are compared relative to body mass, the wide gap in strength performance narrows. Wilmore (1974) showed that relative to body mass, the upper body strength of women was 46% that of men. This is similar to the 44% difference in upper body strength reported between female and male police recruits (Boyce et al. 2009). In the lower body, relative to body mass, women were 92% as strong. When making these strength comparisons relative to lean body mass, Wilmore showed that men were still almost twice as strong as females in the upper body but that females may be stronger (106%) than males in the lower body. These differences were for the average male and female and did not include trained athletes or take body-size differences into account. In a comparison study of a male and female active duty military population, sex differences ranging from 58% to 104% were noted in military-specific strength tasks (i.e., maximal weight lifted to 152 cm height, maximal isometric force on an upright cable pull, and maximal isometric handgrip force) (Vanderburgh et al. 1997). These large differences persisted even when scaled to body mass or lean body mass, but the magnitude of difference was reduced when lean body mass was taken into account. Training responses appear to be more dependent on training experience and initial strength levels than on sex. Women appear to be able to increase upper body strength at the same relative magnitude as men. Over a 12.5-year period, female police officers improved upper body strength by 12.7%, and male police officers improved by 15.8% (Boyce et al. 2009).

Examples of Resistance Training Programs

The following are examples of different types of resistance training programs. Depending on the resistance training experience of the individual, time available for training, and the type of equipment available in the training facility, adjustments to these training programs can be made. If a substitution must be made for a particular exercise, it is important to choose an exercise that recruits the same muscle groups. In addition, the exercise selected should not be too advanced for the training level of the individual.

It is recommended that the training programs be performed in the order listed. Exercises generally proceed from larger muscle mass to smaller muscle mass, but at times the program may call for completing a particular muscle group before proceeding to the next body section. The numbers for each exercise represent the volume and intensity. For example, 1 × 10, 3 × 6-8RM means one warm-up set of 10 repetitions and three work sets of a 6- to 8-repetition maximum. As mentioned earlier, individuals should select a weight that they can lift at least six times but not more than eight times. If a weight cannot be lifted for six repetitions, then it is too heavy and should be reduced. Likewise, if more than eight repetitions can be performed, the resistance is too light and weight should be added.

Training Program 1

Example of a Maximal Strength Program

Training frequency: 4 days per week
Rest period between sets: 2 to 3 min
Rest period between workouts: 72 h between body parts. That is, if the legs are trained on Monday, the chest should be trained at the next session and the leg exercises should not be repeated until Thursday.

Days 1 and 3		Days 2 and 4	
Exercise	**Sets × repetitions**	**Exercise**	**Sets × repetitions**
Squat	1 × 10, 3 × 6-8RM	Bench press	1 × 10, 3 × 6-8RM
Leg extension	3 × 6-8RM	Incline bench press	3 × 6-8RM
Leg curl	3 × 6-8RM	Shoulder press	1 × 10, 3 × 6-8RM
Calf raise	3 × 6-8RM	Upright row	3 × 6-8RM
Lat pull-down	4 × 6-8RM	Triceps push-down	3 × 6-8RM
Seated row	4 × 6-8RM	Sit-up	3 × 20
Biceps curl	4 × 6-8RM		
Bent-knee sit-up	3 × 20		

Training Program 2

Example of a Bodybuilding Program

Training frequency: 4 days per week
Rest period between sets: 1 min
Rest period between workouts: 72 h between body parts

Days 1 and 3		Days 2 and 4	
Exercise	**Sets × repetitions**	**Exercise**	**Sets × repetitions**
Squat	1 × 10, 3 × 10-12RM	Bench press	1 × 10, 3 × 10-12RM
Leg extension	3 × 10-12RM	Incline bench press	3 × 10-12RM
Leg curl	3 × 10-12RM	Incline fly	3 × 10-12RM
Calf raise	3 × 10-12RM	Shoulder press	1 × 10, 3 × 10-12RM
Lat pull-down	3 × 10-12RM	Upright row	3 × 10-12RM
Seated row	3 × 10-12RM	Lateral raise	3 × 10-12RM
Standing barbell biceps curl	3 × 10-12RM	Triceps push-down	3 × 10-12RM
Seated dumbbell biceps curl	3 × 10-12RM	Dip	3 × 12-15
Bent-knee sit-up	3 × 20	Sit-up	3 × 20

Training Program 3

Example of a Muscle-Toning and Strength Program

Training frequency: 3 days per week
Rest period between sets: 1 to 2 min
Rest period between workouts: 48 h between workouts

Day 1		Day 2		Day 3	
Exercise	**Sets × repetitions**	**Exercise**	**Sets × repetitions**	**Exercise**	**Sets × repetitions**
Leg press	3 × 8-10RM	Dumbbell lunge	3 × 8-10RM	Leg press	3 × 8-10RM
Bench press	3 × 8-10RM	Incline bench press	3 × 8-10RM	Bench press	3 × 8-10RM
Shoulder press	3 × 8-10RM	Upright row press	3 × 8-10RM	Shoulder press	3 × 8-10RM
Lat pull-down	3 × 8-10RM	Seated row	3 × 8-10RM	Lat pull-down	3 × 8-10RM
Leg curl	3 × 8-10RM	Leg extension	3 × 8-10RM	Leg curl	3 × 8-10RM
Triceps push-down	3 × 8-10RM	Triceps extension	3 × 8-10RM	Triceps push-down	3 × 8-10RM
Biceps curl	3 × 8-10RM	Biceps curl	3 × 8-10RM	Biceps curl	3 × 8-10RM
Sit-up	3 × 20	Sit-up	3 × 20	Sit-up	3 × 20

Training Program 4

Example of a Circuit Training Program

Training frequency: 2 or 3 days a week
Rest period between sets: 1 to 2 min
Rest period between workouts: 48 h between workouts
Instructions: The circuit is performed continuously from exercise to exercise without rest. When the complete circuit has been performed, a 5 min rest precedes performance of the next circuit.
Number of circuits: Two or three

Exercise	Sets × repetitions
Leg press	1 × 12-15RM
Bench press	1 × 12-15RM
Leg curl	1 × 12-15RM
Shoulder press	1 × 12-15RM
Leg extension	1 × 12-15RM
Lat pull-down	1 × 12-15RM
Triceps push-down	1 × 12-15RM
Biceps curl	1 × 12-15RM
Sit-up	1 × 12-15RM
Cycle ergometer	2 min

Summary

Changes in acute program variables such as intensity, volume, and rest can result in significantly different physiological adaptations. The ability to properly assess the needs of the athlete is important for designing an appropriate resistance training program. There are several different modes of resistance training; each has benefits and limitations. Changes in strength improvements are seen over an athlete's career. A resistance training program can not only stimulate improvements in muscle growth or strength of an individual but can also improve components of fitness (e.g., speed and jump height) that affect the athlete's performance potential. Specific resistance training programs have varying effects on these components of fitness.

REVIEW QUESTIONS

1. Discuss the three components of the needs analysis and why it is important for the development of the athlete's training program.
2. Compare differences in training volume and training velocity between exercise programs focused on maximizing strength versus muscle hypertrophy.
3. What are the benefits and limitations of isokinetic training?
4. Discuss the importance of incorporating Olympic weightlifting exercises into the training programs of strength–power athletes.
5. Discuss why strength improvements can enhance speed and agility performance.
6. What are the potential health benefits associated with resistance exercise?

chapter 9

Power Training

CHAPTER OBJECTIVES

After reading this chapter you should be able to do the following:

- Understand the scientific basis of plyometrics and define a stretch–shortening cycle.
- Explain the relationship between force, velocity, and power.
- Define the rate of force development and explain its importance in athletic performance.
- Explain the importance of ballistic exercises and when they should be incorporated into the training program of competitive athletes.

Many athletic endeavors involve exerting force over a distance as fast as possible. The force can be exerted on an implement such as a baseball bat, a javelin, or a discus or on one's body to move it rapidly. The product of force and velocity is power (Knuttgen and Kraemer 1987). Many strength and conditioning programs focus primarily on improving the force-generating capability of the athlete. The use of traditional resistance training programs that require lifting a heavy load at a slow velocity has generally been considered the primary method of increasing power production (see chapter 8). This is based on the notion that because power is equal to force multiplied by velocity, increasing maximal strength enhances the ability to improve power production. However, to maximize power production, it is imperative to train both the force and velocity components (Kraemer and Newton 2000). This chapter focuses on methods used to generate maximal muscular power, with emphasis on the scientific basis for power development. Another focus is the use of plyometric and ballistic training programs.

Scientific Basis for Power Training

In novice resistance-trained athletes, large increases in strength are common during the beginning stages of training. Improvements in various power components of athletic performance, such as vertical jump height and sprint speed, may also be evident. This is primarily the result of the athlete's ability to generate a greater amount of force. As the athlete becomes stronger and more experienced, the rate of strength development decreases and eventually reaches a plateau. At this stage of an athlete's career, not only are strength improvements harder to achieve, but improving maximal strength does not provide the same stimulus to power performance as it did during the earlier stages of training. In addition, training for maximum force development may have its limitations with regard to improving power performance. An important factor for maximizing power production is exerting as much force as possible in a short period of time. Training for maximal strength using heavy resistance does not appear to enhance the rate of force development (Kraemer and Newton 2000). However, adding plyometric exercises to the training program—or a combination of plyometric and resistance training (using a light resistance as in ballistic training, for example)—may enhance the athlete's ability to increase the rate of force development. This has been demonstrated by a number of studies showing that ballistic training provides a positive stimulus for improving power production even in trained athletes, especially resistance-trained athletes (Newton, Kraemer, and Hakkinen 1999; Wilson et al. 1993).

To maximize power development, a number of components of power need to be trained and emphasized at various stages of the athlete's career. Kraemer and Newton (2000) have described each of these components as a **window of adaptation** (figure 9.1). Each window refers to the magnitude of potential for adaptation. For example, as the athlete's strength level increases, the window of opportunity to improve maximal power production from slow-velocity strength training is reduced. Training must then be aimed at improving performance in the athlete's weakest components, because it is in these components that the athlete has the largest window of opportunity for improvement.

Newton, Kraemer, and Hakkinen (1999) demonstrated this approach in a study on the preseason preparation of a National Collegiate Athletic Association Division I volleyball team. Sixteen athletes with considerable experience in both traditional resistance training and plyometric training participated in this study. They were randomly divided into two groups; the control group continued to perform resistance training as they had previously, while the treatment group performed six sets of six repetitions of the jump squat with a countermovement (ballistic training). At the end of the study (8 weeks), the ballistic training proved to be effective in increasing jumping performance, whereas the subjects performing the traditional resistance training program did not realize any changes in jump performance. No changes in maximal leg strength were seen in either group. The authors suggested that although maximal strength was not improved, the ability of the subjects in the ballistic training group to rapidly contract their muscles while maintaining tension may have contributed to the improved power performance. In addition, these subjects were able to increase their rate of force development, which is thought to be a major contributor to improvements in vertical jump performance (Hakkinen, Komi, and Alen 1985). The window of adaptation with ballistic training appeared to be wide open for these athletes. This study demonstrated the effectiveness of using a mode of training (i.e., window) that had a great potential for adaptation, in comparison with a mode of training with little room for adaptation (slow-velocity resistance training), in causing performance gains in elite athletes with considerable training experience.

Three mechanical properties of the neuromuscular system determine performance (Newton, Cormie, and Kraemer 2012):

- The ability to develop force as rapidly as possible, termed the rate of force development
- The ability of the muscle to produce high force at the end of the eccentric phase and during the early concentric phase of movement
- The ability of the muscle to continue to produce high amounts of force as its velocity of shortening increases

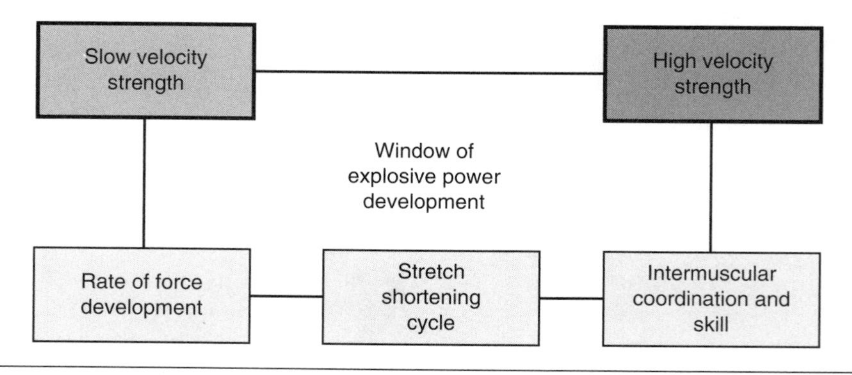

Figure 9.1 Window of adaptation.

Adapted from *Physical Medicine and Rehabilitation Clinics of North America* Vol. 11(2), W.J. Kraemer and R.U. Newton, "Training for muscular power," pg. 361, copyright 2000, with permission of Elsevier.

Several factors contribute to enhancing these mechanical properties of performance. Specific training programs or manipulation of acute program variables within the training program can provide different outcomes or emphases. Most power movements involve a **countermovement** in which the muscles are first stretched and then shortened rapidly. An example is the bending of one's knees before performing a vertical jump. This movement is referred to as a **stretch–shortening** cycle (Komi 2003). The difference in jump height and power expression between a countermovement jump and a **squat jump** (in which the athlete lowers into a squat position and pauses before performing the jump) is related to the greater force output at the start of the upward (concentric) movement. The countermovement jump allows greater forces to be exerted against the ground, a greater increase in the impulse (force × time), and an enhanced acceleration of the body as it leaves the ground (Newton, Cormie, and Kraemer 2012). Other elements reported to contribute to differences in power output between the two types of jumps are recovery of stored elastic energy, activation of the stretch reflex, and muscle–tendon interactions (Bobbert et al. 1996).

Relationship Between Force, Velocity, and Power

As discussed earlier, traditional resistance training programs are very effective in improving strength in high-force, slow-velocity movements (see figure 8.6). However, these strength improvements are not reflected to the same magnitude in the higher velocity range. Exercises using reduced loads performed at a greater velocity result in greater force outputs at the higher velocity ranges. They also result in an improvement in the rate of force development (Hakkinen, Komi, and Alen 1985; Newton, Cormie, and Kraemer 2012). This should not be interpreted as suggesting that training should focus on low-force, high-velocity movements. It is the combination of high force at low velocity and low force at high velocity that should be integrated into the athlete's training regimen. If the trained athlete focused on only one of these aspects of the training model, the other component would potentially decrease in the absence of the appropriate stimulus.

The importance of exerting maximal force as rapidly as possible is the basis for success in strength–power sports. This is referred to as the **maximum rate of force development**. An example of its importance can be seen in American football. In this sport, outcome is often dictated by the athletes (e.g., offensive and defensive linemen) who control the line of scrimmage. If two opposing players of similar size and strength square off against each other, the factor that can determine success may be which athlete can reach peak force more quickly. As the players slam into each other and attempt to extend their arms and control their opponent, the athlete who can generate maximal force more quickly has an advantage. Figure 9.2 shows the advantage of power training (less force, greater velocity) over heavy resistance strength training (maximal force, slow velocity) regarding the rate of force development. Incorporating high-velocity movements into the athlete's training program enhances the rate of force development more than focusing primarily on increasing maximal strength. Although rate of force development is improved by heavy resistance exercise, the magnitude of improvement is superior with higher-velocity exercises. It is prudent to focus on both methods of training, and this is generally accomplished within the annual training program (see chapters 14 and 15).

One of the major problems that has been identified with traditional resistance training exercises is the deceleration of the bar toward the end of the lift (Newton, Cormie, and Kraemer 2012). Deceleration has been reported to occur for 24% of the time spent in the concentric phase, but may increase to 52% if the athlete uses a lighter resistance (~80% of one-repetition maximum [1RM]) (Elliott, Wilson, and Kerr 1989). Thus, even when using lighter resistance, the athlete needs to spend more time decelerating the bar. If the athlete were to be able to accelerate the bar through the full range of motion by actually throwing it as in a bench press throw, or jumping with the weight as in a squat jump, the potential for power development may even be greater. This type of movement is referred to as ballistic resistance training (Newton, Cormie, and Kraemer 2012). Figure 9.3 compares the bar velocities of a traditional bench press movement and a bench press throw. As the athlete moves the bar through its full range of motion, the greater velocity of movement in the bench press throw suggests that it would be advantageous in stimulating power development compared to the traditional bench press movement. Newton and colleagues (1996), comparing the bench press throw to the traditional bench press movement, reported that muscle activation was 19% to 44% higher in the throw versus the press movements, and that deceleration occurred for 40% of the concentric movement during the press.

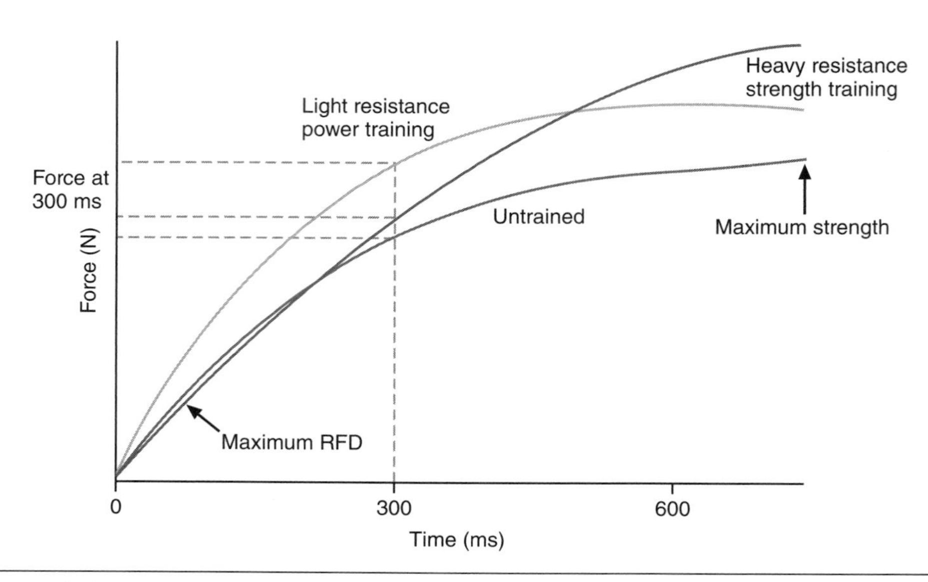

Figure 9.2 Isometric force–time curve indicating maximum strength, maximum rate of force development, and force at 200 ms for (1) pretraining, (2) heavy resistance strength training, and (3) lighter-resistance high-velocity training.

Reprinted, by permission, from R.U. Newton, P. Cormie, W.J. Kraemer, 2012, Power training, National Strength and Condition Association, In *NSCA's guide to program design*, edited by J. Hoffman (Champaign, IL: Human Kinetics), 103; Adapted, by permission, from K. Häkkinen, P.V. Comi, and M. Alen. 1985, "Effect of explosive type strength training on isometric force- and relaxation-time, electromyographic and muscle fibre characteristics of leg extensor muscles," *Acta Physiologica Scandinavica* 125: 587-600.

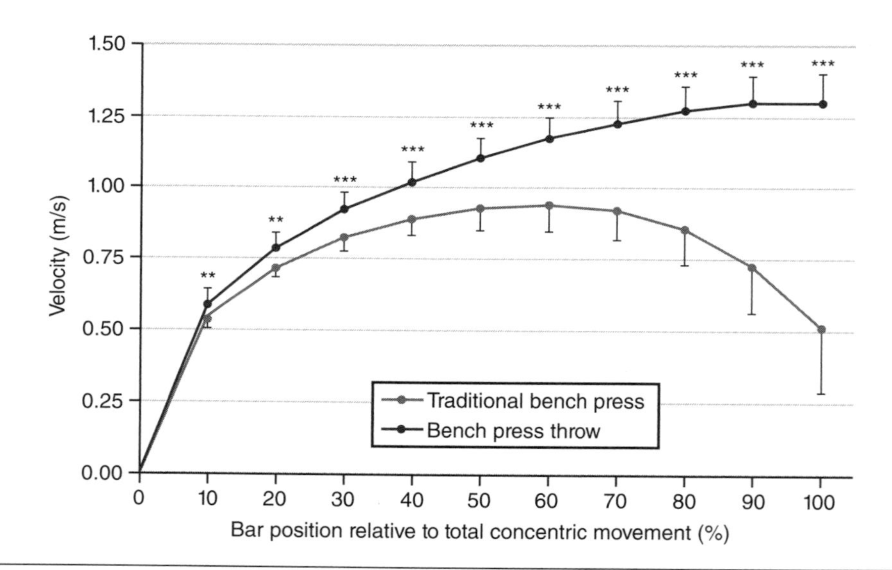

Figure 9.3 Mean (±SD) bar velocity in relation to total concentric bar movement for a traditional bench press, performed as rapidly as possible, and a bench press throw (**$p < 0.01$, ***$p < 0.001$).

Adapted with permission from Newton, et al., 1996. "Kinematics, kinetics, and muscle activation during explosive upper body movements," *Journal of Applied Biomechanics*. 12: 31-43.

The bench press throw may not be a practical exercise, given that attempting to catch the bar once it leaves the hands is not recommended; but several other types of exercises may be very effective. For instance, squat jumps, medicine ball throws, and plyometric drills are all resistance movements that do not require deceleration during the concentric phase of the exercise. Interestingly, technological advances have used hydraulics to allow athletes to use a bench press throw as a legitimate ballistic exercise (figure 9.4).

Figure 9.4 Bench press throw device.

Training Methods for Power Development

Athletes train for power by improving maximal force capability and by enhancing the rate at which a submaximal force can be exerted. However, where the emphasis should be and how it is manipulated is largely determined by the experience and strength level of the athlete. Novice athletes or athletes who are weak would benefit primarily from training programs designed to enhance maximal strength. That may be enough of a stimulus to increase power performance (Cormie, McGuigan, and Newton 2010; Stone, Johnson, and Carter 1979). However, for the experienced and stronger athlete, plyometric drills, ballistic training, and Olympic-style resistance training may provide the necessary stimulus to enhance power performance.

Plyometrics

Plyometrics is a term for exercises that stretch and then shorten the muscle to accelerate the body or limb. As such, plyometrics is often referred to as **stretch–shortening exercise**, which may be the more appropriate term. Plyometrics, or stretch–shortening exercise, has received much attention on the part of coaches since the early 1970s. Track and field coaches were the first to incorporate such training regimens into their programs. This was primarily in response to the superior performances by European Eastern Bloc athletes in international track and field, gymnastics, and weightlifting. It was thought that because these exercises were a staple of the training programs of these athletes, they must contribute to these competitors' success and superiority (Chu 1992).

American track and field coaches quickly began to incorporate many of these exercises into the training programs of athletes, primarily those who jumped, lifted, or threw. Shortly thereafter, coaches in other sports such as volleyball, American football, and basketball began to see the potential applicability of this form of training to their own programs.

Scientific Basis for Plyometric Training

During plyometric, or stretch–shortening, exercises, the muscle is rapidly stretched (**eccentric contraction**), as in a countermovement jump, and then shortened to accelerate the body upward. It has been demonstrated that the stretch–shortening cycle enhances power performance to a greater extent than **concentric** training only (Bobbert et al. 1996; Bosco and Komi 1979; Ettema, Van Soest, and Huijing 1990). Differences of 18% to 20% were shown in a comparison of the countermovement jump (using a **prestretch**) with the squat jump (concentric contraction only) (Bosco et al. 1982). The increased performance seen in the countermovement jump has been attributed to the greater amount of stored **elastic energy**, acquired during the eccentric phase, that can be recruited during the upward movement of the jump (Bosco and Komi 1979). In addition, the prestretch during the countermovement results in a greater neural stimulation (Schmidtbleicher, Gollhofer, and Frick 1988), as well as an increase in the joint moment (a turning effect of an eccentric force, also referred to as torque, at the start of the upward movement) (Kraemer and Newton 2000). The greater joint moment results in a greater **force** exerted against the ground with a subsequent increase in **impulse** (greater force applied over time) and **acceleration** of the body upward. Bobbert and colleagues (1996) have suggested that this latter mechanism may be the primary reason for the greater jump height observed during a countermovement jump, whereas the other mechanisms may play more of a secondary role.

The performance of a stretch–shortening cycle requires a finely coordinated action of **agonist**, **antagonist**, and **synergistic muscle groups**. During the rapid action of the stretch–shortening cycle, the agonist and synergistic muscle groups must apply a great deal of force in a relatively short period of time. To maximize this action, the antagonist muscle groups must be relaxed during the time when the agonists and synergists are active. A novice to stretch–shortening movements needs some training to coordinate these movements. Through training, contraction of the antagonist muscle groups is reduced, which allows for a greater coordination of these muscle groups and produces a more powerful and effective vertical jump (Schmidtbleicher, Gollhofer, and Frick 1988). In addition, during the initial workouts, the EMG (electrical activity of the exercising muscles, indicating extent of activation) of the agonist muscle groups appears to be reduced (Schmidtbleicher, Gollhofer, and Frick 1988). This is likely the result of activation of the **Golgi tendon organ** (a muscle proprioceptor) to protect the muscle from excessive stretch. As the training program continues, the inhibitory effects exhibited by the muscle proprioceptors may be reduced, allowing for improved stretch–shortening performance (Schmidtbleicher, Gollhofer, and Frick 1988).

It has been clearly demonstrated that the use of stretch–shortening exercises provides a distinct advantage for producing greater power than concentric-only movements. The primary question regarding plyometric training is whether plyometric drills should be considered a supplement to the normal training regimen or an alternative way of training. Specifically, is plyometric training as effective in improving power and strength performance as **resistance training**? Does including these drills as part of the overall training program provide any additional benefit for the athlete?

Plyometric training is a training method that is used to enhance power performance. Most plyometric exercises, although not all drills, require the athletes to rapidly accelerate and decelerate their body weight during a dynamic movement. The body weight of the athlete is most often used as the overload, but the use of external objects such as medicine balls also provides a good training stimulus for certain plyometric exercises. To increase intensity of effort or difficulty, experienced athletes may perform plyometric drills wearing weighted vests or holding dumbbells in their hands when performing some of the jumps. Traditional resistance training and plyometric drills are often combined during the power phase in the yearly training program of the strength–power athlete (see chapters 14 and 15).

A number of studies have demonstrated the effectiveness of plyometric training for improving power, generally expressed as increases in vertical jump height (Adams et al. 1992; Bosco et al. 1982; Brown, Mayhew, and Boleach 1986; Ford et al. 1983; Hakkinen and Komi 1985b; Wilson et al. 1993). Traditional resistance training has been shown to improve vertical jump performance as well (Adams et al. 1992; Wilson et al. 1993; Young and Bilby 1993). However, these improvements may be limited in experienced strength-trained individuals (Hakkinen and Komi 1985a) or in athletes who have a high pretraining vertical jump ability (Hoffman, Maresh, et al. 1991). When plyometric drills are combined with a traditional resistance training program, vertical jump performance appears to be enhanced to a significantly greater extent than with either resistance training or plyometric training alone (Adams et al. 1992).

Plyometric Training Program Design

A primary concern with beginning a plyometric training program is the increased potential for injury, because the drills place high forces on the musculoskeletal system. No epidemiological studies have addressed whether plyometric training places the athlete at a greater risk for injury in comparison with other forms of training. However, one should keep in mind that plyometric drills used by athletes comprise the same exercises commonly performed by children in the schoolyard. Hops, skips, and jumps are movements that are typically part of the elementary school physical education curriculum and have been shown to be an effective and safe way of improving physical fitness in youth (Faigenbaum et al. 2007, 2009). Interestingly, evidence supports the use of plyometric training as part of the strength training program of competitive female athletes, and plyometrics may reduce the risk for anterior cruciate ligament injury (see sidebar). As with any training program, certain precautions should be in place to ensure a safe environment for conducting these drills. The following are some recommendations for safe plyometric training:

- Use footwear and landing surfaces with good shock-absorbing qualities.
- Allow for a proper warm-up before beginning the exercise session.
- Use proper progression of drills; master lower-intensity drills before beginning more complex plyometric exercises.
- All boxes used for drills should be stable and should have a nonslip top surface.
- Make sure that space is sufficient for the given drill. For most bounding and running drills, 33 to 44 yd (30-40 m) of straightaway is required, whereas for some of the vertical and depth jumps, only 3 to 4 yd (3-4 m) of space

In Practice

Can Plyometric Training Reduce the Risk for Anterior Cruciate Ligament Injuries in Female Athletes?

A group of sports medicine physicians and sport scientists from South Korea (Yoo et al. 2010) examined the potential beneficial effects of neuromuscular training, both plyometrics and resistance exercise, on anterior cruciate ligament (ACL) injury in female athletes. In a meta-analysis they examined 2215 articles and graded each article based on its methodology. A study was considered relevant if it used a randomized control trial, was either a prospective cohort study or a retrospective case–control study or case series, or was a case series or expert opinion. Only seven studies were deemed eligible for inclusion in the analysis. To determine the effectiveness of the training on injury prevention, the investigators pooled the data and documented the numbers in the trained and untrained groups and the incidence of ACL injury. Five of the seven studies demonstrated efficacy of the training programs, while the other two did not. The meta-analysis revealed that the incidence of ACL injury in athletes performing the plyometrics and resistance training programs was lower (34 incidents among 3999 athletes) compared to no training (123 incidents among 6462 athletes). The odds ratio was reported to be 0.40 with 95% confidence levels of 0.27 to 0.60, meaning that one can state with 95% confidence that female athletes performing a neuromuscular training program have a 0.27 to 0.60 lower risk of suffering an ACL injury than individuals who do not train. Interestingly, plyometrics and strength training were found to be essential in reducing the risk of ACL injuries, while balance training was not.

is needed. For jumping drills, ceiling height should be approximately 4 yd (4 m).

- Select exercises that have a high degree of **specificity** within the athlete's sport to enhance performance gains.
- Ensure that all drills are performed with proper technique.
- Allow for sufficient **recovery** between exercise sessions, and do not perform plyometric drills when fatigued.
- Keep track of all **foot contacts** to maintain the proper **volume** and progression of training.

Similar to the development of the resistance training program, the **exercise prescription** for plyometric training involves controlling several acute program variables. These variables (intensity, volume, **frequency**, and rest) are manipulated based on the ability of the individual athlete, the fitness level of the athlete, and the training phase.

Intensity of Training

In plyometrics, the intensity of training is controlled by the type of exercise performed (Chu 1992). Because most plyometric drills use the athlete's body weight, the complexity of the drill dictates the intensity level. For instance, the intensity for jumping exercises can begin with the two-foot ankle hop as a low-intensity exercise, progress to a more moderate-intensity tuck jump with knees up, and progress even further to a high-intensity drill such as the straight spike jump. In certain situations, increasing or adding weights (e.g., weight of medicine ball) may also increase the intensity of training.

Volume of Training

The volume of training is the total work performed in an exercise session. In plyometric training, volume often refers to the number of foot contacts during each session. Foot contacts provide a means of prescribing and monitoring exercise volume, especially for drills involving jumping. For beginners in the off-season, the volume of training should be between 60 to 100 foot contacts per session. The volume of training increases to 100 to 150 foot contacts for the intermediate athlete in the off-season, and the more advanced athlete may perform 120 to 200 foot contacts during each exercise session in the off-season. Table 9.1 shows sample exercise volumes for beginning, intermediate, and advanced workouts. In drills that involve exaggerated running exercises such as bounding, the volume of training may be measured

Table 9.1 Number of Foot Contacts by Season for Jump Training

	Beginning	Intermediate	Advanced	Intensity
Off-season	60-100	100-150	120-200	Low-moderate
Preseason	100-250	150-300	150-450	Moderate-high
In-season	Sport specific	Sport specific	Sport specific	Moderate

Adapted from Chu 1992.

by distance. During the early stages of the program, distances of 30 m per repetition may be used, with progression up to 100 m during the course of the training program (Chu 1992).

The time to complete a plyometric exercise session should not exceed 20 to 30 min for a beginner. This does not include the time needed for warm-up and cool-down. As the experience level of the athlete increases and a greater number of high-intensity drills are incorporated into the training program, the time needed to complete the workout may increase.

Frequency of Training

Frequency is the number of plyometric training sessions performed per week. To date, research has not addressed the optimum number of weekly training sessions. The athlete's need for recovery has been suggested as a basis for the frequency of plyometric training (Chu 1992). Low-intensity drills such as skipping or front or diagonal cone hops may be performed daily, whereas more complex drills and higher-intensity exercises such as bounding or depth jumps may require a longer recovery period (48-72 h) before the next exercise session. Athletic teams (e.g., American football or track and field) may hold two or three plyometric sessions per week. However, this depends on the time of year. During the off-season, the frequency of plyometric exercise sessions may be relatively high because they are integrated into the off-season strength and conditioning program. However, during the season, coaches and players focus primarily on practices and maintaining strength. Plyometric sessions, although not necessarily eliminated, are substantially reduced in number. Nevertheless, common sense should prevail with the use of plyometric drills during the season. For basketball players who play several games a week and continually scrimmage during practices, plyometric exercises may be more likely to cause injury than enhance power performance. In contrast, track and field athletes who compete in a limited schedule often train through several meets to peak at the more important competitions. Maintaining a high frequency and volume of plyometric sessions during the season may be beneficial for these athletes.

Rest and Recovery

Plyometric training is designed to improve power production, not to serve as a conditioning tool. As such, adequate recovery between each repetition and set is desirable to maximize performance. The quality of each repetition must take precedence over quantity. Similar to resistance training (see chapter 8), a 2 to 3 min recovery period should separate each set to permit adequate phosphagen replenishment. Although less information is available for proper rest intervals between repetitions, the goal is to perform a high-quality repetition; therefore the athlete should rest 5 to 10 s between repetitions, especially for the more intense plyometric drills.

Integration of Plyometric and Resistance Training

An athlete is often asked to perform heavy resistance training and plyometric training together. This combination appears to be more effective in eliciting power performance than either modality alone. However, several recommendations should be followed to maximize the effect of these drills and minimize the risk of injury due to overuse. It is recommended that people avoid performing high-intensity resistance training and high-intensity plyometric training on the same day (Chu 1992). However, some athletes (e.g., track and field athletes who are experienced weightlifters) may benefit from such training (high-intensity lower body resistance exercise immediately followed by a plyometric drill). In this case, coaches need to ensure adequate recovery between the plyometrics and the next high-intensity lower body exercise (Chu 1992). In general, the ideal training program has athletes performing the plyometric exercises on the days they are doing upper body resistance training. If upper body plyometric exercises are to be included, they should be performed on days when the athlete is doing lower body resistance training. Table 9.2

presents a sample training program that integrates strength and plyometric training.

Plyometric Exercises

Plyometric drills are generally categorized by the movement required and the level of intensity. Plyometric exercises consist of jumps, box drills, bounds, and medicine ball drills. Several examples of each of these categories are provided in table 9.3, with a progression from low to high intensity. This is by no means an all-encompassing list of plyometric exercises, but it should suffice to build a good starting base.

Table 9.2 Sample of Integrated Strength and Plyometric Training Program

	Resistance training	Plyometric training
Monday	High-intensity upper body	Low-intensity lower body
Tuesday	High-intensity lower body	
Wednesday		
Thursday	Low-intensity upper body	High-intensity lower body
Friday	Low-intensity lower body	

Plyometric drills do not have to be performed immediately after the resistance exercise session. Sometimes plyometric drills are part of an off-season conditioning program that also includes sprint drills and form running, performed as a team during early morning training sessions (Hoffman et al. 1990). Presently there is a dearth of information on the ideal order of a resistance training and plyometric training program. Future training studies need to address this issue.

Table 9.3 Plyometric Drills

Full descriptions and video clips of how to perform select drills are available in the web resource at www.HumanKinetics.com/PhysiologicalAspectsofSportTrainingAndPerformance.

Drill	Intensity	Starting position	Action
Standing long jump	Low	Stand in semisquat position with feet shoulder-width apart.	With a double-arm swing and countermovement with the legs, jump as far forward as possible.
Squat jump	Low	Stand in squat position with thighs parallel to floor and fingers interlocked behind head.	Jump to maximum height without moving hands. On landing, return to starting position.
Front cone hop	Low	Stand with feet shoulder-width apart at the beginning of a line of cones.	Keeping feet shoulder-width apart, jump over each cone, landing on both feet at the same time. Use a double-arm swing and attempt to stay on the ground for as little time as possible.
Single-leg push-off with box	Low	Stand on the ground in front of a box 6 to 12 in. high. Place heel of one foot on the box near the closest edge.	Push off the top foot to gain as much height as possible by extending through entire leg and foot. Use double-arm action for gaining height and maintaining balance.
Skipping	Low	Stand comfortably.	Lift one leg with knee bent to 90° while lifting the opposite arm with elbow also bent to 90°. Alternate between sides. For added difficulty, push off ground for more upward extension.

(continued)

Table 9.3 *(continued)*

Drill	Intensity	Starting position	Action
Medicine ball throw	Low	Stand with a medicine ball at chest level.	Step forward and throw ball with both arms to a partner at a specified distance. Can also be performed as a sitting chest pass.
Front box jump	Low-moderate	Stand in front of a box 12 to 42 in. high (depending on ability) with feet shoulder-width apart.	Jump up and land with both feet on the box and step down. For a more advanced exercise, hop down and immediately hop back on top. Use a variety of box heights.
Depth jump	Low-moderate	Stand on a box 12 to 42 in. high (the higher the box height, the greater the intensity of the exercise) with toes close to edge and feet shoulder-width apart.	Step from box and drop to the ground with both feet. As soon as there is foot contact, jump explosively as high as possible. Try to have as little ground contact as possible.
Tuck jump with knees up	Moderate	Stand upright and then assume a slight bend in knees with feet shoulder-width apart.	Jump vertically as high as possible, bringing the knees to the chest and grasping them with the hands before returning to floor. Land with knees bent and return to a standing vertical position.
Lateral cone hop	Moderate	Stand with slight bend in knees and feet shoulder-width apart beside a row of 3 to 5 cones stretched out 2 to 3 ft apart.	Jump sideways down the row of cones, landing on both feet. When the row is complete, jump back to starting position.
Double-leg or single-leg zigzag hop	Moderate	Place 6 to 10 cones about 1.5 to 2 ft apart in a zigzag pattern. Begin with slight bend in knees and feet shoulder-width apart.	Jump diagonally over the first cone. On landing, change direction and jump diagonally over each of the remaining cones.
Standing triple jump	Moderate	Stand with feet shoulder-width apart, bending at the knee with a slight forward lean.	Begin with rapid countermovement and jump as far up and forward as possible with both feet, as in the long jump. On landing, make contact with only one foot and immediately jump off. Get maximal distance and land with the opposite foot and take off again. Landing after this jump is with both feet.
Multiple-box jump	Moderate	Stand in front of 3 to 5 boxes 12 to 42 in. high (depending on ability) with feet shoulder-width apart.	Jump onto the first box, then off, and jump onto the next box. Continue to the end of the line, using a double-arm action for gaining height and maintaining balance.
Depth jump to prescribed height	Moderate	Stand on a box 12 to 42 in. high (the higher the box height, the greater the intensity of the exercise) with toes close to edge and feet shoulder-width apart in front of a box of similar height.	Step from box and drop to the ground with both feet. As soon as there is foot contact, jump explosively as high as possible onto the second box. Try to have as little ground contact as possible.
Power skipping	Moderate	Stand comfortably.	With a double-arm action, move forward in a skipping motion, bringing the lead leg as high as possible in an attempt to touch the hands. Try to get as much height as possible when pushing off on back leg. Repetitions should be performed with alternate legs.

Drill	Intensity	Starting position	Action
Alternate-leg bounding	Moderate	Begin with one foot slightly in front of the other with arms at the sides.	Using a rocker step or jogging into the starting position, push off the front leg and drive the knee up and out, trying to get maximal horizontal and vertical distance with either an alternate- or double-arm action. Try to hang in the air for as long as possible. On landing, repeat with opposite leg. Goal is to cover maximal distance with each jump. This is not designed to be a race or sprint.
Pike jump	Moderate-high	Stand with slight bend in knees and feet shoulder-width apart.	Jump up and bring the legs together straight out in front of the body. Flexion should occur only at the hips. Attempt to touch the toes at the peak of the jump. Return to starting position.
Power drop with medicine ball	Moderate-high	Lie supine on the ground with arms outstretched next to a box 12 to 42 in. high, with a partner standing on box holding a medicine ball.	Partner drops ball. The ball is caught and immediately propelled back to the partner.
Split squat	High	Stand upright with feet split front to back as far as possible. The front leg is 90° at the hip and 90° at the knee.	Perform a maximal vertical jump while switching leg positions. As the legs switch, attempt to flex the knee so that the heel of the back foot comes close to the buttocks. Land in the split squat position and jump again.
Single-leg hop	High	Stand with one foot slightly ahead of the other, as in initiating a step, with the arms at the sides.	Using a rocker step or jogging into the starting position, drive the knee of the front leg up and out as far as possible while using a double-arm action. The nonjumping leg is held in a stationary position with the knee flexed during the exercise. The goal is to hang in the air as long as possible. Land with the same leg and repeat.
Multiple-box squat jump	High	Stand in front of 3 to 5 boxes 12 to 42 in. high (depending on ability) in parallel squat position with feet shoulder-width apart and hands behind head or on hips.	Jump onto the first box, maintaining squat position, then jump off and onto the next box. Continue to the end of the line. Keep the hands behind the head or at the hips.
Single-leg depth jump	High	Stand on a box 12 to 18 in. high with toes close to edge and feet shoulder-width apart.	Step from box and drop to the ground, landing with one foot. As soon as there is foot contact, jump explosively as high as possible with that single foot. Try to have as little ground contact as possible.
Single-leg bounding	High	Stand on one foot.	Bound from one foot as far forward as possible, using other leg and arms to cycle in air for balance and increase forward momentum.

Web Resource Preview

Here is a preview of one of the web resource drills. Full drill content can be found at www.HumanKinetics.com/PhysiologicalAspectsofSportTrainingAndPerformance.

Pike Jump

Purpose: Improve lower body power

Type: Moderate- to high-intensity plyometric exercise

Equipment: None

STEPS

1. Athlete stands with slight bend in knees and feet shoulder-width apart *(a)*.
2. Athlete jumps up and brings the legs together straight out in front of the body.
3. Flexion should occur only at the hips.
4. Athlete attempts to touch the toes at the peak of the jump *(b)*.
5. Athlete returns to starting position.
6. The exercise is repeated 6 to 8 times.

TIP

- To increase intensity, you can wear a weighted vest.

a

b

In Practice

Can Plyometrics Performed in the Water Be Effective?

The benefits of plyometrics training are well accepted, but the stress of repeated jumps or ballistic movements may increase the risk for soreness or injury, especially in athletes who perform jumps on a regular basis (e.g., basketball or volleyball players). A group of physical therapists examined whether the use of plyometric drills performed in the water would be efficacious in a group of female volleyball players (Martel et al. 2005). Nineteen female high school volleyball players were recruited for this study. They were randomly assigned to either an aquatic plyometric training program (*n* = 10) or a flexibility program (*n* = 9). Both programs were performed twice per week for 6 weeks. The aquatic plyometric program consisted of power skips, spike approaches, single- and double-leg bounding, continuous jumping for height, squat jumps, and depth jumps. All sessions were performed at a water depth of 122 cm and lasted for 45 min (including warm-up and cool-down performed in the water). All sessions occurred following volleyball practice. Flexibility sessions were led by study investigators and consisted of three sets of 8 to 10 exercises for major upper and lower body muscle groups. Vertical jump height was similar at baseline for both groups; but at the end of 6 weeks, the aquatic plyometric group had improved their vertical jump height by 11%, and the control group (flexibility group) had improved 4%. Results from this study indicate that a plyometric training program performed in an aquatic environment provides significant performance improvements in athletes involved in a ballistic sport. The aquatic environment appears to provide the benefits generally associated with a land-based program, but may be more desirable for reducing the stress of repeated jump training in athletes who perform a high number of jumps as part of their daily routine.

Ballistic Training

Ballistic exercises such as jump squats, bench throws, or medicine ball throws allow the athlete to accelerate a force through a complete range of motion. The benefits of ballistic movements have been demonstrated in several studies. Wilson and colleagues (1993) compared traditional resistance training, ballistic resistance training, and plyometric training in recreational athletes with at least 1 year of resistance training experience. After 10 weeks, the ballistic resistance group showed improvements in a greater number of variables tested—jump height (both countermovement jump [CMJ] and squat jump [SJ]), isokinetic leg extensions, 30 m sprint time, and peak power on a 6 s cycle test—than either the traditional resistance training (improvements in CMJ, SJ, and 6 s cycle test) or plyometric training group (improvements in CMJ only). A closer examination of the effect of each of these training methods on jump performance is presented in figure 9.5.

All three training programs resulted in significant improvements in countermovement jump performance. However, the subjects who performed ballistic training improved to a significantly greater extent than the subjects who performed traditional resistance training but not those who performed plyometric training. In the squat jump, significant pre- to posttraining improvements were realized by the subjects in the traditional resistance training and ballistic training groups, but not the plyometric group. In addition, the ballistic training group improved significantly more than the subjects in the two other groups. The results of this study suggest that ballistic training might be more effective for improving power performance than either traditional resistance training or plyometric training performed separately. The investigators suggested that the use of only body weight as the training load during plyometric drills might pose a disadvantage for optimizing power and sport performance improvements.

Ballistic training, which is a combination of traditional resistance training and plyometric training, may be the most effective method to optimize power and sport performance improvements. Mangine and colleagues (2008) compared 8 weeks of traditional resistance training to a combination of traditional resistance training and ballistic training (jump squats, bench press throws, and ballistic push-ups). All subjects performed a 4 day per week split routine (2 days upper body, 2 days lower body). At the end of 8 weeks, the combined group experienced a 13.6% improvement in jump power, while the traditional resistance training group actually decreased 9.6%. The combined group had a

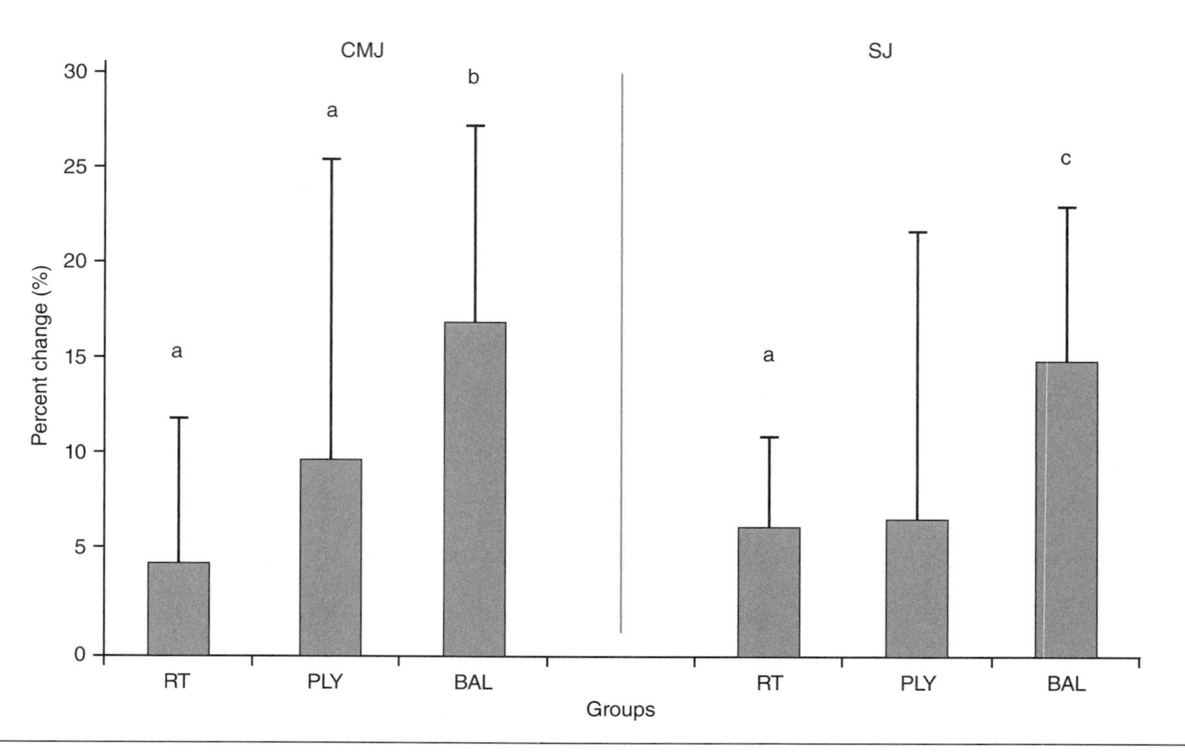

Figure 9.5 Comparison of effect of traditional resistance training, plyometric training, and ballistic training on jump performance. CMJ, countermovement jump; SJ, squat jump; RT, traditional resistance training group; PLY, plyometric training group; BAL, ballistic training group; a = significant pre to post differences; b = significantly different from RT; c = significantly different from both RT and PLY.

Data from Wilson et al. 1993.

6.2% increase in upper body power (measured from ballistic push-up performed on a force plate) while the traditional resistance–trained group improved by 1.4%. The benefits of ballistic exercise training have been shown in a number of other investigations as well (Cormie, McCaulley, and McBride 2007; Cormie, McGuigan, and Newton 2010; Winchester, McBride, et al. 2008). However, ballistic exercises may not provide any additional benefit over traditional resistance training in a weak individual or a novice (Cormie, McGuigan, and Newton 2010).

Athletic performance improvements (jump power, sprint speed) in relatively weak men were similar between subjects performing a traditional resistance training program and those performing a power training program (Cormie, McGuigan, and Newton 2010). The investigators suggested that ballistic training probably should be integrated into the training programs of athletes who have developed a solid strength foundation (squat 1RM/body mass ratio >1.60). For weaker athletes, a focus on basic strength improvement will increase both strength and power. As discussed earlier in the chapter, the experienced strength-trained athlete needs a greater stimulus to make additional improvements in power performance. However, this stimulus may need to be provided for more than 5 weeks. Hoffman, Ratamess, and colleagues (2005) added a jump squat exercise to the training program of competitive football players during the power phase of the off-season conditioning program and compared them to another group of players who did the same training program but without the squat jump exercise. Although significantly greater improvements were noted in 1RM squat in the jump squat group (using 70% of pretraining 1RM squat), no differences were noted between the groups in vertical jump height, 40 yd sprint speed, or agility. The jump squat was performed on a jump squat machine; but the relatively high intensity used for training may have been a disadvantage with respect to stimulating power gains by minimizing velocity of movement compared to lower intensities of training. In addition, a short-duration training program may not be long enough to elicit significant changes in a trained group of athletes.

Ballistic exercises are incorporated into the traditional resistance training program of athletes. These exercises are often integrated in the strength–power, power, and peaking phases of the annual training program. High-velocity ballistic movements (limiting deceleration of the resistance) selectively recruit higher-threshold, fast-twitch motor units, which are also critical to maximal strength development (Kawamori and Haff 2004). It has been shown that peak power outputs are attained with use of a wide range of loads (between 15% and 60% of the athlete's 1RM) for ballistic exercises such as the jump squat and ballistic bench press (Baker, Nance, and Moore 2001a, 2001b; Kawamori and Haff 2004; Mangine et al. 2008; Wilson et al. 1993). The large range in intensity used in these exercises has been attributed to differences in muscle mass recruitment, single joint versus multiple joints, training experience, methods used to measure power output, and strength level of the subjects (Dugan et al. 2004).

Olympic Weightlifting

The snatch, the clean and jerk, and their variants (high pulls, power cleans, power snatch, hang clean, and hang snatch) are generally considered the Olympic weightlifting movements. Similar to what occurs with ballistic exercises, the athlete doing these exercises accelerates the bar through the propulsion phase (second pull) that often results in projection of the bar and or the body into the air (Newton, Cormie, and Kraemer 2012). It has been suggested that the loading that elicits the highest power outputs is between 70% and 80% 1RM in these exercises (Cormie et al. 2007; Kawamori et al. 2005). An important benefit of incorporating the Olympic movements into the training program of strength and power athletes is the similarity of these exercises to athletic movements such as jumping and sprinting (Hori et al. 2005).

A study comparing Olympic weightlifting to traditional powerlifting training in competitive strength–power athletes demonstrated a clear advantage for Olympic lifting in this athletic population (Hoffman, Cooper, Wendell, and Kang 2004). During a 15-week study, both training programs, using a linear periodization model (see chapter 14), resulted in significant improvements in squat strength. However, strength improvements in the Olympic lifting group were 18% greater than in the powerlifting group. In addition, athletes who trained with the Olympic weightlifting exercises experienced a twofold improvement in time for the 40 yd sprint compared to the powerlifting group (0.07 ± 0.14 s and 0.04 ± 0.11 s, respectively). Although these results were not statistically different, the practical differences for a competitive group of athletes were impressive. A significantly greater improvement in vertical jump height was also seen in the Olympic lifting group compared to the powerlifting group. This was later confirmed by others in high school athletes (Channell and Barfield 2008).

Summary

This chapter discusses the scientific basis of power training. In addition, it demonstrates the benefits of plyometric, ballistic, and Olympic exercises in the training programs of athletes interested in maximizing power production. It appears that these exercises may have a greater relevance once the athlete has developed a strength base. These exercises can be successfully integrated into the resistance training programs of strength–power athletes. In addition, as the window of adaptation is reduced for slow-velocity strength (traditional resistance training programs), the combination of resistance training and these other forms of training may provide a new window for adaptation that further enhances power production.

REVIEW QUESTIONS

1. What is the benefit of ballistic training compared to traditional resistance exercises?
2. Describe the stretch–shortening cycle and its importance in power generation.
3. Define the rate of force development and its importance to athletic performance.
4. Provide an example of integrating plyometric exercises into the resistance training program of strength–power athletes.
5. What are the benefits associated with the incorporation of Olympic weightlifting exercises in the resistance training program of strength and power athletes?

chapter 10

Anaerobic Conditioning

CHAPTER OBJECTIVES

After reading this chapter you should be able to do the following:

- Understand the benefits of anaerobic conditioning.
- Describe drills to enhance the anaerobic energy system.
- Explain how technological advances can enhance the anaerobic exercise prescription.
- Differentiate between designing anaerobic conditioning programs for team sport athletes versus track and field athletes.

Physiological adaptations are specific to the exercise program. Thus, to achieve the desired training outcome, the strength and conditioning professional needs to adhere to the basic principle of training specificity. To maximize the training effect and achieve the desired training goal, athletes need to recruit the primary energy system that fuels their sport. Training programs for anaerobic athletes must focus on developing the anaerobic energy system. The specific training adaptations that result from such programs maximize the athlete's ability to perform high-intensity activity with rapid recovery between each exercise bout (i.e., repeated-sprint ability). This chapter focuses on the development of the conditioning program for athletes participating in anaerobic sports. Additional focus is on monitoring anaerobic performance and how technological advances can be used to enhance the sensitivity of these monitoring systems. The final section of the chapter provides examples of anaerobic conditioning drills.

Importance of Anaerobic Conditioning

Part I of this book is devoted to the physiological adaptations to various training programs, including anaerobic training. Table 10.1 lists the significant physiological adaptations generally seen in athletes who train anaerobically. These adaptations help the athlete perform high-intensity activity with rapid recovery between each exercise session. This enables the athlete to perform repeated bouts of exercise with minimal reductions in performance. To develop the most effective program, it is important for coaches to understand the physiological demands that the athlete experiences during competition. However, very little is known about the physical demands and physiological responses of athletes playing team sports such as basketball or American football. This is in contrast to the situation with individual-sport athletes, such as sprinters, who typically perform an isolated event requiring maximal effort. The physiological demands placed on these athletes are easier to understand; consequently, the design of their training programs is more straightforward. This is not to say that developing a training program for an individual-sport athlete is easy. Rather, the **needs analysis** may be clearer, and thus it may be possible to make the **exercise prescription** more specific to the demands of the event.

In team sports such as American football, basketball, or hockey, high-intensity activity (short sprints with various changes in direction) is performed

Table 10.1 Potential Physiological Adaptations to High-Intensity Anaerobic Training

Increase in the transformation of type II fibers to a more glycolytic subtype
Significant elevations in glycolytic enzymes (phosphofructokinase, phosphorylase, lactate dehydrogenase)
Increase in maximum blood lactate concentrations
Reduced blood lactate concentrations during submaximal exercise
Improved buffering capacity

repeatedly for the duration of the game. However, the type of movement patterns and the duration of activity may vary. In addition, different positions on the team may experience different demands and require separate training programs. The goal for coaches developing training programs is to bring each athlete to a level of conditioning that allows for the maintenance of maximum performance during each spurt of activity throughout the game. A vital issue, however, is proper assessment of the activity requirements of each position in each sport. Armed with a thorough understanding of the activity demands of the sport, the coach is able to maximize the effectiveness of the training program through greater **specificity** in the types of exercises and in the **work/rest ratio**.

As already mentioned, limited data are available concerning the physiological demands placed on athletes competing in anaerobic sports such as basketball or football. However, several published studies provide at least a starting base of information to assist in program development. The types of movements, the **intensity** of these movements, the number of consecutive plays, the number of groupings of plays, and the length of rest between plays are all variables that could provide for a more specific exercise prescription.

Recent advances in technology have given sport scientists and coaches an opportunity to understand the movement patterns within a contest, the intensity of the movements (including number and distances of sprints), and the total distance covered during the contest. Use of global positioning systems (GPS) (figure 10.1) has provided a unique advantage with regard to developing sport-specific training programs. GPS technology is based on the use of satellites that are in orbit around the earth. These satellites emit a signal at the speed of light that provides the exact time and its location. The GPS receiver then translates the signal into time and distance using trigonometry. Through calculation of the distance of multiple satellites, the exact position of the receiver can be determined (Townshend, Worrington, and Stewart 2008). Although this technology was originally intended for military use, the development of nondifferential GPS technology during the past decade opened up its potential for civilian use. Some of the earliest publications on potential uses of GPS appeared in 2000 (Schutz and Herren 2000), and it was suggested as a possible resource for sport assessment in 2003 (Larsson 2003). Early issues with GPS technology were the reliability and validity of the device. As the technology of sport-based use of GPS has improved, generally through increasing the frequency of data collection from 1 Hz to 10 Hz, the accuracy, reliability, and validity have been enhanced. The reliability and validity of GPS devices to measure distance traveled and intensity of runs during a field-based team sport have been demonstrated in a number of studies (Coutts and Duffield 2010; Duffield et al. 2010; Gray et al. 2010; Macleod et al. 2009).

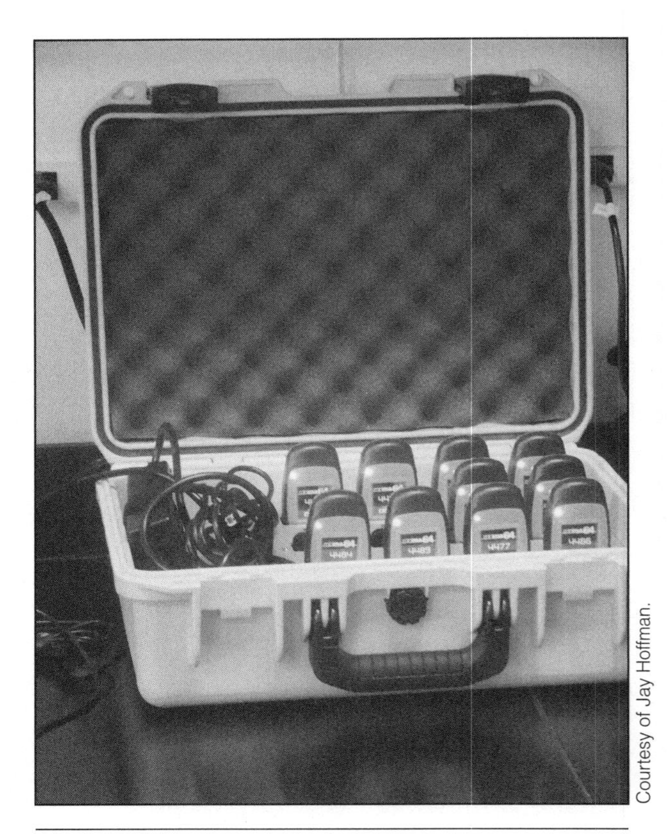

Courtesy of Jay Hoffman.

Figure 10.1 GPS technology.

GPS technology can also be used as an alternative to electronic timing in assessing speed performance. Barbero-Alvarez and colleagues (2010) demonstrated a significant correlation ($r = -0.96$) between speed from a GPS device and electronic timing systems. The benefit for coaches is that they can calculate the running speed of athletes during actual game performance and determine fatigue patterns; these data can provide information regarding the need for player substitution or for developing exercise prescriptions for season preparation. Table 10.2 consists of data collected from elite rugby union players. The analysis separates players by position as well as by half. The game went 83 min, and the players spent 72% of the time either standing or walking, 18.6% jogging, 7.1% cruising or striding, and 2.2% performing high-intensity runs or sprints. Further analysis showed that the majority of high-intensity activity was 4 to 6 s in duration with little difference between positions. The average heart rate during the game was 172 beats/min for backs and 170 beats/min for forwards. The results indicate that competitive rugby union games involve intermittent high-intensity runs for a prolonged duration. This technology provides the coach with the ability to simulate actual game activity during off-season and preseason conditioning sessions. For instance, the variability in these types of runs with varying distances can be easily incorporated into a fartlek training program, or the use of work/rest intervals seen in competition can be the basis for the development of sport-specific interval training programs.

GPS technology can also be used to determine fatigue patterns in competition. Figure 10.2 shows changes in high-intensity running distance by quarter in Australian rules rugby players. Significant decreases in high-intensity run distance were seen between the first and second quarter and between the first and third quarter. High-intensity run distance was significantly lower in the fourth quarter compared to the first three quarters, clearly demonstrating fatigue. The higher-intensity play exhibited in the first quarter was not replicated in the rest of the game. This could have important implications for team success and may be an area of concern that coaches can address. For instance, perhaps better conditioning or more sport-specific conditioning needs to be incorporated into the training program, or perhaps the fatigue is a manifestation of an overtraining syndrome (see chapter 24). The bottom line is that fatigue can be an indication of a need to change substitution patterns or player rotations to keep players fresher throughout the game. Additional analyses are shown in figures 10.3 and 10.4.

The use of technology is always helpful, but technology is not the only method that can be used

Table 10.2 Total Distance and Distance per Designated Speed Zone in Elite Rugby Union Players

		First half	Second half
Total distance (m)	Backs	3318	3542
	Forwards	2931	3236
Standing and walking (0-6 km/h)	Backs	1247	1314
	Forwards	1124	1110
Jogging (6-12 km/h)	Backs	794	1065
	Forwards	722	948
Cruising (12-14 km/h)	Backs	332	330
	Forwards	310	362
Striding (14-18 km/h)	Backs	532	439
	Forwards	479	481
High-intensity runs (18-20 km/h)	Backs	172	120
	Forwards	138	177
Sprints (>20 km/h)	Backs	241	283
	Forwards	157	159

Data from Cunniffe et al. 2009.

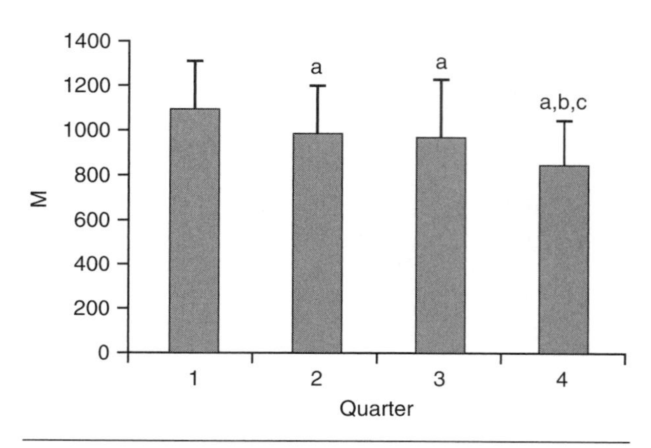

Figure 10.2 High-intensity run distance in Australian rules rugby. a, b, and c = significantly different from 1Q, 2Q, and 3Q, respectively.

Data from Coutts et al. 2010.

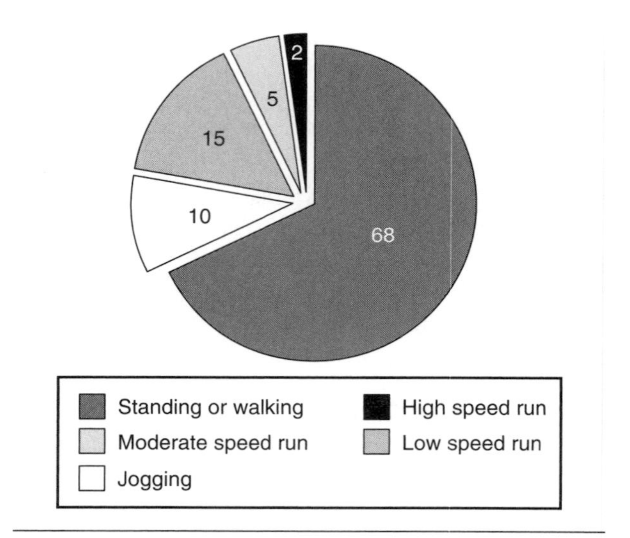

Figure 10.3 Percent time spent in various movement intensities in National Collegiate Athletic Association Division I women's soccer activity.

Research data from J. Hoffman.

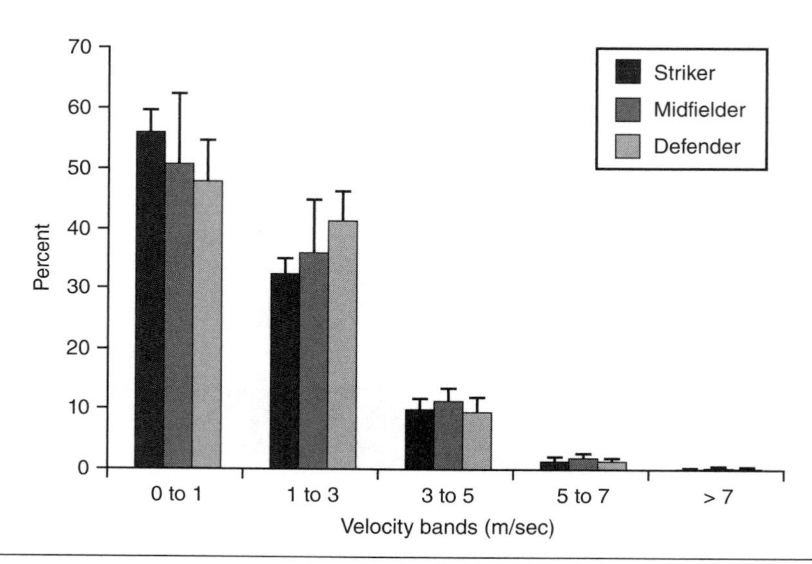

Figure 10.4 Velocity bands in elite women's field hockey.

Data from Gabbett 2010.

to develop appropriate exercise prescriptions. The following section provides an alternative method for developing the anaerobic exercise prescription for two popular anaerobic team sports (basketball and football). This then serves as a basis for the anaerobic exercise prescription.

Basketball

The game of basketball has a continual flow of play, with a smooth transition from offense to defense. All players have shared responsibilities (e.g., rebounding and shooting) that require them to perform similar movements on the basketball court. These movements range from jumps to runs (from a jog to a sprint) and shuffles (backward and side) at various degrees of intensity. A study by McInnes and colleagues (1995) was the first to categorize the movement patterns of a basketball game. In that study, the researchers separated the movement patterns into eight different categories (stand-walk; jog; run; stride-sprint; low, medium, or high shuffle; and jump). Their results illustrated the intermittent nature of basketball by demonstrating 997 ± 183 changes in movement during a 48 min game. This equated to a change of movement every 2 s (players averaged 36.3 min played per game). Shuffle movements (all intensities) were seen in 34.6% of the activity patterns of the game, and running (intensities varying from a jog to a sprint) was observed in 31.2% of all movements. Jumps were reported to occur in 4.6% of all movements, whereas standing or walking was observed during 29.6% of the playing time. Movements characterized as high intensity were recorded once every 21 s of play. When high-intensity shuffles and jumps were considered, only 15% of the actual playing time was spent engaged in activity that could be categorized as high intensity. However, the mean blood lactate concentration during the game was 6.8 ± 2.8 mmol, and heart rates were reported to be at 85% of peak for more than 75% of the actual play and at 95% for 15% of the game. Although the majority of activity in the game appeared to be performed at intensities considered aerobic, some of the physiological measures indicated the highly anaerobic characteristics of the sport.

It should be noted that the subjects in the study were Australian National League players; their style of play may be different from that seen in other organizations (e.g., National Collegiate Athletic Association [NCAA], National Basketball Association). Hoffman and colleagues (1996) analyzed 4 years of basketball at a NCAA Division I institution and reported that speed and other anaerobic performance variables were positively correlated to playing time, whereas aerobic capacity had a significant negative correlation to playing time. Thus, it appears that for basketball, the athlete's training program should place a large emphasis on improving anaerobic conditioning.

An anaerobic exercise prescription specifically for basketball players can be developed. In general, the training of the anaerobic energy system does not start until the preseason training program begins (Hoffman and Maresh 2000). Until this time, the athlete's off-season conditioning program should focus primarily on resistance training, sport-specific skill development (including agility and speed), and playing basketball (scrimmages or summer leagues). Anaerobic conditioning is avoided until the preseason phase to prevent the possibility of **overtraining syndrome** during the season (see chapter 24). During the preseason (approximately 6 weeks long at the collegiate level), the goal is to bring the athlete not necessarily to peak condition, but close to it. Once official basketball practice starts, the team begins practicing with the coaching staff. This period before the competitive season is also referred to as the preseason; however, for the athlete, it is preseason 2. Depending on the specific situation, many basketball coaches devote much of this time to conditioning drills. During this phase of training, it may be wise for the strength and conditioning specialist to continue to leave room for improvement. The problem of **fatigue** or **overreaching** may develop if the athlete peaks too soon.

An example of a preseason anaerobic conditioning program is presented in table 10.3. This program is performed 4 days per week with a progression in

Table 10.3 Example of an Anaerobic Conditioning Program for Basketball

	Day 1	Day 2	Day 3	Day 4
Workout	Intervals	Sprints (distance × repetitions)	Intervals	Sprints (distance × repetitions)
Weeks 1 and 2	3 or 4 laps	400 m × 1 100 m × 2 30 m × 8 Work/rest ratio = 1:4	3 or 4 laps	200 m × 4 or 5 Work/rest ratio = 1:4
Weeks 3 and 4	4 or 5 laps	400 m × 1 100 m × 3 or 4 30 m × 8 to 10 Work/rest ratio = 1:4	4 or 5 laps	200 m × 5 or 6
Weeks 5 and 6	5 or 6 laps	400 m × 2 100 m × 4 or 5 30 m × 10 to 12 Work/rest ratio = 1:3	5 or 6 laps	200 m × 6 or 7 Work/rest ratio = 1:3

both **volume** and intensity. It is important to note how the intensity of the workouts is manipulated via changes in the work/rest ratio during the sprint training. The aim in reducing the work/rest ratio is to improve the recovery time between bouts of high-intensity activity typically seen during a basketball game. In addition, as discussed previously, there is a great deal of variability in the movement patterns and intensity levels of these movements. Thus, to provide for a greater similarity to the game, it may be wise to design the training program with exercises that simulate such changes. For example, interval or fartlek training may become an integral part of the training program. Considering the variation in intensity and length of sprints that these drills can employ, they provide a better opportunity to simulate the changes that may occur in a game of basketball. Specific descriptions of these types of training exercises are presented later.

Football

Football is primarily a game of repeated maximal-intensity bouts of exercise. Every player on the football field has specific responsibilities that may vary considerably (e.g., lineman vs. wide receiver). Thus the physical demands experienced by each player are also different. However, each player must perform these responsibilities at his maximum ability at all times. It has been suggested that the anaerobic energy system is the principal energy system used by the body during a game of football (Gleim, Witman, and Nicholas 1984; Kraemer and Gotshalk 2000). Up to 90% of the energy production during a football game is provided by the **phosphocreatine system**, and the remaining energy production is the result of **glycolysis**.

A football game can be separated into a series of plays. Figure 10.5, *a* and *b,* shows the sequence of series in a NCAA Division III football game. In this particular game, each team had 15 series on offense with an average of 4.5 plays per series (range 1-12). Table 10.4 shows the average number of series and plays seen in nine games (entire season) of Division III football. During each game, an average of 14.4 offensive series per team occurred, with an average of 4.6 plays per series. National Football League (NFL) teams ran approximately one more play per series than reported for the college football teams (between 5.3 to 5.6 plays per series). Each play is reported to last for an average of 5.49 s (ranging from 1.87 to 12.88 s) in college football (Kraemer and Gotshalk 2000), whereas the average NFL play is reported to be 5.0 s in duration (Plisk and Gambetta 1997). Between each play, each team has a maximum of 25 s. However, the time clock does not begin until the referee has set the ball. Thus, the rest interval between plays generally exceeds 25 s in duration. In limited reports, the average time between plays in a college football game was 32.7 s (Kraemer and Gotshalk 2000), whereas in the NFL, the average rest interval between plays has been reported to range between 26.9 and 36.4 s (Plisk and Gambetta 1997). The average time per play and rest time between plays allow for a more precise development of the work/rest ratio needed in the anaerobic exercise prescription. Thus, according to the data, it appears that a work/rest ratio of 1:5 could be used in the off-season conditioning programs for football players performing short-duration sprints that simulate the movement patterns of an actual football game.

An anaerobic exercise prescription specifically for football players can be developed. During the off-season conditioning program, the primary emphasis should be on resistance training. In addition, specific speed and agility exercises are often incorporated into the normal winter conditioning programs (Hoffman et al. 1990). However, these drills are not designed to bring the athletes' anaerobic conditioning to in-season levels. This is reserved for the preseason conditioning program, which emphasizes anaerobic conditioning to maximize the athletes' readiness to compete. This conditioning program begins 6 to 10 weeks before training camp for reasons similar to those discussed with regard to preparing the basketball player. The type of drills and progression of volume and intensity are also similar those shown in table 10.3. However, one can make specific adaptations for the football player. For example, for the college athlete it appears that there are four or five plays per series that last approximately 5 s (see table 10.4). Given that there are three or four series per quarter, one can develop a conditioning program that would simulate a quarter of a football game with realistic work/rest ratios. In addition, incorporation of a range of sprinting distances can simulate the varied runs that are frequently seen in a game. Plisk and Gambetta (1997) have suggested that this type of training program can also incorporate both "successful" (sets of 10 plays or more) and "unsuccessful" (set of four or five plays) series.

Athletes in Anaerobic Individual Sports

In comparison to the conditioning programs for team sports such as football or basketball, the exercise prescription for athletes participating in an individual

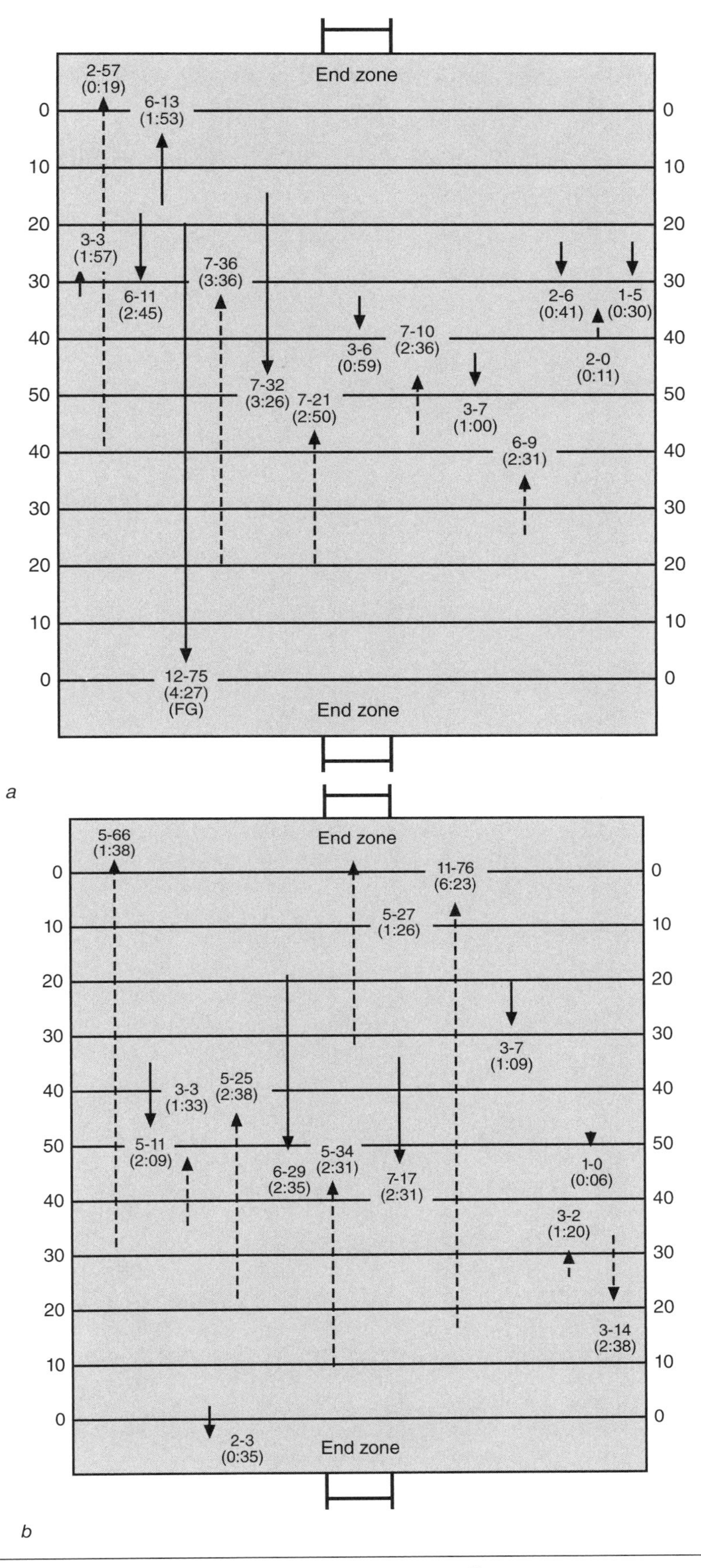

Figure 10.5 *(a)* Drive chart for first half. *(b)* Drive chart for second half.

Table 10.4 Average Number of Series and Plays in NCAA Division III Football (Nine Games Observed)

Observations	Total number
Plays observed	1193
Series observed	259
Series per game	14.4
Plays per series	4.6
Percentage of series 6 plays or greater	31.2%
Percentage of series 10 plays or greater	8.1%

event such as sprinting is easier to develop. Unlike basketball or football players, for whom the movements and intensity of action may vary, athletes such as sprinters have a more straightforward needs analysis. A sprinter is required to run a single sprint at maximum ability during a competition. The sprinter may compete in several different races, but the requirements among them are similar. Any race category that includes distances up to 440 yd (400 m) is considered sprinting (Carr 1999); distances greater than 440 yd (400 m) are considered middle-distance events. The training program for the sprinter focuses primarily on developing power, improving running technique and speed, and increasing speed-endurance. Table 10.5 presents an example of a training program for sprinters.

The training program for the sprinter is unlike that of the basketball or football player, whose primary emphasis during anaerobic conditioning is to prepare for repeated bouts of high-intensity activity with limited rest intervals. In contrast, the sprinter's training program is less concerned with improving fatigue rate and more concerned with the quality of each sprint. Table 10.6 depicts the rest intervals recommended for enhancing speed-endurance for the 400 m sprinter. Notice the long time intervals between sprints. Obviously, the goal for the 400 m sprinter is maximizing the quality of each sprint.

Selection of Rest Intervals for Anaerobic Conditioning

One of the most important components in the development of the anaerobic exercise prescription is the selection of the rest interval. As discussed earlier, the work/rest interval can easily be determined based on the analysis of the sport and positions within the sport. However, that specifies the goal. The question really is, What should the progression be? For example, if the goal of the strength and conditioning professional is to achieve a 1:3 work/rest ratio, this would not necessarily be the starting point of the anaerobic conditioning program. Data actually comparing different rest intervals for anaerobic conditioning are limited. A rest interval can be defined as the time between each sprint and as the time between each workout. The latter has to do primarily with recovery but has important implications for determining the physiological adaptation desired and the time course for that adaptation.

Parra and colleagues (2000) compared 14 consecutive training days of high-intensity sprints to a program comprising the same 14 training days but with a 2-day rest period between exercise sessions (total training time of the second group was 6 weeks). This latter training program consists of a little more than two training sessions per week. Enzymatic changes of skeletal tissue suggested that both programs resulted in significant increases in enzymes related to glycolysis such as phosphofructokinase (107% increase in the shorter program and 68% in the longer-duration program). The shorter training program resulted in significant elevations in creatine kinase, lactate dehydrogenase, and pyruvate kinase; the longer-duration program did not yield similar increases. The shorter program with limited recovery also resulted in a decrease in anaerobic adenosine triphosphate (ATP) consumption, a decrease in glycogen degradation, and no change in exercise performance. The longer program resulted in significant improvements in sprint power performance without any change in anaerobic ATP consumption, glycolysis, or glycogenolysis rate. The results clearly demonstrated that the longer-duration training was able to achieve the desired anaerobic adaptations along with performance improvements. The extra days between training sessions appear to provide the recovery necessary to maximize high-intensity exercise performance.

Table 10.5 Example of a Training Program for a 100 m or 200 m Sprinter

Day 1	Day 2	Day 3	Day 4	Day 5	Day 6
Preseason					
Sprint drills Speed development Turnaround 40 m or 50 m sprints Falling acceleration starts	Interval training	Sprint drills Running technique Block starts and starts to 30 m Stick drill running	15 min jog	Sprint drills Running technique Speed development Stick drill Rollover starts	2 × 300 m, 4 to 6 × 100 m uphill runs
Early season					
Sprint drills Speed development Turnaround 40 m or 50 m sprints	Sprint drills Speed stick running 4 × 60 block starts 4 × 30 flying starts Bounding with a weighted vest	Technique work 4 × 100 m relaxation strides	Sprint drills 4 × 120 m and 6 × 50 m weighted sled pulls	Sprint drills 4 × 20 m starts 2 × 100 m smooth striding	Early competition
Late season					
100 m athletes: 5 × 200 m at 90% of max, 3 min between sprints 200 m athletes: 4 × 300 m at 85% to 90% of max, 10 to 12 min of recovery between sprints	Sprint drills Speed stick drill 2 × 100 m acceleration drill 2 × 120 m strides	6 × 100 m relaxation strides	Sprint drills 3 × 30 m block starts 1 × 150 m at 100% of max	Warm-up only	Most important competition

Data from USA Track and Field 2000.

Table 10.6 Speed–Endurance Training for a 400 m Sprinter

Number of sprints	Distance of each sprint (m)	Recovery time between sprints (min)
10	100	5-10
6	150	5-10
5	200	10
4	300	10
3	350	10
2	450	10

The distance of each sprint per workout can vary. However, the total distance run per workout should be approximately 2.5 times the distance of the athlete's event. Thus, an athlete who is a 400 m sprinter would run 1000 m in sprints per workout. The rest period is long to ensure complete recovery.

Data from USA Track and Field 2000.

Anaerobic Conditioning Exercises

A number of exercises can be used as part of a conditioning program designed to prepare an anaerobic athlete for competition. These types of drills are often thought of as exercises designed to enhance **speed-endurance** and are traditionally used to increase or maintain speed during long-duration sprint events. For drills designed to enhance repeated-sprint ability, it has been shown that short sprints may be more appropriate than shuttle types of runs (Pyne et al. 2008). However, in sports that incorporate intermittent sprints of varying distances and changes in direction, one should employ several types of drills that simulate actual sport performance using the work/rest intervals common to the given sport. Drills like this have also been described as metabolic conditioning, which is a broader term for anaerobic conditioning or anaerobic endurance.

Interval Sprints

Interval sprints are an excellent way to develop anaerobic endurance. This type of drill can be performed on a 400 m track. Typically the athlete sprints the straightaways and jogs or walks the turns (figure 10.6). A 100 m sprint followed by a 100 m

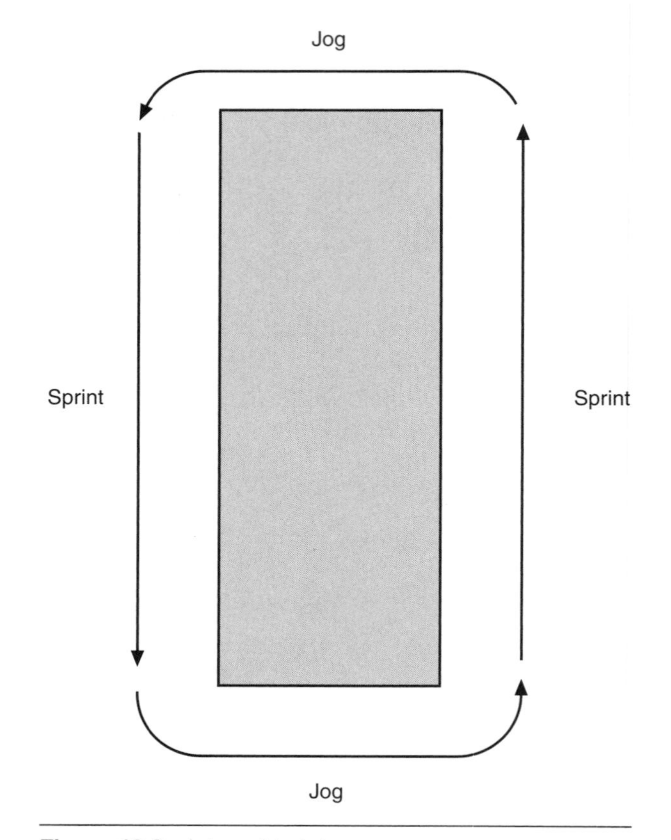

Figure 10.6 Interval training.

In Practice

Comparison Between Interval Training and Repeated-Sprint Training

A group of sport scientists from Germany compared the benefits of interval versus repeated-sprint training in competitive male tennis players (Fernandez-Fernandez et al. 2012). Thirty-one male tennis players competing at the national level were randomly assigned to one of three groups: a control group that maintained their normal training routine, a high-intensity interval training group, and a repeated-sprint training group. The two intervention groups exercised three times per week for 6 weeks. All training sessions were performed on the tennis courts. The high-intensity interval training group performed three sets of three 90 s runs of maximal intensity with a 3 min active recovery. The repeated-sprint group performed three sets of ten 5 s sprints with a 15 s passive rest between each sprint. Between each set all subjects played a 2 on 1 tennis match. Players performed as the single player on one occasion during each workout. Results showed significant improvements in $\dot{V}O_2$peak in both the interval (6.0%) and repeated-sprint (4.9%) groups. These improvements were significantly greater than in the control group. The interval group also experienced significant improvements (28.9%) in tennis-specific endurance; these improvements were significantly greater than those experienced by the repeated-sprint group (14.5%). Repeated-sprint training also resulted in a significant improvement (3.8%) in repeated-sprint ability (e.g., reduction in mean sprint time). Neither training method affected jumping or sprinting ability. The investigators concluded that benefits were seen with both methods of training.

jog is continued for the length of the workout. The length of the workout and of the rest period (jog or walk) depends on both the conditioning and the performance level of the athlete.

Fartlek

Fartlek can be performed on either a track or a cross country course of 2 to 3 mi (about 3 to 5 km). The athlete alternates short bursts of sprinting with jogging. The length of the sprint can be alternated between short and long distances with appropriate adjustments to the rest interval between each sprint. Generally, the same relative work/rest ratio can be maintained for both long and short sprints.

Repetition Sprints

Repetition sprints require the athlete to perform maximum sprints for a given distance. The distance can be either short (e.g., 22-44 yd [20-40 m]) or long (e.g., 110-440 yd [100-400 m]). After a passive rest, the athlete repeats the sprint. The number of repetitions and the work/rest ratio depend on the conditioning level of the athlete.

Repetition Sprints From Flying Starts

Although similar to the repetition sprint drill, **repetition sprints from flying starts** are different in one way. The athlete begins each sprint from a running start and accelerates over 22 yd (20 m) before maintaining the sprint for the required distance. The number of repetitions and the work/rest ratio are again dependent on the conditioning level of the athlete.

Repetitive Relays

Repetitive relays use a group of athletes who form a relay team (figure 10.7). Athlete A sprints to and tags athlete B, who races to athlete C. Athlete C sprints to athlete D, and this process continues for the length of the track. The athlete remains in the position of the runner whom he or she replaced. It is possible to make this drill competitive by having relay teams compete against each other. The number of repetitions depends on the conditioning level of the athletes. The work/rest ratio is controlled by the number of members on the relay team. For instance,

In Practice

Repeated-Sprint Ability Improves Peak Running Speed

Sport scientists from Australia examined the effect of repeated-sprint training in 29 Australian Football League players (Hunter et al. 2011). Subjects were randomly assigned to one of three groups: a control group, a constant-volume group, or a linear increase in volume group. All three groups performed regular football training twice per week during the 4-week study period. Before each football practice, subjects in the constant-volume or linear-volume group completed a high-intensity repeated-sprint training program. The constant-volume group performed an intermittent peak running speed protocol: They began the session by jogging for 15 m (50% maximal intensity), then accelerated to peak speed over 20 m, sustained that velocity for 10 m, decelerated for 20 m, and finally jogged the final 15 m. Each athlete performed 10 sprint repetitions per session. The linear increase group began the protocol using the same 10 repetitions but increased by two repetitions per week (a total of 16 repetitions were performed in the final week). The control group did only the football training. Following the 4-week training period, athletes in the constant-volume and linear-volume groups significantly improved their intermittent peak running speed by 5.1% and 3.7%, respectively. No differences were noted in the control group. A significant improvement in 300 m shuttle time was experienced by the linear-volume group only, and a significant improvement in peak running speed was observed in the constant-volume group only. The investigators concluded that eight sessions of repeated-sprint training improved peak running speed, intermittent peak running speed, and 300 m shuttle time. Interestingly, football practice alone was not sufficient to improve any of these performance variables, highlighting the importance of adding specific sprint training protocols to the preparation of anaerobic athletes.

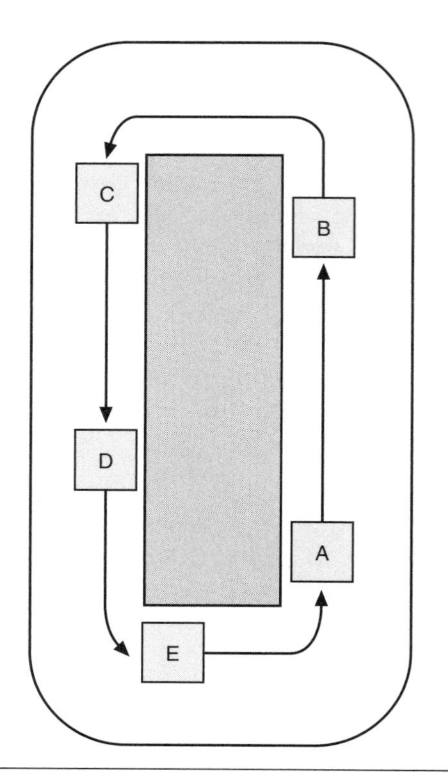

Figure 10.7 Repetitive relays.

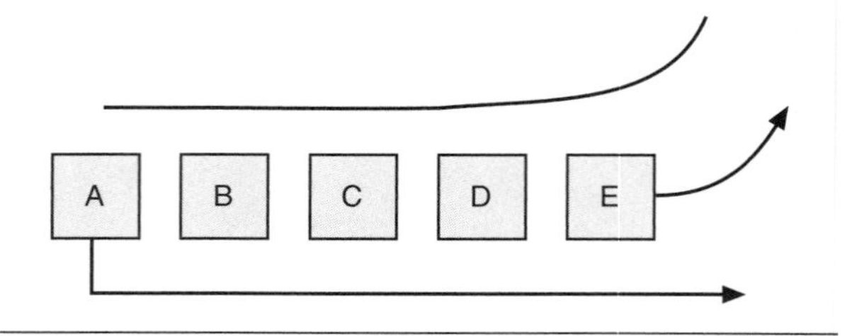

Figure 10.8 Rolling sprints.

if it is assumed that each member of the relay team is similar in speed, having five relay members would result in an approximately 1:4 work/rest ratio.

Rolling Sprints

Rolling sprints are performed with at least four athletes who jog or run slowly one after the other around the track (figure 10.8). On the coach's signal, the last athlete sprints to the front of the line. As the athlete gets to the front, the coach signals again, and the athlete who is now last sprints to the front. This continues for the duration of the run. To increase the intensity of the run, the coach can reduce the time between signals or add more runners to the group.

Summary

Analysis of each sport provides important information regarding the primary energy system that fuels performance. Thus the training program can be specifically developed to stimulate specific adaptations that enhance sport-specific performance. Technological advances have provided very sensitive measures for both analyzing and monitoring training needs and adaptations. Results of these analyses have clearly identified training differences between anaerobic athletes who play team sports such as football or basketball and anaerobic athletes training for a specific event such as a 400 m sprint. In the former situation, emphasis is on preparing the athlete to play a complete game that involves repeated bouts of high-intensity activity with limited rest periods. For the sprinter, emphasis is on speed-endurance so that minimal fatigue occurs toward the later stages of the sprint. One of the primary differences between the training programs of the sprinter and those of the football or basketball player is the emphasis. The sprinter is interested in the quality of each individual sprint, whereas basketball and football players focus primarily on the quantity of high-intensity activity common to their sport.

REVIEW QUESTIONS

1. Specify three physiological adaptations that are common to anaerobic exercise programs.
2. Describe the benefits of using global positioning systems for analyzing athletic performance.
3. Explain the differences between training programs of the basketball player and the 100 m sprinter.
4. Describe the difference between interval and fartlek sprints.
5. Provide an anaerobic training program for an ice hockey player. Be specific about exercises, work/rest ratio, training frequency, and training duration.

chapter 11

Speed and Agility Development

CHAPTER OBJECTIVES

After reading this chapter you should be able to do the following:

- Understand the interaction of stride length and stride rate with regard to speed.
- Develop a training program to enhance speed development.
- Develop a training program to enhance agility performance.
- Understand limitations of speed and agility development.

The importance of both **speed** and **agility** for determining the success of an athlete or team is well acknowledged among coaches of **anaerobic** sports such as American football, basketball, or hockey. Several studies have demonstrated that both speed and agility performance can differentiate between starters and nonstarters in National Collegiate Athletic Association (NCAA) Division I football (Fry and Kraemer 1991; Black and Roundy 1994). In addition, speed and agility have been reported to have a significant positive relationship to playing time in NCAA Division I basketball players (Hoffman et al. 1996). A major question among strength and conditioning professionals is whether speed and agility can be improved during an athlete's career. This chapter discusses speed and agility, exercises designed to improve these aspects of performance, and realistic expectations associated with speed and agility training.

Speed Development

Speed can be defined as the ability to run a specific distance in a particular time (Hoffman and Graham 2012). Good genetics plays a critical role in speed expression in the athlete. Long limbs and a high percentage of fast-twitch muscle fibers provide both a physiological and a biomechanical advantage with respect to speed. Being fast also depends on skill and technique, as well as the rate at which force and power can be generated. It is these extrinsic variables that are the focus of training programs designed to improve speed.

Speed training is part of the periodized training program for athletes who participate in sports in which speed plays a significant role in success (e.g., American football, baseball, basketball, lacrosse, and soccer) (Hoffman and Graham 2012; Little and Williams 2005; Verkhoshansky 1996). Speed can be improved through enhancing the tension and the length–impulse response involved in the rapid exchange between eccentric and concentric muscle action during sprint activity (Bosco and Vittori 1986; Komi 2003). The use of plyometrics or explosive strength training is advantageous to speed development due to the improvements seen in the stretch–shortening cycle (SSC). The SSC is considered a link between power and speed that enables athletes to increase both performance variables (Komi 2003). An enhanced SSC can improve speed, acceleration, and power fairly quickly through better use of elastic energy; but adaptations over prolonged training tend to relate to a reduction in muscle stiffness and enhanced neuromuscular activation (Bosco and Vittori 1986; Komi 2003; Verkhoshansky 1996). This is one of the reasons speed training and power training often take place at the same time within the athlete's periodized training program. This topic is reviewed further in chapters 14 and 15.

Factors in Speed Performance

The ability to accelerate from the start varies considerably among athletes. Some Olympic-caliber athletes can continue to accelerate through the 70 m mark in a 100 m sprint. Although the ability to accelerate is important, the rate at which the acceleration occurs may have even greater importance. This is especially true of the basketball or football player who needs to reach peak velocity as quickly as possible. This is a component of sprinting that the resistance training program (see chapter 8) is aimed at improving.

As long as a runner can still accelerate, he or she will gain speed. Only if the runner begins to decelerate will he or she start to slow down. To maintain running speed, the athlete focuses on improving speed-endurance. Speed-endurance is not a factor in short-distance sprints, such as the 40 yd (37 m) sprint, in which athletes need to accelerate throughout the entire 40 yd. However, it becomes relevant in sprinting distances of 100 m or greater and becomes more of a factor in sprinting success.

Stride Rate and Stride Length

Running speed is the interaction of **stride rate** and **stride length**. The stride contains two steps, or foot strikes. Each step, or foot strike, can be defined as the point of contact of one foot with the ground (figure 11.1). Stride rate is the number of steps taken with each leg during the distance of a run. For example, if a sprinter in a 110 yd (100 m) run took 24 steps with the right leg and 23 steps with the left leg, the total would be 47 strides. If the sprinter ran the race in 11.5 s, the stride frequency would be 4.1 strides per second. Elite sprinters have a stride rate of approximately 5 strides per second (Mero, Komi, and Gregor 1992). As the stride rate increases, the amount of time spent on the ground (**support phase**) decreases, while the time spent in the **flight phase** increases. If stride rate increases and the stride length remains constant, running speed will increase. Similarly, if stride rate remains constant but stride length increases, then running speed will increase. Figure 11.2 shows the effect of changes in stride length and stride rate with running velocity.

As can be seen in figure 11.2, both stride rate and stride length increase as running velocity increases. However, stride length appears to increase only up to velocities of about 8 m/s (Enoka 1994). The contribution of both stride rate and stride length to sprint speed changes at different running velocities. Initial changes in speed are the result of an increase in the runner's stride. As running velocity increases further, increases in both stride length and stride rate contribute to the higher running veloci-

Figure 11.1 A stride in running includes two foot strikes.

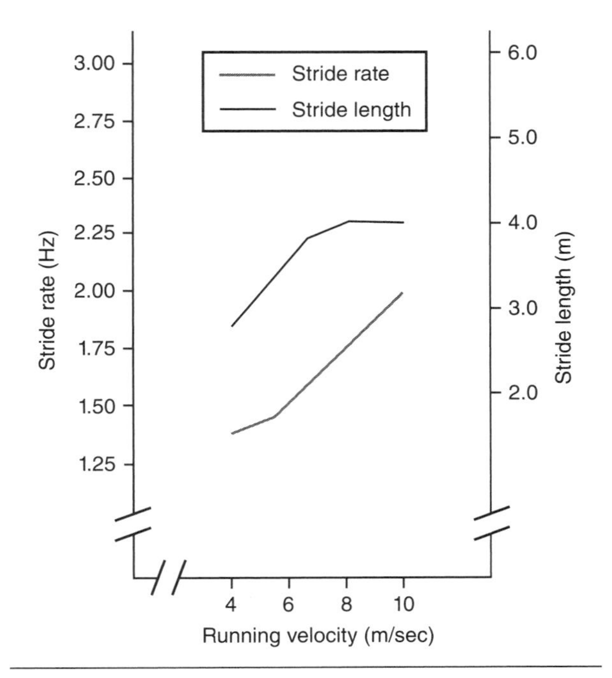

Figure 11.2 Change in stride length and stride rate with running velocity.

Data from Luhtanen and Komi 1978.

ties. However, as speed is increased even further, a slight decrease in stride length is seen along with a sustained increase in stride rate. Thus, stride rate appears to be more important than stride length in determining the runner's maximum velocity (Mero, Komi, and Gregor 1992).

Both stride rate and stride length are variable among individuals. Stride length is dependent on the athlete's height and leg length (Mero, Komi, and Gregor 1992; Plisk 2008). The taller the athlete or the greater the leg length, the longer the stride. Stride rate is also variable, with large differences seen between trained and untrained individuals (Mero, Komi, and Gregor 1992; Plisk 2008). Trained sprinters are able to achieve a greater stride rate than untrained runners and, from a static start, are capable of increasing stride rate for about 27 yd (25 m). Untrained runners appear to reach their maximum stride rate at about 11 to 16 yd (10-15 m) (Plisk 2008). In addition, elite runners can reach their maximum velocity much earlier than untrained runners. Stride rate appears to be very trainable (Plisk 2008). By improving power performance, the athlete can increase acceleration ability by decreasing the ground contact time of each stride and increasing the impulse production on takeoff (Delecluse 1997; Mero, Komi, and Gregor 1992; Plisk 2008). Improvement in running technique may also result in improvement in running speed, especially for individuals with technical flaws in their sprint style.

Muscle Fiber-Type Composition

Skeletal tissue is composed of two types of muscle fibers (see chapter 1 for more detailed discussion). Slow-twitch, or type I fibers, are oxidative fibers whose characteristics are generally associated with low force outputs and slow conduction speeds; the fibers are fatigue resistant. The type I fibers are primarily beneficial to endurance athletes. In contrast, fast-twitch, or type II fibers, are characterized by high force outputs and fast contraction speed but are easily fatigued. A predominance of Type II fibers are common among athletes who are successful in explosive sports that depend on speed and power performance.

As discussed in chapter 1, fiber-type dominance is genetically determined. However, the subtypes of fibers appear to be responsive to a training stimulus. Thus, an appropriately designed training program can stimulate fiber subtype adaptation. Nevertheless, regardless of training stimulus, fiber-type transformation is still limited. Considering that most people are born with an approximately equal number of type I and type II fibers, this limitation is the primary reason why it is not easy to develop elite athletes. Appropriately designed training programs can make a slow athlete faster but cannot transform that athlete into a fast athlete.

Speed is also influenced by the structural architecture of skeletal tissue; muscle thickness, the length of the fascicle, and the pennation angle of the fascicles have been identified as influencing speed performance (Abe, Kumagai, and Brechue 2000; Kumagai et al. 2000). A large pennation angle is associated with an ability to generate greater force compared to a smaller pennation angle, which is associated with greater speed of contraction (Abe, Kumagai, and Brechue 2000; Kumagai et al. 2000). The shorter pennation angle likely allows muscle fibers to shorten faster, enhancing contraction speed and potentially stride rate.

Sprint Technique

Regardless of fiber-type composition, which cannot be coached, the primary interest of the strength and conditioning professional with regard to speed improvement is in improving power, the rate at which force is generated, and running technique. To improve speed, the strength and conditioning professional focuses on the athlete's technique and performance during the start, acceleration, and stride.

The Start

The start is the element of a sprint that coaching may have the greatest influence on. Examination of the athlete's positioning, first step, forward lean, and arm action points toward technique improvements to enhance speed. Athletes who begin a sprint from a two-point stance (e.g., baseball or soccer athletes) should be in an **athletic position** (e.g., feet approximately shoulder-width apart, body mass equally distributed on both feet, and both elbows and knees bent). Before beginning to sprint, the athlete shifts the body mass to the lead leg. Upon initiating movement, the sprinter applies force to the ground with both feet and follows up with an explosive forward action. The leg away from the direction of the sprint should leave the ground first with a fast forward swing, and the arm on that side should also be propelled forward (Dintiman and Ward 2003; Young, McDowell, and Scarlett 2001). When the movement begins from a moving start (either a lower-intensity run, jog, backward run, or side shuffle), upon initiation of the sprint the athlete applies force to the ground with both feet; this is followed by an explosive action with the arm and leg away from the direction of the sprint.

If the athlete begins the sprint from a three- or four-point stance (e.g., football player or sprinter), the body mass should be evenly distributed between the upper and lower body. In a three-point stance, the dominant arm is straight, with fingers planted on the ground, and the nondominant arm is either resting on the thigh or by the side, bent at the elbow approximately 90°. In a four-point stance, both arms are straight with the fingers of both hands planted on the ground. For both stances, the athlete's dominant leg (knee bent to about 125°) should be aligned slightly behind the nondominant-side leg (knee bent to approximately 90°), with the dominant-side toes at the level of the nondominant-side instep or possibly slightly behind), and the head and back should be aligned. Immediately before initiating movement, the athlete shifts the center of gravity above the lead leg. The start begins with a force applied to the ground from both feet, the rear leg moving first with a forward swing. At the same time, the rear arm should move forward (Dintiman and Ward 2003; Young, McDowell, and Scarlett 2001).

Acceleration and Stride

As the athlete accelerates, the body straightens and stride length increases. The foot makes contact with the ground in a dorsiflexed position (toes pointing upward). As the athlete continues the sprint, the head and gaze (i.e., eye direction) should be directed downward, while limiting torso lean. Each stride has two distinct phases: the drive phase and the flight phase. The drive phase involves the time that the lead foot is spent on the ground from landing from the previous stride and pushing off the ground to begin the next stride. As the foot lands from the previous stride, the athlete's center of gravity is at the initial point of contact. As the foot transitions from landing to pushing off the ground, the runner extends the hip, knee, and ankle of that foot. As such, during the drive phase the athlete's center of gravity passes over and in front of the lead foot. When the ball of the lead foot leaves the ground, the drive phase is completed and the flight phase begins (Dintiman and Ward 2003; Mero, Komi, and Gregor 1992).

The flight phase of the stride begins when the ball of the lead foot separates from the ground and continues until it returns to the ground. Keeping the foot dorsiflexed, the athlete flexes the knee and swiftly pulls the heel toward the hip. As the limb is pulled closer to the body's center of gravity and axis of rotation, this provides for a faster swing of the leg. As the heel reaches its maximum height, the athlete drives it forward, raising the dorsiflexed foot above the opposite knee. During the flight phase the foot remains in a dorsiflexed position and returns to the ground via a powerful extension of the hip and knee (Dintiman and Ward 2003; Mero, Komi, and Gregor 1992).

During all phases of the stride, athletes should keep the head in its normal alignment with the trunk while avoiding rotation of the torso and shoulders. The muscles of the head, neck, shoulders, and upper extremities should remain relaxed. The arm swing begins with the lead arm bent to about 70° (palm of the hand parallel to the cheek) and the rear arm bent to approximately 130° (slightly past the hip) (Bosco and Vittori 1986; Mero, Komi, and Gregor 1992). During a sprint the athlete runs with the trunk upright, the head level, and with maximal hip height.

Speed Program Design

The following section provides a blueprint for development of a sprint training program. Further discussion on how speed training can be integrated into the athlete's yearly training program is presented in chapter 15. The sidebar that follows lists some speed drills.

In Practice

Does Ground Reaction Impulse Influence Sprint Acceleration?

Sport scientists from Australia and Japan examined the relationship between ground reaction forces and short-distance (up to 10 m) sprint acceleration from a two-point stance (Kawamori, Nosaka, and Newton 2013). Acceleration during a sprint is determined by the athlete's body mass, gravitational forces, wind or air resistance, and ground reaction forces. The latter three factors are external forces that act on the body, and the ground reaction force is the factor that the athlete has the greatest influence on. To examine this issue, the investigators recruited 30 physically active men who were participating in team sports such as soccer, basketball, rugby, and Australian rules football. All participants performed a 10 m sprint, and ground reaction forces were measured by three force plates that were recessed in the track over the 10 m. Data included relative vertical impulse, relative net horizontal impulse, relative breaking impulse, and relative propulsive impulse. The primary findings demonstrated that 10 m sprint time was correlated weakly with relative net horizontal ($r = -0.52$, $p < 0.01$) and propulsive ($r = -0.66$, $p < 0.01$) impulses, but not with relative resultant vertical ($r = 0.37$, $p > 0.05$) and braking ($r = 0.06$, $p > 0.05$) impulses, suggesting that faster subjects applied ground reaction forces in a more horizontal direction in performing a faster sprint. The resultant impulse (vector addition of vertical, horizontal, and medial-lateral forces) was not correlated with sprint time ($r = 0.21$, $p > 0.05$), indicating that the direction of the impulse (horizontal) is more important for application of force and its relation to sprint performance.

Speed Drills

Full descriptions of how to perform select drills are available in the web resource at www.HumanKinetics.com/PhysiologicalAspectsofSportTrainingAndPerformance. Video clips are also available for most of these drills.

Speed Training Drills

- Kick back
- Arm action (seated)
- Arm action (standing exchange)
- Lean and fall run
- Drop and go
- Jump and go
- Bound and run
- Scramble out
- Two jumps and go
- Push-up start
- Cone or bag jump and sprint

Resisted Speed Training Drill

- Resistive runs

Assisted Speed or Overspeed Training Drills

- Three-person tubing acceleration drill
- Downhill running

Web Resource Preview

Here is a preview of one of the web resource drills. Full drill content can be found at www.HumanKinetics.com/PhysiologicalAspectsofSportTrainingAndPerformance.

Arm Action (Seated)

Purpose: Encourage proper arm movement

Type: Speed training

Equipment: None

STEPS

1. Athlete is seated on the ground with legs extended and toes pointed up.
2. Athlete places one hand at the side of the face, in front of the ear, and the other hand by the hip *(a)*.
3. Athlete initiates the movement by swinging the arms from the shoulder. The arm by the ear is brought down, while the arm by the hip is brought up toward the face *(b)*.
4. This arm movement is performed in a straight line and as rapidly as possible.
5. The movement is repeated 20 times per arm.

VARIATION

- Hold a small weight (1-2 pounds or .45-.9 kilograms) in each hand.

TIPS

- The drill is correct if the butt bounces off the ground. This results from the momentum of the arm movement.
- Keep the knees locked during the whole movement; bent legs will decrease the effectiveness of the arm action.

a

b

Speed Training Exercises

The basic aim of using speed training drills is to improve form and technique during sprinting. The basis of these drills is a focus on specific segments of a sprint and beginning to integrate these segments into the entire sprint action. Speed drills are not a conditioning exercise, and the emphasis in these drills should be on the quality of action. Table 11.1 provides an example of a speed training workout.

Resisted Speed Training

Resisted speed training involves the athlete overcoming a resistance during a sprint activity. Adding resistance has been shown to enhance sprinting speed (Ross et al. 2009; West et al. 2013). These benefits are likely similar to what is seen with resistance training (i.e., enhanced neural function, increased reflex potentiation, and increased type II muscle fiber cross-sectional area) (Linossier, Dormois, Geyssant, et al. 1997; Mero, Komi, and Gregor 1992). However, caution regarding the loading is warranted. Greater resistance may actually damage running technique, negating any potential benefit from resisted running (Lockie, Murphy, and Spinks 2003). The following sidebar lists recommendations for resisted speed training drills.

Table 11.1 Sample Sprint Training Workout

Drill	Reps	Distance and intensity
Arm action (seated)	2-4	10 s
Lean and fall run	2-4	10 yd (9 m)
Push-up start	4-6	20 yd (18 m)
Resistive runs	4-6	20-30 yd (18-27 m)

Recommendations for Resisted Running Drills

Recommended distance: 20 to 40 yd (18 to 37 m)

Sets and repetitions: Three or four sets of four to eight reps

Recovery: 90-120 s

Assisted Speed or Overspeed Training

Assisted speed, or overspeed, training uses downhill sprints or implements to help athletes run faster than they normally can. This training method can increase stride frequency and length more than is generally possible with traditional sprint exercises (Hoffman and Graham 2012). During downhill running, it appears that a downhill slope of 5.8° optimizes sprint speed compared to lesser (2.1°, 3.3°, and 4.7°) and greater (6.9°) slopes (Ebben 2008). One possible negative outcome of overspeed training is the risk of changing running technique (e.g., increase in stride length resulting in a decrease in stride frequency), resulting in slower sprint times. It has been recommended that athletes not exceed 110% of their maximum running speed (Hoffman and Graham 2012). However, precise scientific evidence on appropriate overspeed velocities is lacking.

Agility Development

Agility is the ability to react to changes in direction without loss of speed or accuracy. The expression that someone can "stop on a dime" seems to apply to the ability of an athlete to sprint at maximal velocity and rapidly change direction (either in response to a coach's signal in a practice drill or during an actual competition) without any reduction in speed. In addition, agility often refers to the ability of an athlete to change from one type of movement to another. A defensive back who goes from running backward (backpedaling) to a forward sprint, or a linebacker who goes from a side shuffle to a forward sprint, exemplifies common changes in movement and direction during competition. It requires a combination of strength, power, balance, and coordination to change from a movement performed at a maximal velocity, decelerate as quickly as possible, and accelerate in a new direction, possibly with a new movement. The ability to accomplish this at a high level of precision is an important determinant of an athlete's success, especially in sports such as basketball, football, hockey, and soccer. However, to a large extent this idea is based on empirical evidence (i.e., not scientifically based), since limited scientific data are available on the relationship between agility and athletic performance. Hoffman and colleagues (1996) showed that agility performance was significantly related to playing time in Division I college basketball players. The quicker or more agile the player (as determined by a T-test), the more playing

In Practice

What Are the Benefits of Assisted and Resisted Sprint Training?

Twenty-seven Division I female soccer players participated in a study comparing the effects of a 4-week, 12-session training program of resisted sprint training or assisted sprint training to traditional sprint training on 5 yd (4.5 m) and 15 yd (13.7 m) acceleration and 40 yd (37 m) maximal velocity (Upton 2011). The program took place during the summer preseason training program, with participants having already participated in a 6-week conditioning program (resistance training and running). Athletes continued to perform a 2 day per week resistance training program during the sprint intervention period. Subjects in the assisted sprint group were attached by a shoulder harness to a research assistant who provided the sprint assistance. At 100% stretch of the harness (10 yd [9 m]), the harness provided 45 N of assisted force at 20 yd (18 m); at 200% stretch, the force provided was 90 N. Before the sprint, the harness was stretched to 20 yd. The sprint was initiated by the research assistant. The assisted movement resulted in the participants' performing a supramaximal sprint for 20 yd followed by a 20 yd deceleration to jog. Participants performed 10 assisted sprints with a 3 min recovery between sprints. The resisted sprint training required participants to be attached by a belt to a stationary pole that provided a resistance equal to 12.6% of the participant's body mass. Participants performed 20 yd sprints followed by a 20 yd deceleration and jog. They performed 10 resisted sprints with a 3 min recovery between each sprint. Subjects in the traditional sprint training program performed ten 20 yd sprints with a 3 min rest between each.

The results showed that maximum velocity measured over 40 yd was significantly improved with both assisted ($p < 0.001$) and resisted ($p < 0.05$) sprint training, but remained unchanged ($p > 0.05$) following traditional sprint training. Interestingly, differences were seen between assisted and resisted running in how these improvements occurred. In the assisted running the greatest increases in velocity occurred in the first 5 yd, while in the resisted running group the greatest increase in velocity occurred during the 15 to 25 yd segment of the 40 yd sprint. These results demonstrate the benefit of both modes of sprint training, but one mode may have more application to specific sports than the other. For sports that requite rapid acceleration over a short distance, assisted sprint training would be more appropriate, whereas resisted sprint training would be more beneficial for sports that allow attainment of maximal velocity over a longer distance.

time he had. It has also been suggested that agility is an important component of successful performance in football (Kraemer and Gotshalk 2000) and soccer (Kirkendall 2000).

Plisk (2008) has suggested that the ability to decelerate from a given velocity may be the most important factor determining the athlete's ability to rapidly change direction. He suggested using a progressive, or tiered, approach to enhance agility. First the athlete performs the drill at half speed. On hearing a signal (whistle), the athlete begins to decelerate and, on hearing a second whistle, stops within three steps. The athlete then progresses to running at three-quarters speed and stopping within five steps. Once athletes are successful at this stage, they can perform the drill at full speed using a seven-step braking action. A similar approach can be used for all movements (backward or lateral), with the velocity of movement and braking distance dependent on the athlete's ability.

If agility is one of the variables to be measured as part of a comprehensive testing program for an athlete, the particular agility test used must be part of the athlete's agility program. Agility is often trained within the off-season conditioning program. It is important that the agility drills selected simulate the movements and actions performed during actual competition.

Improving Speed and Agility During an Athlete's Career

An interesting question is whether speed and agility can be improved. Private trainers have made many unsubstantiated claims that they are able to reduce 40 yd sprint time by 0.3 s or more. However, most published studies have seen only limited improvements in both sprint speed and agility. It is likely that large speed gains are related to the conditioning level of the athlete. If the athlete is not in peak condition or has not been performing sprints regularly (which may be the case in the middle of an off-season conditioning program), normal times for a 40 yd or 100 yd sprint may increase remarkably. This may be due to correction of flaws in technique or improvement in speed-endurance. As the athlete begins to train specifically for speed, sprint times will improve but likely only to the point of peak condition of the previous season. Evaluation of a training program should be based on the peak sprint time for the previous season and not from the point at which the training program began (a state of detraining). If not, then the strength and conditioning professional is likely assessing the anaerobic conditioning of the athlete, and not improvements in speed.

Hoffman and colleagues (Hoffman et al. 1990; Hoffman, Fry, et al. 1991; Hoffman, Maresh, et al. 1991) at the University of Connecticut examined the ability of off-season conditioning programs to improve sprint speed and agility in basketball and football players. After 6 months of off-season conditioning, the basketball players showed no improvement in either sprint speed (27 m sprint) or agility (T-test). In a 10-week winter conditioning program for football players, no significant changes in 40 yd sprint speed were observed, despite the incorporation of sprint technique and plyometric training exercises performed twice a week. Considering that these athletes were Division I players, the inability to improve speed or agility during the off-season conditioning programs may be related to the high pretraining level. Other studies have suggested that speed can be improved if high-velocity movements such as stretch–shortening (plyometric) drills are part of the sprint and resistance training program (Delecluse et al. 1995; Rimmer and Sleivert 2000). However, a significant training effect appeared to occur in these studies only in subjects with average speed (Rimmer and Sleivert 2000).

Not enough studies have been performed on the ability to improve sprint speed or agility. From the studies available, it appears that if the athlete has a high pretraining level of speed or agility, the chances of becoming faster or quicker may be limited. If the athlete has only average speed or quickness, or is deficient in a specific component of speed (e.g., explosive strength or running technique), the ability to improve sprint speed or agility is much greater. Similarly to strength, speed and agility may also have a ceiling that limits progress; much of this is determined by genetic potential. As discussed in chapter 1, to run fast it is advantageous to have predominantly **fast-twitch** muscle fibers. Thus, as noted earlier in the chapter, someone who is considered slow may become faster but may not be able to become fast.

A recent study (Hoffman, Ratamess, and Kang 2011), examining 8 years of testing on an NCAA Division III college football team, indicated that speed and agility are difficult to change during the career of these athletes (figure 11.3). When examined across position, no significant changes in speed were noted during the athletes' collegiate football playing career. However, when backs (i.e., wide receivers, quarterbacks, running backs, defensive backs) were separated from linemen (offensive linemen, tight ends, defensive linemen, and linebackers), significant improvements in speed were seen between years 1 and 3 in backs while a trend toward improvement was observed in linemen. Linemen who competed a fifth year were significantly faster than during their first and second years of competition. Thus, it may take 3 or 4 years to see significant improvements in speed in some athletes. Similar to findings for speed, only limited improvements were noted in agility. When all positions were compared, significant decreases in time for the T-drill were noted only between the first and third seasons, while no improvements were noted in the pro-agility drill. Position comparisons revealed no significant improvements in the T-drill for backs, but significant decreases in time for the pro-agility drill were seen for these athletes between years 1 and 3 and between years 2 and 4. In contrast, linemen exhibited no changes in performance in the pro-agility run, but a significant decrease in time for the T-drill between years 1 and 3.

An important issue that needs to be acknowledged is the difference between changes that are statistically significant and those that are practically significant. Statistically, sprint data may not prove the efficacy of a particular training program, especially in elite-level athletes. However, for this caliber of athlete, an improvement of perhaps 0.02 s may be the difference between a medal and fourth place. For example, 0.01 s was the difference between Tyson Gay (9.80 s), who finished fourth, and Justin Gatlin (9.79 s), who received the bronze medal in the 100 m sprint at the 2012 London Olympics. Interestingly, Gay's performance at the 2012 Olympics was faster

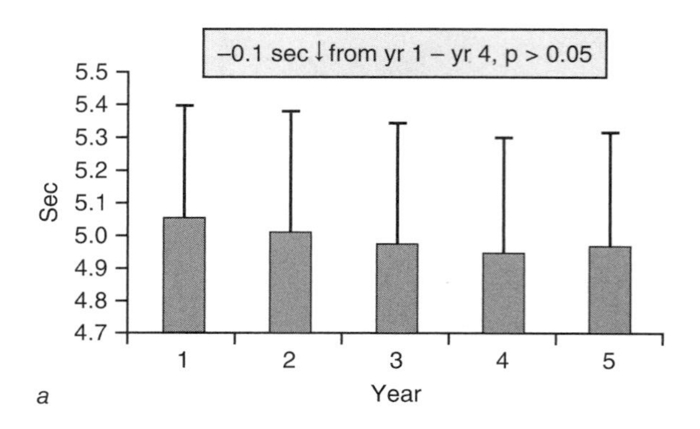

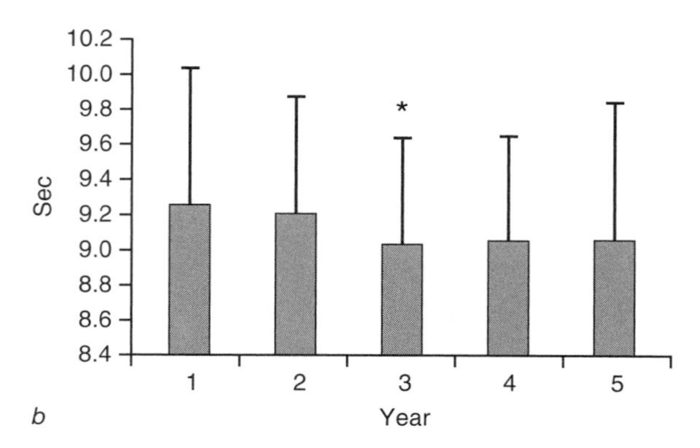

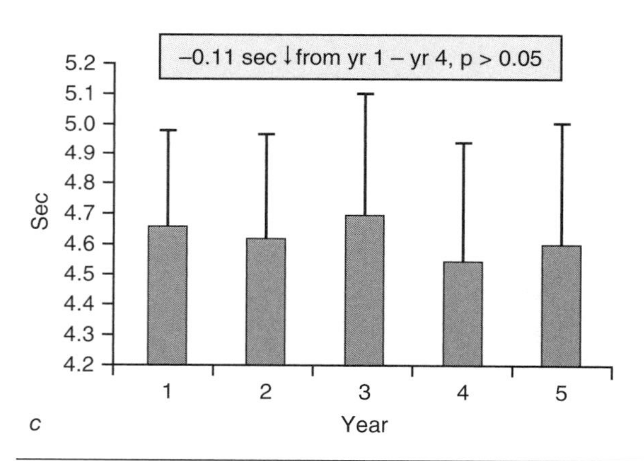

Figure 11.3 Changes in *(a)* 40 yd sprint times and agility, *(b)* T-drill, and *(c)* pro-agility during a college football career. * = Significantly faster compared to year 1.

Data from Hoffman, Ratamess and Kang 2011.

than that of any previous gold medal winner for the Olympic 100 m sprint whose name was not Usain Bolt! Practically, this time holds much significance even though it may not have any statistical significance. Therefore, when interpreting research on the efficacy of training programs, especially speed and agility programs, one must examine the data carefully, including the subject population.

Agility Program Design

Agility training can occur throughout the training year. It can be incorporated into a pre–resistance training warm-up or into a speed and agility program. Often this coincides with a strength–power phase of training. This topic is reviewed in much greater detail in chapter 15. The development of an agility program follows the same basic exercise prescription formats as all training methods: frequency (how often), intensity (how hard), volume (how much), and drill selection. For the greatest "carryover" effect of agility training to actual sport performance, the drills selected should simulate actual sport movements. However, there are also general agility or quickness drills that would benefit athletes in all sports.

- Exercise and drill selection. Agility drills should have as much similarity to the sport requirements as possible. For instance, if the movement patterns of a sport include straight-ahead sprints, lateral shuffles, and backward runs (e.g., basketball), the T-drill may be an appropriate exercise to include. Other sports or positions within sports do not emphasize backward runs (e.g., defensive linemen); in that case the T-drill may not be the most appropriate. However, this does not mean that it could not benefit the athlete.

- Frequency. Frequency refers to the number of training sessions performed during a given unit of time. Improvements in change-of-direction ability have been demonstrated in as few as two or three sessions per week for 4 weeks (Galpin et al. 2008; Walklate et al. 2009). However, duration of training often reaches 10 to 12 weeks (Brown and Khamoui 2012).

- Intensity. The goal of agility training is to enhance the athlete's ability to change direction as quickly as possible; thus these movements need to be performed at maximal or near-maximal effort. The warm-up or skill development should be performed at submaximal speeds.

- Volume. Because intensity in agility exercises does not vary, it is the volume (number of repetitions or exercises) that is manipulated to adjust the difficulty of the program. The optimal volume needed to improve agility performance is not known. However, the use of agility circuits or exercises usually involves about four to six different exercises. The number of repetitions may be determined by time (as many repetitions as can be performed in 1 to 2 min, assuming that a group of athletes is at each station) or can be four or five repetitions per exercise.

• Rest. Recovery periods should be provided between repetitions and drills so that technique can be maintained. A greater work-to-rest ratio (1:6) is typical during the initial stages of training; as the athletes' conditioning level improves, the ratio can be lower (1:4). For example, a drill lasting 10 s would initially use a 1 min recovery period and in the later stages of the program a 40 s recovery. When using an agility circuit, coaches can manipulate the number of athletes per station to achieve the desired rest. A station with seven athletes would have a 1:6 work-to-rest ratio assuming that as one athlete finishes, the next athlete starts. This of course assumes that all athletes at that station are of similar ability.

Agility Drills

A number of drills can be used to improve agility. Many coaches also incorporate specific movements common to their respective sports and use these movements as drills. An example of an agility training program can be seen in table 11.2. The sidebar that follows lists some agility drills.

Table 11.2 Sample Agility Training for a Soccer Player

Drill	Reps
Z-drill	4
T-drill	4
Pro-agility	4
Four-corner drill	3
Ladder drills:	
Quick feet	3
Ickey shuffle	3
In–out shuffle	3

Selected Agility Drills: Speed Drills

Full descriptions and video clips of how to perform all of these drills are available in the web resource at www.HumanKinetics.com/PhysiologicalAspectsofSportTrainingAndPerformance.

Double-leg lateral hopping
Pro-agility
Side shuffle
T-drill
Zigzag drill or Z-drill
Four-corner drill
Quick feet
L-drill
Bag or cone shuffle
Ickey shuffle
In–out shuffle
Snake jump

Summary

Speed training and agility training are important components in the training program of strength–power athletes. Although the ability to improve speed and quickness may be limited by genetic endowment, small changes in performance may still have important practical effects. Speed training should focus on technique, acceleration to maximal velocity, and speed-endurance. Agility training should be specific to movements commonly made by the athlete and should focus on maximizing changes in direction and reaction.

REVIEW QUESTIONS

1. Explain the interaction between stride rate and stride length in the transition from no movement to full sprint.
2. What are the limitations for improving speed and agility, if any?
3. How does muscle architecture influence speed performance?
4. Name and describe five drills for improving running technique.
5. What agility exercises (name at least five drills) could be used to improve agility performance in a basketball player?

chapter 12

Endurance Training

CHAPTER OBJECTIVES

After reading this chapter you should be able to do the following:

- Understand physiological adaptations resulting from endurance training.
- Discuss factors related to endurance performance with specific emphasis on aerobic capacity.
- Explain the lactate threshold and how it can be used to develop the aerobic exercise prescription.
- Develop an exercise prescription for the endurance athlete.
- Discuss whether there is any value of endurance training for the anaerobic athlete.

Endurance training enhances the ability to perform prolonged exercise. Endurance may be specific to either muscular endurance or cardiorespiratory endurance. Muscular endurance involves the ability of a specific muscle or group of muscle fibers to sustain high-intensity, repetitive, or static exercise (Kenny, Wilmore, and Costill 2012). Athletes involved in anaerobic sports such as wrestling or boxing generally train to improve muscle endurance, and this is often accomplished through a specific resistance training program (see chapter 8). Cardiorespiratory endurance refers to the ability to sustain long-duration exercise. Enhancing cardiovascular endurance is the primary goal of distance runners, cyclists, long-distance swimmers, and triathletes. This chapter focuses on the conditioning programs of athletes who are training to improve or maintain cardiorespiratory endurance. In addition, many athletes who participate in anaerobic sports such as American football, basketball, or hockey incorporate cardiorespiratory endurance training into their conditioning programs. This chapter discusses the efficacy of such training for these athletes. For reasons of simplicity, the remainder of the chapter refers to *cardiorespiratory endurance* simply as *endurance*.

Physiological Adaptations to Endurance Training

Endurance training results in physiological adaptations that are different from those seen with either resistance or anaerobic training (see the following sidebar). The effects of endurance training are described in detail in chapters 1 through 5 and are summarized in the following section. The physiological changes common to endurance training programs positively affect the ability of the body to supply adenosine triphosphate (**ATP**) to fuel muscular exercise aerobically. The degree to which these adaptations occur is primarily dependent on the training status of the individual and the person's genetic makeup. People who are deconditioned or are new to endurance training programs experience

Physiological Adaptations Resulting From Endurance Training

Cardiovascular

- Increase in **cardiac output**
- Increase in **stroke volume**
- Increase in blood volume and hemoglobin concentration
- Increase in blood flow to exercising muscles
- Decrease in resting heart rate and blood pressure

Metabolic and Musculoskeletal

- Increase in mitochondrial size and number
- Increase in **oxidative enzymes**
- Increase in **capillary density**
- Increase in reliance on stored fat as an energy source
- Possible increase in myoglobin content

physiological adaptations that reflect a relatively large improvement in endurance performance. However, there are limitations to the extent of these adaptations. The physiological adaptation and performance gains of endurance athletes who are highly experienced and in good condition will not match those seen in the novice or deconditioned individual. The extent of such improvements and the potential for success in endurance sports, as well as in other sports, are largely dependent on genetic makeup.

Factors Relating to Endurance Performance

A number of factors determine the success of an endurance athlete. The extent to which many of these factors can be improved is related to the genetic potential of the individual (e.g., **maximal aerobic capacity** and **fiber type**), whereas other factors (e.g., **lactate threshold** and **exercise economy**) are limited more by the person's training status. Still other factors that influence endurance performance may be more acute (e.g., nutritional and hydration status, rest, and psychological well-being). These acute factors are not within the scope of this chapter and are covered elsewhere in the book.

Maximal Aerobic Capacity

Aerobic capacity, or **$\dot{V}O_2$max**, is widely considered the most objective measure of endurance capacity. $\dot{V}O_2$max is defined as the highest rate of oxygen consumption achieved during maximal exercise. As the **intensity** of aerobic exercise increases, oxygen consumption also rises. As exercise intensity is further elevated, a point is reached where this elevation is not matched by any further increase in oxygen consumption. The point at which this plateau occurs is known as the maximal aerobic capacity, or $\dot{V}O_2$max (figure 12.1).

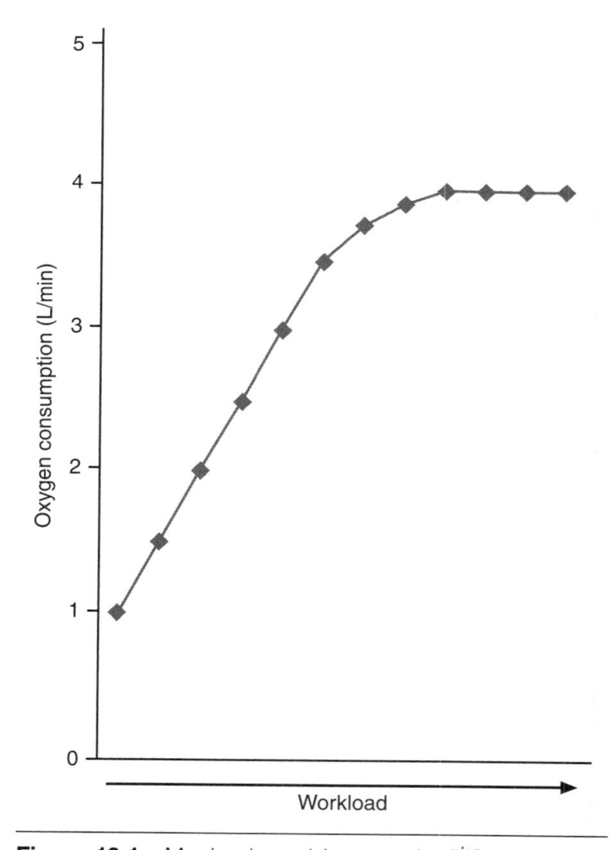

Figure 12.1 Maximal aerobic capacity ($\dot{V}O_2$max).

Table 12.1 provides a range of $\dot{V}O_2max$ values seen in both young (20-29 years) sedentary males and females of average fitness and elite-level male and female distance runners. The differences between the sedentary individuals and the elite-level distance runners can be attributed to both training and genetics. However, it is interesting to note that the $\dot{V}O_2max$ values in the young sedentary population are similar to those in highly trained elite-level basketball players (Hoffman and Maresh 2000). This not only highlights the importance of understanding the **specificity principle** of training, but also indicates that one must take care when interpreting maximal aerobic capacity as an absolute indicator of fitness.

When people begin an endurance training program, their $\dot{V}O_2max$ typically increases. This increase in aerobic capacity can justifiably be interpreted as an increase in fitness level. $\dot{V}O_2max$ values have been reported to increase 15% to 30% over the first 3 months of an endurance training program and may rise as much as 50% within 2 years of training (McArdle, Katch, and Katch 2010). By beginning an endurance training program, previously sedentary individuals could increase their maximal aerobic capacity and reach a level of conditioning that enables them to compete in a marathon. However, their ability to win the race may be limited by factors they are unable to control. As shown in table 12.1, elite male distance runners typically have $\dot{V}O_2max$ values exceeding 70 $ml \cdot kg^{-1} \cdot min^{-1}$ (ranging between 70 and 90 $ml \cdot kg^{-1} \cdot min^{-1}$). As an example of improvements in people who are untrained, consider a sedentary male with an initial $\dot{V}O_2max$ value of 40 $ml \cdot kg^{-1} \cdot min^{-1}$. After 2 years of training, and if one hypothesizes a 50% improvement in maximal aerobic capacity, a $\dot{V}O_2max$ value of 60 $ml \cdot kg^{-1} \cdot min^{-1}$ could result. Even with this remarkable increase in aerobic capacity, he still will likely not be competitive against elite distance runners. With all other factors considered equal (e.g., lactate threshold, exercise economy, nutritional and hydration state, psychological well-being), $\dot{V}O_2max$ is the single most important variable for determining success in distance activities. If the lactate threshold of two people is at 85% of $\dot{V}O_2max$, the one with the higher $\dot{V}O_2max$ has a greater potential for success. The rest of this section discusses the other factors that are critical for the success of endurance athletes.

Muscle Fiber Type

Endurance athletes such as long-distance runners have been reported to possess a higher percentage of **slow-twitch fibers** (approximately 70% type I fibers) than either middle-distance runners or sprinters (Costill, Fink, and Pollock 1976). This is illustrated in figure 12.2. For the endurance athlete, the advantage of a high percentage of slow-twitch fibers is related to the metabolic mechanisms these fibers possess for providing aerobic energy to the exercising muscle. The greater capillary density, **mitochondrial content**, and oxidative enzymes common in type I fibers compared with **type II (fast-twitch) fibers** are more conducive to the metabolic breakdown of stored fat and carbohydrates used for energy during prolonged activity. It would be an advantage if an individual training aerobically were able to convert existing type II fibers to the more beneficial type I fibers. However, there are significant limitations on the ability of an individual to cause any fiber-type transformations, regardless of the training program.

There does not appear to be any convincing evidence that endurance training causes an increase in the percentage of slow-twitch fibers. Although several studies have suggested that endurance training may be able to transform a percentage of type II fibers to the more oxidative type I fibers (Howald et al. 1985; Simoneau et al. 1985), the majority of studies have not shown any transformation between fiber types. However, it is likely that a transformation of fiber subtypes does occur. Thus, a muscle fiber may become either more oxidative or more glycolytic depending on the type of training program being used (see chapter 1).

Despite limitations on fiber-type transformations, significant alterations in the metabolic mechanism can still be accomplished. Changes in capillary density

Table 12.1 Values for Young (20-29 Years) Sedentary Individuals and Elite-Level Endurance Athletes

Test groups	$\dot{V}O_2max$ ($ml \cdot kg^{-1} \cdot min^{-1}$)	
	Males	Females
Sedentary individuals with average fitness	44-51	35-43
Elite-level distance runners	71-90	60-75

Adapted from Martin and Coe 1997.

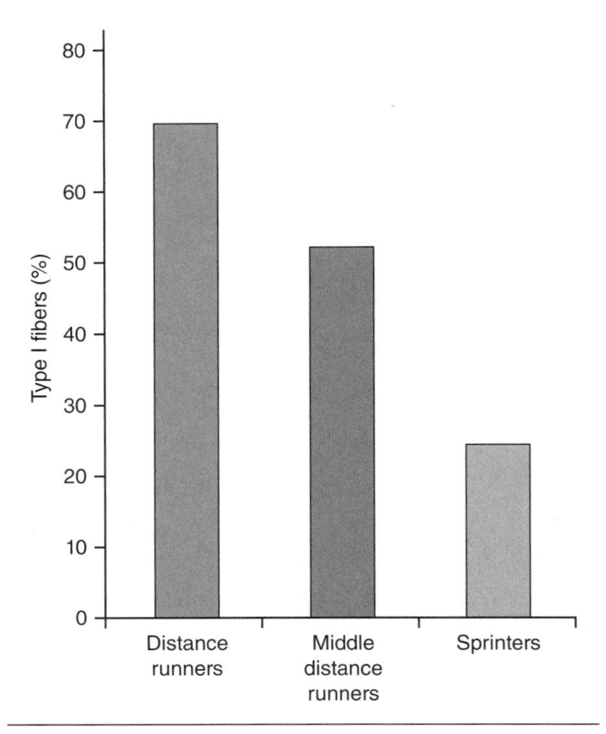

Figure 12.2 Type I fiber composition.

and mitochondrial content are important adaptations to endurance training. Although capillary density is increased with training and a positive relationship ($p < 0.05$) has been seen between capillary density and $\dot{V}O_2$max (Saltin et al. 1977), the primary importance of increased capillary density is that it prolongs the time that oxygen spends within the muscle. That is, the longer the oxygen remains in transit within the muscle (because of a greater capillary network), the greater the opportunity to extract it, even at a high rate of muscle blood flow (Bassett and Howley 2000).

As already mentioned, the high percentage of slow-twitch fibers in the endurance athlete is accompanied by a high concentration of oxidative enzymes in the mitochondria of the cell. Costill, Fink, and Pollock (1976) showed a significantly greater concentration of **succinate dehydrogenase (SDH)** in long-distance runners compared with middle-distance runners and sprinters. In addition, they reported a significant relationship between muscle SDH concentration and $\dot{V}O_2$max ($r = 0.79$, $p < 0.05$). The higher concentration of SDH within the muscle appears to be indicative of the greater oxidative potential of the muscle. Endurance training can increase oxida-

In Practice

Influence of Physiological Factors in Men and Women During a Marathon

Ten male (41.0 ± 11.3 years) and 10 female (42.7 ± 11.7 years) recreational marathon runners were examined by scientists from the University of New Orleans and Louisiana State University before their participation in a marathon (Loftin et al. 1997). All runners were experienced marathoners; the men averaged slightly more than 12 marathons apiece, and the women averaged more than three marathons each. Energy expenditure and physiological factors were examined and compared between sexes. Male runners had a higher $\dot{V}O_2$max (52.6 ± 5.5 ml · kg^{-1} · min^{-1}) than the female runners (41.9 ± 6.6 ml · kg^{-1} · min^{-1}), but ventilatory threshold values were similar between males (76.2 ± 6.1% $\dot{V}O_2$max) and females (75.1 ± 5.1%. $\dot{V}O_2$max). Male runners were heavier (72.4 ± 6.0 kg) than the female runners (60.8 ± 5.7 kg); and body fat, measured by dual-energy X-ray absorptiometry (DEXA), was lower in the men (15.5 ± 6.5%) compared to the women (24.9 ± 5.5%). The men completed the marathon in 220.0 ± 33.2 min, while the women completed the marathon significantly more slowly than the men in 262.5 ± 37.1 min. The runners ran the race close to their ventilatory threshold. The estimated energy expenditure during the marathon was higher among the male runners (2792 ± 235 kcal) compared to the female runners (2436 ± 297 kcal). Examining the influence that these variables had on marathon run performance, the investigators reported that heavier runners, regardless of sex, generally had a slower run time. The bivariate correlation between percent body fat and run time was $r = 0.87$, and the correlation between fat-free mass and run time was $r = -0.50$. Maximal aerobic capacity and ventilatory threshold were also significantly correlated to run time (both correlations were reported as $r = -0.73$). In conclusion, the authors indicated that heavier runners had slower run times and expended a great deal more energy than lighter runners (correlation between body mass and energy expenditure was $r = 0.80$). Decreasing fat mass can have a positive effect on run times for marathon runners.

tive enzyme concentrations almost twofold from untrained levels (Gollnick et al. 1972; Holloszy et al. 1970) as a result of increasing the number and size of the mitochondria within the cell. Although increases in mitochondrial content and capillary density result in a higher $\dot{V}O_2$max, the magnitude of the increase in maximal aerobic capacity after endurance training is not in proportion to the increases in either mitochondrial number or capillary density.

Lactate Threshold

Blood lactate concentration represents the balance between its rate of production and its rate of removal. During light to moderate exercise, the energy demands of the muscles are met sufficiently through adequate oxygen availability. As exercise intensity increases, the muscles are unable to maintain the balance between energy production and energy demand through aerobic metabolism. It is at this point that blood lactate concentrations begin to accumulate. The point at which aerobic metabolism is unable to meet the demands of exercise and the muscles must rely on anaerobic metabolism for energy supply is termed the **onset of blood lactate accumulation (OBLA)**. It is also known as the **anaerobic threshold**, or lactate threshold (Farrell et al. 1979), and is typically reported as a percentage of $\dot{V}O_2$max.

The pattern of lactate threshold is similar for endurance-trained and untrained individuals, the only difference being the percentage of $\dot{V}O_2$max at which lactate threshold occurs. In untrained individuals, blood lactate begins to accumulate at exercise intensities exceeding 55% of their maximal aerobic capacity (Davis et al. 1979). Trained endurance athletes may perform at exercise intensities between 80% and 90% of their $\dot{V}O_2$max (Martin and Coe 1997). Figure 12.3 compares blood lactate responses at different intensities of exercise in trained and untrained endurance subjects. The differences in the lactate thresholds can be attributed to several possible factors relating either to genetic differences or to physiological adaptations. These include a greater percentage of slow-twitch fibers, mitochondrial content, oxidative enzymes, and capillary density in the trained endurance athlete (Gollnick and Saltin 1982; Holloszy and Coyle 1984).

The lactate threshold is trainable and is frequently recognized as an indicator of endurance performance. It is often used as a part of the **exercise prescription**. Exercise intensity is often set at a person's lactate threshold or slightly above it. This is continually adjusted as the individual becomes better conditioned. It is relatively easy to determine the lactate threshold. Typically, blood lactate is plotted against exercise intensity (running speed). The running speed at which a blood lactate level of 4 mmol/L is reached is considered the lactate threshold (Heck et al. 1985) and is typically recommended as the training intensity.

Exercise Economy

The term *exercise economy* is used to refer to the oxygen consumption needed to run at a given velocity. It has been shown that exercise economy may explain some of the variability in distance running performance seen in subjects with similar $\dot{V}O_2$max levels (Bassett and Howley 1997). It has also been shown to be an important factor for predicting performance in athletes running a 10K race (Conley and Krahenbuhl 1980). Figure 12.4 is a comparison in the running economy between two subjects with a similar $\dot{V}O_2$max. Differences in exercise economy have also been seen in cyclists (McCole et al. 1990) and swimmers (Van Handel et al. 1988). Differences in running economy can often be attributed to differences in body position or in performance technique. As technique improves, the energy demand for a

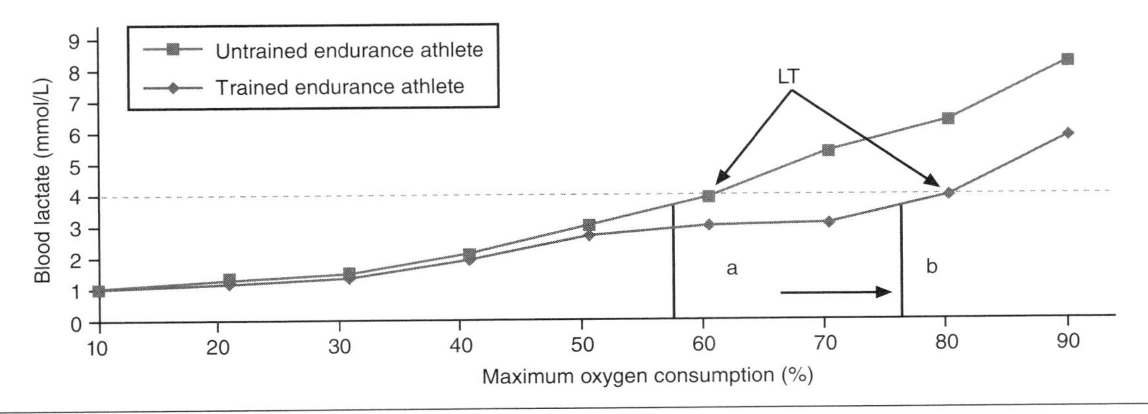

Figure 12.3 Blood lactate response at different intensities of exercise in trained and untrained endurance athletes. LT, lactate threshold.

given exercise velocity is reduced, resulting in a more economical performance. Runners who increase their **stride length** from optimum levels experience a greater energy demand during exercise (Cavanagh and Williams 1982). Changes in stride length and **stride rate** appear to have a significant impact on exercise economy and subsequently on endurance performance. Besides differences in running, cycling, or swimming technique, other factors such as increases in body temperature, wind resistance, and weight (e.g., weight of shoes) also contribute to differences in the economy of exercise between individuals by causing elevations in $\dot{V}O_2max$ (Daniels 1985).

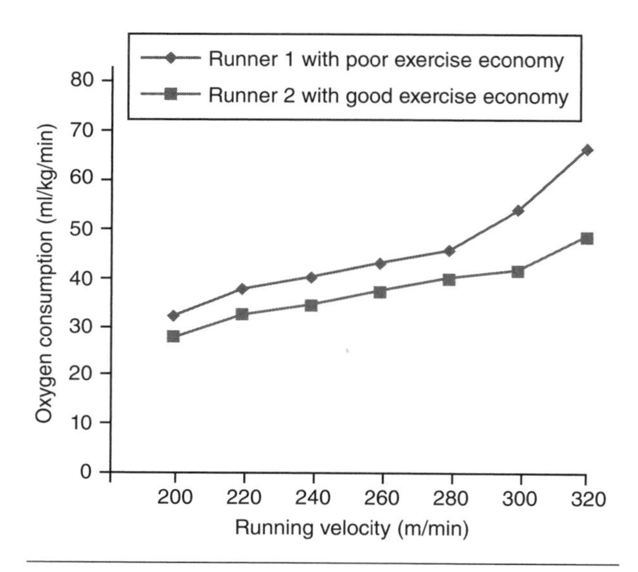

Figure 12.4 Comparison of running economy between two subjects with similar aerobic capacities.

Endurance Exercise Prescription

To design the exercise prescription for the endurance athlete, three primary training variables need to be considered: **training frequency**, **exercise duration**, and training intensity. The mode of exercise depends on the type of endurance sport (e.g., cycling, running, or swimming) that the individual performs. **Cross-training** (performing a different mode of exercise) provides some benefit in the training programs of athletes, and the specifics of this type of training are discussed later in the chapter.

Frequency of Training

How often an athlete should train for endurance is debatable. Research has shown that training twice per week appears to produce changes in $\dot{V}O_2max$ similar to those with training 5 days per week (Fox et al. 1973; McArdle, Katch, and Katch 2010). However, this may be related to the intensity, because as training intensity is reduced, a greater frequency of training may be needed to increase aerobic capacity (McArdle, Katch, and Katch 2010). Frequency may also depend

In Practice

Exercise Economy and Efficiency in Recreational and Elite Cross-Country Skiers

Scientists from Sweden investigated skiing economy and efficiency in recreational and elite skiers (Ainegren et al. 2013). In addition, they examined the influence of age and sex. Considering the difficulty of performing these measures in the snow and outdoors (i.e., air, snow, temperature, humidity, and wind and their effects on grip and glide), the investigators performed their study during roller skiing on a motorized treadmill. A total of 88 subjects volunteered to participate in physiological tests using a free technique or the classic technique. Within each technique the participants were categorized into separate groups according to their performance ability (recreational vs. elite), age, and sex. As hypothesized by the investigators, the elite cross-country skiers were found to have a significantly better skiing economy and efficiency (5-18%) compared to the recreational skiers. Interestingly, the senior elite skiers had greater economy and higher efficiency (4-5%, $p < 0.05$) than the junior skiers. No differences were noted between the sexes. The investigators concluded that in addition to maximal aerobic capacity, skiing economy and efficiency have the greatest effect on performance times in skiers in all categories.

on the training status of the individual. Fewer days of endurance training per week may be needed to maintain aerobic fitness than were needed to reach a given level of performance (Potteiger 2000). The frequency of training, however, is also related to the goals of the training program. If the primary goal is to reduce body fat, then a higher frequency of training will be more beneficial. The greater the frequency of training, the greater the caloric expenditure. It is for this reason that daily exercise is recommended for individuals interested in losing weight (American College of Sports Medicine [ACSM] 2000).

Exercise Duration

The length of time for an exercise session depends on the exercise intensity. If exercise intensity is above the lactate threshold, then the duration will be relatively short because of **fatigue**. If the intensity is mild to moderate, the training could be of much longer duration. An appropriate duration of exercise has not been determined. For deconditioned or sedentary individuals, 5 min per day of low-intensity exercise could elicit a training effect. In contrast, National Collegiate Athletic Association (NCAA) endurance runners average workouts of 8 to 10 mi (13-16 km) a day depending on the phase of training during the season (Kurz et al. 2000). Training sessions for these athletes are obviously much longer. For practical purposes, to produce a training effect it is recommended that most individuals exercise 20 to 30 min per session at an exercise intensity of at least 70% of $\dot{V}O_2$max (ACSM 2013).

One of the concerns for the coach and athlete is the risk-to-benefit relationship that exists when the duration of training is increased. Increasing the duration places the athlete at a higher risk of overtraining (see chapter 24). Because of the high level of conditioning common among distance athletes, coaches may increase the length of each training session to elicit further physiological adaptations, but this may not bring about the desired performance outcomes. Costill and colleagues (1991) failed to see any differences in swimming power, endurance, or performance in collegiate swimmers when training duration was doubled from 1.5 h per day to two daily 1.5 h training sessions.

Training Intensity

Intensity of exercise may be the single most important variable to consider when one is writing the exercise prescription. All of the other acute program variables depend on training intensity. As intensity increases, the duration of exercise becomes shorter; likewise, as intensity is reduced, the duration can be prolonged. As mentioned in the discussion on training frequency, if intensity of exercise is high, the frequency of training needed to elicit a training effect may be reduced. If training intensity is low, a greater frequency may be needed to elicit the desired physiological adaptations and enhance endurance performance. Proper adjustment of exercise intensity is important for achieving the desired training goals. To maximize aerobic capacity, training intensity must create an overload on the physiological processes of the body, resulting in the desired adaptation. However, if exercise intensity is too high, fatigue may occur prematurely, and the training stimulus may be insufficient.

Training intensity can be expressed in several ways. It may be expressed as a relative intensity of a person's $\dot{V}O_2$max or as a percentage of **maximal heart rate**. In addition, the **rating of perceived exertion (RPE)** has been used to prescribe training intensity. The ideal method for determining exercise intensity is to find the velocity of exercise (whether it is running, cycling, or swimming) that corresponds to the athlete's lactate threshold. As described earlier, the relative percentage of $\dot{V}O_2$max that elicits the lactate threshold may vary and depends on the athlete's training status. If the velocity of exercise needed to reach the lactate threshold cannot be calculated, then the other methods may be used to prescribe exercise intensity.

Heart Rate

The use of **heart rate** to prescribe aerobic exercise is well accepted by the exercise science community because of the close relationship between heart rate and oxygen consumption (figure 12.5). Regardless of age, conditioning level, or sex, the relationship between percent $\dot{V}O_2$max and percent maximal heart rate is maintained (ACSM 2013). This relationship is also maintained during different modes of exercise. However, it is important to note that during different forms of exercise, or during exercise that uses a smaller muscle mass (e.g., arm vs. leg), the maximal heart rate may be different. For example, the maximal heart rate during swimming is approximately 13 beats/min lower than that seen during running in trained and untrained men and women (Magel et al. 1975; McArdle et al. 1978). The reason for the lower maximal heart rate during swimming is likely related to the horizontal body position, the cooling effect of water, and the reduced neural stimulation from the smaller active musculature of the upper body (McArdle, Katch, and Katch 2010).

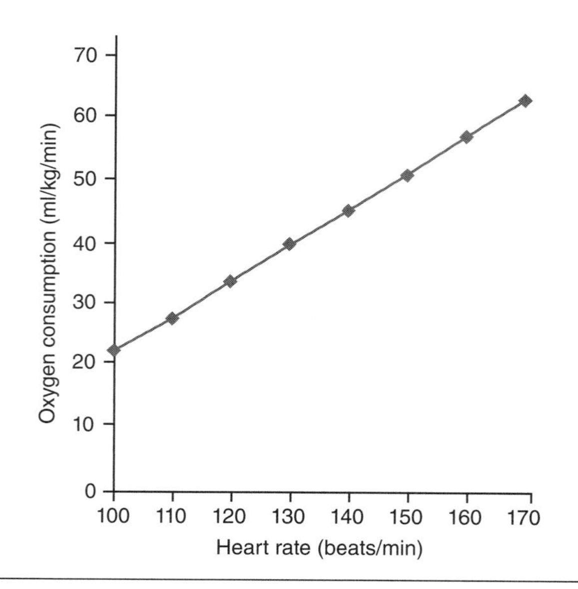

Figure 12.5 Relationship between heart rate and oxygen consumption.

With regard to the endurance exercise prescription for healthy males or females, it is generally acknowledged that exercise should be performed at about 70% of the maximal heart rate to elicit improvements in aerobic capacity. For example, a 20-year-old male whose maximal heart rate is 200 (the formula for predicting maximal heart rate is 220 − age) should run at a pace that elicits a heart rate of 140 beats/min. The use of a percentage of the maximal heart rate is a relatively simple method for determining exercise intensity. An alternate method is to calculate the **heart rate reserve (HRR)**, which provides the functional capacity of the heart. This formula, also known as the Karvonen method, is as effective as using a percentage of the maximal heart rate and is also used with regularity. The HRR can be calculated as follows:

HRR = Maximal heart rate − Resting heart rate

Once the HRR is calculated, the target heart rate can be determined. Target heart rate is given as a range of the functional capacity of the heart plus the resting heart rate:

Target heart rate =
(HRR × Exercise intensity) + Resting heart rate

For example, an exercise prescription for a 20-year-old college student with a resting heart rate of 60 beats/min calls for exercise at 60% to 70% of the functional capacity. After determining the age-predicted maximal heart rate, the HRR is calculated.

HRR = Maximal heart rate − Resting heart rate

HRR = 200 beats/min − 60 beats/min

HRR = 140 beats/min

With the heart rate reserve calculated as 140 beats/min, the next step is to calculate the target heart rate:

Target heart rate (THR) =
(HRR × Exercise intensity) + Resting heart rate

THR = (140 × 0.60) + 60 = 144 beats/min

THR = (140 × 0.70) + 60 = 158 beats/min

Thus, the exercise intensity for this person should range between 144 and 158 beats/min. Endurance training causes physiological adaptations that are expected to decrease the heart rate for an absolute exercise load. To maintain the same exercise stimulus, adjustments must be made to the exercise intensity.

Ratings of Perceived Exertion

Ratings of perceived exertion (RPE) can also be used to prescribe endurance training programs. Borg (1982) created a 15-point rating scale that exercisers can use to rate how they feel in relation to the exercise stress. As energy expenditure and physiological strain increase, the perceived strain on the exerciser also increases. An RPE of 13 or 14, which corresponds to a verbal anchor of "somewhat hard," coincides with a heart rate of about 70% of the maximal heart rate (McArdle, Katch, and Katch 2010). However, recent research suggests that use of the RPE may not be valid for exercise durations of 20 min or longer (Kang et al. 2001).

Endurance Training Programs

How athletes train has changed considerably through the years as the scientific knowledge base regarding the best way to reach peak performance has expanded, and this is certainly true for endurance athletes. Kurz and colleagues (2000) reported that historically, many coaches believed that the longer the race, the more important aerobic training became. In support of this, earlier research had demonstrated that **training volume** (total number of miles run per week) was a significant predictor of marathon run time (Dotan et al. 1983). However, recent thought has refocused training tactics from high volume, low intensity to low volume, high intensity. Still, a single accepted theory of training does not exist; and as might be expected, a number of methods are used to train endurance athletes.

Kurz and colleagues (2000) published a survey on training methods of NCAA Division I distance runners. The survey, which included 14 Division I programs, elicited the training methods employed and the number of weekly sessions during a particular training cycle. These results are adapted into figure 12.6, and the various training methods are listed and described in table 12.2. In the study, the training year was separated into three different phases. The transition phase included the period from May to August; the competition phase ranged from August to October; and the peaking phase was the month of November. In general, the transition phase for an endurance athlete is a period of development in which the volume of training (weekly mileage) is gradually increased as the athlete enters the competitive phase of the season. During the competitive season, greater emphasis is placed on improving the lactate threshold, with training directed toward the final competitions of the peaking phase. During the peaking phase, the athlete focuses all workouts on preparation for the final race. Training volume is tapered, and a greater emphasis is placed on speed work. Typical training volumes for each phase can be seen in figure 12.7.

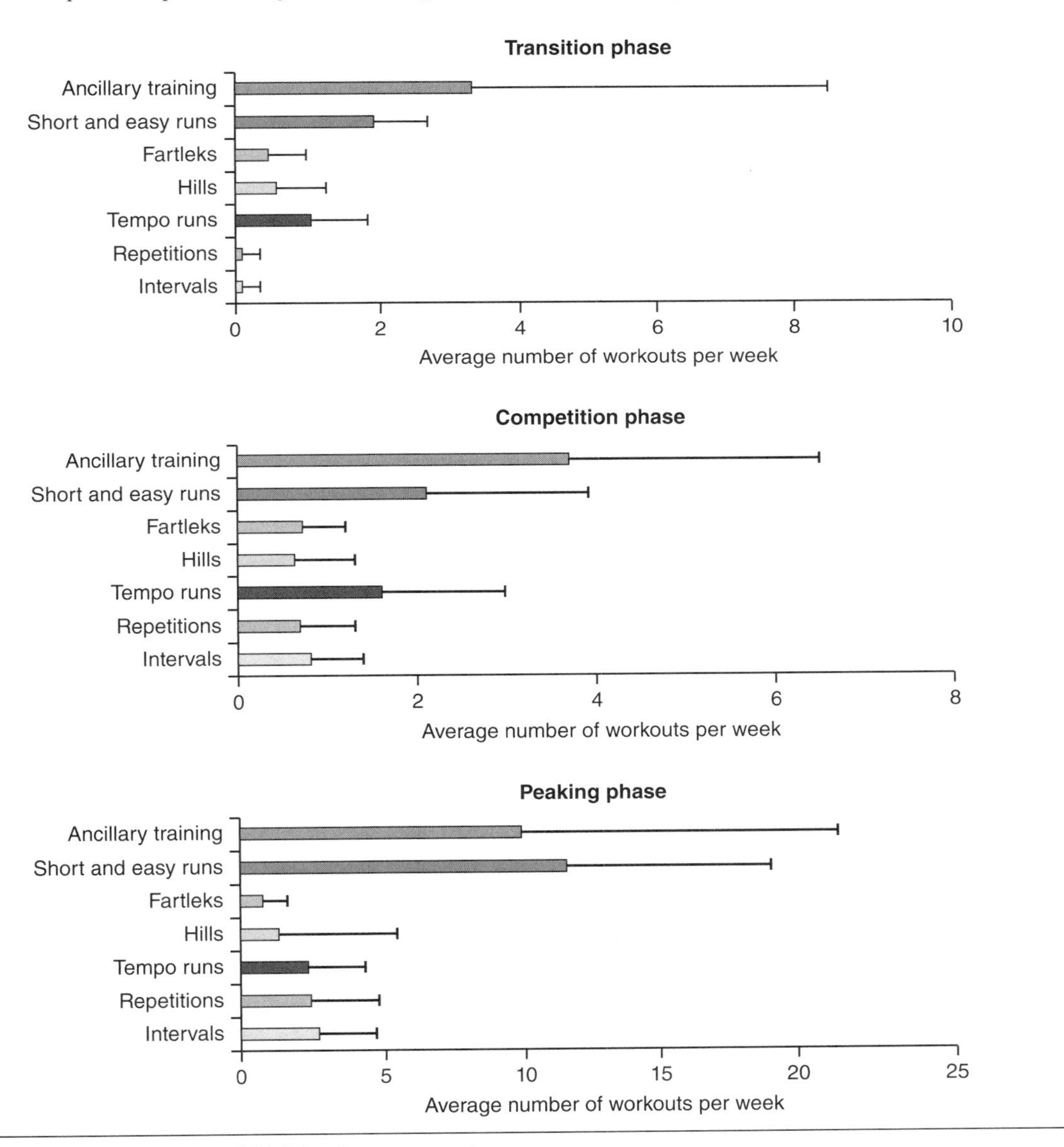

Figure 12.6 Training methods of Division I cross country runners.

Adapted from Kurz et al. 2000.

Table 12.2 Training Methods for Distance Runners

Training method	Description
Intervals	Repeated bouts of high-intensity runs at or exceeding the runners' race pace, interspersed with a recovery period lasting no longer than the time of the high-intensity run.
Repetitions	A series of runs at least 400 m in distance performed faster than race pace with a complete recovery period between each run.
Tempo runs	Training runs performed at a pace 20 to 30 s slower than the normal race pace.
Hills	Runs repeatedly performed on a graded hill at 85% to 90% effort for a time interval between 30 s and 5 min. The recovery period is the jog back down the hill.
Fartleks	These are runs performed at various intensities over various distances.
Short and easy running	As the name implies, this is a run performed at an easy pace over a relatively short distance.
Ancillary training	This comprises all supplementary training, including resistance training, plyometrics, and form drills.

Adapted from Kurz et al. 2000.

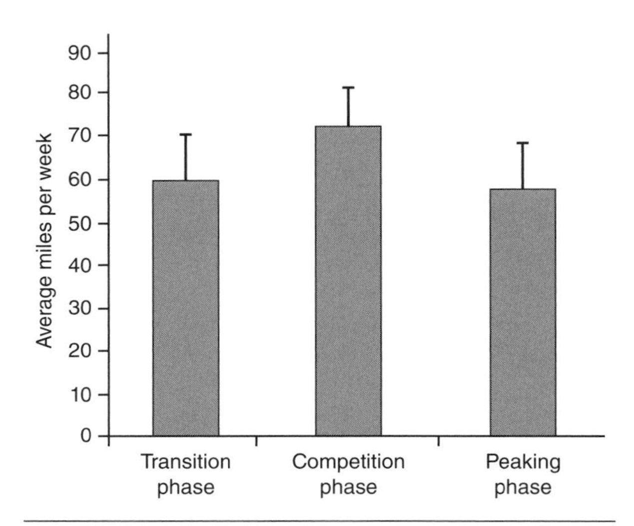

Figure 12.7 Training volume of Division I cross country runners.

Adapted from Kurz et al. 2000.

Triathlon Training

The ultimate endurance event, the triathlon, incorporates swimming, cycling, and running segments performed sequentially. The primary determinant of success is the ability to sustain a high rate of energy expenditure for prolonged periods of time (O'Toole and Douglas 1995). Preparation becomes complicated considering that all three event modes must be trained simultaneously. Triathlons can be of either short or long duration. The Ironman or Olympic triathlon consists of a 3.86K (2.4 mi) swim, a 180.25K (112 mi) bike ride, and a marathon 42.2K (26.2 mi) run. The standard triathlon consists of a 1.5K (~1 mi) swim, a 40K (25 mi) bike ride, and a 10K (6.25 mi) run (Bentley et al. 2002). During shorter-distance triathlons the athlete's $\dot{V}O_2max$ is strongly related to performance, but in the longer events the athlete's aerobic capacity appears to be less important (O'Toole and Douglas 1995). The ability to sustain energy output appears to have a more important role in the longer events. Thus, economy of motion becomes imperative. It is recommended that triathletes who do not have a strong swimming background focus on becoming better swimmers, and that cycling training precede running economy training (O'Toole and Douglas 1995). Swimmers are more efficient (lower energy cost and greater distance per stroke) in the front crawl than triathletes (Toussaint 1990). Although swimmers tend to be taller than triathletes, this height advantage does not appear to be responsible for the greater metabolic efficiency. A greater propulsion force during each stroke seems to play a greater role in distance per stroke (Bentley et al. 2002).

A strategy often used in triathlons and other long-duration endurance events is **drafting**. Drafting is a technique in which two or more competitors perform in an alignment (one behind the other) to reduce the effect of drag. Although this technique is generally associated with cycling, it is also used by swimmers. A lead swimmer can reduce the passive drag of other swimmers between 10% and 26% (Bentley et al. 2002; Chatard, Chollet, and Millet 1998). This provides energy conservation (a 5%-10% decrease is reported in oxygen consumption at both submaximal and maximal velocities) (Bassett et al. 1991; Bentley et al. 2002), and a decrease in blood lactate concentrations is seen as well (Bassett et al. 1991). Drafting in the swimming event may also be beneficial during the subsequent cycling portion of the race. Blood lactate responses are typically higher after the swim event compared to both the cycling and running events (Bentley et al. 2002). This may affect power performance during the cycling stage. Drafting during the cycling stage can result in a significant metabolic savings. The number of cyclists who are in front appears to influence the magnitude of energy savings. Cycling behind a group of eight riders appears to confer a much larger energy savings than drafting behind one to four riders (McCole et al. 1990). In addition, the metabolic cost of alternating cycling positions (going from leader to falling behind and then going back to leader) increases the energy cost of cycling and may affect subsequent run performance (Hausswirth et al. 2001).

Pedaling cadence has an important role in determining energy cost of activity. If the pedaling cadence is too high, the metabolic cost of activity rises, potentially reducing performance. Most studies have indicated that the metabolically optimal pedaling cadence is between 60 and 80 revolutions per minute (rpm) (Bentley et al. 2002). However, others have suggested that the biomechanically optimal pedaling cadence may be closer to 90 rpm (Neptune, Kautz, and Hull 1997; Takaishi, Yasuda, and Moritani 1994). Elite cyclists are able to maintain close to 90 rpm at submaximal intensities during competition, while triathletes pedal closer to 83 rpm (Brisswalter et al. 2000; Lucia, Hoyos, and Chicharro 2001).

The transition from cycling to running interferes with proper running mechanics during the initial stages of the run (Bentley et al. 2002). The running distance or the time of the run at which this effect dissipates is not well understood, but likely relates to the athlete's conditioning level and the drafting that was performed during the cycling stage. Hausswirth and colleagues (1999) showed that triathletes who drafted during the cycling stage were able to run the last 5 km of a short triathlon faster than athletes who did not draft.

In preparation for a triathlon, it is believed that the greatest performance and physiological gains occur when exercise intensity is below the lactate threshold (Esteve-Lanao et al. 2007; Ingham et al. 2008). These studies suggested that 80% of the training load should be performed at exercise intensities that do not elicit significant elevations in blood lactate. When training intensities above the lactate threshold are performed for 30% of the load, no further benefits are noted. Neal and colleagues (2011) examined the effects of training load and training intensity in a 6-month study on Ironman triathletes. They divided the training program into three zones. Zone 1 was below the lactate threshold; zone 2 was above the lactate threshold but below the lactate turnpoint, and zone 3 was above the lactate turnpoint. The lactate turnpoint is the second deflection on lactate threshold tests, signifying that lactate production exceeds maximal clearance rates. The 6-month training period was divided into three 2-month periods. The average volume of training for all three periods combined was 203 ± 71 h with a range of 92 to 266 (8.1 h, 11.0 h, and 9.9 h per week during each 2-month period of training). The second period of training (months 2-4) had the highest total volume. Figure 12.8 shows the percentage of time spent at each training zone during the 6-month training period for swimming, cycling, and running combined. No differences in training intensity were noted between the training modes. The percentage of time spent in zones 1, 2, and 3 across all three training periods was 69 ± 9%, 25 ± 8%, and 6 ± 2%, respectively.

Table 12.3 depicts the physiological changes consequent to the 6-month training program. At the end of the program, modest physiological adaptations were noted in these well-trained athletes. As discussed in part I of this text, expectations for physiological adaptations during relatively short-duration training programs should be modest. However, these results appear to support the work of Esteve-Lanao and coauthors (2007) and Ingham and colleagues (2008), suggesting that too much time in zone 2 may have contributed to the lack of large improvements. It is possible that the greater time spent in zone 2 contributed to a potential overreaching or overtraining phenomenon (see chapter 24).

Training programs are developed such that they are specific to the event (long or short duration) and specific to the training experience of the athlete. An example of a triathlon training program is presented in chapter 15.

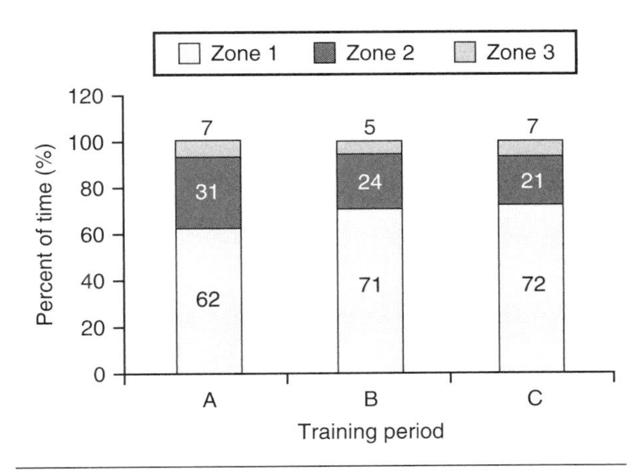

Figure 12.8 Percentage of time spent in each training zone during 6-month training program.

Adapted from Neal et al. 2011.

Table 12.3 Physiological Responses to 6-Month Swimming, Cycling, and Running Program

	Baseline	2 months	4 months	6 months	% change from baseline
Swimming performance					
150 m time at lactate threshold	142 ± 16	141 ± 14	142 ± 14	141 ± 16	0.7
150 m time at lactate turnpoint	130 ± 18	130 ± 15	130 ± 15	127 ± 14	2.3
Cycling performance					
Power at lactate threshold	209 ± 33	213 ± 35	212 ± 29	216 ± 32	3.3
Power at lactate turnpoint	263 ± 34	263 ± 35	267 ± 29	263 ± 33	0.0
Running performance					
Speed at lactate threshold	12.9 ± 1.2	13.6 ± 1.6*	13.4 ± 1.5	13.9 ± 1.6*	7.8
Speed at lactate turnpoint	15.3 ± 1.4	15.3 ± 1.5	15.6 ± 1.2	15.9 ± 1.3*	3.9

*Significantly different from baseline.

Adapted from Neal et al. 2011.

Cross-Training

Athletes often use cross-training (performing exercise different from their usual mode of training) to maintain general fitness. Cross-training is beneficial during periods of injury if athletes are unable to perform their normal exercise routine. As an example, suppose a runner suffers an injury that prevents her from performing her normal training regimen. She can perform an alternative method such as swimming or cycling to maintain fitness levels during the time she is unable to run. Cross-training also prevents boredom and monotony by providing variety in the training program. In addition, the use of different modes of exercise may reduce the likelihood of overtraining because it distributes the stress of training among different muscle groups (Potteiger 2000). Cross-training is also critical for athletes who compete in multiple events. For example, triathletes need to swim, cycle, and run in preparation for their competitions. Although cross-training in one mode appears able to maintain aerobic capacity in other modes, it does not improve aerobic capacity to the same degree as mode-specific training (Foster et al. 1995; Gergley et al. 1984; Magel et al. 1975).

Endurance Training for Anaerobic Athletes

Athletes who play **anaerobic** sports such as football or basketball are often required to perform conditioning drills designed to enhance their aerobic capacity (e.g., long jogs). This may appear strange given that the movement patterns of these sports are performed at predominantly maximal intensities. In addition, aerobic capacity has been reported to have a significant negative correlation to playing time of basketball players (Hoffman et al. 1996). That is, the greater the athletes' aerobic capacity, the less time they actually played. Thus, direct performance benefits from increasing aerobic capacity in these athletes may be questionable and would therefore have important implications for their exercise prescriptions. However, considering the relationship between oxygen supply and skeletal muscle **recovery** (Idstrom et al. 1985), aerobic fitness may be critical for the recovery processes during athletic performance involving repeated high-intensity activities.

Few studies have examined the ability of aerobic capacity to enhance recovery from anaerobic activity. In one study, Koziris and colleagues (1996) examined recreationally trained males and females but were unable to establish any relationship between oxidative metabolism and fatigue during the late stage of a Wingate anaerobic test (30 s of high-intensity cycle ergometry). Similarly, Hoffman, Epstein, Einbinder, and Weinstein (1999) were unable to find any relationship between aerobic capacity and recovery after high-intensity exercise in elite-level basketball players. However, in another study of a large group of subjects with diverse aerobic fitness levels, a relationship between aerobic fitness and recovery during repeated bouts of high-intensity exercise was reported (Hoffman 1997). However, this relationship appears to be limited. A reduction in the rate of fatigue between bouts of high-intensity exercise was observed as aerobic fitness improved (reduced time for a 2000 m run) to values that approached the population mean. In subjects whose level of aerobic fitness neared or exceeded the population mean, no further benefit to fatigue rate was seen (figure 12.9). Although aerobic conditioning appears to provide some advantage for the anaerobic athlete, it is likely that once an average level of aerobic fitness is achieved, the benefits of efforts to increase aerobic capacity further will be minimal. In addition, once an aerobic base is achieved, sport-specific team practices and games are sufficient to maintain aerobic fitness (Hoffman, Fry, et al. 1991).

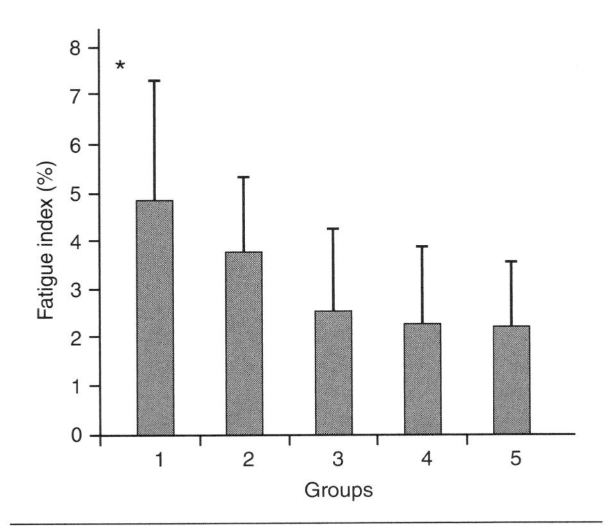

Figure 12.9 Effect of aerobic fitness on fatigue rates during anaerobic exercise. * = Significant change.

Adapted, by permission, from J.R. Hoffman, 1997, "The relationship between aerobic fitness and recovery from high-intensity exercise in infantry soldiers," *Military Medicine* 162:484-488. © Military Medicine: International Journal of AMSUS.

Summary

This chapter reviews the specific adaptations seen after endurance training. Although every athlete who performs endurance exercise experiences these adaptations, the ability of the athlete to be competitive may be limited by genetic factors. There appears to be an upper limit, or ceiling, to improvement of aerobic capacity. In addition, endurance athletes must exercise at or slightly above their lactate threshold to maximize their performance gains. When it is not feasible to calculate lactate threshold, exercise intensity can be given as a percentage of maximal heart rate or as ratings of perceived exertion. Finally, with regard to endurance training for the anaerobic athlete, it appears that once a baseline aerobic fitness level is reached, further concentration on improving aerobic capacity may not be justified.

REVIEW QUESTIONS

1. What are the cardiovascular and metabolic adaptations associated with endurance training?
2. Identify three methods used to prescribe exercise intensity during a run.
3. Explain the importance of the lactate threshold to the endurance athlete.
4. Describe five different training methods used by endurance athletes as part of their training program.
5. Explain the role that endurance training may have in recovery following high-intensity exercise.

chapter 13

Concurrent Training

CHAPTER OBJECTIVES

After reading this chapter you should be able to do the following:

- Understand the limitations associated with performing endurance and resistance exercise.
- Explain issues of compatibility or lack of compatibility between different modes of exercise.
- Discuss the effects of concurrent training on physiological adaptations.
- Determine where potential disruption of optimal performance improvements may occur with the inclusion of different modes of training.
- Understand how training experience may affect performance changes during the competitive season.

For many sports, the reliance on more than one energy system dictates the inclusion of various modes of exercise in the training regimen of the athlete. However, training multiple energy systems and performing various types of training simultaneously, referred to as **concurrent training**, has important implications for the physiological adaptations from such training programs. The preceding chapters have demonstrated that the adaptive responses of the body are specific to the type of training program used. **Endurance training** results in physiological adaptations (e.g., increases in oxidative enzyme activity, capillary density, and mitochondrial content) that are conducive to improving and maintaining prolonged aerobic activities. **Resistance training** produces changes that are often in direct contrast to those seen during endurance training. These adaptations frequently include increases in muscle mass that may parallel decreases in mitochondrial volume density (MacDougall et al. 1979). Such contrasting adaptations due to endurance and resistance training have created a hesitance on the part of both endurance and strength athletes to engage in the other form of training for fear that it may compromise desired training adaptations. This chapter focuses on the compatibility of these forms of training and the effect that concurrent training has on development of **maximal aerobic capacity** and muscle **strength**, growth, and adaptation. In addition, the chapter discusses the use of concurrent training for individuals interested in general fitness and reducing body fat.

Effect of Concurrent Strength and Endurance Training on $\dot{V}O_2$

Hickson, Rosenkoetter, and Brown (1980) were among the first investigators to examine the effects of resistance training programs on aerobic power and short-term endurance. In their study, college-aged males active in recreational sports began a 5 day per week resistance training program. After 10 weeks, the subjects averaged a 38% increase in lower body strength (one-repetition maximum [1RM] squat), but their aerobic capacity was unchanged when expressed relative to body weight. Interestingly, their time to exhaustion in both cycle and treadmill exercise increased significantly, by 47% and 12%, respectively. This was one of the first studies to provide evidence not only contrary to the belief that resistance training may be detrimental to endurance performance but also showing that resistance training may in fact improve endurance performance. Several mechanisms were thought to underlie the improved endurance performance in these subjects. The authors suggested that an improved **glycolytic enzyme capacity** (a potential training adaptation) would provide a greater capacity to resynthesize **adenosine triphosphate (ATP)** as an immediate source of energy. In addition, the neurological adaptations commonly observed during the initial stages of a resistance training program may have resulted in an alteration in motor unit recruitment patterns, providing for an improved **exercise economy**.

Later studies combining endurance and resistance training demonstrated similar effects on aerobic capacity and endurance performance. These studies, using primarily untrained individuals, consistently reported that the combination of resistance and endurance training did not compromise the ability of the subjects to improve their maximal aerobic capacity (Dudley and Djamil 1985; Hunter, Demment, and Miller 1987; Kraemer, Volek et al. 1999; McCarthy et al. 1995). The studies lasted 6 to 12 weeks and typically used a 3 day per week resistance and endurance training program. Several included resistance and endurance training on the same day (McCarthy et al. 1995; Hunter, Demment, and Miller 1987; Kraemer, Volek et al. 1999). However, in some studies, resistance and endurance training were performed on alternate days for a total of 6 consecutive days (Dudley and Djamil 1985). In either scenario, the inclusion of a resistance training program did not impede the ability to improve maximal aerobic capacity (figure 13.1).

Combined endurance and resistance training has produced similar results for untrained subjects and physically active and trained subjects (Bishop et al. 1999; Hennessy and Watson 1994; Hickson et al. 1988). These studies have ranged in duration from 8 to 12 weeks, and the frequency of training has varied from study to study. Bishop and colleagues (1999) examined the effects of adding a 2 day per week resistance training program to the off-season regimen of endurance-trained female athletes. After 12 weeks of concurrent training, no significant differences in endurance performance, **lactate threshold**, or maximal aerobic capacity were reported between subjects who performed resistance training and those who did not. Hennessy and Watson (1994) investigated an 8-week, 5 day per week combined strength (3 days per week) and endurance training (4 days per week) program for anaerobic athletes. When subjects were compared with other athletes similar in training background who performed only endurance training 4 days per week, no differences were seen in aerobic capacity improvements. Other investigators have shown that when a resistance training program was added to the normal training regimen of experienced endurance-trained individuals, not only was aerobic capacity maintained but also short-term endurance was improved (Hickson et al. 1988). Improvements in endurance performance (**time to fatigue**) were

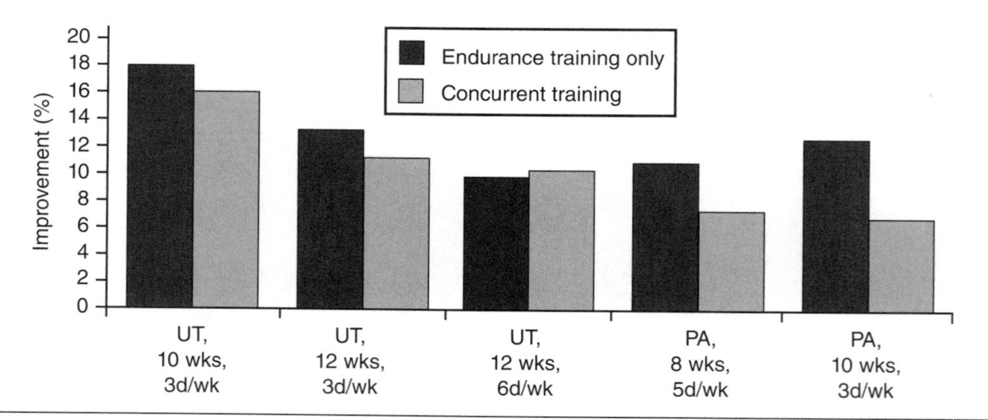

Figure 13.1 Effect of concurrent training on maximal aerobic capacity. UT, untrained subjects; PA, physically active subjects.

Adapted from McCarthy et al. 1995; Kraemer et al. 1999; Hunter et al. 1987; Hennessy and Watson 1994; Dolezal and Potteiger 1998.

seen in both cycling (11% improvement) and treadmill exercise (13% improvement). Although evidence consistently reflects the ability of concurrent training programs to improve maximal aerobic capacity, the benefits of resistance training in individuals with endurance training experience may not always be realized. One study, examining well-trained male cyclists, reported that a 6-week resistance training program (three undulating sessions per week) did not improve performance in a 30 km time trial (Levin et al. 2009). It is possible that the three additional training sessions compromised the athletes' ability to recover.

Some researchers have reported that an added resistance training program improves aerobic capacity in physically active subjects, but the magnitude of this improvement may be less than in subjects with similar training backgrounds who are performing only endurance training (Dolezal and Potteiger 1998). The addition of resistance exercise to the regimen of trained endurance athletes appears to enhance running economy, but has limited effects on $\dot{V}O_2$ kinetics (Millet et al. 2002). To summarize the research on concurrent training and aerobic capacity, it appears that adding resistance training to the aerobic conditioning programs of either previously sedentary individuals or endurance-trained athletes does not compromise improvements in aerobic capacity, and benefits to short-term endurance may also be realized. Resistance and endurance training appear compatible as long as the frequency of the endurance training is not reduced. If resistance training is added at the expense of aerobic exercise, the ability to maintain or improve aerobic capacity may indeed become compromised.

Effect of Concurrent Strength and Endurance Training on Maximal Strength

Results from studies on the effects of concurrently training for both endurance and strength improvements have been inconclusive. Several investigations have suggested that combining endurance and resistance training may compromise the potential for strength gains. However, a number of studies have shown no detrimental effects on strength improvement with the two modes of training combined.

In Practice

How Important Is It for Endurance Athletes to Include Resistance Training in the Training Routine?

The use of resistance training as part of the exercise regimen of endurance athletes is not a well-accepted or common practice, even though evidence is becoming clearer that the inclusion of resistance exercises does not suppress endurance gains as long as it is not at the expense of the aerobic training. However, time is a commodity, and often the coach or athlete has to decide on the priority of various drills in the time available. In any case, it becomes critical for both the athlete and the coach to understand the benefits and importance of incorporating resistance exercises into the weekly training routine. Sport scientists from Italy examined the effects of adding resistance exercises to the training program of master endurance athletes (Piacentini et al. 2013). The use of a maximal strength training program enhanced 1RM strength by 16% and improved running economy by 6% for these endurance runners performing at marathon running pace. The importance of strength training to endurance performance has been endorsed by some of the top track and field coaches within the National Collegiate Athletic Association. At the time of publication, Coach Greg Roy had been the head cross country and track and field coach at the University of Connecticut since 1984. Entering the 2012-2013 season he had a record of 258-41-3 in dual-meet competitions, including nine Big East championship wins, and had produced 31 All-Americans. He had been a big proponent of resistance training for his endurance athletes and had produced some of the strongest endurance athletes known. He is proud to state that one of his top cross country runners could squat 405 lb (184 kg). In addition to the benefits generally associated with resistance training with regard to running economy and reduction of injury risk, improvements in strength and power may give the endurance athlete an advantage during a close race—a greater "kick" at the end of the race (Roy 2013).

These contrasting results are likely related to differences in the training status of the subjects and specific differences in the training program design.

Dudley and Djamil (1985) reported that the combination of endurance and resistance training in previously untrained individuals compromised the magnitude of maximal torque (force acting at a distance from the axis of rotation) increases after 7 weeks. However, this effect was seen primarily at fast speeds of contraction. At slower contraction velocities, strength gains were still observed. In addition, the subjects performed strength and endurance training on alternate days, resulting in a 6 day per week training program. Such a high frequency of training may impede **recovery** and possibly result in **overtraining syndrome** (see chapter 24). A chronic state of fatigue could affect the potential for strength gains. However, because the stimulus that causes overtraining syndrome varies among individuals, training 6 days per week may not always result in fatigue. This was demonstrated in a study by Hunter, Demment, and Miller (1987), who also investigated untrained subjects performing 6 day per week concurrent training. The authors reported no detrimental effects on strength gains when these subjects were compared with a strength training–only group. When concurrent training was performed on the same day (endurance training preceded by resistance training or the reverse), the number of exercise days per week was reduced. When frequency of training is reduced, whether in subjects with limited or no resistance training experience, strength gains do not appear to be compromised (Bell et al. 1991; Gravelle and Blessing 2000; McCarthy et al. 1995).

The effect of adding endurance training to the exercise programs of subjects with experience in strength training does appear to impede the ability to maximize strength performance. Investigations in experienced resistance-trained subjects showed that upper (1RM bench press) and lower (1RM squat) body strength may be compromised when endurance training is performed concurrently with strength

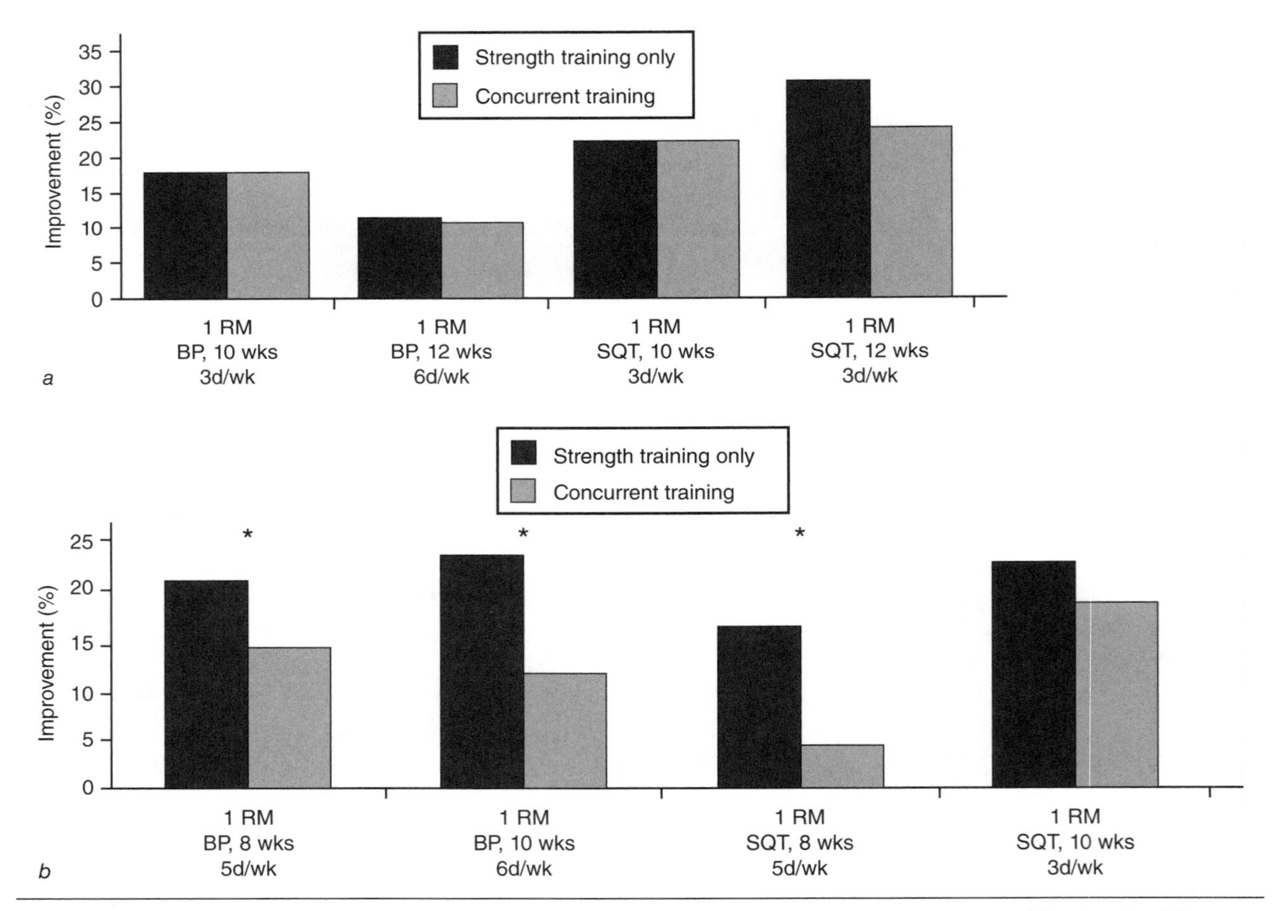

Figure 13.2 Effects of concurrent training on maximal strength in *(a)* untrained subjects and *(b)* trained subjects. BP, bench press; SQT, squat. *Significant difference.

a Adapted from McCarthy et al. 1995; Hunter et al. 1987.

b Adapted from Hennessy and Watson 1994; Delezal and Potteiger 1998.

training (Dolezal and Potteiger 1998; Hennessy and Watson 1994). In both of these studies, subjects performed endurance and resistance training on alternate days, resulting in a total frequency of training of 6 days per week. Thus, the reduction of strength improvement in these subjects may be related to possible chronic fatigue from the high frequency of training. Figure 13.2, *a* and *b,* shows the effect of concurrent training on the percent improvement of strength in previously untrained and trained subjects, respectively.

The cumulative effect of fatigue appears to be a large contributing factor to diminished strength improvements in subjects with some resistance training experience. In studies of novice weightlifters, even a 6 day per week concurrent training program did not compromise strength gains (Hunter, Demment, and Miller 1987). Initial increases in strength are generally the result of neurological adaptations occurring within the first 2 months of training. It is possible that these neurological adaptations are not impeded in the novice lifter even at a higher frequency of training. In subjects with some resistance training experience, the physiological adaptations that contribute to further increases in strength involve muscle structural changes and may be more sensitive to additional modes of training and to fatigue.

Leveritt and Abernethy (1999) studied the acute effects of high-intensity endurance exercise on a resistance training session in recreationally trained subjects, showing that endurance exercise performed 30 min before resistance training reduced strength performance (figure 13.3). The number of repetitions performed during three sets of squat exercises was reduced 13% to 36% when preceded by a high-intensity endurance workout in comparison with the number when no endurance exercise had been performed. In addition, isokinetic testing reduced peak torque 10% to 19% over a range of testing velocities when endurance training preceded the resistance exercise. These results suggest that if endurance training precedes resistance training, the stimulus to the muscle may be reduced, possibly limiting the extent of the physiological adaptations. However, if the strength training session precedes the endurance exercise, strength improvements may not be diminished.

Effect of Sequence of Training on Endurance and Strength Improvements

Studies on the effect of the **sequence of training** on both strength and aerobic capacity improvements are limited. In a study by Collins and Snow (1993) in previously untrained subjects, the sequence of training (whether endurance preceded resistance training or the reverse) did not appear to have any effect on either aerobic capacity or strength improvements. Thus, it appears that in an untrained population, the sequence may not be very important. However, in this study, subjects were required to perform only two sets of 3 to 12 repetitions per exercise using 50% to 90% of their 1RM. Although such a regimen can produce significant strength gains in this subject population, this is not a typical resistance training workout for more experienced resistance-trained subjects (see chapter 8). Further study on the effect of sequence of training using a more experienced subject population would be welcome.

Another concern related to sequence of training is whether concurrent training should be performed on alternate days or whether the resistance and endurance workouts should be performed on the

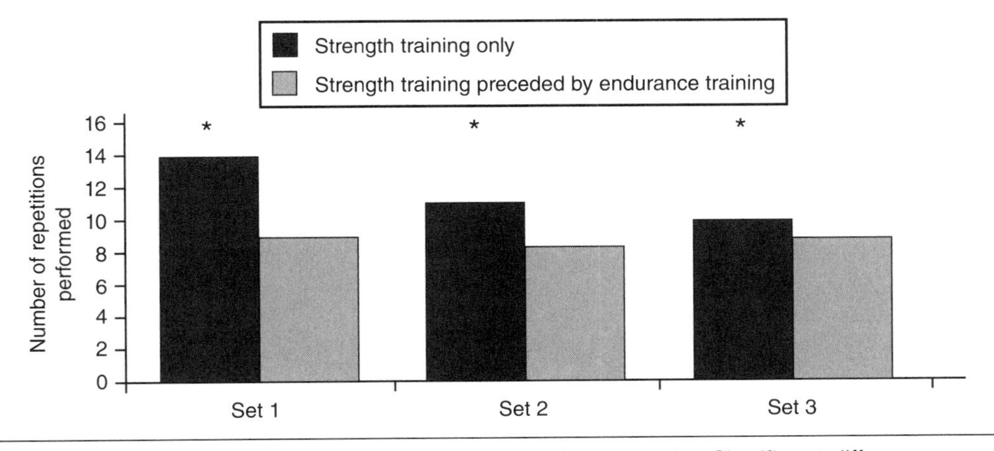

Figure 13.3 Acute effect of endurance training on strength performance. * = Significant difference.

Adapted from Leveritt and Abernathy 1999.

same day. By training on alternate days, the subject may possibly extend the weekly training sessions to 6 days per week. In the untrained subject, this does not appear to impede strength or aerobic capacity improvements (Hunter, Demment, and Miller 1987). However, in more experienced subjects, a high frequency of training (e.g., 6 days per week) may provide insufficient time for recovery between sessions, which compromises strength improvements (Dolezal and Potteiger 1998; Hennessy and Watson 1994). In contrast, when endurance and strength training are performed on the same day, there does not appear to be any detrimental effects on the ability of subjects to increase their strength or aerobic capacity (Bell et al. 1991; Gravelle and Blessing 2000; McCarthy et al. 1995). Although the daily training volume may be greater with reduced weekly training frequencies, the rest between sessions may be sufficient to allow complete recovery.

Sale and colleagues (1990) looked specifically at the question of same-day versus alternate-day training in previously untrained college-aged male subjects. Resistance and endurance training were performed twice per week. Both groups made significant improvements in aerobic capacity, with no significant difference observed between the groups. However, the group that trained 4 days per week (alternate-day training group) made significantly greater strength gains than the group that trained only twice per week (same-day training group) (24% vs. 13%, respectively). Perhaps by training on alternate days but limiting the total number of weekly training sessions, the subjects had a more complete recovery, resulting in significantly greater strength gains. The subjects in this group may have had a better-quality workout because it was not preceded by an endurance training session. Thus, when rest is adequate, significantly greater improvements in strength may be realized, even when the frequency of training is high.

Effect of Concurrent Training on Muscle Growth and Muscle Fiber Characteristics

Many endurance athletes are concerned that the inclusion of a resistance training program may cause physiological changes to the muscle that would be detrimental to endurance performance (e.g., muscle **hypertrophy**, decrease in mitochondrial volume and capillary density). In studies on this particular question, no significant alterations in muscle size or muscle fiber composition were reported in endurance-trained subjects performing resistance exercise for the first time (Bishop et al. 1999; Hickson et al. 1988). It should be noted, however, that these studies were 10 to 12 weeks in duration. It is generally understood that muscle hypertrophy occurs approximately 6 to 8 weeks after the initiation of a resistance training program in previously untrained individuals (see chapter 1 for a review). It is possible that the time course for muscle adaptations in individuals with limited resistance training experience may be longer when they are concurrently performing another mode of training. Based on available evidence, it appears that 10 to 12 weeks of resistance training added to the exercise program of endurance-trained subjects do not cause any significant changes to muscle fiber composition or to fiber cross-sectional area. The effect of training durations exceeding 3 months is not known. However, by manipulating acute program variables, the coach and athlete can specifically focus on the physiological adaptations that are beneficial to the endurance athlete.

The effect of concurrent training on muscle adaptations in other population groups may be different. In previously untrained subjects, 6 weeks of resistance training resulted in a significant increase in the muscle fiber area of both **type I** and **type II** fibers in a group that performed only strength training (Bell, Syrotuik, et al. 2000). These muscles continued to hypertrophy even after 12 weeks of training. In comparison, a group of subjects performing both endurance and resistance training (using the same resistance training program) showed no significant changes in muscle fiber area after 6 weeks of training. However, after 12 weeks, significant increases were observed in the muscle fiber area of type II fibers only. This study, although showing significant increases in muscle fiber area, suggested that combining endurance and strength training might suppress some of the adaptations seen with strength training only. Kraemer, Patton, and colleagues (1995) obtained similar results in physically active subjects. In their study (figure 13.4), significant increases in both type I and type II fibers were seen in both strength- and combined strength- and endurance-trained groups. A significant reduction in the type I fiber area was observed in a group of subjects performing only endurance training. This study demonstrated an apparent benefit of resistance training for the endurance-trained individual. Not only did resistance training not compromise aerobic capacity, but the inclusion of this mode of training appeared to prevent the muscle fiber **atrophy** that may accompany endurance training. As discussed later in this chapter,

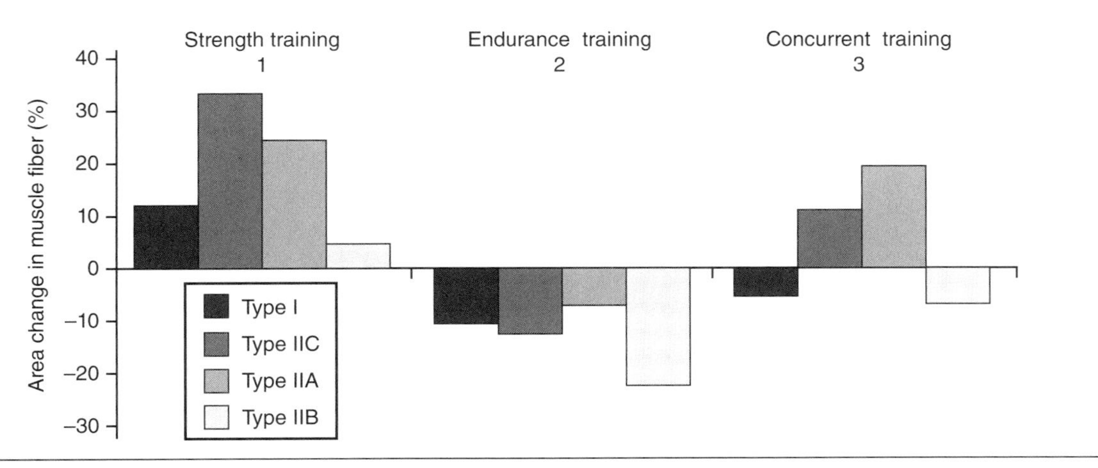

Figure 13.4 Effect of concurrent training on muscle fiber alterations.

Adapted, by permission, from W.J. Kraemer et al., 1995, "Responses of IGF-1 to endogenous increases in growth hormone after heavy-resistance exercise," *Journal of Applied Physiology* 79:1310-1315.

such adaptations may have important implications for individuals interested in increasing their daily energy expenditure as part of a weight reduction program.

Longer-duration concurrent training studies do not appear to support the concept of an **interference effect**. Hakkinen, Alen, and colleagues (2003) randomized men with no previous resistance training experience into either a 2 day per week resistance training–only group or a 2 day resistance training and 2 day endurance training group. The study was 21 weeks in duration, and the resistance training program was periodized from low (50-70% of 1RM) to moderate (60-80% of 1RM) in intensity. The endurance training program consisted of progressive increases in training duration (30 min to 90 min during the examination period) at intensities that varied between above and below the anaerobic threshold. The primary findings of the study were large gains in leg extension maximal strength (21% and 22% in the strength and strength–endurance group, respectively) and significant increases in the cross-sectional area of the knee extensors in type I, type IIa, and type IIb fibers for both groups, with no between-group differences. Thus, in concurrent programs using 2 day per week resistance and endurance training (total weekly workout of 4 days per week), no interference in strength or muscle hypertrophy is seen. Similar results were reported in middle-aged (~40 years) men engaging in a lower-frequency training program (Izquierdo et al. 2005). In this 16-week study, a 2 day per week resistance training program was compared to a 1 day per week resistance and 1 day per week endurance training program. Although strength improvements at week 8 were similar between groups (22% vs. 24% in strength and strength-endurance, respectively), by week 16 the strength improvements were greater in the strength-only group compared to the concurrent training group. Changes in muscle cross-sectional area mirrored changes in strength in that similar increases were seen within the first 8 weeks of training; but by week 16, the strength-only group increased muscle cross-sectional area by 9% while no significant changes were realized by the combined group (4%) or an endurance-only group (1%).

Kraemer, Patton, and colleagues (1995) also demonstrated fiber subtype transformations that were specific to the training program employed. Subjects who performed only strength training (4-day split routine using 2 heavy days [5RM] and 2 moderate days [10RM]) or performed both strength and endurance training (4 days per week using 2 days of prolonged runs [45 min at 80-85% $\dot{V}O_2$max] and 2 days of interval training) saw similar type II fiber subtype transformations (IIa ← IIb) after training. In contrast, subjects who performed only endurance training had a type II muscle fiber subtype transformation from fibers that were more glycolytic to fiber subtypes that were more oxidative (IIc ← IIa ← IIb). In other words, the inclusion of a high-intensity resistance training program appears to reduce the magnitude of the type II fiber transformation to more oxidative fibers.

The effect of concurrent training on muscle fiber subtype changes in elite endurance athletes appears to have the same pattern as seen in studies of recreationally trained individuals. A 16-week study on elite cyclists reported an increase from 26% to 34% in type IIa fibers in the group that performed both resistance and endurance training, while no change in fiber subtype composition was noted in the endurance-only group (Aagard et al. 2011). These changes came at the expense of type IIx fibers. Similar to findings in other studies regarding the interference effect on muscle hypertrophy changes, the endurance training appeared to blunt the growth in muscle cross-sectional area. The blunting of muscle hypertrophy may be a function of training frequency and possibly training level. As the frequency of training increases, which is typically what occurs among elite athletes, skeletal muscle may become more sensitive to the different training stimuli.

Effect of Concurrent Training on Protein Signaling

Resistance and endurance training stimulate specific responses to protein signaling within skeletal muscle (Coffey et al. 2006; Dreyer et al. 2006; Koopman et al. 2007). The inhibition or activation of various signaling pathways determines the magnitude of the anabolic process within skeletal tissue. Resistance training has been shown to stimulate both myofibrillar and muscle mitochondrial protein synthesis in untrained muscle, whereas endurance training appears to stimulate only mitochondrial protein synthesis (Wilkinson et al. 2008). After 10 weeks, resistance training appears to become more specific by stimulating myofibrillar protein synthesis only, while endurance training continues to stimulate only mitochondrial protein signaling. Resistance training appears to heighten the responsiveness of the phosphorylation of key proteins involved in muscle growth. Interestingly, prior training experience may attenuate some of the exercise-specific signaling responses following adaptation to training (Coffey et al. 2006). That is, an endurance-trained athlete may not respond with the same degree of sensitivity as the resistance-trained athlete to a bout of resistance exercise. This may partially explain why resistance training can limit the atrophy of skeletal muscle consequent to endurance exercise, but is unable to compensate to maintain a similar hypertrophy response in the endurance athlete while possibly able to do so in the recreational or untrained individual.

Effect of Concurrent Training on Hormonal Adaptations

Combining endurance and resistance training in previously untrained subjects does not compromise performance gains from either. However, in subjects with resistance training experience, concurrent training may potentially hinder their ability to maximize strength gains, likely related to a reduced adaptive ability of the muscle. As mentioned earlier, the neurological adaptations that lead to initial strength increases are not likely to be affected considering that strength improvements in the novice lifter do not appear to be diminished by concurrent training. Because **anabolic** and **catabolic hormones** are intimately involved in both exercise recovery and muscle growth, it would appear that any difference in the way hormones respond to concurrent training might play a key role in regulating muscle adaptations.

In a study (Kraemer, Volek et al. 1999) addressing the effect of concurrent training on hormonal responses, previously sedentary subjects were placed into one of three training groups: a strength-only group, an endurance-only group, or a group that combined strength and endurance training. Each group exercised 3 days per week. After 12 weeks of training, no significant differences in the resting concentrations of either **testosterone** or **cortisol** were seen among the groups.

In studies of physically active or trained subjects, the hormonal response to these various modes of training appears to be specific to the exercise regimen employed. Kraemer, Patton, and colleagues (1995) examined the compatibility of endurance and strength training in physically active men for 12 weeks. In the group that performed only resistance training, no changes in testosterone concentrations were observed during the course of the program. However, cortisol concentrations were reduced by the eighth week, suggesting that a greater anabolic environment existed for muscle growth (reflected by a greater testosterone/cortisol ratio). In the group that performed only endurance training, no significant changes in testosterone were seen, but significant elevations in cortisol concentrations were reported. In the group that performed both endurance and resistance training, significant increases in both testosterone and cortisol were observed at the end of the training period. However, the relative increase in cortisol far exceeded that of testosterone, reflecting a greater exercise stress from the combination of endurance and resistance training. These results were confirmed by other studies showing a greater catabolic response to combined strength and endurance training versus endurance training only in previously trained individuals (Bell et al. 1997).

The importance of the hormonal environment for muscle growth is discussed in chapter 2. Briefly, changes in the anabolic and catabolic hormonal profile influence cellular changes relating to **protein synthesis** and muscle fiber adaptations. If the catabolic hormone response is elevated, increases in protein degradation may be seen. This was demonstrated by Kraemer, Patton, and colleagues (1995), who showed a decrease in fiber size that corresponded to an increase in cortisol concentrations in subjects performing endurance exercise. The authors further showed that the addition of resistance training to an endurance training program may preserve muscle mass by increasing the anabolic response. This may not only prevent the catabolic effects of the elevated cortisol concentrations but also provide enough of a stimulus to cause slight increases in type IIa muscle fibers. Figure 13.5 shows a diagram of the hormonal response to various modes of training, including concurrent training, and the possible effects on muscle fiber.

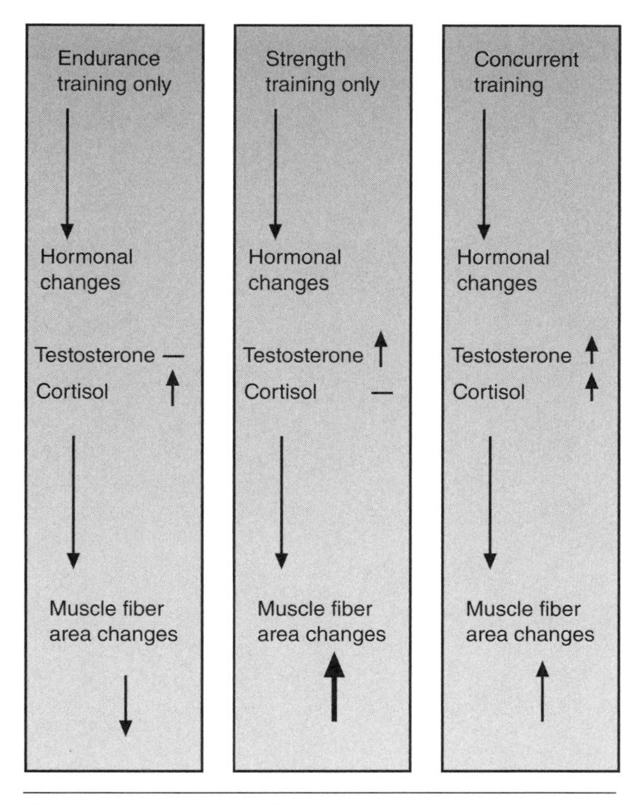

Figure 13.5 Hormonal and muscle fiber response to concurrent training.

Effect of Concurrent Training on Basal Metabolic Rate and Weight Loss

To reduce body weight and alter **body composition**, it is generally accepted that prolonged endurance training is more advantageous for increasing energy output than any other form of training. Endurance training is also thought to play a role in potentiating both **basal metabolic rate (BMR)** and **resting metabolic rate (RMR)** (Ballor and Poehlman 1992). However, results from some studies on the ability of endurance training to increase BMR or RMR have been ambiguous. Some investigations have shown either no change in BMR (Sjodin et al. 1996) or a decrease in RMR (Thompson, Manore, and Thomas 1996) after endurance training. In contrast, resistance training is recognized as the primary mode of training to increase muscular strength, and it also positively alters body composition, primarily by increasing lean tissue. An increase in lean body mass increases the BMR and results in an increase in total energy expenditure. Thus, for individuals interested in weight reduction and body composition changes, the combination of endurance and resistance training appears to provide a more optimal advantage in eliciting a training effect.

Very few studies have examined the effect of concurrent training on BMR and weight loss. Dolezal and Potteiger (1998) were among the first investigators to look at the effects of combined endurance and resistance training on BMR and body composition changes compared with resistance or endurance training alone. They showed that after 10 weeks of concurrent training, subjects were able to significantly increase their BMR and reduce their body fat from preexercise values. Nevertheless, the magnitude of each change was still lower than that seen when subjects performed each mode of training by itself. However, it was clear that by combining the two modes of training, the subjects were able to achieve all the benefits typically derived from performing each mode of exercise alone. The benefits of adding resistance training to an endurance training program or to a diet and endurance training program appear to lie in the ability to prevent the decline in lean tissue resulting from the catabolism consequent to endurance training and from a low-calorie diet (Kraemer, Volek et al. 1999). The combination of resistance and endurance training has been shown to result in significant improvements in lean body

In Practice

What Are the Potential Health Benefits Associated With Concurrent Training?

Much of the discussion surrounding concurrent training focuses on potential negative aspects—how one mode of training may interfere with progression in a fitness component associated with another mode. However, for noncompetitive, physically active individuals, concurrent training may provide an opportunity to experience greater health benefits. Each training component confers specific benefits that can enhance the health of the general population; but by including various modes of training, one expands the health benefits. Takeshima and colleagues (2004) examined the effect of a concurrent resistance and endurance training program in older adults (average age ~ 68 years). Following a 12-week circuit program that included both resistance and endurance exercises, the authors found significant improvements in strength, aerobic capacity, body composition, and lipid profile. For this segment of the population, the primary focus has to be on a broad range of exercise stimuli to provide maximal benefits for health. This is critical for exercise professionals who work with older adults. The inclusion of resistance exercise helps combat osteoporosis and sarcopenia, as well as functional parameters such as gait speed and balance, while endurance training has important benefits for cardiovascular fitness, blood pressure control, lipid profile, and body weight management. For this population it is not a question of which mode of training to use; what becomes more relevant is to ask how to incorporate it all.

mass (2.2%) and decrease in fat mass (1.4%) after 14 weeks in middle-aged women (Fleck, Mattie, and Martensen 2006).

Effects of Combined Sprint and Resistance Training

The exercise science literature has generally considered only endurance and resistance training as the modes of exercise in the context of concurrent training. However, given that resistance and anaerobic endurance training (e.g., sprint and interval training) are two separate forms of training and are performed concurrently by many anaerobic athletes, it seems appropriate to make some mention of this combination. These modes of training appear to complement each other. However, increasing the risk of overtraining would be a major concern. Typically, training programs alter both the intensity and volume for each type of exercise in order to minimize the risk of fatigue (see chapter 14). Wong and colleagues (2010) demonstrated, in professional soccer players, that 8 weeks of concurrent resistance training (four sets of 6RM) and high-intensity interval training (15 s running and 15 s recovery for 16 intervals) during the preseason was effective in increasing strength and power performance and improving aerobic endurance. The mechanisms whereby high-intensity interval training can increase aerobic endurance are discussed in chapter 3. During the competitive season, the emphasis on anaerobic training may compromise the ability of these athletes to improve their strength (Hoffman, Maresh, et al. 1991). Thus, frequency and volume of resistance training are often reduced to twice per week and to four or five exercises per workout, respectively. The ability to enhance strength during the competitive season may be related to the athlete's training experience. Younger, less experienced athletes (i.e., freshman football players) have been shown to increase strength during the season, whereas more experienced resistance-trained players were able only to maintain their strength levels (Baker 2001; Hoffman and Kang 2003; Hoffman, Wendell, et al. 2003). Thus, the goals of the training program are adjusted to accommodate realistic expectations for specific times of the year.

Due to the compatibility of resistance and sprint training, these training programs are often performed in consecutive fashion. That is, an athlete performs the resistance workout and then proceeds to the track or field and begins the sprint workout. Coffey and colleagues (2009) have examined the effect of order of the workouts on sprint performance and muscle protein signaling. Comparisons of a sprint workout (ten 6 s sprints, 54 s recovery) performed before and after a resistance training session revealed no significant difference in mean power generated during the entire set of sprints, but fatigue was greater (19% vs. 11%) when the sprints were performed before compared to after the resistance training. It is possible that the greater fatigue was related simply to a higher power output during the initial sprint—or perhaps the resistance training session potentiated subsequent sprint performance. However, the response of the protein signaling pathway suggested that sprint training performed close in time to resistance training (15 min recovery) may attenuate the signaling pathway responsible for protein synthesis. This is an important consideration for coaches looking to maximize muscle growth and performance. The use of an extended recovery period (>15 min) appears to be warranted, yet little scientific study has been conducted in this area. Furthermore, sprint and resistance training are often integrated in the later stages of the yearly training program. During these phases of training, the primary goals for the strength and conditioning professional and athlete may be directed more toward the conditioning effect than toward muscle hypertrophy. However, the use of longer recovery periods between sprint and resistance training workouts appears to be prudent.

Summary

Concurrent training does not appear to compromise the ability of either trained or untrained individuals to realize gains in aerobic capacity. In addition, the inclusion of resistance training in the programs of endurance athletes may improve their short-term endurance. However, the addition of an endurance training program to the resistance training regimen of physically active or experienced resistance-trained individuals appears to negate the potential for improvement in maximal strength. This does not seem to be the case in previously untrained individuals. Further benefits derived from combining resistance and endurance training include the ability to maintain or improve lean muscle mass and decrease fat mass, suggesting that this is an ideal method of training for people whose primary goal is to reduce body fat percentage.

REVIEW QUESTIONS

1. Explain the effect of endurance training on maximizing muscle strength and size improvements.
2. Discuss the effect that resistance training may have on changes on endurance performance in the endurance athlete.
3. Discuss differences in sequence of training in relation to both strength and endurance improvements.
4. Discuss what is meant by compatibility of training, and provide examples.
5. What benefits might training have for individuals interested in enhancing health and fitness?

chapter 14

Periodization

CHAPTER OBJECTIVES

After reading this chapter you should be able to do the following:

- Explain the general adaptation syndrome and how it applies to the training of athletes.
- Understand the importance of yearly manipulation of training variables to maximize performance at optimal times and minimize the risk of overtraining or overuse injury.
- Differentiate between linear and nonlinear or undulating periodization models.
- Discuss the value of different periodization models and when one would become more applicable than another.
- Learn how to develop periodized training models for team sports, placing emphasis on a season of competition.
- Learn how to develop periodized training models for individual sports or athletes focusing on peaking at a specific time point of the competitive season.

Coaches manipulate training programs at regular time intervals to help athletes achieve optimal performance gains and meet their training objectives. Training goals vary according to the type of sport being played. An athlete who plays American football, or any other sport that places emphasis on performance over a complete season, aims to achieve peak condition at the onset of the season. The athlete is then concerned with maintaining that level of conditioning throughout the season. In contrast, other athletes (e.g., gymnasts, swimmers) may concentrate on a major competition that occurs toward the end of the competitive year. Thus they need to attain peak condition at that finite point in the competitive season. To achieve these varying goals, coaches manipulate training variables (e.g., **intensity**, **volume**). This chapter provides background on **periodization** and discusses the efficacy of periodization and different models of periodized training programs. Much of the published literature has focused on the effect of periodization on strength–power training. However, the aim here is to discuss periodization in relation to all sport disciplines.

Periodization for All Disciplines

Periodized training (development of an annual training program) was reported as far back as the ancient Olympic Games and was employed by coaches throughout the 20th century (Bompa 1999). However, it was not until the later half of the 20th century, when sport scientists began to publish their work, that periodized training received much attention (Bompa 1999). Most of this work was published by scientists from Eastern Bloc countries such as Russia, Romania, and East Germany.

In 1965, Russian sport scientist Dr. Leonid Matveyev published a model of his periodized training program that divided the training year into several different phases and cycles. Most of this program was related to the **general adaptation syndrome (GAS)** developed by Dr. Hans Selye (Stone, O'Bryant, and Garhammer 1981). The GAS suggests that there are three response phases to stressful demands placed on the body. The first phase, referred to as the **alarm phase**, is the initial response to the stimulus (i.e., exercise) and consists of both shock and soreness. An exercise stimulus or a change in the **exercise prescription** frequently results in a reduction in performance. The second phase is an **adaptation** to this new stimulus. The body adapts to the training stimulus or change in the exercise prescription, and an improvement in performance is observed. The third phase is one of **exhaustion**. The body is unable to make any further adaptation to the training stimulus. Unless the stimulus is reduced, a situation leading to chronic fatigue (**overtraining**) may occur. If sufficient recovery is allowed, the body can then make further adaptations, and performance may increase further. The goal of periodization is to avoid or minimize periods of exhaustion and to maintain an effective exercise stimulus, which leads to maximizing athletic potential.

The basic principle of periodization is a shift from an emphasis on high volume and low intensity to an emphasis on low volume and high intensity. This can be seen in a modified version of Matveyev's periodization model (figure 14.1). Matveyev divided his training year into three distinct phases: **preparatory**, **competitive**, and **transitional**. Each phase of the training program relates to a change in the volume and intensity of training. The preparatory phase of training may comprise two subphases known as **general preparation** and **specific preparation**. During the general phase of preparatory training, volume is high and the intensity of exercise is low. The primary purpose of this phase is to prepare the athlete for more intense and sport-specific training in later phases. The second part of this phase is a more specific preparation in which intensity of exercise increases and the volume is reduced. The competitive phase, which consists of all the competitions, may also be divided into subphases. The exhibition contests may be considered part of the **precompetition phase**, and the primary or most important competitions are considered part of the **main competition phase**. The essential difference between these two competition phases is how the volume and intensity of exercise are manipulated. During the precompetition phase, volume of training is reduced as intensity

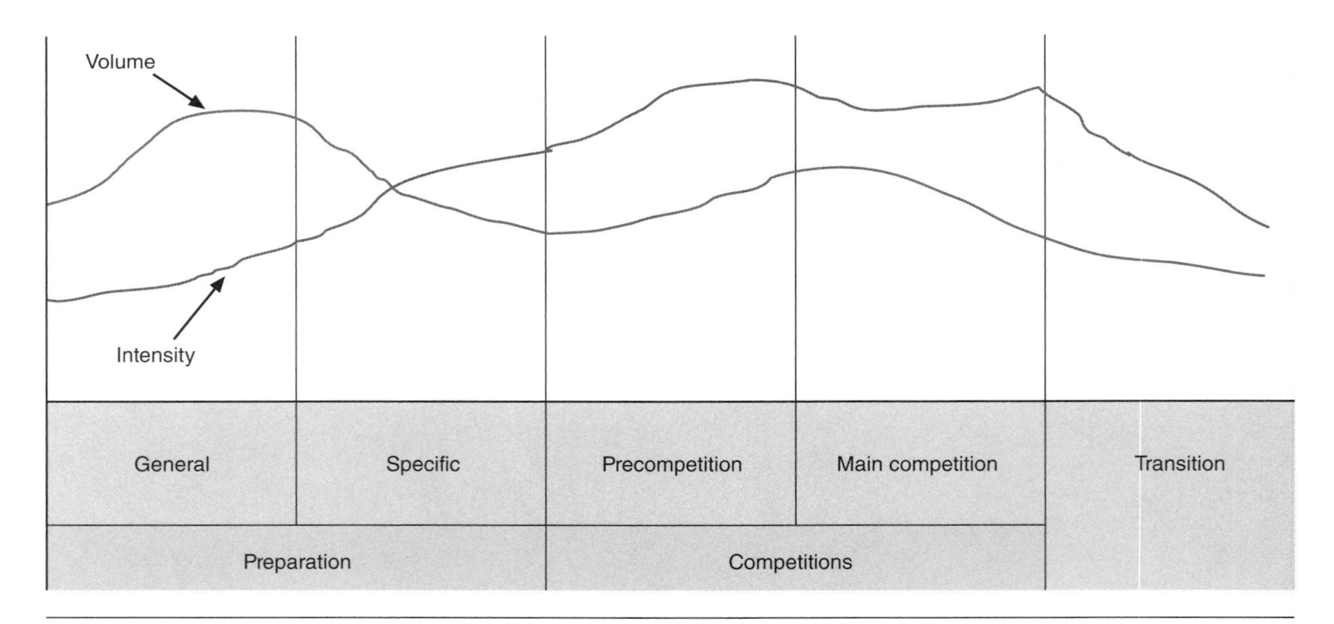

Figure 14.1 Matveyev's model of periodization.
Adapted from Bompa 1999.

steadily increases. As the athlete gets closer to the main competition, training intensity reaches its highest level, and volume falls to its lowest point in the competitive year. The final phase of the year is the transitional phase, when the athlete enters a period of **active rest** in which both volume and intensity of training are substantially reduced.

In the later half of the 20th century, Western scientific literature contained a number of studies examining different combinations of sets and repetitions to produce maximal strength gains (Berger 1962, 1963; O'Shea 1966). However, it was not until the complete domination of the Eastern European countries in the weightlifting events at the Olympic Games of the 1960s and 1970s that the optimal way to develop a training program was reexamined. Stone, O'Bryant, and Garhammer (1981) published a strength training model that adapted the work of Matveyev. Their training program was divided into four different phases, or **mesocycles** (figure 14.2). Each mesocycle may last 2 to 3 months, depending on the athlete. The initial mesocycle is typically called the **preparatory**, or **hypertrophy**, phase. Similar to the first phase of Matveyev's model, this cycle consists of high-volume and low-intensity training. It is designed primarily to increase muscle mass and endurance. The objective is also to help prepare the athlete for the more advanced or intense training seen in the later cycles of the training program. In the next two mesocycles (**strength** and **strength-power**), intensity is elevated and volume is reduced. As the names of these two mesocycles indicate, they are primarily concerned with strength and power development. The final mesocycle of the training year is the **peaking** phase. During this cycle, the athlete prepares for a single contest by reducing volume and increasing intensity in similar fashion to Matveyev's main competition phase. Table 14.1 compares volume and intensity across the various mesocycles in a strength–power athlete.

Athletes who participate in a sport that places importance on an entire season must achieve peak

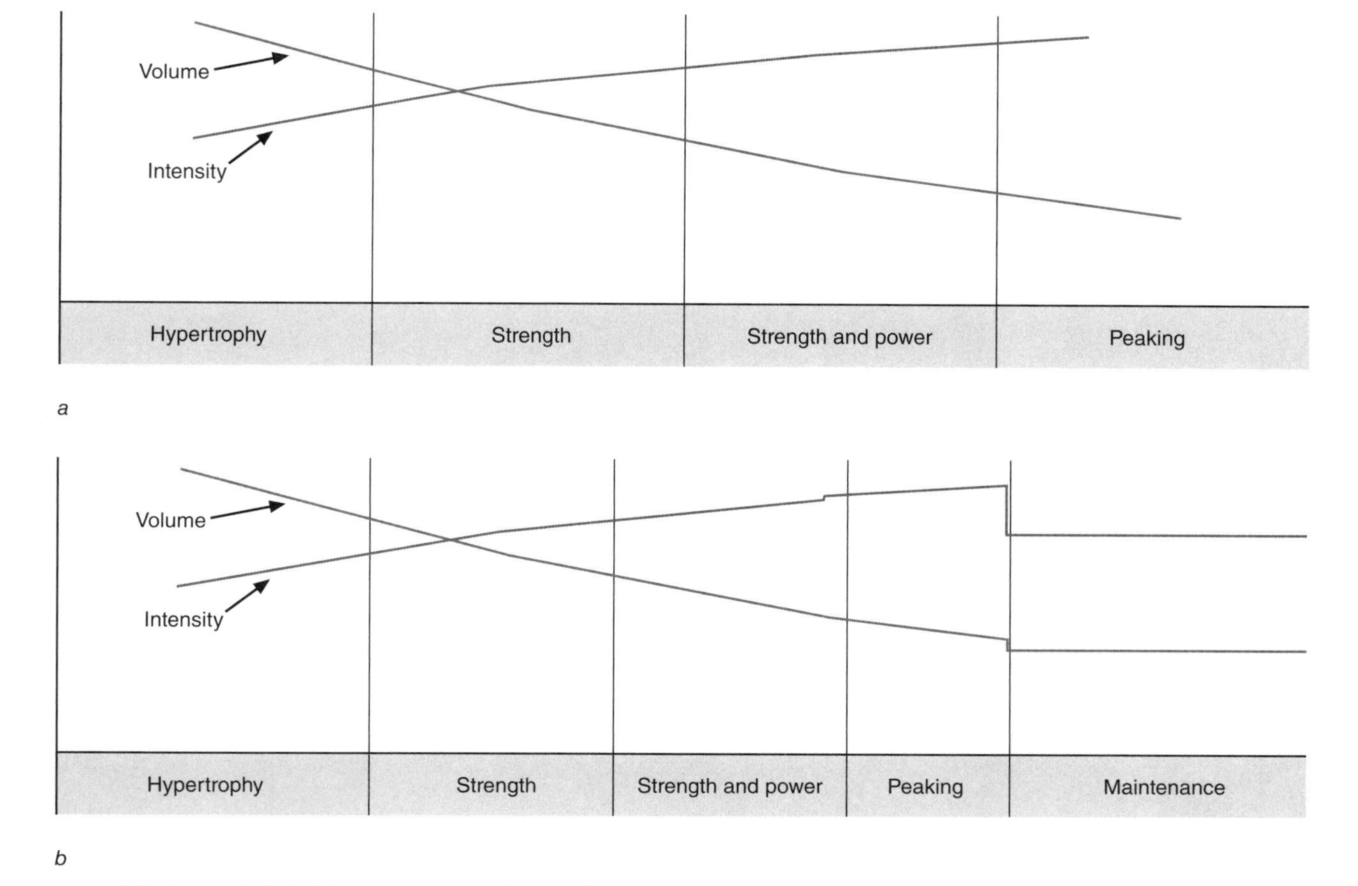

Figure 14.2 Periodization model for a strength–power athlete training for *(a)* a single competition and *(b)* an entire season.

condition by the onset of the competitive year. Thus, after the peaking phase, which may last for several weeks before the preseason, the athlete performs a **maintenance** phase to maintain the strength gains made during the off-season (figure 14.2*b*). During the maintenance phase, the intensity of exercise is reduced to a level similar to what might have been used during the strength phase (6-8RM [repetitions maximum]). The volume of training is also lowered by reduction in the number of assistance exercises performed. These athletes may also have a short peaking phase that precedes a training camp or the start of the season. It is also not uncommon to have short training cycles called **microcycles** that ease the transition from mesocycle to mesocycle.

Models of Periodization

The periodization model that has been the focus of discussion until now consists of uniform changes in training intensity and volume that remain relatively constant throughout each mesocycle. This **linear** model of periodized training is the classic form for designing most periodized training programs. However, **nonlinear**, or **undulating**, periodization models of resistance training are also becoming popular. This model of periodization varies the volume and intensity of training from workout to workout. An example of this model can be seen in table 14.2. Light, moderate, and heavy intensities can be alternated during each week of training. This model has been shown to be as effective as the traditional or linear model of periodization (Baker, Wilson, and Carlyon 1994; Poliquin 1988) and may be appropriate as a training program for sports that place importance on a season of competitions (Fleck and Kraemer 2004). For instance, athletes who play a varied schedule, such as basketball, hockey, baseball, or soccer players, may have several games or competitions in a given week, or their schedule of competitions and travel do not permit a regularly scheduled in-season strength maintenance program. It may be preferable to some coaches and athletes to use relatively light intensities of training preceding or on days of competition. In the undulating periodized program, the athlete can still train at a high intensity but at a more appropriate time of the week.

Efficacy of Periodization

Up to this point, periodization has been discussed as it relates to maximizing athletic potential. However, athletes are unable to maintain peak condition for a prolonged period of time without fatiguing. Manipulating both training volume and intensity allows the athlete to make the necessary physiological adaptations while reaching peak condition at the appropriate time. In addition, the athlete minimizes the risk of developing an overtraining syndrome by altering volume and intensity of training in conjunction with short, appropriately timed **unloading phases** (tapering) after increases in training work.

Table 14.1 Volume and Intensity in a Periodized Strength Training Program

	Volume		Intensity
Mesocycle	Sets	Repetitions	1RM
Hypertrophy	3-5	8-12	60-75%
Strength	3-5	6-8	80-85%
Strength-power	3-5	4-6	85-90%
Peaking	3-5	2-4	>90%

Table 14.2 Example of Nonlinear Periodized Training

	Sets	Repetitions	Rest between sets	Training focus
Day 1	4-5	3-5RM	3 to 4 min	Power
Day 2	4-5	6-8RM	2 to 3 min	Strength
Day 3	4-5	9-12RM	1 min	Hypertrophy

Relatively few studies have examined whether a periodized training program is more effective than a nonperiodized training program. The vast majority of studies have focused primarily on the effects of a periodized resistance training program on maximal strength or on the enhancement of components of athletic performance, such as vertical jump height, speed, or agility (Baker, Wilson, and Carlyon 1994; Hoffman, Ratamess, Klatt, et al. 2009; Kraemer 1997; McGee et al. 1992; O'Bryant, Byrd, and Stone 1988; Stowers et al. 1983; Willoughby 1992, 1993).

Significant increases in strength have been shown in both periodized and nonperiodized resistance training programs. Improvements in upper body strength (as determined primarily by a 1RM bench press) ranged from 8% to 17% from pretest values in individuals performing a nonperiodized resistance training program (Baker, Wilson, and Carlyon 1994; Hoffman, Ratamess, Klatt, et al. 2009; Stowers et al. 1983; Willoughby 1992, 1993). Lower body strength increases (as determined primarily by a 1RM squat) ranged in those same individuals from 20% to 32%. These increases, reported from nonperiodized, multiple-set training programs, reached statistical significance in all studies. Significant improvements were also noted pre- to posttraining in studies that examined single-set nonperiodized training (3-13% and 6-14% increases in upper and lower body strength, respectively) (Kraemer 1997; Stowers et al. 1983). However, the magnitude of improvement was significantly less than that seen with the multiple-set nonperiodized training programs. When comparing multiple-set nonperiodized resistance training programs with periodized resistance training programs, most studies reported significantly greater strength gains in subjects performing a periodized training program. Strength improvements ranged from 8% to 29% in upper body strength and from 11% to 48% in lower body strength in subjects performing periodized resistance training programs (Baker, Wilson, and Carlyon 1994; Hoffman, Ratamess, Klatt, et al. 2009; Kraemer 1997; Stowers et al. 1983; Willoughby 1992, 1993). In addition to greater strength development, subjects performing periodized resistance training also showed a greater improvement in vertical jump power (Stone, O'Bryant, and Garhammer 1981) and vertical jump height (Stowers et al. 1983) than subjects in nonperiodized programs. The results of these studies appear to indicate that periodized resistance training is more effective than nonperiodized training in eliciting strength and motor performance improvements.

However, most of these studies were performed on noncompetitive athletes. A 14-week study, one of the few to examine experienced resistance-trained athletes, failed to show any significant differences between no periodization (6-8RM in traditional power exercises and 3-4RM in Olympic movement exercises) and periodization (a 4-week hypertrophy [9-12RM] phase, a 6-week strength [6-8RM, 3- or 4RM in Olympic movement exercises] phase, and a 4-week power [3-5RM, 1- or 2RM in Olympic movement exercises] phase) (Hoffman, Ratamess, Klatt, et al. 2009). The athletes who maintained the same training program over the 14-week study increased their 1RM bench press by 8.7% and their 1RM squat by 20.7%, while the periodized group improved their 1RM bench press by 8.3% and their 1RM squat by 20.4%. All improvements reached statistical significance, but no differences were noted between the groups. Thus, the benefits of a periodized training program may not be realized over a relatively short term duration of training (~ 15 weeks). However, it is important to appreciate that training programs are developed for the entire year. Thus, the benefits of periodizing the training program are likely more relevant to having the athlete peak at the appropriate time of the year and minimizing the risk of overtraining.

In a critical review of the literature, Fleck (1999) reported that although evidence for the effectiveness of periodized resistance training is convincing, the limited number of studies on this form of training leaves many questions still unanswered. For example, of the few studies that have examined periodized training, only a small percentage addressed motor performance, body composition, or short-term endurance changes. The significant improvements in strength and motor performance reported in these studies do suggest that periodized training is more effective than nonperiodized training. However, this may be largely dependent on the training status of the individual. The magnitude and the rate of strength increases are much greater in untrained individuals than in trained individuals (see chapter 8). Thus, given the rapid strength increases seen in novice lifters, periodized training may not be necessary until a certain strength base has been established. In addition, for the trained athlete, the benefits of periodization may become relevant only for programs consisting of more than several months of continuous training. In short-term programs, clear benefits of manipulating training intensity and volume may not be realized.

It appears that only five studies have compared linear to nonlinear periodized training programs

(Apel, Lacey, and Kell 2011; Baker, Wilson, and Carlyon 1994; Buford et al. 2007; Hoffman, Ratamess, Klatt, et al. 2009; Rhea et al. 2002). Four of these examined recreational lifters exercising 3 days per week for 9 to 12 weeks (Apel, Lacey, and Kell 2011; Baker, Wilson, and Carlyon 1994; Buford et al. 2007; Rhea et al. 2002). Although two of the studies did not demonstrate significant differences in strength gains between linear, nonlinear, and nonperiodized training programs (Baker, Wilson, and Carlyon 1994; Buford et al. 2007), Rhea and colleagues (2002) indicated that a nonlinear training program was more effective in generating strength improvements than the traditional linear resistance training program. In contrast, Apel and colleagues (2011), following 12 weeks of training, found that the traditional linear model was superior. The differences between these studies may be related to training volume. In the studies that showed no significant differences between training paradigms, training volume was equated (Baker, Wilson, and Carlyon 1994; Buford et al. 2007), though Baker and colleagues used a higher volume (five sets per exercise in the core exercises) than Rhea and coauthors (three sets per exercise for all exercises).

However, equating training volume may not be a realistic method of program design. Typically, as intensity increases, volume is reduced. Volume of training during the hypertrophy phase is greater than volume during a strength or strength–power workout. Thus, equating training volume during a high-intensity workout may result in greater fatigue due to the larger number of sets. Hoffman and colleagues (2009) compared linear training, nonlinear training, and no periodization in experienced resistance-trained college football players. This appears to be the only study to have compared these training models in competitive athletes. Figure 14.3 depicts the percent change in upper and lower body strength measures. All three training programs resulted in significant performance increases in both the 1RM squat and bench press, but no significant differences were noted between the groups. However, the nonlinear group had the lowest magnitude of performance improvements in the squat exercise. It was suggested that the single-session lower body power training per week for that group may not have been sufficient to provide an adequate training stimulus. Although limited data exist to support nonlinear training in experienced resistance-trained athletes, the studies have generally used training durations that may not be representative of all training contexts. For instance, in professional athletes whose season may span 6 to 8 months and involves significant travel, the use of a nonlinear model may prove beneficial. However, more research in this area is necessary.

The next sections provide examples of periodized training programs for a strength–power athlete playing a team sport, a strength–power athlete whose primary goal is an isolated competition (e.g., national meet) during the training year, and an endurance athlete. Each program should be viewed as an example only and as a possible guideline for developing a periodized training program to meet an athlete's needs. It is important to remember that one must consider the entire athletic conditioning program while developing the periodization program and not just the resistance training component.

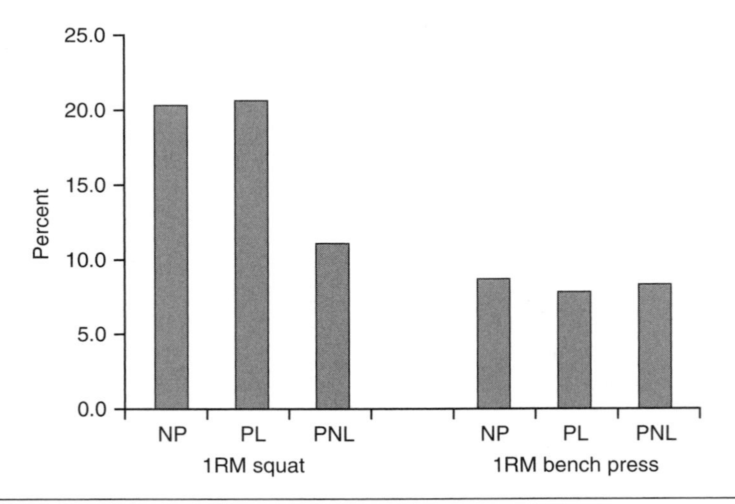

Figure 14.3 Percent change comparison of no periodization (NP), linear periodization (PL), and nonlinear periodization (PNL) in college football players.

Data from Hoffman et al. 2009.

In Practice

Should Recreational Athletes Periodize Their Training Programs?

The primary goal of periodization is to help the athlete reach peak condition or performance at the appropriate time of the year and to decrease the risk of overtraining or overuse injury. The key thought is that the athlete is preparing for a specific competitive goal, whether it is a season of competition or a particular competition (e.g., division or national championships, Olympics). Recreational athletes are most likely exercising for the physical fitness benefits and may not have any specific target date for reaching any training goals. Their training goal may in fact be to remain physically fit, or they can focus on a specific goal such as muscle hypertrophy (e.g., bodybuilding) or muscle strength improvement. In that case the training program may be specific to the training goal without any manipulation of intensity or volume. Although many of the studies on the efficacy of periodization were conducted with recreational athletes, none appear to have focused on whether periodization can be effectively used to enhance an individual's health and fitness profile. Given the importance of periodization to peak athletic performance and prevention of overtraining syndrome, one can make the argument that periodizing does not apply to the recreational athlete. Although periodizing a recreational athlete's training program is not necessarily wrong and may in fact provide different stimuli, preventing boredom or monotony, it may not have the same effect as in competitive athletes. The recreational athlete does not exercise at the intensity or volume common among competitive athletes, and the risk of overuse is not as great a concern. Though not much evidence exists to suggest that periodizing a recreational training program is advantageous, it is very unlikely that periodization would negate any training effects.

Periodized Training Program for a Strength–Power Athlete in a Team Sport

The training program of an athlete preparing for a season of competition involves bringing the athlete to peak condition for a specific time period. A periodized training program that has only one peaking phase is referred to as a **monocycle** (Bompa 1999). An example of an annual periodized training program for a strength–power athlete playing a team sport (e.g., American football) is shown in figure 14.4. After the competitive season, the athlete generally goes through a transitional period. During this time the athlete undergoes an active rest in which there are no formal workouts, and the only activity that the athlete may participate in is low-intensity and low-volume recreational sports such as jogging, swimming, basketball, or any of the racket sports. For the collegiate or high school football player, this period of time generally coincides with exams and winter vacation. Once the athlete returns to school, the off-season conditioning program usually begins. As seen in figure 14.4, the initial mesocycle is the preparatory period, a resistance training program that prepares the athlete for the more strenuous training of the subsequent phases. This mesocycle may last 6 to 8 weeks. In addition, the intensity and volume of training during this phase are beneficial to athletes needing to add muscle mass; this phase is also referred to as the hypertrophy phase. During this phase, the athlete also performs some form of conditioning activity to maintain aerobic capacity. This may include normal endurance-type activities such as jogging, cycling, and swimming; or the athlete may play recreational basketball, volleyball, or some other sport for 20 to 30 min. This conditioning activity should be performed two or three times each week. The mesocycle may conclude with an unloading phase that significantly reduces the intensity and volume of training to prepare the body for the next phase of training.

During the next mesocycle, the strength phase, the intensity of training is increased and the volume reduced. The primary emphasis during this mesocycle, which may last 6 to 8 weeks, is increasing maximal strength. The athlete maintains the endurance–conditioning program begun during the preparatory phase. To allow the athlete to adequately recover from this training cycle, another unloading phase may be added. The athlete can then proceed to the next mesocycle or, depending on the team, possibly to spring football.

If there is spring football, the training program goes in a different direction, resembling a maintenance

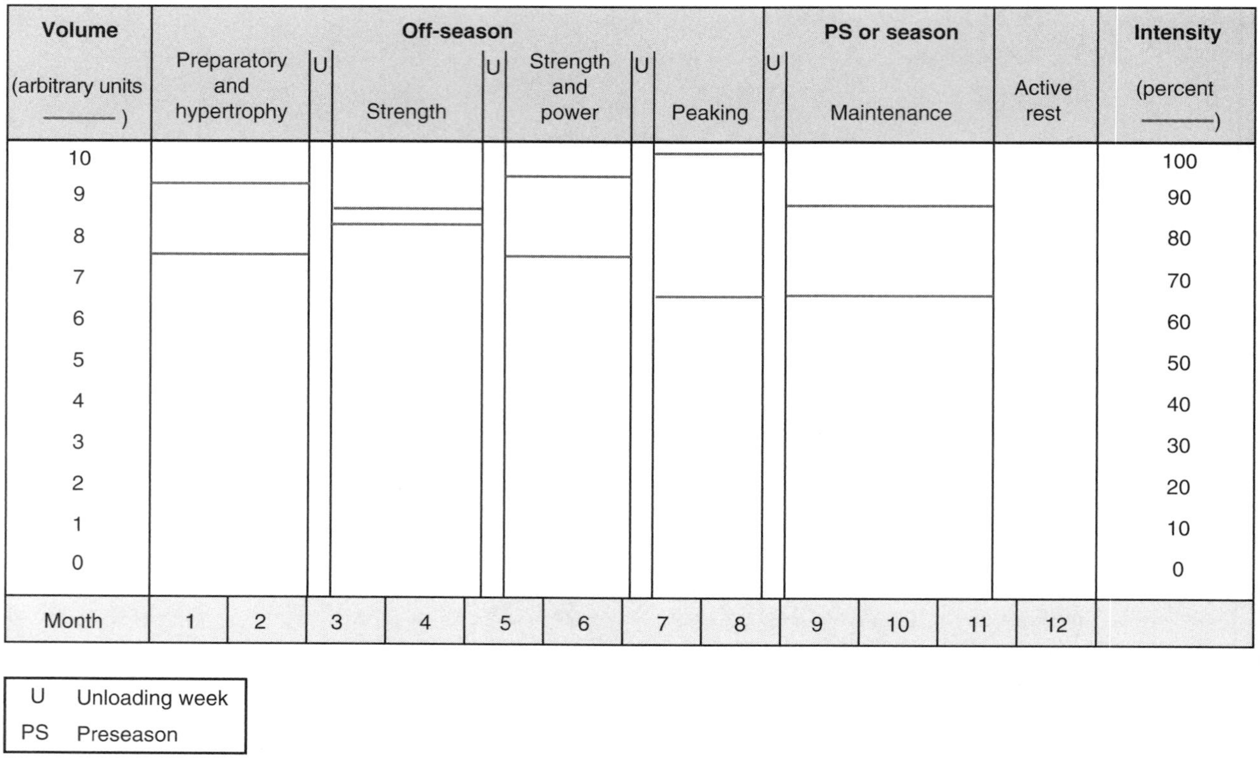

Volume (arbitrary units ———)
Off-season
Preparatory and hypertrophy | U | Strength | U | Strength and power | U | Peaking | U
PS or season
Maintenance
Active rest
Intensity (percent ———)
10 9 8 7 6 5 4 3 2 1 0
100 90 80 70 60 50 40 30 20 10 0
Month 1 2 3 4 5 6 7 8 9 10 11 12

U Unloading week
PS Preseason

Figure 14.4 Periodized training program for a strength–power athlete playing a team sport (e.g., American football).

program typically used during the regular competitive season. Training volume is significantly reduced. However, intensity may remain similar to that during the previous strength phase. At the end of spring football, a week of active rest may precede the continuation of the conditioning program.

The next mesocycle (either immediately following the strength phase or after spring football) is the strength–power phase. Olympic exercises (e.g., power cleans, push presses, high pulls), if not already part of the training regimen, may be incorporated into this phase of training. The exercises have a greater specificity to the movements on the field of play. The intensity of exercise is further elevated, and the volume of training (related to the reduced number of repetitions, as the number of sets per exercise might remain constant) is decreased. During this phase, stretch–shortening exercises may also be included in the program. In addition, sport-specific conditioning, agility, and speed training can be integrated into the 2 or 3 day per week running program. This mesocycle should also last between 6 and 8 weeks.

The next phase of training, which precedes training camp, is of shorter duration (4-6 weeks) and is designed to bring the athlete to peak strength and condition for the start of the football season. During this peaking phase, training intensity is further elevated, and the volume of resistance training is again reduced. This is accomplished via a reduction in the number of assistance exercises in the resistance training program. By this phase of training, the athlete should be concentrating primarily on getting into the proper physiological condition to play football. The conditioning program emphasizes **anaerobic** training (e.g., intervals, both long and short sprints, and agility exercises). The stretch–shortening exercises incorporated into the previous mesocycle should be included in this phase also.

The preseason period, which lasts until the start of the regular season, begins when the athlete reports to training camp. During this period and for the remainder of the competitive season, the resistance training program may be reduced to a 2 day per week maintenance program. The maintenance phase generally comprises the core exercises plus several assistance exercises that work the antagonist muscle groups. Training intensity is similar to what was used during the strength phase, and training volume may be similar to what was used during the peaking phase. Similarly, sport-specific conditioning needs to be maintained at a reduced volume and intensity. Some form of anaerobic training (e.g., sprints or intervals) should be continued 2 or 3 days per week to maintain the athlete's peak condition. This can be easily incorporated into the practice schedule. Practices (in which conditioning drills are included as part of the practice routine) and games have been shown to effectively maintain conditioning level during the competitive season of an anaerobic sport (Hoffman, Fry, et al. 1991).

Periodized Training Program for a Strength–Power Athlete Preparing for a Specific Event

Figure 14.5 presents an example of an annual periodized training program for a strength–power athlete preparing to peak for a single event. This program is also considered a monocycle because the athlete is preparing to reach peak condition only once during the competitive year. As shown in the previous example, it is easy to divide the year into precise mesocycles for an athlete participating in a sport that has a well-defined off-season, preseason, and season. However, to prepare an athlete to peak for a specific event that often occurs at the end of the competitive season, precise control of the training variables is required. Unlike the situation with team sports, in which there is some room for maneuverability, a mistake in the training prescription for an athlete preparing for a single competition could lead to an undesirable outcome. The athlete may not reach peak condition by the time of the contest, or may peak too early and possibly overtrain in an attempt to maintain a high performance level for an extended period of time.

During the initial phase of training, the program is similar to what is typically seen for the strength–power athlete participating in a team sport. However, the competitive phase may be long, with many of the earlier competitions considered of lesser importance. In this instance the athlete trains through these early season competitions, preferring instead to peak for the more important competition at the end of the competitive year. During this competition period,

Volume (arbitrary units ——) | Preparation | Competition | Intensity (percent ——)

Hypertrophy | U | Strength | U | Strength and power | U | Peaking | U | Active rest

Volume: 10, 9, 8, 7, 6, 5, 4, 3, 2, 1, 0

Intensity: 100, 90, 80, 70, 60, 50, 40, 30, 20, 10, 0

Month	1	2	3	4	5	6	7	8	9	10	11	12

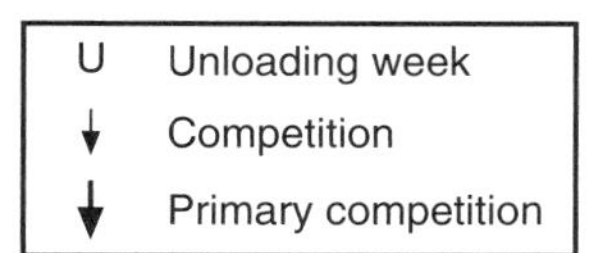

Figure 14.5 Periodized training program for a strength–power athlete preparing to peak for a single event.

the major difference between this athlete and the athlete participating in a team sport is the absence of a maintenance phase. The athlete preparing for a single competition may have several mesocycles during the early to middle part of the competitive year and then enter a peaking phase to maximize performance before the major competition.

Many sports (e.g., track and field) have both an indoor and an outdoor season. In this situation there are two competitive seasons, each having a competition that the athlete primarily focuses on. The training program for two distinct competitive phases is called a **bi-cycle** (Bompa 1999). When athletes compete in three or more competitions in a single year (e.g., gymnastics or boxing), these training plans are referred to as a **tri-cycle** or **multi-cycle**. Figure 14.6 shows an example of a bi-cycle periodized training program for a strength–power athlete.

A bi-cycle consists of two monocycles that are linked through a short unloading period. The approach to each monocycle is similar to the approach previously discussed for an athlete preparing for a single competition. However, if greater importance is placed on the competition during the second monocycle, then the volume of training will be higher in the preparatory phase during the first monocycle. In this scenario, the condition of the athlete is slightly lower during the first monocycle in comparison with where it should be during the second monocycle. In tri-cycle programs, the last competition is generally the most important. Thus the highest volume of training is seen during the preparatory phase of the first monocycle. In addition, the preparatory phases of the second and third monocycles are relatively short compared with the preparatory phase of the first monocycle. Changes in intensity of training are similar during each monocycle. The challenge of multi-cycle training programs is the reduction in the preparatory phase of training. The higher intensities of training performed more frequently during the year place the athlete at a greater risk of overtraining. It has therefore been recommended that tri-cycle or

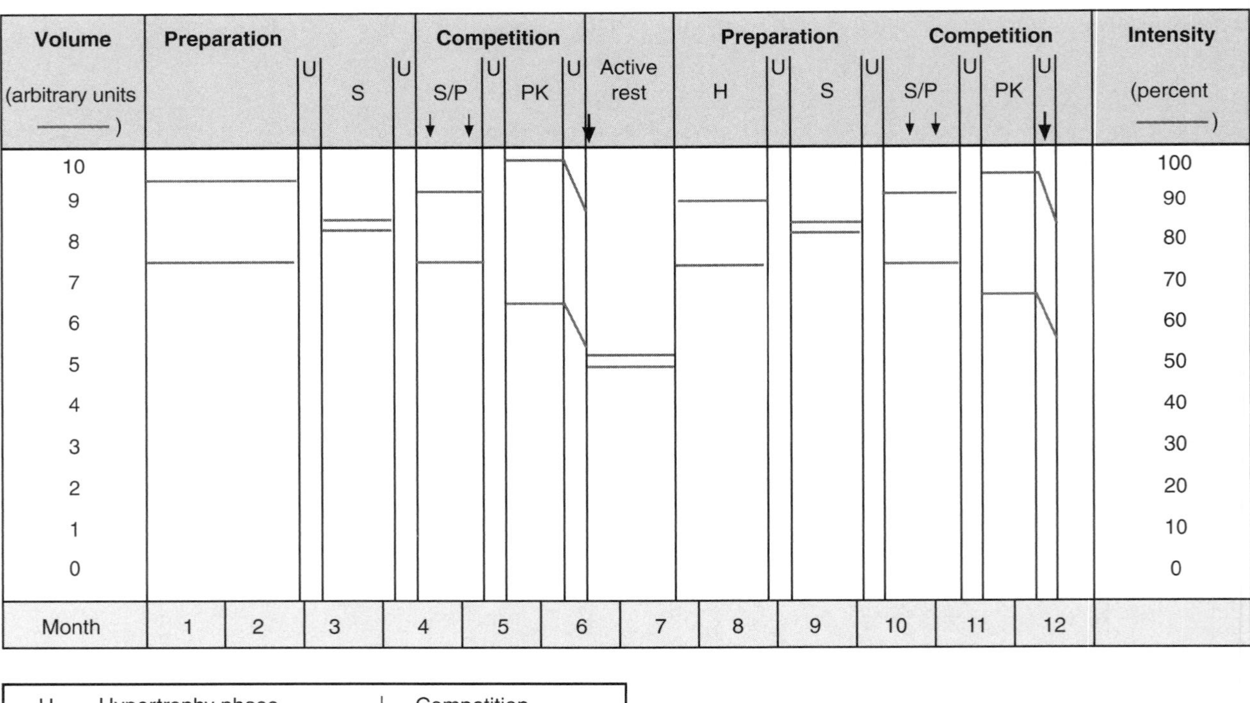

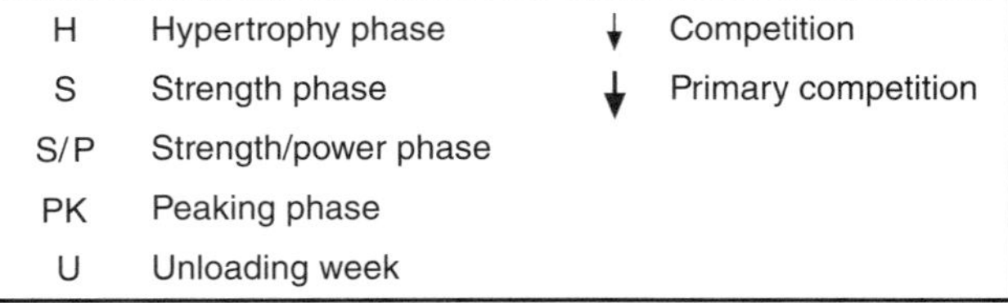

Figure 14.6 Bi-cycle periodized training program for a strength–power athlete.

In Practice

Periodization for the Tactical Athlete

The approach to military, police, and fire and rescue personnel has begun to change so as to prepare these individuals for their tactical position similarly to how athletes are prepared for competition, hence the term *tactical athletes*. In fact, the Israeli military has an entire base dedicated to combat fitness that develops periodized training programs for the various military units in all branches of the service. These programs begin in basic training, carry the soldiers through their advanced training, and are maintained throughout their military service. The American military does not have an established military unit dedicated to combat fitness but has made tremendous efforts in hiring sport performance professionals to work as civilian contractors within the various military installations around the world.

Working in the tactical community presents several challenges. The first challenge is the broad array of military and police occupations, requiring performance of a clear needs analysis (see chapter 8) for each profession. Secondly, for many in the tactical community there is no "season" and no "off-season." These individuals have to be ready on a daily basis as if every day was "game day," which is quite a challenge. For soldiers who are preparing for deployment, the development of a periodization program may be a bit simpler, as the performance professional can use the deployment as the "season" and the preparation for deployment as the "off-season." However, military and fire and rescue personnel, as well as special operations units, may have to perform at a moment's notice; thus their training goals may be quite different. These individuals may prefer a nonlinear routine to allow them to focus on several different training goals, or they may decide to focus on one specific training goal and vary only the type of exercise. With the latter type of nonperiodized training program, it may become important for the tactical athlete to incorporate several unloading periods to avoid overtraining.

multi-cycle periodized training programs be limited to more advanced athletes (Bompa 1999). The thought is that the experience and ability of these athletes give them a better opportunity to adapt to this highly stressful training program.

Periodized Training Program for an Endurance Athlete

The primary difference between a training program for an endurance athlete and others is the high volume of training that is maintained throughout the preparatory and competitive phases (figure 14.7). If volume of training is not sufficient, athletic performance may be affected. The volume of training is always higher than training intensity in endurance sports even when there are several monocycles during the year. Bompa (1999) has suggested that periodization for an endurance sport is accomplished in three main phases: **aerobic endurance**, aerobic endurance plus **specific endurance**, and finally specific endurance. Aerobic endurance development, which is the enhancement of the athlete's cardiorespiratory system, occurs during the transition and early preparatory phases. Training is performed at a moderate intensity, and the volume steadily increases as the athlete progresses. During the preparatory phase, the endurance athlete continues to emphasize aerobic endurance but also begins to perform some higher-intensity training (e.g., long and medium interval training). The volume of training continues to increase as the athlete completes the preparatory phase. As the athlete enters the competition phase, the emphasis on specific endurance training is greater. Intensity of training is high, and the athlete often exercises at intensities that exceed racing intensity. Training volume is also high. Finally, as the athlete enters the unloading (tapering) phase, training intensity is reduced to a much greater extent than is volume. The greater reduction in training intensity during the unloading period is viewed as having primary importance because of the potential impact that prolonged high-intensity training has on development of the overtraining syndrome (Bompa 1999). Training volume may also be reduced, but not to the extent typically seen in the strength–power athlete.

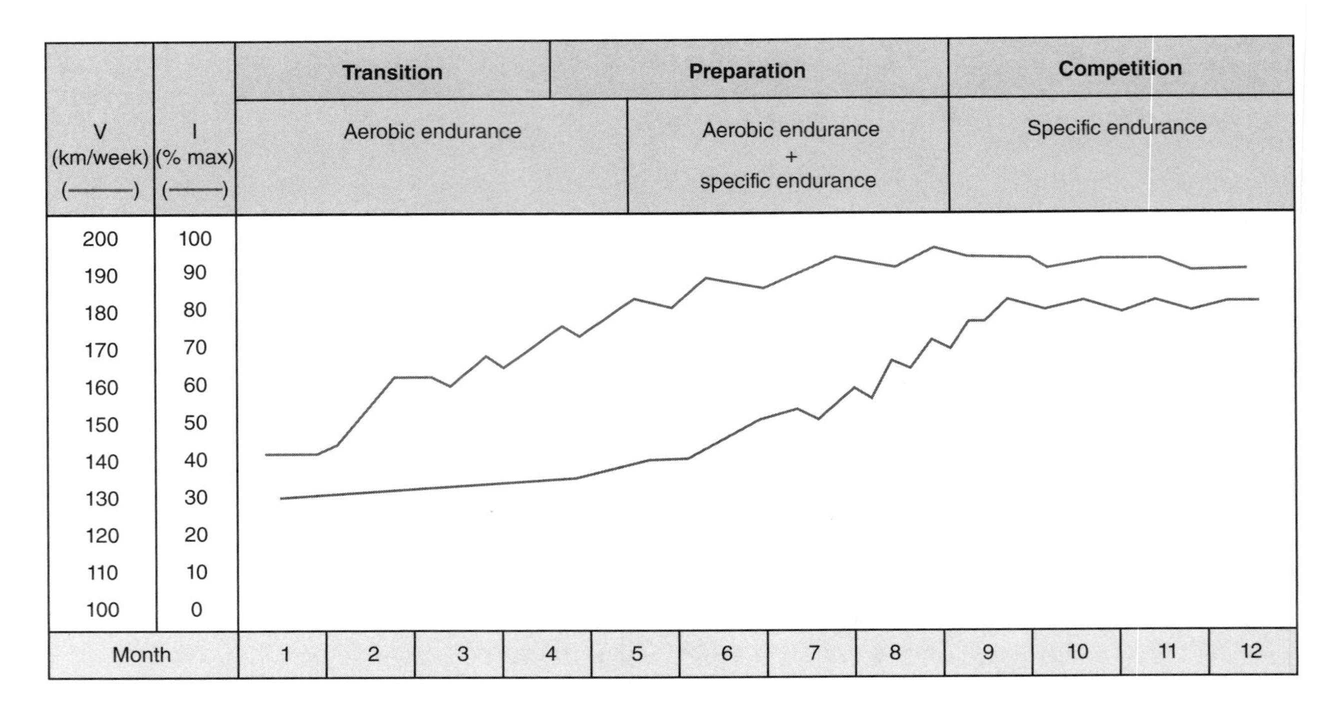

Figure 14.7 Periodized training program for an endurance athlete.
Adapted from Bompa 1999.

Summary

The goal of periodization is to maximize the potential of the athlete to reach peak condition by manipulating both training volume and training intensity. Through the proper manipulation of these training variables, not only does the athlete peak at the appropriate time, but also the risk of overtraining is reduced because the periodized program is designed to ease the strain on the athlete. The limited number of studies on the efficacy of periodized training have demonstrated it to be more effective than nonperiodized training programs for maximizing athletic performance. However, this may be more of a factor for the experienced athlete than for those with less experience.

REVIEW QUESTIONS

1. For a periodized training program, explain the general relationship between training intensity and training volume.
2. Explain the differences between linear and nonlinear training and when each type of program may be appropriate.
3. Explain how Selye's general adaptation syndrome applies to training adaptations for athletes.
4. Describe the primary differences in the training program of a college football player focusing on success during the entire season and that of a shot-putter focusing on conference championships at the end of the season.
5. Describe alterations in training programs for athletes who participate in spring football.

chapter 15

Program Development and Implementation

CHAPTER OBJECTIVES

After reading this chapter you should be able to do the following:

- Understand how to integrate various training modes into a year-long training program.
- Evaluate specific aspects of a sport and know how to design a training program to meet those specific needs.
- Differentiate between off-season and in-season training programs and develop realistic training expectations and goals.

The most challenging aspect of the job for the strength and conditioning coach or fitness professional is developing and implementing the yearly training program. The goal in developing this plan is to maximize performance at the appropriate time of the year and minimize the risk for overtraining. Although this is the primary focus of chapter 14 in its review and explanation of periodization, that chapter does not provide specifics on how to incorporate all fitness components needed to prepare the athlete for competition. Chapter 14 discusses the theory behind periodization but does not detail the practical aspects of developing the year-long training program. The present chapter provides examples of specific training programs to emphasize important aspects of program development and implementation.

The training program developed is specific to each sport and ideally specific to the individual athlete. As discussed in the earlier chapters of this part of the book, a number of training methods give the strength and conditioning professional the tools that need to be used at the appropriate time in the yearly training cycle. In addition, each sport has nuances that must be considered with regard to the development of the yearly training program. For example, both American football and basketball are considered strength–power sports. For such sports, plyometric exercises are often incorporated into the yearly training program; however, the coach needs to consider the volume of jumps performed by basketball players as they play pickup games in the off-season. Adding plyometric exercises to the training program for basketball players may increase the risk for overtraining. This chapter focuses on **training program development and implementation**. Although it presents topics and program examples discussed in previous chapters, the emphasis here is on how the various training modes are integrated in the overall training scheme.

Training Sessions

The use of a warm-up routine before any training session is strongly recommended. The benefits of a warm-up are detailed in chapter 7. The importance of the warm-up relates primarily to enhancing subsequent athletic performance but also to reducing the risk for injury during the training session. The warm-up should be dynamic, should use exercise-specific movement patterns, and should not be fatiguing. Table 15.1 presents an example of a dynamic warm-up.

Table 15.1 Sample Dynamic Warm-Up

Exercise	Repetitions
Jog	2
High knee walk	2
Stepping trunk twist	2
Glute kick	2
Side lunge	2
Side shuffle (switch sides every 10 yd)	2
Carioca	2
Backward run	2
Power skip	2
Two to four bounds that accelerate into a run for 30 yd	2

All exercises are performed for 30 yd (27 m).

Off-Season Training Program

During the development of the off-season training program, the strength and conditioning professional needs to ensure that the athletes are physically ready for each phase of training. As any new exercise is introduced during any phase of training, it is important that technique be emphasized first. Increasing the intensity of the exercise (i.e., increasing training load) should not receive emphasis until the athlete has demonstrated successful technique. During the off-season, the initial focus is often on muscle size and strength and power development. Chapter 8 provides details on specific guidelines for the development of the resistance training program, including guidelines for exercise order. However, strength and conditioning professionals should also be concerned that the athlete does not become deconditioned during the off-season by only focusing on the resistance training program. They should maintain some conditioning activity even if it is recreational athletics such as basketball or racquetball or another recreational activity. With competitive athletes, recreational tournaments such as basketball or volleyball can provide a competitive, low-stress, fun atmosphere that allows team members to stay conditioned, remain competitive, and develop camaraderie.

Before the onset of the off-season training program, the strength and conditioning professional often performs an **assessment** to help set the training goals for the team and individual athletes. When working with a team, the strength and conditioning professional provides the framework for the specific training goals of each training phase. However, to optimize the training stimulus and maximize performance benefits, it is also necessary to focus on the individual athlete's strengths and weaknesses. This should be based on assessment results and in consultation with both the sport coach and the athlete. Importantly, analysis of the assessment should be relative to the age of the athlete and how the athlete's results compare to normative data for similar athletes in the sport. This affords a more educated approach to the development of the training program best suited to the given athlete. Thus, the goals of the off-season training program should be specific for each athlete. For example, ballistic exercises (e.g., squat jumps or bench press throws) may be more effective for an experienced strength–power athlete than for a novice. This highlights the need to examine which training component has the greatest window of adaptation (or the most potential for improvement), identified through careful evaluation of the testing results (see chapter 16).

Identifying the fitness component with the greatest potential for growth is a logical method of determining the athlete's training goals. However, other fitness components cannot be ignored; if the others are not maintained, the athlete can actually see them decrease. Identifying a window of adaptation simply sets the goals for the component that has the greatest potential for improvement.

Once the training goals are established, the primary focus at the onset of the off-season training program is in the weight room. The resistance training program generally begins with the preparatory or hypertrophy phase, consisting of high-volume (greater number of repetitions performed per set) and low-intensity (loads with a low percentage of the athlete's one-repetition maximum [1RM]) training. During this phase, the primary purpose is to prepare athletes for the higher-intensity lifting in subsequent stages of training and to focus on increasing muscle size. Significant increases in strength can occur relatively quickly (~7 weeks) even in experienced athletes during this phase, but these increases are likely a return to baseline strength values from the previous season (Hoffman, Ratamess, Klatt, et al.

2009). Rapid improvements in strength are generally seen in individuals who have been detrained (Staron et al. 1991). The magnitude of the strength decrement is likely related to the length of time from the end of the season to the onset of the off-season conditioning program. In collegiate athletes, this may be up to 6 weeks. Considering the rapidity with which strength levels return, a prolonged recovery time between the end of the previous season and the beginning of the off-season conditioning provides an opportunity for complete recovery (both physical and mental) and leads to a greater desire to participate in the off-season conditioning program. For training goals focused on increasing muscle size, a longer duration of time at the hypertrophy phase may be recommended.

Figure 15.1 provides a general example of an 8-month off-season training program for a strength–power athlete who plays a fall sport. In this example there are two periodization cycles; the cycles are separated by an event such as spring football that will interrupt the training program. This is not an ideal way of training athletes, as they stop training in the middle of the cycle. In this case the athletes should maintain strength through the inclusion of a maintenance phase. Some programs, however, may want to train through spring practice and use the opportunity for technique work and sport-specific conditioning. In these circumstances the athletes may train through this period without any alteration to the training program. In any case, it is not incorrect to repeat training cycles. It may also be prudent to use an off-loading period (1 week) between some of the training phases to promote recovery.

A number of resistance training programs can be used to train athletes. Several examples are provided in this chapter. Regardless of the type of program developed, the key point is whether the strength and conditioning professional can defend its use through an evidence-based approach. That is, any program should be supportable by scientific method (e.g., evidence presented in all chapters within this part of the book). Table 15.2 depicts a 15-week resistance training program for soccer. Soccer is not considered a strength–power sport per se, but soccer players benefit tremendously from resistance training.

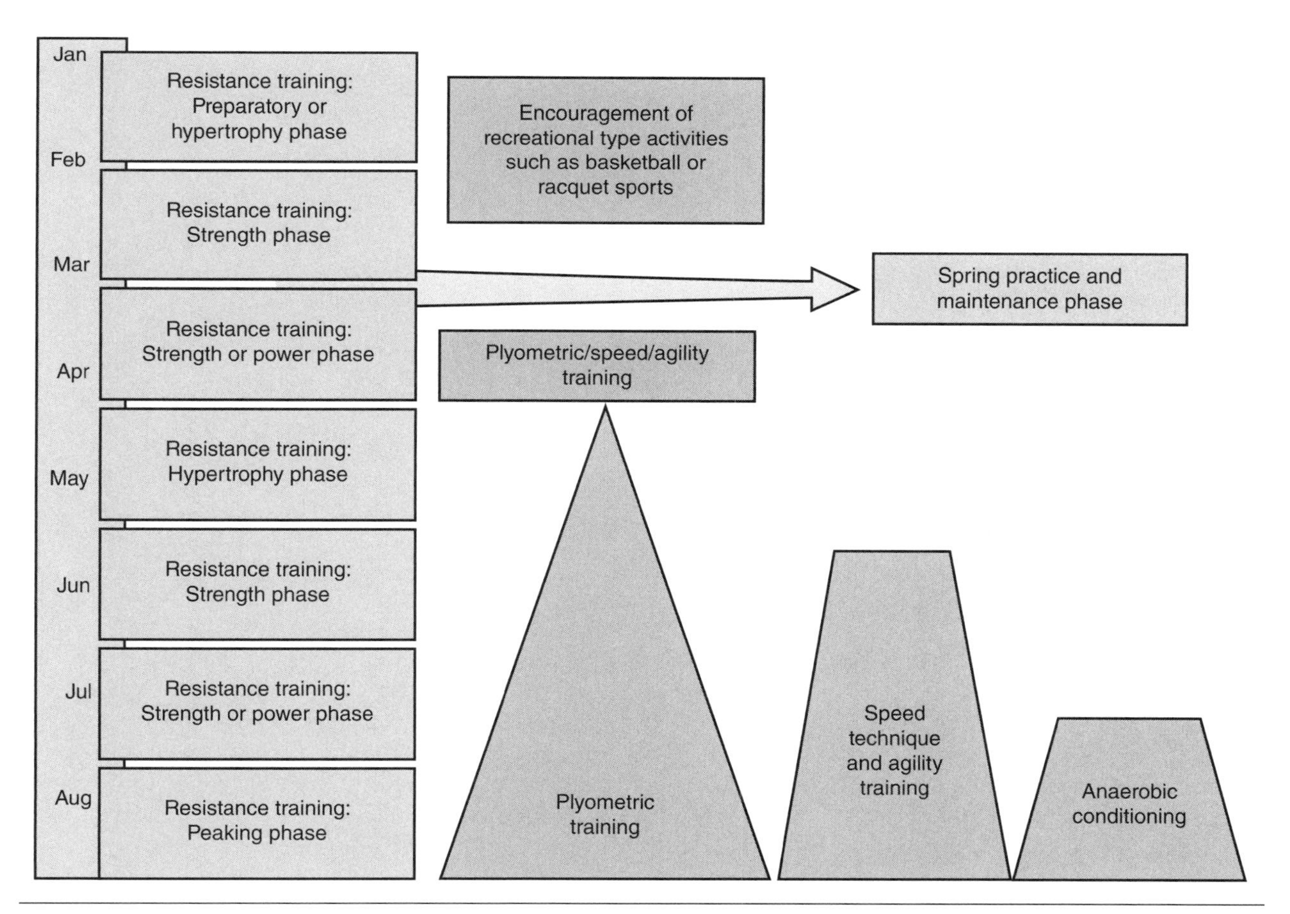

Figure 15.1 Example of an off-season training program for a strength–power sport such as American football.

Table 15.2 Fifteen-Week Soccer Program (Training 3 Days per Week)

	Sets × reps				
Exercise	**Weeks 1-5**	**Weeks 6-10**		**Weeks 11-15**	
Monday and Thursday					
Squat	4 × 8-10	Hang pull	4 × 4-6	Hang pull	4 × 3-5
Romanian deadlift (RDL)	3 × 6-8	Squat	4 × 6-8	Squat	4 × 4-6
Lunge	3 × 8-10	RDL	4 × 4-6	Dumbbell box jump	3 × 5
Calf raise	3 × 8-10	Lunge	3 × 6-8	Lateral box jump	3 × 5
Lat pull-down	3 × 8-10	Calf raise	3 × 6-8	RDL	3 × 4-6
Seated row	3 × 8-10	Lat pull-down	3 × 6-8	Calf raise	3 × 6-8
Biceps curl	3 × 8-10	Seated row	3 × 6-8	Lat pull-down	3 × 6-8
Glute-ham	3 × 10	Biceps curl	3 × 6-8	Biceps curl	3 × 6-8
		Glute-ham	3 × 6-8	Glute-ham	3 × 6-8
Tuesday and Friday					
Bench press	4 × 8-10	Bench press	4 × 6-8	Bench press	4 × 4-6
Dumbbell shoulder press	3 × 8-10	Dumbbell inclined press	3 × 6-8	Dumbbell in-clined press	3 × 6-8
Upright row	3 × 8-10	Shoulder press	3 × 6-8	Shoulder press	3 × 4-6
Dumbbell shrug	3 × 8-10	Lateral and front raise (compound set)	3 × 6-8	Lateral and front raise (compound set)	3 × 6-8
Triceps push-down	3 × 8-10	Triceps push-down	3 × 6-8	Triceps push-down	3 × 6-8
Neck exercise	3 × 8-10	Neck exercise	3 × 6-8	Neck exer-cise	3 × 6-8
Abdominal exercise	3 sets				
Intensity (% 1RM)	75-80%	80-85%		85-90%	
Rest interval between sets	1 to 2 min	2 to 3 min		3 to 4 min	

Preparatory–Hypertrophy Phase

The initial training phase in a periodized training cycle is the preparatory–hypertrophy phase. Table 15.3 provides an example of the resistance training program used for a strength–power athlete during this phase. This phase is generally 4 to 8 weeks in length, depending on the athlete's goals. As discussed earlier, during this training cycle athletes may also maintain their conditioning by participating in basketball or racket sports, but the primary focus is on the resistance exercise program.

If one of the training goals is to reduce body fat, appropriate adjustments to the training program must be made. Greater emphasis could be placed on endurance training, but evidence suggests that high-intensity training is an effective stimulus for increasing fat oxidation (Tremblay, Simoneau, and Bouchard 1994). The addition of endurance training is potentially problematic for the resistance-trained athlete, as a greater emphasis on aerobic endurance activity will likely attenuate maximal strength and

Table 15.3 Example of a Split-Routine Resistance Training Program (4 Days per Week): Preparatory–Hypertrophy Phase for a Strength–Power Athlete

Exercise	Sets × repetitions	Exercise	Sets × repetitions
Days 1 and 3		Days 2 and 4	
Squat	1, 4 × 8-12	Bench press	1, 4 × 8-12
Leg extension	3 × 8-12	Inclined bench press	3 × 8-12
Leg curl	3 × 8-12	Inclined fly	3 × 8-12
Standing calf raise	3 × 8-12	Seated shoulder press	1, 4 × 8-12
Lat pull-down	1, 4 × 8-12	Upright row	3 × 8-12
Bent-over dumbbell row	1, 4 × 8-12	Compound set: lateral and front raises	3 × 8-12
Seated dumbbell curl	3 × 8-12	Triceps push-down	3 × 8-12
Standing hammer curl	3 × 8-12	Triceps extension	3 × 8-12
Hyperextension	3 × 8-12	Abdominal crunch	3 × 20
Sit-up	3 × 20		

Rest: 1 min between each set, 72 h between days 1 and 3 and between days 2 and 4. Days 1 and 2 and days 3 and 4 can be performed consecutively. For instance, this split-routine training program can be performed on Monday, Tuesday, Thursday, and Friday (4 days per week). The 1, 4 in the sets represents 1 warm-up set and 4 work sets. A work set is a set where the athlete uses the required training load.

Adapted from Hoffman, Brown and Smith 2012.

power gains (Kraemer, Patton, et al. 1995). Considering the beneficial effects associated with high-intensity training and fat oxidation, as well as the compatibility between these two modes of training (resistance and high-intensity running), this combination of exercise modes may better accommodate the desired training outcomes than the combination of endurance and resistance training.

The training program suggested in table 15.3 is a 4-day split routine, in which each body part is trained twice per week. During this routine the athlete trains the legs, back, and biceps on Mondays and Thursdays while training the chest, shoulders, and triceps on Tuesdays or Fridays. The order of routines is not important and can easily be reversed. Often the exercises for the back and biceps involve a "pulling" motion (i.e., elbow flexion), whereas the exercises for the chest, shoulders, and triceps involve a pushing motion (elbow extension). The athlete trains with the same relative intensity during each workout. However, some variability in training volume is to be expected, since the number of repetitions per session may differ to account for recovery from the preceding session. To maximize the training stimulus, the coach and athlete should follow the progressive overload principle. That is, once the athlete achieves the desired training goal (e.g., maximum number of repetitions per session), the resistance should be elevated during the next workout. Although the relative intensity remains the same, training volume actually increases (load × number of repetitions). This method of program prescription, as discussed in chapter 14, is described as traditional or classic periodization and is also referred to as linear periodization.

Some strength and conditioning professionals incorporate an alternative training paradigm in which the focus of each workout can be different. For instance, in a single week the athlete can do a hypertrophy session, a strength session, and a power session. This method of training has been referred to as undulating or nonlinear periodization (Kraemer and Fleck 2007). Table 15.4 provides an example of a 3-day nonlinear off-season training program that can be implemented for athletes during the off-season. The efficacy of this method has recently been called into question, as a study by Hoffman, Ratamess, Klatt, and colleagues (2009) reported that during a 15-week off-season conditioning program in collegiate football players, the traditional periodized training model was more effective than nonlinear training programs. This study is discussed in much greater detail in chapter 14.

In determining the appropriate type of periodized program, one important consideration is the experience level of the athlete. For the experienced

Table 15.4 Undulating Program for Competitive Strength–Power Athletes

Exercise	Sets × reps	Exercise	Sets × reps	Exercise	Sets × reps
Monday Training goal: Power Intensity: 85% to 90% 1RM Rest: 3 to 4 min between sets		Wednesday Training goal: Strength Intensity: 80% to 85% 1RM Rest: 2 to 3 min between sets		Friday Training goal: Hypertrophy Intensity: 75% to 80% 1RM Rest: 1 to 2 min between sets	
Squat	1, 4 × 3-5RM	DB lunge	1, 4 × 6-8RM	Front squat	3 × 10-12RM
Bench press	1, 4 × 3-5RM	Inclined bench press	1, 4 × 6-8RM	DB bench press	3 × 10-12RM
Push press	1, 4 × 3-5RM	Seated DB shoulder press	1, 4 × 6-8RM	Seated shoulder press	3 × 10-12RM
High pull (clean grip)	3 × 3-5RM	Upright row	3 × 6-8RM	Compound set: shrug and lateral raise	3 × 10-12RM
Lat pull-down	3 × 3-5RM	Seated row	3 × 6-8RM	DB bent-over row	3 × 10-12RM
Triceps push-down	3 × 6-8RM	Triceps extension	3 × 6-8RM	Compound set: triceps push-down and extension	3 × 10-12RM
Standing biceps curl	3 × 3-5RM	DB seated biceps curl	3 × 6-8RM	DB hammer curl	3 × 10-12RM
Sit-up	3 × 20RM	Crunch	3 × 20RM	Back extension	3 × 8-10RM

DB, dumbbell. The 1, 4 in the sets represents 1 warm-up set and 4 work sets. A work set is a set where the athlete uses the required training load.

Adapted from Hoffman, Brown and Smith 2012.

resistance-trained athlete, the inclusion of assistance exercises (e.g., inclined bench press, inclined fly) is important for providing the necessary stimulus to generate training adaptation (Hoffman et al. 1990). If the strength and conditioning professional opts to use the nonlinear method of training, the workout often involves training the entire body each workout. This limits the number of assistance exercises that can be used. The linear training model may be more appropriate for the experienced strength–power athlete.

Strength Phase

The next phase of the off-season training program focuses on strength development. During the strength training phase, the intensity of training is higher while volume is lower than in the preparatory–hypertrophy phase. During this mesocycle, additional exercises (primarily multiple-joint, structural movement exercises) can be incorporated into the program to increase the training stimulus. An example of a strength training program can be seen in table 15.5.

Olympic-style lifting movements (e.g., high pulls) are often included during this phase. Although these exercises could also be used in the initial phase of training, inexperienced or novice athletes may benefit from first developing a strength base and proper technique with traditional powerlifting exercises. Incorporating these exercises into the later phases of training provides a degree of variation in the program that prevents monotony. This training phase often lasts between 4 and 6 weeks.

During the strength training phase, the strength and conditioning professional can begin to incorporate additional training modes. Plyometric exercises or speed and agility drills are often integrated into this phase. As with the resistance training program, the plyometric exercises should progress from low to high intensity. Low-intensity plyometric exercises generally require two-legged jumps from ground level. As the athlete moves to single-leg jumps, bounds, jumps from height (depth jumps), and repeated jumps onto boxes, the intensity is increased. The volume of plyometric training also needs to be considered. If working with basketball or volleyball athletes, the strength and conditioning professional should be aware that they are likely expected to play pickup games during the off-season. The jumps that the athletes perform during this recreational play will contribute to overall fatigue. Adding plyometric exercises to the training programs of these athletes without accounting for their recreational participation can increase the risk for **overtraining syndrome**. A sample plyometric training program for strength–power athletes is shown in table 15.6.

Table 15.5 Example of a Split-Routine Resistance Training Program: Strength Phase for Strength–Power Athletes (4 Days per Week)

Exercise	Sets × reps	Exercise	Sets × reps
Days 1 and 3		Days 2 and 4	
Squat	1, 4 × 6-8	High pull (clean grip on day 2 and snatch grip on day 4)	1, 4 × 6-8
Deadlift	3 × 6-8	Bench press	1, 4 × 6-8
Leg curl	3 × 6-8	Inclined DB bench press	3 × 6-8
Standing calf raise	3 × 6-8	Inclined DB fly	3 × 6-8
Lat pull-down	1, 4 × 6-8	Seated DB shoulder press	1, 4 × 6-8
Bent-over row	1, 4 × 6-8	Front raise on day 2 and lateral raise on day 4	3 × 6-8
DB biceps curl	3 × 6-8	Triceps push-down	3 × 6-8
Seated hammer curl	3 × 6-8	Triceps DB extension	3 × 6-8
Hyperextension	3 × 8-12	Sit-up	3 × 20
Crunch	3 × 20		

Rest: 3 min between each set, 72 h between days 1 and 3 and between days 2 and 4. Days 1 and 2 and days 3 and 4 can be performed consecutively. For instance, this split-routine training program can be performed on Monday, Tuesday, Thursday, and Friday (4 days per week). DB, dumbbell. The 1, 4 in the sets represents 1 warm-up set and 4 work sets. A work set is a set where the athlete is using the required training load.

Adapted from Hoffman, Brown and Smith 2012.

Table 15.6 Example of a Plyometric Training Program for a Strength–Power Athlete

Exercise	Sets	Repetitions	Notes
Squat jump	2	5	Can use weighted vests to increase intensity.
Front box jump (24 in.)	3	5	Can jump with weighted vest or dumbbells in hand to increase intensity. Intensity can also be increased by increasing jump box height.
Single-leg zigzag hop	3	5	Use cones in a zigzag pattern. Can increase intensity by using weighted vest.
Multiple bag or box side jump	3	5	Use bags and boxes to jump laterally in both directions. Can increase resistance by changing height of bags or boxes or using a weighted vest.
Medicine ball throw	3	10	Upper body plyometric drill. Can increase weight of ball to increase intensity.

All drills should be performed at maximal effort, with a 2 to 3 min rest between each set. Athletes should understand that the quality of the repetitions is important. Training frequency should be twice per week.

Strength and Power

The next phase of the off-season training program is the power or strength–power phase. During this mesocycle, exercise intensity continues to rise, while training volume is lower than during the strength phase. This phase places greater emphasis on Olympic and ballistic exercises. Improvements in power appear to be enhanced by multiple-joint, compound structural movements that combine speed of movement and high force output (Cormie, McGuigan, and Newton 2011; Hoffman, Cooper, Wendell, and Kang, 2004; Newton, Cormie, and Kramer 2012). Exercises such as the snatch, power clean, and push press are often substituted for the traditional powerlifting exercises or assistance exercises. Table 15.7 provides an example of a 4-day split-routine resistance training program for the strength–power mesocycle. Hoffman and colleagues (2004) have indicated that Olympic exercises can enhance speed and power development in experienced resistance-trained athletes during their off-season training program to a greater extent than traditional powerlifting exercises. Speed and agility training is often included in this phase. Due to the compatibility between these training modes and resistance training (see chapter 13), no change to the resistance training program is necessary.

Speed and Agility

The use of speed and agility training during the strength–power mesocycle has several purposes. The obvious one is to enhance speed and agility development; another may be to enhance team camaraderie. A **circuit training program** incorporated as a team event fosters teamwork and team spirit while improving speed and agility performance. What the speed and agility program is not designed to do is enhance conditioning. Thus the work-to-rest interval is relatively longer than one would expect when goals include conditioning.

The **work-to-rest ratio** for an exercise that enhances anaerobic capacity may be 1:4 (e.g., a 10 s sprint would be followed by a 40 s rest); however, when the focus is on the quality per repetition, the work-to-rest ratio should be lengthened to perhaps 1:8. It is the quality of the work that is important, not the quantity. As the athlete moves into the later stages of the off-season training program, the speed and agility work becomes more up-tempo to contribute to aspects of anaerobic conditioning. However, a longer work-to-rest ratio still requires 100% effort for each drill.

Table 15.8 shows an example of a sprint technique and agility drill training session for athletes participating in the strength–power phase. These

Table 15.7 Example of a Strength–Power Mesocycle (4 Day per Week Split Routine)

Days 1 and 3		Days 2 and 4	
Exercise	**Sets × reps**	**Exercise**	**Sets × reps**
Squat	1, 4 × 4-6	Power clean	1, 4 × 4-6
Snatch	4 × 4-6	Bench press	1, 4 × 4-6
Leg curl	3 × 4-6	Inclined bench press	3 × 4-6
Lat pull-down	1, 4 × 4-6	Push press, push jerk	1, 4 × 4-6
Seated row	1, 4 × 4-6	High pull (snatch grip)	3 × 4-6
Dumbbell biceps curl	4 × 4-6	Triceps push-down	3 × 4-6
Hyperextension	3 × 8-12	Triceps extension	3 × 4-6
Crunch	3 × 20	Abdominal or core routine	3 × 20

Rest: 3 min between each set, 72 h between days 1 and 3 and between days 2 and 4. Days 1 and 2 and days 3 and 4 can be performed consecutively. For instance, this split-routine training program can be performed on Monday, Tuesday, Thursday, and Friday (4 days per week). The 1, 4 in the sets represents 1 warm-up set and 4 work sets. A work set is a set where the athlete uses the required training load.

Adapted from Hoffman, Brown and Smith 2012.

Table 15.8 Sprint Technique and Agility Drill Circuit Training

Sprint technique drills	Sets	Repetitions	Agility drills	Sets	Repetitions
Standing arm exchange	2	10	Pro-agility	3	3
Lean and fall run	2	5	T-drill	3	3
Drop and go	2	5	Edgren side shuffle	3	3
Scramble out	2	5	L-drill	3	3
Light sled pulls	2	5	Zigzag drill	3	3
Assisted speed	2	5	Agility ladder	3	3

drills can be performed in a circuit fashion twice per week. Similar circuit-type training can also be used for plyometric drills. These circuits are commonly used for American football, basketball, baseball, and other strength–power team sports. As the tempo for these drills increases, they require greater levels of cardiorespiratory endurance but still benefit power development.

Implementation of a speed technique and agility circuit involves several **administrative considerations**. The initial concern is the number of drills and the number of assistants that can help administer each drill. This obviously is the determining influence on the number of drills that can be incorporated. After deciding on the number of drills, the strength and conditioning professional uses the number of athletes participating in the workout to divide them among the drills. For example, with 54 players and six coaches to assist in the drills, a total of six drills can be used with nine athletes per drill. Athletes can be randomly assigned to each drill area (it is recommended that athletes of similar ability be grouped together, perhaps separated by position). They move from one drill to the next, generally in a clockwise fashion. Since the speed technique is lower in intensity, it is recommended that the athletes first complete the speed technique circuit and then perform the agility circuit. Administratively, coaches remain at their stations and take on the responsibility of administering an agility drill.

The agility drills selected should have as much movement specificity to the sport as possible. Although table 15.8 provides the exercise prescription in terms of sets and repetitions, the exercise prescription can also specify time (in minutes) per station. This becomes especially relevant if there are time constraints associated with the practice. If the time allotment per station was 2 min, with nine athletes per station and average performance of 5 s per drill, a total of 24 repetitions could be performed, but only 2 or 3 repetitions per athlete. If the time was extended to 5 min, then a total of 60 repetitions could be performed (6 or 7 repetitions per athlete). Nine athletes per station would also result in a 1:9 work-to-rest ratio. By reducing the number of athletes per station and increasing the number of drills, it would be possible to increase the number of repetitions per time allotted and decrease the work-to-rest ratio.

Peaking

The final mesocycle of the off-season training program is the peaking phase. During this phase training intensity is at its highest, while training volume is reduced even further. For some athletes this phase provides the opportunity to make final preparations for the most important competition (e.g., track and field or swimming), for other athletes (basketball or soccer) to prepare for the start of preseason training camp. For the team sport athlete preparing for preseason training camp, the lower volume of resistance training provides more time to focus on speed, agility, and anaerobic conditioning exercises. In the final 2 or 3 weeks of the strength–power phase and throughout the peaking phase, these athletes emphasize improving their anaerobic conditioning in preparation to play a competitive season. An example of a resistance training program for a peaking mesocycle is shown in table 15.9; table 15.10 provides an example of a 12-week anaerobic conditioning program including speed, technique, and agility for athletes preparing for a competitive season.

Table 15.9 Example of a Peaking Mesocycle (4 Day per Week Split Routine)

Monday		Tuesday	
Exercise	**Sets × reps**	**Exercise**	**Sets × reps**
Squat	5 × 1-3	Bench press	5 × 1-3
Box jump	5 × 5	Power clean	3 × 1-3
Deadlift	5 × 1-3	Push press	5 × 1-3
Bent-over row	5 × 3-5	High pulls (snatch grip)	5 × 1-3
Hyperextension	4 × 10	Abdominal muscles	4 × 25
Thursday		**Friday**	
Squat	5 × 1-3	Bench press	4 × 1-3
Front squat	3 × 1-3	Snatch	5 × 1-3
RDL	5 × 3-5	Inclined bench press	3 × 5
Lat pull-down	5 × 3-5	Push jerk	4 × 1-3
Chest hammer curl	4 × 4-6	High pull (clean grip)	4 × 3
Hyperextension	4 × 10	Crunches	4 × 20

The daily goal is to achieve a three-rep max with each core lift. The athlete may fall short of three reps on the last sets (i.e., do one or two instead). Before reaching the workout session weight, the athlete may need to perform two or three warm-up sets. The rest period between sets should be 3 to 5 min. RDL = Romanian deadlift.

Table 15.10 Agility, Sprint Technique, and Conditioning Program

	Monday	Tuesday	Wednesday	Thursday	Friday	Saturday
Week 1		Agility and sprint technique		Agility and sprint technique		
Week 2		Agility and sprint technique		Agility and sprint technique		
Week 3		Agility and sprint technique		Agility and sprint technique		
Week 4		Agility and sprint technique		Agility and sprint technique		
Week 5		Agility and sprint technique		Agility and sprint technique		
Week 6		Agility and sprint technique	2 × 200, 5 × 40 yd sprints	Agility and sprint technique	1 line drill, 2 intervals	
Week 7		Agility and sprint technique	4 × 200, 6 × 40 yd sprints	Agility and sprint technique	1 line drill, 3 intervals	
Week 8	6 × 10 yd, 3 intervals	Agility and sprint technique	4 × 200, 6 × 40 yd sprints	Agility and sprint technique	1 line drill, 3 intervals	4 × 200, 4 × 100, 4 × 60 yd sprints
Week 9	8 × 10 yd, 3 intervals	Agility and sprint technique	5 × 200, 8 × 40 yd sprints	Agility and sprint technique	2 line drills, 3 intervals	4 × 200, 4 × 100, 4 × 60 yd sprints
Week 10	8 × 10 yd, 4 intervals	Agility and sprint technique	6 × 200, 8 × 40 yd sprints	Agility and sprint technique	2 line drills, 4 intervals	5 × 200, 5 × 100, 5 × 60 yd sprints
Week 11	10 × 10 yd, 4 intervals	Agility and sprint technique	7 × 200, 10 × 40 yd sprints	Agility and sprint technique	3 line drills, 4 intervals	5 × 200, 5 × 100, 5 × 60 yd sprints
Week 12	10 × 10 yd, 5 intervals	8 × 200, 10 × 40 yd	3 line drills, 4 intervals	Rest	Rest	Report to camp

In Practice

What Is the Correlation Between Injury and Training Intensity in College Athletes?

Scientists from Northwest Missouri State University examined 411 student-athletes (both men and women) participating in 16 sports from a Division II university (Vetter and Symonds 2010). All athletes completed a survey providing demographic information, training frequency, training intensity, and injury incidence. The frequency of training during the competitive season was similar between men (5.56 ± 0.90 days/week) and women (5.10 ± 1.22 days/week), as were the hours spent training per day (2.51 ± 1.09 h/day and 2.93 ± 1.25 h/day, respectively). The frequency of training during the off-season was also similar between men (4.98 ± 1.29 days/week) and women (4.31 ± 1.47 days/week), as were the hours spent training per day (2.13 ± 1.09 h/day and 2.36 ± 1.52 h/day, respectively). Male athletes spent 2.32 ± 1.02 days/week training at a moderate intensity and 2.22 ± 1.02 days/week training at a vigorous intensity. Female athletes spent 2.20 ± 1.47 days/week training at a moderate intensity and 2.48 ± 1.48 days/week training at a vigorous intensity. Men reported 1.62 ± 0.49 and 1.44 ± 0.50 acute and chronic injuries during the past year, respectively. Women reported 1.54 ± 0.50 and 1.48 ± 0.63 acute and chronic injuries during the past year, respectively. Significant negative correlations were seen in the female athletes between mental and physical exhaustion during the off-season and chronic injury (r = −0.379 and −0.328, respectively). Significant negative correlations were noted in the male athletes between frequency of vigorous exercise intensity and mental exhaustion during the competitive season and acute injury (r = −0.221 and −0.315, respectively). The authors concluded that high frequency of workout days with an exercise intensity ranging from moderate to vigorous increases the risk for injury. Thus, to avoid overuse injuries and overtraining, it is recommended that coaches provide some "down time" during the yearly training cycle. This should occur between the end of the competitive season and the start of the off-season.

Anaerobic Conditioning

The goal of the anaerobic conditioning program is to prepare the athlete for a season of competition. The conditioning program comprises both short- and long-distance sprints, as well as interval training. Chapter 10 presents the basics of developing the anaerobic conditioning program, as well as examples of various exercises used to condition athletes. Similar to what occurs with other modes of training, the volume for the conditioning program often increases weekly, while intensity is controlled by altering the work-to-rest ratio. As preseason training camp nears, the time between each sprint should decrease. Toward the end of the training cycle, the work-to-rest ratio should simulate that typical of actual performance. To provide an additional overload that can stimulate further physiological adaptations, the strength and conditioning professional may opt to further decrease the work-to-rest ratio to a level that is below that seen during competition.

Some sports appear to be more conducive for their athletes to maintain a high level of conditioning during the off-season by playing their respective sport (e.g., basketball players scrimmaging). If the strength and conditioning professional does not understand this, there is the potential for making a mistake in the exercise prescription that puts these athletes at risk for overtraining. In general, strength and conditioning professionals should plan for 6 to 8 weeks of anaerobic conditioning to prepare athletes for competition (Hoffman 2003; Hoffman and Maresh 2000).

The preseason period for many sports is approximately 4 to 6 weeks. Thus the goal of the anaerobic conditioning program during the strength–power and peaking mesocycles is to bring the athlete as close to peak condition as possible, but not necessarily to peak condition. This not only provides an opportunity for further adaptation but also potentially reduces the risk of fatigue or overreaching that may develop if the athlete peaks too soon. Ideally, the strength and conditioning professional and the team's coaching staff communicate regarding the development of the yearly training program. If they do not, errors in developing the exercise prescription could have negative consequences on the athlete's preparation.

Anaerobic Conditioning for Athletes Participating in Individual Sports

The primary difference in the anaerobic conditioning program between athletes who participate in an individual sport (such as sprint-distance running events) and team sport athletes preparing for a season of competition is the focus on decreasing fatigue rate over the course of a single sprint, rather than preparing the athlete for repeated high-intensity activity over a long time. Table 15.11 provides an example of how rest intervals can be manipulated to enhance speed-endurance for the 400 m sprinter. The long time intervals between sprints are not intended to improve fatigue rates but to maximize the quality of each sprint.

Competitive Season (Maintenance Phase)

During the competitive season, the primary focus is on actual sport performance. Although conditioning may still occur during practice, one should take the intensity of practice into consideration when planning the volume of conditioning drills. It is prudent to account for the duration and frequency of sprinting that occurs during practice to determine whether specific conditioning drills should be performed during or following practice or whether they would pose a greater risk for overtraining. Many coaches are using GPS devices (see chapter 10) to monitor the volume of training during practice. This provides a sensitive index for monitoring training stresses. Strength and conditioning professionals may decide to use additional conditioning exercises for players who are not in the regular playing rotation. This provides those athletes with the physiological stimulus necessary for them to maintain their level of conditioning and be prepared for an increase in playing time.

In the in-season resistance training program, the primary goal is to maintain the strength and power gains made during the off-season. This phase of training is generally referred to as the maintenance phase. Training in the maintenance phase is typically performed twice per week. Several studies have demonstrated that strength and power gains can be maintained with a 2 day per week in-season resistance training program (Hoffman and Kang 2003; Hoffman, Maresh, et al. 1991; Hoffman, Wendell, et al. 2003). The volume and intensity of training during the maintenance phase are similar to those used during the strength mesocycle, and only the core exercises are performed (see table 15.12). Although the goal of this phase is to maintain strength and power gains, several studies have shown that strength improvements can also be seen during this phase of training in young athletes with limited resistance training experience (Hoffman, Maresh, et al. 1991; Hoffman, Wendell, et al. 2003).

Training Program Considerations for Aerobic Endurance Athletes

The training program for the endurance athlete is periodized; however, instead of preparing for a season of competition, the endurance athlete is typically preparing for the most important competitions that occur at the end of the season. Although

Table 15.11 Speed–Endurance Training for 400 m Sprinters

Number of sprints	Distance of each sprint (m)	Recovery time between sprints (min)
10	100	5–10
6	150	5–10
5	200	10
4	300	10
3	350	10
2	450	10

The distances of the sprints can vary. However, the total distance run per workout should be approximately 2.5 times the distance of the athlete's event. Thus, 400 m sprinters should run 1000 m in sprints per workout. The length of the rest period should ensure complete recovery.

Adapted from *USA Track and Field Coaching Manual* 2000.

endurance athletes generally do not perform speed or agility training, they often use resistance training to support specific training goals. The following sections discuss specific training strategies for athletes competing in endurance events.

Long Distance: Marathon

The basis for developing the endurance training program is discussed in detail in chapter 12. Heart rate is often used to determine training intensity (Snyder 2012). The training program for someone planning to compete in a marathon includes high-intensity runs and easy distance runs to enhance recovery. It is important that the preparatory phase provide sufficient runs at low to moderate intensity to prepare the athlete for the more intense runs in later stages of the training cycle. In general, the preparatory phase should last 4 to 8 weeks depending on the initial conditioning level of the athlete (Snyder 2012). As the athlete progresses toward the race, the volume of training (i.e., distance run) increases. Rasmussen and colleagues (2013) have suggested that runners preparing for a marathon run a minimum of 30 km (18.6 mi) per week to reduce the risk of a running-related injury. These investigators were unable to find any significant differences in the injury rate in runners who ran 30 to 60 km per week compared to those who ran more than 60 km per week. However, runners who ran less than 30 km per week had a twofold greater risk for a running-related injury than runners who used a higher training volume. This was consistent with other studies reporting greater injury rates in runners performing less than 40 km per week compared to those who exceeded that volume threshold (Van Middelkoop et al. 2008).

A large part of the training preparation for the marathon is to increase the lactate or anaerobic threshold. By increasing the anaerobic threshold, runners can run faster for longer distances. They accomplish this through hill running, long speed intervals, and fartlek runs. These speed days should be at the beginning of the training week when the runner is less fatigued. The tempo or slower distance runs can be performed in the later part of the week. Table 15.13 provides an example of a 24-week workout to prepare for a marathon.

Table 15.12 In-Season Resistance Training Maintenance Program for the Strength–Power Athlete

Exercise	Sets	Reps
Squat	4	6-8RM
Power clean	4	4-6RM
Bench press	4	6-8RM
Push press	4	4-6RM

Table 15.13 Sample 24-Week Training Program for a Marathon

Weeks 1 to 6						
Monday	**Tuesday**	**Wednesday**	**Thursday**	**Friday**	**Saturday**	**Sunday**
Rest	3 mi tempo run	Ancillary training and 2 mi tempo run	4 mi tempo run	4 mi easy run	Ancillary training and 2 mi hill run	5 mi easy run
Weeks 7 to 12						
Rest	Fartleks: (60 min) • 12 min warm-up run • 5 min hard:2.5 min easy (2) • 4 min hard:2 min easy (2) • 3 min hard:1.5 min easy (2) • 2 min hard:1 min easy (2) • 5 min cool-down	Ancillary training and 6 mi easy run	Hill runs or intervals	6 to 8 mi tempo run	Ancillary training	10 to 12 mi easy run

(continued)

Table 15.13 *(continued)*

Monday	Tuesday	Wednesday	Thursday	Friday	Saturday	Sunday
Weeks 13 to 18						
Rest	Hill running or fartlek or both (60 min)	Ancillary training and 5 to 7 mi easy run	Intervals: 3 or 4 × 10 min at high intensity with a 3 min run at low intensity	6 to 8 mi tempo run	Ancillary training	12 to 15 mi easy run
Week 19 (recovery week—reduce training volume)						
Rest	5 mi tempo run	Rest	5 mi tempo run	Rest	10 mi easy run or rest	Rest
Week 20						
Rest	Intervals: • 10 × 400 m at race pace with 200 m recovery jogs • 5 min recovery jog • 2 × 800 m sprint with 400 m recovery jog	Ancillary training and 3 mi easy run	12 to 15 mi easy run	Rest	Ancillary training and 3 mi easy run	8 mi race pace
Week 21						
Rest	Long run, 20 to 22 mi at moderate intensity	Ancillary training and 3 mi easy run	Intervals: • 8 × 200 m at race pace with 100 m recovery jog • 1 mi at race pace with 400 m recovery jog • 8 × 200 m at race pace with 100 m recovery jog	Rest	Ancillary training and 3 mi easy run	5 mi tempo run
Week 22						
Rest	Intervals: 3 × 15 min at race pace with 5 min recovery at low intensity	Ancillary training and 3 mi easy run	12 mi easy run	Rest	Ancillary training and 3 mi easy run	5 mi tempo run
Week 23 (start of taper)						
Rest	2 × 30 to 40 min at race pace with a 10 min recovery at low intensity	Ancillary training and 3 mi easy run	Intervals: 3 × 15 min at race pace with 5 min recovery at low intensity	5 mi tempo run	Ancillary training and 3 mi easy run	Rest
Week 24 (final taper)						
10 to 12 mi easy run	3 × 5 min runs at race pace with 5 min recovery jogs	4 mi easy run	Ancillary training and 3 mi easy run	Rest	2 to 3 mi easy run	Race

Competitive runners also often participate in both shorter-distance races (half-marathon is quite popular, as well as 5K runs) and ultra-endurance events. As with training for the marathon, athletes interested in participating in shorter races still need to develop their aerobic base with low-intensity runs for the first 6 weeks of training. The inclusion of tempo runs and hill training is also highly recommended (Poston 2005). A major difference between the training program of athletes training for a half-marathon versus the marathon is the lower training volumes per week. Further, these athletes may spend more time on speed sessions during each week of training to prepare for the shorter-duration event.

Recreational athletes who prepare for a 5K race should be able to do so in as few as 6 to 8 weeks (Hoffman, Brown, and Smith 2012). Training for the short-duration races is more focused on intensity than on volume. In longer-duration "short events" such as a 10 km race, the volume of training is greater than that for the 5K run. For competitive runners, one or two training runs per week should be longer than the race distance. For novice runners the longer training runs may not be as necessary, as their primary goal is just to finish the race. Competitive runners will perform speed workouts to complement their distance training. That is, the total distance of sprints performed approaches the race distance. For example, when training for a 5K event, the runner can use 12 × 400 m sprints or 6 × 800 m. Training should progress to this volume of work.

Triathlon

Training for a triathlon incorporates swim, bike, and running workouts. Training volume for each exercise mode is dependent on the distance of the triathlon. In any case, the training program includes long- and short-distance runs, rides, or swims, as well as intervals and fartleks. Table 15.14 provides a sample 10-week triathlon training program for an Olympic-distance triathlon, which includes a 1.5K (0.9 mi) swim, 40K (25 mi) bike, and 10K (6.2 mi) run. In addition to the Olympic-distance triathlon are the sprint-distance triathlon (0.75K swim, 20K bike, 6K run), long-distance or half-Ironman triathlon (1.9K swim, 90K bike, 21.1K run), and the ultra-distance or Ironman triathlon (3.8K swim, 180.2K bike, 42.2K run).

In Practice

Speed-Endurance Is a Powerful Stimulus for Performance Improvements

Speed–endurance training is a typical method of training to delay fatigue in a long sprint. Iaia and Bangsbo (2010) examined the benefit of this method of training to enhance performance of both sprinters and athletes in short-duration endurance events such as 40K cycling or 10K running races. They suggested that speed–endurance training is very effective in improving performance in short-duration endurance events. One of the benefits of speed–endurance training in endurance athletes is that it can reduce the training duration, which may decrease the risk for overuse injury or overtraining. Iaia and Bangsbo (2010) suggested that the total amount of training can be reduced by 30% with the inclusion of speed–endurance training sessions. Although it is questionable whether speed–endurance training can improve maximal aerobic capacity, it appears to be able to reduce energy expenditure during exercise. In addition, it has been suggested that speed–endurance training is a potent stimulus for decreasing the rate of muscle glycogenolysis during submaximal exercise. It is well acknowledged that this form of training has an important role in improving an athlete's ability to maintain high-intensity sprinting for a prolonged duration, but it also appears to be able to enhance performance in short-duration endurance events. To develop speed-endurance, it is suggested that exercise bouts of 10 to 40 s at near-maximal effort be separated by 1 to 5 min. To maintain speed-endurance, the duration of the exercise bout may be longer (5-90 s), but the interval between bouts is shorter than during development.

Table 15.14 Example of a 10-Week Triathlon Training Cycle

Week	Monday	Tuesday	Wednesday	Thursday	Friday	Saturday	Sunday
1	Rest	300 yd swim 3 mi tempo run	Ancillary training 3 mi easy run	8 mi bike	300 yd swim 2 mi tempo run	Ancillary training	10 mi bike
2	Rest	300 yd swim 3 mi tempo run	Ancillary training 4 mi easy run	10 mi bike	500 yd swim 3 mi tempo run	Ancillary training	12 mi bike
3	Rest	500 yd swim 4 mi tempo run	Ancillary training 5 mi easy run	12 mi bike	500 yd swim 3.5 mi tempo run	Ancillary training 10 mi bike	12 mi bike
4	Rest	500 yd swim Intervals: 8 × 400 m, 200 m recovery jogs	Ancillary training 5 mi easy run	12 mi bike interval (1 mi high intensity and 1 mi moderate intensity)	500 yd swim 3.5 mi tempo run	Ancillary training 12 mi bike ride at race pace	15 mi easy bike ride
5	Rest	500 yd swim Intervals: 8 × 400 m, 200 m recovery jogs	Ancillary training 5 mi easy run	12 mi bike interval (1 mi high intensity and 1 mi moderate intensity)	750 yd swim 3.5 mi tempo run	Ancillary training 12 mi bike ride at race pace	15 mi easy bike ride
6	Rest	750 yd swim Intervals: 10 × 400 m, 200 m recovery jogs	Ancillary training 5 mi easy run	14 mi bike interval (1 mi high intensity and 1 mi moderate intensity)	750 yd swim 3.5 mi tempo run	Ancillary training 14 mi bike ride at race pace	18 mi easy bike ride
7	Rest	750 yd swim Intervals: 10 × 400 m, 200 m recovery jogs	Ancillary training 6 mi easy run	16 mi bike interval (1 mi high intensity and 1 mi moderate intensity)	1000 yd swim 4 mi tempo run	Ancillary training 16 mi bike ride at race pace	20 mi easy bike ride
8	Rest	1000 yd swim Intervals: 10 × 400 m, 200 m recovery jogs	Ancillary training 6 mi easy run	16 mi bike interval (1 mi high intensity and 1 mi moderate intensity)	1250 yd swim 4 mi tempo run	Ancillary training 16 mi bike ride at race pace	20 mi easy bike ride
9	Rest	1250 yd swim 6 mi run race pace	Ancillary training 7 mi easy run	18 mi bike interval (1 mi high intensity and 1 mi moderate intensity)	1500 yd swim 5 mi tempo run	Ancillary training 20 mi bike ride at race pace	22 mi easy bike ride
10	Rest	1000 yd swim 4 mi tempo run	Ancillary training 10 mi easy run	1000 yd swim 15 mi easy bike ride	Rest	3 mi easy run and 5 mi easy bike	Race

Summary

A multitude of program combinations can be used to train athletes. Determining if one program is more effective than another is quite difficult, primarily due to the large variability in the training response of athletes. To ensure that a training program is indeed appropriate, it is highly recommended that strength and conditioning professionals develop programs based on sound scientific evidence. This gives them the ability to defend their sport-specific training programs. It is also imperative that the coach or fitness trainer teach the correct techniques in all exercises to maximize the success of their athletes.

REVIEW QUESTIONS

1. What are the training considerations for the experienced strength–power athlete compared to a novice lifter?
2. Explain concerns that the strength and conditioning professional should have with regard to the training program of athletes participating in a jumping sport (e.g., basketball or volleyball).
3. Provide an example of a circuit plyometric training program for a group of 60 athletes.
4. For an athlete focused on losing body fat, what adjustments could be made to the training program to accomplish this goal?
5. What differences are seen between the anaerobic conditioning programs of a volleyball player and a basketball player?

chapter 16

Athletic Performance Testing and Normative Data

CHAPTER OBJECTIVES

After reading this chapter you should be able to do the following:

- Develop a testing protocol based on the specific physiological requirements of a sport.
- Understand what factors can influence assessment results.
- Understand the importance of reliability and validity of all performance tests.
- Select specific tests to assess various physiological components of interest.
- Compare assessment outcomes to normative data.

The growing knowledge base on physiological, biomechanical, and psychological responses to exercise and various training programs has given coaches a greater understanding of how to maximize their players' athletic performance. As a result, the development of optimal training programs over the last few years has grown immensely. In addition, exercise scientists have focused their research on **athletic profile** development. The objective for these scientists is to determine the type of athlete that would have the greatest potential to achieve success in a particular sport. The development of an athletic profile for each sport not only helps coaches in the selection process for their teams but also provides standards for both athletes and coaches in setting their training goals.

Developing an athletic profile requires a detailed battery of testing to thoroughly analyze all the components of athletic performance (**strength**, **anaerobic power**, **speed**, **agility**, **maximal aerobic capacity** and **endurance**, and **body composition**). Test results help determine the relevance and importance of each fitness component to a particular sport and permit the appropriate emphasis on the given variable in the athlete's training program. In addition, a sport-specific athletic profile helps establish standards for predicting potential success in the given sport. Athletes and coaches alike can use these standards as a motivational tool when establishing personal training goals by comparing their results to normative data from similar athletic populations. Performance testing can also be used to provide baseline data for individual exercise program prescription, provide feedback in the evaluation of a training program, and provide information concerning the extent of recovery after injury. This chapter focuses on factors that affect performance testing and provides examples of both laboratory and field tests used to assess various athletic components.

Factors Affecting Performance Testing

In interpreting performance tests, it is important to recognize that test results may be influenced by several factors, such as body size, **muscle fiber type**, the **training status** of the athlete, the **specificity** and **relevance** of the test to the sport and training program, and the **validity** and **reliability** of the test.

Body Size

The athlete's size may have a large influence on the results of some performance tests. For instance, in strength testing, absolute strength has a positive correlation to body size, whereas a negative correlation is seen between body size and the strength/mass ratio (Hoffman, Maresh, and Armstrong 1992). Thus, in an evaluation of American football players, linemen would be expected to have greater absolute strength and players in the skill positions (e.g., running backs) greater relative strength.

Fiber-Type Composition

For a given muscle size and architecture, considerable variability is seen in **force production**, **contraction velocity**, and **fatigue rates**. This variability may be related to the inherent contractile properties of the muscle, which are determined in part by the fiber-type composition (percentage of fibers that are **fast twitch** vs. **slow twitch**). Athletes with a higher percentage of fast-twitch fibers appear to have a greater force capability and faster contraction velocity (see chapter 1 for a review). Thus, athletes who have a greater percentage of fast-twitch fibers are likely to be associated with anaerobic sports such as sprinting, basketball, or football. On the other hand, athletes whose fiber-type composition is primarily slow twitch have a slower rate of fatigue but a lower force capability and a slower contraction velocity. Athletes with predominantly slow-twitch fibers would find success in endurance sports. As discussed earlier in this book, the ability of athletes to significantly alter their fiber-type composition through training is extremely limited. Therefore, with regard to speed or agility, there might be limitations to the extent of improvement that can be realized in certain athletes based on their physiological limitations. As noted earlier in the book, it may be possible to make a slow athlete faster, but it is highly unlikely that a slow athlete can be made into a fast athlete.

Training Status of the Athlete

As mentioned in earlier chapters, the more experienced the athlete, the smaller the potential to achieve performance gains. During the early stages of a training program, significant improvements in performance can be seen. As the duration of training increases, the rate of performance improvement declines. At some point during the athlete's career, perhaps after several years of training, further performance improvements are harder to achieve and a plateau appears to have been reached. And not only is the athlete's training experience a factor in determining performance improvement expectations; but also, athletes with a high ability level, regardless of training status, may not experience significant performance improvements even when participating in a training program for the first time (Hoffman et al. 1990; Hoffman, Maresh, et al. 1991). Thus, to properly evaluate the effectiveness of a training program or to set realistic performance goals, it is imperative that the coach have a clear understanding of the training status of the athlete.

Recognizing the athlete's experience level is also essential for interpreting performance results. For instance, in a 1-year investigation of elite weightlifters, small increases in strength were observed, but these increases did not reach statistical significance (Hakkinen, Komi, et al. 1987). Although the athletes and the strength and conditioning professionals could not see a statistical change, practically speaking they could rate the training program a success. In a group of elite athletes, training improvements are so difficult to achieve that even small improvements can mean the difference between winning and losing. When interpreting test results, especially in an elite athletic population, practical significance should take precedence over statistical significance (Hoffman 2006).

Test Selection

Test selection is based on the importance or relevance of each fitness component within a sport. This in turn is usually based on the needs analysis of the sport. Generally, assessments are selected from a battery of tests that may include strength (both upper and lower body), power, speed, and agility. Cardiorespiratory endurance, body composition, and flexibility are also components that may be part of a performance testing protocol. Once the fitness components for assessment have been determined, the next step is to ensure that the tests selected are reliable, valid, specific, and relevant to the sport being assessed.

Specificity and Relevance of the Test

To provide any significant information about the athlete's performance, it is imperative that each test be specific to the athlete's training program. In strength testing, for instance, when the training and the testing use the same or a similar mode of exercise (e.g., squats), the testing results more accurately reflect the magnitude of strength improvements. If training and testing use different training modes (e.g., machines vs. free weights) or exercises (e.g., squats vs. leg presses), the testing may not demonstrate actual strength improvement. The importance of this is highlighted in a study by Pipes in 1978. He examined two groups of subjects for 10 weeks. All subjects were tested on the leg press and squat exercise; however, one group had trained with only the leg press, while the other group had trained with the squat. The group that had trained with the leg press increased their strength by 27% in that exercise; however, when they were tested on the squat exercise, the strength increase was only 7.5%. For the group that had trained with the squat exercise, a strength gain of 28.9% was seen in that exercise, but only a 7.5% increase was seen in the leg press exercise. It appears that strength testing using a mode of exercise that is different from the one used in training (but uses similar muscle groups) may reflect 25% of the magnitude of strength gains. In further support of Pipes' study, Fry and colleagues (1991) showed that a 21% increase in squat strength resulted in only an 8% increase in leg extension strength.

When selecting tests, it is also important that each test have relevance to the given sport. Selected tests should provide both the athlete and coach with information concerning the athlete's ability to succeed in the particular sport. For example, the Wingate anaerobic test is the most widely accepted power test available. However, because it is performed on a cycle ergometer, its relevance to many sports has been questioned. As a result, efforts have been made to develop more specific anaerobic power tests that have a greater relevance to sports consisting primarily of running or jumping movements (Hoffman et al. 2000; Ross et al. 2009). An example of a sport-specific anaerobic power test is the vertical jump test. The athlete can perform it on a force plate or while attached to an accelerometer for the sports of basketball and volleyball.

Validity and Reliability of the Test

Two of the most important characteristics of a test are validity and reliability. Validity refers to the degree to which a test measures what it is intended to measure or claims to measure. Reliability refers to the ability of a test to produce consistent and repeatable results. Tests with proven reliability can reflect even slight changes in performance when one is evaluating a conditioning program. If a test is unreliable, differences in testing may reflect only the variation of the test and not the effectiveness of the training program.

Administrative Considerations for Assessment

Several administrative concerns must be addressed to minimize the risk associated with assessment protocols and to maximize the knowledge attained with appropriately designed testing schedules and testing sequences. In addition to the administrative concerns, the ability to communicate the results from the assessment to both coaches and athletes is critical in accurately developing an appropriate exercise prescription or setting training goals.

Safety Considerations

Before people participate in any performance assessment, regardless of level of competition, it is essential that they be medically cleared by the team physician or, if appropriate, their own personal physician. This prevents people from participating in an assessment that may increase risk for injury in cases of an existing medical limitation or contraindication. It is the responsibility of the strength and conditioning or fitness professional to ensure that medical clearance has been obtained, and it is strongly recommended that the requirement of attaining medical clearance for each participant be included in the manual for standard operating procedures.

Timing of Assessment

Each evaluation period needs to have a specific purpose and goal. These are generally determined by the time of the training year. The testing period may focus on determining training needs (goal development) and baseline performance measures, while some assessment periods measure the effectiveness of the training program or evaluate the readiness of athletes to compete.

The effectiveness of a training program is usually evaluated at the beginning and conclusion of the training cycle. However, if the goal of the assessment period is to assess the physical readiness of the athletes to participate in a competitive season, testing should occur as the athletes report to training camp. Some consideration should be given to providing both competitive and recreational athletes without prior experience in specific exercises (i.e., strength tests) sufficient time to learn how to perform each test. This reduces the risk of injury and provides for a more accurate assessment.

An example of a testing protocol can be seen in figure 16.1, which provides a testing schedule for collegiate American football players. The competitive season for these athletes lasts from September to November-December. The first testing session typically occurs before the onset of the off-season (winter) workouts. This testing period guides the exercise prescription and establishes training goals. The next testing period usually occurs at the end of winter workouts. This testing session generally serves to evaluate the winter conditioning program, check the athletes' progress, and adjust training goals. The third and final assessment period occurs at the start of training camp and gives coaches a final evaluation of the effectiveness of the summer training program and a measure of the athlete's readiness to participate in the upcoming competitive season.

Testing Sequence

An important administrative concern for the strength and conditioning professional with regard to the assessment protocol is the order in which the testing battery is performed. The least fatiguing tests should always be performed first. Assessments requiring high-skill movements, such as agility drills, should be performed before any fatiguing tests. Any performance test that fatigues the athlete will confound the results of subsequent assessments. For example, a repeated sprint test such as the line drill or 300 yd shuttle run that precedes a strength assessment may result in a significant decrease in strength. In general, the more fatiguing assessments (e.g., 300 yd shuttle run, line drill, or 1.5 mi run) are performed last in the testing sequence.

A number of administrative factors can influence testing sequence. For instance, based on the number of athletes being tested and the number of strength and conditioning professionals who can assist in the testing program, the head strength and conditioning coach can determine the number of testing stations and the number of athletes to be rotated through each station. In addition, the length of the testing period (e.g., 2 h, 1 day), provides a window of time

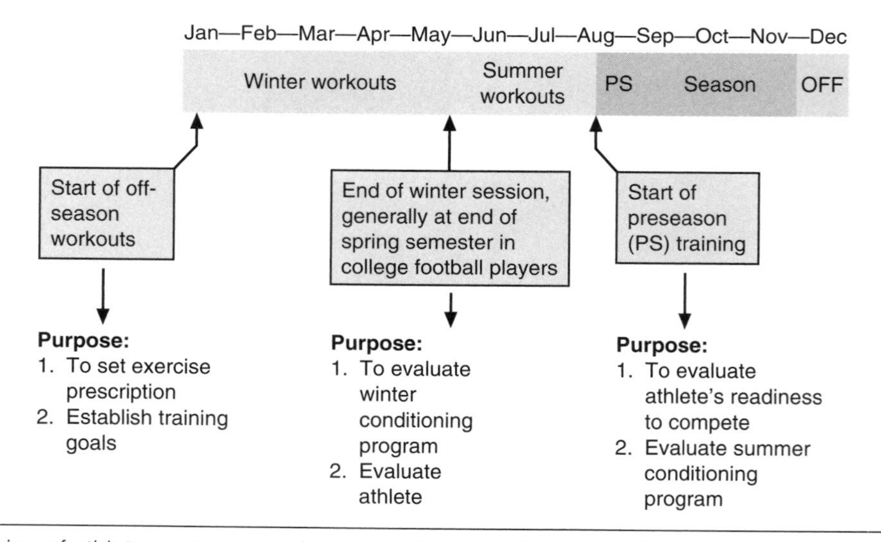

Figure 16.1 Timing of athlete assessments for a collegiate American football team.

Adapted, by permission, from J.R. Hoffman, 2006, *Norms for fitness, performance, and health* (Champaign, IL: Human Kinetics), 9.

that the strength and conditioning staff has to complete the testing program. This helps to determine the number of assessments that can be performed. Ideally, all athletes being assessed will perform the tests in the same sequence. If testing is performed over an extended time period (i.e., over 2 days), the most fatiguing tests should always be performed last. Time constraints, though, may make the ideal testing sequence unrealistic. Testing a large number of athletes may require simultaneous use of several different stations. Athletes often rotate through the stations within a set period of time. Some athletes may perform a 40 yd sprint, followed by strength tests. Other athletes perform their strength tests before the sprint and agility tests. Either of these testing sequences may not affect the results; however, other measures could potentially be dependent on testing order. For instance, a maximal squat test may potentiate subsequent vertical jump performance. If some of the athletes are tested in the vertical jump before the squat and others in the reverse order, the potentiating effect from the maximal squat may provide a significant advantage in jump height and power (Hoffman, Ratamess, Faigenbaum, et al. 2007).

Delineation of Test Results

At the conclusion of the testing program, the results must be communicated to the athlete and, when appropriate, to the sport coach. Individual athletes' performance results can be compared to their previous results to evaluate progress or can be compared to normative values of athletes playing in the same sport and position. This provides a measure of the athlete's potential and also may be a way to determine training goals.

Tests for Needs Assessment and Program Evaluation

The remainder of this chapter covers tests commonly used for each of the performance variables. It is not meant to be an all-inclusive discussion of tests but focuses on those that are widely accepted and used.

In Practice

Effect of Age on Performance Measures in Professional Baseball Players

A collaborative study from the University of Central Florida and several professional baseball teams examined the effect of age on changes in performance variables in professional baseball players (Mangine et al. 2013). Data from 1157 professional baseball players who were on a major league or minor league roster were examined retrospectively. The players were divided into cohorts based on age. The first group consisted of professional players who were adolescents (16-19 years, *n* = 82); the second group consisted of players who were college-aged athletes (20-22 years, *n* = 285); the third through sixth groups were divided into 3-year segments (23-25 years, *n* = 364; 26-28 years, *n* = 206; 29-31 years, *n* = 112; and 32-34 years, *n* = 63, respectively). The seventh and final group comprised athletes older than 35 years (*n* = 63). All performance assessments were conducted as part of the teams' normal spring training camp assessment routine. Assessments included lower body power (vertical jump test), speed (10 yd sprint), agility (pro-agility), grip strength, 300 yd shuttle, and body composition (skinfold measures). No differences were noted in 10 yd sprint speed or pro-agility time in any age group. Grip strength was significantly elevated in the age group 26 to 28 years compared to the adolescent and college-aged players. However, the highest grip strength was noted in players 29 to 31 years, which was significantly greater than the value for adolescent (15.0%) or college-aged (10.0%) players. Professional baseball players 23 to 25 years had significantly greater (8.3%) jump power than adolescent players. Players reached peak jump power between the ages of 26 and 28. Interestingly, position players (nonpitchers) who had continued playing into their mid-30s were able to maintain their power output. Older pitchers (35+ years) were not, suggesting that they rely on greater skill as they age versus maintaining their power capability. The authors concluded that age-related differences suggest the importance of strength and conditioning programs that can extend the length of the athlete's playing career.

Strength

When selecting a strength test, it is vital that the test be familiar to the athlete and specific to the athlete's resistance training program. As mentioned earlier, this ensures a clear understanding of the effectiveness of the conditioning program. In addition, using an exercise that is part of the athlete's conditioning program ensures proper technique, reduces the potential of injury during testing, and allows for proper selection of the resistance attempted during the test. If initial testing occurs before the onset of a conditioning program, which might be the case for freshman athletes on the first day of practice, the exercises used to assess strength may be novel from the point of view of the athletes. It would still be appropriate to use these tests as long as the exercises will be part of their resistance training program and the same tests are used to reassess them at the conclusion of the training program.

If strength tests are to be part of an evaluation to predict potential performance, then the strength test should incorporate similar movement patterns and involve the same muscle mass routinely recruited during actual sport performance (Hoffman, Maresh, and Armstrong 1992). Strength testing generally involves the use of a multijoint, large muscle mass exercise to evaluate an athlete's adherence to a resistance training program or to make comparisons between athletes. However, there are times when using an exercise that recruits a smaller muscle mass for an isolated joint action provides additional information. For example, comparing muscle groups of bilateral limbs (e.g., right knee flexors vs. left knee flexors) or agonist versus antagonist muscle groups (e.g., knee flexors vs. knee extensors) may indicate a weakness that can predispose the athlete to injury.

The mode of exercise used to assess strength is dependent on the goals of the testing program. Generally, strength testing is performed with either **dynamic constant-resistance** exercises or an **isokinetic testing** device. Isokinetic testing devices are designed to measure joint movements at a constant velocity. The force exerted by a moving body segment is met with an equal and opposite resistance that is constantly altered as the body segment moves through its full range of motion. This force exerted by the body segment to produce rotation about its axis is referred to as **torque** and is expressed in newton-meters. The test–retest reliability of isokinetic devices has been well established (Farrel and Richards 1986). Since isokinetic devices permit the evaluation of only single-joint unilateral movement, their role in strength evaluation is limited primarily to evaluating the athlete's potential for muscle injury as a result of either a bilateral deficit or an agonist–antagonist muscle imbalance (Hoffman, Maresh, and Armstrong 1992).

Isokinetic Testing

Peak torque measures are frequently used to establish a ratio between agonist and antagonist muscle groups and make bilateral comparisons. They are a reliable indicator of muscle ability in both healthy and injured knee joints. The hamstring-to-quadriceps ratio (H:Q) is the most prevalent agonist-to-antagonist relationship reported for the lower body. A 3:5 ratio of H:Q strength has been commonly accepted as the normal strength proportion between these two muscle groups (Hoffman, Maresh, and Armstrong 1992). However, ratios of 2:3 (Fry and Powell 1987) and 3:4 (Knapik et al. 1991) are also acceptable strength ratios. The H:Q ratio can also be reported as a percentage. That is, a 2:3 H:Q ratio means that the hamstring muscle group is 67% as strong as the quadriceps.

When the H:Q is compared among athletes, a large variation is often seen between sports and at different testing velocities. Ratios have been reported to range from 52% in college endurance runners (Worrell et al. 1991) to greater than 100% in National Collegiate Athletic Association (NCAA) Division I basketball players (Hoffman, Fry, et al. 1991). This wide variation of H:Q likely reflects the specific demands of each sport as well as the specific speed of contraction.

Differences in H:Q may be a function of participation in a resistance training program. A lower H:Q observed at slow velocities of muscle contraction may reflect the type of strength training regimen employed. For example, performing the squat exercise may disproportionately increase knee extensor strength at a slow velocity of movement (e.g., 60°/s), causing a reduction in the H:Q (Fry and Powell 1987). In addition, the type of athlete being tested appears to influence the H:Q. Athletes involved in high-intensity anaerobic activities such as basketball, football, or hockey appear to have greater knee flexor strength, resulting in a high H:Q at all speeds of contraction (Hoffman, Fry, et al. 1991; Hoffman, Maresh, and Armstrong 1992; Housh et al. 1988).

Isokinetic testing at different velocities of muscle contraction has merit. However, testing at fast velocities of contraction in the lower limb may have added importance, considering the greater power development of the knee flexors (hamstring muscle group) at fast speeds of contraction (Read and Bellamy 1990). Knapik and colleagues (1991) reported that female athletes with an H:Q less than 75% at 180°/s were

1.6 times more likely to be injured than athletes with stronger knee flexors. However, that same study also indicated that the H:Q may not be an independent predictor of injury when bilateral comparisons are also considered.

Initial studies on bilateral strength differences reported that a 10% difference in the strength between the right and left knee flexors indicated an enhanced likelihood of muscle injury (Burkett 1970). However, subsequent studies have suggested that strength deficits between bilateral muscle groups may be able to reach 15% before a significant increase in the injury rate is seen (Knapik et al. 1991). Athletes with strength imbalances greater than 15% had a 2.6 times greater incidence of muscle injury (Knapik et al. 1991). Still, others have been unable to conclusively determine that bilateral deficits can be used as a significant predictor of lower leg injury (Worrell et al. 1991). Although bilateral deficits in the lower extremity exceeding 10% to 15% may be a cause for concern, normative strength balances are still uncertain.

In some athletes, bilateral deficits are seen between muscles of the shoulder, elbow, and wrist because of activity patterns that rely predominantly on unilateral arm action (e.g., tennis, baseball pitching) (Cook et al. 1987; Ellenbecker 1991). Strength differences approaching 20% in the upper limb have been seen in tennis players and baseball pitchers. These large bilateral strength differences may be compounded by the non–weight-bearing requirements of the upper body musculature. Whether this large bilateral strength difference negatively affects performance or increases the risk of injury in these athletes is still not fully understood.

Athletes who rely on unilateral arm action during performance also show significant differences in agonist and antagonist strength balance (shoulder internal–external rotators) (Cook et al. 1987; Ellenbecker 1991; McMaster, Long, and Caiozzo 1991). These imbalances have been attributed to the emphasis placed on shoulder adduction and internal rotation during these activities (e.g., tennis, baseball pitching, or water polo). These movements are characterized by a significant weakness in the external rotators and by significantly stronger internal rotators in the dominant shoulder. Since the external rotators are important for shoulder joint stability, any imbalance may lead to injury in athletes with shoulder capsule laxity (McMaster, Long, and Caiozzo 1991).

In Practice

Is the Isometric Midthigh Pull a Valid Measure for Jump and Sprint Acceleration Performance?

An international collaboration among sport scientists from the United Kingdom and Australia examined the relationship between an isometric measure of force development and dynamic performance (West et al. 2011). Thirty-nine experienced resistance-trained professional rugby players volunteered for the study. All participants performed an isometric midthigh pull while standing on a force platform. Participants stood in a position that was similar to the body position during the second pull of a power clean (e.g., slight forward lean with straight back, with shoulders positioned in line with the bar). The knee angle was approximately 120° to 130°, and the bar height was adjusted to accommodate the different heights of the participants. On performance of the midthigh pull, both peak force and the rate of force development were determined. In addition to the midthigh pull, participants performed a countermovement vertical jump and a 10 m sprint. Absolute peak force from the midthigh pull was not related to dynamic performance; however, when reported as relative to body mass, it was significantly correlated to both 10 m sprint time ($r = -0.37$) and countermovement jump height ($r = 0.45$). Significant correlations were also noted between the peak rate of force development and 10 m sprint time ($r = -0.66$) and countermovement jump height ($r = 0.39$). Force generated at 100 ms was inversely related to 10 m sprint time ($r = -0.54$); when expressed relative to body mass, the correlation increased to $r = -0.68$. The investigators concluded that measures of maximal force development and rate of force development from an isometric measure are related to jump and sprint acceleration performance.

Dynamic Constant-Resistance Testing

The primary advantage of dynamic constant-resistance testing versus isokinetic testing is the ability to simulate actual sport movement or recruit a larger muscle mass. In addition, most athletes use dynamic constant-resistance exercises as part of their resistance training program. A controversy frequently discussed in connection with testing maximal strength is whether to test for a one-repetition maximum (1RM) or to predict maximal strength from the number of repetitions performed with a submaximal load. Often the motivation for using the latter method is the time difference between achieving a maximal lift versus predicting maximal strength from one set. When testing large groups of athletes (as is often the case in strength testing for a football team), the time factor is an important and valid consideration. In addition, the possible risk of injury with performance of a maximal lift may also be considered a drawback to using the 1RM as a strength test.

Several studies have shown that the use of submaximal loads to predict maximal strength is highly valid ($r > 0.90$) (Landers 1985; Mayhew, Ball, and Bowen 1992; Mayhew et al. 1999; Shaver 1970). In fact, many teams in the National Football League (NFL) routinely use submaximal testing to estimate the maximal strength of their players. However, several studies have also reported that the number of repetitions performed at selected percentages of 1RM is variable between exercises, and the variance within an exercise is also large (Hoeger et al. 1987; Hoeger, Hopkins, Barette, et al. 1990). A recent examination of four published submaximal equations to predict maximal strength showed that all four significantly underestimated or overestimated maximal strength performance in strength-trained athletes (Ware et al. 1995). In addition, a closer examination of the NFL's 225 lb bench press test to assess maximal upper body strength showed it to be accurate as long as the number of repetitions was below 10 (Mayhew et al. 1999). If more than 10 repetitions were performed, the equation became less valid and tended to underestimate actual strength ability. This may be a major concern with testing of athletes who can bench press more than 300 to 315 lb (remember from chapter 8 that more than 10 repetitions can usually be performed with exercises at percentages less than 75% of a subject's 1RM).

The other concern about using 1RM tests to assess maximal strength is the possible increased risk of injury. However, no published reports have determined whether a greater risk of injury exists with proper execution of a 1RM test versus a submaximal strength test. If proper safety precautions are used (e.g., spotters, proper technique, and qualified supervision), the possibility of injury is minimized. In addition, the resistance selected during a 1RM is generally based on the training experience of the athlete. That is, based on the training history of the athlete, both the athlete and coach should have a general idea of what the maximum strength level is for a particular exercise. Another factor that should be considered is that when a competitive athlete performs a maximal number of repetitions with an absolute resistance, the last repetition is a RM. Although it may be an 8RM, a 14RM, or even a 25RM, in each circumstance the athlete is performing the RM with a potentially fatigued muscle. A fatigued muscle may be more susceptible to injury because of changes in neuromuscular recruitment patterns or factors relating to muscle–tendon stress. Thus, in testing of an experienced strength-trained athlete, the risk of injury may be lower with a 1RM versus a maximal number of repetitions with a submaximal load. A protocol for assessing a 1RM is presented in the sidebar. Remember to use proper safety precautions.

Protocol for Testing Maximal Strength (1RM)

The athlete should do the following:

1. Perform a warm-up set of 10 repetitions at a resistance that is equivalent to approximately 50% of the expected 1RM.
2. Perform another warm-up set of 5 repetitions at a resistance that is equivalent to approximately 75% of the expected 1RM.
3. Rest 3 to 5 min.
4. Perform 1 repetition with a resistance that is equivalent to approximately 90% to 95% of the expected 1RM.
5. Rest 3 to 5 min.
6. Attempt 1RM lift.
7. Rest 3 to 5 min.
8. If attempt was successful, increase resistance and attempt a new 1RM.
9. Continue this protocol until failure.

Adapted, by permission, from J.R. Hoffman, 2006, *Norms for fitness, performance, and health* (Champaign, IL: Human Kinetics), 34.

Submaximal tests to predict maximal strength have been demonstrated to be valid (correlation coefficients > 0.90) (Landers 1985; Mayhew, Ball, and Bowen 1992; Mayhew et al. 1999). However, the validity of these equations is dependent on the number of repetitions performed. If more than 10 repetitions of an exercise are performed, the equations lose their validity and tend to underestimate actual strength levels (Mayhew et al. 1999). Thus, if the decision is to use a submaximal test to predict maximal strength, it is important to ensure that the load tested is relative to the strength level of the athlete. To predict strength in an American football team, the strength and conditioning professional may opt to use loads specific to the player's position. For instance, linemen may perform as many repetitions as possible in the bench press exercise using 330 lb (150 kg) while linebackers perform as many repetitions as possible with 300 lb (136 kg), and so on. This time-efficient method gives athletes a better opportunity to produce a valid test.

As mentioned earlier, the number of repetitions performed at selected percentages of the 1RM is quite variable among exercises, and the variance within an exercise is also quite large (Hoeger, Hopkins, Barette, et al. 1990). Table 16.1 provides formulas used to predict a 1RM.

The bench press, squat, and power clean are widely used dynamic constant-resistance tests for upper body strength, lower body strength, and explosive power, respectively. These tests have been demonstrated to have strong test–retest reliability ($r > 0.90$) (Hoffman et al. 1990; Hoffman, Fry, et al. 1991). In tables 16.2 through 16.4, strength characteristics for various high school and collegiate sports are shown.

Table 16.1 Equations for Predicting 1RM Strength

Equation	Reference
Repetition weight / [1.0278 – 0.0278(reps)]	Brzycki 1993
[0.033(reps)] (repetition weight) + (repetition weight)	Epley 1985
(100) (repetition weight) / [101.3 – 2.67123(reps)]	Landers 1985

Adapted from Hoffman 2006.

Table 16.2 Percentile Values of High School and College Football Players in the 1RM Bench Press, Squat, and Power Clean Exercises

	High school 14 to 15 years			High school 16 to 18 years			NCAA Division III		NCAA Division I		
% rank	BP	SQT	PC	BP	SQT	PC	BP	SQT	BP	SQT	PC
90	245	400	215	265	449	250	345	455	370	500	300
80	210	340	201	245	408	235	315	425	345	455	280
70	195	325	190	235	376	225	299	405	325	430	270
60	185	310	183	225	355	213	290	385	315	405	261
50	170	295	173	210	324	203	275	365	300	395	252
40	165	275	165	205	302	192	265	345	285	375	242
30	155	255	161	194	285	175	255	335	270	355	232
20	145	240	153	175	270	163	235	315	255	330	220
10	130	215	141	155	245	144	224	275	240	300	205
Mean	180	296	177	211	336	200	280	368	301	395	252
SD	45	73	32	44	84	43	50	71	53	77	38
N	191	149	160	298	210	246	580	550	1189	1074	1017

BP, bench press; SQT, squat exercise; PC, power clean.

Adapted from Hoffman 2006.

Table 16.3 Percentile Values of the 1RM Bench Press, Squat, and Power Clean Exercises (lb) in Various NCAA Division I Female Collegiate Athletes

	Basketball			Softball			Swimming		Volleyball	
% rank	BP	SQT	PC	BP	SQT	PC	BP	SQT	BP	SQT
90	124	178	130	117	184	122	116	145	113	185
80	119	160	124	108	170	115	109	135	108	171
70	115	147	117	104	148	106	106	129	104	165
60	112	135	112	99	139	100	101	120	100	153
50	106	129	110	95	126	94	97	116	98	143
40	102	115	103	90	120	93	94	112	96	136
30	96	112	96	85	112	88	93	104	90	126
20	88	101	88	80	94	80	88	101	85	112
10	82	81	77	69	76	71	78	97	79	98
Mean	105	130	106	94	130	97	98	118	97	144
SD	18	42	20	18	42	20	15	19	14	33
N	120	86	85	105	97	80	42	35	67	62

BP, bench press; SQT, squat exercise; PC, power clean.

Adapted, by permission, from J.R. Hoffman, 2006, *Norms for fitness, performance and health* (Champaign, IL: Human Kinetics), 37.

Table 16.4 Percentile Values of the 1RM Bench Press, Squat, and Power Clean Exercises (lb) in NCAA Division I Male Baseball and Basketball Athletes

	Baseball			Basketball		
% rank	BP	SQT	PC	BP	SQT	PC
90	273	365	265	269	315	250
80	260	324	239	250	305	235
70	247	310	225	240	295	230
60	239	293	216	230	280	220
50	225	270	206	225	265	215
40	218	265	200	216	245	205
30	203	247	190	210	225	195
20	194	237	182	195	195	180
10	175	218	162	185	166	162
Mean	227	281	210	225	251	209
SD	41	57	36	33	57	34
N	170	176	149	142	131	122

BP, bench press; SQT, squat exercise; PC, power clean.

Adapted, by permission, from J.R. Hoffman, 2006, *Norms for fitness, performance and health* (Champaign, IL: Human Kinetics), 38.

Anaerobic Power and Anaerobic Fitness

Anaerobic power can be assessed in both laboratory and field settings. In the laboratory, a variety of tests have been used over the years to assess anaerobic power. These tests have included sprints on a motorized treadmill (Cunningham and Faulkner 1969; Falk et al. 1996; Ross et al. 2009), repeated jumps on a force plate or contact mat (Bosco, Mognoni, and Luhtanen 1983), and maximal-effort cycling tests with exercise durations ranging from 7 to 120 s against various braking forces (Ayalon, Inbar, and Bar-Or 1974; Katch et al. 1977; Sargeant, Hoinville, and Young 1981). Other laboratory tests for assessing anaerobic power have used isokinetic testing devices (single isolated movements for **peak power** and multiple repetitions to assess anaerobic fatigue) (Thorstensson and Karlsson 1976) and sprints performed with a vertical elevation (Margaria, Aghemo, and Rovelli 1966). Some of these tests assessed peak power (highest power output attained during the test), and others evaluated **mean power** (average power output of entire test). Another variable reported was fatigue rate, which involved assessment of the ability of the athlete to maintain power output through the duration of the test.

Laboratory Assessments

The laboratory-based anaerobic test most commonly used today is the Wingate anaerobic test (Bar-Or 1987). This is a 30 s cycling or arm-cranking test performed at a maximal effort against a resistance relative to the subject's body weight. The Wingate anaerobic test (WAnT) was developed at the Wingate Institute in Israel (Ayalon, Inbar, and Bar-Or 1974). Of the available laboratory-based anaerobic power tests, the WAnT has the most extensive research base to date. Test–retest reliability has been shown consistently ($r > 0.90$) (Bar-Or 1987).

The WAnT provides assessments of an individual's peak power, mean power, and **fatigue index**. Peak power is the highest mechanical power output achieved at any stage of the test and represents the explosive capability of the athlete's lower body. Most laboratories report peak power as the highest power output achieved over a 5 s interval; however, some laboratories use a 3 s interval or lower. Mean power is the average power output during the 30 s test. This measure provides an assessment of the athlete's anaerobic endurance, or the ability to maintain high power outputs over a long duration. The fatigue index is often determined by dividing the lowest 5 s power segment (generally the last 5 s of the test) by peak power. Although it is not clear whether the fatigue index is a good indicator of anaerobic fitness, it does appear to correlate highly with the percentage of fast-twitch fibers (Bar-Or et al. 1980). Typically, a higher fatigue index is seen in athletes with a greater percentage of fast-twitch fibers. Athletes who are endurance trained generally have a lower fatigue index. Technological advances with the WAnT have given sport scientists the flexibility to alter the methodology of the assessment by varying the duration of the test or the resistance that individuals pedal against. Some laboratories have used repeated trials of shorter duration (10-20 s) or have performed a longer, 60 s test. Each configuration is developed to meet the specific needs of the assessment. Figure 16.2 depicts a typical performance diagram produced from a 30 s WAnT.

Although the WAnT is often acknowledged as the standard of laboratory-based anaerobic power measurements, it has not achieved widespread acceptance among coaches as a performance test for their athletes. This may be related to questions concerning muscle and activity pattern specificity as well as accessibility to laboratories with such testing capabilities. For example, anaerobic power assessment for a basketball player may be more specific if performed as a vertical jump power test. These tests

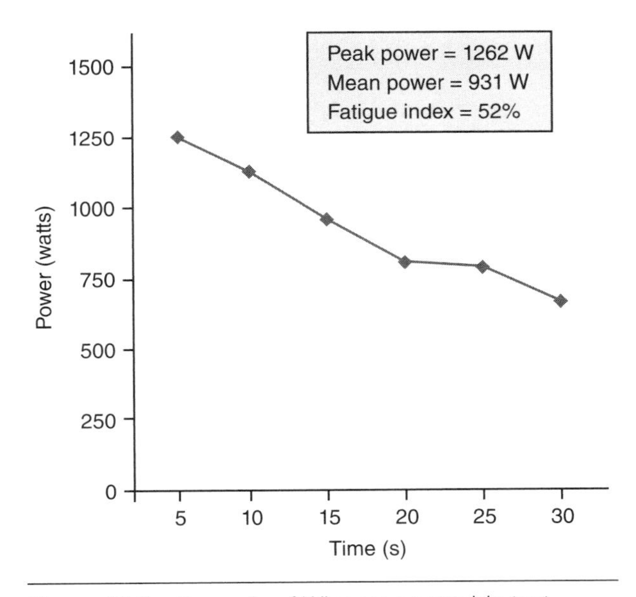

Figure 16.2 Example of Wingate anaerobic test.

Data from a football athlete collected at the Human Performance Laboratory, The College of New Jersey.

generally require the athlete to perform repeated countermovement jumps on a force plate or contact mat. The flight time of each jump is recorded (moment subject breaks contact with the mat until contact is made on landing). The time in flight is used to calculate the change in the body's center of gravity (Bosco, Mognoni, and Luhtanen 1983). Using body weight and the calculated jump height, mechanical work is calculated. Finally, using both mechanical work and length of contact time between jumps, anaerobic power can be determined. A vertical jump anaerobic power test appears to be more sport specific, especially for sports such as basketball and volleyball (Hoffman et al. 2000).

Field Tests

In testing large groups of individuals, administrative concerns (equipment availability and time constraints—only one subject can be tested at a time) may prevent wide adoption of any of the previously mentioned tests. As a result, many coaches have searched for a field-based test that provides assessments similar to those obtained with laboratory measures. The vertical jump is a popular field test for anaerobic power. A few field tests can be used to evaluate anaerobic fitness. Some of the more popular field assessments are discussed in the following sections.

Vertical Jump

The vertical jump is a popular field test to assess anaerobic power. Its popularity is likely related to the relative ease in performing the assessment and the ability to provide a sport-specific power measure for athletes participating in sports that involve jumping. The only negative aspect of the vertical jump test is that it measures only jump height. To provide a more accurate assessment of power, a formula can be used to estimate power output from the vertical jump test (Harman et al. 1991). Power outputs are recorded in watts (W), and the absolute watts generated divided by the athlete's body mass (in kilograms) provides the relative power output (W/kg). The equations to calculate peak and mean power are as follows:

$$\text{Peak power (W)} = 61.9 \times \text{Jump height (cm)} + 36 \times \text{Body mass (kg)} + 1822$$

$$\text{Mean power (W)} = 21.2 \times \text{Jump height (cm)} + 23 \times \text{Body mass (kg)} - 1393$$

Tables 16.5 through 16.12 provide percentile rankings for vertical jump height and power in various population groups.

Table 16.5 Percentile Ranks for Vertical Jump Heights in Youth

% rank	7-8	9-10	11-12	13-14 Male	13-14 Female	15-16 Male	15-16 Female	17-18 Male
90	9.6	11.5	16.5	21.0	17.0	27.0	18.5	28.2
80	9.3	11.0	14.3	20.0	16.0	24.0	17.5	26.0
70	8.7	10.4	12.3	19.0	16.0	22.5	16.9	25.0
60	8.1	9.9	11.8	18.4	15.0	22.0	16.0	23.8
50	8.0	9.5	10.5	17.0	14.5	20.5	15.5	22.0
40	7.7	9.0	10.0	16.0	14.0	20.0	14.9	20.2
30	7.5	8.6	9.6	15.0	14.0	20.0	14.1	19.4
20	7.1	7.8	8.8	13.8	13.5	17.0	13.2	18.6
10	6.9	7.0	6.8	12.3	13.0	17.0	10.0	18.0
Mean	8.1	9.3	11.2	16.8	14.6	20.9	15.2	22.6
SD	1.0	1.7	3.5	3.4	1.5	3.4	2.7	3.8
N	26	67	74	42	19	29	16	27

Adapted, by permission, from J.R. Hoffman, 2006, *Norms for fitness, performance and health* (Champaign, IL: Human Kinetics), 59-60.

Table 16.6 Percentile Ranks for Vertical Jump (No Step) in Football

% rank	High school (14 years)	High school (15 years)	High school (16 years)	High school (17 years)	High school (18 years)	NCAA D III	NCAA D I
90	27.6	27.5	28.5	30.0	30.0	30.5	33.5
80	25.5	26.0	27.0	28.0	28.0	29.0	31.5
70	24.0	25.0	25.5	26.5	26.5	28.0	30.0
60	23.5	24.0	25.0	26.0	26.0	27.0	29.0
50	22.3	23.0	24.0	25.0	25.0	26.5	28.0
40	21.9	22.0	23.5	23.0	24.0	25.5	27.0
30	21.2	21.1	22.0	22.0	23.0	24.5	25.5
20	19.3	20.0	20.5	20.5	22.0	23.0	24.0
10	17.7	18.0	18.0	19.0	21.0	21.5	21.5
Mean	22.6	23.0	23.8	24.8	25.2	26.1	27.6
SD	3.6	3.9	3.7	4.3	3.7	3.7	4.4
N	30	86	79	64	39	524	1495

Adapted, by permission, from J.R. Hoffman, 2006, *Norms for fitness, performance and health* (Champaign, IL: Human Kinetics), 60.

Table 16.7 Percentile Ranks for Vertical Jump (No Step) in Basketball Players

% rank	High school (14 years)	High school (15 years)	High school (16 years)	High school (17 years)	NCAA D I male	NCAA D I female	NBA
90	25.6	27.1	29.0	28.3	30.5	21.6	31.2
80	23.4	25.0	27.5	26.5	30.0	20.1	29.5
70	22.5	24.0	25.7	24.5	28.5	19.7	28.4
60	21.6	23.0	24.7	24.0	28.0	18.5	27.5
50	21.0	23.0	24.0	24.0	27.5	18.0	27.0
40	20.9	22.0	23.0	23.5	26.8	17.5	26.2
30	20.3	21.5	22.4	22.9	26.0	16.5	24.6
20	18.0	20.5	20.9	21.6	25.5	15.9	23.6
10	15.4	20.0	19.5	21.0	24.5	14.5	22.4
Mean	21.0	23.1	24.0	24.0	27.7	18.0	26.7
SD	3.1	3.0	3.9	2.3	2.4	2.5	3.3
N	21	87	58	22	138	118	40

Adapted, by permission, from J.R. Hoffman, 2006, *Norms for fitness, performance and health* (Champaign, IL: Human Kinetics), 61.

Table 16.8 Vertical Jump Heights (in.) for College Football Players Participating in the NFL Combine

% rank	DL	LB	DB	OL	QB	RB	TE	WR
90	38.0	37.4	40.0	33.5	36.5	38.7	38.4	41.0
80	35.5	36.5	38.5	32.5	35.5	37.2	37.0	39.1
70	34.5	35.5	37.5	31.5	35.5	35.5	36.0	38.4
60	33.5	34.6	37.0	30.0	34.5	34.0	35.1	36.7
50	32.0	33.5	36.0	29.5	32.8	33.5	34.5	36.0
40	31.5	33.0	35.5	28.5	31.5	32.5	33.9	35.5
30	30.5	32.5	35.0	27.5	30.8	32.2	32.8	34.5
20	30.0	31.0	34.2	27.0	30.0	31.5	31.0	33.5
10	28.3	30.0	33.1	25.9	29.0	30.8	30.6	32.0
Mean	32.7	33.7	36.3	29.5	32.6	34.1	34.2	36.5
SD	3.5	3.1	2.6	3.0	3.7	3.0	3.0	3.3
N	65	41	81	96	24	37	31	63

DL, defensive linemen; LB, linebackers; DB, defensive backs; OL, offensive linemen; QB, quarterbacks; RB, running backs; TE, tight ends; WR, wide receivers.

Adapted from Hoffman 2006.

Table 16.9 Percentile Ranks for Vertical Jump (No Step) in NCAA Division I Female Athletes

% rank	Volleyball	Softball	Swimming
90	20.0	18.5	19.9
80	18.9	17.0	18.0
70	18.0	16.0	17.4
60	17.5	15.0	16.1
50	17.0	14.5	15.0
40	16.7	14.0	14.5
30	16.5	13.0	13.0
20	16.0	12.0	12.5
10	15.5	11.0	11.6
Mean	17.3	14.6	15.3
SD	2.1	2.9	3.0
N	90	118	40

Adapted, by permission from J.R. Hoffman, 2006, *Norms for fitness, performance and health* (Champaign, IL: Human Kinetics), 62.

Table 16.10 Percentile Ranks for Vertical Jump Height and Power in Professional Baseball Players (Minor and Major League Combined)

% rank	VJ height (in)	VJ mean power (W)	VJ mean power (W/kg)
90	32.0	2643	26.5
80	30.0	2509	25.5
70	29.0	2429	24.8
60	28.0	2342	24.3
50	27.0	2275	23.7
40	26.5	2195	23.1
30	25.5	2121	22.8
20	24.0	2038	21.9
10	22.0	1922	20.8
Mean	27.1	2276	23.7
SD	3.9	293	2.3
N	855	718	718

Mean power was calculated using the Harman formula.

Table 16.11 Percentile Ranks for Vertical Jump Power in NCAA Division I Women's Basketball and Soccer

	Women's soccer	Women's basketball			
% rank	VJ mean power (W)	VJ peak power (W)	VJ peak power (W/kg)	VJ mean power (W)	VJ mean power (W/kg)
90	978	1882	24.5	1005	31.1
80	874	1800	23.5	967	12.7
70	804	1749	22.6	913	12.3
60	778	1699	22.3	884	11.8
50	739	1660	21.9	849	11.4
40	701	1588	21.3	816	10.8
30	689	1498	20.7	784	10.6
20	675	1418	19.9	761	10.2
10	601	1294	19.0	644	9.5
Mean	751	1620	21.8	845	11.4
SD	195	233	2.1	133	1.3
N	29	92	92	92	92

Power outputs were calculated using an accelerometer attached to the athlete's waist.

Table 16.12 Percentile Ranks for Vertical Jump Power in Professional Basketball Players in the NBA

% rank	VJ peak power (W)	VJ peak power (W/kg)	VJ mean tower (W)	VJ mean power (W/kg)
90	3675	31.2	1690	15.2
80	3395	30.3	1632	15.1
70	3090	28.7	1567	14.8
60	2956	27.8	1538	14.4
50	2870	27.0	1480	13.9
40	2754	26.4	1419	13.2
30	2553	24.9	1368	13.0
20	2508	23.9	1325	12.4
10	2377	23.2	1202	11.7
Mean	2897	27.0	1460	13.7
SD	493	3.5	200	1.5
N	68	68	68	68

Power outputs were calculated using an accelerometer attached to the athlete's waist.

300-Yard Shuttle Run

The 300 yd shuttle run is a field test used to assess anaerobic capacity. Following a warm-up, the athlete lines up at the starting point. The athlete begins the test by sprinting to a point 25 yd (23 m) away and returns to the starting line. A total of six round trips are performed (12 × 25 yd = 300 yd, or 274 m). As the athlete crosses the line on the final sprint, the time is recorded to the nearest 0.1 s, and a 5 min rest interval begins. Following the 5 min rest interval, another 300 yd shuttle is performed. The average of the two times is recorded. Table 16.13 shows percentile ranks for the 300 yd shuttle for professional baseball players.

Line Drill

The line drill is another field test used to assess anaerobic fitness (see figure 16.3 for an example) (Hoffman et al. 2000; Hoffman and Kaminsky 2000). To perform this test in an outdoor facility, cones are placed at the starting point and 10, 20, 30, and 40 yd (9, 18, 27, 36 m) away. The athlete sprints from the starting line to the 10 yd marker and back, then pivots and sprints to the 20 yd marker and returns to the starting line, and continues this pattern to the 30 yd and 40 yd markers. A total of three trials are often used with a 2 min rest period between each. All sprint times are recorded, and the fastest time is reported. To calculate a fatigue index, the fastest score is divided by the slowest score. When the line drill is performed outdoors, the total distance run is ~200 yd; when it is performed on a basketball court (sprint to foul line and back, half-court and back, far foul line and back, and full court and back), the distance is approximately 155 yd. Percentage ranks for sprint times and fatigue rates are depicted in table 16.14. Times may vary depending on running surface.

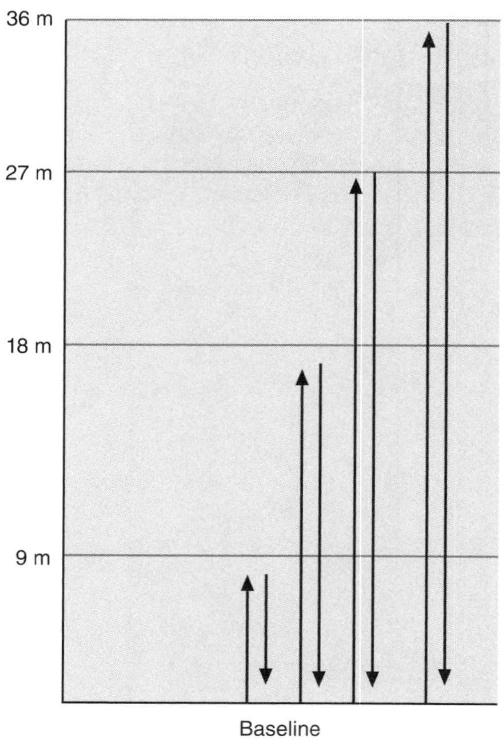

Figure 16.3 Line drill.

Adapted from Seminick 1994.

Table 16.13 Percentile Ranks for the 300 yd Shuttle in Professional Baseball Players (Minor and Major Leagues)

% rank	300 yd shuttle (s)
90	49.0
80	50.0
70	51.0
60	52.0
50	52.0
40	53.0
30	51.0
20	55.0
10	57.0
Mean	52.7
SD	3.38
N	737

Maximal Aerobic Capacity and Aerobic Endurance

Athletes who excel in endurance sports such as cross-country skiing, running, swimming, or cycling generally have a large aerobic capacity (figure 16.4). Although many factors determine aerobic performance (e.g., capillary density, mitochondrial number, muscle fiber type), the $\dot{V}O_2max$ of the

Table 16.14 Percentile Ranks for the Line Drill and Fatigue Index in NCAA Division III Football and Division I Women's Basketball and Soccer

	Football (200 yd on turf)		Women's soccer (200 yd on track)		Women's basketball (155 yd on court)	
% rank	Line drill peak time (s)	Fatigue index (%)	Line drill peak time (s)	Fatigue index (%)	Line drill peak time (s)	Fatigue index (%)
90	34.0	.930	36.6	.943	28.6	.990
80	34.7	.900	36.9	.928	29.4	.990
70	35.2	.880	37.3	.918	29.8	.980
60	35.9	.870	37.6	.911	30.2	.980
50	36.5	.855	38.0	.908	30.7	.980
40	37.0	.840	38.2	.901	31.0	.970
30	37.9	.830	38.6	.889	31.2	.960
20	39.4	.810	38.9	.874	31.6	.958
10	41.8	.780	39.2	.851	32.5	.939
Mean	37.1	.853	36.4	.869	30.2	.957
SD	3.2	.063	7.5	.180	3.8	.112
N	596	590	51	51	78	78

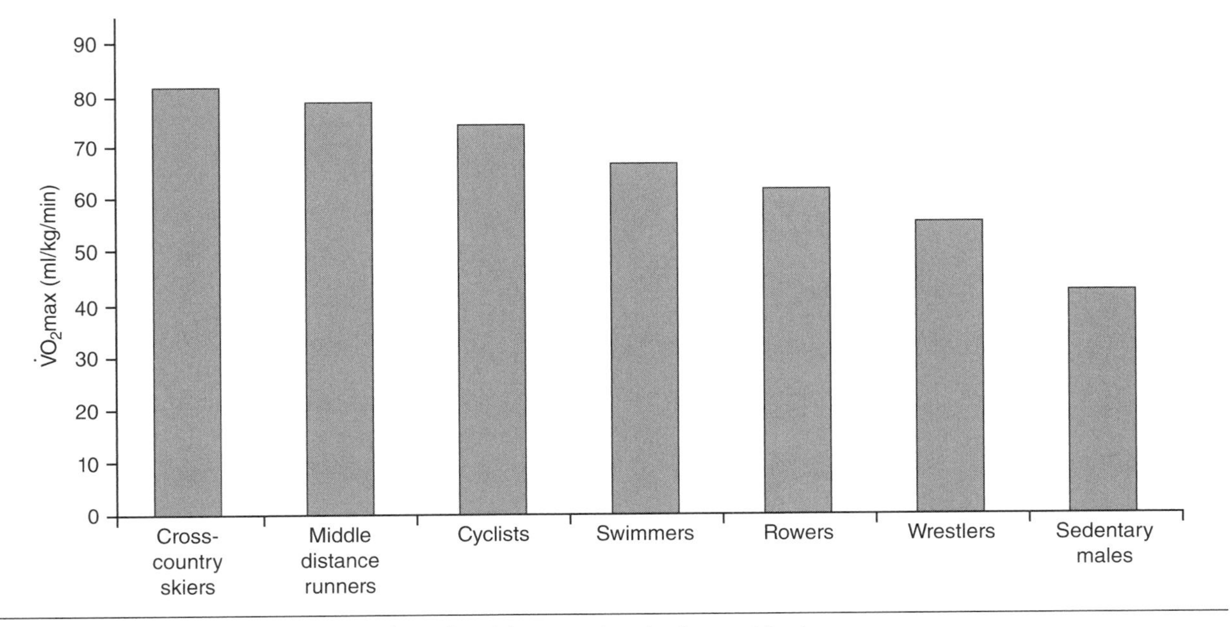

Figure 16.4 Maximal oxygen uptake in male athletes and sedentary subjects.

Adapted from Saltin and Åstrand 1967.

athlete provides important information concerning the capacity of the aerobic energy system. Maximal aerobic capacity can be determined either through direct measurement of oxygen consumption ($\dot{V}O_2$) during exercise to exhaustion or through submaximal exercise tests. Directly measuring $\dot{V}O_2$ during a graded exercise test to exhaustion on a treadmill is considered the standard. The premise underlying determination of an individual's maximal aerobic capacity is discussed in chapter 12. Maximal aerobic capacity can be determined on a treadmill or cycle ergometer or through tethered swimming. The choice of the exercise modality to use depends on the sport that the athlete plays. However, if specificity is not an issue, the treadmill produces the best results. In a study of triathletes, the $\dot{V}O_2$max results from tethered swimming and cycle ergometry were 13% to 18% and 3% to 6% lower, respectively, than values obtained from treadmill running (O'Toole, Douglas, and Hiller 1989).

Figures 16.5, 16.6, and 16.7 show popular treadmill and cycle ergometer testing protocols for assessing maximal aerobic capacity. Many protocols have been developed, and some are population specific. For instance, some exercise protocols are designed primarily for cardiac subjects whereas others are designed primarily for athletes. The main differences between the two are the starting point (elevation and speed of the treadmill) and the increments for each stage of exercise (increases in elevation or speed).

Before the onset of a maximal exercise test, subjects should be allowed to warm up for a minimum of 5 min or until they feel ready to proceed. Generally, the warm-up is performed at 0% grade at a speed the subject considers comfortable. After the warm-up, the subject is attached to the breathing apparatus and the testing begins. The test ends when the subject indicates that he or she has reached exhaustion, or when the subject has met at least two of the first three of the following criteria. These criteria deter-

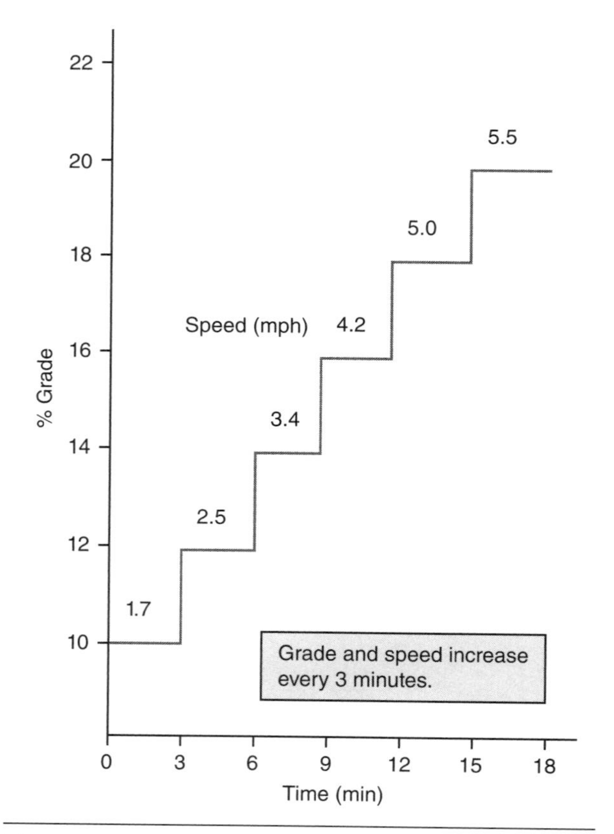

Figure 16.5 Bruce treadmill protocol for assessing maximal oxygen consumption.

Adapted from Bruce et al. 1973.

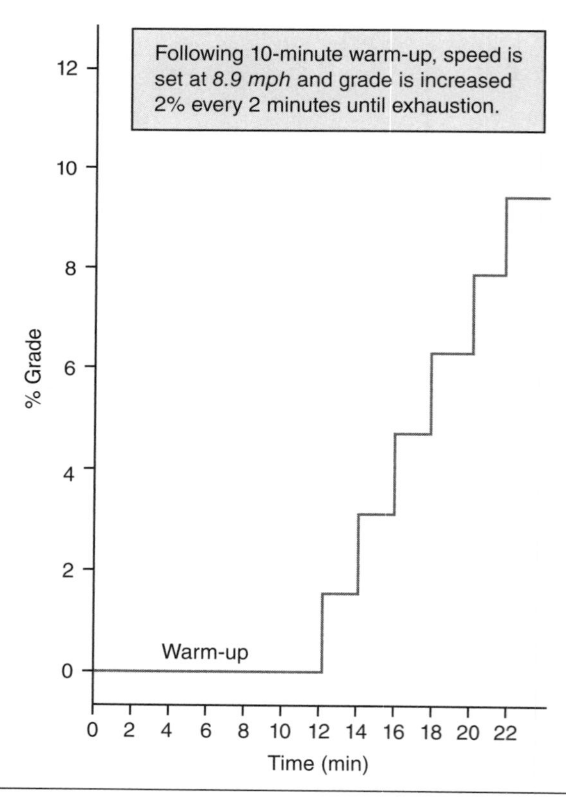

Figure 16.6 Costill and Fox treadmill protocol for assessing maximal oxygen consumption.

Adapted from Costill and Fox 1969.

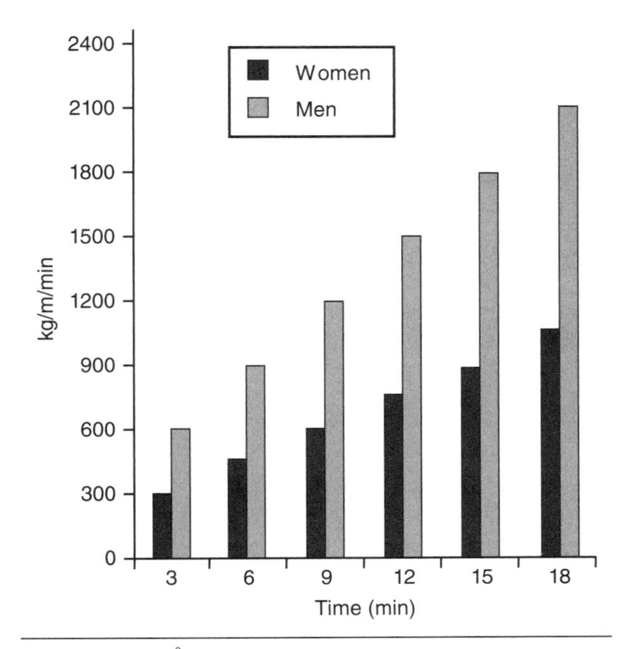

Figure 16.7 Åstrand cycle ergometer protocol for assessing maximal oxygen consumption. Men start at 600 kg/m/min and increase 300 kg/m/min every 3 min. Women start at 300 kg/m/min and increase 150 kg/m/min every 3 min. Pedaling speed is set at 50 rpm.

Adapted from Åstrand 1965.

mine whether $\dot{V}O_2$max has been reached. The fourth criterion can be assessed only once the test has been completed and is used to confirm that $\dot{V}O_2$max has been attained.

- Increase in oxygen uptake no greater than 150 ml/min despite an increase in exercise intensity (plateau criterion)
- Attainment of age-predicted maximal heart rate
- Respiratory exchange ratio ($\dot{V}CO_2/\dot{V}O_2$) greater than 1.10
- Plasma lactate concentration of at least 8 mmol/L 4 min after exercise (confirms that $\dot{V}O_2$max has been reached)

Considering the costs associated with the equipment, space, and personnel needed to directly measure oxygen consumption, this method of testing is generally reserved for research or clinical settings. When direct measurement of $\dot{V}O_2$max is not possible, a variety of submaximal and maximal exercise tests are available to predict $\dot{V}O_2$max. These tests are generally performed in a controlled environment and are administered on an individual basis by an exercise technician. The validity of these tests has been well established and is based on several assumptions. The assumptions listed next apply to using a submaximal exercise test to predict $\dot{V}O_2$max (American College of Sports Medicine 2000). If any of these assumptions are not met, the validity of the test may be reduced.

- A steady-state heart rate is obtained for each stage of exercise.
- A linear relationship exists between heart rate and intensity of exercise.
- The maximal heart rate for a given age is consistent.
- The efficiency of exercise (e.g., $\dot{V}O_2$ for the intensity of exercise) is the same for everyone.

Submaximal testing can be performed on either a cycle ergometer or a treadmill. Generally, a submaximal test uses an end point of 85% of age-predicted maximal heart rate. With a treadmill, the speed and grade of the final stage can be used to estimate $\dot{V}O_2$max (figure 16.8). The benefit of using a treadmill relates primarily to the familiarity that most individuals have with either walking or running versus riding on a cycle ergometer. However, cycle ergometers may still be a more popular mode of testing because of the ease of performing other measures (e.g., blood pressure and electrocardiogram) during the test and the non–weight-bearing character of the test. The relatively lower cost of a cycle ergometer compared with a treadmill and greater safety (e.g., reduced chance of subject's tripping or falling while cycling compared with running on a treadmill) are other reasons that may contribute to a greater use of submaximal cycle ergometer testing. Table 16.15 provides the protocol for the popular YMCA submaximal cycle ergometer test. The initial workload is 150 kgm/min (0.5 kp). Each stage of the test is 3 min in duration, and workload for the second stage is dependent on the heart rate in the last minute of the first stage. The heart rate measured during the last minute in each stage is plotted against work rate. The line generated from the plotted points is extrapolated to the individual's age-predicted maximal heart rate, and a perpendicular line is dropped to the *x*-axis to determine the work rate that would have been achieved if the person had worked to maximum. An individual's $\dot{V}O_2$max can then be calculated with the formula:

$$\dot{V}O_2\text{max (ml/min)} = (\text{kgm/min} \cdot 2\ \text{ml/kg/m}) + 3.5\ \text{ml/kgmin} \cdot \text{kg})$$

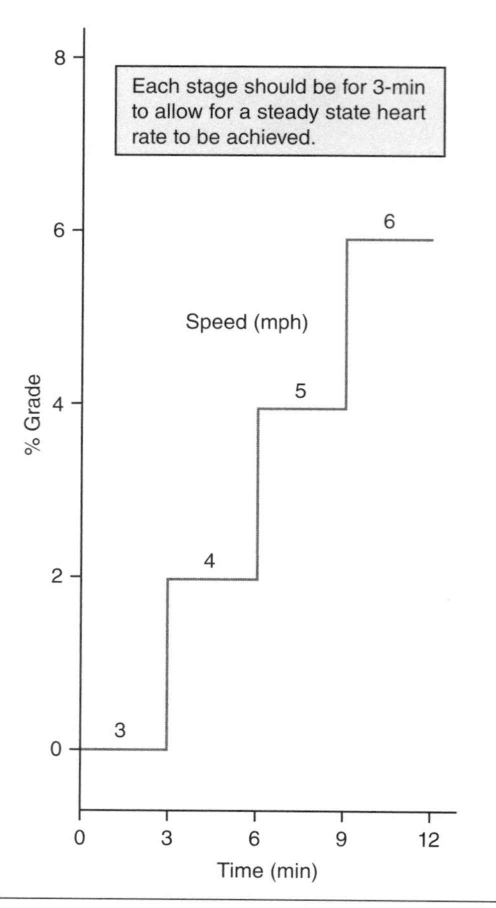

Figure 16.8 Submaximal treadmill protocol to estimate aerobic fitness. To estimate $\dot{V}O_2max$, the following formula may be appropriate (from Ebbeling et al. 1991): $\dot{V}O_2max$ (ml · kg^{-1} · min^{-1}) = 15.1 + (21.8 × speed in mph) – (0.327 × heart rate) – (0.263 × speed in mph × age) + (0.00504 × heart rate × age) + (5.98 × sex). For sex, 0 = females and 1 = males. This formula reportedly predicts $\dot{V}O_2max$ within 4.85 ml · kg^{-1} · min^{-1} of actual $\dot{V}O_2max$.

When large groups of subjects are being tested, it may be more feasible with regard to time to administer a field test. Field tests measuring the time to run 1.0 to 1.5 mi or the distance that can be run in 12 min have been used to estimate aerobic fitness. Popular tests include the Cooper 12 min run and the 1.5 mi test for time. The goal of the Cooper test is for the individual to run as great a distance as possible in the allotted time period (12 min). To estimate the athlete's $\dot{V}O_2max$ for the 12 min run, the following formula can be used:

$$\dot{V}O_2max\ (ml \cdot kg^{-1} \cdot min^{-1}) = (0.0268 \times \text{Distance covered in meters}) - 11.3$$

Considering that the distance for a single lap on most oval tracks is 400 m, an athlete running six laps, or 2400 m during the 12 min run, would have a predicted $\dot{V}O_2max$ of 53.0 ml · kg^{-1} · min^{-1} [(0.0268 × 2400) – 11.3]. A potential flaw of this test is that it may be quite difficult to estimate distance run, especially if the athlete did not complete a set fraction of a lap. Administratively, it may be easier to have athletes run a given distance. This would allow the test administrator to call out the times of each runner as he or she completes the six laps. Aerobic capacity can then be estimated for the 1.5 mi run with the following formula:

$$\dot{V}O_2max\ (ml \cdot kg^{-1} \cdot min^{-1}) = 3.5 + 483 / (\text{Time in minutes to run 1.5 mi})$$

If an athlete ran 1.5 mi in 10.0 min, the $\dot{V}O_2max$ would be calculated as 51.8 ml · kg^{-1} · min^{-1} [3.5 + (483/10.0)].

Percentile values for maximal aerobic power in the general population can be seen in table 16.16.

Table 16.15 Protocol for YMCA Submaximal Cycle Ergometer Test

Stage 1	Heart rate response	Stage 2	Stage 3	Stage 4
150 kgm/min (0.5 kp)	Heart rate <80	750 kgm/min (2.5 kp)	900 kgm/min (3.0 kp)	1050 kgm/min (3.5 kp)
	Heart rate 80 to 89	600 kgm/min (2.0 kp)	750 kgm/min (2.5 kp)	900 kgm/min (3.0 kp)
	Heart rate 90 to 100	450 kgm/min (1.5 kp)	600 kgm/min (2.0 kp)	750 kgm/min (2.5 kp)
	Heart rate >100	300 kgm/min (1.0 kp)	450 kgm/min (1.5 kp)	600 kgm/min (2.0 kp)

Adapted from ACSM, 2014, *Guidelines for exercise testing and prescription,* 9th ed., (Philadelphia, PA: Lippincott, Williams, and Wilkins), 82.

Table 16.16 Percentile Values for Maximal Aerobic Power (ml · kg^{-1} · min^{-1})

Men	Age (years)				
Percentile	**20-29**	**30-39**	**40-49**	**50-59**	**60+**
90	51.4	50.4	48.2	45.3	42.5
80	48.2	46.8	44.1	41.0	38.1
70	46.8	44.6	41.8	38.5	35.3
60	44.2	42.4	39.9	36.7	33.6
50	42.5	41.0	38.1	35.2	31.8
40	41.0	38.9	36.7	33.8	30.2
30	39.5	37.4	35.1	32.3	28.7
20	37.1	35.4	33.0	30.2	26.5
10	34.5	32.5	30.9	28.0	23.1
Women					
90	44.2	41.0	39.5	35.2	35.2
80	41.0	38.6	36.3	32.3	31.2
70	38.1	36.7	33.8	30.9	29.4
60	36.7	34.6	32.3	29.4	27.2
50	35.2	33.8	30.9	28.2	25.8
40	33.8	32.3	29.5	26.9	24.5
30	32.3	30.5	28.3	25.5	23.8
20	30.6	28.7	26.5	24.3	22.8
10	28.4	26.5	25.1	22.3	20.8

Reprinted, by permission, from J.T. Cramer and J.W. Coburn, 2004, Fitness testing protocols and norms. In *NSCA's essentials of personal training*, National Strength and Condition Association, edited by R.W. Earle and T.R. Baechle (Champaign, Ill: Human Kinetics), 217-264.

Speed

Speed is the ability to perform a movement in as short a time as possible. It is a relatively easy variable to measure, requiring only the use of a stopwatch and a track. For programs with large training budgets, electronic timers are available and are becoming more popular. The major issue with using a stopwatch is the potential for measurement error. Even under ideal conditions with an experienced tester, stopwatch times may be approximately 0.2 s faster than electronically measured times because of the tester's reaction-time delays in pressing the start and stop buttons as the athlete begins and ends the sprint (Hoffman 2006).

The 40 yd sprint is the most popular distance used to assess the speed of athletes, probably because of the familiarity that most coaches have with sprint times associated with this distance. The 40 yd sprint has achieved tremendous popularity among football coaches and, given the number of strength and conditioning coaches who have a football background, has become a staple for most athletic testing programs. The justification for the 40 yd distance is not entirely clear. It may have originated as an arbitrary distance that over time has become well accepted.

Coaches of several other sports have used longer or shorter sprint distances depending on the specific requirements of their sport. Some strength and conditioning coaches for basketball have begun to use a 30 yd sprint to assess speed since this is the approximate length of a basketball court. However, the use of this particular distance in basketball has not gained the popularity of the 40 yd sprint (Latin, Berg, and Baechle 1994). Baseball, on the other hand, has used the 60 yd sprint, most likely because this is the distance between bases (e.g., first to third or second to home).

Table 16.17 Percentile Ranks for 40 yd Sprint Times in American Football Players

% rank	High school (14-15 years)	High school (16-18 years)	High school (14-15 years) E	High school (16-18 years) E	NCAA D III team	NCAA D III backs	NCAA D III linemen	NCAA D I	NCAA D I E
90	4.89	4.70	5.10	4.99	4.60	4.55	4.78	4.58	4.75
80	5.00	4.80	5.23	5.11	4.72	4.63	4.92	4.67	4.84
70	5.10	4.89	5.28	5.25	4.80	4.69	4.97	4.73	4.92
60	5.20	4.97	5.33	5.33	4.87	4.74	5.05	4.80	5.01
50	5.30	5.09	5.48	5.42	4.94	4.80	5.13	4.87	5.10
40	5.40	5.20	5.55	5.48	5.02	4.85	5.25	4.93	5.18
30	5.51	5.30	5.65	5.64	5.13	4.90	5.35	5.02	5.32
20	6.00	5.46	5.87	5.73	5.27	4.99	5.50	5.18	5.48
10	6.20	5.76	6.41	5.84	5.49	5.12	5.74	5.33	5.70
Mean	5.43	5.16	5.58	5.44	5.01	4.81	5.20	4.92	5.17
SD	0.54	0.46	0.51	0.34	0.36	0.22	0.37	0.32	0.37
N	96	173	89	120	599	305	304	757	608

E, electronic timing; other times were from handheld stopwatches.

Adapted, by permission, from J.R. Hoffman, 2006, *Norms for fitness, performance and health* (Champaign, IL: Human Kinetics), 109.

Table 16.18 Times for 40 yd Sprint for College Football Players Participating in the NFL Combine

% rank	DL	LB	DB	OL	QB	RB	TE	WR
90	4.71	4.57	4.40	5.04	4.59	4.38	4.63	4.40
80	4.77	4.62	4.44	5.11	4.68	4.47	4.78	4.45
70	4.84	4.65	4.47	5.19	4.75	4.55	4.80	4.48
60	4.88	4.71	4.51	5.22	4.79	4.58	4.80	4.50
50	4.92	4.75	4.53	5.28	4.80	4.59	4.87	4.53
40	4.96	4.77	4.56	5.31	4.82	4.63	4.96	4.55
30	5.03	4.80	4.60	5.36	4.86	4.65	4.99	4.58
20	5.08	4.87	4.62	5.43	4.92	4.71	5.02	4.63
10	5.20	4.91	4.66	5.55	5.02	4.86	5.06	4.68
Mean	4.93	4.74	4.54	5.29	4.81	4.62	4.88	4.54
SD	0.18	0.13	0.11	0.19	0.15	0.19	0.14	0.10
N	65	41	77	101	27	33	30	61

DL, defensive linemen; LB, linebackers; DB, defensive backs; OL, offensive linemen; QB, quarterbacks; RB, running backs; TE, tight ends; WR, wide receivers.

Adapted from Hoffman 2006.

Table 16.19 Percentile Ranks for 30 and 60 yd Sprint Times in Male High School Baseball and Basketball Players

% rank	Baseball (14-15 years) 30 yd	Baseball (16-18 years) 30 yd	Basketball (14-15 years) 30 yd	Basketball (16-18 years) 30 yd
90	3.86	3.78	4.07	3.91
80	3.90	3.85	4.17	3.96
70	3.99	3.89	4.23	4.00
60	4.00	3.90	4.28	4.11
50	4.11	3.91	4.31	4.19
40	4.20	3.99	4.38	4.26
30	4.25	4.00	4.44	4.31
20	4.30	4.09	4.53	4.34
10	4.45	4.20	4.75	4.47
Mean	4.12	3.95	4.35	4.19
SD	0.21	0.16	0.25	0.22
N	45	27	150	109

Reprinted by permission, from J.R. Hoffman, 2006, *Norms for fitness, performance and health* (Champaign, IL: Human Kinetics), 110.

Tables 16.17 through 16.20 provide percentile rank sprint times for high school, collegiate, and professional football players. Sprint times for scholastic baseball and basketball players are also included.

Agility

Agility refers to the ability to change direction rapidly and with accuracy. It is a testing variable commonly measured during athletic performance testing. Like speed, it is a relatively easy performance variable to measure. All that is needed are a stopwatch and cones. A variety of agility tests can be selected; however, agility testing provides more relevant information if the test selected incorporates movements that are similar to those that the athlete performs during competition and if the test is part of the athlete's training program. Two popular examples of agility tests are the T-test and the Edgren side-step test. The protocols for these tests are outlined in figures 16.9 and 16.10, respectively. The pro-agility test is also a very popular agility assessment, and steps for performing this measure are presented in chapter 11.

Table 16.20 Percentile Ranks for 40 yd Sprint Times in Male Youth

% rank	12 to 13 years	14 to 15 years	16 to 18 years
90	5.41	5.02	4.76
80	5.63	5.15	4.85
70	5.77	5.24	4.90
60	5.84	5.32	4.98
50	5.97	5.46	5.10
40	6.08	5.54	5.13
30	6.25	5.78	5.21
20	6.32	6.02	5.30
10	6.64	6.08	5.46
Mean	5.99	5.54	5.09
SD	0.39	0.43	0.28
N	28	92	94

Reprinted by permission, from J.R. Hoffman, 2006, *Norms for fitness, performance and health* (Champaign, IL: Human Kinetics), 110.

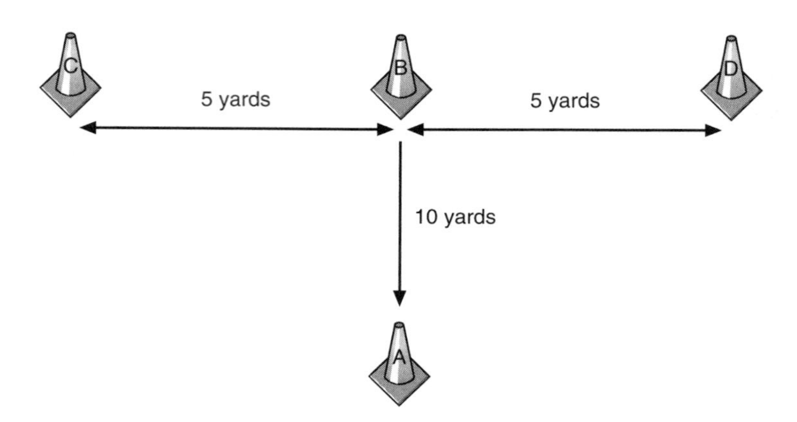

Figure 16.9 T-test. The T-test requires that four cones be arranged in a T formation. The first two cones (A and B) are set 10 yd apart. Cones C and D are 5 yd from either side of cone B. The athlete begins by sprinting from cone A to cone B and touches its base with the hand. The athlete then shuffles to cone C, facing forward at all times without crossing feet, and touches the base of that cone with the hand. The athlete next shuffles to cone D and touches the base, back to cone B and touches the base, and then performs a backward run to cone A.

Adapted from Seminick 1990.

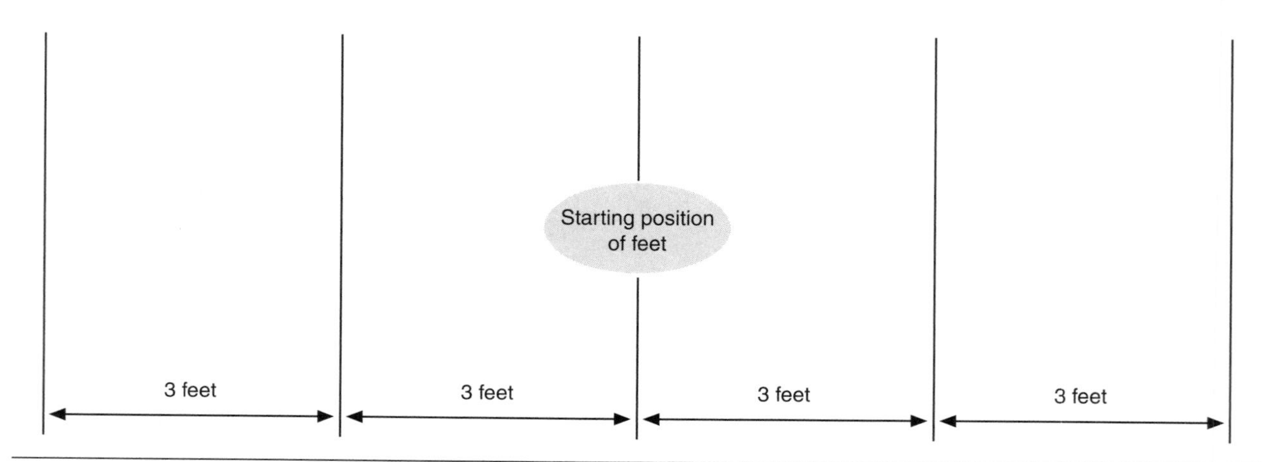

Figure 16.10 Edgren side-step test. For the Edgren side-step test, the test administrator marks five lines 3 ft (about 1 m) apart on the floor. The athlete begins astride the center line and then sidesteps to the right until the right foot has touched or crossed the outside line. The athlete next sidesteps to the left until the left foot has touched or crossed the outside line. The athlete sidesteps back and forth to the outside lines as rapidly as possible for 10 s. The number of lines crossed in 10 s is recorded. Any time the athlete crosses feet, a single line or point is subtracted from the total score.

Adapted from Harmon et al. 2000.

Tables 16.21 through 16.25 provide normative data on T-test and pro-agility test results for various high school, collegiate, and professional athletes.

Body Composition

Body composition refers to the proportion of body weight that is fat and the proportion that is lean tissue. The range in body fat percentage varies among athletes (figure 16.11), related primarily to the specific demands of each sport. Endurance athletes or gymnasts are generally on the very lean side, whereas some football players (primarily linemen) may be borderline obese. Standards of body composition for men and women are shown in table 16.26. Some football players, despite superior athletic skills in other performance variables, may be in the lower 20th percentile of the population in body fat. A study reported that the body fat percentage of NFL

Table 16.21 Percentile Ranks for the T-Test in NCAA Division III College Football Players

% rank	Team	DB	DL	LB	OL	RB	WR	QB/TE
90	8.40	8.16	8.47	8.45	8.89	8.18	8.33	8.37
80	8.63	8.37	8.78	8.78	9.38	8.54	8.52	8.57
70	8.82	8.60	8.98	8.95	9.49	8.75	8.57	8.73
60	8.96	8.77	9.38	9.00	9.65	8.89	8.62	8.89
50	9.07	8.89	9.58	9.05	9.79	8.98	8.76	9.10
40	9.23	8.93	9.67	9.13	9.93	9.05	8.87	9.21
30	9.45	9.06	9.86	9.19	10.12	9.15	9.09	9.25
20	9.68	9.21	9.94	9.37	10.50	9.38	9.16	9.33
10	10.06	9.43	10.25	9.72	10.71	9.85	9.44	9.41
Mean	9.17	8.82	9.40	9.07	9.86	8.97	8.82	8.98
SD	0.69	0.50	0.68	0.51	0.68	0.54	0.37	0.38
N	438	42	30	37	31	24	29	21

DB, defensive backs; DL, defensive linemen; LB, linebackers; OL, offensive linemen; QB, quarterbacks; RB, running backs; WR, wide receivers; TE, tight ends.

Adapted from Hoffman 2006.

Table 16.22 Percentile Ranks for the T-Test in Elite High School Soccer Players

% rank	Team
90	9.90
80	10.01
70	10.08
60	10.13
50	10.18
40	10.37
30	10.53
20	10.67
10	10.90
Mean	10.30
SD	0.42
N	40

Reprinted, by permission, from J.R. Hoffman, 2006, *Norms for fitness, performance and health* (Champaign, IL: Human Kinetics), 113.

offensive linemen may exceed 25% (Snow, Millard-Stafford, and Rosskopf 1998). Despite body fat percentages that can be categorized as borderline obese, specific needs of the position (e.g., extreme physical contact) require that a higher body fat percentage be deemed acceptable. A concern for the athlete's long-term health is warranted once his playing career is over. Body composition can be assessed in a number of ways. Methods vary in terms of complexity, cost, and accuracy. The following sections briefly describe these methods.

Dual-Energy X-Ray Absorptiometry

The new gold standard for body composition assessment is dual-energy X-ray absorptiometry (DEXA). DEXA assessment provides regional and total body measurements of lean and fat tissue, bone density, and bone mineral content. The reliability and validity of DEXA for body composition assessment have been established at low, moderate, and high levels of body fat and in both athletic and nonathletic populations (Fornetti et al. 1999; Visser et al. 1999). An advantage of the DEXA assessment is that it uses a three-compartment model (fat mass, lean tissue mass, and bone density) to determine body composition. This is superior to the more common two-compartment model (fat and lean tissue mass), and appears to result in a more accurate measurement of body composition by eliminating sources of error seen during

Table 16.23 Percentile Ranks for the Pro-Agility Test in NCAA Division I College Athletes

% rank	Women's volleyball	Women's basketball	Women's softball	Men's basketball	Men's baseball	Men's football
90	4.75	7.65	4.88	4.22	4.25	4.21
80	4.84	4.82	4.96	4.29	4.36	4.31
70	4.91	4.86	5.03	4.35	4.41	4.38
60	4.98	4.94	5.10	4.39	4.46	4.44
50	5.01	5.06	5.17	4.41	4.50	4.52
40	5.08	5.10	5.24	4.44	4.55	4.59
30	5.17	5.14	5.33	4.48	4.61	4.66
20	5.23	5.23	5.40	4.51	4.69	4.76
10	5.32	5.36	5.55	4.61	4.76	4.89
Mean	5.03	5.02	5.19	4.41	4.53	4.54
SD	0.20	0.26	0.26	0.18	0.23	0.27
N	81	128	118	97	165	869

Reprinted, by permission, from J.R. Hoffman, 2006, *Norms for fitness, performance and health* (Champaign, IL: Human Kinetics), 113.

Table 16.24 Pro-Agility Times for College Football Players Participating in the NFL Combine

	Pro-agility							
% rank	DL	LB	DB	OL	QB	RB	TE	WR
90	4.21	4.08	3.90	4.48	4.07	4.07	4.18	3.96
80	4.30	4.16	4.07	4.54	4.12	4.16	4.23	4.06
70	4.36	4.21	4.12	4.60	4.17	4.21	4.27	4.07
60	4.41	4.23	4.15	4.64	4.19	4.24	4.31	4.10
50	4.45	4.28	4.18	4.71	4.27	4.30	4.35	4.15
40	4.52	4.31	4.20	4.77	4.33	4.34	4.39	4.20
30	4.56	4.50	4.21	4.84	4.36	4.38	4.42	4.23
20	4.68	4.52	4.25	4.92	4.40	4.44	4.45	4.26
10	4.75	4.57	4.34	4.99	4.49	4.50	4.59	4.33
Mean	4.47	4.32	4.16	4.76	4.27	4.30	4.35	4.15
SD	0.21	0.17	0.14	0.46	0.15	0.15	0.13	0.14
N	56	15	40	75	25	29	28	50

DL, defensive linemen; LB, linebackers; DB, defensive backs; OL, offensive linemen; QB, quarterbacks; RB, running backs; TE, tight ends; WR, wide receivers.

Adapted from Hoffman 2006.

Table 16.25 Pro-Agility Times for NCAA Division III College Football Players

% rank	Team	Backs	Linemen
90	4.28	4.21	4.40
80	4.35	4.28	4.51
70	4.44	4.32	4.56
60	4.50	4.39	4.71
50	4.56	4.44	4.85
40	4.63	4.50	4.94
30	4.75	4.55	4.99
20	4.93	4.59	5.18
10	5.15	4.66	5.37
Mean	4.64	4.43	4.84
SD	0.35	0.17	0.37
N	151	79	76

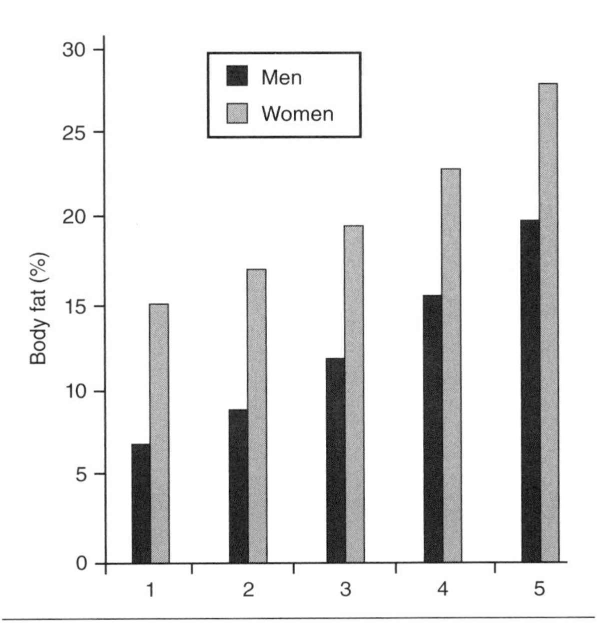

Figure 16.11 Body fat descriptive data for male and female athletes in various sports. 1 = Gymnastics, bodybuilders, wrestlers, endurance runners; 2 = soccer players, men's basketball players, male and female track and field athletes; 3 = baseball players, skiers, speed skaters, Olympic-style weightlifters; 4 = women's basketball players, football skill position players, hockey players, tennis players, volleyball players, softball players; 5 = football linemen, female shot putters.

Adapted from Harman et al. 2000.

Table 16.26 Body Composition Standards for Men and Women

	Males by age			Females by age		
Percentile	**20-29**	**30-39**	**40-49**	**20-29**	**30-39**	**40-49**
90	7.1	11.3	13.6	14.5	15.5	18.5
80	9.4	13.9	16.3	17.1	18.0	21.3
70	11.8	15.9	18.1	19.0	20.0	23.5
60	14.1	17.5	19.6	20.6	21.6	24.9
50	15.9	19.0	21.1	22.1	23.1	26.4
40	17.4	20.5	22.5	23.7	24.9	28.1
30	19.5	22.3	24.1	25.4	27.0	30.1
20	22.4	24.2	26.1	27.7	29.3	32.1
10	25.9	27.3	28.9	32.1	32.8	35.0

Adapted from American College of Sports Medicine 2000.

estimation of body density (e.g., residual volume). Limitations of DEXA measurements are the costs of purchasing and operating the machine. In addition, because the DEXA is an X-ray device, radiological boards of various states in America require physician prescription and operation by a licensed X-ray technician. These add costs to the use of DEXA and also limit potential use of this assessment technique due to lack of needed personnel.

Hydrostatic Weighing

For years, **hydrostatic weighing** was considered the gold standard of body composition analysis. However, due to the complexity of this test and the space required to perform it, human performance labs are using DEXA scans as the primary means of assessing body composition. The determination of body composition by hydrostatic weighing is based on the premise that a volume of water is displaced when an individual is submerged. As the body is immersed under water, it is buoyed by a counterforce equal to the weight of the water displaced. The loss of mass in water, corrected by the density of water, allows body density to be calculated by the following formula:

$$\text{Body density} = \text{Body mass in air} \,/\, \{[(\text{Body mass in air} - \text{Body mass in } H_2O) \,/\, \text{Density of } H_2O] - \text{Residual volume}\}$$

When body density is calculated, several assumptions are made that may increase the error in atypical populations. Hydrostatic measurement is assumed to provide an estimation of body fat within 2.5% of the true value (Gettman 1993). To accurately assess body density, it is necessary to calculate lung **residual volume**. Accurate measure of residual volume is important to reduce error; unless the residual volume is accounted for, the formula for calculating body density will not account for possible air in the intestines or what remains in the lungs following maximal expiration. Residual volume can be measured directly or predicted through various formulas. Once body density is determined, **percent body fat** can be calculated using various equations. Since body density is variable and is affected by aging, growth and maturation, sex, and ethnicity, an equation that is specific to a population group is often used. Table 16.27 provides population-specific formulas for conversion of body density. The following is an equation commonly used to derive percent body fat from body density:

$$\%\ \text{Body fat} = (495 \,/\, \text{Body density}) - 450$$

Plethysmography

For people who are not comfortable being fully submerged in a hydrostatic tank or pool, an alternative method of assessing body composition can be performed. An air displacement method known as plethysmography has been found to be highly reliable in a number of subject populations (Anderson 2007; Davis et al. 2007). Plethysmography takes place within a closed chamber that measures body volume by changes in pressure. This method has been shown to be a valid measure of body composition (Pineau, Guihard-Costa, and Bocquet 2007; Radley et al. 2003). However, some have suggested that it may overestimate body fat percentage in comparison to DEXA (Pineau, Guihard-Costa, and Bocquet 2007; Radley et al. 2003). Although the use of air displacement provides a valid assessment of body fat percent-

Table 16.27 Population-Specific Formulas for Conversion of Body Density to % Body Fat

Population	Age	Sex	% body fat (%BF) formula
White	17-19	Males	%BF = 4.99/D – 4.55
		Females	%BF = 5.05/D – 4.62
	20-80	Males	%BF = 4.95/D – 4.50
		Females	%BF = 5.01/D – 4.57
Black	18-32	Males	%BF = 4.37/D – 3.93
	24-79	Females	%BF = 4.85/D – 4.39

Adapted from Heyward and Stolarczyk 1996.

age, the calculation may be higher than that from DEXA measures, making comparisons between these modalities difficult to perform.

Skinfold Measurements

Skinfold measurements are the most popular method used to assess body composition. They take significantly less time to complete than the other methods that have been discussed. The principle behind skinfold measurements is that the amount of subcutaneous fat is proportional to the amount of body fat. By measuring skinfold thickness at various sites on the body, one can calculate percent body fat using a regression equation. However, because subcutaneous fat to total amount of body fat varies according to age, sex, and ethnicity (Lohman 1981), it is necessary to select the appropriate regression equation. In addition, regression equations vary in the number of skinfold sites needed. Even when the appropriate regression equation is used, there may be a 3% to 4% error in the body fat percentage attained from skinfold measurements (Lohman 1981). Tables 16.28 and 16.29 provide several examples of commonly used regression equations and a description of skinfold sites.

Table 16.28 Examples of Commonly Used Regression Equations for Computing Body Fat Percentage From Skinfold Measurements and Description of Skinfold Sites

Sites	Sex and age	Formula
Durnin and Womersley 1974		
Biceps, triceps, subscapular, and suprailiac	**Males (age in years)**	
	17-19	$D = 1.1620 - 0.0630 \times (\log \Sigma \text{ skinfolds})$
	20-29	$D = 1.1631 - 0.0632 \times (\log \Sigma \text{ skinfolds})$
	30-39	$D = 1.1422 - 0.0544 \times (\log \Sigma \text{ skinfolds})$
	Females (age in years)	
	17-19	$D = 1.1549 - 0.0678 \times (\log \Sigma \text{ skinfolds})$
	20-29	$D = 1.1599 - 0.0717 \times (\log \Sigma \text{ skinfolds})$
	30-39	$D = 1.1423 - 0.0632 \times (\log \Sigma \text{ skinfolds})$
Jackson and Pollock 1985		
Seven sites		
Chest, midaxillary, triceps, subscapular, abdomen, suprailiac, and thigh	Males	$D = 1.112 - 0.00043499\ (\Sigma \text{ 7 skinfolds}) + 0.00000055\ (\Sigma \text{ skinfolds})^2 - 0.00028826\ (\text{age})$
	Females	$D = 1.097 - 0.00046971\ (\Sigma \text{ 7 skinfolds}) + 0.00000056\ (\Sigma \text{ skinfolds})^2 - 0.00012828\ (\text{age})$
Three sites		
Chest, abdomen, and thigh	Males	$D = 1.10938 - 0.0008267\ (\Sigma \text{ 3 skinfolds}) + 0.0000016\ (\Sigma \text{ skinfolds})^2 - 0.0002574\ (\text{age})$
Triceps, suprailiac, and thigh	Females	$D = 1.1099421 - 0.0009929\ (\Sigma \text{ 3 skinfolds}) + 0.0000023\ (\Sigma \text{ skinfolds})^2 - 0.0001392\ (\text{age})$

D, body density.

Reprinted, by permission, from NSCA, 2011, Athlete test and program evaluation, J.R. Hoffman. In *NSCA's guide to program design*, edited by J.R. Hoffman (Champaign, IL: Human Kinetics), 48.

Table 16.29 Description of Skinfold Sites

Location	Method
Abdominal	Horizontal fold, 2 cm to the right of the umbilicus
Biceps	Vertical fold on the anterior aspect of the arm over the belly of the biceps muscle
Chest–pectoral	Diagonal fold, 1/2 the distance between the anterior axillary line and the nipple (men), or 1/3 the distance between the anterior axillary line and the nipple (women)
Midaxillary	Horizontal fold on the midaxillary line at the level of the xiphoid process of the sternum
Subscapular	Diagonal fold at a 45° angle, 1 to 2 cm below the inferior angle of the scapula
Suprailiac	Diagonal fold in line with the natural angle of the iliac crest taken in the anterior axillary line
Thigh	Vertical fold on the anterior midline of the thigh midway between the proximal border of the patella and the inguinal crease
Triceps	Vertical fold on the posterior midline of the upper arm midway between the acromion process of the scapula and the inferior part of the olecranon process of the elbow

Table 16.30 Percentile Ranks of Body Fat % (Skinfold) in College Football and Professional Baseball Players

Percentage	Team	Backs	Linemen	Professional Baseball
90	11.0	10.3	14.2	7.6
80	12.5	11.3	15.8	9.4
70	13.9	12.2	17.0	10.9
60	14.9	13.0	18.8	12.3
50	16.1	13.8	20.0	13.3
40	17.4	14.7	21.5	14.1
30	19.5	15.4	23.3	15.2
20	21.8	16.5	25.1	16.5
10	25.1	18.7	27.0	18.1
Mean	17.0	14.1	20.2	13.1
SD	5.2	3.4	5.0	4.1
N	379	195	187	989

Bioelectrical Impedance

Bioelectrical impedance analysis is another popular means of estimating body composition. It is similar to skinfold measures with regard to accuracy, and it may be easier to use because it eliminates potential error among testers. The principle underlying bioelectrical impedance is the relationship between total body water and lean body mass. Since water is an excellent conductor of electricity, a greater resistance to an electrical current passing through the body indicates a higher percentage of body fat. Likewise, if resistance is minimal, a higher percentage of lean tissue is present.

In assessment of body composition by bioelectrical impedance, any change in body fluid can have a significant effect on body fat calculation. If one chooses to use bioelectrical impedance as an assessment tool, it is strongly recommended that subjects refrain from drinking or eating within 4 h of the measurement; void completely before the measurement; and refrain from ingesting alcohol, caffeine, or any diuretic agent before assessment (Hoffman 2006). Table 16.30 depicts percentile ranks for NCAA

Division III college football players and professional baseball players (minor and major league players).

Summary

Performance variables are commonly analyzed as part of an athlete's performance testing routine. The use of these performance tests provides an opportunity to assess the athlete's adherence to an off-season training program, evaluate a rehabilitation program, and possibly indicate performance potential by comparing test results to existing performance standards. It is important that the tests selected have both validity and reliability in order for them to have any significant meaning for the coach or athlete. Finally, to minimize measurement error, the testing protocol should follow the same format over repeated test dates. In addition, several performance variables (e.g., speed, agility, and body composition) may be influenced by the test supervisor. To reduce variability between testing sessions, the same tester should perform the same tests over time.

REVIEW QUESTIONS

1. What is the importance of validity and reliability in an assessment?
2. Provide examples of four factors that can influence performance results in an assessment.
3. Provide support for the use of maximal and submaximal strength tests for competitive athletes.
4. Develop a testing battery for a women's intercollegiate volleyball team and justify your answer.
5. Compare methods of body composition analysis and describe the benefits and limitations of each.

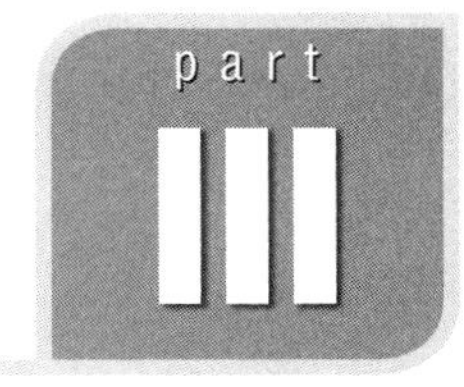

NUTRITION, FLUID REGULATION, AND NUTRITIONAL SUPPLEMENTATION

This section focuses on the basics of sports nutrition, hydration, sport supplementation, and performance-enhancing drugs. The sports nutrition chapter covers macronutrients, micronutrients, and vitamins and minerals in detail, as well as the issue of nutrient timing. Precompetition and postcompetition meals are another specific focus. The importance of hydration to both physiological function and performance is discussed in the chapter on hydration. Another chapter discusses the popularity of sport supplementation, with specific attention to the efficacy of popular supplements that are used today. The chapter on performance-enhancing drugs provides an in-depth discussion on the frequency of use, efficacy, and risks associated with athletes using these substances.

chapter 17

Sports Nutrition

CHAPTER OBJECTIVES

After reading this chapter you should be able to do the following:

- Understand the classes of macronutrients and micronutrients.
- Discuss the importance of amino acids and differentiate between essential and nonessential amino acids.
- Explain the importance of nutrient utilization during sport performance, including the effect of nutrient timing.
- Understand the importance of protein intake and how the daily requirements for protein differ among strength–power and endurance athletes.
- Discuss importance of preexercise, during-exercise, and postexercise meals, including providing a discussion on the glycemic index.

The importance of proper nutrition for athletic performance has been clearly documented over the past 25 to 30 years. For athletes to maintain a high level of training, they need to have an energy intake that equals their high energy expenditures. For the average person, a **recommended dietary allowance (RDA)** has been established to provide standards for promoting and maintaining good health. However, nutritional requirements are much greater for athletes than for the average population. Depending on the needs of the athlete (e.g., size, male vs. female, requirements of the sport), energy intake may be three- to fourfold higher than that recommended for the average individual. The nutritional needs of the athlete have been the focus of much research concerning what to eat, when to eat, and which nutritional **supplements** to take in order to maximize athletic performance. For example, a recommended balance of 55% to 60% **carbohydrate**, 30% **fat**, and 10% to 15% **protein** appears to provide a sufficient dietary composition for most people. However, many athletes are concerned that their specific nutritional needs may not be met unless they alter the recommended balance with respect to protein or carbohydrate intake. This has led to many studies examining the proper diet for athletes and the efficacy of protein and carbohydrate supplementation.

Sport nutrition covers a broad range of topics that may be specific to the individual athlete (e.g., the needs of a marathon runner may be different from the needs of a wrestler). To thoroughly review each topic area would be beyond the scope of this chapter. Thus, the chapter provides a brief review of the nutritional classes, with additional focus on topics considered to be common areas of concern for the general athletic population.

Classes of Nutrients and Their Functions

Six classes of nutrients are required for the energy and health needs of the individual: carbohydrates, fats, proteins, **vitamins**, **minerals**, and **water**. Carbohydrates, fats, and proteins are the principal compounds that make up our food and provide energy for our bodies. Vitamins and minerals play an important role in energy production and are also involved in bone health and immune function. However, they provide no direct source of energy. Water may be the most important nutrient available. It is needed for nutrient transport, waste removal, body cooling, and most other body reactions. Water is discussed more thoroughly in chapter 18.

Carbohydrates

There are several types of carbohydrate, all of which are treated differently by the body. **Monosaccharides** are the simplest form of carbohydrate. They consist of a one-unit sugar molecule such as **glucose**, fructose, or galactose. **Disaccharides** are composed of two-unit sugar molecules such as **sucrose** (table sugar), maltose (grain sugar), or lactose (milk sugar). Each of these carbohydrates can be broken down to its simpler form through the process of digestion. Both mono- and disaccharides are considered simple sugars and are a good source of quick energy. Sucrose consists of a glucose and a fructose molecule, maltose consists of two glucose molecules, and lactose consists of a glucose and a galactose molecule. However, for galactose to be used as energy, it must undergo a secondary conversion to glucose. So, to increase blood glucose levels quickly, a glass of milk may not be ideal. Simple carbohydrates are found in natural foods such as fruits (fresh, dried, and juices) and vegetables, as well as in processed foods such as candies and soft drinks.

Carbohydrates that contain more than two monosaccharides are termed **polysaccharides** and are known as complex carbohydrates. Common polysaccharides include **starch** and **glycogen**, which is made up primarily of chains of glucose molecules. The bonds that bind the monosaccharides of the complex carbohydrate may be either digestible (such as those found in potatoes, pasta, bread, and beans) or indigestible. Indigestible polysaccharides are known as **fiber** and are common in some grains, fruits, and vegetables.

The consumption of simple carbohydrates results in a relatively fast rise in blood glucose. This results in an **insulin** response to move the glucose from the blood into the muscles, where it can be used for immediate energy or stored for later use. Carbohydrates that are not used immediately are stored in the muscles and liver as glycogen. Having full glycogen storage depots is critical for fueling athletic performance. However, if the body's glycogen stores are completely full, the excess carbohydrate is converted to fat and stored in **adipose** sites around the body. The benefit of consuming complex carbohydrates such as starchy foods is that the time required for complete digestion is slower, which results in a more gradual increase of blood glucose. Further discussion of glycogen utilization for athletic performance appears later in the chapter.

Fats

Fat is a highly concentrated fuel that has limited water solubility. Fats, also referred to as **lipids**, exist in the body in several forms. The most common form of lipid is **triglyceride**, which is composed of three **fatty acids** and a **glycerol** molecule. Another common lipid is cholesterol. Although these lipids are frequently linked to heart disease, they also serve the following important functions in the body:

- Provide up to 70% of total energy during the resting state
- Support and cushion vital organs
- Make up essential components of cell membranes and nerve fibers
- Serve as a precursor for steroid hormones
- Store and transport **fat-soluble** vitamins
- Serve as an insulator to preserve body heat

The basic unit of fat is the fatty acid, which is also the part of fat that is used for energy production. Fatty acids occur in one of two forms: saturated and unsaturated. The difference between these two types of fatty acids is the binding between each molecule. Saturated fatty acids have no double bonds, meaning that each carbon atom of the fat molecule is saturated with its full complement of hydrogen atoms. Unsaturated fatty acids have at least one (monounsaturated) or more (polyunsaturated) double bonds and tend to be liquid at room temperature. Saturated fatty acids tend to be solid at room temperature and are generally derived from fat of animal origin. Exceptions do exist; examples of common oils high in saturated fats are palm kernel oil and coconut oil, which are liquid at room temperature. The consumption of saturated fats is also associated with a greater risk for cardiovascular disease. Polyunsaturated fatty acids are primarily found in vegetable oils and are the preferred source of fat for lowering the risk of cardiovascular disease.

Proteins

Proteins are nitrogen-containing substances that are formed by **amino acids**. Proteins serve as the major structural component of muscle and other tissues in the body. In addition, they are used to produce hormones, enzymes, and **hemoglobin**. Proteins can also be used as energy; however, they are not the primary choice as an energy source. For proteins to be used by the body, they must be broken down into their simplest form, amino acids. Twenty amino acids have been identified that are needed for human growth and metabolism (table 17.1). Eleven of these amino acids are **nonessential**, meaning that our body can synthesize them and they do not need to be consumed in the diet. The remaining nine amino acids are **essential**, which means that our body cannot produce them and they must be consumed in the diet. Absence of any of these essential amino acids from the diet prevents the production of the proteins that are made up of those amino acids. As a result, the ability for tissue to grow, to be repaired, or to be maintained is compromised.

Table 17.1 Essential and Nonessential Amino Acids

Essential	Nonessential
Histidine (in children only)	Alanine
Isoleucine	Arginine
Leucine	Asparagine
Lysine	Aspartic acid
Methionine	Cysteine
Phenylalanine	Glutamic acid
Threonine	Glutamine
Tryptophan	Glycine
Valine	Histidine (in adults only)
	Proline
	Serine
	Tyrosine

Histidine is not synthesized in children, making it an essential amino acid for that population group. However, in adults it is synthesized, making it a nonessential amino acid for that population.

If the protein portion of a food contains all of the essential amino acids, the protein is called a complete protein. Meats, fish, eggs, and milk are the best sources of complete proteins. Proteins from plant and grain sources do not supply all of the essential amino acids and are referred to as incomplete proteins. Thus, for vegetarians to receive all of the essential amino acids, they need to combine proteins from several different plant and grain sources.

Vitamins

Vitamins are needed by cells to perform specific functions that promote growth and maintain health, including enabling cells to use carbohydrates, fats, and proteins for energy. Vitamins are classified as either **water soluble** or fat soluble. There are four fat-soluble vitamins: vitamins A, D, E, and K. Once absorbed, they are bound to lipids and transported throughout the body. Excess fat-soluble vitamins are stored within the fat depots of the body, and excessive intake of fat-soluble vitamins can cause toxic accumulations. The remaining vitamins are water soluble and, once absorbed, are transported throughout the body in water. In general, any excess of water-soluble vitamins is excreted in the urine. Table 17.2 provides a list of the various vitamins and their major functions, dietary sources, symptoms of deficiency, and RDAs.

Most of these vitamins are important for athletic performance. Their relevance is likely related to their roles in energy metabolism and muscle growth. Many athletes, concerned that their high-intensity workouts require a greater vitamin intake, have spent significant dollars purchasing vitamin supplements. However, research has been unable to support any need for vitamin supplementation for athletes involved in either power or endurance sports as long as a vitamin deficiency does not preexist (Benardot 2000; Singh, Moses, and Deuster 1992; Telford et al. 1992). It appears that the higher caloric intakes of these athletes more than compensate for any increases in vitamin requirements due to high-intensity training. However, vitamin supplementation may prove to be beneficial for athletes who do not have a well-balanced diet or are on a calorie-restricted diet.

A topic receiving a lot of attention is the efficacy of both vitamins C and E as **antioxidants**, which prevent the cellular damage that occurs after the release of metabolically generated oxygen **free radicals**. It appears that a free radical is produced during periods of oxidative stress when univalent oxygen intermediates leak from the electron transport chain within the mitochondria during oxidative phosphorylation

Table 17.2 Vitamin Functions and Requirements

Vitamin	Dietary sources	Functions	Symptoms of deficiency	RDA*
A (retinol; β-carotene can also be made into vitamin A)	Foods of animal origin (e.g., liver, egg yolk, butter, and milk); β-carotene is found in dark green leafy vegetables and yellow- and orange-pigmented fruits and vegetables	Rhodopsin synthesis (necessary for eyesight), health of skin and soft tissue membranes, bone development, reproduction, and immune system; β-carotene is a powerful antioxidant	Rhodopsin deficiency, night blindness, frequent infections, poor growth, and skin disorders	800 mg in women 1000 mg in men
B_1 (thiamine)	Whole grains, legumes, and milk	Carbohydrate and amino acid metabolism, necessary for growth	Beriberi (muscle weakness including cardiac), confusion, and depression	1.0 mg in women 1.2 mg in men
B_2 (riboflavin)	Dairy products, meats, green leafy vegetables, and enriched and whole-grain products	Energy metabolism, vision (especially in bright light), and health of the skin	Bright light sensitivity; skin rashes, especially near corners of the mouth	1.1 mg in women 1.3 mg in men
B_3 (niacin)	Dairy products, meats, poultry, fish, and enriched and whole-grain products	Energy metabolism, nerve and digestive system function, health of skin	Pellagra (diarrhea, dermatitis, mental confusion, and weakness)	14 mg in women 16 mg in men
B_5 (pantothenic acid)	Found in most foods	Energy metabolism through its involvement as a structural part of coenzyme A (a compound involved in energy metabolism)	Neuromuscular dysfunction and fatigue	5 mg
B_6 (pyridoxine)	Meats, poultry, fish, green leafy vegetables, and whole-grain products	Involved in amino acid metabolism	Poor tissue repair and retarded growth, irritability, nausea, dermatitis	1.3 mg
B_{12} (cobalamin)	Meats, poultry, fish, eggs, and dairy products	Red blood cell production, amino acid metabolism, and nerve cell maintenance	Anemia, fatigue, and confusion	2.4 µg
Folic acid (folate)	Green leafy vegetables, legumes	Red blood cell production, maintenance of gastrointestinal (GI) tract health	Anemia, GI discomfort (e.g., diarrhea, constipation)	400 µg
C (ascorbic acid)	Citrus fruits, tomatoes, and green vegetables	Collagen formation, improved resistance to infection, protein metabolism, powerful antioxidant	Scurvy, poor wound healing, muscle soreness, bleeding gums	75 mg in women 900 mg in men

Vitamin	Dietary sources	Functions	Symptoms of deficiency	RDA*
D (cholecalciferol)	All dairy products, eggs, green vegetables, and fish oil	Promotes calcium and phosphorus absorption and bone mineralization	Rickets, weak bones, joint pain	5 μg
E (tocopherol)	Oils of vegetable origin, green vegetables, nuts, seeds, and whole-grain foods	Powerful antioxidant that protects cells from oxidative damage, also protects vitamin A from oxidative damage	Premature red cell death, possible role in muscular dystrophy	8 mg in women 10 mg in men
H (biotin)	Available in most foods	Energy metabolism and glycogen synthesis	Mental and muscle dysfunction, fatigue, and nausea	30 μg
K (phylloquinone)	Green vegetables, milk, liver; also made from bacteria in the gut	Involved in the synthesis of clotting factors	Excessive bleeding due to poor blood clotting	70 to 140 μg

*RDA based on 1989 RDA and 1998 DRI (dietary reference intakes) from the Food and Nutrition Board, Institute of Medicine, National Academy of Sciences.

(Kanter 1995). The free radicals are highly reactive and are thought to increase the rate of **fatigue** and contribute to the muscle damage seen after exercise. Although several different foods are known to have antioxidant properties, vitamins C and E have been given the most attention for their antioxidant action.

Vitamin E is generally considered the most important antioxidant in biological systems because of its association with the cell membrane (Bjorneboe, Bjorneboe, and Drevon 1990). Much of the damage from free radicals is characterized by disruption of the cell membrane. Vitamin E is thought to use several mechanisms in its role as an antioxidant. It can act directly on the oxygen-derived free radical by scavenging the potentially damaging singlet oxygen or act indirectly by protecting **β-carotene** and sparing **selenium** usage (Kanter 1995). Selenium is a mineral and is a component of glutathione peroxidase, an antioxidant enzyme. Although β-carotene and selenium cannot be substituted for one another, they complement each other to provide protection against the free radicals. β-Carotene, whose antioxidant ability is well documented, is the most widely distributed carotenoid compound (Bendich 1989). It also appears to have a direct effect by scavenging singlet oxygen radicals. Vitamin C also serves as an important antioxidant and free radical scavenger (Kanter 1995).

Studies have yielded promising results regarding the ability of vitamin supplementation and antioxidant activity to reduce markers of lipid peroxidation after exercise. Studies by Cannon and colleagues (1990) and Meydani (1992) demonstrated that vitamin E supplementation reduced concentrations of malondialdehyde (MDA, a skeletal muscle marker of lipid peroxidation) after exercise, indicating the benefit of vitamin E for reducing exercise-induced lipid peroxidation. Vitamin C supplementation also appears to have antioxidant ability. Several studies have reported reduced muscle soreness and a greater recovery from exercise in subjects taking vitamin C supplements compared with subjects who did not supplement (Jakeman and Maxwell 1993; Kaminski and Boal 1992). Although these studies attributed the differences to the antioxidant activity of vitamin C, they failed to measure any lipid peroxidation markers. Research on a combination of antioxidant vitamin mixtures has also demonstrated reduced MDA activity (Kanter, Nolte, and Holloszy 1993). However, it is not known whether a combination mixture provides any further protection than supplementing with only one vitamin. In addition, it is still unclear whether the protective mechanism that vitamins E and C appear to provide can be duplicated in all types of physical activity. Kanter (1995) suggested that vitamin supplements are an effective antioxidant during high metabolic activity. However, during exercise with a high mechanical stress (e.g., downhill running or resistance training), the protective mechanisms of these antioxidants may be diminished. Thus, the type of stress imposed (metabolic or mechanical) may determine the magnitude of effect that these vitamins have as antioxidants. Nevertheless, the benefits demonstrated in a number of studies do suggest a positive effect from supplementing with these vitamins during periods of high-intensity training. In addition, vitamin C supplementation (600 mg per day) has been shown to reduce the incidence of upper respiratory tract infections in marathon runners (Peters et al. 1993).

Minerals

Minerals are inorganic substances needed for normal cell function. Minerals are found everywhere in the body, and the total mineral content of the body is approximately 4% of an individual's body weight. They can act by themselves or function in combination with other minerals or various organic compounds. Minerals that are required by the body at a level greater than 100 mg per day are defined as **macrominerals**. Minerals that are needed in smaller amounts are known as **trace elements**, or **microminerals**. Table 17.3 provides a list of essential macrominerals and trace elements and their major functions, dietary sources, symptoms of deficiency, and RDAs.

Calcium and **phosphorus** are the most abundant minerals in the body. They constitute approximately 40% and 22% of the total mineral content of the body,

In Practice

Vitamin D Status and Athletic Performance

Vitamin D is necessary for maintaining bone health through its role of increasing calcium absorption and bone deposition (Holick 2004). Normal vitamin D concentrations are generally reported to be greater than 30 ng/ml, while levels between 20 and 30 ng/ml are deemed insufficient, and levels <20 ng/ml are categorized as deficient for all age groups worldwide (Holick 2004). The past few years have seen a growing relationship between vitamin D deficiency and a prevalence of immune dysfunction and inflammation (Cannell et al. 2008). In contrast to the situation with other vitamins, sufficient vitamin D levels may be difficult to obtain through the diet alone. The best source of dietary vitamin D is oily fish; the other dietary sources such as milk, juices, and other food groups reach adequacy via fortification with vitamin D (Holick 2004). The principal source of vitamin D production is believed to be sunlight (Cannell et al. 2008). However, this source may become deficient due to a number of factors, including place of residence, season of the year, time of day, skin pigmentation, sunscreen use, age, and the extent of clothing covering the body (Cannell et al. 2008; Holick 2007). Even individuals who reside in warm, sunny climates may find themselves deficient in vitamin D because of inadequate light exposure due to the use of sunscreen, clothing, or avoidance of sunlight (Constantini et al. 2010).

A concern that athletes deficient in vitamin D may increase their susceptibility to immune or inflammatory disease (or both) and increase risk to bone health motivated investigators to examine vitamin D deficiency or insufficiency and the prevalence of overuse injuries during an academic year in collegiate student-athletes (Halliday et al. 2011). Male and female athletes from the University of Wyoming were examined during an academic year. Athletes who trained or competed outdoors (e.g., cross country, football, soccer, track and field) were classified as outdoor athletes; athletes who trained or competed indoors (e.g., basketball, swimming, wrestling) were classified as indoor athletes. Vitamin D concentrations were shown to change during the academic year, averaging 49.0 ± 16.6 ng/ml, 30.5 ± 9.4 ng/ml, and 41.9 ± 14.6 ng/ml during the fall, winter, and spring, respectively. With 40 ng/ml used to define optimal status, 75.6%, 15.2%, and 36.0% of the athletes were classified as being at an optimal level for the respective seasons. Multivitamin intake and tanning bed use were significantly correlated to vitamin D concentrations (r = 0.39 and r = 0.48, respectively). An interesting finding was the significant correlation seen between vitamin D concentrations in the spring and frequency of illness (r = 0.40). These results indicate that insufficiency in vitamin D status increases the risk for illness, and that consideration should be given to supplementing vitamin D, especially during the winter months.

Table 17.3 Mineral Functions and Requirements

Mineral	Dietary sources	Functions	Symptoms of deficiency	RDA*
Calcium	All dairy products, green leafy vegetables, and legumes	Bone and tooth formation, blood clotting, muscle activity, and nerve function	Bone fragility, stunting of growth in children	1200 mg
Chloride	Table salt	Component of digestive enzymes	Cramps, lethargy	750 mg
Chromium	Whole-grain foods and meat	Involvement in blood glucose control and glucose metabolism	Unknown	50 to 200 mg
Copper	Meats and most drinking water	Hemoglobin and melanin production, also a component of several enzymes	Anemia and loss of energy	1.5 to 3.0 mg
Fluoride	Fluoridated water and seafood	Provides extra strength in teeth by creating decay-resistant enamel	Possible risk of developing dental cavities	3.0 mg in women 4.0 mg in men
Iodine	Iodized salt and seafood	Thyroid hormone production and maintenance of normal metabolic rate	Fatigue, decrease in metabolic rate; a serious deficiency may result in goiter (thyroid gland enlargement)	150 mg
Iron	Meats, poultry, fish, legumes, and dried fruits	Component of hemoglobin, adenosine triphosphate production in electron transport system	Anemia, decreased oxygen transport, fatigue	15 mg in women 10 mg in men
Magnesium	Nuts, legumes, whole grains, dark green leafy vegetables, and seafood	Involved in bone strength, protein synthesis, muscle and nerve function	Nervous system irritability and muscle weakness	320 mg in women 420 mg in men
Manganese	Whole-grain wheats, nuts, seeds, legumes, and fruits	Hemoglobin synthesis, bone and cartilage growth, and activation of several enzymes	Associated skeletal problems (osteoarthritis, osteoporosis, increased risk for fracture)	2.5 to 5.0 mg
Molybdenum	Legumes and whole-grain products	Component of enzymes	Unknown	75 to 250 mg
Phosphorus	Present in all foods of animal origin and legumes	Bone and tooth formation, energy transfer, and maintenance of body acid–base balance	May see loss of cellular function	700 mg

(continued)

Table 17.3 *(continued)*

Mineral	Dietary sources	Functions	Symptoms of deficiency	RDA*
Potassium	Meats, poultry, dairy products, fruits, vegetables, and legumes	Muscle and nerve function	Muscle weakness and abnormal electrocardiogram	Not established
Selenium	Food content based on selenium content of soil and water where food was grown	Cellular antioxidant; component of many enzymes	May be associated with an increased risk for cancer, possible cardiac dysfunction	55 mg
Sodium	Table salt and most processed foods	Important to body fluid regulation, nerve and muscle function	Nausea, vomiting, exhaustion, and dizziness	Not established; about 2500 mg
Sulfur	Present in most foods	Component of hormones, vitamins, and proteins	Unknown	Not established
Zinc	Meats, poultry, and fish	Component of several enzymes, necessary for protein metabolism and CO_2 transport	Deficient protein metabolism and CO_2 transport	12 mg in women 15 mg in men

*RDA based on 1989 RDA and 1998 DRI (dietary reference intakes) from the Food and Nutrition Board, Institute of Medicine, National Academy of Sciences.

respectively. They both are essential for bone health, and calcium plays a critical role in muscle function (see chapter 1). However, supplementing any of these minerals in the normal diet does not appear to have any ergogenic effect on athletic performance. The only mineral that is frequently supplemented is **iron**. Iron is a trace element, found in relatively small quantities in the body, that has an important role in oxygen transportation. Iron is required for the formation of hemoglobin and myoglobin. Hemoglobin is found in red blood cells, and myoglobin is found within the cytoplasm of muscle. Both of these molecules bind oxygen and store it until needed.

Iron deficiency is prevalent in the world today, and it is more common in women than men because of menstruation and pregnancy. The primary problem associated with iron deficiency is **anemia**, a condition in which hemoglobin concentrations are low. As a result, the oxygen-carrying capacity of the blood is reduced, causing fatigue, headaches, and other problems. Poor diet and prolonged endurance exercise may also contribute to iron deficiency anemia. Several studies have shown that prolonged endurance exercise, or high-intensity training over several weeks in previously untrained subjects, causes iron levels to fall (Magazanik et al. 1988; Pattini, Schena, and Guidi 1990). Endurance athletes tend to have a lower hemoglobin concentration than the normal population. This is commonly referred to as sports anemia. However, comparisons between athletes and untrained controls were unable to equivocally show that athletes are at a greater risk for iron deficiency (Clarkson 1991). Iron supplementation in athletes who are iron deficient improves performance, particularly endurance performance. Iron supplementation in athletes without any iron deficiency does not have any ergogenic benefit.

Water

Water is second only to oxygen in its importance for maintaining life. It constitutes about 60% of a man's and 50% of a woman's total body weight. Water is critical for dissipating body heat, transporting nutrients in the blood, and removing metabolic waste. Water is also crucial for maintaining athletic performance. Levels of dehydration of only 2% of an athlete's body weight can have significant detrimental effects on athletic performance (Hoffman, Stavsky, and Falk 1995; Hoffman et al. 2012). Chapter 18 provides detailed discussion on the importance of hydration.

In Practice

Does Zinc Supplementation Enhance the Hormonal Response During Exercise?

The use of the mineral zinc as a supplement for elite athletes received much attention during the BALCO investigation on performance-enhancing drugs. A supplement known as **ZMA** (**z**inc monomethionine and aspartate [30 mg], **m**agnesium **a**spartate [450 mg], and vitamin B_6 [10.5 mg]), developed by Victor Conte, founder of BALCO, was reputed to enhance muscle strength performance and testosterone concentrations. However, there has been no scientific evidence to support the ergogenic effect of this specific combination of minerals. Koehler and colleagues (2009) examined the acute effect of ZMA on testosterone secretion patterns. The investigators provided 14 men who had a daily zinc intake between 11.9 and 23.2 mg with ZMA. Serum zinc concentrations were significantly elevated, but no change in total or free testosterone concentrations were observed. However, this may have been a function of the acute nature of the study, the baseline levels of zinc, or both. Others have examined extended use of zinc supplementation in zinc-deficient men and demonstrated that 3 to 6 months of supplementation increased resting testosterone concentrations (Prasad et al. 1996). Still other research showed that when zinc was provided for 4 weeks to wrestlers performing exhaustive exercise, testosterone concentrations were maintained in those athletes given physiological doses of zinc compared to athletes not given zinc (Kilic et al. 2006). Thus, zinc may be part of the mechanism involved in maintaining normal testosterone concentrations, and may play an important role in preventing hypogonadal effects of overtraining.

Nutrient Utilization in Athletic Performance

Although all nutrients work together so that an athlete can maintain a high level of performance, several types are of particular value. Carbohydrates, fats, and proteins play critical roles in any athlete's sustenance.

Carbohydrate Utilization in Athletic Performance

Carbohydrate is the critical fuel needed to sustain exercise. The production of adenosine triphosphate (ATP) during events lasting more than several seconds is dependent on the availability of muscle glycogen or blood glucose. Although protein and fat can also contribute to energy production, they can do so only in the presence of carbohydrate. The oxidation of fat during prolonged exercise depends on Krebs cycle intermediates that are produced from the breakdown of carbohydrate during the process of **glycolysis**. When both muscle and liver glycogen stores are depleted, the ability to provide a sufficient amount of the Krebs cycle intermediates is drastically reduced, thus limiting exercise performance even in the presence of plentiful fat and protein supplies. Marathon runners frequently feel the effects of depleting their glycogen reserves when they "hit the wall" during a race. It is at this point that their ability to sustain exercise at their race pace is drastically reduced. This places a large emphasis on maximizing storage of carbohydrate before training and competition and replenishing it following exercise and competition (Ivy et al. 1988; Sherman et al. 1981).

During the early stages of exercise, muscle glycogen is the primary energy source. As exercise duration becomes prolonged, blood glucose makes a greater contribution. As glucose uptake by the active muscle increases, the liver needs to increase the rate at which it breaks down stored glycogen. Unfortunately, the glycogen content of the liver is limited, and the liver cannot rapidly produce glucose from other substrates (e.g., amino acids), so blood glucose levels decrease. Fatigue quickly sets in as glycogen stores in the muscle become reduced.

The body does not have the ability to use glycogen reserves from inactive muscles to fuel active muscles. This is related to the lack of the enzyme **phosphatase** within skeletal muscle. This enzyme is present in the liver and to a small extent in the kidneys, thereby permitting glucose to leave those sites and to be transported to exercising muscle. This process is depicted in figure 17.1.

Glucose is metabolized through the process of glycolysis. As the glucose molecule enters the

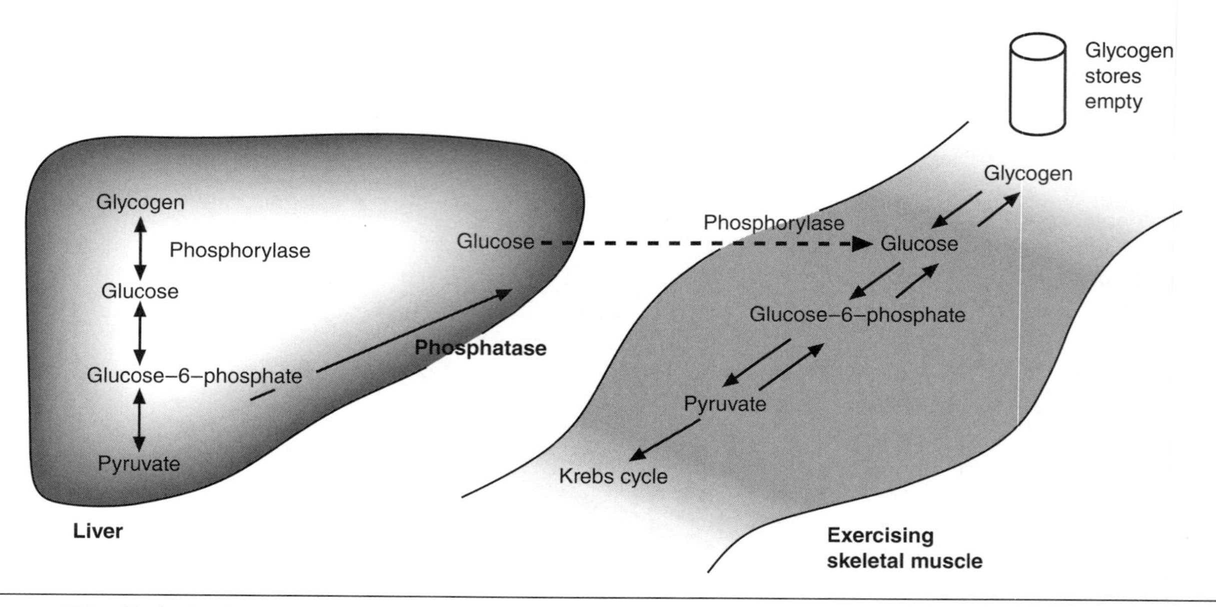

Figure 17.1 Carbohydrate utilization during exercise.

cell, it can be used as immediate energy or stored within the muscle as glycogen. The process during which glucose molecules are polymerized (linked together) to form glycogen depends on the enzyme **glycogen synthetase**. The process of metabolizing glycogen back to glucose is regulated and limited by the enzyme **phosphorylase**. The activity of this enzyme is influenced to a great extent by the hormone epinephrine (see chapter 2). In the first reaction, glucose is phosphorylated from an ATP molecule to glucose-6-phosphate. In muscle tissue, the phosphorylation of the glucose molecule traps it within the muscle. However, in the liver and to some extent in the kidneys, the enzyme phosphatase can split the phosphate group from glucose, making it available to leave the cell and be transported throughout the body.

At the onset of exercise, muscle glycogen is the primary source of carbohydrate used for energy. The rate of muscle glycogen depletion depends on several factors, including exercise intensity, physical condition, mode of exercise, environmental temperature, and preexercise diet (Costill 1988). As exercise intensity increases, the oxygen uptake may not meet the demands of the exercising muscle. At this point, there is a greater reliance on carbohydrate for energy. This is emphasized in figure 17.2, which shows greater muscle glycogen use at increasingly higher intensities of exercise. The rate of glycogen use may be 40-fold greater during sprints than during walking (Costill 1988). When glycogen stores are depressed to very low levels, the intensity of exercise must be lowered to match fuel availability.

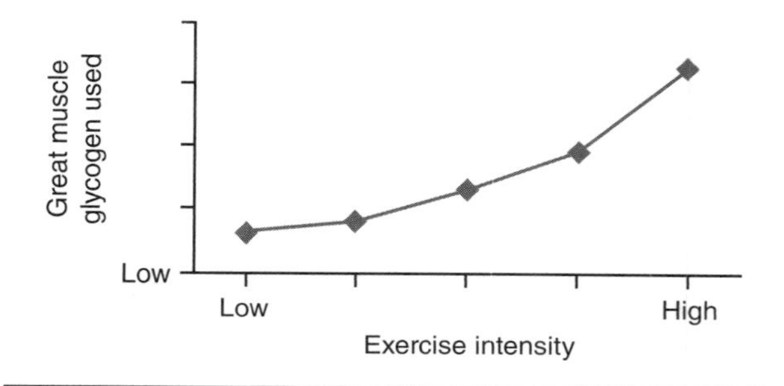

Figure 17.2 Muscle glycogen use at increasing intensity of exercise.

Adapted from Costill 1988.

Several chapters of this text discuss the effects of the environment, exercise economy, or mode of exercise on the metabolic demands of activity. Any factor that makes the athlete exercise at a greater percentage of his or her maximal capacity causes an increase in glycogen utilization, resulting in an earlier onset of fatigue. In addition, the time to exhaustion during prolonged endurance exercise is proportional to the preexercise glycogen content of the muscle. Bergstrom and colleagues (1967) showed that subjects exercising at 75% of their $\dot{V}O_2$max under normal conditions (muscle glycogen content 100 mmol/kg wet weight) could exercise for 115 min before becoming fatigued. However, if muscle glycogen content was reduced to 35 mmol/kg wet weight, exercise duration decreased to only 60 min. When subjects were provided with a carbohydrate-rich diet for 3 days, muscle glycogen content increased to 200 mmol/kg wet weight and subsequent exercise performance was prolonged to 170 min. The results of this study are depicted in figure 17.3. It should be noted that glycogen depletion and hypoglycemia cause fatigue in exercise that lasts more than 60 min. Fatigue occurring during athletic events of shorter duration is likely the result of other factors, such as an accumulation of lactic acid and hydrogen ions within the muscle fiber (Costill 1988).

During periods of starvation or inadequate carbohydrate consumption, the body can catabolize muscle or liver protein to form glucose through the process of **gluconeogenesis**. However, it does not appear that a significant amount of carbohydrate can be generated in this manner. Thus, the glycogen storage capability of the muscle and liver depends primarily on the dietary consumption of carbohydrate.

The study by Bergstrom and colleagues (1967) showed that normal muscle glycogen content was 100 mmol/kg wet weight for subjects fed a normal percentage of carbohydrate (55% of total calories). However, when subjects were fed a carbohydrate-rich diet (60-70% of total calories) for 3 days before exercise, muscle glycogen content was doubled. This procedure is known as **glycogen loading** and is a widely used practice among endurance athletes before competition.

A number of studies have shown that a combination of diet and exercise can result in a supercompensation with respect to glycogen replenishment.

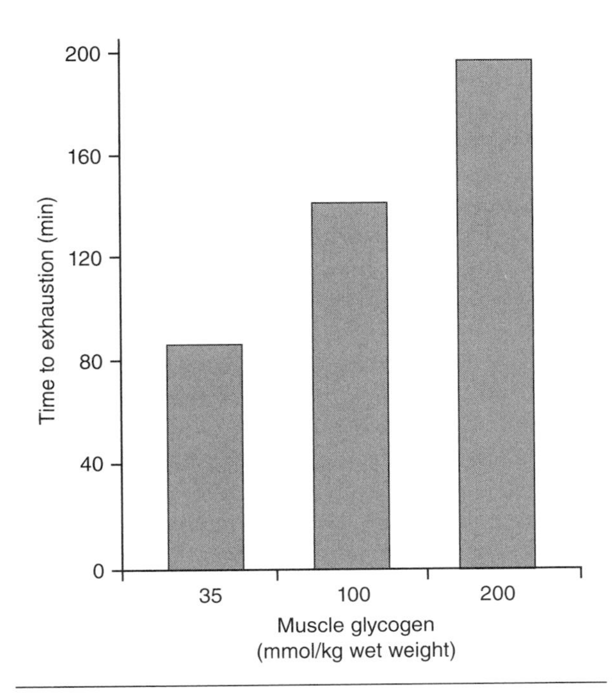

Figure 17.3 Relationship between muscle glycogen content and duration of exercise.

Data from Bergstrom et al. 1967.

The idea behind glycogen loading arose during the mid-1960s when scientists were looking for a way to maximize the amount of glycogen within the muscle. Researchers suggested that endurance athletes should deplete existing glycogen stores about 1 week before competition by performing an exhausting training routine and then consume a low-carbohydrate diet for the next 3 days (Hawley et al. 1997). The depleted glycogen stores cause an increase in the enzyme glycogen synthetase, which is responsible for glycogen synthesis. The athlete begins to consume a diet rich in carbohydrate 3 days before competition. Because glycogen synthetase concentrations are increased, the high carbohydrate intake results in a supercompensation and a greater muscle glycogen content. It is important that the exhausting exercise recruit the same musculature that is recruited during the competition. In addition, after the bout of exhausting exercise, the volume and intensity of exercise should be very low for the remainder of the week. Such a regimen may increase muscle glycogen content twofold.

Other studies have shown that similar increases in muscle glycogen content can be attained without the need for exhausting exercise and the subsequent low-carbohydrate, high-fat, and protein diet (Sherman et al. 1981). Reducing training intensity and consuming a normal diet (55% carbohydrate) until 3 days before competition, and then performing only a daily 10 to 15 min warm-up in combination with consuming a high-carbohydrate diet, caused muscle glycogen content to reach about 200 mmol/kg wet weight.

Fat Utilization in Athletic Performance

The two primary fuels that are used to provide energy for muscular activity are carbohydrates and fats. As discussed earlier, the amount of carbohydrate available for energy use is limited. Fat, however, has unlimited availability. During light to moderate exercise, the energy needs of the muscle are met by triglycerides from within the muscle itself and by free fatty acids. Free fatty acids are released from adipose sites around the body and bind to the protein **albumin** in the blood for transport to the active muscle. At the beginning of exercise, energy is used equally from carbohydrate and fat resources. However, as duration of exercise increases, a greater reliance on fat utilization occurs. When exercise duration exceeds more than 1 h, the carbohydrate reserves become limited and eventually depleted. The utilization of stored fat increases such that stored fat may supply more than 90% of the total energy required by the end of the exercise session (Jeukendrup 2003). During prolonged (2 h) low-intensity exercise (25% $\dot{V}O_2max$), there seems to be little change in substrate availability and oxidation (Romijn et al. 1993). Lipolysis in peripheral adipocytes appears to provide the predominant source of energy during exercise of this duration and intensity. As exercise intensity increases (~65% $\dot{V}O_2max$), a greater contribution comes from intramuscular fat stores, but peripheral breakdown is still maintained (Achten, Gleeson, and Jeukendrup 2002; Romijn et al. 1993). This likely reflects the greater catecholamine response occurring at this exercise intensity (see chapter 2). As exercise duration increases to 85% $\dot{V}O_2max$, the contribution of fat oxidation as an energy substrate is significantly reduced (Romijn et al. 1993). Figure 17.4 shows increases in free fatty acid uptake by the muscle, reflecting a greater utilization of fat as an energy source as duration of exercise increases. A greater reliance on fat is maintained as long as the intensity of exercise remains light to moderate. As intensity increases, the added energy requirements are met by a greater use of blood glucose and muscle glycogen stores.

The increase in **lipolysis** (breakdown of triglycerides into free fatty acids) is a result of sympathetic neural stimulation, which is activated by reduced blood glucose levels, accompanied by a decrease in insulin and an increase in glucagon concentrations. The sympathetic hormone norepinephrine binds to its receptor site on the adipose tissue, stimulating activation of the enzyme lipase, which regulates the breakdown of triglycerides into free fatty acids. On liberation, the glycerol molecule diffuses out of the adipose tissue and is used for gluconeogenesis in the liver (Bjorntorp 1991). The free fatty acids bind to albumin and are transported to the muscle or other end organs.

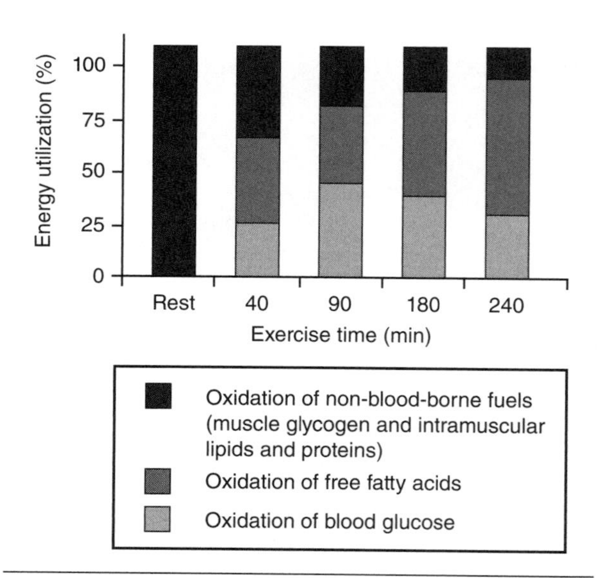

Figure 17.4 Relationship between duration of exercise and substrate utilization.

Data from Ahlborg et al. 1974.

Endurance-trained athletes appear to use fat for a greater percentage of their energy requirements during exercise (Saltin and Åstrand 1993). This is helpful because it allows greater conservation of both muscle and liver glycogen stores. The mechanisms that bring about the greater utilization of lipids during training are related to the specific physiological adaptations resulting from endurance training. These adaptations lead to a greater concentration of enzymes involved in lipolysis and **β-oxidation** (metabolism of fatty acids for energy) and parallel other adaptations that enhance the delivery system to the muscle (e.g., greater capillary density). Thus, the mechanisms involved in both production (breakdown of triglycerides to free fatty acids) and delivery (transport of free fatty acids from the adipocyte to the exercising muscle) are enhanced through training, providing endurance athletes with a greater concentration of lipids at their disposal for use as energy.

Protein Utilization in Athletic Performance

In general, proteins are not used to any appreciable extent as a fuel during exercise. The primary role of dietary proteins is for use in the various anabolic processes of the body. However, if an insufficient amount of carbohydrate is consumed or if exercise is prolonged, then protein can become a major source of energy. There is no storage depot of protein in the body as there is for carbohydrate (muscle and liver glycogen) and fat (adipose tissue). This is the reason for the atrophy of skeletal muscle mass in people on calorie-reduced diets. When protein is used, it is metabolized for energy primarily from the breakdown of skeletal muscle and liver proteins. During this catabolic process, protein is degraded into its amino acid components. In the liver, the nitrogen is removed from the amino acid molecule through the process of **deamination** and then excreted as urea. In the muscle, the nitrogen is removed from the amino acid molecule through the enzymatic process of **transamination** and then attached to other compounds. In either process, the carbon skeleton can be further metabolized for use as energy. When the body excretes more nitrogen than it consumes, it is said to be in a negative nitrogen balance, indicating that catabolic processes are occurring within the muscle. On the other hand, if nitrogen intake exceeds nitrogen excretion, the body is in a positive nitrogen balance and anabolic processes are likely to be occurring.

Not all proteins in the body can be used to supply energy. Typically, the proteins from connective tissue and nerves are fixed, meaning that they are unable to be metabolized and used as energy (McArdle et al. 2010). Proteins from skeletal muscle, however, are easier to degrade and can be used as an energy source when carbohydrate stores are near depletion. The **glucose–alanine cycle** is the mechanism that provides amino acids as substrates for energy use (Felig and Wahren 1971). The amino acids that have been metabolized in the muscle are converted to glutamate and then to alanine. Alanine is then transported to the liver, where it is deaminated. The carbon skeleton is converted to glucose through the process of gluconeogenesis and then released back into the circulation, where it is transported to the exercising muscle. Ahlborg and colleagues (1974) have shown that after 4 h of light exercise, the glucose–alanine cycle may account for 45% of the total glucose released from the liver. If exercise intensity is increased, the percentage of alanine-derived glucose is greater. Thus, depending on an individual's carbohydrate reserve and both the duration and intensity of training, the reliance on protein as an energy source can become more vital. This magnifies the importance of maintaining a high carbohydrate intake. In addition, for athletes whose training programs focus on muscle size and strength development, the use of a restricted-carbohydrate diet may have severe negative consequences on achievement of their goals.

In general, the protein requirement for an adult is 0.8 g/kg body weight. However, many athletes and coaches believe that high-intensity training results in a greater protein requirement. The issue of protein needs for athletes has been the subject of much debate within the scientific community.

Strength–Power Athletes

The belief that strength and power athletes require more protein than the normal sedentary population stems from the notion that additional protein or amino acids available to the exercising muscle promote protein synthesis. Resistance exercise has been shown to be a potent stimulator of muscle protein synthesis, resulting in protein accretion greater than protein degradation (Biolo et al. 1995; Phillips et al. 1997). Muscle protein synthesis was shown to increase 112% from resting levels at 3 h postexercise, while muscle protein degradation increased 31% from resting levels within the same time frame. When amino acids are ingested or infused following resistance exercise, muscle protein synthesis is enhanced to a greater extent than is seen after resistance exercise in a fasted state (Biolo et al. 1997; Tipton, Ferrando, et al. 1999). Figure 17.5 shows the effect of resistance training, amino acid infusion, and the combination of the two on muscle protein synthesis. Amino acid infusion during resistance exercise has been demonstrated to increase protein synthesis between 50% and 100% more than performance of resistance exercise only (Biolo et al. 1997). Other investigators have reported that the combination of oral ingestion of amino acids and resistance exercise may produce an even greater increase (3.5-fold) in muscle protein synthesis (Miller et al. 2003). Although resistance exercise and protein intake can each increase muscle protein synthesis, the combination of the two is clearly superior in eliciting significant gains.

There does appear to be compelling evidence that strength and power athletes have a greater daily protein requirement than other segments of the population. In studies examining high versus low daily protein intakes, higher protein consumption was associated with greater gains in protein synthesis, muscle size, and body mass. Within 4 weeks of protein supplementation (3.3 vs. 1.3 $g \cdot kg^{-1} \cdot day^{-1}$) of subjects in a resistance training program, significantly greater gains were seen in protein synthesis and body mass in the subjects with the greater protein intake (Fern, Bielinski, and Schutz 1991). Similarly, Lemon

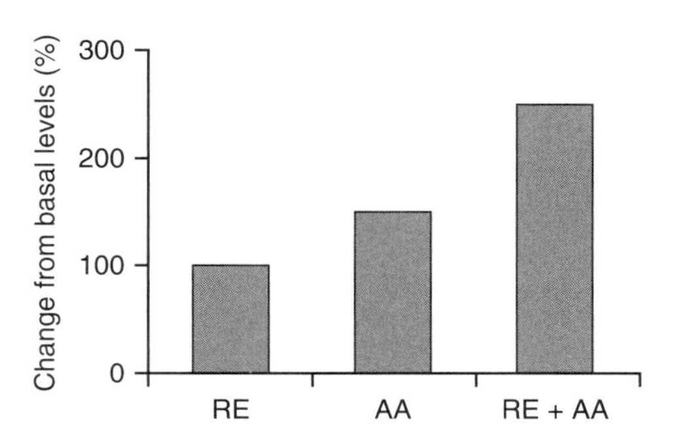

Figure 17.5 Rates of protein synthesis after resistance exercise (RE), amino acid infusion (AA), and resistance exercise + amino acid infusion (RE+AA), expressed as a percent change from basal.

Reprinted, by permission, from J.R. Hoffman, 2007, "Protein intake: effect of timing," *Strength and Conditioning Journal* 29: 26-34; Data from Biolo et al. 1995, 1997.

and colleagues (1992) reported a greater protein synthesis in novice resistance-trained individuals with protein intakes of 2.62 versus 0.99 $g \cdot kg^{-1} \cdot day^{-1}$. However, no significant differences were observed between the two groups in muscle size or strength. This may be related to the novelty of the training stimulus in these subjects. It is generally accepted that strength gains in previously untrained individuals are attributable to neural adaptations and not to any muscle structural changes. Hoffman, Ratamess, Kang, and colleagues (2007), examining competitive strength–power athletes (collegiate football players), reported that a group consuming 2.0 $g \cdot kg^{-1} \cdot day^{-1}$ for 12 weeks during the off-season conditioning program showed greater gains in squat strength than a group consuming 1.2 $g \cdot kg^{-1} \cdot day^{-1}$.

In studies of strength-trained individuals, higher protein intakes have generally been shown to have a positive effect on muscle protein synthesis and size gains (Lemon 1995). A comparison of two groups of bodybuilders ingesting a low-calorie diet with differing protein intakes (0.8 vs. 1.6 $g \cdot kg^{-1} \cdot day^{-1}$) showed that the bodybuilders consuming the higher protein intake were in a positive nitrogen balance and those consuming the RDA for protein were in a constant negative nitrogen balance (Walberg et al. 1988). Tarnopolsky and colleagues (1992) showed that strength-trained individuals needed a protein intake equivalent to 1.76 $g \cdot kg^{-1} \cdot day^{-1}$ to maintain a positive nitrogen balance, whereas sedentary controls needed 0.89 $g \cdot kg^{-1} \cdot day^{-1}$. This is consistent with other studies demonstrating that protein intakes between 1.4 and 2.4 $g \cdot kg^{-1} \cdot day^{-1}$ keep resistance-trained athletes in a positive nitrogen balance (Lemon 1995). However, there seems to be a limit to the benefit that can be derived from an increased protein intake. When protein intakes exceed 2.0 $g \cdot kg^{-1} \cdot day^{-1}$, no further increases in protein synthesis appear to occur (Fern, Bielinski, and Schutz 1991; Lemon et al. 1992; Tarnopolsky et al. 1992). In a comparison of three different daily protein intakes (1.2 $g \cdot kg^{-1} \cdot day^{-1}$, 1.7 $g \cdot kg^{-1} \cdot day^{-1}$, and 2.4 $g \cdot kg^{-1} \cdot day^{-1}$) in strength–power athletes, no significant strength or lean body mass differences were noted between the groups (Hoffman, Ratamess, Kang, et al. 2006). However, the greatest gains in strength in the bench press and squat exercises (between 3.0 kg and 8.4 kg) were seen in the highest (2.4 $g \cdot kg^{-1} \cdot day^{-1}$) daily protein intake group compared to the other groups. Daily protein intake recommendations for strength-trained athletes range from 1.2 to 1.7 $g \cdot kg^{-1} \cdot day^{-1}$ (Rodriguez, DiMarco, and Langley 2009). However, higher daily protein intakes (~2.0 $g \cdot kg^{-1} \cdot day^{-1}$) appear to be tolerated in healthy athletes and may provide performance benefits. Additional research on daily protein intakes in competitive strength-trained athletes still appears to be warranted.

Endurance Athletes

The importance of protein intake for the endurance athlete may be similar to that for the strength–power athlete. Although protein is not a major fuel source for exercising muscles, an increased reliance on protein as an energy source is observed during prolonged endurance exercise. To prevent a significant loss in lean tissue, endurance athletes have begun to increase the protein content of their diet to replace the protein lost during exercise (Lemon 1995). Although the goal for endurance athletes is not necessarily to maximize muscle size and strength, loss of lean tissue can have a significant detrimental effect on endurance performance. Therefore, these athletes need to maintain muscle mass. Several studies have determined that protein intake for endurance athletes should be 1.2 to 1.4 $g \cdot kg^{-1} \cdot day^{-1}$ to ensure a positive nitrogen balance (Rodriguez, DiMarco, and Langley 2009).

Timing of Nutritional Intake

Athletes are concerned about being properly fueled before exercise or competition. Although most realize the importance of eating, many do not understand the proper timing of the preexercise or preevent meal and what this meal should consist of. Over the last few years, a greater understanding of timing has been attained from studies on the effects of food and fluid consumption before and during exercise and competition.

Nutrient Timing

One of the more interesting areas of research within sport nutrition is nutrient timing. With regard to protein, the time of ingestion appears to have a significant impact on maximizing skeletal muscle adaptation during resistance training programs (Cribb and Hayes 2006; Esmarck et al. 2001), and may also be beneficial in enhancing muscular recovery in trained athletes following an acute exercise session (Hoffman, Ratamess, Tranchina, et al. 2010). These effects appear to be related to an enhanced delivery of amino acids to exercising muscle that provides an immediate availability of nutrients at the workout's conclusion. Stimulating greater muscle protein synthesis immediately following the workout is thought to enhance the adaptation due to a heightened muscle sensitivity from the exercise stimulus. In support of this hypothesis, several studies have demonstrated that when essential amino acids are consumed immediately before a workout, the rate of delivery and uptake of these amino acids to skeletal muscle is significantly greater than it is when these nutrients are consumed following the workout (Tipton, Ferrando et al. 1999; Tipton et al. 2001). This issue is discussed in greater detail in chapter 19.

Evidence is quite convincing regarding the importance of ingesting protein immediately before or after a training session (or both) (Cribb and Hayes 2006; Esmarck et al. 2001; Hoffman, Ratamess, Tranchina, et al. 2010). The benefits associated with the feedings surrounding the workout may indicate that skeletal muscle is more sensitive to protein feedings immediately before and following an acute resistance exercise session than to feedings at other times of the day. Adding further support, Hulmi and colleagues (2009) showed that protein consumption before and immediately after a resistance exercise session can increase messenger RNA (mRNA) expression, suggesting higher cell activation, and prevent a postexercise decrease in myogenin mRNA expression. Simply speaking, this study provided evidence that protein ingestion surrounding the training session directly stimulated muscle protein synthesis, which can potentially enhance recovery from the training session and promote muscle adaptation.

Another often-raised question is whether the benefits of protein ingestion are better obtained from a supplement or whether similar benefits can be achieved from a food source. Interestingly, the latter appears to be the case. Evidence indicates that milk consumption can stimulate amino acid uptake by skeletal muscle and result in an increase in net muscle protein synthesis (Elliot et al. 2006; Lun et al. 2012). Whole milk appears to be more beneficial than fat-free milk, unless the quantity of the fat-free milk consumed makes it similar in caloric value to whole milk (Elliot et al. 2006). Whole milk and isocaloric fat-free milk ingested 1 h following a resistance exercise session stimulated significant elevations in amino acid uptake—80% and 85% greater, respectively, than the isocaloric fat-free milk. These results indicate that a food source such as milk is suitable for ingestion during recovery from both resistance and endurance exercise; and a food source may be an effective and cheaper alternative to protein supplements. The primary benefit of taking a postworkout drink as a supplement is the convenience; a protein drink also provides a large amount of protein that is quickly absorbed at the time when the exercising tissue is at a heightened sensitivity. However, milk from a milk machine or refrigerator in the workout facility can offer the same benefits.

If a carbohydrate is added to fat-free milk, the benefits associated with whole milk consumption and protein synthesis may also be seen. Kammer and colleagues (2009) reported that when cereal was consumed with fat-free milk during 2 h of cycling exercise at 60% to 65% $\dot{V}O_2max$, significant increases in the insulin response, as well as activation of protein signaling, was seen compared to a carbohydrate-electrolyte drink. The addition of a carbohydrate with protein stimulates greater uptake of both carbohydrates and amino acids as a result of the insulin response. This is further emphasized in a study by Lun and colleagues (2012) who reported that chocolate milk can enhance muscle protein synthesis by stimulating protein signaling pathways to a significantly greater extent than a carbohydrate drink following a 45 min run at 65% $\dot{V}O_2max$. The chocolate milk increased muscle glycogen content to the same magnitude as the carbohydrate only drink.

Amount of Protein Consumed per Ingestion

Evidence on the benefits of nutrient timing has been quite convincing. Another question regarding protein intake relates to the amount to consume per ingestion. What is the maximum effective dose that one can consume postexercise? Moore and colleagues (2009) examined postexercise protein drinks containing 0, 5, 10, 20, or 40 g protein. Protein was ingested following an acute bout of leg extension exercise. Whole-body leucine oxidation was measured over 4 h. The results indicated increasing muscle protein synthesis with increased protein intake up to 20 g. No further increase in protein synthesis was seen with the higher dose. Whether these results would have differed if the investigators had used a multijoint structural exercise such as the squat is an interesting and legitimate question. However, this study was the first to examine an important issue and set the stage for future investigations.

Nutrient Timing and Carbohydrate Ingestion

A major emphasis in the nutritional strategy of competitive athletes is the maintenance and replenishment of carbohydrate stores. As discussed earlier, carbohydrates are the primary fuel used by athletes during training and competition. Given that athletes may be competing in several events per day, the ability to maintain sufficient energy stores can make the difference between success and failure. Thus, what is needed is a strategy focused on maximizing carbohydrate storage and consuming sufficient carbohydrate between competitions and training sessions. The type of carbohydrate consumed also has an important effect on the rate of absorption. Therefore careful selection of food in relation to the time of exercise or competition becomes critical for maximizing energy availability and avoiding a hypoglycemic response that can impede performance.

To help people determine the correct types of carbohydrate to consume, the glycemic index classifies foods based on their acute glycemic impact (Jenkins et al. 1981). Foods with a high glycemic index are digested quickly and raise blood glucose levels fairly rapidly. Examples of foods with a high glycemic index are baked potato, rice cakes, waffles, and instant rice. Foods with a lower glycemic index take longer to be digested. Examples of such foods include nuts, fruits, dairy products, and pasta.

Precompetition Meal

It is generally accepted that athletes benefit greatly from having a meal before practice or competition as opposed to performing in a fasted state (Rodriguez, DiMarco, and Langley 2009). The guidelines for such a meal, as recommended by the American College of Sports Medicine, the American Dietetic Association, and the Dietitians of Canada in their joint position stand, are listed at the end of this paragraph. Ideally, the meal should be eaten approximately 3 to 4 h before the event and should comprise 200 to 300 g carbohydrate (Schabort et al. 1999; Sherman et al. 1989). There is concern regarding a hypoglycemic response and premature fatigue the closer the meal is to a practice or competition (Foster, Costill, and Fink 1979). However, eating close to the time of competition may not be as detrimental as once thought (Alberici et al. 1993; Devlin, Calles-Escandon, and Horton 1986; Horowitz and Coyle 1993). The meal before competition contributes very little to the glycogen content of the muscle, but it helps to ensure adequate blood glucose levels and prevent feelings of hunger. A liquid meal may be a wise choice if the meal is to be eaten close to the time of competition, as such a meal may be less likely to cause gastric discomfort. According to recommendations set forth by Rodriguez and colleagues (2009), the precompetition meal should be

- sufficient in fluid to maintain hydration,
- low in fat and fiber to facilitate gastric emptying and reduce risk of gastrointestinal distress,
- high in carbohydrate to maintain blood glucose and maximize filling of glycogen stores,
- moderate in protein intake,
- made up of foods familiar to the athlete, and
- eaten 3 to 4 h before the game or practice.

Food Supplements During Competition

The importance of carbohydrate supplementation during competition has begun to attain full acceptance, and its efficacy has scientific support. Several studies have shown that providing carbohydrate during exercise maintains blood glucose levels and improves exercise performance (Coggan and Coyle 1989; Currell and Jeukendrup 2008; Jeukendrup et al. 1997). Figure 17.6 shows the benefits of carbohydrate feedings (3 g/kg body mass of a liquid comprising 85% glucose polymers and 15% sucrose in a 50% solution) versus placebo on duration of cycling exercise performed at 70% of the subjects' $\dot{V}O_2$max (Coggan and Coyle 1989). In this study, subjects consumed either the carbohydrate drink or placebo at minute 135 of exercise. The exercise time to fatigue averaged 21% longer in subjects who drank the carbohydrate mixture. In addition, the carbohydrate supplement

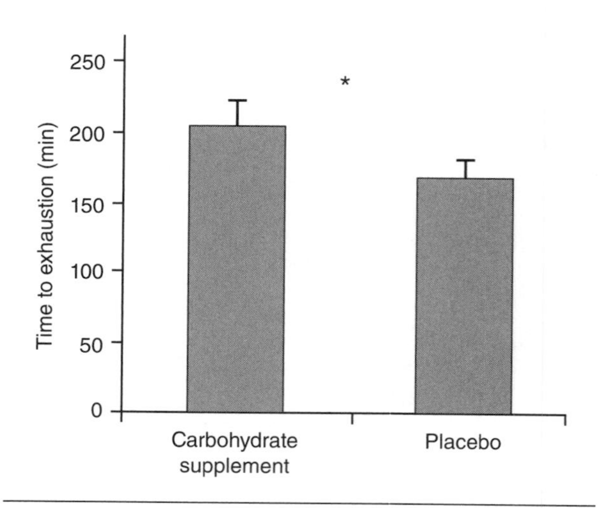

Figure 17.6 Effect of carbohydrate supplementation on time to exhaustion during cycling exercise. * = Significant difference ($p < 0.05$) between the groups.

Adapted from data of Coggan and Coyle 1989.

did not cause any elevation in plasma insulin concentrations.

Later research demonstrated that providing carbohydrate at 15 to 20 min intervals during the first 2 h of activity may be more beneficial than waiting until after 2 h of exercise to provide the supplement (McConnell et al. 1997). This is especially relevant in situations in which the athlete has not carbohydrate loaded or consumed a precompetition meal. The carbohydrate should be primarily glucose; however, a combination of glucose and fructose may be used without gastric problems (Coggan and Coyle 1991). Currell and Jeukendrup (2008) reported that compared with glucose alone, the combination of glucose and fructose in a 2:1 ratio, ingested at a rate of 1.8 g/min, resulted in an 8% improvement in cycling performance in trained cyclists. The carbohydrate consumed during training or competition can be in liquid, solid, or gel form, as long as fluid intake is adequate (Rodriguez, DiMarco, and Langley 2009).

The effect of carbohydrate feeding during competitions involving high-intensity, intermittent exercise (e.g., football, basketball, or hockey) has not been investigated to the same extent as it has in endurance sports. However, speculation would suggest that if athletes enter the competition with an inadequate glycogen supply, they will benefit from carbohydrate supplementation during the event. This may be relevant for athletes participating in tournaments conducted over several days during which glycogen replenishment is incomplete.

Postcompetition Meal

The timing of the postexercise or postcompetition meal is important. It is generally recommended that the postexercise meal be eaten within 2 to 4 h (Ivy et al. 1988; Volek, Houseknecht, and Kraemer 1997). However, the closer the meal is to the conclusion of the exercise or competition, the greater the opportunity to maximize glycogen loading. Ivy and colleagues (1988) showed that when carbohydrates were given immediately after exercise, muscle glycogen content at 6 h postexercise was significantly higher than when ingestion was delayed by 2 h.

The type of carbohydrate may also be important. It is generally recommended that carbohydrates with a high **glycemic index** be consumed following exercise or competition. Postexercise carbohydrate meals with a high glycemic index result in a higher muscle glycogen content than a similar amount of carbohydrate with a lower glycemic index (Burke, Collier, and Hargreaves 1993).

The primary role of protein consumption postexercise is muscle repair and other anabolic processes within the muscle. It does not appear to have any effect on muscle glycogen replenishment. However, there is evidence that a combined protein and carbohydrate supplement provided immediately postexercise may enhance the anabolic processes after resistance exercise (Roy et al. 2000; Stearns et al. 2010). It is thought that the interaction of these two nutrients leads to a more anabolic state because of the combined influence of the hormone insulin (involved in both glucose uptake and protein synthesis) and the exercise stimulus. Roy and colleagues (2000) examined the effect of a combined carbohydrate/protein (CHO/PRO) or carbohydrate-only (CHO) postexercise supplement, which was provided immediately after and 1 h after resistance exercise, on whole-body protein synthesis. Subjects in the CHO/PRO group consumed a supplement of carbohydrate (~66%), protein (~23%), and fat (~12%); subjects in the CHO group consumed a supplement of 56% sucrose and 44% glucose polymer from corn syrup solids. A third group of subjects was given a placebo (PL). The results depicted in figure 17.7 show that both CHO/PRO and CHO supplementation resulted in significantly greater protein synthesis (NOLD) than did PL. No significant difference in protein synthesis was seen between CHO/PRO and CHO. It appears that the hyperinsulinemia after carbohydrate consumption caused an increase in protein synthesis, likely by enhancing amino acid uptake within the muscle.

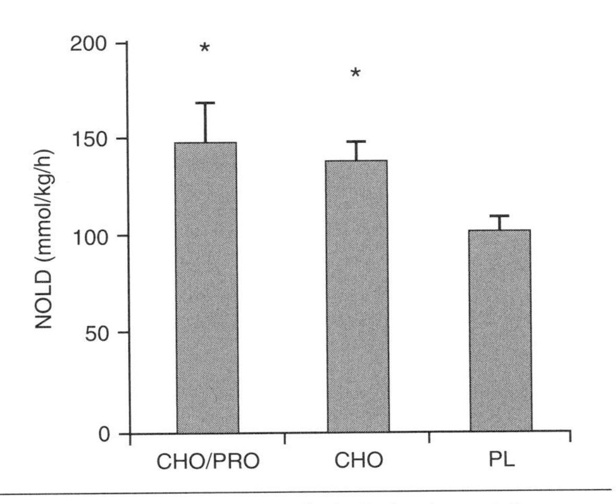

Figure 17.7 Comparison of whole-body protein synthesis after varied postexercise supplements. * = Significantly better than PL (placebo).

Adapted from Roy et al. 1997.

Summary

Six classes of nutrients are required for the energy and health needs of the body. It appears that to maintain a high level of performance, athletes need to increase their energy intake to a level that equals their high energy expenditures. An increase in caloric expenditure does not appear to necessitate any change in the vitamin or mineral requirements of athletes as long as they are not under a restricted diet regimen. The primary difference between the composition of the athlete's diet and that of noncompetitive contemporaries appears to be the amount of protein required. The RDA for the normal adult population is 0.8 g/kg body weight. Competitive athletes appear to need approximately twice that amount to enhance muscle protein synthesis.

A primary strategy for many athletes is to maximize muscle glycogen content. Several loading strategies have been employed to enhance the athlete's ability to delay fatigue, including providing carbohydrate supplementation during and immediately after exercise performance. Such supplementation appears to prolong exercise duration as well as to enhance muscle glycogen replenishment after an exercise session. In addition, a combined carbohydrate and protein meal immediately after exercise may not only enhance glycogen replenishment but also increase protein synthesis, which has tremendous relevance for the strength–power athlete.

REVIEW QUESTIONS

1. What is an essential amino acid and how does it differ from a nonessential amino acid?
2. Explain the physiological reason why nonexercising muscle cannot provide glucose to exercising muscle but glycogen stores within the liver can be used to fuel exercise in all active muscles in the body.
3. Discuss the importance of protein timing with regard to muscle recovery.
4. Compare protein intake needs between endurance and strength–power athletes.
5. Explain the importance of the glycemic index and its role in determining pre- and postexercise food intake.
6. What is the importance of the glucose–alanine shuttle?

chapter 18

Hydration

CHAPTER OBJECTIVES

After reading this chapter you should be able to do the following:

- Explain water balance at rest and during exercise.
- Define euhydration, dehydration, hypohydration, and rehydration.
- Describe how hypohydration affects physiological function and athletic performance.
- Discuss osmolality and how it affects the drinking response.
- Explain physiological effects and potential warning signs of dehydration.
- Understand strategies associated with maintaining normal fluid balance.

As mentioned in chapter 17, water is essential for human life. It provides the medium for all biochemical reactions that occur within the body and is essential for preserving blood volume. It therefore plays a critical role in maintaining cardiovascular function and **thermoregulation**. Water constitutes approximately 60% of body weight and about 72% of lean body mass. Approximately two-thirds of the water in the body is found within the cells and is referred to as **intracellular fluid**. The remaining one-third is found in various compartments outside of the cell and is known as **extracellular fluid**. The extracellular fluid includes the fluid surrounding the cells (**interstitial fluid**), blood **plasma**, lymph, and other bodily fluids. Water is second only to oxygen as a necessity for maintaining life. Although the body can withstand a 40% loss in body mass from starvation, a 9% to 12% loss of body mass from fluid loss can be fatal.

Considering the role that water plays in physiological function, it is no surprise that water also has a significant role in maintaining exercise performance. During exercise, the metabolic rate may increase 5 to 20 times over its resting level. This results in a large increase in body heat that must be dissipated to maintain thermal homeostasis. If exercise is performed in a hot environment, the need for heat dissipation is further magnified. Heat dissipation is primarily accomplished through **evaporative cooling (sweating)**. Evaporative cooling may result in a large loss in body water because of sweat loss (this is reviewed more thoroughly in chapter 21). If fluid intake does not match body water loss, the body is said to be **dehydrating**, meaning that a body water deficit is occurring. A state of **hypohydration** (referring to an existing body water deficit) has significant effects on both physiological function and the ability to perform exercise. This chapter focuses on the effects of hypohydration on both physiological function and exercise performance. Fluid replacement during exercise is also discussed.

Water Balance at Rest and During Exercise

Under normal conditions, body water is generally in balance (referred to as **euhydration**). Water intake occurs through fluid intake (accounting for approximately 60% of daily water intake), food consumption (approximately 30%), and metabolic processes within the body. Fluid loss occurs through excretion from the kidneys and large intestine and by evaporation of water from the skin surface or from the respiratory tract. During resting conditions, the primary avenue of water loss (approximately 60%) is through excretion from the kidneys (urination). Water loss as a result of evaporation from both the skin and respiratory tract accounts for approximately 35% of the total water loss. The remaining 5% is the result of water loss in the feces.

Water loss is accelerated during exercise because of the increase in metabolic heat production. Sweat rate can vary greatly and depends on environmental conditions (ambient temperature, humidity, and wind velocity), clothing (insulation and moisture permeability), and the intensity of physical activity (Sawka and Pandolf 1990). Sweat rates generally range between 1.0 and 1.5 L/h, which is equivalent to about a 2% decrease in body water per hour in a 155 lb (70 kg) man. The highest sweat rate reported in the literature is 3.7 L/h, which was measured for Alberto Salazar during preparation for the 1984 Olympic marathon (Armstrong et al. 1986).

The primary problem during exercise in the heat is countering the large fluid loss from sweating through sufficient **rehydration**, which is related to the reliance on a thirst sensation to begin fluid intake. Studies examining ad libitum fluid intakes (the ability to drink at will) have shown that it is common for exercising individuals to voluntarily dehydrate despite the availability of adequate amounts of fluids (Armstrong et al. 1986; Hubbard et al. 1984). Thirst does not appear to be perceived until an individual has incurred a body water deficit of approximately 2% (Rothstein, Adolph, and Wells 1947). In addition, voluntary **dehydration** may impair the ability to rehydrate during later stages of exercise. This impairment may be related to the magnitude of hypohydration. Neufer, Young, and Sawka (1989) reported that **gastric emptying** may be reduced by approximately 20% to 25% in hypohydrated individuals (5% loss of body weight) performing moderate exercise in the heat (102 °F [39 °C]). However, at body water deficits of lower magnitudes (below 3% body weight) during exercise at a moderate intensity, gastric emptying or intestinal absorption may not be impaired (Ryan et al. 1998). Nevertheless, it appears to be prudent practice to force hydration during the early stages of exercise to prevent any compromise in fluid availability to the body at later stages.

During dehydration, water is lost from both intracellular and extracellular spaces. At body water deficits of ~3% body weight, water is lost primarily from the interstitial spaces. As the body water deficit increases in magnitude, a greater percentage of the lost body fluid comes from intracellular spaces. This may be related to the **glycogen** depletion occurring within the cell during prolonged exercise. Approximately 3 to 4 g water is bound to each gram of glycogen (Olsson and Saltin 1970). As exercise increases in duration, the large intracellular fluid loss may be partly related to the water released during the breakdown of glycogen (Costill et al. 1981). In addition, when body water deficits become very low, the body redistributes water from both intracellular and extracellular spaces to maintain the water content of the organs necessary to maintain life (e.g., brain and liver) (Nose, Morimoto, and Ogura 1983).

Effects of Hypohydration on Physiological Function

A body water deficit has significant implications for both cardiovascular and thermoregulatory function. Hypohydration results in a reduction in plasma volume. As a consequence, less blood is available to both exercising muscle and the skin. Decreases in plasma volume are also associated with a reduction in **stroke volume** (Nadel, Fortney, and Wenger 1980). This is likely related to a lowering of central venous pressure, which decreases cardiac filling pressure (Kirsch, von Ameln, and Wicke 1981). To compensate, heart rate increases to maintain normal blood flow. However, depending on the magnitude of the body water deficit, the increase in heart rate may not be sufficient to fully compensate for the lower stroke volume. For example, if the magnitude of dehydration exceeds 2% of one's body weight and is accompanied by a moderate to severe thermal stress, an increase in heart rate does not appear to fully compensate for the decrease in stroke volume. As a result, **cardiac output** is also reduced (Nadel, Fortney, and Wenger 1980; Sawka, Knowlton, and Critz 1979). However, it does appear that cardiac output can be maintained at higher degrees of dehydration if the hypohydration occurs in the absence of a thermal strain (Sproles et al. 1976).

A body water deficit also impairs thermoregulation, the severity of which depends on the magnitude of hypohydration. Core temperature, an indicator of thermal strain, is elevated relative to the degree of hypohydration (Sawka, Young, Francesconi, et al. 1985). As the body water deficit increases, the ability of an individual to dissipate heat is reduced. The reduction in blood volume and an increase in blood displacement to the peripheral vasculature, especially during exercise in the heat, causes a reduction in venous return and subsequent cardiac output (Nadel, Fortney, and Wenger 1980; Sawka, Knowlton, and Critz 1979). As a result, the ability to dissipate heat is reduced, reflected by a lower cutaneous blood flow for a given core temperature. Increases in core temperature are also associated with decreases in both sweating rate and blood flow to the skin (Sawka and Pandolf 1990). Similar to changes in core temperature, these changes are dependent on the magnitude of the body water deficit.

The physiological mechanisms that reduce sweat rate and cutaneous blood flow are not entirely understood but are likely related to increases in plasma **osmolality** and decreases in plasma volume (Sawka and Pandolf 1990). Hyperosmolality has been shown to increase the threshold temperature for sweating and vasodilation, even in the absence of a fall in blood volume (Fortney et al. 1984; Sawka, Young, Francesconi, et al. 1985). The delay in the sweating response because of increases in plasma osmolality is likely related to both central and peripheral mechanisms (Sawka and Pandolf 1990). **Osmoreceptors** within the hypothalamus, sensitive to changes in osmolality, are thought to be the central mechanism responsible for the delay in the sweating response; osmotic pressure changes at the sweat gland have been suggested as a peripheral mechanism responsible for the reduced sweating response. These changes in the osmotic pressure gradient between the plasma and sweat gland reduce the fluid available to the sweat gland. In addition, the reduction in blood volume causes a decrease in peripheral blood flow, reducing the ability to dissipate heat. A significant relationship (r = 0.53-0.75) between changes in blood volume and changes in sweat rate has been reported during exercise in the heat (Sawka, Young, Francesconi, et al. 1985). It is thought that atrial baroreceptors sensitive to changes in blood volume provide afferent input to the hypothalamus to reduce fluid loss in sweat (Sawka and Pandolf 1990). Although the ability of the body to thermoregulate is reduced, the importance of conserving fluid balance for essential organs needs to take precedence.

The effect of body water deficit on muscle glycogen resynthesis has also been examined (Neufer et al. 1991). The potential for a dysfunction in glycogen resynthesis is related to the association between water and glycogen previously discussed. It has been hypothesized that hypohydration impairs muscle glycogen resynthesis because of a permissive role that water may play in this process. However, Neufer and colleagues (1991) were unable to see any difference in glycogen resynthesis between subjects euhydrated or hypohydrated to 5% of their body weight immediately after 2 h of exhaustive exercise or 15 h postexercise (figure 18.1), despite significant differences between the groups in muscle water content. Thus, it does not appear that a body water deficit causes any decline in muscle glycogen resynthesis.

Table 18.1 summarizes the physiological effects and potential warning signs of dehydration. Body water deficits greater than 5% of body weight compromise health. If the water deficit exceeds 9% of body weight, the physiological systems of the body may be impaired to the extent that life may be threatened.

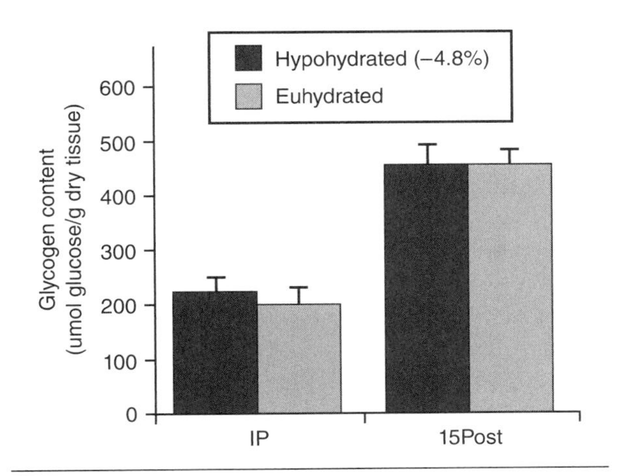

Figure 18.1 Effect of hypohydration on glycogen resynthesis. IP, immediately postexercise; 15Post, 15 min postexercise.

Adapted from Neufer et al. 1991.

Table 18.1 Summary of the Physiological Effects and Potential Warning Signs of Dehydration

Body weight loss (%)	Physiological change	Warning signs
0-2	↑ Core temperature	None
2-4	↓ Plasma volume ↓ Muscle water ↓ Stroke volume ↓ Blood flow to the skin and muscle ↑ Heart rate	Thirst, verbal complaints, and some discomfort
4-6	↓ Sweat rate	Flushed skin; apathy; clear loss of muscle endurance; impatience; muscle spasms; muscle cramps; tingling sensation in arms, back, and neck
6-8	↑ Urine acidity ↑ Protein in urine ↓ Blood flow to kidney	"Cotton mouth," headache, dizziness, shortness of breath, indistinct speech
8-12		Swollen tongue, spasticity, delirium

Adapted from Armstrong 1988.

Electrolyte Balance During Exercise

In addition to fluid loss as a result of sweating during exercise, a large volume of sweat leads to a loss of **electrolytes** from the body. Sweat is primarily made up of water (approximately 99%) and includes a number of minerals, such as potassium (K^-), sodium (Na^+), chloride (Cl^-), magnesium (Mg^{2+}), and calcium (Ca^{2+}). The concentrations of these minerals in sweat are much lower than in plasma or other body fluids (table 18.2). Na^+ and Cl^- are considered the predominant ions in sweat and account for its salty taste.

Electrolyte loss is variable among individuals and depends on sweat rate, physical condition, and state of heat acclimatization (Maughan and Noakes 1991). During exercise, the large loss of Na^+ and Cl^- in sweat results in the release of the hormone aldosterone (for a review of the fluid regulatory hormones, refer back to chapter 2), which acts on the kidneys to increase retention of these electrolytes. As the plasma concentration of these ions increases, plasma osmolality increases as well. This activates osmoreceptors in the hypothalamus, triggering a thirst sensation and stimulating the individual to drink. Electrolyte losses in the sweat cause these ions to be redistributed among body tissues. The resulting change in the electrolyte balance between intracellular and extracellular compartments may affect exercise performance by altering the membrane potential of the motor unit (nerve and muscle fibers that it innervates), possibly leading to performance decrements (Sjogaard 1986).

Effects of Hypohydration on Performance

The physiological changes that result from a body water deficit may have profound implications for exercise performance. Based on the changes to both cardiovascular and thermoregulatory function during dehydration, one could assume that exercise performance, especially endurance performance, would be negatively affected. Investigations on the effect of moderate to severe levels of hypohydration on aerobic activity have indeed shown significant performance decrements; their extent appears to be related to the magnitude of the body water deficit, the environmental conditions in which the exercise was performed, and the duration of exercise.

Aerobic Performance

Endurance performance appears to be more adversely affected when exercise is performed in the heat while dehydrated (Sawka 1992). In temperate environments, **maximal aerobic power** appears to be maintained if the level of dehydration does

Table 18.2 Concentrations of Major Electrolytes in Plasma, Sweat, and Muscle

	Electrolyte concentrations (mEq/L)		
Electrolyte	**Plasma**	**Sweat**	**Muscle**
Sodium (Na^+)	137-144	40-80	10
Chloride (Cl^-)	100-108	30-70	2
Calcium (Ca^{2+})	4.4-5.2	3-4	0-2
Potassium (K^-)	3.5-4.9	4-8	148
Magnesium (Mg^{2+})	1.5-2.1	1-4	30-40

Adapted from Maughan 1991.

not exceed 3% of body weight (Armstrong, Costill, and Fink 1985; Caldwell, Ahonen, and Nousiainen 1984). At body water deficits greater than 3% of body weight, significant reductions in aerobic power are seen (Buskirk, Iampietro, and Bass 1958; Caldwell, Ahonen, and Nousiainen 1984; Webster, Rutt, and Weltman 1990). If exercise is performed in the heat, the combined dehydration and thermal stress can result in performance decrements occurring at a lower level of dehydration. Craig and Cummings (1966) reported that subjects exercising in a hot environment (115 °F [46 °C]) experienced a significant (10%) decline in aerobic power at only a 2% body weight deficit.

The effect of hypohydration on physical work capacity appears to be more pronounced. Similar to the situation with aerobic power, declines in physical work capacity are related to the magnitude of the body water deficit and the environmental temperature. However, performance decrements appear to occur at a lower level of dehydration. Armstrong, Costill, and Fink (1985) reported a 3.3% ($p > 0.05$) increase in the time recorded for a 1500 m run at a body water deficit of 1.9% body weight. Although this difference was not statistically significant, it may have much practical significance. During a race of relatively short duration (e.g., 1500 m), the range in performance times among competitors would not be expected to be very large. Thus, a 3% difference in time may be the difference between winning and losing the race. During competitions of greater duration (e.g., 5000 or 10,000 m), low levels of dehydration (1.6% and 2.1%, respectively) were shown to result in a significant decline in performance (6.7% and 6.3%, respectively) (Armstrong, Costill, and Fink 1985). These results are depicted in figure 18.2.

Other investigators examining similar levels of dehydration have reported even greater performance

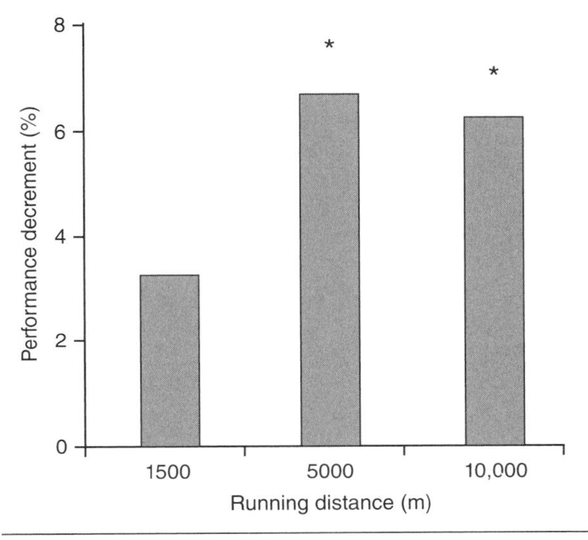

Figure 18.2 Effect of dehydration on changes in running times for 1500 m, 5000 m, and 10,000 m races. * = Significant decline in performance.

Data from Armstrong et al. 1985.

decrements. A 31% difference ($p < 0.05$) in the time to **fatigue** was seen in cyclists dehydrated to 1.8% of body weight who performed at 90% $\dot{V}O_2max$ (Walsh et al. 1994). The difference in the magnitude of performance decrements between the study just discussed and this investigation may relate to the method of dehydration. The subjects in the running study were dehydrated with a diuretic, whereas the subjects in the cycling study were fluid depleted from 60 min of cycling exercise at 70% $\dot{V}O_2max$, which was performed immediately before the high-intensity bout of exercise. The accumulated fatigue from the preceding exercise session likely resulted in the large differences in performance between the studies, suggesting that the method of dehydration may also have a significant effect on the extent of performance declines.

As the level of dehydration increases, a greater decline in physical work capacity is seen. Caldwell, Ahonen, and Nousiainen (1984) reported more than a threefold drop in cycling performance (7 W decline vs. 23 W decline) as the level of dehydration went from 2% to 4% in a temperate environment. Dengel and colleagues (1992) reported a 3.7% ($p > 0.05$) decline in cycling duration in subjects with a body water deficit of 3.3%. As the level of hypohydration reached 5.6%, fatigue occurred 6.4% ($p > 0.05$) earlier than under euhydrated conditions. Performance decrements are even further magnified when exercise is performed in the heat. Craig and Cummings (1966) reported that subjects dehydrated to 2% of their body weight reached fatigue 22% faster when walking (3.5 mph [5.6 km/h]) to exhaustion in a hot environment (115 °F [46 °C]). When the level of dehydration was further reduced to 4% of their body weight, the subjects' time to fatigue was 48% faster. It seems fairly clear that the significant impairments to both the cardiovascular and thermoregulatory systems from a body water deficit directly affect the ability of the athlete to perform aerobic exercise.

Anaerobic Performance

The effect of hypohydration on **anaerobic power** performance is less clear. Studies addressing the effect of dehydration on anaerobic exercise performance have been conducted primarily on athletes who reduce weight quickly in order to compete at a given weight class. Most of this research has been done in wrestlers or other athletes who participate in high-intensity, short-duration athletic events. These studies have shown that strength, anaerobic power, and anaerobic capacity are not affected by varying levels of hypohydration as long as the duration of the high-intensity activity is less than 40 s (Houston, Marin, et al. 1981; Jacobs 1980; Park, Roemmick, and Horswill 1990; Viitasalo et al. 1987). Jacobs (1980) reported no significant differences in either peak or mean power even as the magnitude of dehydration increased to 5%. Power outputs tended to increase ($p = 0.06$) when examined relative to body weight (figure 18.3). As the duration of anaerobic exercise increases (either sustained or intermittent), decreases in anaerobic power and capacity have been reported (Hickner et al. 1991; Horswill et al. 1990; Webster, Rutt, and Weltman 1990). In examining the effect of dehydration on simulated or actual sport performance, Klinzing and Karpowicz (1986) reported significantly slower performance times in wrestlers performing a standardized series of wrestling drills. These studies suggest that for sports in which strength and power performance are emphasized over a relatively short duration (<40 s), rapid weight loss may result in an increase of relative strength and power, offering a competitive advantage to the athlete. However, as duration of anaerobic activity increases, the strain of hypohydration appears to affect performance.

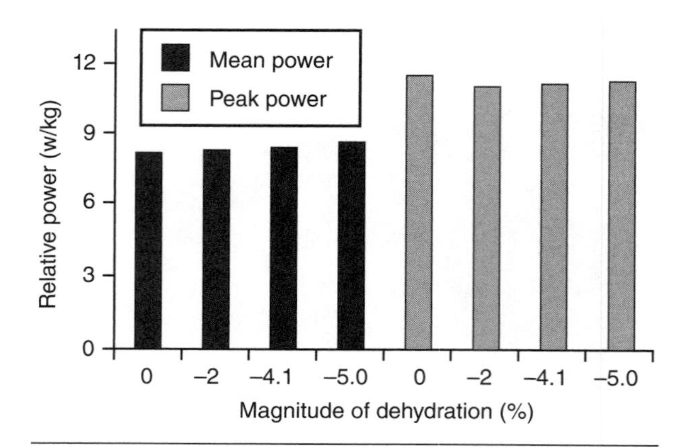

Figure 18.3 Effect of differing magnitudes of dehydration on anaerobic power outputs relative to body weight.

Adapted from Jacobs 1980.

Much less is known about the effects of dehydration in sports that rely on intermittent bouts of high-intensity exercise over a relatively long duration. In such sports (e.g., basketball, American football, and hockey), athletes do not voluntarily dehydrate to compete in a lighter weight class. Rather, dehydration may occur because of inadequate fluid intake during game performance. When anaerobic power needs to be maintained over a moderate duration of time (e.g., 30-60 min), a body water deficit may hinder performance. One study addressed the effect of fluid restriction on anaerobic power and skill performance during a simulated basketball game (Hoffman, Stavsky, and Falk 1995). Fluid restriction during the 40 min game resulted in an average body water deficit of 1.9%. Although this magnitude of dehydration did not result in any significant changes in vertical jump height, anaerobic power, or shooting performance, an 8% difference in shooting percentage observed ($p > 0.05$) would result in a possible 6-point deficit (average number of shots in a game/calculated shooting %) during a game (figure 18.4). Practically

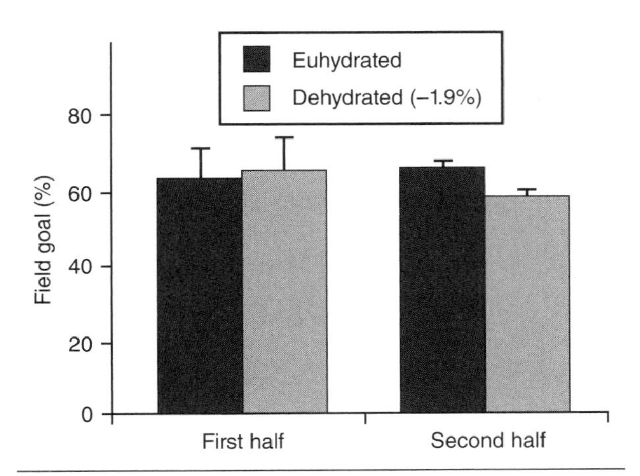

Figure 18.4 Effect of dehydration on basketball shooting performance.

Adapted, by permission, from J.R. Hoffman, H. Stavsky, and B. Falk, 1995, "The effect of water restriction on anaerobic power and vertical jumping height in basketball players," *International Journal of Sports Medicine* 16: 214-218.

speaking, this could clearly affect the outcome of a game. The results of this study suggest that even at low levels of dehydration, at which the athlete may not perceive thirst, performance in activities that require fine motor control may be impaired. Previous research has shown that motor unit recruitment patterns and muscle contraction capabilities may be reduced because of elevated body temperatures and electrolyte imbalances occurring with combined exercise and hydration stress (Sjogaard 1986). In support of these studies, Hoffman and colleagues (2012) reported that a 2.3% loss in body mass due to fluid restriction during a competitive basketball game in women resulted in impaired shooting percentage and reaction time, but no change in jump power compared to that in a contest in which the women were permitted to drink.

Fluid Replacement During Exercise

The decline in physiological function resulting from a body fluid imbalance and the subsequent decrease in athletic performance emphasize the need to rehydrate during exercise. If athletes rely on thirst as an indicator to drink, they will undoubtedly perform in a dehydrated condition. As discussed previously, the thirst sensation does not occur until body water deficit reaches approximately 2%. Even at this magnitude of hypohydration, physiological changes are apparent and performance decrements can be seen. Although the need for fluid replenishment is obvious, in many athletic events the volume and frequency of fluid consumption may be limited by the rules of the competition (e.g., rest periods or time-outs) or by availability (e.g., distance between drinking stations on a race course). In addition, even when fluid is available, athletes seldom consume more than 0.5 L/h (Noakes 1993). Thus, athletes need to adopt several strategies to maintain a normal body fluid balance. The following are recommendations for maintaining fluid balance.

General Strategies

- Ingested fluids should be between 59 °F and 72 °F (15 °C and 22 °C) and flavored to enhance palatability to promote maximal fluid replacement.
- Athletes should weigh themselves after each practice and game and replace all lost fluid. They should rehydrate 1 L for each kilogram of body weight lost.
- Changes in body weight, urine color, and urine specific gravity are indicators of hydration status and should be routinely examined by the athlete, coaches, and trainers (Armstrong et al. 1998).

Preexercise or Precompetition Hydration Strategies

- Athletes should drink 0.5 L water 2 h before exercise to promote adequate hydration and allow for excretion of excess fluid.
- They should then drink an additional 200 to 300 ml fluid 10 to 20 min before exercise.

During-Exercise Hydration Strategies

- During exercise, forced hydration should be practiced (drinking even when the thirst sensation is absent) to prevent physiological and performance impairments.
- Fluids should be readily available at shorter intervals during a road race, and athletes should be instructed to replace fluids during every time-out or stoppage of play during athletic contests.
- Athletes should optimize, not maximize, fluid intake without overhydrating. Athletes who know their sweat rate can match fluid loss with fluid intake (see sidebar for method of determining sweat rate).

- To optimize gastric emptying, about 0.5 L fluid should be in the stomach at all times (Convertino et al. 1996). If the fluid contains carbohydrate, the carbohydrate concentration should be between 4% and 8% (Murray et al. 1997). This provides for optimal intestinal absorption.
- Electrolyte drinks have not been proven to delay fatigue during exercise less than 3 to 4 h in duration. However, glucose–electrolyte drinks may delay fatigue in exercise of shorter durations.

Postexercise Hydration Strategies

- Fluids consumed following exercise can include or not include carbohydrate and electrolytes. However, water alone may not be the optimal rehydration beverage at this time because it decreases osmolality (a measure of solute concentration), limits the drive to drink, and may increase urine output (Hoffman and Maresh 2011). Salt (sodium or sodium chloride) in a rehydration beverage or food can conserve fluid volume and increase the urge to drink (Armstrong and Maresh 1996). For the athlete who must rehydrate between training or competitive sessions without meal consumption, it is important to consider choosing a rehydration beverage with electrolytes.
- As discussed in connection with general hydration strategies, the volume of fluid consumed during recovery should meet or exceed the volume of sweat that was lost during exercise. Otherwise, the athlete will remain somewhat dehydrated because of urinary fluid losses.
- Because replacement of expended glycogen stores is also a postexercise goal, the rehydration solution should contain carbohydrate. Carbohydrate may also help to improve the intestinal absorption of sodium and water (Murray 1987).
- Some team physicians use an intravenous fluid infusion to rehydrate athletes at halftime of a contest. However, scientific studies assessing the efficacy of this practice reported no performance differences between infusion and normal oral rehydration techniques when the rehydration and rest periods ranged from 20 to 75 min between exercise sessions (Casa et al. 2000; Castellani et al. 1997). During a 15 min halftime period, this rehydration strategy may still be debatable.

Determination of Sweat Rate

1. Get dry nude body weight (e.g., 65.3 kg).
2. Mimic event as much as possible.
3. Do not hydrate or urinate.
4. Get dry nude body weight again (e.g., 64.2 kg).
5. Calculate exercise time (e.g., 30 min).
6. Determine sweat rate per hour (in liters).

Example: 65.3 − 64.2 = 1.1 kg × (30/60 min) = 2.2 L/h

Fluid Temperature, Fluid Palatability, and Consumption

The temperature of water and its palatability are important factors for stimulating fluid intake. Several studies have demonstrated that as water temperature is increased, fluid consumption is significantly reduced (Armstrong, Costill, and Fink 1985; Szlyk et al. 1990). The highest water consumption appears to occur when water temperatures range from 59 °F to 72 °F (15 °C to 22 °C) (Hubbard, Szlyk, and Armstrong 1990). Szlyk and colleagues (1990) demonstrated that subjects walking on a treadmill (3.0 mph [4.8 km/h] at a 5% grade) for 6 h at 104 °F (40 °C) drank significantly more water when it was chilled to 59 °F (15 °C) than when its temperature was 104 °F (40 °C) (figure 18.5). These results were shown in two types of subjects: avid drinkers (subjects who were able to maintain body weight loss within 2% of initial body weight when given cold water ad libitum) and reluctant drinkers (subjects who lost more than 2% of initial body weight even though they were provided water ad libitum).

It also appears that water palatability is important for enhancing the drive to drink. If water has a color, odor, or taste that negatively affects human senses, individuals may refuse to drink (Hubbard, Szlyk, and Armstrong 1990). However, water consumption is enhanced if water is flavored, regardless of its temperature (Hubbard et al. 1984; Szlyk et al. 1989). Cherry, raspberry, and citrus appear to be the most

In Practice

Hydration and Sweating Responses During Preseason American Football Practices

Investigators from the University of Connecticut examined the hydration practices of high school American football players during preseason training camp (Yeargin et al. 2010). Twenty-five high school players volunteered and were divided into two separate groups, younger (*n* = 13; 14 or 15 years of age) and older (*n* = 12; 16 or 17 years); all were starters on either the varsity, junior varsity, or freshmen football team. The players were examined for 10 days during the summer preseason training camp. Practices occurred once per day on days 1 through 5 and days 8 through 10. Two practices occurred on days 6 and 7. Practices lasted on average slightly less than 3 h per session and consisted of football drills, contact, and conditioning. Sweat rates varied across training camp. The sweat rate was highest on days 1 through 4 and lowest on days 5 through 7, without any clear pattern of change. Sweat rate was lower in the younger (0.6 ± 0.4 L/h) than in the older players (0.8 ± 0.3 L/h). Using body mass and surface area as covariates did not explain the differences. The total average sweat lost in practice ranged from 1.5 ± 0.7 L to 2.9 ± 1.2 L on warm days and 1.0 ± 0.5 L to 2.1 ± 0.9 L on cool days. The amount of fluid consumed was similar between younger (16.8 ± 3.4 ml/kg) and older players (16.8 ± 4.5 ml/kg). Given that fluid consumption was similar, the percentage of fluid replacement differed between age groups (younger participants at 81% and older participants at 68%). Although the players replaced most of the sweat loss during practices, they still remained mildly hypohydrated throughout the preseason practices, indicating that their rehydration habits were inadequate. The investigators concluded that the high school athletes tended to underestimate their fluid losses and overestimate their rehydration. Thus coaches and health professionals need to educate athletes on appropriate hydration habits during and after practice.

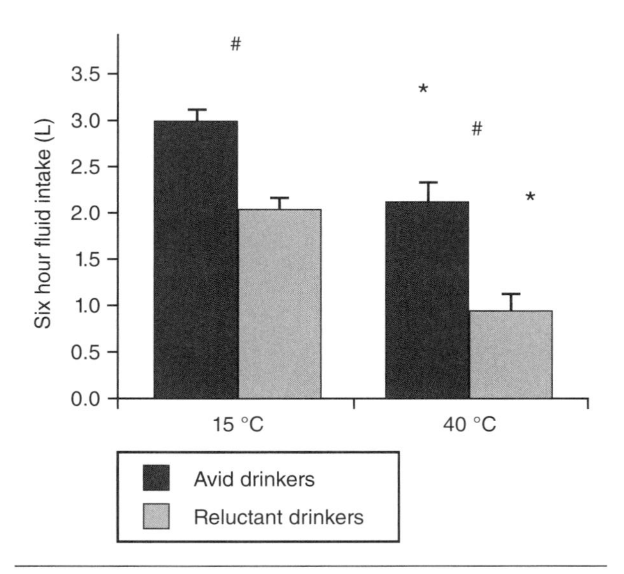

Figure 18.5 Effect of water temperature on drinking. * = Significant difference from 15 °C. # = Significant difference between avid and reluctant drinkers.

Data from Szlyk et al. 1990.

popular flavors for increasing fluid consumption (Hubbard, Szlyk, and Armstrong 1990).

Fluid Consumption and Gastric Emptying

Once the athlete or active individual has established a pattern of replacing lost fluids, the next major concern with respect to ensuring the availability of ingested fluids is gastric emptying. The rate of gastric emptying is determined primarily by the volume and composition of the fluid consumed (Maughan and Noakes 1991). Gastric emptying occurs at a higher rate when the volume of fluid in the stomach is high and falls rapidly as fluid volume decreases (Leiper and Maughan 1988). Although this highlights the importance of drinking repeatedly during exercise, the gastric distension that may occur with a large fluid consumption may also give a sensation of fullness, affecting the athlete's drinking behavior. The primary objective, though, is for the athlete to match fluid intake to fluid loss. Through trial and error, athletes find the appropriate fluid volume and the rate of intake that does not impair their performance. Gastric emptying also appears to be greater during exercise than at rest (Neufer et al. 1986). This is likely related to the increased mechanical movement of fluid within the stomach during exercise. Athletes could possibly consume a greater volume of fluid during exercise without experiencing feelings of fullness because of the greater emptying rates during activity.

The composition of ingested fluids has a major effect on the rate of gastric emptying. Both caloric content and osmolality of the drink can affect the ability of fluid to empty from the stomach. Besides replacement of lost fluids, the proposed effect of many commercial sports drinks is to spare muscle glycogen and replace lost electrolytes. With regard to the caloric composition of the drink, as the caloric content rises, the rate of gastric emptying declines (Murray 1987). The average gastric emptying rate for fluids containing calories ranges from less than 5 ml/min to 20 ml/min. The gastric emptying rate for solutions of plain water or saline is at the higher end of the scale (Brener, Hendrix, and McHugh 1983; Costill and Saltin 1974). Similarly, drinks of high osmolality empty from the stomach at a slower rate than drinks of low osmolality (Murray 1987), likely related to the positive relationship between osmolality and caloric content. In addition, the ingestion of glucose solutions of high concentrations may exacerbate plasma volume loss because of a movement of water into the gut resulting from the high osmolality of such concentrated solutions (Maughan, Fenn, and Leiper 1989).

Glucose polymers appear to promote the emptying of glucose–electrolyte solutions from the stomach by reducing the osmolality of the solution while maintaining the total glucose content (Maughan 1991). A number of studies have demonstrated that glucose–polymer solutions have a faster emptying rate than solutions containing free glucose in similar concentrations (Foster, Costill, and Fink 1980; Sole and Noakes 1989). However, results have not been consistent and have often contrasted. For instance, Foster, Costill, and Fink (1980) showed a significantly greater emptying rate with a 5% glucose–polymer solution than with a 5% solution of free glucose, but no differences between glucose–polymer solutions and free-glucose solutions of 10%, 20%, and 40% concentrations. In contrast, Sole and Noakes (1989) showed that glucose–polymer solutions of higher concentrations (15%) had significantly greater emptying rates than similar free-glucose solutions, but they did not see any differences when comparing solutions of lower concentrations (5% and 10%). Nevertheless, these studies and others on glucose–polymer solutions have consistently reported a faster emptying rate with the glucose–polymer drink even if the differences were not statistically significant. However, no polymer solution has shown an emptying rate as fast as that of plain water.

Electrolyte Replacement Drinks

The primary purpose of electrolyte drinks is to replace electrolytes lost in sweat. Sodium may be lost at a rate of 75 mmol/h during exercise. As discussed earlier, this may contribute to the reduced sweat rate caused by the increase in electrolyte concentrations in the plasma, resulting in a change in the osmotic gradient between the circulation and the sweat gland. Through replacement of the lost fluid with plain water, the plasma **osmolality** is lowered and sweat rate returns to its normal level (Senay 1979). However, complete restoration of extracellular fluid is not possible unless the lost sodium is restored first (Takamata et al. 1994).

There does not seem to be compelling evidence to support the use of electrolyte drinks during exercise less than 3 to 4 h in duration. However, in events that are prolonged (>4 h), the ingestion of electrolyte drinks appears to prevent **hyponatremia**. Hyponatremia is a condition of low plasma Na^+ (between 117 mmol/L and 128 mmol/L) that is associated with disorientation, confusion, and grand mal seizures (Noakes et al. 1990). It is brought on by rehydrating with large amounts of water (>2 L/h) for a prolonged period of time. Also referred to as *water intoxication,* hyponatremia was implicated in the death of endurance athletes during several prolonged events. Almond and colleagues (2005), investigating runners competing in the Boston Marathon, reported that 13% of the runners were hyponatremic at the end of the race. These runners consumed on average >3 L of water during the race and drank at every mile mark. Their run times were >4 h, and by the end of the race they had actually gained weight. Consuming electrolyte drinks during an event appears to prevent this relatively rare condition.

Most of the studies involving glucose and electrolyte replacements have examined athletes participating in prolonged aerobic activity. The effect of electrolyte supplementation during athletic contests involving repeated high-intensity, short-duration activities (e.g., football, basketball, and hockey) is unknown. Considering the possible implications of electrolyte loss on motor unit recruitment patterns and muscle contractile capabilities (Sjogaard 1986), it appears that future research should address the efficacy of electrolyte drinks in these high-intensity sports.

Role of the Alanine–Glutamine Dipeptide in Fluid Absorption

Glutamine infusion or ingestion during acute dehydration stress has been reported to enhance fluid and electrolyte absorption resulting from intestinal disorders (Nath et al. 1992; Silva et al. 1998; Van Loon et al. 1996), but its effects may not be consistent (Li et al. 2006). This is possibly related to stability issues of glutamine in the gut. The ability of glutamine in combination with alanine to enhance electrolyte and fluid absorption appears to be greater than that of glutamine alone. This is likely related to an increased stability of the molecule and an enhanced rate of absorption via specific ion transporters within intestinal epithelia (Lima et al. 2002). The greater stability of the alanine–glutamine dipeptide appears to be maintained at a low pH (Furst 2001). This may have potential implications for athletes during competition.

One study on the potential ergogenic benefit of the alanine–glutamine dipeptide during exercise showed that acute ingestion (0.05 g/kg and 0.2 g/kg) enhanced fluid uptake and reduced the magnitude of performance decrement during exercise to exhaustion under hypohydrated conditions (Hoffman, Ratamess, Kang, et al. 2010). Furthermore, the alanine–glutamine dipeptide was significantly more effective than water alone. This has important implications during athletic performance in that dehydration can play a critical role in the outcome of a contest. Another study examined the efficacy of the alanine–glutamine dipeptide (1 g/500 ml fluid ingested) in women's collegiate basketball (Hoffman et al. 2012). As discussed earlier in the chapter, a 2.3% loss of body mass occurred during a game in which players were not permitted to rehydrate. They were able to maintain jump power throughout the game, but basketball shooting performance and reaction time were significantly impaired. In rehydration trials, the alanine–glutamine dipeptide maintained basketball shooting accuracy to a better extent than water alone (figure 18.6) and enhanced visual reaction time. Players rehydrating with the alanine–glutamine dipeptide were able to respond to a visual stimulus more quickly than when dehydrated (figure 18.7). No significant differences in visual reaction time were observed in subjects ingesting water compared to subjects in the dehydrated condition. This dipeptide, which may enhance fluid and electrolyte uptake from the gut, may have important implications for activities that rely on fine motor control.

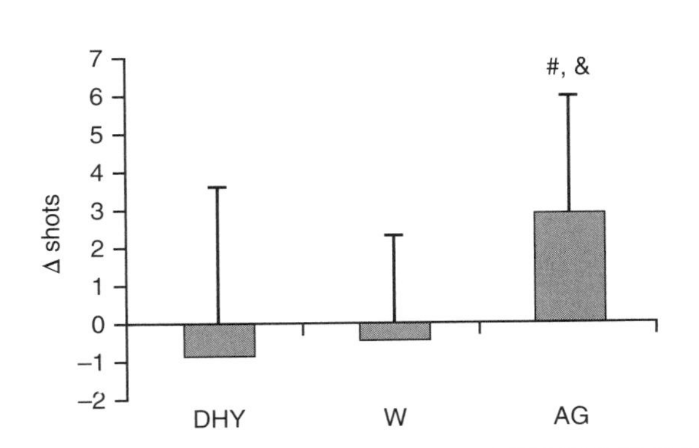

Figure 18.6 Field goal shooting. # = Significantly different from DHY. & = Significantly different from W. DHY, dehydrated; W, water only; AG, alanine–glutamine dipeptide. All data are presented as mean ± SD.

Adapted from J.R. Hoffman et al., 2012, "L-alanyl-L-glutamine ingestion maintains performance during a competitive basketball game," *Journal of the International Society of Sports Nutrition* 9: 4.

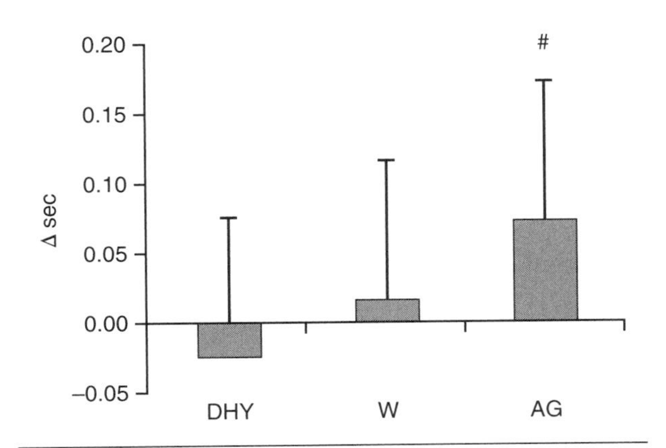

Figure 18.7 Change in visual reaction time. # = Significantly different from DHY. DHY, dehydrated; W, water only; AG, alanine–glutamine dipeptide. All data are presented as mean ± SD.

Adapted from J.R. Hoffman et al., 2012, "L-alanyl-L-glutamine ingestion maintains performance during a competitive basketball game." *Journal of the International Society of Sports Nutrition* 9: 4.

In Practice

Performance Benefits From Intravenous Fluid Intake and Glycerol

Although the efficacy of intravenous (IV) fluid intake has not been well established, it remains a part of the rehydration protocol of many team physicians. A survey of National Football League teams revealed that 75% of the teams reported regularly administering 1.5 L normal saline to up to 20 players 2 h before a game (Fitzsimmons, Tucker, and Martins 2011). Part of the issue associated with IV fluid delivery is the relatively quick diuresis associated with this method of rehydration (Van Rosendal et al. 2012). However, when glycerol is provided as part of the rehydration protocol, it acts as an osmotic agent and promotes fluid retention (Van Rosendal et al. 2010). Glycerol use before exercise is thought to delay dehydration and progression of the performance decrements associated with reduced body fluid balance (Van Rosendal et al. 2010).

Van Rosendal and colleagues (2012) examined the effect of combined IV and glycerol on rehydration and performance. Nine endurance athletes were recruited to perform a 40 km cycling time trial while dehydrated by 4% of their body mass. The participants performed four IV fluid and oral glycerol protocols. Once they achieved their required body mass reduction, they were randomized to one of four rehydration regimens: (1) 100% oral rehydration with a carbohydrate–electrolyte sports drink (64%) and distilled water (36%); (2) 100% oral rehydration with the carbohydrate–electrolyte sports drink (64%), distilled water (36%), and 1.5 g/kg glycerol; (3) oral and IV rehydration—50% sports drink (88% carbohydrate–electrolyte content with 12% distilled water) and 50% IV fluids (0.9% NaCl); (4) oral and IV rehydration—50% sports drink (88% carbohydrate–electrolyte content with 1.5 g/kg glycerol and 12% distilled water) and 50% IV fluids (0.9% NaCl). Participants were rehydrated with 150% of the fluid lost. Following the rehydration protocol, participants performed the 40 km time trial. Compared to oral rehydration, the oral glycerol improved time to complete the trial by 3.7%. The IV rehydration protocol improved time by 3.5% compared to oral rehydration, and IV with glycerol improved time by 4.1%. Although all three rehydration protocols had significantly greater time to completion than oral rehydration only, no further significant differences between trials were noted. Plasma volume restoration was highest for IV intake with oral glycerol, followed by IV alone, oral glycerol, and then oral ($p < 0.01$ for all of these comparisons). No differences in heart rate, tympanic or skin temperatures, sweat rate, blood lactate concentration, thermal stress, or rating of perceived exertion (RPE) values were seen between groups. Thus, it appears that combining IV fluid intake with oral glycerol resulted in the greatest fluid retention, but it did not improve exercise performance compared with either modality alone. The results of this study support the use of a combined IV rehydration protocol with oral consumption of glycerol for athletes who need a large volume of fluid in a short period of time.

Summary

Hypohydration has detrimental effects on both physiological and thermoregulatory systems. These effects are seen beginning at low levels of hypohydration (1% body weight) and increase as the magnitude of the fluid deficit grows. Decreases in athletic performance have been reported primarily in aerobic events. However, several studies have indicated that performance decrements may also be seen during high-intensity anaerobic activity exceeding 40 s in duration or in contests that comprise repeated bouts of high-intensity activity of shorter durations. The significant decline in both physiological function and athletic performance highlights the importance of maintaining a normal fluid balance. Fluid palatability and fluid temperature are factors that enhance fluid consumption. Although significant decreases in electrolytes are seen during hypohydration, the advantages of electrolyte replacement drinks are not realized until exercise durations exceed 3 to 4 h.

REVIEW QUESTIONS

1. Discuss the difference between dehydration and hypohydration.
2. Define hyponatremia and the potential dangers associated with it.
3. What type of strategy can be employed to ensure that athletes maintain normal fluid balance?
4. Describe the cardiovascular effects associated with hypohydration and how a body fluid deficit can affect thermoregulation.
5. What type of performance decrements can be seen at 2% hypohydration and 5% hypohydration?

chapter 19

Dietary Supplementation

CHAPTER OBJECTIVES

After reading this chapter you should be able to do the following:

- Understand the purpose of the Dietary Supplement Health and Education Act.
- Describe the purpose and efficacy of dietary supplements designed to improve muscle growth and increase muscle strength and power.
- Describe physiological differences between consumption of amino acids and whole proteins and explain how these differences may relate to timing of their actions.
- Explain the purpose and efficacy of dietary supplements designed to enhance energy, focus, and alertness.
- Understand the popularity of energy drinks, their efficacy to enhance performance and reduce body fat, and any health risks associated with their consumption.
- Describe some new dietary supplements that have been proposed to be efficacious for different athletic populations.

More than $28 billion a year is spent in the United States on thousands of different dietary supplements (Cohen 2012). The dietary supplement industry focuses on almost every segment of the population, but the prevalence of dietary supplement use by competitive athletes appears to be quite high. Reports from 2004 indicated that between 65% and 89% of intercollegiate athletes were using some type of dietary supplement (Froiland et al. 2004; Herbold, Visconti, et al. 2004). These percentage ranges are consistent with what is seen in dietary supplement use patterns among elite athletes worldwide (Heikkinen et al. 2011; Lun et al. 2012; Sato et al. 2012). However, the use of dietary supplements is not limited to mature, competitive athletes. According to one study, dietary supplement use in adolescent populations ranged from 23% to 32% (Briefel and Johnson 2004); others have indicated that use patterns among adolescents may even be higher, with numbers ranging from 42% to 74% of students surveyed (Bell et al. 2004; Herbold, Vazquez, et al. 2004; Hoffman, Faigenbaum, et al. 2008). Understanding dietary supplements becomes an integral part of the education of coaches, trainers, and medical professionals who advise and consult with athletes, clients, or patients. Considering the multitude of dietary ingredients that are promoted as supplements, it would be nearly impossible in a single chapter to provide an in-depth discussion of each supplement. Therefore, it is the purpose of this chapter to present an overview of the more popular supplements used among athletic populations. The supplements are separated based on their expected physiological role. Thus, supplements developed for muscle growth, strength, and power are grouped together. Other sections focus on intra- and intercellular buffers and on caffeine and energy drinks. A final section covers supplements that do not fall precisely into any of these categories but are becoming popular or may have ergogenic potential.

Dietary Supplement Regulation

The dietary supplement industry for many years was unregulated, meaning that this industry could develop, market, and sell supplements without evidence of efficacy or safety. In 1994, the Dietary Supplement Health and Education Act (DSHEA) was enacted to provide a framework for regulation of the dietary supplement industry. It required nutritional companies to provide more consumer information about supplements, reaffirmed that supplements are a category of food, and provided the Food and Drug Administration (FDA) with the ability to establish good manufacturing process (GMP) requirements for supplement companies. This law requires the food and supplement industry to educate consumers so that they can determine the risk versus reward relating to its products. It also places the onus on the manufacturers to ensure that their products are safe and are properly labeled. Specific requirements include the following:

- The ingredients composing a supplement must be safe.
- The ingredients in the supplement must be effective, meaning that the product does what the company says it does (enforcing truthful label claims).
- The product must be manufactured in a manner that is consistent with good manufacturing process (ensuring quality).

Although government control was intended to "clean up" the industry, a number of companies have intentionally or perhaps unintentionally sold contaminated supplements and may currently be doing so. This is often the defense of professional athletes who test positive for a banned substance and claim that they took a contaminated "supplement." Although some of these claims have been bogus, there are more than enough documented cases of intended or unintended adulteration of products to create concern. As this became an international issue, the International Olympic Committee (IOC) sponsored a study to examine the prevalence of contaminated supplements. Geyer and colleagues (2004) examined 634 nonhormonal dietary supplements purchased in 13 different countries from 215 different companies. A total of 94 samples (14.8%) contained testosterone precursors that were not listed on the label. Most of the products that produced a positive test were purchased in the United States and Germany, but the country with the greatest relative risk for purchase of a product containing a banned substance was the Netherlands (25.8%). Martello, Felli, and Chiarotti (2007) reported that 12.5% of the nutritional supplements in their analysis tested positive for substances not declared on the label (androgens and stimulants). Products may be contaminated because of an intention on the part of the manufacturer to increase the potency of their product. In other cases the problem is one of cross-contamination, resulting from the contamination of the raw materials or a lack of hygiene in the machinery used during the production process (De Hon and Coumans 2007).

In response to these issues, the FDA has continued to set standards and guidelines for manufacturing, labeling, and storage operations for dietary supplements. The enactment of the GMP Final Rule (Title 21, Code of Federal Regulations Part 111) requires that a dietary supplement label list what the supplement contains and state that it is not contaminated with any harmful or undesirable substances such as pesticides or heavy metals. In addition, it requires the company not only to identify each ingredient within the supplement but also to provide the strength or concentration of each ingredient. The current GMP (cGMP) was designed to enhance consumer confidence by having the federal government work cohesively with industry in ensuring the quality and safety of products (Crowley and Fitzgerald 2006). However, the ruling applies only to companies that market the final product. It does not impose the same requirements on manufacturers of dietary ingredients, which are often based outside of the United States.

The evidence supporting the benefits of many products is clear. However, the risk of using a contaminated product is a concern for many competitive athletes who are tested as part of the screening process for banned substances. Several third-party certification laboratories have arisen to provide assurances that supplements do not contain any banned ingredients. These laboratories, accredited and certified by the World Anti-Doping Agency, test supplements sent directly from the manufacturing company to certify that they contain no banned ingredients. This provides athletes "peace of mind" that the supplement they are ingesting will not result in a positive drug test. However, the laboratories do not certify that the supplement is efficacious. With regard to that, education is still at a premium.

Dietary Supplements for Muscle Growth and Strength–Power Development

This section focuses on dietary supplements that are popular or have been shown to be efficacious for people interested in promoting muscle growth and increasing muscle strength and power.

Protein and Essential Amino Acids

The importance of dietary protein is discussed in depth in chapter 17. As a dietary supplement, protein and amino acids appear to be second only to vitamins in athletic and nonathletic populations (Hoffman, Faigenbaum, et al. 2008). As discussed in chapter 17, protein needs are dependent on the type of athlete. Strength–power athletes have a greater daily protein requirement than endurance athletes. With regard to protein and amino acid supplementation, the leading dietetic and sports medicine organizations generally take a conservative approach (Rodriguez, DiMarco, and Langley 2009). However, these organizations do acknowledge the role of protein and amino acids in optimizing the training response and enhancing recovery. The largest benefit supporting the use of protein and amino acid supplementation relates to the importance of the timing of ingestion. The most convenient and efficient method for providing preevent (prepractice) or postevent (postpractice) protein may be supplementation.

The combination of resistance exercise and protein ingestion has a more positive effect on net protein balance better than either resistance training or protein intake alone (Biolo et al. 1995; Phillips et al. 1997; Tipton et al. 2001). Although resistance exercise results in protein degradation, it also has been shown to stimulate muscle protein synthesis to a greater magnitude causing a net gain in protein accretion (Biolo et al. 1995; Phillips et al. 1997). This is demonstrated in figure 19.1, which compares protein fractional synthesis rate (FSR) to protein fractional breakdown rate (FBR) in skeletal muscle following resistance exercise in a fasted state. At

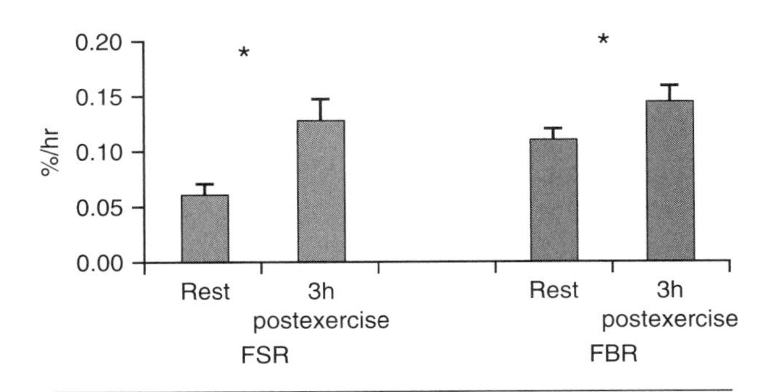

Figure 19.1 Fractional synthetic rate (FSR) and fractional breakdown rate (FBR) at 3 h after resistance exercise in a fasted state. * = Significantly different from resting value.

Reprinted, by permission, from J.R. Hoffman, 2007, "Protein intake: effect of timing," *Strength and Conditioning Journal* 29: 26-34; Data from Phillips et al. 1997.

3 h postexercise, the FSR was elevated 112% from rest, while FBR was elevated 31% from resting levels during the same time period. Protein synthesis and breakdown increased simultaneously; however, protein synthesis increased to a greater extent than breakdown. A strong correlation ($r = 0.88$) has been reported between FSR and FBR, suggesting that in the fasted state, amino acid uptake from the circulation via protein degradation provides the necessary amino acids for protein synthesis (Phillips et al. 1997).

When protein is ingested following resistance exercise, muscle protein synthesis is enhanced to a greater extent than after resistance exercise alone (Biolo et al. 1997; Tipton, Ferrando, et al. 1999). After a resistance training session, protein synthesis is elevated by about 100% from resting concentrations (Biolo et al. 1995). After protein ingestion that is not preceded by exercise, muscle protein synthesis is elevated by approximately 150%; but protein ingestion in combination with a bout of resistance exercise may elevate muscle protein synthesis by more than 200% (Biolo et al. 1997). To enhance the anabolic potential (i.e., the effects that occur when protein accretion exceeds protein degradation) of the workout, evidence is quite convincing on the importance of combining protein ingestion with resistance exercise. However, it is important to differentiate between aspects of study design in these investigations and actual protein use by competitive athletes. Several issues pertaining to the results of these studies have been raised. These well-controlled studies infused amino acids directly into the body, a method of ingestion that is not comparable to how most athletes consume their protein. In addition, evidence indicates that between 20% and 90% of the amino acids ingested normally (oral ingestion of capsules or powders) are removed from the circulation as they initially pass through the liver (Cortiella et al. 1988; Matthews, Marano, and Campbell 1993a, 1993b), which is further exacerbated by exercise (Halseth et al. 1997; Williams et al. 1996). Nevertheless, subsequent investigations demonstrated comparable changes in net muscle protein balance (synthesis – degradation) between oral ingestion and infused essential amino acids following resistance exercise (Tipton, Ferrando et al. 1999), indicating that oral consumption of protein, which is typical in people ingesting protein supplements, is efficacious in enhancing the anabolic response to resistance exercise.

As discussed in chapter 17, the need for protein appears to be greater for the strength–power athlete than for the endurance-trained athlete or the sedentary population (Lemon et al. 1992; Tarnopolsky et al. 1992). An increase in protein ingestion is thought to enhance the recovery and remodeling processes within skeletal tissue that has been damaged during resistance exercise (Tipton et al. 2004). Recent investigations have demonstrated that protein ingestion following resistance exercise can decrease the extent of muscle damage, attenuate force decrements, and enhance recovery (Hoffman, Ratamess, Tranchina, et al. 2010; Kraemer, Ratamess, et al. 2006; Ratamess et al. 2003). Given that this greater protein requirement for athletes, especially strength–power athletes, has been accepted (Rodriguez, DiMarco, and Langley 2009), the focus of many research endeavors is the timing of protein consumption in relation to the workout.

Acute Protein Intake Pre– or Post–Resistance Exercise or Both

The time at which protein is ingested appears to be important in maximizing the anabolic response from the training session (Anderson et al. 2005; Cribb and Hayes 2006; Esmarck et al. 2001; Tipton et al. 2001). In one study, untrained subjects were provided 6 g essential amino acids (0.65 g histidine, 0.60 g isoleucine, 1.12 g leucine, 0.93 g lysine, 0.19 g methionine, 0.93 g phenylalanine, 0.88 g threonine, and 0.70 g valine) with 35 g sucrose following a resistance training workout (Rasmussen, Tipton, et al. 2000). The investigators reported no difference in net muscle protein synthesis whether the supplement was consumed at 1 or 3 h postworkout. However, when this same combination of essential amino acids and carbohydrate was infused immediately before exercise, the increase in muscle protein synthesis was significantly greater than when it was infused immediately after exercise (Tipton, Ferrando et al. 1999). The preexercise infusion resulted in a 46% increase in amino acid concentration within skeletal muscle immediately postexercise and an 86% elevation 1 h after exercise (Tipton, Ferrando et al. 1999). These values were significantly higher than those in subjects provided the amino acid and carbohydrate blend immediately following exercise. By 3 h postexercise, muscle amino acid concentrations were still 65% above resting levels in subjects who received the amino acid and carbohydrate blend before the workout.

It appears that one of the primary benefits of amino acid ingestion before exercise is related to an increased rate of delivery and subsequent uptake by skeletal muscle during exercise. Tipton and colleagues (2001) demonstrated that a 2.6-fold greater increase in the rate of phenylalanine delivery to skeletal muscle was seen when amino acids were consumed before resistance exercise compared to afterward (see figure 19.2). In this study the investigators examined amino acid uptake by skeletal muscle for a 3 h period (rest, exercise, and postexercise). Amino acid uptake was reported to be 160% greater for the total 3 h period in subjects who ingested amino acids before exercise compared to those consuming them immediately postexercise. An enhanced uptake of amino acids by skeletal muscle is assumed to correspond to greater muscle protein synthesis. Thus, the combination of

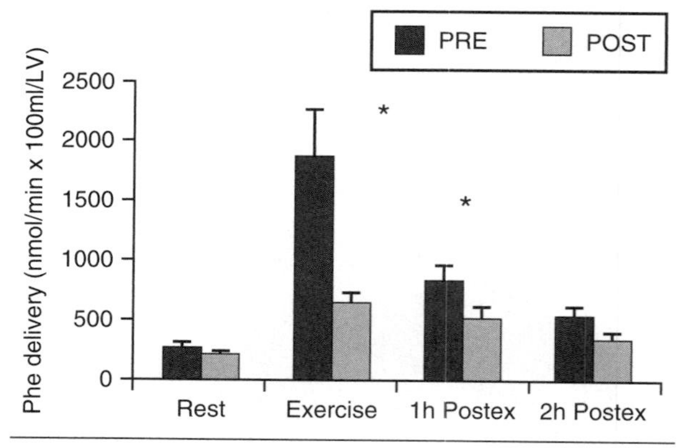

Figure 19.2 Mean delivery of phenylalanine to the leg for protein supplementation provided before (PRE) and immediately following (POST) exercise. Phe, phenylalanine; LV, leg volume; Postex, postexercise. * = Significant difference between PRE and POST.

Reprinted, by permission, from J.R. Hoffman, 2007, "Protein intake: Effect of timing," *Strength and Conditioning Journal* 29: 26-34; Data from Eipton et al. 2001.

amino acids and carbohydrate provided before the onset of exercise appears to be a potent stimulator of amino acid delivery via increased blood flow to exercising muscle and subsequent muscle uptake, resulting in greater protein synthesis than consumption of this supplement blend postexercise. Although the response of phenylalanine is depicted in figure 19.2, similar responses have been reported for other essential amino acids (Tipton, Ferrando et al. 1999).

The amino acid blend used in many of these investigations was based on the availability of each of these essential amino acids in proportion to their necessity for the synthesis of muscle protein (Borsheim et al. 2002). Several studies have determined that only the essential amino acids are necessary for stimulating protein synthesis (Tipton, Ferrando et al. 1999; Tipton, Gurkin et al. 1999). In one investigation, subjects were provided 40 g (21.4 g essential and 18.6 g nonessential) of amino acids (Tipton, Gurkin et al. 1999). The increase in muscle protein synthesis, however, was in proportion to only the essential amino acid composition of the supplement. In further support of the idea that only the essential amino acids are important in stimulating muscle protein synthesis, Borsheim and colleagues (2002) compared a 6 g mixed amino acid (3 g essential and 3 g nonessential) to a 6 g essential amino acid supplement and reported a dose–response effect relative to the concentration of essential amino acids. There is little to no evidence to support the use of nonessential amino acids as part of any nutritional supplement.

Interestingly, a proportional increase in the availability of all of the essential amino acids from ingestion of a supplement with an amino acid composition similar to that of muscle does not truly occur (Borsheim et al. 2002). Differences in the clearance rates of individual amino acids following ingestion likely result in an uptake by muscle that differs from the ingested mixture (Borsheim et al. 2002). Leucine and isoleucine appear to have a more potent effect than the other amino acids on muscle protein synthesis. When leucine was added to a whole protein (whey) and carbohydrate supplement blend, the net protein balance was significantly greater than that following either the protein and carbohydrate–only blend or carbohydrate alone (Koopman et al. 2005).

Differences in Whey and Casein Protein Ingestion With Respect to Protein Accretion

The two most common whole proteins used in dietary supplements are casein and whey. These proteins from bovine milk have different digestive properties. Casein, accounting for 80% of the protein in milk, forms a gel or clot in the stomach following ingestion that makes it slow to digest. As a result, casein provides a sustained but slow release of amino acids into the bloodstream, sometimes lasting for several hours (Boirie et al. 1997). Whey protein accounts for the other 20% of the protein in bovine milk and contains high amounts of the essential and branched chain amino acids (Hoffman and Falvo 2004). Whey protein is the translucent liquid part of milk that remains following the cheese manufacturing process (coagulation and curd removal); as a result, it is absorbed into the body much more quickly than casein.

Boirie and colleagues (1997) examined differences between casein and whey protein supplementation. They demonstrated that a 30 g feeding of casein had significantly different effects on postprandial protein gain than whey ingestion. Whey protein ingestion resulted in a rapid appearance of amino acids in the plasma; casein ingestion resulted in a slower rate of absorption, producing a slow but sustained elevation in plasma amino acid concentrations. Whey protein ingestion resulted in a peak increase in protein synthesis of 68%, while casein ingestion stimulated a peak increase in protein synthesis of 31%. Analysis of postprandial leucine balance at 7 h postingestion showed that casein ingestion resulted in a significantly higher leucine balance, and no change from baseline was seen at that time point following whey consumption. Providing further support, Tipton and colleagues (2004) also suggested that the differences in digestive properties between whey and casein result in a fast and slow increase in muscle protein synthesis, respectively. However, ingestion of 20 g of these proteins at 1 h following resistance exercise resulted in a similar net muscle protein synthesis over a 5 h examination period. The results suggested that although whey protein can stimulate a rapid increase in protein synthesis, a large part of this protein is oxidized (used as fuel), while casein may result in a greater protein accretion over a longer duration. In a comparison of single versus multiple ingestions of whey protein (total protein consumed was equivalent) over 4 h, the repeated pattern of ingestion resulted in a greater net leucine oxidation than a single feeding of either casein or whey (Dangin et al. 2002). Even though casein and whey are both complete proteins, their amino acid composition is different. Whey protein contains a much greater leucine content than casein. Given that leucine has an important role in muscle protein metabolism, the benefits of whey protein consumption may be seen in its enhanced absorption capability and the resultant increase in protein synthesis. This may be especially relevant for enhancing muscle remodeling and recovery with regard to a potential window of adaptation (immediately before, after, or both before and after a workout) in which the heightened sensitivity of the muscle causes it to respond differently than at other times (Esmark et al. 2001; Cribb and Hayes 2006; Hoffman, Ratamess, Tranchina, et al. 2010).

Both casein and whey are effective in stimulating muscle protein synthesis. However, differences in the digestive properties of these proteins appear to stimulate a different pattern of protein synthesis. Whey protein ingestion results in a greater acute response compared to a more gradual rise in protein synthesis following a feeding of casein. Although the total net muscle protein synthesis appears to be similar between these proteins, it is possible that the acute elevation seen following whey protein ingestion postexercise provides a greater window of opportunity for enhancing the recovery and remodeling of skeletal muscle.

Does Muscle Protein Synthesis Differ With Ingestion of Amino Acids Versus Whole Protein?

Ingestion of either essential amino acids or whole proteins stimulates muscle protein synthesis. A question, though, is whether there is an advantage to consuming one type of protein versus the other. Present knowledge indicates that amino acid consumption (specifically essential amino acids) before exercise stimulates greater muscle protein synthesis than amino acid ingestion immediately afterward (Tipton et al. 2001) or 1 to 3 h postexercise (Rasmussen et al. 2000). However, no benefits are seen for stimulation of protein synthesis with preexercise ingestion of whey protein compared to a 1 h postexercise ingestion (Tipton et al. 2007). Although preexercise ingestion of essential amino acids appears to be beneficial to postexercise protein synthesis, this same effect is not seen with whole protein consumption. These differences are not well understood but likely relate to differences in absorption rates and subsequent delivery to exercising muscle. As shown by Tipton and colleagues (2007), arterial amino acid concentrations are approximately 100% higher than resting levels following ingestion of essential amino acids but only 30% following whey protein ingestion. This also corresponds to a greater amino acid availability to active muscle. It is possible that the addition of carbohydrate to the amino acid blend (no carbohydrate was included with the whey protein) influenced the response of muscle to amino acid ingestion by stimulating a greater insulin response, resulting in greater amino acid uptake by muscle.

Importance of Carbohydrate and Protein Combinations for Muscle Protein Synthesis

The addition of carbohydrate to a protein supplement is based on a desire to stimulate an insulin response. As discussed in chapter 2, insulin has a critical role in regulating glucose uptake by tissue. In addition, exercise enhances skeletal muscle responsiveness to glucose by causing a greater sensitivity of muscle to the effects of insulin (Mikines et al. 1988; Richter et al. 1989). Thus, the inclusion of carbohydrate in a recovery supplement can enhance the uptake of amino acids by stimulating an insulin response (Biolo, Fleming, and Wolfe 1995). Although carbohydrate alone has only a minor effect on improvements in muscle protein balance following exercise (Borsheim et al. 2004; Roy et al. 1997), the combination of carbohydrate and protein or amino acids in a supplement results in more effective protein uptake and an enhanced rate of muscle protein synthesis. In a study comparing (a) carbohydrate only, (b) carbohydrate and protein, and (c) carbohydrate, protein, and leucine combinations on muscle protein synthesis rate following resistance exercise, the combination of carbohydrate and protein was superior to carbohydrate only in stimulating whole-body protein balance (Koopman et al. 2005). The inclusion of leucine with the carbohydrate–protein combination provided an even greater stimulus to muscle protein synthesis compared to the carbohydrate–protein mixture alone. The role of leucine in stimulating muscle protein stimulation is becoming very clear, and its mechanism of action is thought to be related to its ability to potentiate the signaling process at the translational level for enhancing protein synthesis (Kimball and Jefferson 2004).

Investigations of the timing of a carbohydrate and protein supplement ingestion are equivocal. The variability among the studies may relate to the type of protein used—whether it was a whole protein or an amino acid blend. No differences in muscle protein synthesis were observed when a carbohydrate (35 g sucrose) and amino acid (6 g) supplement was provided at 1 or 3 h post–resistance exercise (Rasmussen et al. 2000). However, when a supplement consisting of 10 g protein (primarily casein), 8 g carbohydrate (sucrose), and 3 g lipid (milk fat) was provided immediately postexercise, a significant difference in muscle protein synthesis was seen compared to that with the same supplement provided 3 h following exercise (Levenhagen et al. 2001). In this latter investigation, plasma glucose, insulin, and amino acid concentrations were similar between the two supplement periods; however, glucose and amino acid uptake by exercising skeletal muscle was greater when the supplement was provided immediately after exercise. Thus, skeletal muscle appears to show the highest responsiveness and potential for greatest adaptation in a relatively short time (within 1 h) following exercise.

Protein Timing and Training Studies

Protein ingestion taking place close in time to the workout (e.g., immediately before or within an hour afterward) has been shown to significantly enhance muscle protein synthesis and muscle protein accretion when compared to ingestion times outside of that window. Hulmi and colleagues (2009) have suggested that protein consumption before and immediately after a resistance exercise session can accelerate muscle adaptation by increasing messenger RNA (mRNA) expression and prevent a postexercise decrease in myogenin mRNA expression. Evidence is indicating that the timing of the protein supplement may take on greater importance in stimulating muscle adaptations that occur during prolonged training.

Several investigations have demonstrated that protein ingestion before and immediately after a resistance exercise session is a potent stimulus for improving muscle growth and performance compared to carbohydrate-only supplements in young (approximately 19-23 years) and previously trained (Hoffman, Ratamess, Kang, et al. 2007) and untrained individuals (Anderson et al. 2005; Willoughby, Stout, and Wilborn 2007). In contrast, an investigation on the effect of protein timing in untrained elderly men reported no change in muscle mass or strength following 12 weeks of protein supplementation and resistance training even though ingestion occurred immediately before and after exercise (Candow et al. 2006). The differences between these studies may be related to differences in the endocrine response to a resistance exercise session between young and older men (Kraemer, Hakkinen, et al. 1999). Although both young and older adults respond positively to resistance training programs, it is possible that lower concentrations of testosterone common in older men result in a decreased sensitivity to the timing of protein intake.

Interestingly, one of the first studies on the effects of protein timing and muscle hypertrophy was published by Esmarck and colleagues (2001) in an examination of older adults (74.1 ± 1 years). Participants consumed a liquid protein supplement (10 g protein, 7 g carbohydrate, and 3 g fat) immediately after or 2 h following each resistance training session (three times per week) for 12 weeks. Results showed that muscle cross-sectional area and individual muscle fiber area were significantly increased in the subjects who consumed the supplement immediately following each workout, but was not altered in the participants who ingested it 2 h following the workout. In a subsequent study, Cribb and Hayes (2006) investigated the effects of a protein (40 g whey isolate) and carbohydrate (43 g glucose) supplement in young (21-24 years), recreationally trained male bodybuilders. Participants consumed the supplement blend immediately before and after each resistance training session or in the morning and evening. When the supplement was consumed immediately before and after each workout, participants experienced significantly greater gains in lean body mass, increases in the cross-sectional area of type II fibers (see figure 19.3), increases in contractile protein content (see figure 19.4), and increases in strength compared to participants consuming the supplement in the morning and evening. When pre- and postingestion times were compared to morning and evening ingestion times in competitive athletes, the effect of protein timing was not as clear.

Hoffman, Ratamess, Tranchina, and colleagues (2009) provided a 42 g protein supplement twice per day (either pre- and postworkout or morning and evening) to strength–power athletes (university

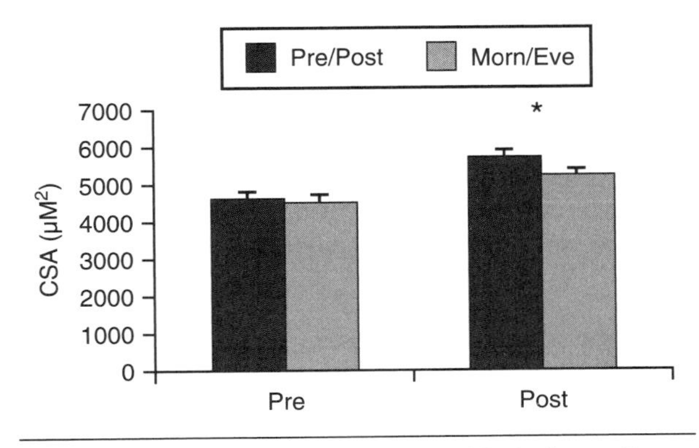

Figure 19.3 Cross-sectional area (CSA) of type IIa fibers. * = Significant difference between pre- and postworkout and morning and evening.

Reprinted, by permission, from J.R. Hoffman, 2007, "Protein intake: Effect of timing," *Strength and Conditioning Journal* 29: 26-34; Data from Cribb and Hayers 2006.

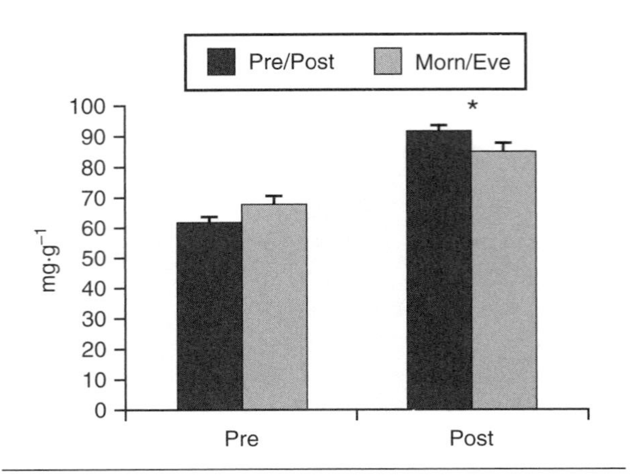

Figure 19.4 Contractile protein content. * = Significant difference between pre- and postworkout and morning and evening.

Reprinted, by permission, from J.R. Hoffman, 2007, "Protein intake: Effect of timing," *Strength and Conditioning Journal* 29: 26-34; Data from Cribb and Hayers 2006.

football players). A third group of participants did not take any supplement and served as the control, meaning that they did the same workout as the other two groups but were not provided a protein supplement. All three groups of athletes studied significantly increased their upper and lower body strength and power, with no between-group differences seen in any of the strength or power measures. All three groups had a daily protein intake that exceeded 1.6 g/kg per day. Based on the results of this study, if dietary protein intake is at or exceeds recommended levels for a strength–power athlete (1.6 g/kg), additional protein from a supplement, regardless of when it is ingested, may not result in further performance gains. In addition, all three study groups were in a positive nitrogen balance, indicating that their protein intakes were sufficient to meet their protein needs. Although results did not support the importance of protein timing, this study, as with all investigations, should be interpreted in the appropriate context. The study was only 10 weeks in duration, which may not have been long enough to permit teasing out differences in nutrient timing in experienced strength-trained athletes. Previous studies had examined either untrained or recreationally trained subjects only. Another factor that potentially contributed to the results is that the protein supplement did not contain any carbohydrate. This may have delayed nutrient absorption, and subjects may have missed the window of adaptation. Additional research focusing on the effects of nutrient timing in experienced athletes is still warranted.

Creatine

Creatine is one of the dietary supplements most commonly used by strength–power athletes and has likely been the most widely researched dietary supplement over the past 20 years. This section focuses on the physiology of creatine, its mechanism of action, and its efficacy.

Physiology of Creatine

Creatine is a nitrogenous organic compound that is synthesized endogenously from the amino acids arginine, glycine, and methionine. Creatine synthesis occurs primarily in the liver and in smaller amounts in both the kidneys and pancreas. Creatine is then transported from its site of synthesis to skeletal muscle or other organs via the circulatory system. Creatine is found naturally in meat and fish, and dietary consumption of these foods results in greater creatine concentrations. The vast majority (98%) of creatine is stored within skeletal muscle in either its free form (~40%) or its phosphorylated form (~60%) (Heymsfield et al. 1983). Small amounts of creatine are also stored in the heart, brain, and testes. It is the phosphorylated form of creatine, referred to as phosphocreatine (PC), that is used to fuel high-intensity exercise.

The PC molecule acts as a substrate in the formation of adenosine triphosphate (ATP) by rephosphorylating adenosine diphosphate (ADP). This is a critical chemical reaction that occurs during short-duration, high-intensity exercise and is dependent on the enzyme creatine kinase and the availability of PC within the muscle. This topic is reviewed in greater detail in chapter 3. As PC concentrations decrease, the ability to perform or sustain high-intensity exercise becomes compromised. During short-duration, high-intensity activity (e.g., 50-100 m sprint), the energy needed to perform the activity is derived primarily through the hydrolysis of PC (Gaitanos et al. 1993; Hirvonen et al. 1987). As duration of high-intensity exercise increases, the contribution of PC as the primary source of energy is reduced. For example, during short-duration, maximal-intensity exercise such as an eight-repetition maximum (RM) squat or a 100 m sprint, PC concentrations within the muscle are reduced between 35% and 57% from resting levels (Gaitanos et al. 1993; Boobis 1987). As duration of maximal-effort exercise increases to 30 s (e.g., a 200 m sprint), PC concentrations in the muscle are reduced between 64% and 80% of resting concentrations (Bogdanis et al. 1996; Boobis 1987; Cheetham et al. 1986). During high-intensity, repetitive exercise (common to sports such as American football, basketball, or ice hockey), PC concentrations within muscle approach depletion (McCartney et al. 1986). As these concentrations decrease, the ability to perform maximal exercise is reduced (Hirvonen et al. 1992); however, if PC concentrations in muscle remain elevated, the ability to maintain high-intensity exercise improves. This is the physiological basis for creatine supplementation.

Muscle creatine concentrations have been shown to increase by approximately 20% following supplementation (Febbraio et al. 1995; Hultman et al. 1996). However, there appears to be a **ceiling effect**. Once muscle creatine concentrations are saturated, further intake will not result in any further elevation of the muscle creatine content. It appears that the upper limit of creatine concentrations in skeletal muscle is approximately 150 to 160 mmol/kg dry weight (Balsom, Soderlund, and Ekblom 1994; Greenhaff 1995). This makes it possible to determine an accurate dosing protocol and prevents dosing according to the philosophy common among athletes who use nutritional supplements that "if a little is good, then more must be better."

Creatine Dosing

To rapidly increase muscle creatine concentrations to their maximal levels, athletes typically use a 20 to 25 g daily dose for 5 days, or 0.3 g/kg body mass if they dose relative to their body mass (Hultman et al. 1996). This is referred to as the loading phase. A maintenance dose of 2 to 5 g/day or 0.03 to 0.075 $g \cdot kg^{-1} \cdot day^{-1}$ is often used following the loading phase. If the athlete bypasses the initial loading dose, muscle creatine content can still reach maximal levels with smaller doses. The only difference is that it will take longer to reach that same muscle creatine concentration (~30 days vs. 5 days). As long as the maintenance dose is taken, muscle creatine concentrations will remain elevated. Once creatine supplementation ceases, muscle creatine content returns to baseline levels within approximately 4 weeks (Febbraio et al. 1995; Hultman et al. 1996).

Efficacy of Creatine Supplementation

Creatine is one of the most widely studied ergogenic aids to date. The ergogenic benefits seen in studies on creatine use have been consistent (Bemben et al. 2001; Brenner, Walberg-Rankin, and Sebolt 2000; Eckerson et al. 2004; Haff et al. 2000; Hoffman, Ratamess, Kang, Mangine, et al. 2006; Kirksey et al. 1999; Kreider et al. 1998; Lehmkuhl et al. 2003; Pearson et al. 1999; Volek et al. 1999). The efficacy associated with creatine use has been reported primarily during prolonged supplementation periods (4-12 weeks); however, a number of studies have used creatine supplementation protocols of varying durations with different dosing regimens and have met with varying results. The vast majority of short-duration creatine dosing regimens (3-6 days) have not resulted in significant performance improvements during a single, acute bout of explosive exercise (Cooke, Grandjean, and Barnes 1995; Dawson et al. 1995; Mujika et al. 1996; Odland et al. 1997; Snow et al. 1998). However, one study using a 20 g loading dose for 3 days in National Collegiate Athletic Association (NCAA) Division I strength–power athletes reported significant improvements in repeat sprint cycle performance (Ziegenfuss et al. 2002). Considering that muscle creatine concentrations are significantly elevated within the first 2 to 3 days of a loading phase, it is likely that the differences between this latter study and the others are related to the exercise protocol used. The investigation by Ziegenfuss and colleagues (2002) showing a positive effect employed a multiple-set high-intensity sprint protocol (6 × 10 s loaded sprints) with a 60 s rest interval between each sprint. In contrast, the other studies appear to have used an isolated bout of exercise or a longer rest interval that would provide for greater muscle phosphagen restoration. Other investigations demonstrating the efficacy of short-duration (5-6 days) creatine supplementation have generally used a loading dose (20 g/day) with a more appropriate testing protocol (e.g., repeated sprint or prolonged high-intensity exercise) (Cottrell, Coast, and Herb 2002; Cox et al. 2002; Mujica et al. 2000; Preen et al. 2001; Stout et al. 2000). Only two known investigations have demonstrated ergogenic benefits from a low-dose creatine supplementation regimen. Burke and colleagues (2000) provided 7.7 g/day (0.1 g/kg) for 28 days and showed efficacy with a low-dose, relatively short supplement duration; Hoffman, Stout, and colleagues (2005) demonstrated that even a low-dose (6 g/day), short-duration (6 days) supplementation regimen in active, college-aged men can improve fatigue rate, although no improvements were reported in maximal power performance.

The magnitude of strength improvements observed during long-duration (4-12 weeks) supplementation periods has generally been two- to threefold greater in athletes supplementing with creatine compared to a placebo (figure 19.5). One of the clear benefits of creatine supplementation is seen in a more effective or higher-quality workout. With improved workout quality, the training stimulus to the muscle will ultimately yield greater physiological adaptation resulting in improved performance.

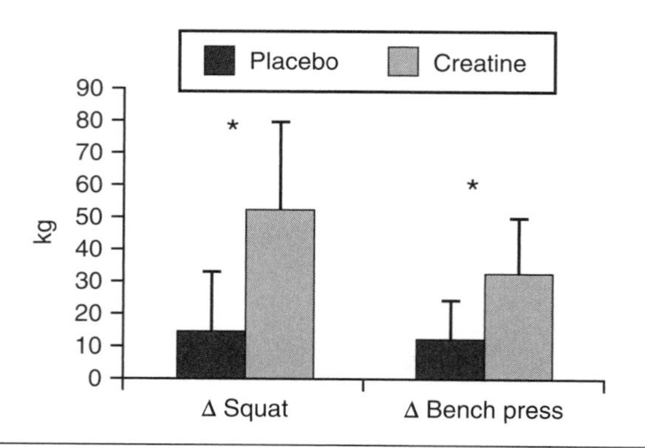

Figure 19.5 Effect of creatine supplementation on muscle strength gains in collegiate football players. * = Significant difference between creatine and placebo.

Adapted, by permission, from J.R. Hoffman et al., 2006, "Effect of creatine and ß-alanine supplementation on performance and endocrine responses in strength/power athletes," *International Journal of Sport Nutrition and Exercise Metabolism* 16: 430-446.

Effect of Creatine Supplementation on Body Mass Changes

Creatine supplementation has also been generally associated with increases in body weight. During prolonged supplementation, increases in body mass are primarily seen as fat-free mass. Initial increases in weight gain appear to occur relatively quickly as individuals supplement with creatine; this is likely related to an increase in total body water. As creatine content within skeletal muscle increases, the resulting intracellular osmotic gradient causes water to fill the cell (Volek and Kraemer 1996). As supplement duration increases, additional increases in lean body mass are likely related to the greater stimulus to the muscle from higher-quality training sessions that result in an increased synthesis of muscle contractile proteins (Balsom et al. 1993; Bessman and Savabi 1990).

Creatine Supplementation and Skeletal Muscle Adaptation

Improvements in lean body mass subsequent to creatine supplementation are likely caused by

In Practice

Can Creatine Supplementation Enhance Cognitive Performance in Older Adults?

A study from the University of Chichester in England examined the effect of creatine monohydrate ingestion on short- and long-term memory in older adults (McMorris et al. 2007). The hypothesis was based on studies showing that creatine supplementation is able to increase creatine concentrations in the brain, as well as studies showing positive responses in cognitive performance in younger adults supplementing with creatine (McMorris et al. 2006). Older men and women (76.4 ± 8.5 years) were recruited. Participants were provided either creatine (20 g/day) or an equivalent amount of placebo for 5 days and tested before and following supplementation. Participants performed verbal and spatial short-term memory tests (forward and backward recall). In addition, they performed a long-term memory test. This test required the participant to examine 10 photographs of individuals with their occupation written below the photograph. The participants waited for 1 h and were shown 20 photographs, including the 10 they had seen earlier. They were asked to match the face with the person's occupation. Results indicated that creatine supplementation in older adults was able to enhance both short-term (forward but not backward recall) and long-term memory. The investigators concluded that short-term creatine supplementation can enhance cognitive function in an older population.

physiological adaptations that occur within skeletal muscle as a result of the training stimulus, which are augmented by creatine use. Volek and colleagues (1999) examined the effects of creatine supplementation during a 12-week periodized resistance training program and reported significantly greater increase in the cross-sectional area of type I, type IIa, and type IIab fibers in participants supplementing with creatine compared to participants who consumed a placebo. These results were confirmed in a subsequent investigation by Willoughby and Rosene (2001). Figure 19.6 depicts the combined effect that creatine supplementation and resistance training can have on amplifying the response of myosin heavy chain protein synthesis (these are the microfilaments composing muscle that are involved in muscle contraction). Creatine ingestion appears to be effective in eliciting muscle morphological changes, most likely indirectly by giving the athlete the ability to sustain a higher-quality workout, stimulating a greater physiological response.

Evidence indicates that creatine may have a direct effect on changing skeletal muscle morphology. Creatine has been shown to increase satellite cell proliferation and differentiation when added to a culture of satellite cells (Vierck et al. 2003). Satellite cells are muscle stem cells found between the basal lamina of the myofiber and the plasma membrane. They are thought to be able to regenerate muscle fibers (Jankowski, Deasy, and Huard 2002). When myogenic stem cells are activated, they proliferate, differentiate, and fuse with existing myofibers within muscle, resulting in muscle hypertrophy. The fusion appears to be necessary for muscle differentiation (Dayton and Hathaway 1989). Creatine supplementation appears to have both direct and indirect actions on stimulating muscle morphological adaptation.

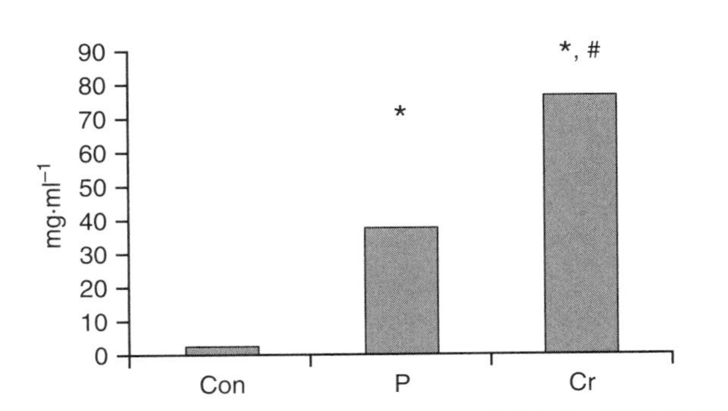

Figure 19.6 Creatine supplementation and change in myofibrillar protein content. Con, untrained control; P, placebo; Cr, creatine; * = significantly different ($p < 0.05$) from Con; # = significantly different from P.

Reprinted, by permission, from J.R. Hoffman, 2010, "Creatine and ß-alanine supplementation in strength/power athletes," *Current Topics in Nutraceutical Research* 8: 19-32; Adapted from Willoughby and Rosene 2001.

Creatine Supplementation: Responders Versus Nonresponders

As discussed in chapter 6, individuals vary greatly in their responses to training programs. Variability appears to be a trait associated with creatine use as well. It has been suggested that between 20% and 30% of subjects who ingest creatine may not respond to the supplement (Greenhaff et al. 1994; Syrotuik and Bell 2004). Some evidence suggests that for creatine to provide an ergogenic benefit, muscle creatine concentrations need to increase by 20 mmol/kg dry weight (Greenhaff 1997). People who supplement with creatine at 20 g/day for 5 days but show an increase of muscle creatine content less than 10 mmol/kg dry weight are deemed nonresponders. This has been confirmed by other investigators as well (Syrotuik and Bell 2004). Syrotuik and Bell (2004) reported that the majority of creatine users were quasi-responders; individuals whose muscle creatine concentrations increased more than 10 mmol/kg dry weight but less than 20 mmol/kg dry weight. These individuals had greater strength scores than the nonresponders but slightly lower scores than the responders.

The sensitivity of the response to creatine ingestion may depend on the individual's initial muscle creatine content (Syrotuik and Bell 2004; Ekblom 1996; Harris, Soderlund, and Hultman 1992). People who begin a supplementation regimen with low initial muscle creatine content likely benefit the most from creatine supplementation, while those with high initial levels benefit the least. While evidence supports the influence of baseline muscle creatine content as the most important factor determining the effectiveness of creatine supplementation, muscle fiber composition appears also to contribute to the response. Syrotuik and Bell (2004) reported that responders had the highest percentage of fast-twitch muscle fibers, followed by the quasi-responders; the nonresponders had the lowest percentage. Individuals with predominantly fast-twitch fibers and with low initial levels of muscle creatine appear to be most responsive to creatine supplementation.

Adverse Responses Associated With Creatine Supplementation

Adverse responses or side effects are an undesired response, with potential debilitating effects from supplement or drug ingestion. A number of anecdotal reports have linked creatine supplementation to gastrointestinal, cardiovascular, and muscular problems, with muscle cramps the most frequently mentioned adverse response. Controlled prospective studies have been unable to support any of the alleged adverse effects associated with creatine use. Supplementation protocols of 10 to 12 weeks have not documented any difference in adverse events in individuals supplementing with creatine versus a placebo (Hoffman, Ratamess, Kang, Mangine, et al. 2006; Kreider et al. 1998; Volek et al. 1999). Creatine use for even longer durations has not changed the risk associated with its use. According to a retrospective investigation on current and former competitive athletes who had used creatine for up to 4 years, an occasional gastrointestinal issue occurring during the loading phase was common and was described primarily as "gas" to mild diarrhea (Schilling et al. 2001). Another major concern raised about creatine supplementation was a possible strain on kidney function, due to the high nitrogen content of creatine, and reported increases in creatinine excretion during short-term ingestion (Harris, Soderlund, and Hultman 1992). However, several studies on this specific issue have reported no renal dysfunction during either short-term (5 days) or long-term (up to 5 years) creatine use (Poortsmans and Francaux 1999; Poortsmans et al. 1997).

As creatine concentrations within the muscle cell increase, an osmotic gradient is created that draws water into the cell through the process of osmosis. This has created a concern that creatine supplementation can increase the risk for heat illness, especially in athletes performing in a hot environment (Terjung et al. 2000). It is thought that the osmotic gradient will exacerbate the plasma volume loss generally seen during exercise in the heat (Lopez et al. 2009). However, several well-controlled scientific studies have refuted the alleged risk associated with creatine supplementation and dehydration or reduced thermoregulation (Lopez et al. 2009; Mendel et al. 2005; Vogel et al. 2000; Watson et al. 2006; Wright, Grandjean, and Pascoe 2007). One study examined the effect of creatine supplementation and fluid regulation by dehydrating subjects to 2.5% and 4.0% of their body weight following a 5-day creatine loading protocol (Vogel et al. 2000). After 75 min of exercise on a cycle ergometer, subjects showed a −7% and a −9% loss in plasma volume, respectively. Creatine ingestion in these subjects did not affect their hydration status, and no increase in muscle cramping was reported. In a critical analysis of the research examining creatine supplementation and exercise in the heat, Lopez and colleagues (2009) concluded that creatine supplementation neither hinders nor negatively affects athletes' ability to exercise in the heat or their body fluid balance.

What Type of Creatine Is Most Effective?

The marketing of creatine supplements by sport nutrition companies has created a bit of confusion regarding the efficacy of various forms of creatine. Creatine was introduced to the market based on research using its monohydrate form. Given an understanding of its physiological role, the next most important question relates to its ability to be absorbed when ingested orally. Several studies have demonstrated that the intestinal absorption of creatine supplied in the monohydrate form is nearly 100% (Chanutin 1926; Deldicque et al. 2008). For any other form to demonstrate greater efficacy, it would need to have enhanced absorption ability. One study compared plasma creatine concentrations and the pharmacokinetics of creatine absorption following ingestion of equal concentrations of creatine monohydrate, tri-creatine citrate, and creatine pyruvate (Jäger et al. 2007). The authors reported slightly altered kinetics of plasma creatine absorption between the three creatine forms, with the highest plasma concentrations seen for creatine pyruvate. However, they did not believe that there were any physiological differences in the bioavailability of these forms of creatine since the absorption of creatine monohydrate is practically 100% to begin with. Effervescent powders containing di-creatine citrate, found in a number of creatine supplements, have also been shown to have bioavailability similar to that of creatine monohydrate (Dash and Sawhney 2002; Ganguly, Jayappa, and Dash 2003).

Another form of creatine touted by sport nutrition marketers is creatine ethyl ester. Marketing claims suggest that the esterification of creatine can enhance creatine absorption, delay its degradation, and increase its bioavailability. The ability to enhance absorption over that of a molecule that already is nearly 100% absorbed was just discussed. More concerning, though, is that the creatine ethyl ester molecule is rapidly degraded to creatinine in the gut, likely related to the low pH (Mold et al. 1955). In a study comparing creatine ethyl ester to creatine monohydrate and placebo following 5 days of ingestion (Spillane et al. 2009), the creatine monohydrate and placebo groups experienced significantly greater gains in body mass and lean body mass than the creatine ethyl ester group. In addition, no significant increase in serum and muscle creatine content was observed in the group supplementing with the creatine ethyl ester molecule, suggesting that a large portion of the supplement was degraded within the gut following ingestion.

β-Hydroxy-β-Methylbutyrate

β-Hydroxy-β-methylbutyrate (HMB) is a derivative of the amino acid leucine and its metabolite α-ketoisocaproate. It is believed that HMB has both anabolic and lipolytic effects. Studies of HMB administration demonstrated possible anti-catabolic properties of the supplement in both animal and human subjects (Sapir et al. 1974; Tischler, Desautels, and Goldberg 1982). This is reflected by decreased muscle damage and enhanced recovery periods after HMB administration during times of high muscular stress (Nissen and Abumrad 1997). Wilson and colleagues (2012) demonstrated that prolonged administration of HMB to rats (equivalent to 16% of the animal's life span) resulted in a blunting of the negative changes associated with aging (e.g., body composition and muscle cross-sectional area).

Studies examining HMB administration and performance in human subjects have been limited. The first study of HMB supplementation in humans reported significant strength and lean body mass increases in previously untrained subjects after 4 weeks of resistance training (Nissen et al. 1996). Panton and colleagues (2000) provided a single daily 3.0 g dose of HMB to untrained men and women for 4 weeks. These investigators reported significant gains in upper body strength, as well as a trend toward improvements in lean body mass ($p = 0.08$) and a decrease in body fat % ($p = 0.08$). In addition, a lower creatine kinase concentration in the group ingesting the supplement compared to the placebo suggested that membrane integrity was maintained to a greater extent in the group supplementing with HMB. Jowko and colleagues (2001), using a similar dosing regimen in untrained men, reported similar results (i.e., increases in lean body mass and strength) following 3 weeks of resistance training. Others, though, were unable to duplicate these results after 8 weeks of resistance training (Gallagher et al. 2000a). Although the results of this latter study did not reach statistical significance, overall improvement of 1RM strength in 10 exercises increased 32.5% in the placebo group and between 43.5% and 45.5% in two groups taking different doses of HMB ($p > 0.05$). No adverse effects on hepatic enzyme function, lipid profile, renal function, or the immune system were noted after 8 weeks of HMB supplementation (Gallagher et al. 2000b). Kraemer, Hatfield, and colleagues (2009), in a 12-week study on recreationally active men, reported that ingestion of a proprietary amino acid blend including 1.5 g calcium HMB resulted in significantly greater gains in lean body mass and strength (1RM bench press and 1RM squat) than the placebo. In addition, they reported that HMB was able to attenuate muscle damage during training and reduce protein degradation to a greater degree than was seen in the placebo group.

As already mentioned, it has been suggested that HMB has anti-catabolic properties that may lead to an enhanced recovery from exercise. This has been attributed primarily to reduced enzyme markers of muscle damage and enhanced recovery during periods of high muscular stress following HMB administration (Knitter et al. 2000; Nissen et al. 1996; Panton et al. 2000). Specifically, these studies reported significantly lower creatine kinase and lactate dehydrogenase concentrations following a prolonged run (20 km) in endurance-trained athletes (Knitter et al. 2000) and in a resistance training program (Kraemer, Hatfield, et al. 2009; Nissen et al. 1996). The subjects in these studies supplemented with 3 g/day HMB for periods lasting 3 to 6 weeks. In studies of short duration (~2 weeks), these effects may not be realized (Hoffman, Cooper, Wendell, Im, et al. 2004; Nunan, Howatson, and van Someren 2010). Hoffman, Cooper, Wendell, Im and colleagues (2004) examined the effects of 10-day supplementation on collegiate football players during preseason training camp. No significant differences were noted between subjects supplementing with HMB and subjects consuming a placebo in any performance measure or in any markers of stress and fatigue. There does not appear to be any benefit from acute ingestion of HMB, even when it is consumed either immediately before or immediately following a workout (Wilson et al. 2009).

The ergogenic effects of HMB in trained athletic populations are less clear. Studies using resistance-trained or competitive athletes were unable to duplicate the results seen in a recreationally trained population using similar supplementation schedules (Hoffman, Cooper, Wendell, Im et al. 2004; Kreider et al. 1999; O'Connor and Crowe 2003; Paddon-Jones,

Keech, and Jenkins 2001; Ransone et al. 2003; Slater et al. 2001). These studies do not support the efficacy of HMB supplementation in resistance-trained athletes. Additional research appears warranted to further explore the effects of HMB supplementation for prolonged periods and during various types of physical stress.

Dietary Supplements for Intracellular and Intercellular Buffering

This section focuses on supplements purported to enhance the athlete's ability to sustain high-intensity exercise for a prolonged time. Although improved buffering capacity is an expected and desired adaptation resulting from anaerobic exercise training, enhancing buffering ability through nutritional intervention can have positive performance effects.

β-Alanine

β-alanine has become one of the most popular sport supplements used by strength and power athletes worldwide. The popularity of β-alanine is related to its ability to enhance intramuscular buffering capacity and attenuate fatigue during high intensity exercise. This section will focus on the physiology of β-alanine and examine the value of the research published during the past decade on its efficacy.

Physiology of β-Alanine

β-alanine is a nonessential, nonproteogenic amino acid (i.e., does not stimulate protein synthesis) that is synthesized in the liver (Matthews and Traut 1987). It is also present in many of the foods we eat, such as beef, chicken, and turkey. When consumed within the diet, β-alanine is a part of dipeptide molecules such as anserine, balenine, or carnosine. Carnosine is the principal histidine-containing dipeptide found in human skeletal tissue. However, when carnosine is consumed in the diet, it is quickly metabolized into β-alanine and histidine and cannot be taken up as an intact molecule by skeletal muscle (Bauer and Schulz 1994). β-Alanine is then taken up into skeletal muscle for the resynthesis of carnosine. Skeletal muscle has a relatively low concentration of β-alanine (Skaper, Das, and Marshall 1973) but high concentrations of histidine and carnosine synthetase, the enzyme responsible for carnosine synthesis. Thus it is β-alanine that is the rate-limiting step in carnosine synthesis. The ergogenic properties of β-alanine alone are limited, and it appears to be effective only as a precursor for the synthesis of carnosine (Dunnett and Harris 1999). Carnosine appears to have a role in enhancing performance of strength and power athletes by acting as an intracellular buffer during high-intensity exercise. It has been estimated that carnosine can contribute up to 40% of the capacity of skeletal muscle to buffer H^+ produced during high-intensity exercise (Harris et al. 2006; Hill et al. 2007).

Carnosine is present in both type I and type II muscle fibers, but in much higher concentrations in fast-twitch skeletal tissue (Harris et al. 2006; Hill et al. 2007). Histidine-containing dipeptides (e.g., anserine, balenine, and carnosine) are found in great concentrations in several mammalian species that perform highly anaerobic activities. Whales, despite the need to breathe air for oxygen in order to spend prolonged time under water, have concentrations of 400 mmol/kg dry muscle of histidine-containing dipeptides; thoroughbred race horses and greyhound race dogs have concentrations of 110 mmol/kg dry muscle and 90 mmol/kg dry muscle, respectively (Abe 2000; Harris et al. 1990). In contrast, human sprinters have muscle carnosine concentrations ranging between 17 and 25 mmol/kg dry muscle (Harris et al. 1990). As might be expected, the muscle carnosine content of sprinters is significantly higher than that typically found in endurance athletes, untrained individuals, and the elderly (Harris et al. 2006; Suzuki et al. 2002).

Muscle carnosine concentrations appear to be affected by participation in high-intensity anaerobic sports and training. In a comparison of various athletic populations, Parkhouse and colleagues (1985) reported that highly trained anaerobic athletes have a greater buffering capacity and a significantly greater muscle carnosine content than endurance athletes and untrained subjects. In a study comparing trained bodybuilders and untrained control subjects, the carnosine concentration in the vastus lateralis was significantly greater in the bodybuilders than the control subjects (Tallon et al. 2005). These differences could not be explained by differences in the muscle size between the groups. It appears that the high-volume and moderate- to high-intensity resistance training program of bodybuilders can stimulate endogenous changes in muscle carnosine concentrations. Others have reported a positive relationship between muscle carnosine concentration and mean power output from a 30 s maximal sprint on a cycle ergometer (Suzuki et al. 2002). This would suggest that the variability of muscle carnosine concentrations between individuals influences high-intensity exercise performance. However, it does not indicate a cause-and-effect relationship between training and muscle carnosine concentrations.

Only one study has reported a significant elevation in muscle carnosine concentrations from 8 weeks of high-intensity training (Suzuki et al. 2004). The majority of long-duration training studies (4-16 weeks) have been unable to support these results (Kendrick et al. 2008, 2009; Mannion, Jakeman, and Willan 1994). The results from the study by Tallon and coauthors (2005) may have been tainted, as the bodybuilders examined in that study self-admitted to using anabolic steroids. Carnosine synthesis has been shown to be upregulated by circulating testosterone concentrations (Penafiel et al. 2004). The role that exogenous androgen administration has in increasing muscle carnosine concentrations is not clear. Although there is limited support related to a training effect on increasing muscle carnosine concentrations, further research may be warranted to ascertain the mechanism for higher muscle carnosine concentrations in anaerobic athletes. Genetic endowment may be a factor, but as sport scientists increase their use of noninvasive technology (e.g., magnetic resonance spectroscopy) to examine muscle carnosine content, our ability to advance understanding of specific training adaptations will improve.

β-*Alanine Dosing Scheme*

β-Alanine is ingested in either a pill form or as a powder mixed in fluids. Increases in muscle carnosine concentrations appear to be dependent on the daily dose. One of the initial studies examined the effect of three different dosing regimens (10, 20, and 40 mg/kg body weight) on changes in plasma β-alanine concentrations (Harris et al. 2006). The two highest doses resulted in the greatest increase in plasma β-alanine concentrations (see figure 19.7) but were also associated with uncomfortable side effects (e.g., paresthesia; a tingling-like sensation felt in the skin). The lowest dose (10 mg/kg), which was equivalent to an approximate 800 mg dose, also resulted in an elevation in plasma β-alanine concentrations. However, this occurred without the adverse effects seen at the higher dosing levels. Plasma β-alanine concentrations peaked between 30 and 40 min following ingestion of the lower dose. Its half-life (time at which a 50% reduction in peak concentration occurred) was 25 min, and plasma β-alanine concentrations returned to baseline levels by 3 h postingestion. This kinetic profile suggests that the appropriate dosing regimen is 800 mg β-alanine taken every 3 to 4 h, that is, a daily dosing regimen between 4.8 g and 6.4 g per day.

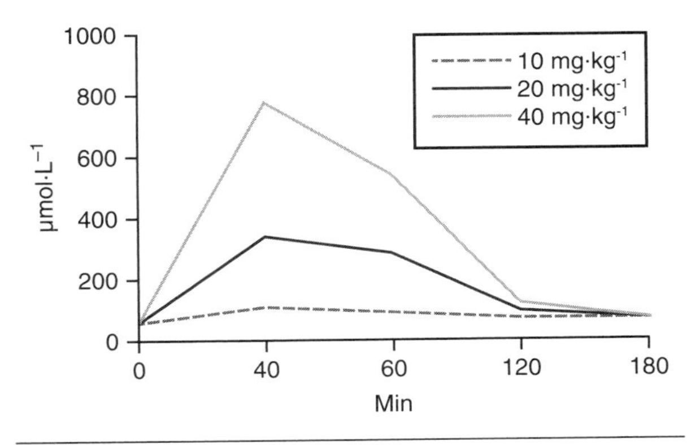

Figure 19.7 Plasma β-alanine concentrations.

Reprinted, by permission, from J.R. Hoffman, 2010, "Creatine and β-alanine supplementation in strength/power athletes," *Current Topics in Nutraceutical Research* 8: 19-32; Adapted from Harris et al. 2006.

Hill and colleagues (2007) examined β-alanine supplementation for 10 weeks. During the initial week of supplementation the participants consumed 4.0 g/day; they then consumed 6.4 g/day for an additional 9 weeks. Muscle carnosine concentrations were measured at weeks 0, 4, and 10. In addition, subjects performed a cycling to exhaustion test at 110% of maximal power. Within 4 weeks of supplementation, muscle carnosine concentrations increased by 58%; and by week 10, they increased an additional 15%. Changes in muscle carnosine concentrations corresponded to a 13% and 16% increase in total work done, respectively. Harris and colleagues (2007) reported significant elevations (37%) in muscle carnosine concentrations in untrained subjects following only 2 weeks of supplementation. Technology to deliver supplement in a slow-release pattern was examined by Derave and colleagues (2007). The benefit of such technology is that the slower release of β-alanine can reduce the adverse effects without affecting muscle carnosine elevations or performance improvements. One of the more interesting aspects of β-alanine ingestion is the notion that the greater the dose, the greater the increase in muscle carnosine elevation. Whether a ceiling effect exists regarding muscle carnosine content and β-alanine supplementation is not clear. The use of time-release capsules may allow for a greater daily dosing, resulting in greater muscle carnosine content than ingestion of powder or regular capsules. Information on the washout period (length of time to return to baseline levels) for muscle carnosine concentrations following cessation of β-alanine supplementation is not clear. One study provided 4.8 g/day of β-alanine to subjects for 6 weeks (Baguet et al. 2009). Following a 3-week period of no supplementation, muscle carnosine content decreased by 30%. By 9 weeks of cessation, muscle carnosine concentrations had returned to baseline levels. The investigators also suggested that the washout period may be dependent on the initial increase in muscle carnosine content. High responders (subjects with a greater accumulation of muscle carnosine content) required a greater washout time to return to baseline levels (~15 weeks) than low responders, whose muscle carnosine concentrations returned to baseline levels within 6 weeks.

Efficacy of β-Alanine Supplementation

β-Alanine appears to be most effective for athletes competing in repetitive or prolonged high-intensity activity. Studies examining this specific activity have been consistent in providing evidence to support the ergogenic benefit of β-alanine supplementation (Derave et al. 2007; Hill et al. 2007; Hoffman, Ratamess, Kang, Mangine, et al. 2006; Hoffman, Ratamess, Faigenbaum et al. 2008; Hoffman, Ratamess, Ross, Kang, et al. 2008; Stout et al. 2006, 2007; Van Thienen et al. 2009). Studies in untrained subjects have demonstrated significant improvements in fatigue (Stout et al. 2006, 2007). In a study examining untrained young men, Stout and colleagues (2006) provided 6.4 g/day (four servings, 1.6 g per serving) of β-alanine for 6 days and then 3.2 g/day for an additional 22 days. Before and after supplementation, they had participants perform an incremental cycle ergometry test to determine the physical working capacity at fatigue threshold (PWC_{FT}). The PWC_{FT}, determined from bipolar surface electromyography recorded from the vastus lateralis muscle, provides a measure of the highest level of exercise intensity a person can maintain without signs of fatigue. The results demonstrated that 4 weeks of β-alanine supplementation resulted in a significant increase (9%) in PWC_{FT} compared to no change in the placebo group. In a follow-up study on young untrained women, Stout and colleagues (2007) reported that 28 days of β-alanine supplementation was able to a significantly enhance PWC_{FT} (12.6%), ventilatory threshold (13.9%), and time to exhaustion (2.5%) during a graded exercise cycle ergometry test compared to values in a placebo group.

The benefits of β-alanine supplementation have also been demonstrated in experienced strength–power athletes. Collegiate football players were provided 4.5 g/day of β-alanine 2 weeks before the onset of preseason training camp (Hoffman, Ratamess, Faigenbaum et al. 2008). Supplementation continued for an additional 2 weeks during preseason practices. Following 2 weeks of supplementation, no differences were noted in sprint times or fatigue rates during performance of repeated line drills (an approximate 30-35 s shuttle run performed three times with 2 min rest between each sprint) compared to values in a control group. Although no significant differences were observed in peak power, mean power, and total work in a 60 s Wingate anaerobic power test, a trend ($p = 0.07$) toward a reduced rate of fatigue was observed in the athletes consuming the supplement versus the placebo. As supplementation continued 2 weeks into training camp, examination of the athletes' training logs revealed no significant difference in training intensity in either the bench press or squat exercise, but a trend ($p = 0.09$) toward a higher (9.2 %) volume of training (bench press and squat combined) was observed in the athletes supplementing with β-alanine compared to placebo. In addition, subjective feelings of fatigue during camp were significantly lower in athletes using the supplement compared to the placebo. As discussed earlier, 2 weeks of β-alanine supplementation can significantly elevate muscle carnosine concentrations, but not to the same magnitude as seen with longer-duration supplement periods. The trend toward an improved fatigue rate during the 60 s maximal-intensity bout of exercise suggests that even during suboptimal β-alanine ingestion durations, improvements in intracellular buffering capacity can be seen. When β-alanine supplementation is prolonged (4 weeks), significant delays in fatigue were seen during repeated isokinetic bouts (five sets) of exercise in competitive 400 m sprinters, but no improvements were noted in 400 m race time (Derave et al. 2007). Although improved buffering capacities are associated with β-alanine ingestion and result in delaying fatigue, an enhanced buffering capacity may not cause significant improvements in times in sprints lasting between 30 and 60 s. However, during situations of prolonged fatigue, an improved buffering capacity may affect sprint performance. Van Thienen and colleagues (2009) demonstrated that after 8 weeks of β-alanine ingestion in trained cyclists, 30 s sprint performance improved following a 110 min time trial.

The efficacy of β-alanine supplementation for resistance training has been demonstrated in several studies. Hoffman, Ratamess, Ross, Kang, and colleagues (2008) provided β-alanine at 4.8 g/day for 4 weeks to experienced resistance-trained athletes. Although no significant differences were seen in squat strength, improvements were observed between the groups following the 4-week supplementation period; the total number of repetitions performed in the squat exercise per workout (expressed as the difference between workouts performed at week 0 and week 4) was significantly greater in the supplement

In Practice

Can Significant Performance Effects Be Seen With β-Alanine Supplementation in Older Adults?

One of the most pressing concerns for an aging population is the loss of skeletal mass and muscular function, which is known as sarcopenia. Sarcopenia is one of the deleterious effects associated with aging and occurs in both type I and type II muscle fibers (Doherty 2003). A significant reduction in muscle carnosine content is also seen with aging (Tallon et al. 2007). This has led to the hypothesis that β-alanine supplementation can enhance performance capacity in older adults by increasing muscle carnosine concentrations (del Favero, Roschel, Solis, et al. 2012). Eighteen participants were randomly assigned to a β-alanine supplement group (65 ± 4 years) or placebo (64 ± 7 years). Participants were provided 3.2 g/day of either β-alanine or placebo for 12 weeks. Maximal aerobic capacity, ventilatory anaerobic threshold, and time to exhaustion were measured in all subjects before and after supplementation. Sit-to-stand and quality of life measures were also used. Following 12 weeks of supplementation, muscle carnosine concentrations were increased by 85.4% in the β-alanine group compared to a 7.2% increase in the placebo group. Significant improvements were noted in the time to exhaustion (36% improvement in the β-alanine group and only a 8.6% improvement in the placebo group). No differences were noted in sit-to-stand time or in any quality of life measure. The main effects from this study demonstrated that muscle carnosine concentrations can be significantly elevated with ingestion of a daily dose of β-alanine, and the increase in muscle carnosine content was related to performance improvements in the time to exhaustion test. This study demonstrates the potential positive benefits of dietary supplement intervention on delaying the decline in muscle function in older adults.

group than the placebo group (9.0 ± 4.1 and 0.3 ± 7.8, respectively). A significant difference in mean power was also seen between the groups. The lack of strength improvement is consistent with findings from other studies of similar or longer durations (4-10 weeks) in which strength performance did not improve (Kendrick et al. 2008, 2009). These results are likely related to the physiological mechanism that is affected by elevations in muscle carnosine concentrations and the duration of the study. Delaying fatigue through an enhanced intracellular buffering system would not have a direct effect on strength improvement. However, the improved training volume, seen as an increase in the number of repetitions performed at a given exercise intensity, enhances the quality of the training session (Hoffman, Ratamess, Kang, Mangine, et al. 2006; Hoffman, Ratamess, Faigenbaum, et al. 2008; Hoffman, Ratamess, Ross, Kang, et al. 2008). That should enhance the training stimulus to the muscle and stimulate further adaptation. However, in trained athletes, this effect may be greater during longer training durations.

The greater volume of work (i.e., more repetitions) performed during the resistance training program of experienced strength–power athletes supplementing with β-alanine has also been shown to stimulate significant decreases in fat mass and increases in lean body mass (Hoffman, Ratamess, Kang, Mangine, et al. 2006). Interestingly, the higher training volumes associated with β-alanine supplementation provide a stimulus that would simulate the effects of a resistance training program designed to enhance muscle hypertrophy. Higher volumes of training typify the training program of bodybuilders, and the resulting improvements in lean muscle mass and decreases in fat mass are desirable outcomes for individuals whose training goals include an improvement in body composition.

Adverse Effects Associated With β-Alanine Supplementation

The only reported adverse effect associated with β-alanine supplementation is paresthesia. Paresthesia is a sensation of numbing or tingling in the skin. It generally occurs with ingestion of high doses of β-alanine but often disappears within 1 h following ingestion (Harris et al. 2006). When β-alanine is combined with a carbohydrate and electrolyte drink, this effect appears to be negligible (Hoffman, Ratamess, Kang, Mangine, et al. 2006). The use of time-release capsules reduces the risk for paresthesia significantly, as the slower release of β-alanine reduces the magnitude of plasma β-alanine concentrations. Although studies examining prolonged (greater than 15 weeks) supplementation durations have not been performed, the fact that β-alanine is an amino acid with an important physiological role in the body, consumed in dosages that are similar to that consumed in the regular diet, suggests that it is likely a very safe supplement to use (Artioli et al. 2010).

Creatine and β-Alanine Combination

Two of the more popular supplements used by strength–power athletes are creatine and β-alanine. In one of the first studies to examine the efficacy of these two supplements combined, Hoffman, Ratamess, Kang, Mangine, and colleagues (2006) looked at whether this combination would provide a significant benefit to strength–power athletes. In a 10-week study on collegiate football players, athletes were randomly divided into three groups: placebo, creatine only (10.5 g/day), or creatine (10.5 g/day) plus β-alanine (3.2 g/day). The investigators demonstrated that the combination of creatine and β-alanine significantly improved the quality of the workout (i.e., significantly greater number of repetitions) compared to creatine alone (see figure 19.8). The greater training volume was associated with significantly greater gains in lean body mass and decreases in fat mass. No differences were noted between creatine and creatine plus β-alanine in strength gains in 1RM bench press and 1RM squat, but improvements in both supplement groups were significantly greater than in the placebo group. A similar investigation of previously untrained men randomly assigned subjects to one of four groups: creatine and β-alanine, creatine only, β-alanine only, or placebo (Stout et al. 2006). Significant improvements in all three supplement groups were reported, but no additive benefits were noted in physical work capacity in the combined creatine and β-alanine group.

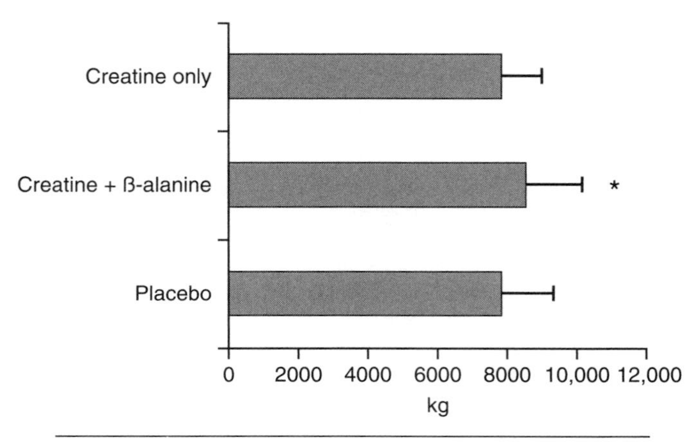

Figure 19.8 Comparison of effects of creatine and creatine plus β-alanine on training volume in the squat exercise. * = Significantly different from P and C.

Reprinted, by permission, from J.R. Hoffman, 2010, "Creatine and β-alanine supplementation in strength/power athletes," *Current Topics in Nutraceutical Research* 8: 19-32; Adapted from Hoffman 2006.

Sodium Bicarbonate

During high-intensity exercise, accumulation of hydrogen ions (H^+), primarily through increases in lactic acid within the muscle, lowers muscle pH. The increase in intracellular acidosis results in a decrease in force production, possibly through the impairment of excitation–contraction coupling and cross-bridge cycling and the inhibition of glycolytic enzymes (e.g., phosphofructokinase) needed for the resynthesis of ATP (Heigenhauser and Jones 1991). With the idea of countering this high level of intracellular acidosis, athletes have attempted to augment the bicarbonate buffering system of the body by ingesting **sodium bicarbonate**.

There has been much variability in studies examining the efficacy of **bicarbonate loading** for improving sprint and power performances. While some investigations have shown improved total work capacity (McNaughton et al. 1999) and a reduced fatigue rate during maximal-effort runs (Costill et al. 1984; Goldfinch, McNaughton, and Davies 1988; Wilkes, Gledhill, and Smyth 1983) or cycling exercise (Edge, Bishop, and Goodman 2006), an impressive number of studies have also failed to show any efficacy of bicarbonate loading in activity of similar duration and intensity (Gaitanos et al. 1991; Horswill et al. 1988; Katz et al. 1984; Kozak-Collins, Burke, and Schoene 1994; Tiryaki and Atterbom 1995). Several researchers have also examined the ability of sodium bicarbonate to delay fatigue during a resistance exercise session (Portington et al. 1998; Webster et al. 1993). These studies were unable to demonstrate any effect of sodium bicarbonate ingestion on the total number of repetitions performed during four or five sets of the leg press exercise (subjects were asked to perform approximately 12 repetitions at 70% to 85% of 1RM). It should be noted, however, that none of the subjects were competitive weightlifters or bodybuilders. Whether this supplement may be beneficial for competitive weightlifters is not known.

The variability of these studies may be related to differences in the metabolic demands of the training stresses or to differences in the dosing regimen. It appears that a dose of 0.3 g/kg sodium bicarbonate is necessary to have any beneficial effect during high-intensity performance between 1.0 and 7.5 min (Heigenhauser and Jones 1991). In addition, many of these studies examined noncompetitive athletes. As has often been demonstrated, there is a large degree of variability in the efficacy of ergogenic agents in the noncompetitive subject population. Several minor side effects are associated with sodium bicarbonate use, primarily gastrointestinal disturbances such as nausea, diarrhea, and bloating. Taking sodium bicarbonate with large volumes of fluid reduces the incidence of these disturbances (Heigenhauser and Jones 1991). Although no long-term side effects are associated with its use, one of the physiological effects of sodium bicarbonate ingestion is lowered plasma potassium concentrations. This could potentially result in cardiac arrhythmias (Heigenhauser and Jones 1991).

Sodium Citrate

Another alkalizing agent that can buffer increases in H^+ is sodium citrate. It appears that sodium citrate can increase blood pH without the gastrointestinal distress commonly seen with sodium bicarbonate (Van Someren et al. 1998). It is believed that once it is in the blood, sodium citrate is metabolized into bicarbonate, thus increasing extracellular pH (Tiryaki and Atterbom 1995). It appears that sodium citrate regulates intramuscular pH during high-intensity exercise via the same mechanism as sodium bicarbonate. However, in contrast to sodium bicarbonate, most studies have demonstrated that sodium citrate does not have any ergogenic benefit during short-duration, high-intensity exercise (Cox and Jenkins 1994; Van Someren et al. 1988). Linossier, Dormois, Bregere, and colleagues (1997) demonstrated that sodium citrate ingestion during maximal exercise lasting between 2 and 15 min does have an ergogenic benefit. Others have reported that a dose approximating 100 mg/kg body mass ingested 60 to 90 min before exercise can improve (about 20% greater) muscle endurance during maximal isometric knee extensions (Hausswirth et al. 1995).

Sodium citrate ingestion also appears to be ergogenic for the endurance athlete. Subjects in a study by Oopik and colleagues (2003) ingested 0.5 g sodium citrate per kilogram body mass 2 h before a 5K run time trial. The time required to complete the run was significantly faster in the sodium citrate group (1153 ± 74 s) than in the placebo group (1184 ± 91 s). In addition, lactate concentrations following the run were significantly higher in the sodium citrate group (11.9 ± 3.0 mmol/L) than in the placebo group (9.8 ± 2.8 mmol/L). Similar results were reported during a 3 km run in subjects using the same dosing regimen (Shave et al. 2001). The only concern mentioned was the gastrointestinal distress experienced by several of the runners.

Dietary Supplements for Energy

This section deals with supplements that are popular or have been demonstrated to be efficacious for individuals interested in enhancing energy, focus, and alertness. These supplements are popular with athletic and nonathletic populations.

Caffeine

Caffeine is one of the most widely used drugs in the world today. It is found in coffee, tea, soft drinks, chocolate, and various other foods. Caffeine is a central nervous system (CNS) stimulant, and its effects are similar to but weaker than those associated with amphetamines.

Mechanism of Action

Caffeine is used as an ergogenic aid by both aerobic and anaerobic athletes. However, the mechanism of action for these two groups of athletes may be quite different. In the aerobic athlete, caffeine is thought to prolong endurance exercise. The mechanisms proposed to cause this effect involve an increase in fat oxidation by mobilization of free fatty acids from adipose tissue or intramuscular fat stores (Acheson et al. 1980; Cox et al. 2002; Dulloo et al. 1989; Kovacs, Stegen, and Brouns 1998; Spriet 1995). The greater use of fat as a primary energy source slows glycogen depletion and delays fatigue. However, a number of researchers have questioned this mechanism (Graham 2001; Graham et al. 2000; Greer, Friars, and Graham 2000; Laurent et al. 2000).

During short-duration, high-intensity exercise, the primary ergogenic effect attributed to caffeine supplementation is enhanced power production. A number of possible mechanisms have been suggested to explain caffeine's effect on strength–power performance. These mechanisms include action on both the CNS and neuromuscular systems. One of the more significant effects on the CNS is caffeine's action as an adenosine antagonist, possibly delaying fatigue by binding to adenosine receptors and reducing adenosine's inhibitory effects (Astorino and Roberson 2010; Goldstein, Ziegenfuss, et al. 2010). In a meta-analysis, Warren and colleagues (2010) suggested that caffeine's effect on the CNS is the most likely source of improvement in strength–power performance through enhancement of muscle activation (motor unit recruitment). In addition, evidence supports caffeine as an analgesic, lowering pain and ratings of perceived exertion (Doherty and Smith 2005; Gliottoni and Motl 2008; Motl, O'Connor, and Dishman 2003). There is also evidence suggesting that caffeine produces an enhanced excitation–contraction coupling, affecting neuromuscular transmission and mobilization of intracellular calcium ions from the sarcoplasmic reticulum (Tarnopolsky 1994). Additionally, caffeine is thought to enhance the kinetics of glycolytic regulatory enzymes such as phosphorylase (Spriet 1995). The exact mechanism explaining caffeine's effect on strength–power performance is not clear at this time. However, evidence supports both central and peripheral factors as contributing to this effect.

Effect of Caffeine on Endurance Performance

Initial studies examining the effect of caffeine supplementation on endurance performance reported a 21 min improvement in the time to exhaustion (from 75 min in a placebo trial to 96 min in the caffeine trial) during cycling at 80% $\dot{V}O_2max$ (Costill, Dalsky, and Fink 1978). These results were confirmed by a number of additional studies demonstrating the ergogenic effect of caffeine during prolonged endurance activity (Essig, Costill, and Van Handel 1980; Graham and Spriet 1995; Ivy et al. 1979; Spriet et al. 1992). These studies showed that caffeine in doses ranging from 3 to 9 mg/kg (equivalent to approximately 1.5-3.5 cups of automatic drip coffee in a 70 kg person) produces a significant ergogenic effect.

Caffeine ingestion from a food source (coffee consumption) and anhydrous caffeine have both been shown to have significant performance effects; however, the extent of performance improvements appears to be greater when caffeine is ingested in tablet form (Graham, Hibbert, and Sathasivam 1998). When caffeine is provided as a pure caffeine supplement, the ergogenic benefit for improved endurance performance has been reported to range between 28% and 43% (Graham, Hibbert, and Sathasivam 1998; Graham and Spriet 1995). However, when caffeine is provided in a food source such as coffee (either caffeinated or added to decaffeinated coffee), the ergogenic benefit for endurance exercise may not be seen (Casal and Leon 1985; Graham, Hibbert, and Sathasivam 1998) or may be seen only at a reduced rate (Costill, Dalsky, and Fink 1978; Trice and Haymes 1995; Wiles et al. 1992). Graham and colleagues (1998) suggested that although the bioavailability of caffeine is the same whether it is consumed in a food source or in an anhydrous form, some compound in coffee antagonizes the action of caffeine. It appears that the ergogenic benefit is realized only when other nutritional ingredients are added to coffee. Hoffman and colleagues (2007) have shown that a nutritionally enriched coffee product containing 450 mg caffeine, 1200 mg garcinia cambogia (50% hydroxycitric acid), 360 mg citrus aurantium extract (6%), and 225 µg chromium polynicotinate can improve endurance performance 29% in endurance-trained subjects (figure 19.9).

Effect of Caffeine on Strength Performance

The evidence supporting caffeine alone as an ergogenic aid in strength, power, and anaerobic activities appears to be inconclusive. When caffeine is ingested in dosages of 5 to 6 mg/kg body weight, significant increases in acute strength and power performance, as well as increases in training volume, have been reported (Astorino et al. 2010, 2011; Bazzucchi et al. 2011; Duncan and Oxford 2011; Goldstein, Ziegenfuss, et al. 2010). A significant increase in maximal bench press strength has been observed in resistance-trained women following caffeine ingestion (Goldstein, Jacobs, et al. 2010). Others have shown that acute caffeine ingestion can improve peak torque and power in the first of two bouts of isokinetic knee extension–flexion exercise (Astorino et al. 2010) and

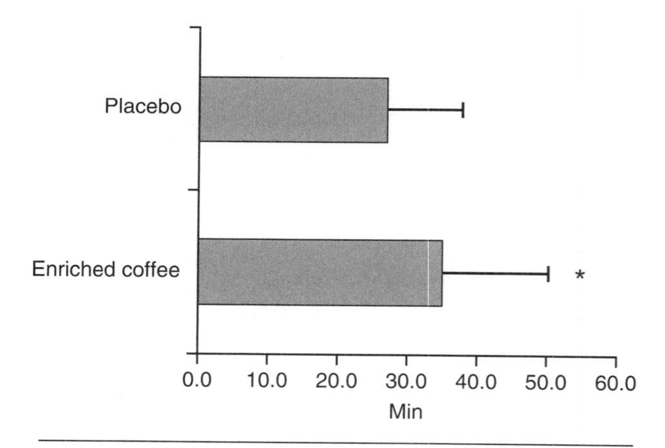

Figure 19.9 Time to exhaustion: * = enriched coffee significantly ($p < 0.05$) different from placebo. Data are reported mean ± SD.

Adapted, by permission, from J.R. Hoffman et al., 2007, Effect of nutritionally enriched coffee consumption on aerobic and anaerobic exercise performance," *Journal of Strength and Conditioning Research* 21: 456-459; Adapted from Hoffman et al. 2007.

increase contraction velocity during elbow flexion (Bazzucchi et al. 2011). Increases in training volume (repetitions performed) have also been demonstrated following acute caffeine ingestion in the first two sets of the leg press to exhaustion (Astorino et al. 2011), knee extension-flexion (Astorino et al. 2010), and bench press to exhaustion at 60% 1RM (Duncan and Oxford 2011). In contrast, no changes in 1RM in the bench press and leg extension exercises were seen in untrained male subjects following ingestion of 400 mg caffeine (Hendrix et al. 2010). Similarly, a 6 g/kg ingestion of caffeine was unable to enhance strength performance during maximal-effort isometric plantar flexion exercise (Fimland et al. 2010).

Effect of Caffeine on Sprint, Agility, and Reaction Time

Caffeine ingestion has been demonstrated to have an ergogenic effect on repeated sprint performance during both running and cycling modes of exercise (Fukuda et al. 2010; Lee, Lin, and Cheng 2011; Paton, Lowe, and Irvine 2010; Pontifex et al. 2010; Schneiker et al. 2006; Stuart et al. 2005). Acute caffeine intake before exercise has been reported to improve repeated sprint performance for subjects performing five sets of 6 × 20 m sprints (Pontifex et al. 2010), and repeated 30 s sprints on a cycle ergometer (Paton, Lowe, and Irvine 2010). Schneiker and colleagues (2006) reported significant improvements in sprint performance during two 36 min simulated game halves, with each half including 18 × 2 min blocks of 4 s sprints with 120 s recovery (both active and passive), following caffeine ingestion. This study simulated the physiologic demands of a team sport and suggests that caffeine may be able to maintain sprint performance during a competitive contest. In a similar study, Stuart and coauthors (2005) also observed greater repeated sprint performance during a simulated rugby contest following caffeine ingestion. Others have shown that acute ingestion of a proprietary blend of caffeine, creatine, and amino acids ingested 30 min before treadmill sprint drills resulted in an improvement in anaerobic running capacity at 110%, 105%, and 100% of peak velocity (Fukuda et al. 2010). Caffeine also appears to augment performance above levels seen with other supplements. Lee and colleagues (2011) showed that following a 5-day loading dose of creatine that increased repeated sprint performance, the addition of an acute intake of caffeine (6 mg/kg) was able to stimulate further improvements in power outputs during intermittent high-intensity sprints on a cycle ergometer.

Acute ingestion of caffeine also appears to have beneficial effects on agility performance and reaction time (Duvnjak-Zaknich et al. 2011; Stuart et al. 2005). Caffeine ingestion appeared to benefit competitive athletes by improving reactive agility in both fresh and fatigue situations (Duvnjak-Zaknich et al. 2011), including a rugby game (Stuart et al. 2005). However, this may be limited to trained individuals, as no ergogenic effect was observed in untrained individuals performing the pro-agility test following caffeine supplementation (Lorino et al. 2006).

Dose Response

There appears to be a dose response with caffeine supplementation in strength–power performance. The normal dose in most studies showing a positive effect of caffeine supplementation is 5 to 6 mg/kg body mass. This would mean that the average dose for a person weighing 175 lb (80 kg) would be approximately 400 mg caffeine. For comparison, a generic cup of drip coffee contains between 110 and 150 mg caffeine per 8 oz. (Higgins, Tuttle, and Higgins 2010). A 12 oz. Coca-Cola or Pepsi contains between 30 and 40 mg caffeine. Energy drinks typically contain 75 to 80 mg caffeine per 8 oz.; however, some contain as much as 174 mg per serving (Hoffman 2010b; Schneider and Benjamin 2011). No significant effects have been seen with low caffeine doses (up to 2.1 mg/kg body mass) in studies examining multiple sprints, grip strength, and repetitions in the bench press exercise (Bellar, Kamimori, and Glickman 2011; Campbell et al. 2010; Dragoo et al. 2011).

Caffeine Ingestion and Risk of a Positive Drug Test

Currently the World Anti-Doping Agency (WADA) lists caffeine as part of its monitoring program, meaning that caffeine levels are still tested for and reported following urine testing but that caffeine is not banned from use. Caffeine was removed from the WADA banned list in 2004. However, the NCAA still lists caffeine as a "drug subject to restrictions" if concentrations in urine exceed 15 μg/ml. The NCAA does not list an equivalent dosage required to test positive, since there are a number of confounding factors, but states that the cutoff is set so that moderate caffeine consumption will not cause a positive test. Van der Merwe and colleagues (1988) found that approximately 17.5 mg/kg caffeine, which would be equivalent to 8 to 10 cups of coffee, produced a urinary caffeine level of 14 μg/ml, still below the NCAA cutoff level of 15 μg/ml. In addition to caffeine, guarana seed extract is included in the stimulant group banned by the NCAA. Taurine, L-theanine, and green tea extract, are also listed as "Impermissible" by the NCAA, which means institutions may not supply these to their student-athletes. Intercollegiate athletes should check with their coaches and medical staff before consuming energy drinks. Athletes competing under the rules of the U.S. Anti-Doping Agency should also check this organization's rules before consuming energy drinks.

Energy Drinks

Research has indicated that energy drinks are the most popular supplement besides multivitamins in the American adolescent and young adult population (Froiland et al. 2004; Hoffman, Faigenbaum, et al. 2008). More than 30% of all American male and female adolescents use these supplements on a regular basis. Energy drinks are also reported to be the most popular supplement (41.7% of the 403 athletes surveyed) among young (17.7 ± 2.0 years) elite British athletes (Petroczi et al. 2008). The primary reason for their use is thought to be related to a desire to reduce or control body fat (Bell et al. 2004; Dodge and Jaccard 2006; Froiland et al. 2004; Hoffman, Faigenbaum, et al. 2008). However, many competitive athletes also use these drinks for their potential ergogenic effect. The basic active ingredient in energy drinks is caffeine; and although ergogenic benefits have been seen with caffeine supplementation in doses ranging from 3 to 9 mg/kg (equivalent to approximately 1.5-3.5 cups of automatic drip coffee in a 70 kg person), there appears to be a difference in ergogenic potential when caffeine is ingested in a food source (coffee or sports drink) versus its anhydrous form. Although both forms have been shown to have an ergogenic effect, the magnitude of performance improvements appears to be greater when caffeine is ingested in tablet form (Graham, Hibbert, and Sathasivam 1998). To maximize the effectiveness of caffeine in an energy drink, supplement companies often add several other ingredients to enhance the stimulatory potential of caffeine. The following sections briefly review the efficacy of energy drinks in relation to performance improvements and metabolic enhancement. Issues relating to the safety of energy drink consumption are also discussed.

Energy Drinks and Weight Loss

Caffeine alone, as discussed earlier in this chapter, is effective in enhancing lipolysis and fat oxidation and in reducing glycogen breakdown. However, when caffeine is combined with other thermogenic agents, its effectiveness in enhancing fat utilization is magnified (Diepvens, Westerterp, and Westerterp-Plantenga 2007; Haller, Jacob, and Benowitz 2004). In the past, caffeine was often combined with ephedra, and the combination was shown to be an effective supplement for increasing metabolic rate and stimulating fat loss (Boozer et al. 2002; Greenway et al. 2004). However, as a result of the 2004 FDA ban on ephedrine alkaloids, efforts were made to explore alternative therapeutic means to combat obesity. Synephrine, considered a mild stimulant, is thought to contribute to appetite suppression and increased metabolic rate and lipolysis (Fugh-Berman and Myers 2004), and has been used in many energy drinks in place of ephedrine. To maximize its effectiveness as a weight loss supplement, synephrine is often combined with other herbal products (Haller, Benowitz, and Jacob 2005). These products may include yohimbine, yerba mate extract, hordenine, phenylethylamine, and methyl tetradecylthioacetic acid, all of which have been shown to play a role in enhancing lipolysis and increasing energy expenditure (Andersen and Fogh 2001; Barwell et al. 1989; Galitzky et al. 1988). Many of these ingredients have multiple effects; for instance, phenylethylamine is an endogenous neuroamine that is often included in weight loss supplements to enhance mood. Several studies have shown that phenylethylamine can relieve depression and improve mood in clinical populations (Grimsby et al. 1997; Sabelli et al. 1996). Whether these ingredients can enhance mood in an apparently healthy population is not well established. Hoffman, Kang, Ratamess, and colleagues (2008) examined the efficacy of an energy drink containing phenylethylamine, among other ingredients, and despite showing greater fat utilization, they were unable to demonstrate any significant effect on mood state.

Studies on the effect of prolonged consumption of energy drinks on weight loss have yielded some promising results. In a clinical examination, Boozer and colleagues (2002) reported significant decreases in body mass and body fat, with positive alterations to lipid profiles, following 6 months of use of an ephedrine and caffeine supplement. Providing further support, another study, examining a combination of ephedra, caffeine, omega-3 fatty acids, and several vitamins for 9 months in women, showed significant decreases in body mass and body fat and improvements in various metabolic indices such as insulin sensitivity and lipid profiles (Hackman et al. 2006). A beneficial effect of energy drinks in short-duration studies and without the use of ephedra compounds has also been reported. Roberts and colleagues (2008) demonstrated that consuming an energy drink containing 200 mg caffeine, guarana extract, green tea leaf extract, glucuronolactone, ginger extract, and taurine for 28 days resulted in a significant decrease in body fat and body mass in healthy college-age participants. The benefits associated with multi-ingredient energy drink formulations have been observed in the absence of any other manipulation of diet or exercise habits. Nevertheless, there is no evidence to suggest that the use of an energy drink alone, without any dietary or exercise intervention, is a recommended method of weight and body fat management.

Energy Drinks and Athletic Performance

Energy drinks containing caffeine and other synergistic ingredients appear to have positive effects on athletic performance. Several studies have demonstrated that ingestion of an energy drink before exercise can delay fatigue and improve the quality of a resistance training workout (Gonzalez et al. 2011; Hoffman, Ratamess, Ross, Shanklin, et al. 2008; Ratamess, Hoffman, et al. 2007). The combination of 450 mg caffeine, 1200 mg garcinia cambogia (50% hydroxycitric acid), 360 mg citrus aurantium extract (6%), and 225 μg chromium polynicotinate in an enriched coffee drink significantly enhanced time to exhaustion during cycle ergometer exercise by 29% compared to the value in subjects consuming decaffeinated coffee (Hoffman, Kang, et al. 2007). However, in the same study, no difference in anaerobic power performance was noted between subjects consuming the supplement and those consuming the placebo (decaffeinated coffee). These results were similar to findings in another study, which showed that an energy supplement (Red Bull energy drink) increased upper body muscle endurance but had no effect on power performance during repeat Wingate anaerobic power tests (Forbes et al. 2007). Improvements in training volume (defined as total number of sets × repetitions in a resistance training workout) have been seen following ingestion of an energy drink (Gonzalez et al. 2011; Hoffman, Ratamess, Ross, Shanklin, et al. 2008). The combination of caffeine, L-taurine, glucuronolactone, and branched chain amino acids consumed 10 min before a resistance exercise session enhanced acute exercise performance by increasing the number of repetitions and the total volume of exercise performed during the session. The greater volume of training augmented both the growth hormone and insulin response to exercise (Hoffman, Ratamess, Ross, Shanklin, et al. 2008), indicating that consumption of this preexercise energy supplement enhanced the anabolic response to the training session (figure 19.10).

Energy drinks improve endurance performance and the quality of a resistance exercise workout; however, many athletes use these drinks primarily for their stimulatory effect, specifically to enhance focus, alertness, and reaction time. The data supporting this effect are limited, but they provide evidence to support many of the empirical claims made by athletes. The popular energy drink Red Bull has been shown to enhance cognitive performance through improved choice reaction time, concentration, and memory, reflecting improved alertness (Alford, Cox, and Westcott 2001). Others have shown that an energy drink containing caffeine and several herbal and botanical compounds (including evodiamine, N-acetyl-L-tyrosine, hordenine, 5-hydroxytryptophan, potassium citrate, N-methyl tyramine, sulbutiamine, vinpocetine, yohimbine HCl, and St. John's wort extract) can increase focus, alertness, and reaction time. In addition, improvements in reaction time to visual and auditory stimuli were noted. Interestingly, despite a significant improvement in reaction ability, no significant improvements were observed in anaerobic power performance as assessed by repeated Wingate anaerobic power tests. Thus, energy drinks may have a greater role in focusing an athlete and improving exercise duration but less effect on strength or power performance.

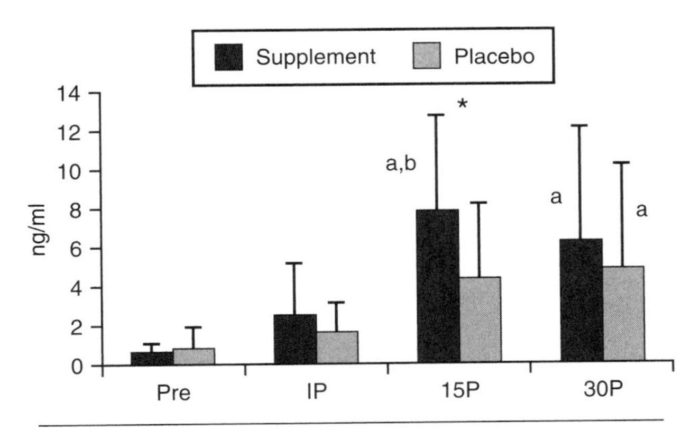

Figure 19.10 Increases in workout training volume after ingestion of an energy drink result in significant elevations in the serum growth hormone response postexercise (mean ± SD) at various time points. * = Significantly different between the supplement and placebo exercise sessions; a = significantly different from Pre; b = significantly different from IP. Pre, preexercise; IP, immediately postexercise; P, postexercise.

Adapted, by permission, from J.R. Hoffman et al, 2008, "Effect of a pre-exercise 'high-energy' supplement drink on the acute hormonal response to resistance exercise," *Journal of Strength and Conditioning Research* 22: 874-882.

Safety Issues Related to Energy Drink Consumption

As mentioned earlier, caffeine is a mild stimulant and is commonly found in coffee, tea, chocolate, and soft drinks. The concentration of caffeine in these products ranges from 40 to 150 mg (Lieberman 2003). In contrast, the top-selling energy drinks have caffeine levels that range from 75 to 174 mg per serving; in some of the higher-caffeine energy drinks, levels may exceed 500 mg per serving (Reissig, Strain, and Griffiths 2009). The adverse effects seen with caffeine in these doses include insomnia, nervousness, headache, and tachycardia (Clauson et al. 2003). However, the blood pressure response to energy drink consumption appears to be inconclusive. Several studies have reported significant elevations in systolic blood pressure (Hoffman, Kang, et al. 2006; Taylor et al. 2007), while others have shown no change (Alford, Cox and Westcott 2001; Dalbo et al. 2008; Hoffman, Kang, Ratamess, et al. 2008; Roberts et al. 2007). Differences between the studies are not clear, but they are likely related to differences in the combinations of ingredients generally associated with these energy drinks. These studies have consistently shown no alterations in diastolic blood pressure. The ingredients usually added to energy drinks, such as guarana, ginseng, and taurine, are in concentrations far below the amounts associated with adverse events (Clauson et al. 2003). However, a concern would apply to energy drinks that contain ephedra alkaloids or other β-agonists such as citrus aurantium (e.g., synephrine) that may present a higher risk for an exaggerated sympathetic response.

Several case reports have indicated that energy drinks may increase the risk for ventricular tachycardia (Nagajothi et al. 2008) or myocardial ischemia (Berger and Alford 2009). Others, though, report no association between caffeine consumption and cardiac conduction abnormalities (Katan and Schouten 2005). A greater risk for cardiac events may be present if energy drinks are consumed with alcohol. Individuals who are predisposed to cardiac arrhythmias may be at increased risk for a significant adverse event if they combine alcohol with an energy drink (Wiklund et al. 2009).

A concern that energy drinks can increase the risk for dehydration has been raised based on evidence that caffeine can induce dieresis and natriuresis (Riesenhuber et al. 2006). However, in several well-designed studies, caffeine consumption did not impair hydration, exacerbate dehydration, or impair thermoregulation (Del Coso, Estevez, and Mora-Rodriguez 2009; Grandjean et al. 2000). In addition, it has been surmised that caffeine does not reduce exercise–heat tolerance or increase the risk for hyperthermia (Armstrong, Casa, Maresh, et al. 2007). However, concern should be directed toward energy drinks that contain ephedra or other β-agonistic compounds. This combination of ingredients may pose a risk for both athletes and nonathletes during high intensity activity performed in the heat. Several well-documented heat deaths of professional athletes who were using ephedra contributed to the banning of this herbal ingredient in 2004. It may be prudent to advise against the use of energy drinks that contain these ingredients in individuals who are poorly conditioned or overweight and exercise in the heat.

Additional concerns recently raised about energy drinks pertain to issues relating to dependence, withdrawal, and tolerance (Reissig, Strain, and Griffiths 2009). Although many of these issues have been studied in connection with caffeine, direct studies in connection with energy drink consumption are limited. There is considerable debate about whether caffeine can produce a dependence syndrome similar to that associated with narcotics. A few studies have suggested that habitual caffeine users may fulfill diagnostic criteria for substance dependence (Hughes et al. 1998; Oberstar, Bernstein, and Thuras 2002); however, evidence documenting such behaviors in individuals consuming energy drinks is lacking. Symptoms of caffeine withdrawal include headache, tiredness or fatigue, sleepiness, and irritability. Whether these symptoms are also associated with cessation of energy drink consumption is not known. Another concern, though, is tolerance. For athletes who use energy drinks on a regular basis, the issue of tolerance may have important implications as the competitive season progresses. Although high caffeine ingestion has been associated with tolerance (Reissig, Strain, and Griffiths 2009), no studies to date have examined the issue of tolerance in relation to energy drinks.

L-Carnitine

L-carnitine is synthesized from the amino acids lysine and methionine and is responsible for the transport of fatty acids from the cytosol into the mitochondria to be oxidized for energy (Kerner and Hoppel 2000). Consequently it is used by many as a weight loss supplement. The supporting physiology relates to its mechanism of action. Fatty acids are first bound to coenzyme A (CoA) via a thioester bond in the cytosol of the cell. This reaction is catalyzed by fatty acyl-CoA enzyme synthetase. The acyl group on CoA is then transferred to carnitine to form acyl-carnitine, which is then transported into the mitochondrial matrix for oxidation. The role of carnitine in lipid oxidation has generated interest in its efficacy as a dietary supplement, primarily to enhance exercise performance by increasing fat utilization and sparing muscle glycogen. However, studies on L-carnitine's role in enhancing lipid oxidation have not shown clear efficacy in either human or rat models (Aoki et al. 2004; Arenas et al. 1991; Brandsch and Eder 2002). Although some investigators have reported enhanced fatty acid oxidation following 3 weeks of L-carnitine supplementation (Bacurau et al. 2003), most studies have been unable to demonstrate elevated muscle carnitine levels after supplementation (Arenas et al. 1991; Barnett et al. 1994). As such, this likely limits the ability of carnitine to enhance intracellular fatty acid transport. This may be related in part to limits on the amount of carnitine that can be absorbed through oral supplementation (Hultman, Cederblad, and Harper 1991), or potentially related to limits on the amount of fat that can be transported into the mitochondria through the carnitine system due to feedback regulators within skeletal muscle such as malonyl CoA, which is a product of metabolism.

Interestingly, several studies have suggested that L-carnitine may enhance recovery from exercise (Giamberardino et al. 1996; Spiering et al. 2007; Volek et al. 2002). Decreases in pain and muscle damage (Giamberardino et al. 1996), decreases in markers of metabolic stress (Spiering et al. 2007), and an enhanced recovery (Volek et al. 2002) have been demonstrated following high-intensity resistance exercise in untrained or recreationally trained individuals. Proposed mechanisms involve enhancing blood flow regulation through an enhanced vasodilatory effect that reduces the magnitude of exercise-induced hypoxia (Hulsmann and Dubelaar 1988, 1992). In addition, Kraemer, Spiering, and colleagues (2006) indicated that L-carnitine supplementation (2 g/day for 3 weeks) upregulates androgen receptors and increases insulin-like growth factor (IGF) binding proteins that preserve IGF-1 concentrations (Kraemer et al. 2003). These endocrine adaptations from the supplement can have an important role in the enhanced recovery seen following high-intensity exercise.

Up to 3 g daily L-carnitine supplementation (for 3 weeks) appears to be well tolerated in healthy volunteers, with no subjective, hematologic, or metabolic adverse events reported (Rubin et al. 2001). As with most supplements, though, this information should not be extrapolated to apply to greater doses or more prolonged supplementation periods.

Popular Dietary Supplements That May Have Ergogenic Potential

The sport supplement industry is an ever-growing industry. This section focuses on relatively new supplements that have made ergogenic claims and have started to appear in the scientific literature.

Betaine

Betaine is a trimethyl derivative of the amino acid glycine. It is a significant component of many foods, including wheat, spinach, beets, and shellfish (Zeisel et al. 2003). It is estimated that the daily intake of betaine in the human diet ranges from an average of 1 g/day to a high of 2.5 g/day in individuals who consume a diet high in whole wheat and shellfish (Craig 2004). In addition, betaine can be synthesized in the body through the oxidation of choline-containing compounds (Craig 2004). The physiological functions attributed to betaine include its action as an osmoprotectant (Eklund et al. 2005). That is, it protects the cell against dehydration by acting as an osmolyte, thereby increasing the water retention of cells. Other studies have indicated that betaine supplementation may lower plasma homocysteine concentrations (Olthof and Verhoef 2005; Olthof et al. 2003) and reduce inflammation (Detopoulou et al. 2008), perhaps decreasing cardiovascular disease risk. In addition, it has been suggested that betaine acts as a methyl donor by providing a methyl group to guanidinoacetate via methionine that can synthesize creatine in skeletal muscle (du Vigneaud et al. 1946).

Animal studies have reported that betaine supplementation can increase muscle creatine concentrations (Zahn et al. 2006). However, no known studies have examined changes in muscle creatine concentrations in humans supplementing with betaine. It is thought that the donation of methyl groups from betaine occurs via a series of enzymatic reactions in the mitochondria of liver and kidney cells (Delgado-Reyes, Wallig, and Garrow 2001). Betaine donates a methyl group to homocysteine to form methionine. Methionine is converted to S-adenosylmethionine (SAM), which acts as a methyl donor contributing to the synthesis of creatine as well as a number of other proteins (Craig 2004). Dietary betaine has been shown to increase serum methionine, transmethylation rate, and methionine oxidation in healthy men (Storch, Wagner, and Young 1991); and animals injected with betaine have shown a dose–response increase in red blood cell SAM (Wise et al. 1997). A study by del Favero, Roschel, Artioli, and colleagues (2012) was unable to support the ability of betaine to increase muscle creatine concentrations following 2 g/day ingestion for 10 days in untrained subjects. In addition, no performance changes were noted. Thus further research on the precise mechanism of action appears warranted. Despite some questions regarding the physiological mechanisms involving betaine, some ergogenic effect appears to be associated with its ingestion.

Armstrong and colleagues (2008) examined the effect of acute betaine ingestion following a dehydration protocol and prolonged treadmill running (75 min at 65% of $\dot{V}O_2max$) in the heat. Following the treadmill running, subjects performed a sprint to exhaustion. The investigators were unable to report any ergogenic benefit with regard to time to exhaustion in the sprint test. In addition, they reported a greater loss in plasma volume in subjects consuming fluids with betaine than in subjects consuming fluids that did not contain betaine. In an examination on the efficacy of betaine supplementation for strength and power performance, Lee and colleagues (2010) examined 14 days of betaine ingestion in recreationally trained men. They found no significant changes in repetitions performed in the squat or bench press exercise, but they did find significant improvements in bench press throw power, isometric bench press force, vertical jump power, and isometric squat force. A recent study suggested that 2 weeks of betaine ingestion could significantly improve muscle lower body endurance by increasing the number of repetitions performed in the squat exercise, and also improve the quality of the workout by increasing the number of repetitions performed at 90% of the subject's maximal mean and peak power outputs (Hoffman, Ratamess, Kang, et al. 2009). These improvements occurred within 1 week of supplementation. This effect was not seen in the upper body measure or in other measures of anaerobic power (Wingate test, vertical jump test, or bench press throw). Previous research suggested that betaine supplementation may enhance mood in a clinical population suffering from motor neuron disease (Liversedge 1956). However, Hoffman and colleagues (2009) were unable to provide any support for improved mood or reduction in soreness ratings with 2 weeks of betaine ingestion. Additional research is needed to further explore the potential benefits of betaine on mood. A subsequent study by the same research team confirmed the earlier results showing that 2 weeks of betaine ingestion did not alter subjective feelings of fatigue or energy (Hoffman et al. 2011).

Clinical studies have not reported any adverse effects associated with betaine supplementation. Human and animal studies have demonstrated no toxic or carcinogenic effects in 52- and 104-week trials, respectively. Research on the efficacy of betaine supplementation is limited. Published studies have indicated that there are some potential benefits following 2 weeks of betaine supplementation in active college males. However, additional research is warranted to determine the rate of muscle creatine synthesis from betaine supplementation and to compare muscle creatine synthesis kinetics with creatine supplementation versus betaine supplementation.

Choline

Choline is an essential nutrient that has an important function in synthesis of the neurotransmitter acetylcholine. Neurons are unable to synthesize choline and rely on dietary intake to produce sufficient acetylcholine (Amenta and Tayebati 2008). Acetylcholine is critical for many physiological functions; and any deficiency could result in a multitude of physiological problems involving, for example, cell membrane signaling (phospholipids), lipid transport (lipoproteins), and methyl group metabolism (homocysteine reduction). One of the more interesting findings from the research is the benefit of choline supplementation for memory and cognition improvements (Buchman et al. 2001; Canal et al. 1991; DiPerri et al. 1991; Gossell-Williams et al. 2006).

The importance of enhancing neurotransmitter function has interesting implications for athletic performance. If choline can improve neurotransmitter concentration, it likely has a potential ergogenic role in athletic events that involve reaction time and power, even at times when plasma choline concentrations are normal. Choline provided as phosphatidylcholine is 12-fold more effective than inorganic choline salts in increasing serum concentrations and maintaining elevated concentrations for a longer duration (12 h vs. 30 min) (Hirsch, Growdon, and Wurtman 1978; Wurtman, Hirsch, and Growdon 1977). Thus, most supplement studies provide choline as phosphatidylcholine or L-alpha glycerylphosphorylcholine (alpha-GPC), a water-soluble form lacking the hydrophobic tail groups. Phosphatidylserine is also a phospholipid, found in the membrane of organs with high metabolic activity such as brain, heart, lung, liver, and skeletal muscle (Blokland et al. 1999; Starks et al. 2008). Phosphatidylserine has been shown to reduce inflammation (Huynh, Fadok, and Henson 2002; Monteleone et al. 1990) and to act as an antioxidant (Dacaranhe and Terao 2001; Lactorraca et al. 2003). These results have led to further study on the efficacy of phosphatidylserine to enhance recovery from exercise. Daily ingestion of 750 mg phosphatidylserine for 7 days before exhaustive exercise resulted in significant improvements in sprint performance (Kingsley et al. 2005) and an increase in time to exhaustion during intermittent cycling exercise (Kingsley, Miller, et al. 2006). Starks and colleagues (2008) reported a lowered stress response (decrease in cortisol concentrations) to moderate-intensity cycling exercise (65-85% $\dot{V}O_2max$) following 10 days of supplementation with 600 mg phosphatidylserine. However, Kingsley, Kilduff, and colleagues (2006) were unable to support improved recovery in individuals performing an acute bout of eccentric exercise (downhill running).

The type of activity that could benefit from choline supplementation may be prolonged endurance events. Exercise that reduces plasma choline supplementation (e.g., marathon running) may benefit from choline supplementation. However, researchers have been unable to provide evidence to support this hypothesis (Deuster et al. 2002; Warber et al. 2000). This may be related to the inability of prolonged exercise to actually deplete plasma choline concentrations (Deuster et al. 2002). As discussed earlier, the role of choline and neurotransmitter formation suggests that choline supplementation could potentially influence reaction time, especially during fatiguing events. Hoffman, Ratamess, Gonzalez, and colleagues (2010) gave subjects a daily supplement containing 150 mg α-glycerophosphocholine, 125 mg choline bitartrate, and 50 mg phosphatidylserine for 4 weeks. Following the initial ingestion, the group that consumed the supplement maintained reaction time to both visual and auditory stimuli following a high-intensity bout of exhaustive exercise, while subjects consuming a placebo experienced significant reductions. In addition, acute ingestion of the supplement resulted in maintained focus and alertness following exhaustive exercise, while subjects consuming a placebo experienced significant declines. Interestingly, after 4 weeks of supplementation, both groups exhibited significant declines in reaction performance. It appeared that some habituation had occurred during the 4-week ingestion period. Further research in this area seems warranted.

Quercetin

Polyphenolic compounds or polyphenols occur naturally in plant foods. There are two main classes of polyphenols; flavonoids and phenolic acids. Flavonoids are present in most fruits and vegetables (Manach et al. 2005) and are associated with a lower incidence of heart disease and certain types of cancers (Boots, Haenen, and Bast 2008). Flavonoids also are antioxidants (Boots, Haenen, and Bast 2008) and reportedly have anti-inflammatory effects (Manach et al. 2005). In addition, flavonoids may stimulate muscle mitochondrial biogenesis (Davis et al. 2009). These physiological effects suggest that flavonoids may have potential ergogenic implications for the endurance athlete.

Quercetin is the most abundant flavonoid, with broad biological effects (Chun, Chung, and Song 2007). Humans can absorb significant amounts of quercetin without any adverse effects (Harwood et al. 2007). Quercetin has a half-life ranging between 3.5 and 28 h (Manach et al. 2005; Moon et al. 2008). This is important with regard to ingestion times surrounding exercise. Nieman and colleagues (2009) provided trained cyclists 1000 mg quercetin for 14 days before and during a 3-day period of intensified exercise and for 7 days after the training period (total supplement duration was 24 days). The immune and inflammatory responses were significantly lowered in the supplement group, but no performance differences were noted. Despite the lack of any performance difference, the results indicate that quercetin may be effective in countering exercise-induced inflammation. In a study on untrained males, 2 weeks of quercetin supplementation (1000 mg/day) resulted in significant improvements in a 12 min time trial on a treadmill and a nonsignificant increase (*p* ranging from 0.08 to 0.192) in mRNA expression of four genes related to skeletal muscle mitochondrial biogenesis (Nieman et al. 2010). The differences between the studies may have to do with differences in training experience. Quercetin may have greater relevance to performance improvements in the untrained individual but greater relevance to enhancement of recovery in the trained athlete. This is an interesting supplement on which future research will yield more information.

Summary

The use of ergogenic aids is prevalent among athletes at all levels and in most sports. Many athletes appear to begin their supplementation routines at the high school level. This chapter reviews a number of the more popular supplements in use today. It was beyond the scope of the chapter to provide background on every supplement available for purchase. However, it is strongly recommended that readers have a clear understanding of the physiology governing potential ergogenic effects and know the potential risks associated with the ingredients in a supplement. In work with competitive athletes, it is imperative to know whether ingestion of a supplement can lead to a positive drug test. Students and professionals already employed in the field are encouraged to critically examine the research on supplementation. For the most part, studies are not conducted with the population that the supplement is intended for. Most of the research is performed on recreationally trained subjects, which results in a great deal of variability in the outcomes. Consequently, a study concluding that a particular agent does not confer any ergogenic benefit may unfortunately not be applicable to the population that is actually using the supplement.

REVIEW QUESTIONS

1. What are the physiological differences between whey and casein protein? Discuss how these proteins can be used to maximize muscle remodeling.
2. Describe the difference between essential and nonessential amino acids with regard to protein synthesis.
3. Provide the physiological rationale for the ability of creatine and β-alanine to enhance the training response in strength–power athletes.
4. What are the risks associated with creatine and β-alanine supplementation, if any? Discuss each supplement individually.
5. Discuss nutrient intervention to enhance buffering capacity during high-intensity anaerobic exercise.
6. What is the potential ergogenic role of caffeine? Is it legal to use this product at any concentration? Explain your answer.
7. What is the frequency of use of energy drinks among various population groups, and what are the primary reasons for their use?
8. What are the potential mechanisms that may support betaine's role as an ergogenic aid?
9. Describe the potential role that choline may have as an ergogenic aid.

chapter 20

Performance-Enhancing Drugs

CHAPTER OBJECTIVES

After reading this chapter you should be able to do the following:

- Understand how exogenous use of testosterone can enhance muscle growth and athletic performance and whether the use of testosterone precursors has similar efficacy.
- Understand the benefits associated with the clinical use of androgens.
- Describe the risks associated with exogenous testosterone administration.
- Understand whether exogenous administration of growth hormone is ergogenic and what potential risks are associated with its use.
- Determine the efficacy of central nervous system stimulants.
- Understand the physiological outcomes and potential risks associated with blood doping and erythropoietin use.

The use of performance-enhancing drugs by athletes has received a lot of attention in the media over the past decade. The publicity has incorrectly given the impression that performance-enhancing drugs are a relatively new phenomenon. For the past half-century, drug use by athletes has been an issue. This has led to both medical and scientific focus on the efficacy and dangers of these compounds. The medical risks linked with these drugs, as well as associated ethical considerations, have led the major sport governing bodies to institute measures to prevent their use. As part of educational and awareness programs, much effort has focused on combating use of these drugs through better understanding of their efficacy and risks. Many of these drugs are dual use, meaning that they are often a part of treatment options for various diseases and illnesses, but their physiological effects have an ergogenic potential (potential to improve athletic performance). This chapter focuses on some of the more popular performance-enhancing drugs used by strength–power athletes, bodybuilders, and endurance athletes. It covers their physiological roles, their efficacy, and risks associated with their use.

Anabolic Steroids

The use of anabolic steroids has received a tremendous amount of attention for the past 10-plus years. However, competitive athletes have used anabolic steroids for many decades. Anabolic steroids are synthetic derivatives of the male sex hormone testosterone. They have both anabolic (referring to increase in muscle mass and strength) and androgenic (referring to development of male secondary sex characteristics including hair growth) properties. However, these compounds are unable to confer an isolated physiological effect and generally have varying degrees of anabolic or androgenic (masculinizing) effects. Given the diversity of these compounds, the term *androgens* is believed to be more appropriate than *anabolic steroids* (Hoffman, Kraemer, et al. 2009). The physiological actions of testosterone are reviewed in chapter 2.

The physiological changes regulated by testosterone have made it one of the drugs of choice for strength–power athletes or athletes interested in increasing muscle mass. However, testosterone itself is a very poor ergogenic aid. Rapid degradation is seen when testosterone is given through either oral or parenteral (injectable) administration (Hoffman, Kraemer et al. 2009). Thus it became necessary to chemically modify testosterone to retard the degradation process, thereby maintaining effective blood concentrations for longer periods of time and achieving androgenic and anabolic effects at lower concentrations of the drug (Hoffman, Kraemer, et al. 2009). Anabolic steroid use through either oral or parenteral administration became possible after these modifications. Examples of commonly used oral and injectable anabolic steroids are listed in table 20.1.

Slight biochemical modifications to the testosterone molecule can alter the biological activity of the compound. Adding a methyl or ethyl group to the molecule (specifically at the 17-α-carbon position) can extend the half-life of the compound when it is administered orally (avoiding degradation during its first pass through the liver), but causes greater changes in the blood lipid profile (decrease in high-density lipoprotein cholesterol) (Dimick et al. 1961). Another biological modification to testosterone is linking an ester (carboxylic acid group) to the 17-β hydroxyl group. The ester molecules differ in size, shape, and function, which affects the rate at which the androgen is released from the tissue (Hoffman, Kraemer, et al. 2009). The larger the ester, the slower the release into the circulation. The ester also increases fat solubility and reduces solubility in water. As a result, the degradation rate is very slow (dependent on the size of the ester). However, while the compound is attached to the ester, it has no biological activity. Only when the ester is detached (via the esterase enzyme) does the molecule become active. Examples of common testosterone esters include testosterone propionate, testosterone cypionate, and testosterone enanthate.

Effect of Androgens on Size and Strength Gains

After early reports that anabolic steroids were widely used by Eastern European athletes and were becoming increasingly prevalent in other countries, scientific and medical communities began investigating their effect on strength and muscle mass. Initial studies were unable to see any significant differences in strength or body mass gains in subjects taking anabolic steroids compared with subjects taking a placebo (Fahey and Brown 1973; Fowler, Gardner, and Egstrom 1965; Golding, Freydinger, and Fishel 1974; Loughton and Ruhling 1977; Stromme, Meen, and Aakvaag 1974). As a result, the scientific and medical community at the time suggested that anabolic steroids had little influence on athletic performance. This was contrary to the anecdotal reports emanating from gyms and caused a bit of a credibility gap between the researchers and athletes. However, examination of these studies revealed several methodological flaws. Several of the

Table 20.1 Commonly Used Orally and Parenterally Administered Anabolic Steroids

Oral compounds		Parenteral compounds	
Trade names	**Generic names**	**Trade names**	**Generic names**
Dianabol	Methandrostenolone	Deca-Durabolin	Nandrolone decanoate
Anavar	Oxandrolone	Delatestryl	Testosterone enanthate
Anadrol	Oxymetholone	Depo-Testosterone	Testosterone cypionate
Winstrol	Stanozolol	Durabolin	Nandrolone phenylpropionate
Maxibolin	Ethylestrenol	Primobolan Depot	Methenolone enanthate
Halotestin	Fluoxymesterone	Parabolan	Trenbolone acetate

investigations used a physiological dosage instead of the suprapharmacologic dosages typically used by androgen users. In essence, the subjects were shutting down their own endogenous production and replacing it with an exogenous source. Another flaw concerned the method of strength assessment. In some of these studies, performance was assessed by a mode of exercise that was different from the training stimulus. This lack of specificity may have masked any possible training effects. In addition, several of these studies used subjects who had minimal resistance training experience. As discussed earlier in this book, subjects with limited training experience have a large potential for strength gains, negating any need for a performance-enhancing supplement. When androgens are administered to experienced resistance-trained athletes in the dosages commonly used, results appear to confirm the anecdotal claims of superior performance and size improvements.

Examining the efficacy of androgens in athletic populations is difficult, as most institutional review boards, especially in the United States, view the administration of androgens in nonclinical settings as unethical. The majority of studies examining androgen administration to both clinical populations and experienced resistance-trained athletes have reported significant strength and body mass gains (Alen, Hakkinen, and Komi 1984; Alen, Reinila, and Vihko 1985; Bhasin et al. 1997; Boone et al. 1990; Friedl et al. 1991; Hervey et al. 1981; O'Shea 1971; Stamford and Moffatt 1974; Ward 1973). The magnitude of performance (muscle size and strength) gains appears to be dose related. Comparisons of 100 and 300 mg/week testosterone enanthate or 100 and 300 mg/week nandrolone, administered for 6 weeks in physically active men, showed that the largest gains in body weight and isokinetic muscle strength were seen in most cases with the larger doses (Friedl et al. 1991). Bhasin and colleagues (1997) examined the effects of a high dose (600 mg/week) of testosterone enanthate in resistance-trained men and reported that the combination of androgen administration and resistance training produced greater increases in lean body mass and muscle strength than training with a placebo or androgen administration alone with no resistance training. Maximum squat and bench press strength increased 38% and 22%, respectively, in the combination group compared to ~20% and 10% in the placebo plus resistance training and androgen administration-only groups. Additional studies administering 25, 50, 125, 300, or 600 mg/week of testosterone enanthate (with no resistance training) to resistance-trained men over 20 weeks showed that increases in lean body mass, type I and II muscle fiber area, thigh muscle volume, and strength were greatest at the higher doses, that is, at least 125 mg/week (Bhasin et al. 2001; Sinha-Hikim et al. 2002; Woodhouse et al. 2003).

The use of androgens by experienced strength–power athletes or bodybuilders, in whom small performance or muscle hypertrophy gains are quite difficult to achieve, appears to provide a significant performance advantage. Kuipers and colleagues (1993), administering 100 mg/week of nandrolone decanoate or placebo to bodybuilders for 8 weeks, reported that muscle fiber area increased substantially only in the bodybuilders using steroids. Another study, on the effect of a suprapharmacologic dose of androgens compared to a more therapeutic dose (200 mg/week of nandrolone decanoate), reported a 12.6% increase in the deltoid muscle fiber area in the bodybuilders, whereas administration of the therapeutic dose did not increase fiber area (Hartgens et al. 2002). Another investigative team administered 3.5 mg/kg body mass of testosterone enanthate or a placebo during a 12-week periodized resistance training program and reported greater increases in body mass and bench press strength (22% vs. 9%) in the androgen-treated group compared to the placebo group (Giorgi, Weatherby, and Murphy 1999). Others have indicated that the gains in strength may be two- to threefold higher (figure 20.1) in athletes using androgens than in similarly trained athletes who are not using them (Hervey et al. 1981; Ward 1973). These studies were of relatively short duration (5-12 weeks). The strength differences observed were remarkable considering the relatively short training regimen. Studies on longer durations of androgen administration are limited. Alen and colleagues (1985) examined competitive powerlifters during 26 weeks of androgen administration; body mass increased more than 5 kg in the group using the drugs, while the control group, consisting of both powerlifters and bodybuilders, did not achieve any increase in body mass during the same time period. A case study by Alen and Hakkinen (1985) reported on changes in an elite bodybuilder who was followed for 1 year of self-administration of anabolic steroids. During the phases when the bodybuilder took anabolic steroids, strength levels increased. During cycles in which the athlete trained drug free, strength levels decreased. Although well-controlled studies using similar subject populations are limited, the results reported in this study are consistent with many of the anecdotal reports of these athletes.

In addition to the greater gains in strength, power, and size experienced by athletes using androgens, another reason for use may be related to the potential

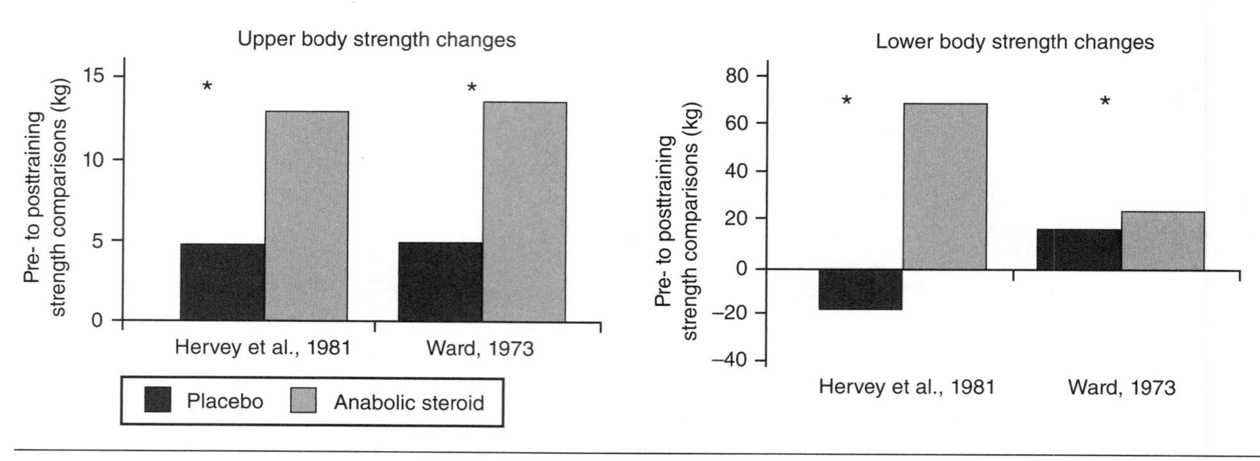

Figure 20.1 Effect of short-term anabolic steroid administration on strength gains. * = Significant difference between groups.

Adapted from Hervey et al. 1981; Ward 1973.

for increased recovery. The principal mechanisms responsible for increasing strength and lean body mass are enhancement of protein synthesis and inhibition of the potential catabolic effects associated with high-intensity training. The anticatabolic activity of androgen administration is reflected by changes in the testosterone-to-cortisol ratio (T/C ratio) (Rozenek et al. 1990), but more importantly may have a direct effect in suppressing catabolic activity by blocking the catabolic signaling pattern in muscle (Zhao et al. 2004). These changes can give the athlete the ability not only to maintain a higher intensity and volume of training but also to enhance the recovery processes between exercise sessions. If the athlete can train harder and for a longer duration, the stimulus presented to the muscle will result in a greater physiological adaptation but also allow the athlete to return to a high level of play much faster. This may have been one of the motivations for the apparent rampant use of androgens by Major League Baseball players during the 1990s and the early part of the 21st century.

Typical Use of Androgens

Exogenous androgen administration in physiological dosages (i.e., in which the amount of androgen administered is similar to the concentrations produced by endogenous sources such as the testes and adrenal glands) serves only to shut down the production of androgens in the body through a negative feedback mechanism. Such dosing would only replace the body's own endogenous production. Thus, the athlete needs to take the drug in concentrations greater than those that the body can normally produce by itself. Forbes (1985) was one of the first scientists to develop a dose–response curve of the effect of androgens. He reported that the total dose of androgens ingested appears to have a logarithmic relationship to increases in lean body mass (figure 20.2). Athletes typically use androgens in a **stacking** fashion, in which several different drugs are administered simultaneously. The rationale for stacking is that the potency of one anabolic agent may be enhanced when it is consumed at the same time as another anabolic agent (Hoffman, Kraemer, et al. 2009).

The most popular way of using androgens is through oral and parenteral (injectable) means; however, the vast majority of individuals (77%) self-administer androgens via an injection (Cohen et al. 2007). This is related to the user's belief that self-injecting is a safer or healthier method of administration and that it may have a better effect (Cohen et al. 2007). Most users take androgens in a cyclic pattern, meaning that they use the drugs for several weeks or months and alternate these cycles with periods of discontinued use. Often the drugs are administered in a pyramid (step-up) pattern in which dosages are steadily increased over several weeks. Toward the end of the cycle the user "steps down" to reduce the likelihood of negative side effects. At this point, some people discontinue use of the drug or may initiate another cycle of different drugs (i.e., drugs that may restart endogenous testosterone production to prevent the undesirable drop in testosterone concentrations that follows the removal of the pharmaceutical agents). Although the length of each cycle is quite variable (ranging from 1 to 728 weeks), the median cycle length is reported to be 11 weeks (Cohen et al.

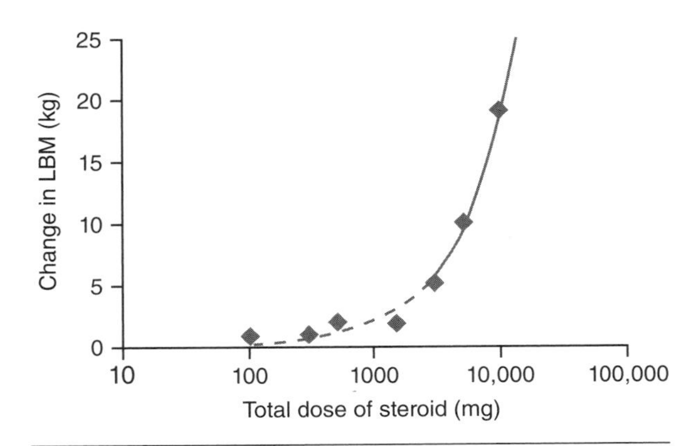

Figure 20.2 Dose–response curve for anabolic steroids and changes in lean body mass (LBM).

Adapted from Forbes 1985.

2007). Recent surveys have indicated that the typical nonmedical use pattern is 4 to 6 months over a year (Cohen et al. 2007; Parkinson and Evans 2006). A typical steroid regimen involves an average of 3.1 agents, with a typical cycle ranging from 5 to 10 weeks (Perry et al. 2005). The dose is reported to vary between 5 and 29 times greater than normal physiological replacement doses (Perry et al. 2005). One study showed that nearly 50% of individuals who self-administer androgens exceed 1000 mg testosterone or its equivalent per week (Parkinson and Evans 2006). However, this number may be exaggerated, as Cohen and colleagues (2007) suggested that the number of androgen users who self-administer more than 1000 mg testosterone or its equivalent per week may be closer to 10%. In any case, as suggested previously, higher pharmacologic dosages appear necessary to elicit the gains that athletes desire. Unfortunately, these results accentuate the athlete's philosophy that if a low dose is effective, then more must be better.

Demographics of Androgen Users

Due to heightened media exposure, there is a strong belief among the general population that the use of performance-enhancing drugs, specifically anabolic steroids and growth hormone, is rampant among professional athletes today. However, the available scientific evidence from the past few years indicates that illegal performance-enhancing drug use among competitive athletes is declining. In the past 20 years, National Collegiate Athletic Association (NCAA) surveys have reported that the number of collegiate athletes who self-admit to androgen usage has declined (Anderson, Albrecht, and McKeag 1993; NCAA 2001). These trends were apparent in all sports, including football, in which androgen use among those athletes was reduced by approximately 50% (Anderson, Albrecht, and McKeag 1993; NCAA 2001). Use patterns among professional and Olympic-caliber athletes are not known, as professional sport organizations in the United States do not release any of their drug-testing results to the general public. Thus most information emanating from professional sport has been based on innuendo and hearsay.

There is limited scientific evidence concerning the prevalence of the nonmedical use of androgens in the United States, and evidence regarding the use of testosterone precursors is even more sparse. However, considering that these precursors are now classified similarly to androgens, an understanding of the prevalence of androgen use may shed some light on the prevalence of use of these drugs. A 2007 report indicated that since 1993, the lifetime use of androgens for nonmedical reasons had remained at a consistent 1% in the college-aged population (McCabe et al. 2007). According to a more recent estimate, more than 800,000 individuals in the United States had used androgens during their lifetime (Hoffman, Kraemer, et al. 2009). However, most surveys on the nonmedical use of androgens have focused on collegiate and adolescent students and athletes. Information concerning adult use is generally limited to surveys of individuals who are self-administering androgens.

In the adult population of androgen users, the median age is 29 years, with nearly half of these people holding at least a bachelor's degree and more than 5% of self-admitted users holding a terminal degree (e.g., JD, MD, or PhD) (Cohen et al. 2007). Most adult users of androgens in the United States are Caucasian (88.5%) and are employed as professionals with yearly income exceeding that of the general population (Cohen et al. 2007). The primary reason for drug use among the general population of androgen users appears to be related to increases in strength and muscle mass and wanting to "look good" (Cohen et al. 2007). Other motivations for drug use include reduction of body fat, improvement in mood, and attraction of sexual partners. Interestingly, of 1955 androgen-using males surveyed, bodybuilding and sport performance were either not motivation for androgen use or of little importance (Cohen et al. 2007). Although recent media reports have focused on performance-enhancing drug use in professional athletes and youth, the majority of adults who self-administer androgens for nonmedical purposes appear to be intelligent, economically stable Caucasian men who are not competitive athletes.

Androgen use among African American collegiate athletes (1.1%) appears to be as common as in Caucasian student-athletes (1.1%) (Green et al. 2001).

A concern that was first noted on NCAA performance-enhancing drug surveys was the change in the age of initial androgen use among collegiate athletes who self-admitted to using these drugs. During the initial years of the survey, the majority of college athletes using these banned drugs did so toward the end of their college careers. Presumably, this was to enhance their chances of playing at the next level (i.e., professional sport); however, recent publications of the survey began to show a decrease in the age of initial androgen use. More than 40% of college athletes who admit to using androgens indicated that they first began using these drugs in high school (NCAA 2001). There have also been reports that androgen use has trickled down to the middle school level (Faigenbaum et al. 1998; Stilger and Yesalis 1999).

Early studies on performance-enhancing drug use in adolescents suggested that androgen use at the secondary level ranged from 6% to 11% in males. During the past 10 to 15 years, the use of androgens among male adolescents appears also to have been on the decline, with self-reported use ranging from 1.6% to 5.4% (Hoffman, Faigenbaum, et al. 2008). Studies showing a higher incidence (>6% self-admitting) have specifically examined high school football players. However, comparisons of androgen use among adolescent athletes and nonathletes have been inconclusive. The pattern of performance-enhancing drug use among adolescents does appear to increase as students move through high school, with a recent study indicating that 6% of high school male 12th graders admitted to using androgens (Hoffman, Faigenbaum, et al. 2008).

One of the biggest changes in androgen use patterns has been in the prevalence among female athletes and adolescents. The prevalence of androgen use among males has generally been reported to be three- to fourfold greater than in females, with frequency-of-use patterns in females varying between 1.2% and 1.7% (DuRant, Escobedo, and Heath 1995; DuRant et al. 1997). However, in contrast to the declining use reported among male adolescents, a greater frequency of androgen use in female adolescents has been reported, ranging from 2.0% (Miller et al. 2005) to 2.9% (Irving et al. 2002). Although these reports are alarming, several recent studies have suggested that the trend toward a greater frequency of androgen use among female adolescents may have been overstated or may at least be declining (Dodge and Jaccard 2006; Hoffman, Faigenbaum, et al. 2008).

Clinical Applications of Androgen Administration

The use of androgens by competitive athletes violates the basic ethics that govern competitive sport. However, the physiological benefits associated with testosterone administration have given clinicians a potential clinical tool to combat diseases and illnesses that are associated with muscle wasting and that limit functional performance. Exogenously administered testosterone is an approved method of treatment for hypogonadism in adult men (Bhasin et al. 2006). The goal of testosterone therapy is to raise testosterone levels into the midnormal range for healthy young men, correct symptoms of androgen deficiency, induce and maintain secondary sex characteristics, and maintain sexual function. In addition, there has been a concerted effort to determine the benefits of androgen use in patients suffering from chronic disease such as human immunodeficiency virus (HIV) infection with wasting syndrome, chronic obstructive pulmonary disease (COPD), and end-stage renal disease (Basaria, Wahlstrom, and Dobs 2001; Hoffman, Kraemer, et al. 2009). Strong scientific evidence indicates that the use of androgens to maintain or improve muscle mass is effective in improving functional performance and quality of life in these individuals (Bhasin et al. 2000; Casaburi et al. 2004; Grinspoon et al. 2000; Johansen, Mulligan, and Schambelan 1999; Knapp et al. 2008). In addition, substantial pharmaceutical effort has been directed toward exploring the anabolic effects of androgens on skeletal muscle for the prevention and treatment of sarcopenia and ameliorating functional limitations associated with aging (Bhasin et al. 2006b).

Medical Issues Associated With Androgen Use

Potential and documented adverse effects associated with androgen use are listed in table 20.2. Early medical and scientific publications (from the 1960s through the 1980s) were consistent in minimizing the efficacy and highlighting the risks associated with androgen use (Darden 1983; Fahey and Brown 1973; Fowler, Gardner, and Egstrom 1965; Golding, Freydinger, and Fishel 1974). This approach unfortunately proved to be ineffective. It caused athletes to lose trust in the physician's knowledge of androgens and forced them to seek advice from friends, Internet sites, or drug suppliers (Pope et al. 2004; Hoffman and Ratamess 2006). Recent literature has suggested that the medical issues associated with androgen,

Table 20.2 Potential Adverse Effects Associated With Androgen Use

Potential adverse effects of high doses of androgens

• Behavioral and psychiatric side effects
  ◦ Depression
  ◦ Hypomania and mania
• Cardiovascular complications
  ◦ Lowering of high-density lipoprotein cholesterol
  ◦ Increased low-density lipoprotein cholesterol
  ◦ Sudden cardiac death
  ◦ Myocardial hypertrophy and dysfunction
• Prolonged suppression of hypothalamic–pituitary–testicular axis
• Hepatic dysfunction, neoplasms, and peliosis hepatis
• Gynecomastia

Potential adverse effects of intramuscular injections
• Local infection and abscess • Systemic infection, including HIV and hepatitis C virus (HCV)
Unique adverse effects associated with androgen use in women
• Hirsutism • Clitoral enlargement • Change of body habitus, including widening of upper body torso • Breast atrophy • Menstrual irregularity
Unique adverse effects associated with androgen use in children
• Premature epiphyseal fusion and growth retardation • Premature virilization in boys and masculinization in girls

Adapted from J.R. Hoffman et al., 2009, "Position stand on androgen and human growth hormone use," *Journal of Strength and Conditioning Research.* 23: S1-S59.

although real, may be somewhat overstated (Berning, Adams, and Stamford 2004; Hoffman and Ratamess 2006; Sturmi and Diorio 1998; Street, Antonio, and Cudlipp 1996). Many of the side effects of androgen administration are reversible upon cessation. It is also important to note that there are differences in the side effects associated with androgen use under medical supervision versus abuse (i.e., the polypharmacy that may accompany androgen administration).

Anecdotally, it appears that a disproportionate magnitude of adverse events associated with androgen use is seen in bodybuilders (who may also be consuming several other drugs that relieve some side effects but potentiate other risk factors as well, for example diuretics, thyroid hormones, insulin, and anti-estrogens) compared to strength–power athletes. The mind-set and motivation of these two types of athletes can be quite different. Strength–power athletes who self-administer androgens do so to prepare for a season of competition. They typically cycle the drug to help them reach peak condition at a specific time of the training year. In contrast, bodybuilders use androgens to maximize muscle growth and definition. Their success is predicated on their aesthetic appearance. Thus many of these athletes use androgens excessively for several years without cycling off or perhaps minimize the length of off cycles depending on their competition schedule. Recent research has indicated that these athletes exhibit behavior consistent with substance dependence disorder (Perry et al. 2005). A potential for muscle dysmorphia may be seen in athletes who become obsessed with their appearance. Although the medical issues associated with androgens may be different between strength–power athletes and bodybuilders, the scientific literature generally does not differentiate between the two. The following sections discuss adverse effects on specific physiological systems. It is important to note that many of the side effects connected with androgens could be potentiated by the use of additional performance-enhancing or recreational drugs.

Psychiatric and Behavioral Effects

There have been a number of anecdotal reports of rage in androgen users, referred to as "roid rage."

However, studies of behavioral changes from androgen use have been inconsistent with regard to anger or measures of aggressiveness (Daly et al. 2003; Kouri et al. 1995; Pope, Kouri, and Hudson 2000; Tricker et al. 1996). This inconsistency with respect to behavioral or psychiatric changes may be related to concurrent use of other substances such as alcohol or psychoactive drugs or to a preexisting personality or psychiatric disorder (Bahrke and Yesalis 1994). No controlled trial of androgen administration has demonstrated significant change in aggression at physiologic replacement doses of testosterone. In fact, testosterone replacement in healthy androgen-deficient men has been reported to improve positive aspects and attenuate negative aspects of mood (Wang et al. 1996). Only a small number of subjects (less than 5%) in controlled trials have demonstrated marked increases in aggression, and this appears to occur only with the use of supraphysiologic doses; the vast majority of subjects administered androgens showed little or no change in behavior (Daly et al. 2003; Kouri et al. 1995; Pope, Kouri, and Hudson 2000; Tricker et al. 1996). It is possible that high doses of androgens might provoke rage reactions in a subset of individuals with preexisting psychopathology (Hoffman, Kraemer, et al. 2009).

Mood disorders such as mania, hypomania, or major depression appear to be a behavior change associated with androgen use (Malone et al. 1995; Pope and Katz 1994; Porcerelli and Sandler 1998). Major depression has been reported during periods of use but may become more prevalent following cessation (Malone et al. 1995; Pope and Katz 1994). Interestingly, a higher proportion of female athletes appear to report symptoms of hypomania and depression and a preoccupation with their physique during and following androgen use (Gruber and Pope 2000).

Cardiovascular Effects

Androgens may have a deleterious effect on blood lipid profiles (Bhasin et al. 1996; Jockenhovel et al. 1999). The effects of androgens on plasma lipids appear to be dependent on the dose, the route of administration (oral or injectable), and whether the androgen is aromatizable or not. Parenteral administration of replacement doses is associated with a small decrease in high-density lipoprotein (HDL) cholesterol levels with little to no effect on total cholesterol, low-density lipoprotein (LDL) concentration, or triglyceride concentrations (Bhasin et al. 2006a; Whitsel et al. 2001). On the other hand, supraphysiologic doses of testosterone, even when administered via injection, cause a significant decrease in HDL concentrations (Bhasin et al. 1996); and orally administered 17-α-alkylated, nonaromatizable androgens appear to result in an even greater reduction in plasma HDL and increase in LDL than does parenterally administered testosterone (Jockenhovel et al. 1999).

Increases in left ventricular mass have been reported among users of androgenic steroids (Dhar et al. 2005; Karila et al. 2003; Payne, Kotwinski, and Montgomery 2004). Since many androgen users resistance train, it is not clear whether the left ventricular hypertrophy reported in powerlifters is a consequence of resistance training or androgen use or both (Hoffman, Kraemer, et al. 2009). Interestingly, one study used echocardiography to assess left ventricular mass and wall thickness among male powerlifters and bodybuilders who were currently using androgens, ex-users who had not used androgens for over 12 months, and weightlifters who had never used androgens. The population that was currently using androgens had higher left ventricular muscle mass than nonusers or previous users (Urhausen, Albers, and Kindermann 2004). The E/A ratio (ventricular filling ratio) was reduced in the powerlifters using androgens, suggesting altered diastolic function (Stolt et al. 1999).

There have been several case reports of sudden death among power athletes who were abusing androgens (Dickerman et al. 1995; Di Paolo et al. 2007; Fineschi et al. 2007; Hausmann, Hammer, and Betz 1998; Luke et al. 1990). Several of these deaths were associated with nonthrombotic myocardial infarction, leading to speculation that androgens might induce coronary vasospasm (Payne, Kotwinski, and Montgomery 2004). However, most of these case reports were anecdotal, and a causative relationship between androgen use and the risk of sudden death has not been established. Strength–power athletes who are using androgens often have short QT intervals but increased QT dispersion, in contrast to endurance athletes with similar LV mass who have long QT intervals but do not have increased QT dispersion (Stolt et al. 1999). QT interval dispersion has been used as a noninvasive marker of susceptibility to arrhythmias (Pye, Quinn, and Cobbe 1994); however, it is unknown whether this would predispose a resistance-trained athlete who takes large doses of androgens to ventricular arrhythmias.

Hypothalamic–Pituitary–Testicular Axis Suppression

Androgen administration suppresses endogenous pituitary luteinizing hormone (LH) and follicle-stimulating hormone (FSH) secretion and, indirectly, testicular testosterone via the negative feedback mechanism (see chapter 2). Prolonged suppression of the hypothalamic–pituitary–testicular axis may result in infertility (Lloyd, Powell, and Murdoch 1996). After discontinuation of exogenous androgen use, circulating testosterone concentrations may fall to very low levels as the effects of exogenous testosterone wear off and the endogenous axis has yet not recovered. It may take up to a year for LH and FSH secretion patterns to return to normal (Garavik et al. 2011). During this period, people may experience troublesome symptoms of androgen deficiency, including loss of sexual desire and function, depressed mood, and hot flushes.

Liver Toxicity

Elevations of liver enzymes, cholestatic jaundice, hepatic neoplasms, and peliosis hepatis have been reported with the use of oral 17-α-alkylated androgenic steroids (Cabasso 1994; Kosaka et al. 1996; Pavlatos et al. 2001; Socas et al. 2005; Soe, Soe, and Gluud 1992), but not with parenterally administered testosterone or its esters (Calof et al. 2005). Furthermore, it is not clear whether elevations in the enzymes aspartate aminotransferase, alanine aminotransferase, and creatine kinase during androgen administration are the result of liver dysfunction or of muscle injury resulting from resistance exercise. The risk of hepatic dysfunction during androgen administration has probably been overstated (Dickerman et al. 1999; Pertusi, Dickerman, and McConathy 2001), especially with regard to parenteral administration.

Gynecomastia

Breast tenderness and breast enlargement in men, referred to as gynecomastia, are frequently associated with the use of aromatizable androgenic steroids (Strauss and Yesalis 1991). The occurrence of gynecomastia has been suggested to be as high as 54% in androgen users (Hoffman, Kraemer, et al. 2009; Strauss and Yesalis 1991). It is not uncommon for athletes to use an aromatase inhibitor or an estrogen antagonist in combination with androgenic steroids to prevent breast enlargement.

Intramuscular Injection Risks

The majority of androgen users administer androgens by intramuscular route; 13% of those who use intramuscular injections reported unsafe injection practices (Parkinson and Evans 2006). Parenteral self-administration increases the risk of infection, muscle abscess, and even sepsis (Evans 1997). Transmission of HIV infection and hepatitis has also been reported, presumably because of needle sharing or the use of improperly sterilized needles and syringes (Hoffman, Kraemer, et al. 2009).

Medical Issues Among Women

Women taking androgens may undergo masculinization and experience hirsutism, deepening of the voice, enlargement of the clitoris, widening of the upper torso, decreased breast size, menstrual irregularities, and male pattern baldness (Derman 1995; Pavlatos et al. 2001). Some of these adverse effects may not be reversible. In addition, epidemiologic studies have reported an association of elevated testosterone concentrations in women with increased risk of insulin resistance and diabetes mellitus (Ding et al. 2006).

Medical Issues Among Children and Adolescents

In addition to the adverse effects associated with adult abuse, the use of androgens in pre- or peripubertal adolescents may result in premature epiphyseal fusion, limiting attainment of full adult height (Casavant et al. 2007; Vandenberg et al. 2007). Boys may undergo premature or more accelerated pubertal changes, while girls may experience virilization.

Androgen Abuse and Mortality

There is little evidence to establish a causative role in increased mortality directly related to androgen use. However, there have been some anecdotal reports of increased mortality among androgen users. One study investigated mortality and underlying causes of mortality among 62 powerlifters who had achieved the top five positions in weightlifting competitions in the 82.5 to 125.0 kg weight categories during the period from 1977 to 1982 (Parssinen et al. 2000). The investigators made the assumption that these powerlifters were using androgens. Age-matched individuals from the general population were used as a control group. Thirteen percent of the powerlifters and 3% of the control group died during this period. Suicides, myocardial infarction, hepatic coma, and non-Hodgkin's lymphoma contributed to the deaths among powerlifters. In this small sample, the risk of death was 4.6 times higher in the powerlifters than in the control population. However, since these were larger men, it is not known whether androgen use or their larger body mass contributed to the greater mortality rates. A study comparing toxicology findings and manner of death between androgen users and users of heroin or amphetamines reported that users of androgens were more likely to become involved in incidents leading to violent death and to die at younger ages (Petersson et al. 2006).

Legal Issues Associated With Androgens

Although use of androgens has always required a physician's prescription, they were not always listed as a controlled substance. Consequent to mounting pressure regarding androgen use among American adolescents, the U.S. Congress in 1990 amended the Controlled Substances Act to include androgens. This was known as the Anabolic Steroid Control Act and reclassified androgens as a Schedule III substance. The impact of the legislation was to make it a crime to use these drugs for nonmedical purposes. Other Schedule III substances are weak opioids such as codeine and Vicodin, barbiturates, amphetamines, and methamphetamines. By 2004, an amended version of the Anabolic Steroid Control Act modified the definition of androgens to include 26 additional compounds that included designer androgens, such as tetrahydrogestrinone (THG) and prohormones like androstenedione.

Testosterone Precursors

As described in chapter 2, the rate-limiting step for testosterone synthesis is the side-chain cleavage of cholesterol to form pregnenolone. The conversion of pregnenolone to testosterone occurs through one of two pathways; these produce various intermediary compounds or precursors that are steps in the synthesis of testosterone. Since these precursors result in a conversion to testosterone, it is thought that their use as an ergogenic aid will result in an increase in endogenously produced testosterone. The basis for the use of testosterone precursors as an ergogenic aid evolved from a study in which a threefold increase in testosterone was seen in healthy women given 100 mg of either androstenedione or dehydroepiandrosterone (DHEA) (Mahesh and Greenblatt 1962). Given that females synthesize very small amounts of the male hormone testosterone, results suggesting that a precursor that gets converted to testosterone can increase testosterone concentrations in females were not surprising.

Athletes looking for a competitive advantage began to supplement with these testosterone precursors (androstenedione, androstenediol, and DHEA) based on the premise that even in healthy athletes, ingestion would lead to increases in the endogenous production of testosterone and achieve body composition and performance changes similar to those with anabolic steroids. However, these precursors have only relatively weak androgenic properties on their own, with androstenedione and DHEA having only 1/5 and 1/10 the biological activity of testosterone, respectively (Migeon 1972). Even if a large dose is ingested, these substances (similarly to orally ingested testosterone derivatives) are likely in large part to be rapidly removed from circulation by hepatic uptake and elimination. To enhance the anabolic potential of these testosterone precursors, the 19th carbon atom from androstenedione and androstenediol is sometimes removed to form norandrostenedione and norandrostenediol, respectively. The benefit of these precursors is that they are both converted to nandrolone, which is reported to have greater anabolic potency than testosterone itself (Sundaram et al. 1995). Based on a small sample of studies, testosterone precursors were officially listed as controlled substances in the 2004 Anabolic Steroid Control Act, and a physician's prescription was then required for use. As a result, testosterone precursors have been labeled illegal performance-enhancing drugs and have been banned by all sport governing bodies.

Strength and Body Composition Changes

Studies examining the efficacy of testosterone precursors have produced varying results. The majority have shown that androstenedione (in dosages varying between 100 and 300 mg per day) or DHEA (in dosages varying between 50 and 100 mg per day) supplementation for study durations varying between 8 and 12 weeks has no significant benefit on strength or body composition compared to placebo in untrained or recreationally trained men (Brown et al. 1999; King et al. 1999; Brown, Vukovich, Reifenrath, et al. 2000; Broeder et al. 2000; Brown and McKenzie 2006). However, these studies suffered from a number of methodological flaws. The use of untrained individuals to determine the efficacy of a supplement to improve strength and lean tissue accruement is highly questionable. As discussed earlier in this book, the window of adaptation for untrained subjects is quite large. This subject population can make large strength gains in a relatively short period of time, regardless of supplementation. In general, untrained subjects are not the population that is typically used to determine the efficacy of nutritional supplements. Furthermore, they are not a group for whom supplements of any kind would be recommended. This problem is similar to the one seen in the early studies on the efficacy of androgens, in which untrained subjects were provided anabolic steroids and showed no differences when compared to a control group (Hoffman, Kraemer, et al. 2009). When these drugs were provided to experienced resistance-trained subjects in doses commonly used by athletic populations, their efficacy became quite evident.

Only a limited number of studies have examined the use of testosterone precursors in experienced resistance-trained individuals. Van Gammeren and colleagues (2001, 2002), using both low (156 mg per day) and high (344 mg per day) doses of both norandrostenedione and norandrostenediol for 8 weeks, were unable to demonstrate any strength or body composition improvements in men (19-35 years) with at least 1 or 6 years, respectively, of resistance training experience.

Another problem in many of these studies was the large disparity in subject age. Subjects ranged from the third to the seventh decade. The significance of this was highlighted by the work of Brown, Vukovich, Martini, and colleagues (2000), who demonstrated that 28% of the variability in the testosterone response to androstenedione supplementation could be explained by the subjects' resting testosterone concentrations. Although no performance measures were used in that study, it gave impetus to examining the role that testosterone precursors may have in elevating endogenous testosterone production in men with low resting testosterone levels.

Endogenous Testosterone Synthesis

Jasuja and colleagues (2005), examining hypogonadal men (clinically diagnosed low resting testosterone levels), demonstrated that 1500 mg androstenedione per day can significantly elevate both total and free testosterone concentrations for 8 h following supplement administration. In addition, it was observed that these subjects significantly elevated strength and lean body mass following 12 weeks of supplementation. The investigators used a dosing pattern fivefold greater than that used in any other study. In addition, the design of the study was highly questionable considering that no control group was used and subjects were not permitted to exercise during the 12 weeks. It is unknown whether a group of similar subjects who trained during the 12-week study would have achieved similar strength and lean body mass gains. Possibly a control group participating in a resistance training program would have achieved similar strength and muscle skeletal adaptations.

The vast majority of studies using a double-blind crossover design (subjects taking both the supplement and the placebo with appropriate wash-out periods between dosing regimens) have been unable to support an elevation in either total or free testosterone concentrations from an oral dose of androstenedione or DHEA (Brown et al. 1999; King et al. 1999; Wallace et al. 1999; Rasmussen, Volpi, et al. 2000; Brown, Vukovich, Martini, et al. 2000). These studies used doses that varied between 100 and 300 mg per day of androstenedione and 50 and 150 mg per day of DHEA. Some investigations have shown that ingestion of testosterone precursors can significantly elevate resting testosterone concentrations, but these elevations may be dose dependent. Leder and colleagues (2000) reported significantly greater elevations in testosterone when 300 mg of androstenedione was provided for 7 days versus 100 mg per day in untrained men between the ages of 20 and 40. In a similar subject population (30- to 59-year-old men), Brown and colleagues showed similar results (37% increase in testosterone concentrations) following 4 weeks of 300 mg of androstenedione ingestion. When this dosing pattern was provided to younger men only (acute administration), no changes in testosterone concentrations were observed (King et al. 1999; Brown, Vukovich, Reifenrath, et al. 2000). Interestingly, 200 mg doses in college-aged physical education students resulted in an acute elevation in testosterone concentrations (Earnest et al. 2000), but not in resistance-trained young men (Ballantyne et al. 2000). This may be a function of study duration. Others have shown that the elevations in testosterone levels may be transient, in that testosterone concentration may increase after 4 weeks of supplementation but return to baseline levels after 12 weeks (Broeder et al. 2000). The results of these studies appear to suggest that the ability of testosterone precursors to elevate testosterone concentrations in apparently healthy men is inconsistent at best.

Mode of Ingestion

The primary method of intake of testosterone precursors has been through oral means. It is likely that much of the testosterone synthesized from the ingestion of these precursors is metabolized quite rapidly in the liver. In the research on sublingual delivery (an alternative means of delivery), the ability to increase testosterone production has been consistent (Brown et al. 2002; Brown and McKenzie 2006). It has been demonstrated that the use of sublingual delivery significantly elevates free testosterone concentrations 30 min following ingestion and that these concentrations remain elevated for 180 min. The time course of the elevation of total testosterone may be a bit delayed (peaking at ~60 min postingestion).

Reviews have generally concluded that the use of testosterone precursors falls far short of providing the anabolic effects generally associated with androgens (Broeder 2003; Brown, Vukovich, and King 2006). However, it is important to acknowledge that these precursors have not been examined in highly trained athletes. Small changes in a training regimen can have a profound influence on elite-level athletic performance improvement. What is important to keep in mind is that differences between success and lack of success at the elite level of athletic competition are quite small. For the competitive athlete, these small changes may have important practical implications that would never be confirmed by normal statistical analysis. It is these small changes that drive many competitive athletes to look for a competitive edge.

Regarding the potential health risk of these supplements, it is clear from the research available that the use of testosterone precursors does not provide the same benefit as androgens but may be associated with similar side effects. The aromatization (conversion of androstenedione to estradiol and estrone) of these compounds may result in the same adverse responses as are associated with androgens. In addition, the consistent decline in HDL concentrations observed in the investigations previously discussed suggests that the precursors adversely affect lipid profiles.

Selective Androgen Receptor Modulators

Although the benefits associated with medically prescribed androgens have been documented, the adverse effects remain a concern. This has stimulated the pharmaceutical industry to develop an androgen that eliminates or minimizes the side effects associated with androgen administration. In response, selective androgen receptor modulators (SARMs) were developed. These synthetic drugs were designed to produce selective tissue-specific anabolic actions in muscle and bone with minimal androgenic actions in other peripheral tissues. SARMs were first reported in 1998, and currently several classes exist, including quinolines, tricyclics, bridged tricyclics, bicyclics, aryl propionamides, and tetrahydroquinolines (Omwancha and Brown 2006; Segal, Narayanan, and Dalton 2006). SARMs bind to the androgen receptor with high affinity, thereby strengthening the anabolic properties; yet they are not subject to aromatase or 5α-reductase activity, so their androgenic properties are low (Omwancha and Brown 2006; Segal, Narayanan, and Dalton 2006). In addition, they have favorable pharmacokinetic properties and great potential for biochemical modifications, suggesting that they may have greater use in a clinical population (Segal, Narayanan, and Dalton. 2006). Although this drug is not commonly used among competitive athletes, it is likely to become a drug of choice in the near future.

Masking Agents

To avoid a positive drug test, athletes may take substances to hide their use of androgens and other performance-enhancing drugs. These are known as **masking agents** and are banned by all sport governing bodies. A variety of substances are used. For the most part, these substances are not ergogenic. Diuretics, sulfonamides, and epitestosterone are drugs that have been used by athletes to avoid a positive drug test. Athletes have also used tampering methods, such as urine substitution, catheterization, or consumption of adulterants (Jaffee et al. 2007).

Diuretics have been used to dilute urine and mask drug use. Their physiological action is to block sodium reabsorption in the kidneys and induce fluid and electrolyte loss in urine. They are generally used to treat diseases such as hypertension, congestive heart failure, or kidney problems. The athlete uses diuretics to reduce the concentration of drugs in the urine by increasing dieresis (Hoffman, Kraemer, et al. 2009). A variety of diuretics affect different areas of the kidney. Interestingly, ingestion of a diuretic is reported to be more ergolytic (causing a decrease in performance) than ergogenic, specifically during endurance events (Barr 1999). Sulfonamides, such as probenecid, have been used in the past to decrease the excretion rate of androgens metabolites (i.e., alter the way that testosterone is metabolized). However, these are no longer effective, as current drug tests can detect androgens in the urine despite the use of sulfonamides. Athletes have also used epitestosterone as a masking agent (Aguilera, Hatton, and Catlin 2002). Epitestosterone is a 17α epimer of testosterone found in the urine in concentrations similar to those of testosterone. The physiological role of epitestosterone is not well understood, but it is used to detect exogenously administered androgens. In general, the ratio between testosterone and epitestosterone

In Practice

Can Tea Consumption Alter a Drug Test for Androgen Abuse?

The glucuronidation of testosterone occurs naturally in the body and is the way in which testosterone is metabolized to epitestosterone. Several enzymes are involved in the glucuronidation of testosterone, but the enzyme UDP-glucuronosyltransferase (UGT2B17) is considered the most important. Jenkinson and colleagues (2012) examined the effect of both dietary green tea and white tea consumption on the ability to suppress UGT2B17-mediated testosterone glucuronidation. Any inhibition of this enzyme could potentially raise circulating levels of testosterone and affect the T:E ratio. The authors' hypothesis was based on the known effect that consuming tea, specifically the catechins, has to inhibit the production of several enzymes. Both green and white tea extracts were added to testosterone solutions. To analyze testosterone glucuronidation, the investigators compared UGT2B17 assays of testosterone glucuronidation with detection of unglucuronidated testosterone using high-performance liquid chromatography. The addition of both white and green tea extracts to the testosterone solution resulted in an inhibition of the glucuronidation of testosterone by impeding or inhibiting the UGT2B17 enzyme. This has potentially important implications for testing androgen use in competitive athletic populations. Inhibition of testosterone glucuronidation could increase the risk of a positive drug test by raising testosterone concentrations and potentially increasing the T:E ratio. This is potentially very concerning in that the investigation provided evidence that consumption of a popular drink may influence drug tests; the results also suggest the potential impact that everyday food or beverages may have in influencing drug tests.

(T:E) is 1:1. However, to account for outliers, a 4:1 ratio is considered to be within legal (negative drug test) limits. Athletes who are using androgens may self-administer epitestosterone (in doses similar to those in their testosterone regimen) to reduce the ratio to within legal limits. Reports from athletes have indicated that epitestosterone injections 1 h before testing resulted in a passed drug test. Although epitestosterone administration can be detected (Aguilera, Hatton, and Catlin 2002), it can result in an athlete's passing a drug test unless it is specifically tested for.

Human Chorionic Gonadotropin

Human chorionic gonadotropin (hCG) is a dimeric glycoprotein hormone found in the placenta of pregnant women (Handelsman 2006). In males, hCG acts similarly to LH, as it has specific target receptors on Leydig cells, stimulating testosterone secretion (Kicman, Brooks, and Cowan 1991). Some athletes use hCG at the end of an androgen cycle to help restart the body's own endogenous production of testosterone; others stack it with androgens to stabilize the T:E ratio (Kicman, Brooks, and Cowan 1991). Single doses of hCG (3000-5000 IU) have resulted in substantial elevations in testosterone but no significant change in the T:E ratio (Cowan et al., 1991). A single 6000 IU dose of hCG has been reported to result in a 50% elevation in plasma testosterone level 2 h after injection (Saez and Forest 1979). The response appears biphasic in that peak elevations in plasma testosterone may be observed 3 to 4 days after administration (Kicman, Brooks, and Cowan 1991). About 20% to 30% of administered hCG is excreted in urine within 6 days (Kicman, Brooks, and Cowan 1991). Detection of hCG in the urine results in a positive drug test. Considering the significant elevations in testosterone secretion patterns following hCG injection, side effects associated with androgen administration may be seen at higher doses. Athletes using hCG often inject doses of 1000 to 7000 IU of hCG every 5 days in 3- to 4-week cycles, although others have used greater quantities for cycles extending beyond 8 weeks (Llewellyn 2007).

Anti-Estrogens

Anti-estrogens are drugs used clinically to treat a variety of diseases but used primarily in treatment plans for breast cancer (Baumann and Castiglione-Gertsch 2007; Steiner, Terplan, and Paulson 2005).

Physiologically, they inhibit the effects of estrogen by inhibiting the enzyme aromatase or by blocking estrogen receptor action (Handelsman 2006). Similar to SARMs, selective estrogen receptor modulators (SERMs) have been developed to antagonize estrogen actions in specific tissues (Handelsman 2006). Athletes may administer anti-estrogens to reduce the aromatizing effects from androgen use and often at the end of an exogenous cycle of androgens to increase the body's own production of testosterone (Handelsman 2006). The use of anti-estrogens may also reduce several undesirable side effects (i.e., gynecomastia, water retention, and other health risks) associated with androgen use that are caused by aromatization into estradiol and other estrogens.

The two specific categories of anti-estrogens are aromatase inhibitors and receptor blockers. There are a number of aromatase inhibitors; aromasin is thought to be one of the most effective among athletes (Llewellyn 2007). SERMs and receptor blockers antagonize estrogen receptors. Clomid (clomiphene citrate) and Nolvadex (tamoxifen citrate) are among the more popular estrogen receptor blockers (Hoffman, Kraemer, et al. 2009). Clomid is a popular drug among male bodybuilders (50-100 mg/day) and is frequently used for 4 to 6 weeks upon termination of an androgen cycle. Aromatase inhibitors, SERMs, and other anti-estrogens such as Clomid are prescription drugs banned by most national and international sport governing bodies.

Growth Hormone

The use of growth hormone (GH) as a performance-enhancing drug has become more prevalent in the last few years, in part due to the level of sophistication of detection of androgen use among athletes. The anabolic effects of GH (see chapter 2 for a review of the physiological role of endogenously secreted GH)

In Practice

A Potential New Method to Detect Exogenous Growth Hormone Use

Testing for exogenous GH use has proven very challenging. One proposed method included measuring the proportions of different isoforms of GH. Considering that the 22 kDa form has the greatest concentration in the circulation and is the isoform used during GH therapy, any exogenous administration would lower all of the isoforms of GH, making detection possible (Wu et al. 1999). However, the window for assessing GH abuse with this test would be between only 12 and 24 h (Velloso et al. 2013). Other proposals for detecting GH abuse have focused on measuring the markers of recombination GH administration. Both IGF-1 and N-terminal propeptide of collagen type III (PIIINP) have been shown to provide a sensitive and specific test for identifying individuals taking exogenous GH (Powrie et al. 2007). The validity of the test has also been established (Erotokritou-Mulligan et al. 2008), and it was used successfully to detect and disqualify two powerlifters at the 2012 Paralympics in London (Velloso et al. 2013).

Velloso and colleagues (2013) compared the effects of 2 weeks of recombinant GH administration in 16 resistance-trained individuals in a placebo-controlled study on serum concentrations of IGF-1 and PIIINP. In addition, these investigators were interested in determining the effects of an acute bout of resistance exercise on these markers. Results indicated that GH administration can significantly increase IGF-1 and PIIINP concentrations. However, IGF-1 concentrations return to baseline following 1 week of no administration, but PIIINP concentrations remained significantly elevated. In addition, an acute bout of resistance exercise (six sets of 10 repetitions of the leg press exercise at 80% of maximal strength) did not significantly change IGF-1 or PIIINP concentrations from resting values. The authors concluded that exogenous administration of GH can elevate both IGF-1 and PIIINP concentrations before and immediately following an acute resistance exercise session, but that exercise by itself did not alter the resting concentrations of these variables. When these results were applied to the GH-2000 discriminant function formula (see Wallace et al. 1999), four of the eight subjects given GH were identified as users, and no false positives were noted. The authors concluded that this test had high specificity and good sensitivity for the detection of GH abuse among competitive athletes.

and limited detection capability are the motivations suggested for the surge in its use (Hoffman, Kraemer, et al. 2009; Saugy et al. 2008). As an ergogenic aid it has become popular because of two specific physiological effects: increase in lean tissue (anabolic effect) and decrease in body fat (metabolic effect). Growth hormone is a peptide hormone secreted by the pituitary gland in pulsatile fashion, leading to largely fluctuating concentrations in the circulation. Because it is a peptide hormone, its half-life is very short (~20 min) (Van Helder, Goode, and Radomski 1984). Thus the chance for detection is very small, but it needs to be administered in high doses (10-25 IU per injection) at three or four times per week to provide any effect (Saugy et al. 2008).

Clinical Role

Medical administration of GH has important clinical roles. It is primarily administered to promote physiologic and psychological well-being and altered body composition in adults with GH deficiency and muscle wasting due to HIV/AIDS, as well as in patients diagnosed with short bowel syndrome. In addition, GH is administered to promote linear growth in short children. Other Federal Drug Administration (FDA)-approved indications for GH include chronic kidney disease, Turner syndrome, Prader-Willi syndrome, *SHOX* gene haploinsufficiency, and Noonan syndrome. The most common efficacy outcome of GH in infants, children, and adolescents is an increase in linear growth.

Efficacy of Growth Hormone Use Among Competitive Athletes

The medical use of GH was initially undertaken in children and adolescents with subnormal growth and adults with GH deficiency. It has now become very popular with aging adults as an elixir for returning a youthful image and with athletes for enhancing physical performance. Growth hormone was first prepared in the 1940s using extracts from humans or monkeys (Knobil and Hotchkiss 1964; Li and Papkoff 1956). Monkeys were the only species that was shown to demonstrate efficacy in humans, but the amount was only available in small quantities. Only in the 1980s with advances in technology was recombinant human GH synthesized; this provided an unlimited supply for medical use but also created an opportunity for athletes to experiment.

Only a limited number of studies have examined the efficacy of GH as an ergogenic aid. A meta-analysis (Liu et al. 2008) reported that most people (the study included more than 300 users of GH) received the drug for an average of 20 days and that a number had received it as only a single injection. The subjects were primarily young men (~27 years) and were recreational, not elite, athletes. The average dose was 36 $\mu g \cdot kg^{-1} \cdot day^{-1}$, approximately 5- to 10-fold the therapeutic dose in adults with GH deficiency. Lean body mass increased an average of 2.1 kg in the GH-treated groups compared to those not treated, and a small, not statistically significant decrease in fat mass (−0.9 kg) was seen. Body weight did not change significantly. Only two studies appropriately evaluated change in strength (Deysigg et al. 1993; Zmuda et al. 1997). These were the longest trials, 42 and 84 days in duration. Neither of these studies was able to demonstrate efficacy of GH administration for strength improvements. A recent study on GH administration (19 $\mu g \cdot kg^{-1} \cdot day^{-1}$) in subjects with normal GH levels showed a significant increase in strength and power following a single week of use (Graham et al. 2008).

The effect of GH may be more metabolic than anabolic. Healy and colleagues (2006), using 0.2 $U \cdot kg^{-1} \cdot day^{-1}$ of GH or placebo for 4 weeks, reported significant increases in plasma levels of glycerol and free fatty acids before and after exercise, indicating greater fat oxidation in the subjects receiving GH. This was consistent with other studies showing an increase in lipolysis following GH administration in normal subjects (Lange et al. 2002; Moller et al. 1990).

The adverse events reported in these studies mirrored those often seen in adults administered with clinical doses of GH. These adverse events include soft tissue edema, joint pain, carpal tunnel syndrome, and excessive sweating. Many of these side effects are thought to be related to GH's effect on fluid retention (Birzniece et al. 2011). At higher doses, which may be used by athletes, the adverse effects may include diabetes, negative effects on cardiac function, acromegaly, and additional metabolic disorders and myopathies (Birzniece et al. 2011).

Thyroid Drugs

The thyroid gland secretes two hormones, triiodothyronine (T3) and thyroxine (T4), that function in virtually all cells of the human body and are involved in metabolism and energy expenditure (Wexler and Sharretts 2007). Thyroid drugs (primarily sodium levothyroxine) are typically used to treat thyroid insufficiency or hypothyroidism (Escobar-Morreale et al. 2005; Wiersinga 2001). Bodybuilders have used thyroid drugs to potentially enhance anabolic growth processes and enhance metabolic rate. Thyroid drugs used by athletes include Cytomel, Triacana, and Synthroid in supraclinical doses (Llewellyn 2007). Unsupervised use of thyroid drugs can disrupt the hypothalamic–pituitary–thyroid axis and produce negative side effects: bone and skeletal muscle catabolism, heart palpitations, agitation, shortness of breath, irregular heartbeat, sweating, nausea, irritability, tremors, restlessness, and headache (Clarke and Kabadi 2004; Escobar-Morreale et al. 2005). There is a paucity of research on the ergogenic effects of thyroid drugs.

Central Nervous System Stimulants

Stimulants increase central nervous system (CNS) activity, primarily through their interaction with β-receptors on nerve endings. Their ergogenicity is related to enhanced mental acuity, alertness, physical energy, thermogenesis, and exercise performance (e.g., muscle strength, endurance, and improved reaction time) (Avois et al. 2006). However, side effects such as nervousness, anxiety, heart palpations, headaches, nausea, cardiomyopathy, high blood pressure, and in some rare cases a stroke may occur (Hoffman, Kraemer, et al. 2009). Most stimulants are banned substances, but caffeine, pseudoephedrine, synephrine, and ephedrine and methylephedrine (both in concentrations <10 μg/ml) are not prohibited. Amphetamines release stores of norepinephrine, serotonin, and dopamine from nerve endings and prevent reuptake, which leads to increased amounts of dopamine and norepinephrine in synaptic clefts (Avois et al. 2006). Amphetamines are absorbed from the small intestine, and plasma concentrations reach their peak 1 to 2 h following ingestion (Avois et al. 2006). The effects of stimulants may begin to be seen 10 to 40 min after ingestion and may last for up to 6 h. Amphetamine metabolites are excreted in the urine, where they can be detected (for up to 4 days following use) via drug testing.

Ephedrine

Ephedra, also called ma huang, is a plant that contains ephedrine. Ephedrine is a β-agonist that is reported to be relatively popular among bodybuilders (Gruber and Pope 1998; Tseng et al. 2003) but also used by weightlifters (Gruber and Pope 1998). It is thought to have a strong thermogenic quality desired by bodybuilders in order to reduce body fat. It is often used as a stacking agent with caffeine to enhance the thermogenic effect. Similar to caffeine, ephedrine increases fat oxidation and spares muscle glycogen. Clinically, it has been used to treat respiratory problems and is commonly present in pharmaceuticals such as bronchodilators, antihistamines, decongestants, and weight loss products.

Findings on performance changes with ephedrine are inconclusive. Ergogenic effects—improved power during cycling (Gill et al. 2000) and improved time in a 1500 m run (Hodges et al. 2006)—have been reported with both ephedrine and pseudoephedrine, but the vast majority of studies on performance have not shown any benefit (Bell, Jacobs, and Ellerington 2001; Chester, Reilly, and Mottram 2003; Chu et al. 2002; Gillies et al. 1996; Jacobs, Pasternak, and Bell 2003; Swain et al. 1997). Caffeine–ephedrine stacking has shown consistent results with regard to improving endurance performance (Bell and Jacobs 1999; Bell, Jacobs, and Zamecnik 1998; Bell, Jacobs, et al. 2000; Bell, Jacobs, and Ellerington 2001); however, it is also associated with higher blood pressure and heart rates during exercise (Bell, Jacobs, and Ellerington 2001; Haller, Jacob, and Benowitz 2004). Adverse effects (vomiting and nausea) following exercise have also been reported in 25% of the subjects ingesting a mixture of 5 mg/kg caffeine and 1 mg/kg ephedrine (Bell and Jacobs 1999). A subsequent study by the same research team showed that use of a lower dose (4 mg/kg caffeine and 0.8 mg/kg ephedrine) could provide similar ergogenic benefits without any of the previously observed side effects (Bell, Jacobs, et al. 2000). The caffeine–ephedrine mixture appears to have greater benefit than either supplement taken alone.

In 2004, the FDA banned all products containing ephedra or ephedrine after determining that they posed an unreasonable risk to users. This was based on a report published by the Rand Institute (Shekelle et al. 2003) indicating that 16,000 adverse events were linked to the use of ephedra-containing dietary supplements. The use of ephedra-containing dietary supplements or ephedrine plus caffeine is associated with two to three times the risk of nausea, vomiting, psychiatric symptoms such as anxiety and change in mood, autonomic hyperactivity, and palpitations, and in a few cases death (Shekelle et al. 2003). As a result of the many adverse effects, ephedrine use has been banned by most sport governing bodies.

However, ephedrine alkaloids are found in common cold medicines; thus competitive athletes need to be aware that consumption of these products can result in a failed drug test.

Citrus Aurantium

Citrus aurantium is from a fruit otherwise known as bitter orange and is commonly used as an Asian herbal medicine to treat digestive problems (Fugh-Berman and Myers 2004). It is also a mild stimulant and is thought to contribute to appetite suppression, increased metabolic rate, and lipolysis (Fugh-Berman and Myers 2004). Citrus aurantium contains synephrine, a sympathomimetic agent, which has been suggested to stimulate specific adrenergic receptors that stimulate fat metabolism without any of the negative side effects generally associated with compounds that stimulate the other adrenergic receptors (Carpene et al. 1999). Synephrine, an active component of citrus aurantium, is thought to interact with β-3 receptors to increase lipolysis and minimize the cardiovascular effect typical of adrenergic amines (Carpene et al. 1999). Although synephrine has been shown to stimulate peripheral α-1 receptors, resulting in vasoconstriction and elevations in blood pressure (Brown et al. 1988), other research has shown that citrus aurantium ingested alone has no effect on blood pressure (Haller, Benowitz, and Jacob 2005); however, when combined with other herbal products, it may cause significant elevations in systolic blood pressure (Haller, Benowitz, and Jacob 2005; Hoffman, Kang, et al. 2006). In addition, when citrus aurantium is combined with caffeine and other herbal products, significant improvements in time to fatigue have been reported (Hoffman, Kang, Ratamess, et al. 2007). Synephrine has been on the NCAA's list of banned performance-enhancing drugs but not that of the World Anti-Doping Association or of the U.S. Anti-Doping Association. It is recommended that athletes be extremely cautious in using this product, as it may produce a positive drug test.

Modafinil

Modafinil is a psychostimulant that has been used clinically to treat narcolepsy and idiopathic hypersomnia (Lyons and French 1991; Repantis et al. 2010). In healthy individuals it has been shown to be very effective in maintaining alertness and cognitive performance in situations of sleep deprivation (Repantis et al. 2010). Acute modafinil ingestion (4 mg/kg) appears to have an ergogenic effect in healthy individuals by significantly increasing time to exhaustion compared to values in placebo and control trials (Jacobs and Bell 2004). When compared to a high dose of caffeine (600 mg per day), modafinil ingestion (at 100, 200, and 400 mg doses per day) produced similar performance-enhancing (alertness and reaction time) effects (Wesenten et al. 2002). However, the amount of caffeine used was threefold greater than the amount typical of over-the-counter caffeine tablets, whereas the modafinil dosages were within the prescribed range. Modafinil is a banned substance, and athletes who test positive for this drug are suspended.

Salbutamol

Salbutamol is a β-2 agonist that is frequently used for the prevention of exercise-induced asthma (EIA). It is currently approved by the World Anti-Doping Association (WADA) for use via inhalation as long as the athlete receives a therapeutic use exemption (TUE). Regarding the ergogenic effects of salbutamol, recent research does not provide any evidence supporting a change in substrate oxidation from a 4 mg dose of salbutamol (Arlettaz et al. 2009) or in endurance activity following 200 μg, 400 μg, and 800 μg dosages (Sporer, Sheel, and McKenzie 2008) compared to placebo. Based on available evidence, salbutamol does not have any ergogenic effect in athletes using this medication for treatment of EIA.

Clenbuterol

Clenbuterol is a β_2-agonist that is generally used to reverse bronchial constriction. In recent years athletes have begun to use it as an ergogenic aid to increase lean muscle tissue and reduce subcutaneous fat (Prather et al. 1995). This is based on studies in a rodent model demonstrating that clenbuterol can increase muscle protein synthesis (MacRae et al. 1988; Reeds et al. 1986). Clenbuterol is banned in competition by WADA. Though studies in humans are limited, several findings have indicated an ergogenic potential of β_2-agonists for strength improvements (Maltin et al. 1993). Athletes typically use clenbuterol in doses that are twice the recommended amounts administered for clinical purposes, in a cyclic fashion (3 weeks on alternated with 3 weeks off, with a 2-day-on, 2-day-off cycle during the "on" week) (Prather et al. 1995). It is believed that this cycling regimen avoids β_2-receptor downregulation (Di Pasquali 1992). In contrast to the inhalation route often used for relieving bronchial constriction, athletes consume clenbuterol in capsule form. A number of potential side effects of clenbuterol have been suggested, for example transient tachycardia, hyperthermia, tremors, dizziness, palpitations, and malaise (Kierzkowska, Stanczyk, and Kasprzak 2005); and recent evidence has suggested altered cardiac function (Sleeper, Kearns, and McKeever 2002) and inhibited longitudinal growth in bones (Kitaura, Tsunekawa, and Kraemer 2002). However, these latter studies were conducted with equine and rodent models, respectively. Further research on human studies is warranted.

Site Enhancement Drugs

Site enhancement drugs are primarily used by bodybuilders. These drugs, for example Synthol, Nolotil, and Caverject, cause a temporary increase in the size of the muscle into which they are injected (Llewellyn 2007). Esiclene was a drug that was popular because it led to swelling and inflammation when injected locally. However, other drugs have become more popular among bodybuilders for local site enhancement. Synthol is composed of medium-chain triglycerides, lidocaine, and benzyl alcohol and is injected intramuscularly, where it lodges between the fascicles. Repeated injections lead to greater volume within the muscles. Evidence suggests that bodybuilders inject 1 to 3 ml every day or every other day for 2 to 3 weeks (Llewellyn 2007). Scientifically, little is known about these drugs and their potential side effects.

Blood Doping

Over the past 20 years, many anecdotal reports have emanated from the Olympic Games, various world championships, and other athletic events about endurance athletes involved in blood doping. Blood doping entails removing a certain volume of the person's blood, freezing the red blood cells, and reinfusing them in a saline solution at a later date once the red blood cells in the body have returned to normal concentrations. The goal of blood doping is to increase the number of red blood cells per unit volume of blood, with the intent of delivering more oxygen to exercising muscles and increasing maximal oxygen consumption. As a result, endurance performance is improved.

Studies on the ergogenic benefit of blood doping have consistently demonstrated a postinfusion increase in $\dot{V}O_2$max and improved endurance performance (Buick et al. 1980; Robertson et al. 1984; Sawka et al. 1987; Thomson et al. 1983). $\dot{V}O_2$max has been reported to be 5% to 13% greater in subjects reinfused with two to four units of blood (each unit is ~450 ml) compared with control subjects. Spriet (1991) analyzed several studies that examined the effect of blood doping on race performance times. His analysis revealed that reinfusion of two units of blood may improve run times in 2, 6, and 10K races by approximately 7, 30, and 68 s, respectively. Figure 20.3 shows improvements in run times for various race distances after blood doping. The data from these studies show convincingly that blood doping enhances both $\dot{V}O_2$max and endurance performance.

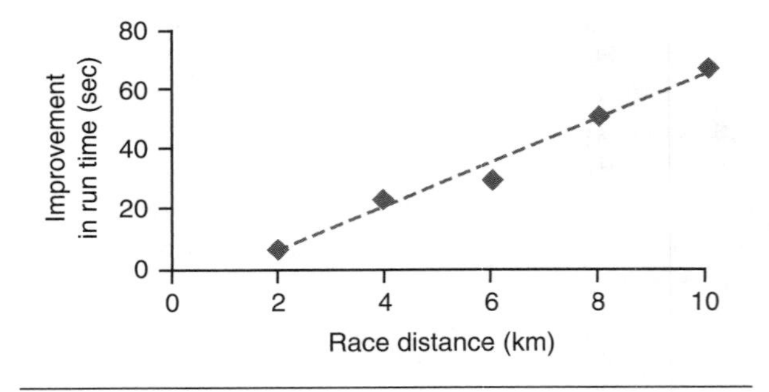

Figure 20.3 Effect of blood doping on run times.
Data from Goforth et al. 1982; Spriet 1991; Williams et al. 1981.

A major concern for athletes who participate in blood doping is the medical risks they may encounter. As with any transfusion, potential risks include hepatitis and acquired immuno deficiency syndrome (AIDS). In addition, athletes may decide to infuse blood that is not their own. Certain athletes reportedly used homologous transfusions (e.g., transfusions from a family member) during the 1984 Olympics (Spriet 1991). This type of practice could place athletes at a much higher risk of contracting the diseases mentioned. In addition, reinfusion of blood may result in an overload of the cardiovascular system, causing an increase in the viscosity of the blood. A more viscous blood could lead to blood clotting and potentially to heart failure.

Erythropoietin

For many endurance athletes, an alternative to blood transfusions is the administration of erythropoietin. Erythropoietin (EPO) is a hormone produced by the kidneys that stimulates red blood cell production. It is this hormone that is responsible for increases in red blood cell volume during exposure to altitude (see chapter 23). Studies of EPO in athletes have been limited. Over a 6-week period, EPO injections increased hemoglobin concentrations 10%, aerobic capacity 6% to 8%, and time to exhaustion up to 17% (Ekblom and Berglund 1991). Four months before the study, 7 of the 15 subjects had reinfused red blood cells (undergone blood doping). Interestingly, the researchers reported that the increases in aerobic capacity and run time to exhaustion were almost identical. A 15-week study on healthy males used a treatment protocol of 5000 IU of EPO every other day for the first 2 weeks, three injections on 3 consecutive days in week 3, and one injection per week for the next 12 weeks (Lundby et al. 2007). A 9.8% increase in red blood cell volume was seen at week 5, a 5.9% increase at week 11, and a 7.8% increase at week 13. The increase in red blood cells was accompanied by decreases in plasma volume to nearly the same magnitude, keeping blood volume relatively the same. Hemoglobin concentrations increased 20% by week 13. The decrease in plasma volume was thought to be mediated by a downregulation of the renin–angiotension–aldosterone axis.

The risks of EPO appear to be serious. The deaths of a number of racing cyclists have been related to EPO administration (Gareau et al. 1996). However, this allegation has not been confirmed. The primary risk associated with EPO is its lack of predictability compared with red blood cell infusion. Once EPO has been injected into the body, the stimulus for producing red blood cells is no longer under control. The risks are similar to those with blood doping. The athlete may be at a greater risk for increased viscosity of the blood, possibly leading to blood clotting and heart failure.

β-Blockers

β-Blockers are a class of drugs that block the β-adrenergic receptors, preventing the catecholamines (e.g., norepinephrine and epinephrine) from binding. β-Blockers are generally prescribed by cardiologists to treat a wide variety of cardiovascular diseases, including hypertension. The ergogenic benefit of these drugs may reside in their ability to reduce anxiety and tremors during performance. Thus, athletes who rely on steady, controlled movements during performance (e.g., archers or marksmen) would appear to benefit from these drugs. In addition, it has been suggested that β-blockers improve physiological adaptations to endurance training by causing an upregulation of β-receptors (Williams 1991). This may result in an exaggerated response to sympathetic discharge (e.g., improved contractile response) during intense exercise once the drugs are discontinued.

Several studies have shown that β-blockers improve both slow and fast shooting accuracy (Antal and Good 1980; Kruse et al. 1986; Tesch 1985). In addition, the dose taken appears to have significant effects on the magnitude of improvement. In shooters given β-blockers in two different doses (80 vs. 40 mg oxprenolol), the group taking the higher dosage shot with greater accuracy (Antal and Good 1980). However, in certain sports, some degree of anxiety may be important. Tesch (1985) reported that bowlers whose performance was improved during blockade with oxprenolol had significantly greater heart rates before, during, and after competition compared with the subjects whose performance did not improve while on β-blockers.

β-Blockers may also have an ergolytic effect (decrease in performance). Studies have shown that β-blockers impair the cardiovascular response to exercise by reducing oxygen and substrate delivery to exercising muscles (Williams 1991). Risks associated with these drugs include bronchospasm in individuals with asthma, light-headedness (due to decreases in blood pressure), increased fatigue, and hypoglycemia in type 2 diabetics (β-blockers can increase insulin secretion).

Summary

The use of performance-enhancing drugs is prevalent among athletes at all levels and in most sports. This chapter reviews a number of the more common drugs in use today. Several have been well researched, whereas only scant information is available about others. Many of the problems associated with studying performance-enhancing drugs are related to the difficulty of conducting studies with dosages that are commonly used by athletes. Most institutional review boards do not allow research with substances in dosages that exceed the FDA's recommendations or may potentially place the subject at risk. In addition, many researchers do not have access to the elite-level athletes who are typically using the drugs. Most of the research is performed on recreationally trained subjects, which leads to a great deal of variability in the outcomes. Consequently, studies have concluded that a particular agent may not have any ergogenic benefit, but this unfortunately may not be applicable to the population that is actually using the drug.

REVIEW QUESTIONS

1. What is the clinical role of androgen administration, and how does dosing differ in competitive athletes who use these drugs?
2. What are the potential adverse effects associated with androgen abuse? Do they differ between the sexes?
3. Describe and provide examples of masking agents.
4. What is the clinical role of growth hormone administration?
5. What ergogenic role does exogenous growth hormone have in an athletic population? Make sure to discuss the efficacy of its role in this population.
6. What type of performance-enhancing drugs might be popular among endurance athletes and why?
7. Provide examples of three performance-enhancing drugs that are used to enhance alertness and focus.

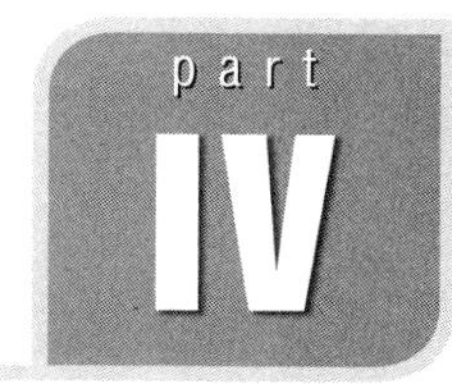

ENVIRONMENTAL FACTORS

This part of the book provides an in-depth review of the physiological stresses associated with various environmental conditions (heat, cold, and altitude). How these environmental stresses influence performance is also discussed. Each chapter considers how athletes can adapt to various environmental conditions, as well as the dangers and risks associated with each of these conditions. The chapter on altitude focuses on how athletes use altitude to stimulate physiological changes that can enhance athletic performance.

chapter 21

Heat

CHAPTER OBJECTIVES

After reading this chapter you should be able to do the following:

- Describe how the body dissipates heat.
- Understand the physiological stresses associated with hyperthermia.
- Describe how exercise is affected by a hyperthermic environment.
- Understand uncompensable heat stress.
- Understand the etiology of heat acclimatization.
- Discuss the heat illnesses.

During exercise, the **metabolic rate** may increase to a level 5 to 15 times higher than that at rest to provide energy for muscle contraction. This higher metabolic rate creates a large heat production (approximately 80% of all energy that is metabolized is transformed to heat) that needs to be dissipated in order to maintain heat balance within the body. This places a large burden on the **thermoregulatory system** to defend the **core temperature** of the body. When exercise is performed in the heat, the strain experienced by the thermoregulatory system is further magnified. The extent of the physiological stress depends on several factors, including the individual's heat **acclimatization**, level of physical fitness, and **hydration** state. These factors affect the athlete's ability to perform. Under severe environmental conditions, exercising in the heat poses a significant risk to health and well-being.

This chapter reviews the physiological response to exercise in the heat. Further discussion focuses on the factors that influence physiological strain during exercise in the heat and their potential impact on athletic performance. In addition, the heat illnesses are reviewed and the mechanisms used to monitor **heat stress** are examined. Finally, the chapter discusses several necessary precautions the athlete should take to minimize the risk of heat injury.

Physiological Response to Exercise in the Heat

During exercise in the heat, the core temperature of the body increases rapidly. This correlates to the increased metabolic heat that is generated during exercise in hot conditions. Until heat loss equals heat production, core temperature will continue to rise. Once heat loss equals heat production, a steady-state core temperature is achieved (Sawka et al. 1993). The primary goal of the thermoregulatory system during exercise in the heat is to remove the heat from the body. If **hyperthermia** (high body temperature) is not controlled, body temperature can rise to a level that places the individual in grave danger. Thermal signals in the hypothalamus sense increases in the core temperature of the body. The thermal regulatory center in the hypothalamus interprets these signals as an imbalance in temperature **homeostasis** and activates a defense mechanism to return core temperature to normal levels. Sweat glands are stimulated to increase sweat production, and the smooth muscle in the arterioles of the skin is relaxed to allow for **vasodilation** and an increase in blood flow to the periphery. The brain also diverts blood flow from internal organs to the skin to expedite **heat dissipation**.

How the Body Dissipates Heat

Body heat can be dissipated to the environment though **evaporative** or **nonevaporative** means. Evaporative heat loss involves the evaporation of water (the result of sweating) from the body. Evaporation accounts for approximately 85% to 90% of the heat dissipation during exercise in a hot, dry environment (Adams et al. 1975). When the exercise environment is hot and wet (humidity above 50-70%), evaporative heat loss is reduced and a greater amount of heat is stored in the body. In hot, wet conditions, the skin becomes hot and reddish in color, which represents the increased blood that has been diverted to the periphery. The body must rely more on nonevaporative mechanisms to dissipate heat. Nonevaporative heat loss is the result of the combined effects of **conduction**, **radiation**, and **convection**. Conduction is heat exchange between two solid surfaces that are in direct contact. The rate of conductance depends on the temperature difference between the two surfaces. Conduction may have limited value during exercise in the heat because of the limited surface area that is in direct contact with the ground. It accounts for less than 2% of heat loss in most situations (Armstrong 2000). Conduction is easily illustrated by the example of a person camping out during the winter months. When the camper sleeps on snow-covered ground, significant heat exchange occurs between the person's body and the ground. In the morning, the camper wakes up lying in a depression that was created as body heat melted the packed snow beneath the tent floor.

Radiation is the transfer of energy waves that are emitted by one object and absorbed by another (Armstrong 2000). Convection is heat exchange that occurs between a surface and a fluid medium. Both air and body fluids can dissipate heat through convective means. As mentioned previously, evaporative heat loss is severely reduced during exercise in a hot, wet environment. The body relies more on radiation and convection to dissipate heat under these conditions. Figure 21.1 shows the possible ways in which the body regulates heat exchange during exercise.

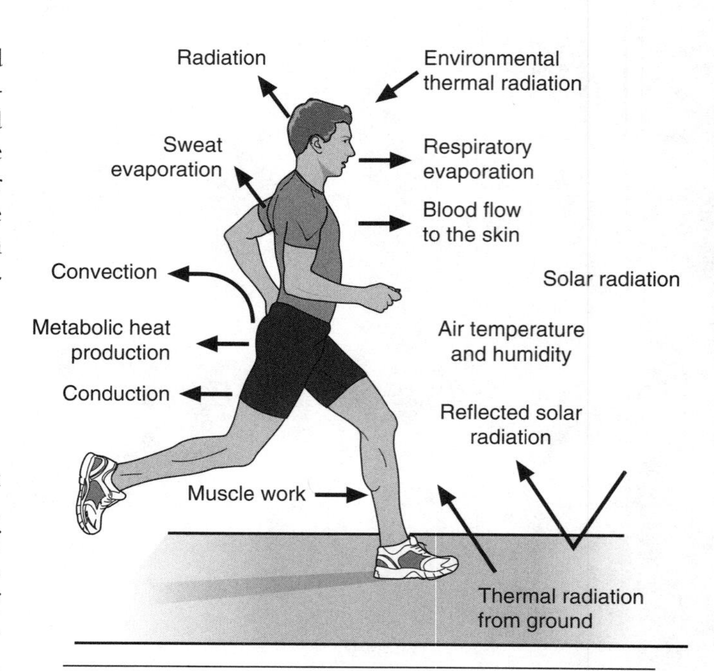

Figure 21.1 Regulation of heat exchange.

Unlike evaporative cooling, which always dissipates heat, the ability of radiation and convection to lower body heat depends on the ambient temperature. Skin temperatures generally range from 93 to 98 °F (34 to 37 °C); if air temperature is 100 °F (38 °C) or greater, heat will likely be added to the body. During conditions in which both ambient temperature (>100 °F [38 °C]) and humidity are high, the ability to dissipate heat through either evaporative or nonevaporative mechanisms becomes limited. In such conditions, the athlete has difficulty dissipating body heat and is at an increased risk for heat illness.

Cardiovascular Response to Exercise in the Heat

During exercise in the heat, the volume of blood that is diverted to the skin may be as high as 7 L/min (Rowell 1986). This increase in blood flow to the periphery enhances heat dissipation through convection and radiation (a higher blood flow to the periphery increases skin temperature). As blood flow to the skin is increased, the blood vessels of the skin become engorged, creating blood pools in the skin (Sawka et al. 1993). The large volume of blood that remains in the periphery causes a reduced **venous return** and less cardiac filling. The reduced cardiac filling causes a cardiovascular strain, reflected by a reduced stroke volume. As a result, **cardiac output** is lowered. To compensate, the heart rate must increase to maintain cardiac output. In addition, blood flow from splanchnic and renal areas is reduced to compensate for the greater blood flow diverted to the exercising muscles and the periphery (Rowell 1986). The effect of exercise in the heat on blood distribution is depicted in figure 21.2.

During exercise in the heat, **sweat rate** is increased to enhance evaporative cooling. Sweat rates vary from athlete to athlete; a rate of 1 L/h is common (Sawka et al. 1993). The highest sweat rate reported in the literature is 3.7 L/h, measured for Alberto Salazar (Armstrong et al. 1986). As sweat loss accumulates, a decrease in blood volume and an increase in plasma tonicity are seen (Sawka and Pandolf 1990). The decrease in blood volume results in a reduction in skin blood flow and a decrease in sweat output. As a result, the ability of the body to dissipate heat is reduced. Consequently, core temperature rises and, in combination with a reduced blood volume, increases both cardiovascular strain and the risk of heat illness. The importance of maintaining proper hydration levels during exercise in the heat cannot be overestimated. Dehydration exacerbates the physiological strain during exercise in a hot environment. The importance of hydration is reviewed in chapter 18.

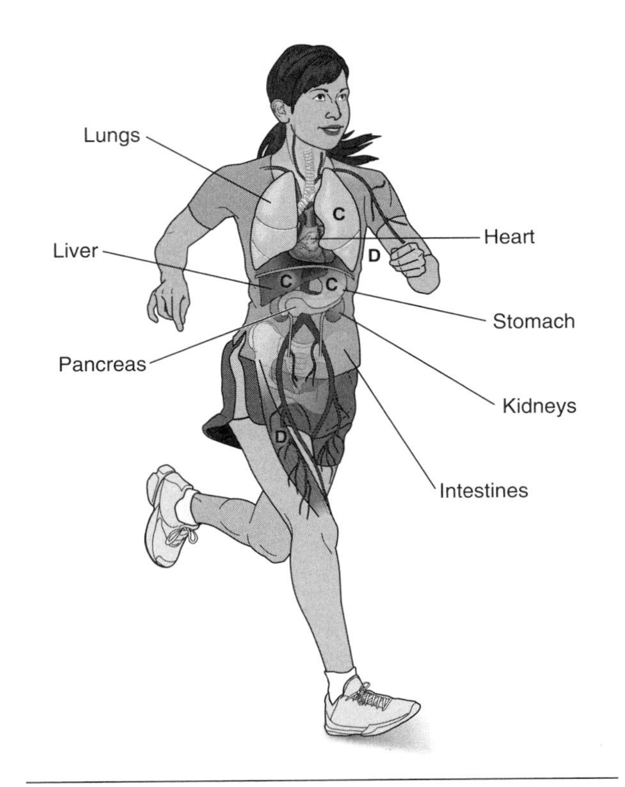

Figure 21.2 Distribution of blood during exercise in the heat. C, blood vessels constricted (smaller); D, blood vessels dilated (larger).

In Practice

Do Sweating Patterns Differ Between Male and Female Athletes?

A study from the United Kingdom produced a whole-body sweat map in females and compared it to previously reported sweat patterns in male athletes (Smith and Havenith 2012). Elite and subelite male (23 ± 3 years) and female (21 ± 1 years) runners were recruited to participate. Sweat was collected at regional locations on the body using absorbent materials applied directly to the skin for 5 min durations. All participants performed a 60 min run: 30 min at 60% of their $\dot{V}O_2$max and 30 min at 75% of their $\dot{V}O_2$max. Exercise sessions were performed in a climate-controlled room at 25.7 °C ± 0.4 °C, 45% ± 7% relative humidity, and a 2 m/s frontal air velocity. Male runners had a significant higher gross sweat loss than females at both exercise intensities (458 ± 115 $g \cdot m^{-2} \cdot h^{-1}$ and 272 ± 103 $g \cdot m^{-2} \cdot h^{-1}$, respectively). Regional sweat rates during the lower exercise intensity for women were highest at the central upper back, heels, and dorsal foot and between the breasts (223, 161, 139, and 139 $g \cdot m^{-2} \cdot h^{-1}$, respectively). Sweat rates increased at all regions as exercise intensity was elevated, with the exception of the feet, ankles, and lateral lower breast. Regional sweat rates were higher for men, but both sexes experienced greater sweating responses on the anterior aspect than the posterior aspect of the body, with the greatest sweat rates seen on the central and lower back and the lowest sweat rates toward the extremities. In conclusion, the investigators reported that in a mild thermal environment, endurance-trained women demonstrated large regional variation in sweat rates with a consistent pattern of distribution. Men had an overall greater sweat rate, likely due to greater metabolic heat production, but both sexes showed similar "high" and "low" sweat distribution patterns. Interestingly, the variation of regional sweat rates in the women could not be explained by differences in regional skin temperature, nor did they correspond with sweat gland densities. These latter points were not consistent with other reports in the scientific literature and provide additional impetus to further exploration of issues related to regional sweat gland densities and sweat rates.

Uncompensable Heat Stress

In certain professions (e.g., firefighters, soldiers engaged in a chemical or biological battlefield), workers are required to wear protective clothing. Often these clothing ensembles are worn in hot and humid environments, which results in an uncompensable heat stress that minimizes the evaporative cooling. **Uncompensable heat stress** exists when the evaporative cooling requirements exceed the environment's cooling capacity (Givoni and Goldman 1972). The additional heat stress created by protective clothing can reduce the time to reach a body temperature of 38.5 °C in a 37 °C environment for a worker performing moderate work, from 90 min (when wearing normal cotton, single-layer working gear) to only 20 min with fully encapsulating clothing that has an impermeable outer layer (Havenith 1999). In addition, the wearing of protective clothing has been shown to reduce the physiological strain a person can tolerate (Montain et al. 1994). These investigators reported that the magnitude of reduction in physiological tolerance (e.g., the time one is able to spend in the heat) with full-encapsulation protective clothing was similar between desert and tropical climates. It appears that high aerobic fitness and habitual exercise provide a significant benefit for increasing exercise–heat tolerance (Cheung and McLellan 1998). Individuals who were highly fit (defined as those engaged in a regular exercise program and having a $\dot{V}O_2$ max >55 ml · kg^{-1} · min^{-1}) were able to tolerate 110 min in 40 °C, 30% relative humidity while wearing a protective suit, while individuals who were moderately fit ($\dot{V}O_2$ max <50 ml · kg^{-1} · min^{-1}) were able to tolerate only 88 min in those environmental conditions. Interestingly, fluid replacement does not alter the rate of heat storage during uncompensable heat stress, but may increase the heat storage capacity during light exercise exceeding 60 min (McLellan and Cheung 2000).

A system of evaluating the insulating effect of clothing has been developed for understanding the effect of various layers of clothing on the ability to reduce heat gain or loss. The unit typically used for measuring clothing insulation is the **Clo** unit, which is reported as $m^2K \cdot W^{-1}$ (watts per square meter per kelvin) (1 Clo = 0.155 as $m^2K \cdot W^{-1}$). A naked person has a Clo value of 0.0, and a person wearing a typical business suit has a Clo value of 1.0. An overall Clo value is calculated by summing the Clo values for all clothing layers. In addition, evaporative heat transfer is affected by the thickness of the clothing ensemble. The moisture permeability index (i_m) describes the efficiency of fabrics in transferring moisture that the ambient air can evaporate (Havenith 1999; Woodcock 1962). An i_m of 0 represents an impermeable layer of clothing, whereas an i_m of 1 indicates that all of the moisture can be evaporated. A typical i_m value for most permeable clothing ensembles in "still air" is about 0.5 (Goldman 2007). For tactical personnel (i.e., military, police, and fire and rescue), the protective clothing ensembles often include several layers of material. For combat personnel, this may include their battle dress uniform (BDU), a ballistic protection vest, a combat vest, and a backpack. If engaged in a combat arena exposed to chemical or biological agents, they don their chemical protection ensemble to provide protection to the entire body. The physiological strain associated with uncompensable heat stress, as discussed earlier, reduces the tactical athlete's ability to function. Several investigations have examined the ability to reduce uncompensable heat stress by providing some cooling mechanism to the ensemble. Ice vests, liquid delivery systems, and cold-air devices have been shown to be effective in reducing the physiological strain associated with uncompensable heat stress (Heled, Epstein, and Moran 2004; Kenny et al. 2011; Pimental et al. 1987).

Many sports also involve elaborate equipment ensembles that can increase physiological strain. For instance, football players may wear undergarments, shoulder pads, pants, girdle with padding, jersey, and a helmet. However, depending on the practice (e.g., shorts only), the ensemble may change. Depending on the equipment worn and the fabric structure, clothing insulation has been reported to range from a 1.15 Clo to a 1.50 Clo, and the permeability index ranged from 0.37 to 0.40 i_m (McCullough and Kenny 2003). A recent study by Armstrong and colleagues (2010) examined two uniform ensembles: A partial uniform ensemble included compression shorts, socks, shoes, gloves, T-shirt, jersey, and pants with knee and thigh pads, while the full uniform included all this plus shoulder pads and helmet. The investigators also used a control condition in which subjects wore compression shorts, athletic shorts, socks, and shoes. The exercise protocol included 80 min of treadmill walking and box lifts in a hot environment (33 °C, 48-49% relative humidity). Results demonstrated a reduced time to exhaustion, increased rate of core temperature elevation, and a correlation between lean body mass and increase in core temperature in the full uniform condition, but not in the partial and control uniform conditions. Depending on the environmental conditions, coaches, trainers, and medical personnel can ease physiological strain by limiting the equipment worn.

In Practice

What Is the Most Valid Device for Assessing Body Temperature During Outdoor Exercise in the Heat?

A multisite study examined the validity of several commonly used devices to measure body temperature during outdoor exercise (Casa et al. 2007). Twenty-five men and women (26.5 ± 5.3 years) volunteered for this study. All subjects were physically active but were not competitive athletes. Body temperature was measured in several ways: (1) rectal thermometer; (2) having subjects swallow an ingestible thermistor to measure gastrointestinal temperature; (3) oral digital thermometer; (4) forehead skin temperature using a forehead sticker; (5) aural temperature using a tympanic ear thermometer; (6) axillary temperature; and (7) temporal artery temperature measured by a temporal artery scanner. The rectal temperature assessed by a rectal probe was considered the gold standard. All participants performed all measurements in euhydrated conditions. Participants remained active throughout the measurement period by playing team sports such as soccer and Ultimate Frisbee. Temperature was assessed every hour during the 3 h exercise period and every 20 min during the 1 h recovery period. Short rest periods (5-10 min) were provided each hour to provide time for temperature measures. Compared to rectal temperature, gastrointestinal temperature was the only measurement that provided accurate core body temperature. The other measures—oral, axillary, aural, temporal artery, and forehead temperatures—were reported to be significantly different and invalid for assessing hyperthermia in individuals exercising outdoors. This study has very important implications for medical venues during outdoor sporting events, in which core temperature measures influence triage and treatment decisions.

Heat and Performance

Exercising in the heat has significant effects on athletic performance. However, the magnitude and direction of the effect (degree of improvement or decrement) depend on the extent of hyperthermia as well as the mode of activity being performed.

Aerobic Exercise in a Hot Environment

When exercise is performed in the heat, the **maximal aerobic capacity ($\dot{V}O_2max$)** of the athlete has been reported to be lowered (Sawka, Young, Cadarette, et al. 1985; Smolander et al. 1986). $\dot{V}O_2max$ appears to decrease regardless of the individual's state of heat acclimatization or physical fitness (Sawka, Young, Cadarette, et al. 1985). Not only is **aerobic power** reduced during exercise in the heat, but the time to reach exhaustion decreases also (MacDougall et al. 1974).

The physiological mechanisms responsible for these performance decrements are likely related to the increased blood flow diverted to the peripheral vasculature. As mentioned in a previous section, cardiac output may decrease because of the reduction in venous return. Cardiac output during exercise in the heat has been reported to be 1.2 L/min below that seen during exercise in a temperate environment (Rowell 1986). Dilation of the peripheral vascular beds may also result in a diversion of some blood from the exercising muscles to the skin (Sawka and Young 2000). This would also contribute to the reduced cardiac output available to the muscles, affecting the metabolism of muscle contraction.

During submaximal exercise in the heat, an increase in core temperature causes a shift in metabolism from primarily aerobic to **anaerobic** in the exercising muscles and the liver (Dimri et al. 1980; Young et al. 1985). Using postexercise oxygen uptake measures, Dimri and colleagues (1980) showed that as ambient temperature increases, metabolic rate also increases, and a larger percentage of energy use is derived from anaerobic sources. Young and colleagues (1985) provided further evidence of a greater reliance on anaerobic metabolism by demonstrating a greater lactate concentration in both muscle and blood after exercise in the heat. The greater reliance on anaerobic metabolism indicates an increased reliance on carbohydrate stores within the body to fuel exercise. This likely plays a significant role in the quicker rate of **fatigue** seen during exercise in the heat.

Anaerobic Exercise in a Hot Environment

The relatively few studies on anaerobic exercise in the heat have generally shown either no change (Dotan and Bar-Or 1980; Stanley et al. 1994) or an increase in strength or power performance (Falk et al. 1998; Sargeant 1987). Sargeant (1987), using a warm-water (111 °F [44 °C]) bath immersion, reported significantly greater maximal power performance; however, the increase in maximal power was accompanied by an increase in the rate of fatigue. In contrast, Falk and colleagues (1998), using a study design that simulated a competitive situation (bouts of high-intensity exercise [five 15 s Wingate anaerobic tests] interspersed with relatively short **recovery** periods [30 s between each bout]) in the heat (95 °F [35 °C], 30% relative humidity), reported an increase in power performance but no difference in fatigue rate. The improved power production during exercise in the heat may be related to a warm-up effect. The higher ambient temperatures likely cause a greater increase in muscle temperature. An increase in muscle temperature of 9 °F (5 °C) has been hypothesized to cause a 10% increase in power development (Binkhorst, Hoofd, and Vissers 1977) in some studies, whereas others have suggested that for each 1.8 °F (1 °C) rise in muscle temperature, a 4% improvement in power performance may be seen (Sargeant 1987). The higher power performances in the heat may be related to an increase in speed of muscle contraction or a faster metabolic rate (Falk et al. 1998).

In the study by Falk and colleagues, a high ambient temperature did not appear to affect recovery from exercise compared with a thermoneutral condition (72 °F [22 °C]), 40% relative humidity) (figure 21.3). In that study, subjects performed a series of five bouts of 15 s of the Wingate anaerobic test (WAnT) separated by 30 s of active recovery (cycling with no resistance). The 3 min 15 s exercise–rest period was followed by 60 min of passive recovery in the heat. After the recovery period, another series of five 15 s WAnTs was performed without any observed decrement in performance. The ability to maintain anaerobic power performance in a hot environment may have been related to the ability to keep the subjects euhydrated during the recovery period.

Heat Acclimatization

During initial exposure to the heat, the exercising individual experiences weakness, dizziness, flushed skin, and other signs and symptoms that represent heat stress. However, after several days of heat exposure, a reduction in these symptoms occurs and the person's heat tolerance improves. This is a result of a number of physiological adaptations that improve the thermoregulatory function of the body. When such adaptation occurs in a naturally hot environment, it is termed *acclimatization*. However, if it is accomplished in an artificial environment, as when one exercises in a heat chamber in the middle of the winter, it is termed *acclimation*. Acclimatization and acclimation produce similar physiological adaptations; thus, acclimatization is the inclusive term used to refer to both types of adaptation (adaptation occurring in either a natural or laboratory setting).

Acclimatization to heat occurs after repeated exposures to a heat stress that is sufficient to raise core body temperature and bring about moderate

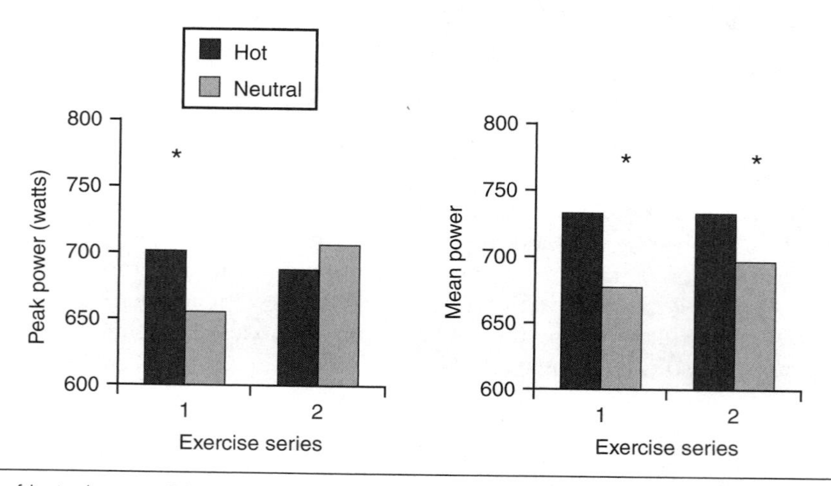

Figure 21.3 Effect of hot, dry conditions on anaerobic power performance. * = Significant difference between thermal conditions.

Data from Falk et al. 1998.

to profuse sweating (Wenger 1988). This is best accomplished through exercising in the heat at an intensity ranging from 50% to 95% of $\dot{V}O_2max$ for approximately 1 h a day. Heat acclimatization appears to occur in two stages. After a few days of heat exposure, several adaptations are noted, resulting primarily from a reduction in the cardiovascular strain (e.g., reduced heart rate) at a relative level of exercise in the heat (Armstrong and Maresh 1991; Wenger 1988). The reduced cardiovascular strain is also associated with an increase in both plasma volume and exercise tolerance and a decrease in core body temperature and perceived level of exertion. As exposure to the heat is prolonged (up to 14 days), further adaptations may be observed. An increase in sweat rate and sweat sensitivity (i.e., sweat loss per degree rise of core body temperature), as well as a decrease in **electrolyte** losses in both sweat and urine, may also occur. However, these latter adaptations may depend on the environmental conditions (Armstrong et al. 1987). Table 21.1 summarizes the adaptations commonly seen during heat acclimatization.

Sweat rate is different when heat acclimatization occurs in a dry versus a humid environment. During dry heat exposure, little to no change in sweat rate is seen; however, during acclimatization in hot, humid conditions, increases in sweat rate do occur (Armstrong and Maresh 1991; Wenger 1988). Under hot, dry conditions, there is a high rate of evaporative cooling. However, in a hot, humid environment, in which ambient water vapor pressure is high, the body needs to increase the wetted skin area to achieve a cooling effect similar to that experienced in hot, dry conditions. The body accomplishes this either by increasing sweating in areas that did not produce much sweat before acclimatization or by increasing the sweating intensity (more profuse sweating) (Wenger 1988). A greater sweat rate at body surface areas that already produce a large sweat concentration would have a minimal effect on whole-body cooling, because much of the increase in sweat production would be wasted in dripping. In order to maximize efficiency of adaptation, the body selectively increases sweating in areas that previously had poor sweat rates.

Questions about heat acclimatization frequently concern whether one can become acclimatized without exercising (passive heat acclimatization) and whether a high aerobic capacity provides inherent protection during exercise in the heat. Individuals native to a hot or tropical climate are reported to have a lowered core temperature and a more efficient sweating response to exercise (Wenger 1988). However, for people not native to such a climate, heat exposure without exercise also induces a heat acclimatization response that is reflected by an improved ability to dissipate heat. Yet to maximize heat acclimatization, the athlete must exercise in the heat. High-intensity exercise has been thought to be necessary to increase exercise heat tolerance (Gisolfi and Robinson 1969; Shvartz et al. 1977). However, ample evidence demonstrates comparable heat tolerance when exercise is performed at mild to moderate exercise intensities (>40% of $\dot{V}O_2max$) (Armstrong and Maresh 1991; Houmard et al. 1990). Full heat acclimatization is seen after 14 consecutive days of exercising in the heat, and partial heat acclimatization can be seen after 4 days of exercising in the heat even at a mild exercise intensity (47% of $\dot{V}O_2max$) (Hoffman et al. 1994).

Table 21.1 Physiological Adaptations to Heat Acclimatization[1]

Adaptation	Days of heat acclimatization													
	1	2	3	4	5	6	7	8	9	10	11	12	13	14
Heart rate decrease			—	—	—	—								
Plasma volume expansion			—	—	—	—								
Rectal temperature decrease					—	—	—	—						
Perceived exertion decrease			—	—	—	—								
Sweat Na^+ and Cl^- concentration decrease[2]					—	—	—	—	—	—				
Sweat rate increase								—	—	—	—	—	—	—
Renal Na^+ and Cl^- concentration decrease			—	—	—	—	—							

[1]Point at which approximately 95% of the adaptation occurs.

[2]During consumption of a low NaCl diet.

Adapted from Armstrong and Dziados 1986.

This ability to partially heat acclimatize after only a few days of exercise in the heat has an important practical benefit. As an example, consider a football team based in the northern or eastern United States. The team's next game is to be played in a warm climate (e.g., Florida or Arizona) in the middle of November. The players' ability to tolerate the heat will be a strong factor affecting the outcome of the game. This poses a tremendous advantage for the opponent. However, several possible approaches may be tried to partially acclimatize the athletes to enhance their heat tolerance for the game. With only a week to prepare, it is not possible to attain full heat acclimatization. The first potential solution involves flying the team to the opposing city early in the week to practice. Another potential solution is to practice indoors (if an indoor facility is available) and increase the heat during practice to simulate the weather expected at the game. If this is not possible, perhaps the players could be heat acclimated in a laboratory facility that has a heat chamber or some other facility where room temperature can be controlled. Within several days of exercising in the heat, the players will have a greater tolerance. This minimizes the advantage that the warm-weather team would have had.

A high level of physical fitness may provide some benefit for tolerating a heat stress but does not appear to substitute for exercising in the heat to produce heat acclimatization. Most researchers agree that a high level of physical fitness improves the physiological responses to exercise in the heat (Armstrong and Pandolf 1988; Armstrong and Maresh 1991). Although a high fitness level is not as beneficial as full heat acclimatization, it may enhance the rate at which full or partial heat acclimatization can be reached (Pandolf, Burse, and Goldman 1977). Several studies have shown that intense training for 2 weeks in a temperate environment does not produce the same heat tolerance level as exercising in the heat (Gisolfi and Cohen 1979; Strydom and Williams 1969). However, an interesting study by Armstrong and colleagues (1994) showed that exercise heat tolerance could be maintained in highly trained endurance athletes through the winter months as long as rigorous training was continued. Other studies suggest that the heat tolerance of highly trained endurance athletes may improve further if they undergo a heat acclimation regimen (Gisolfi 1973; Piwonka et al. 1965).

Heat Illnesses

Heat stress places an added burden on the cardiovascular system to maintain thermoregulation. However, when exercise is performed in the heat, the control of hemodynamic stability to maintain exercise performance may take precedence over thermoregulation, even to an extent that places the individual at great risk of heat illness (Hubbard 1990). In other words, during exercise, blood flow to active muscles is not diverted to peripheral tissues to reduce the risk of heat injury. Four heat illnesses are commonly seen among athletes: **heat cramps**, **heat exhaustion**, **exertional heatstroke**, and **heat syncope**. Although some have suggested that these heat illnesses lie on a continuum, it is possible to exhibit symptoms of any of them without manifesting symptoms of the others (Armstrong and Maresh 1993). The incidence of heat illnesses is reduced when the athlete becomes acclimatized to the heat (Armstrong and Maresh 1991).

Heat Cramps

During prolonged or repetitive exercise, the athlete may experience painful contractions within the exercising muscles. These heat cramps are believed to be the result of a large loss of Na^+ and Cl^- in sweat and the replacement of sweat loss with dilute fluid, pure water, or both (Armstrong 2000). Heat cramps should not be confused with exercise-induced muscle cramps. Heat cramps are usually observed in the large muscles of the extremities or in the abdomen. They usually begin as a weak tingling sensation and progress to a localized contraction of several muscle fibers that appears to wander over the muscle as adjacent motor units become activated. Unlike exercise-induced muscle cramps, which appear to affect the entire muscle, heat cramps do not typically affect the entire muscle mass.

Although the exact mechanisms that induce heat cramps are not clear, it is thought that the loss of NaCl causes a shift in the intracellular–extracellular NaCl and water ratio. This in turn results in an alteration of the electrical properties of the muscle membranes and could alter muscle contraction or relaxation (Armstrong 2000). The importance of electrolytes in both the development and treatment of heat cramps is becoming clearer. A significant reduction of NaCl in the urine of individuals with heat cramps has been reported (Leithead and Gunn 1964) and appears to suggest that the kidneys are reabsorbing sodium because of the whole-body electrolyte imbalance. Further evidence of the involvement of electrolytes was presented in several reports showing a rapid and permanent relief of heat cramps with the consumption of fluid–electrolyte beverages (Armstrong and Maresh 1993; Bergeron 1996).

The basic treatment for heat cramps is to restore homeostatic equilibrium by replacing both water and electrolytes. Intravenous (IV) solutions are effective and result in rapid relief. Oral NaCl solutions—two 10-grain salt tablets dissolved in 1 L of water—are another common treatment. Preventive measures may include consuming salt in the diet or isotonic electrolyte solutions during exercise.

Heat Exhaustion

Heat exhaustion is the most common form of heat illnesses. It is considered a volume depletion problem and primarily occurs when cardiac output is reduced, resulting in an inability of the cardiovascular system to meet the demands of both the exercising muscle and peripheral tissues. It is defined as an inability to continue exercise in the heat (Armstrong and Maresh 1993; Hubbard and Armstrong 1989). Symptoms include various combinations of nausea, vomiting, irritability, headache, anxiety, diarrhea, chills, piloerection, hyperventilation, tachycardia, hypotension, and heat sensations in the head and upper torso (Armstrong et al. 1987). Orthostatic intolerance (dizziness) and syncope (fainting) may also occur. Rectal temperature is usually less than 40 °C. The signs and symptoms of heat exhaustion are quite variable and may differ under different exercise–heat scenarios (Armstrong and Maresh 1993).

Since heat exhaustion is the result of volume depletion (plasma) resulting in inadequate cardiovascular compensation, simply rehydrating often results in complete recovery within 15 to 30 min (Hubbard and Armstrong 1988). In most cases, fluids can be replenished orally; however, in severe cases of heat exhaustion in which the subject is unconscious or is lacking in mental acuity, IV solutions may be the preferred method of fluid replenishment.

During heat exhaustion, volume depletion can be attributed to either a salt depletion or water depletion. Often it is a combination of the two forms, which have similar signs and symptoms. However, the treatments for these two forms of volume depletion are different. Water depletion occurs quite rapidly and may be the result of a single exercise session. In such a case, fluid replenishment, which may include only drinking plain water, is effective in restoring the body to its normal euhydrated state. Salt depletion is generally the result of a prolonged electrolyte deficit (approximately 3-5 days) brought about by excessive water loss in sweat that could not be replenished through dietary salt intake (Armstrong 2000). In a salt depletion situation, it is imperative that an appropriate amount of NaCl be included with fluid intake. Armstrong (2000) suggested that salt and water losses can be estimated as 2 g NaCl and 1.5 L water per hour of continuous moderate to heavy exercise. Appropriate adjustments to the IV or oral solution should be made accordingly.

Exertional Heatstroke

Unlike heat exhaustion, in which volume depletion leads to a progressive fatigue and an inability to continue exercise, heatstroke is one of the few illnesses that can threaten the life of a healthy athlete. A rectal temperature greater than 102 °F (39 °C) and an increase in serum enzymes (e.g., ALT, AST, CPK, and LDH), indicative of cellular damage, are characteristics that differentiate heat exhaustion from the more severe and life-threatening exertional heatstroke. In addition, during heat exhaustion, spontaneous body cooling may still be present or perhaps variable in a severe case; however, during heatstroke, spontaneous body cooling is absent (athlete will have hot, dry skin). Another distinguishing characteristic of heatstroke is the bizarre behavior exhibited by the victim. A person who has exertional heatstroke may be disoriented with a complete loss of mental acuity or, in a most severe case, may become comatose.

The mortality rate or the extent of multisystem tissue damage is closely related to the patient's core temperature at the point of collapse and the time elapsed between the patient's collapse and the initiation of cooling. If cooling therapy is begun rapidly, the risk of death or disability is dramatically reduced (Costrini 1990). The method of cooling also affects the recovery rate of the heatstroke patient.

The primary goal of cooling therapy is to lower the core body temperature to a safe and non–life-threatening level (99 °F [37 °C]). A number of different cooling methods have been used to lower core body temperature, including ice packs placed at the groin, armpits, and neck; ice packs covering the body; air and water sprays at both warm and cool temperatures; and ice-water or cool-water immersion. Although concern has been raised that ice- or cool-water baths may cause a shivering-induced heat production or vasoconstriction of peripheral blood vessels, these issues are minimal or nonexistent in the vast majority of heatstroke cases (Armstrong et al. 1996). The rapid cooling observed with immersion was demonstrated to be the most effective cooling technique, and the benefits were seen without any of these concerns becoming an issue (Casa et al. 2007).

Heat Syncope

Heat syncope (fainting) is rare in a conditioned athlete. In most instances, it occurs when individuals stand for a prolonged period of time in the heat or exercise for a prolonged period in an upright position. Heat syncope is caused by a pooling of blood in the vasculature of the limbs and skin because of excessive ambient temperatures. In response to the hot environment, the cutaneous vesicles of the skin dilate to allow for greater cooling. The increase in vasodilation reduces the volume of blood that is returned to the heart, which decreases cardiac output and lowers blood pressure. Blood flow to the brain is therefore reduced, resulting in a syncopic episode.

The medical diagnosis of heat syncope is based on a fainting spell in the absence of an elevated rectal temperature (Armstrong 2000). Before the syncopic episode, the patient may experience nausea, weakness, tunnel vision, or vertigo. Treatment for heat syncope is to replace any fluid and electrolyte deficits and have the patient lie in a horizontal position with the feet elevated. The horizontal position allows for a greater venous return to the heart. Subsequently, cardiac output and blood pressure increase, resulting in a return of normal blood volume to the brain.

Monitoring Heat Stress

When exercise is performed in the heat, it is imperative to take the necessary precautions to minimize the athlete's risk of heat illness. One of the first steps is to monitor the environment. Weather reports that provide both temperature and humidity for the geographical region of concern help assess the risk of heat illness during training or competition (figure 21.4). The risk of heat illness is determined by the intersection of temperature and relative humidity. If the intersection is in the "moderate risk" zone, the athlete should be monitored for signs and symptoms of heat exhaustion or exertional heatstroke (table 21.2). If the intersection lies in the "high risk" to "very high risk" zones, the exercise plan needs to be revised. It may be necessary to postpone exercise until weather conditions become less dangerous or, at the very least, reduce the distance or duration of the workout (Armstrong 2000).

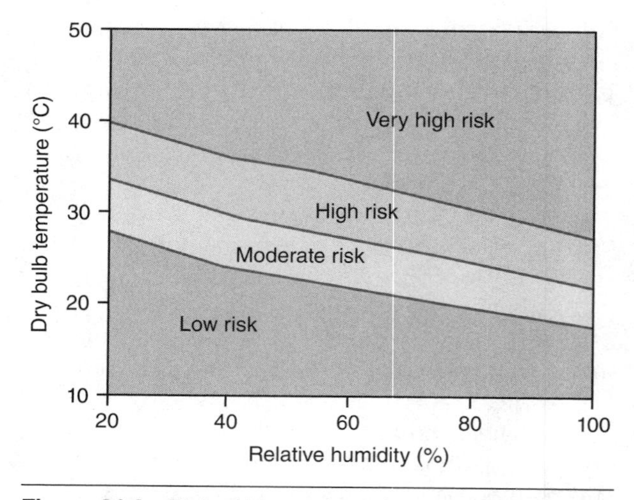

Figure 21.4 Risk of heat exhaustion or heatstroke in hot environments.

Adapted from Armstrong et al. 1995.

Table 21.2 Warning Signs for Heat Exhaustion and Exertional Heatstroke

Heat exhaustion	Exertional heatstroke
Headache, irritability	Headache
Tingling or numbness in head, neck, back, or limbs	Unconsciousness, coma*
Chills or shivering	Loss of mental clarity
Great fatigue*	Bizarre behavior*
Rapid, weak pulse	Rapid, strong pulse
Pale, moist, cool skin	Hot, red skin*
Dizziness*	Profuse sweating in most cases
Vomiting, nausea	Fainting
Dehydration*	

*Important; these occur in most cases.

Adapted from Armstrong 2000.

Heat Stress Indices

To quantify the strain of the thermal environment, a concerted effort has been made to develop an index that provides a measure of risk. This type of index can be used to minimize heat-related injuries and make appropriate adjustments to training schedules. The heat stress index is a single value that integrates the effects of the basic parameters of a thermal environment that will vary with the thermal strain experienced by the individual (Parsons 2003). A number of heat stress indexes have been developed, but many of these have proven to be too complicated for significant use by personnel in the field.

One of the more widely accepted indexes is the wet bulb globe temperature. The wet bulb globe temperature (WBGT) is a heat stress index that can be obtained at the site of activity and used to indicate the level of environmental stress (Armstrong 2000). The WBGT quantifies perceived heat stress imposed through all four heat exchange pathways. It consists of a dry bulb temperature (measuring ambient air temperature), a wet bulb temperature (dry bulb thermometer measured under a water-saturated cloth wick), and a black globe temperature (dry bulb thermometer placed inside a black metal sphere). It provides both ambient temperature and relative humidity at the site of activity. The black globe temperature ensures full absorbance of radiation and constant exposed surface area, regardless of the location of the radiative heat source. The importance of this apparatus over a regular thermometer should not be underestimated. Ambient temperature alone is not sufficient to determine heat stress. The dry bulb temperature accounts for only 10% of the heat stress index, whereas the wet bulb temperature (indication of relative humidity) accounts for 70% (Yaglou and Minard 1957). WBGT index is as follows:

$$\text{WBGT} = (0.7 \text{ temp wet bulb}) + (0.2 \text{ temp black globe}) + (0.1 \text{ temp dry bulb})$$

In the absence of a significant radiant load, the equation can be

$$\text{WGBT} = (0.7 \text{ temp wet bulb}) + (0.3 \text{ temp dry bulb}).$$

The result of this index corresponds to the four categories seen in figure 21.4. A WBGT index above 82 °F (28 °C) places the athlete at a very high risk. A WBGT index between 73 and 82 °F (22-28 °C) places the athlete at a high risk. A WBGT index between 64 and 73 °F (18-22 °C) places the athlete at a moderate risk, and a WBGT index below 64 °F (18 °C) is considered a low risk for heat illness. Although a heat stress index of 63 °F (17 °C) is considered a low risk for heat illness, there is no guarantee that the athlete will not experience heat exhaustion or exertional heatstroke. A number of factors (e.g., sleep deprivation, hydration status, and diet) may interact to increase the individual's risk of heat illness, even at a relatively low heat stress index (Armstrong, De Luca, and Hubbard 1990; Epstein 1990).

An inherent limitation of the WBGT is its limited applicability across a broad range of scenarios and environments, as measuring globe temperature is inconvenient (Moran and Pandolf 1999). The black globe temperature is measured by a temperature sensor placed in the center of a thin copper matte-black globe (diameter, 150 mm), which in many circumstances is cumbersome and impractical. Moran, Shitzer, and Pandolf (1998) also reported that many heat strain indexes were valid only under certain conditions. They suggested that some of the difficulty related to the number of parameters measured, which contributed to increased error. The authors developed a physiological strain index to provide a measure of the combined strain of the cardiovascular and thermoregulatory systems from workloads performed in thermal environments. Their results showed that rectal temperature (T_{re}) and heart rate (HR) were valid indicators of the physiological strain experienced by individuals performing in the heat. This is the physiological strain index (PSI) equation:

$$\text{PSI} = 5(T_{ret} - T_{re0}) \times (39.5 - T_{re0})^{-1} + 5(HR_t - HR_0) \times (180 - HR_0)^{-1}$$

where T_{re0} and HR_0 are initial T_{re} and HR, and T_{ret} and HR_t are simultaneous measurements taken at any time. T_{re} and HR, which depict the combined loads of the cardiovascular and thermoregulatory systems, were assigned the same weight constant of 5. The index was scaled to a range of 0 to 10 (no or little strain to very high strain). Moran and colleagues (2003) later combined the physiological strain index with an environmental stress index (ESI) to provide a simple yet complete index for work–rest cycles. It was suggested that the ESI could replace the WBGT as a quick and simple index for understanding the heat strain of present environmental conditions. The ESI was developed from fast-reading meteorological response sensors of ambient temperature (T_a), relative humidity (RH), and global radiation (GR), which require only a few seconds to reach equilibrium. The ESI is calculated as follows:

$$\text{ESI} = 0.63T_a - 0.03\text{RH} + 0.002\text{SR} + 0.0054(T_a\,\text{RH}) - 0.073(0.1 + \text{SR})^{-1}$$

To integrate the two indexes, only the heart rate measures of the PSI were used. The integration appeared to provide a simple method of categorizing the strain of the environment that would assist in reducing the risk of heat injuries.

Summary

Exercise performed in the heat places emphasis on removing heat and maintaining the core temperature of the body. This strains the cardiovascular system, which must distribute blood to the exercising muscles and to the peripheral blood vessels to help cool the body. The adjustments that the body needs to make may have a negative effect on exercise performance (primarily aerobic) and may also place the athlete at a greater risk of heat illness. If the body is unable to control increases in core temperature, the resulting hyperthermia may have lethal consequences. Although heat acclimatization causes physiological adaptations that improve heat tolerance, it will not prevent heat illnesses if the heat stress is severe. Regardless of the level of heat acclimatization, an athlete who is not properly prepared for exercising in the heat may still be susceptible to heat illnesses. The following recommendations may help minimize the risk of heat illnesses for individuals preparing to exercise in the heat:

- Before exercise, determine the heat stress index and make appropriate adjustments to the workout.
- Wear lightweight, loose-fitting clothing.
- Exercise at cooler times of the day.
- Avoid long warm-ups on hot, humid days.
- Ensure proper hydration before exercise (body weight should be within 1% of average body weight).
- Ensure proper hydration during exercise by replacing fluids lost through sweating (drink 1 L water for every kilogram of body weight lost during exercise).
- Ensure proper salt intake at meals.
- Know the signs and symptoms of heat illnesses.

REVIEW QUESTIONS

1. Explain the difference between heat acclimatization and heat acclimation.
2. Describe how the body dissipates heat.
3. How does heat dissipation differ in a hot and dry environment compared to a hot and wet environment?
4. What is uncompensable heat stress?
5. Describe the effect that a hot environment has on endurance performance.
6. What are the physiological adaptations associated with heat acclimatization?
7. Name five warning signs for both heat exhaustion and heatstroke.

chapter

22

Cold

CHAPTER OBJECTIVES

After reading this chapter you should be able to do the following:

- Understand the factors related to heat loss during a cold stress.
- Discuss the physiological response to exercise in the cold.
- Understand the role of body composition and thermal balance.
- Determine physiological adaptations to cold stress.
- Discuss the effect of hypothermic conditions on exercise performance.
- Understand injuries related to cold exposure.

In contrast to the situation with heat illnesses, many cold-induced injuries can be prevented through behavioral regulation during exercise. The ability of humans to insulate themselves from a harsh environment serves as a protective measure against the dangers of the cold. Technological advancements in the design of clothing allow individuals to insulate themselves in a way that permits ease of movement while causing only minimal increases in energy costs. Nevertheless, athletes may still be subjected to **cold stress** either by self-imposed actions (e.g., not dressing appropriately) or by sudden changes in environmental conditions during an event. Under such circumstances, the physiological response of the individual is altered to combat the specific challenge and maintain **thermal homeostasis**. Such changes can have implications for both exercise performance and the risk of cold-related injuries. This chapter reviews factors that contribute to heat loss, physiological adaptations to the cold, potential performance effects of exercising in the cold, and the types of injuries caused by cold exposure.

Cold Stress: Factors Contributing to Heat Loss

Exposure to cold, whether it is cold air or cold water, results in a transfer of heat from the body to the environment. The heat transfer progresses from the core of the body to the skin and from the skin to the environment. Most heat loss from the skin occurs through **conductive** or **convective** mechanisms (see chapter 21 for a review of these heat transfer modes). As ambient temperature drops below the **core temperature** of the body, a gradient develops that results in a loss of body heat. The difference between the core temperature and the environment can be further magnified by wind speed. Wind accelerates body-heat loss by removing warm air trapped in insulative clothing, increasing evaporative cooling when insulative material is wet, and increasing evaporative cooling directly from the skin when the skin is wet (Hamlet 1988; Armstrong 2000). The term ***windchill*** is used to refer to the combined effects of cold ambient temperature and air movement. A windchill index (figure 22.1) has been developed to reflect the relative risk for freezing injuries to tissue, but several investigators have pointed to limitations of this index (Danielsson 1996; Kaufman and Bothe 1986; Sawka and Young 2000). The windchill index may overestimate the effect of increasing wind speed and underestimate the effect of lowering skin temperature (Danielsson 1996). In addition, the index estimates only the risk for freezing of exposed skin, not well-insulated skin (Kaufman and Bothe 1986). Thus, someone who is well insulated is not at as high a risk in a cold, windy environment as might be predicted by the index. However, the index may still have merit if any part of the body is exposed (e.g., fingers or face) and therefore at risk of freezing.

When the body is in cold water, the individual is at a much greater risk of cold injury than in any other environmental condition. The transfer of body heat through conductive mechanisms is approximately 25 times greater in the water than it is in the air (Toner and McArdle 1988). During immersion in icy water, total body cooling may exceed 6 °C/h. At this rate of heat loss, death could occur within 45 min to 3 h (Hayward and Eckerson 1984). Even when exercise is performed in cold water, the **metabolic heat** generated is often not sufficient to compensate for the rapid body cooling. This is witnessed in swimming, when heat loss is augmented by both conductive and convective means (Nadel et al. 1974). The debilitating effects of a cold, wet environment are frequently seen during exercise performed in the rain or during partial water immersion, as the insulative ability of clothing is compromised when it becomes wet. This facilitates heat loss through conductive, convective, and **evaporative** means.

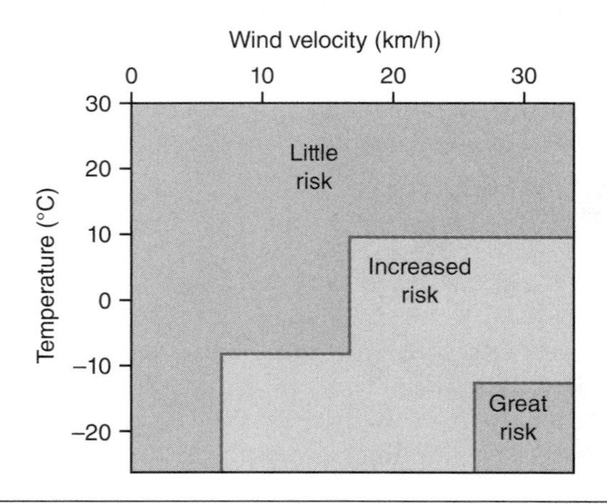

Figure 22.1 Windchill chart: risk of freezing tissue.
Adapted from Armstrong 2000.

Physiological Responses to Exercise in the Cold

Exercise in the cold, whether conditions are cold and dry or cold and wet, results in physiological responses that focus on bringing the body back to thermal homeostasis. During exposure to a cold environment, core body temperature falls because of the rapid transfer of heat from the core to the periphery. To defend core temperature, the peripheral blood vessels constrict. **Vasoconstriction** appears to occur when skin temperature falls below 95 °F (35 °C) and becomes maximal when skin temperature drops below 88 °F (31 °C) (Veicteinas, Ferretti, and Rennie 1982). The defense of core temperature is also accomplished by increasing metabolic heat production, primarily through a **shivering** response of the muscles. However, a portion of the increased **thermogenesis** during cold exposure may be the result of an increase in metabolism that is not related to muscular contraction (Toner and McArdle 1988).

In certain animals, an increase in metabolic heat production in the cold occurs without the need for muscle activity. This source of heat is thought to result from an elevated rate in aerobic metabolism of brown adipose tissue (Toner and McArdle 1988). Although the metabolism of brown fat may also be seen in human infants during a cold stress, its contribution to thermogenesis in adults has not been shown to be significant. Nonshivering thermogenesis in adults appears to result primarily from an increase in metabolic heat production stimulated by elevated catecholamines, glucocorticoids, and thyroid hormones (Toner and McArdle 1988). However, the actual contribution of this potential source of increased metabolic heat has not been determined.

Shivering accounts for much of the increased metabolism during a cold stress. After several minutes of cold exposure, the muscles of the torso begin to rhythmically contract, followed by the muscles of the extremities (Horvath 1981). Seventy percent of the total energy liberated during shivering is liberated as heat, with the remainder generating external force (Sawka and Young 2000). As the cold stress becomes more severe, the extent of shivering increases. Shivering has been reported to cause an increase in oxygen uptake to a level that corresponds to 15% of the individual's $\dot{V}O_2max$ (Young et al. 1986). Although maximal shivering is difficult to quantify, increases in oxygen consumption approaching 46% of the subject's $\dot{V}O_2max$ have been reported during immersion in 54 °F (12 °C) water (Golden et al. 1979).

The vasoconstrictor response to cold exposure helps to reduce heat loss and defend core temperature. However, this results in a significant decline in skin and muscle temperatures. Cold-induced vasoconstriction has pronounced effects on exposed skin, especially the hands and fingers, making them particularly susceptible to cold injury and a loss of dexterity (Brajkovic and Ducharme 2003). A physiological effect often reported during cold exposure is a return of blood flow and temporary rewarming that occurs in the fingers and the forearm. This phenomenon was described by Sir Thomas Lewis in 1930 as a "hunting" response (Lewis 1930). It is more commonly referred to as **cold-induced vasodilation** (CIVD), and it serves to modulate the effects of vasoconstriction (O'Brien 2005). It involves periodic oscillations of skin temperature following the initial decline during cold exposure. CIVD provides a protective effect in maintaining tissue integrity and minimizing the risk of cold injury (Cheung 2010) and may be beneficial for improving dexterity and tactile sensitivity during exposure to the cold (Daanen 2003). The mechanisms for CIVD have not been well defined, but several hypotheses have been suggested. These include (a) an increase in vasodilatory substances in the blood (e.g., nitric oxide) that act on cutaneous blood vessels; (b) norepinephrine modulating CIVD; or (c) fatigue resulting from sustained contraction of arteriovenous anastomoses during cold, causing a release of vasoconstrictory tone and resulting in an increase in blood flow and skin temperature (Cheung 2010).

Effect of Exercise on Maintaining Thermal Balance

Although shivering contributes significantly to metabolic heat production, the best way to maintain thermal homeostasis is through exercise. The increase in physical activity during a cold stress generates enough metabolic heat that shivering may not be needed. Claremont and colleagues (1975) showed that core temperature could be maintained within 0.9 °F (0.5 °C) when exercise is performed at ambient temperatures ranging between 32 and 95 °F (0 and 35 °C). When the individual is properly attired, exercise in temperatures as low as −22 °F (−30 °C) may be sustained without significant changes in core temperature (Toner and McArdle 1988). Clearly, physical activity is the greatest contributor to metabolic heat production in the cold.

Although it has been proven that exercise has a positive effect on maintaining thermal homeostasis, a number of published reports have suggested that prolonged exercise in cold water may increase the risk of **hypothermia** (Centers for Disease Control and Prevention 1983; Danzl, Pozos, and Hamlet 1995; Pugh 1966). The risks of exercise in cold water are well acknowledged. As previously mentioned, the heat loss through conductive, convective, and evaporative mechanisms exceeds the ability of the body to generate metabolic heat through exercise. During exercise in the water, heat loss may be augmented by a number of factors. The dangers of exercising in cold water were demonstrated by Hayward, Eckerson, and Collis (1975), who reported significantly lower core temperatures in subjects exercising in cold water compared to subjects who remained still. The lower core temperatures of the exercising subjects suggested that they were at a much greater risk for hypothermia than the subjects who did not exercise. The following are possible mechanisms for the increased risk of hypothermia during exercise in cold water (Toner and McArdle 1988):

- Increased heat transfer from the core to the periphery because of increased blood flow
- Increased heat production in the extremities versus the trunk compared with that in a nonexercise condition
- Reduced insulatory benefit of the boundary layer of the skin–water interface if movement of the upper and lower extremities increases
- Increased effective surface area for heat transfer provided by the redistribution of blood from the trunk core to the extremity

A combination of these four mechanisms results in a change in the heat gain–heat loss dynamics of the extremities where the ratio of surface area to mass is high, causing significant reduction in core temperature.

During exercise in cold weather, some concern has focused on a **thermoregulatory fatigue** that could blunt the shivering response and reduce vasoconstriction. A study by Young and colleagues (1998) showed that prolonged exercise in the cold accompanied by sleep deprivation and a negative energy balance resulted in a lowered shivering threshold. However, the conditions experienced by these subjects are not commonly encountered by exercising or competing athletes. An additional concern is the effect of exercise in a temperate environment on the thermoregulatory responses during subsequent cold exposure. The thermoregulatory response during exercise in a temperate environment focuses on dissipating heat.

In Practice

Can Performing Exercise Before Putting on a Dry Immersion Suit Reduce the Rate of Cooling?

Scientists from Norway were interested in determining whether performing exercise before putting on an immersion suit could delay the rate of body cooling (Faerevik and Reinertsen 2012). This issue is important because various types of workers are required to perform their job duties in cold water for prolonged periods of time. Two groups of male workers were examined; one group exercised before putting on the immersion suit, while the other group performed their normal routine before putting the suit on. The exercise protocol consisted of a 20 min ride on a cycle ergometer at 40% of the age-predicted maximal heart rate. Participants were then immersed in 5 °C water. The temperature in the laboratory was 10 °C. The time in the water was 140 min. These conditions were designed to simulate the water temperature of the operational area of the Sea King helicopter in Norway. Exercise before water immersion resulted in elevations in both skin and rectal temperatures at the onset of cold-water immersion. However, this group of participants experienced a faster rate of cooling within the first 10 min of immersion, resulting in significant decreases in both rectal and skin temperature. Overall the rate of cooling did not differ between groups; participants in the preimmersion exercise group cooled 0.34 ± 0.11 °C/h, while the participants who did not exercise cooled 0.31 ± 0.05 °C/h. Results indicated that wearing a dry immersion suit eliminated long-term differences in core cooling between participants who were prewarmed compared to those who entered the water in a normothermic state.

During exposure to a cold environment, there may be a **thermoregulatory lag** in switching from heat dissipation to heat conservation (Castellani et al. 1999).

Part of this hypothesis has been generated from research demonstrating that the increased perfusion of blood to the active muscle during exercise remains elevated for an extended duration (Thoden et al. 1994). Castellani and colleagues (1999) examined this question in a group of subjects who rested for 2 h at 40 °F (4.4 °C) after either a passive heat exposure or a 1 h exercise period on a cycle ergometer at 55% peak O_2 uptake in 95 °F (35 °C) water. The results showed a greater cooling effect for the exercise group. This greater cooling effect was not attributed to a fatigued shivering response but predominantly to a thermoregulatory lag (i.e., an inability of the subjects to change from a situation of heat dissipation to one of heat conservation). The researchers suggested that the postexercise **hyperemia** was still evident in the previously active muscles, which increased convective heat transfer during cold exposure. In addition, a redistribution of body heat from the core to the periphery was thought to be occurring, related to the increased blood flow during and after exercise. For thermoregulatory fatigue to occur, exercise needs to be of much greater duration. Thermoregulatory fatigue appears to set in during exercise exceeding 4 h in severe cold and wet conditions (Thompson and Hayward 1996). These results have important implications for people exercising in a temperate environment who are quickly exposed to the cold, as well as for individuals exercising in the cold who fail to insulate themselves properly after the conclusion of exercise.

Role of Body Composition and Thermal Balance

It is rare to find a situation in which a high percentage of body fat may be beneficial. However, there appears to be one such situation. During cold exposure, individuals with a higher percentage of body fat appear to have a greater ability to maintain their core body temperature than their leaner counterparts. A number of reports have demonstrated a positive linear relationship between body fat composition and core temperature during cold exposures (Toner et al. 1986; Toner and McArdle 1988). In addition, people with a high body fat percentage shiver less than leaner individuals (Toner et al. 1986). The high levels of subcutaneous fat apparently have a greater insulatory ability, which limits the transfer of heat from the core of the body to the periphery by decreasing the rate of heat conduction.

Considering that women have more body fat than men do, it might stand to reason that they could tolerate cold stress better than men can. However, it does not appear that women have a thermoregulatory advantage over men when it comes to cold exposure. Women have a greater body surface area and a smaller total body mass than men do (when men and women of equivalent body fat are compared). The larger surface area in women provides for an increased thermal gradient, causing a greater heat loss through convective mechanisms (Sawka and Young 2000). Thus, a greater total heat loss is seen in women than in men when body composition is controlled.

Acclimatization to the Cold

The degree to which people can acclimatize to the cold has not been studied as extensively as heat **acclimatization**. It does appear that when individuals are chronically exposed to the cold, some degree of physiological adaptation may occur. Two potential adaptations have been suggested that may contribute to a greater cold tolerance in individuals with prolonged cold exposure. The first adaptation is related to increased metabolic heat production through an exaggerated shivering response or potentially through a nonshivering mechanism (Sawka and Young 2000). Huttunen, Hirvonen, and Kinnula (1981) reported that Finnish workers who were chronically exposed to a cold environment retained brown adipose tissue. This theory is not well accepted, however, because of the alleged inability of humans to retain brown fat as adults. The second possible adaptation may be an enhanced sympathetic response, which results in more rapid cutaneous vasoconstriction (Sawka and Young 2000). When compared with heat acclimatization, the physiological adaptations that have been proposed to result from chronic cold exposure appear to develop at a much slower pace, and they may not be as effective in preventing injury.

In Practice

What Are the Long-Term Effects of Cold-Air Exposure?

Concerns about the effects of exercise during long-term exposure to hypothermic conditions have been raised by sports medicine specialists in Norway (Sue-Chu 2012). Winter sport athletes may compete in a cold and dry environment for years, often at exercise intensities exceeding 80% of their maximal aerobic capacity (endurance athletes). A concern was that prolonged exposure for a number of years under these environmental stresses may result in a greater prevalence of respiratory symptoms and airway hyperresponsiveness. Only a handful of studies have examined the prolonged effects of training under these conditions. One study showed an accelerated decline in lung function, including evidence of airflow limitation during exercise, in elite cross-country skiers at the end of a 9- to 12-year observation period (Verges et al. 2004). Sue-Chu (2012) suggested that repeated exposure to cold air may result in epithelial injury and mucosal inflammation in both the proximal and distal airways. In the proximal airway, a twofold increase in neutrophil infiltration was reported in adolescent skiers with 7 years of experience, with similar responses in mast cell infiltration and even greater responses in eosinophil accumulation. These inflammatory responses appear to be different from those in asthmatics, as they are not associated with respiratory symptoms of hyperresponsiveness. It is hypothesized that these inflammatory responses may be associated with healing, repair, or remodeling of the proximal airway. Similarly, a greater inflammatory response is seen in the distal airway. Bronchoalveolar lavage has indicated significant increases in the inflammatory response as detected by increases in lymphocyte counts. In addition, changes in mucosal morphology, including a decrease in the number of ciliated cells and a thicker epithelium, are suggestive of a remodeling or repair of the airway. It is not known at this time whether these changes are reversible, or what the long-term risk is to athletes competing for years under these environmental stresses and experiencing these morphological changes.

Exercise Performance and the Cold

To prepare for exercise in a cold environment, athletes can wear insulative clothing that will often negate the harsh conditions. Insulative clothing not only prevents the athlete's core body temperature from decreasing but may even elevate it. However, athletes may forego proper attire because bulky insulative clothing impedes the fluidity of movement. Athletes sometimes experience a sudden change in the weather in the middle of a training session or competition, leaving them unprepared to combat changed environmental conditions. Thus they are exposed to the harshness of the environment, and the body is forced to battle the elements to maintain core body temperature. The physiological changes that occur during cold exposure may have important implications for the subsequent performance capability of the athlete.

Aerobic Exercise in the Cold

During prolonged endurance exercise, oxygen consumption appears to be inversely related to the ambient temperature (Beelen and Sargeant 1991; Claremont et al. 1975; Galloway and Maughan 1997). At similar relative exercise intensities, a greater oxygen uptake is seen as the environmental conditions become colder. Galloway and Maughan (1997) reported that when exercise was performed at 39 °F (4 °C) compared to a more temperate environment, increases in oxygen uptake were accompanied by increases in **minute ventilation** and carbohydrate utilization (figure 22.2). These changes resulted in a decrease in the mechanical efficiency of exercise, as reflected by the significantly shorter time to fatigue (81.4 min) during exercise at 39 °F (4 °C) versus exercise performed at 52 °F (11 °C) (93.5 min to fatigue). However, exercise performed in the cold conditions still appeared to be less exhausting than exercise in a more temperate environment (81.2 min and 51.6 min at 70 °F [21°C] and 88 °F [31 °C], respectively). The changes in mechanical efficiency that resulted in these performance decrements appear to be related more to a drop in skin and muscle temperature than to changes in core body temperature. Core temperature of the subjects exercising at 39 °F (4 °C) increased to 100 °F (38 °C) during exercise, indicating that these subjects were not experiencing any cold stress. Thus, changes in the efficiency of exercise may occur when temperature changes are present only at the periphery and do not necessar-

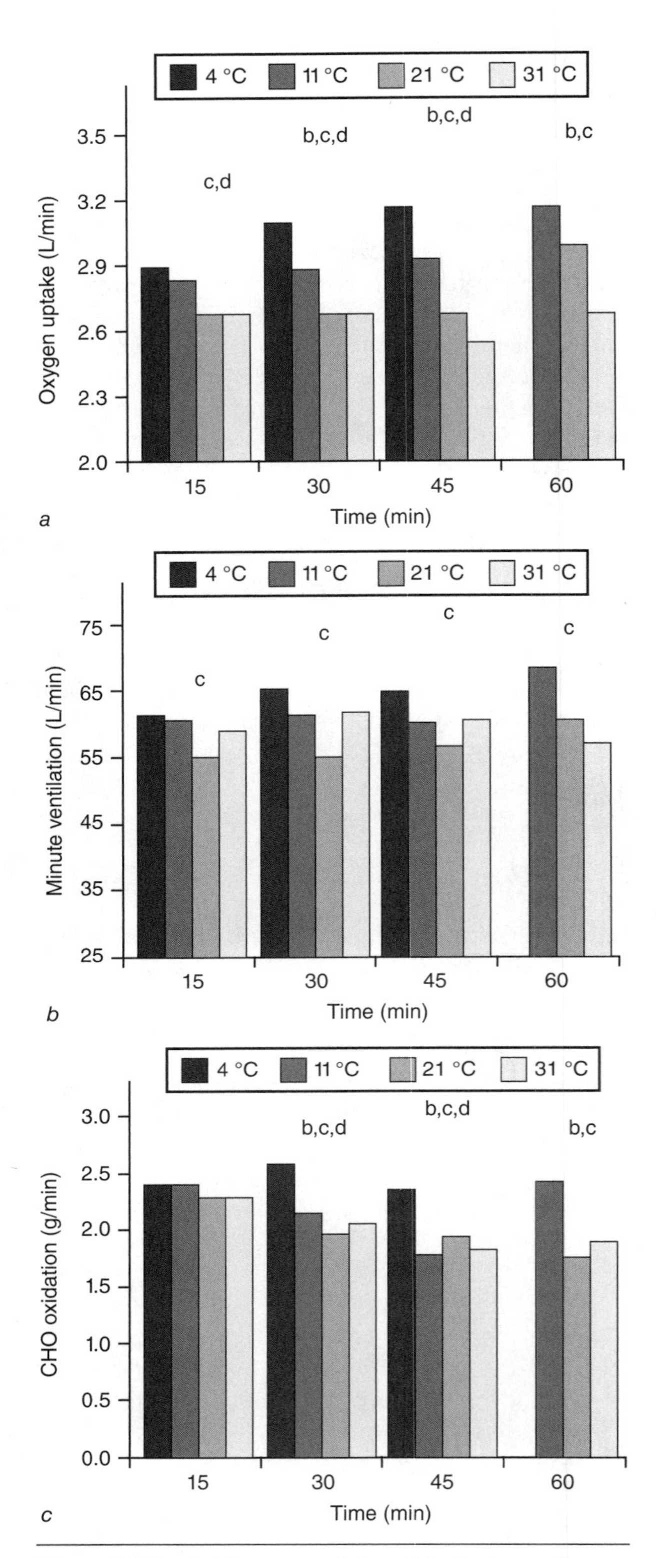

Figure 22.2 *(a)* Oxygen uptake, *(b)* minute ventilation, and *(c)* carbohydrate (CHO) oxidation during exercise. b = Significant difference between 4 and 11 °C; c = significant difference between 4 and 21 °C; d = significant difference between 4° and 31 °C.

Adapted from Galloway and Maughan 1997.

ily reflect a significant threat to the subject's thermal homeostasis.

Not all studies of prolonged endurance exercise in the cold have been consistent in reporting increases in oxygen consumption. Others have suggested that oxygen uptake during exercise in the cold may be lower than or even equal to oxygen consumption during exercise performed in warmer conditions (Young 1990). This is likely related to changes in the core body temperature. If core temperature is significantly lowered, the effect on exercise performance may become more pronounced. Studies examining $\dot{V}O_2max$ and the cold have reported that maximal oxygen consumption may not be affected unless core temperature is decreased to a level that reflects a severe cold stress (Horvath 1981; Bergh and Ekblom 1979). Bergh and Ekblom (1979) suggested that $\dot{V}O_2max$ is not significantly lowered until core temperature is reduced by at least 0.9 °F (0.5 °C). A reduction in the maximal aerobic capacity of an individual during a severe cold stress is likely related to a change in myocardial contractility and limits on attainment of maximal heart rate (Sawka and Young 2000).

Increases in oxygen consumption during endurance exercise in the cold may also be related to the greater oxygen requirement of the shivering muscles. However, this may be more relevant when exercise is performed at a low intensity of training. At higher intensities, the increase in exercise metabolism is sufficient to prevent shivering, and oxygen consumption with this type of exercise stimulus may be similar between cold and temperate environments (Sawka and Young 2000).

Anaerobic Exercise in the Cold

The ability to generate maximal **strength** and **power** is related to muscle temperature. When exercise is performed in a cold environment, muscle temperatures may decline if the muscle is not sufficiently warmed. As muscle temperature declines, significant decrements in performance can be seen. Davies and Young (1983a) reported a 43% decrease in jump power and a 32% decrease in cycling power when muscle temperatures were reduced by more than 14 °F (8 °C). In addition, cooling of the muscle results in an increase in the time to peak force (Davies, Mecrow, and White 1982; Davies and Young 1983a). The extent of muscle power and force decrements appears to be related to the degree of muscle temperature reduction. Power outputs have been reported to decrease 3% to 6% for every 1.8 °F (1 °C) reduction in muscle temperature (Bergh and Ekblom 1979; Sargeant 1987).

The magnitude of force and power decrements during cold exposure may also be related to the velocity of movement. In a study examining the effect of cold immersion on muscle strength (Howard et al. 1994), subjects who were immersed for 45 min in cold water (54 °F [12 °C]) showed significant decrements in both peak torque and average power during the leg extension exercise when compared to a thermoneutral environment. However, these reductions in peak torque and power occurred only at contraction speeds greater than 180°/s. At slower velocities of joint movement (0°/s and 30°/s), no significant differences were observed (figure 22.3). These results were consistent with other studies demonstrating minimal effects on isometric strength after cold exposure (Binkhorst, Hoofd, and Vissers 1977; Bergh and Ekblom 1979). The performance decrements seen at the higher velocities of contraction may be caused by a variety of factors related to the cold exposure, including decreased rate of cross-bridge formation (Godt and Lindly 1982; Stein, Gordon, and Shriver 1982), reduced nerve conduction velocity (Montgomery and MacDonald 1990), and changed motor unit recruitment patterns (Rome 1990). As muscle temperatures decrease, the rate of muscle enzyme activity may also decrease, resulting in a reduced ability to replenish high-energy phosphates (Ferretti 1992). Most of these mechanisms are speculative, and further research is needed to provide a better understanding of muscle performance decrements during exercise in the cold.

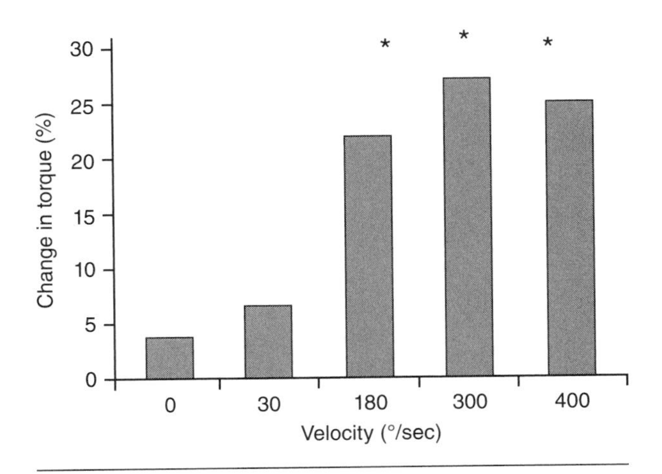

Figure 22.3 Effect of cold immersion on peak torque. * = Significant difference from values in a temperate climate.

Adapted from Howard et al. 1994.

Manual Dexterity

One of the major issues related to performance in a cold weather environment is the effect it has on manual dexterity. This has important relevance to tactical athletes (i.e., soldiers, police, fire and rescue) or any other profession that requires precision of movement with fingers and hands. These changes can occur even when core temperature is normal. When skin temperatures fall to 20 °C, pain sensation increases; and when skin temperature declines to 15 °C, a decrease in manual dexterity becomes apparent (Heus, Daanen, and Havenith 1995; Enander 1987). The colder temperatures result in the cooling of superficial tissues in the fingers and hands accompanied by decreases in joint mobility. As temperature drops further (below 7 °C), tactile sensitivity is reduced (Kenefick et al. 2008). The duration of cold exposure experienced by uncovered skin is related to the magnitude of performance loss. Longer exposures result in greater declines in performance as muscles and nerves continue to cool. Immersion of the hands and forearms in chilled water (10 °C) for as few as 5 min can lower manual dexterity by 20% to 50% (Cheung et al. 2003). Interestingly, O'Brien and colleagues (2011) have reported that finger temperature and thermal comfort can be maintained if the face is covered during cold weather activity. They suggested that wearing a balaclava may improve one's comfort in the cold, but that prolonged exposure of uncovered hands will still result in performance decrements.

Medical Concerns

Most injuries related to cold ambient temperatures are the result of prolonged exposure without proper clothing. Cold injuries among athletes engaged in competitive or recreational sports are rare. Individuals performing physical activity in the cold are able to maintain body temperature by wearing layers of clothing as insulation. However, athletes participating in outdoor winter sports (e.g., cross-country skiing, speed skating, American football) may refrain from wearing bulky insulative clothing in order to maintain freedom of motion. This not only alters the physiological response to the physical activity but also places athletes at an increased risk of cold injuries.

Cold injuries may occur during all seasons and can result from unexpected cold-water immersions or unexpected storms that catch athletes without proper attire. Armstrong (2000) reported on the risk of cold injuries in marathon runners performing in cool weather (39-50 °F [4-10 °C]) or in races in which a sudden drop in temperature or a change in weather conditions occurs midway through the race. During such races, the runner loses a great deal of body heat through convection, radiation, and evaporation. The body cools rapidly, and it is more difficult to maintain core body temperature as fatigue sets in during the later stages of the race. A greater risk of cold injury may arise in races that begin under mild conditions with runners seemingly dressed appropriately. With sudden changes in the weather (cold and rain) presenting challenges that the runner is ill prepared for (lack of appropriate clothing), the athlete becomes more susceptible to cold injury.

Several injuries are common to prolonged cold exposure. Hypothermia and **frostbite** are the most dangerous of the cold injuries. **Immersion foot** and **chilblain** are also common but pose much less threat to survival.

Hypothermia

Hypothermia is defined as a lowering of the core body temperature below 95 °F (35 °C) (Hamlet 1988; Ward, Milledge, and West 1995). It can be classified according to both core temperature and length of exposure. Core temperature between 90 to 95 °F (32 to 35 °C) is considered mild hypothermia, whereas core temperature below 90 °F (32 °C) is considered severe. When exposure to the cold is of relatively short duration but the cold stress exceeds the ability of the body to maintain core temperature despite maximum heat production, the hypothermia is classified as **acute hypothermia**. When prolonged activity in the cold is accompanied by exhaustion and depletion of energy reserves, as might occur during mountaineering, the ability to maintain core temperature is diminished and the ensuing hypothermia is classified as **subacute** (Ward, Milledge, and West 1995). In acute hypothermia, the body still has the capability to produce heat, but the cold stress far exceeds its ability to maintain warmth, whereas in subacute hypothermia, the body is able to maintain core temperature but is unable to maintain heat production because of prolonged activity and exhaustion. This is the primary difference between the two forms of hypothermia. During a mild cold stress over a prolonged period of time (days or weeks), the thermoregulatory response is not overwhelmed but is unable to maintain sufficient body heat. This gradual decrease in body temperature is known as **chronic hypothermia** and is commonly seen in an elderly population (Ward, Milledge, and West 1995).

Clinical features of hypothermia are shown in table 22.1. During mild hypothermia, the individual begins to shiver uncontrollably and skin color appears grayish. Speech patterns may become slurred and slow,

Table 22.1 Clinical Features of Hypothermia

Temperature °C	°F	Clinical features
37.0	98.6	Normal body temperature
36.0	96.8	Increase in metabolic rate to compensate for heat loss
35.0	95.0	Maximal shivering
34.0	93.2	Lowest temperature compatible with continuous exercise
31.0-33.0	87.8-91.4	Severe hypothermia, retrograde amnesia, clouded consciousness
28.0-30.0	82.4-86.0	Progressive loss of consciousness, muscular rigidity, slow respiration and pulse, ventricular fibrillation if heart is irritated
27.0	80.6	Cessation of voluntary motion; person appears to be dead, pupils are nonreactive to light, and deep tendon and superficial reflexes are absent
26.0	78.8	Victim seldom conscious
25.0	77.0	Ventricular fibrillation may develop spontaneously
21.0-24.0	69.8-75.2	Pulmonary edema develops
20.0	68.0	Cardiac standstill
17.0	62.6	No measurable brain waves

Adapted from Ward et al. 1995; Armstrong 2000.

and the person's attitude and demeanor toward the task at hand may become negative and irritable. Muscular coordination becomes impaired, and an increase in fatigue becomes evident (Ward, Milledge, and West 1995). As core temperature decreases below 90 °F (32 °C), mental acuity is diminished. Thinking is slowed, memory deteriorates, and the individual becomes confused. The slurred speech and unresponsive behavior resemble the signs in a stroke victim. As core temperature continues to decrease, the individual becomes progressively unresponsive and eventually lapses into a coma.

The principal plan of management for cases of hypothermia is to prevent any further heat loss and restore body temperature to normal. The patient should be removed from the cold environment and placed in a shelter out of the wind, rain, or snow. For mild hypothermia, a slow rewarming of the body is considered acceptable; however, for severe hypothermia, a rapid, active rewarming should be performed (Ward, Milledge, and West 1995).

For a case of mild hypothermia, surface rewarming is the correct approach. Wet clothing should be replaced with dry; and the body should be covered with material that has some insulating ability (e.g., sleeping bag or blankets), or hot water bottles should be placed at the victim's underarms or groin. When no dry clothing is available, the wet clothing may be wrung out and put back on. It is important to cover the clothing with additional insulating material to prevent further heat loss. Rewarming can also be accomplished through conduction from close body-to-body contact. Warm fluids should be given either orally (when victim is alert) or possibly through an IV. In addition, the victim should inhale warm air.

During severe hypothermia, the victim should be evacuated as quickly as possible. It is important to remember that in extreme cases of hypothermia, the patient may appear dead. Core temperature must be elevated to normal levels before this possibility can be considered (Ward, Milledge, and West 1995). Rewarming should be started using the techniques discussed for mild hypothermia. Once the patient has been transported to an appropriate facility, warm IV fluids should be provided, and airway warming via an endotracheal tube is recommended (Hamlet 1988). In addition, internal organ warming is generally accomplished through gastric and peritoneal lavage (i.e., irrigation with a warm saline solution). Rewarming must proceed with caution, and the patient must be treated gently to avoid **ventricular fibrillation**.

Frostbite

As ambient temperature drops toward the freezing point, the skin becomes numb and loses its sense of touch and pain at about 50 °F (10 °C) (Hamlet 1988). This is accompanied by a transient general vasoconstriction. As ambient temperature drops below freezing, the skin actually freezes. Several injuries are associated with freezing of the skin. Their extent depends on the environmental temperature, wind velocity, and duration of exposure. Exposure resulting in freezing of only the superficial layers of the skin is referred to as **frostnip**. Frostnip is not associated with any subsequent damage or tissue loss and is not considered a serious cold injury. During frostnip, the skin may appear red and scaly. Sensation in the affected areas may be lost, but the skin remains pliable. Once the affected areas are rewarmed, their appearance may be similar to that of first-degree sunburn.

Superficial frostbite is the result of freezing of the skin and subcutaneous tissues. During this injury, the skin becomes white and frozen, but the deep underlying tissues remain pliable. Rewarming should be rapid. After rewarming, the skin swells and becomes mottled blue or purple. Within several days, the affected areas may become gangrenous.

If freezing involves some of the deeper structures (muscle, bone, and tendons), the more serious **deep frostbite** has occurred. The affected area is insensitive and becomes hard and fixed over joints (nonpliable). The color of the skin may be grayish purple or white marble. Actual crystallization of tissue fluids in the skin or subcutaneous tissues occurs. Although the tissue is frozen, movement in the affected body part may still occur because the tendons are not as sensitive to the cold as other tissues (Ward, Milledge, and West 1995), and the primary muscle groups are at a distance from the area of injury. However, the affected areas will still become gangrenous, and a permanent loss of tissue is almost inevitable (Ward, Milledge, and West 1995).

The mechanisms of frostbite injuries involve changes of blood flow to the affected area and the freezing of tissue (Foray 1992). During cold exposure, the unprotected body part has a reduced oxygen supply because of an increase in blood viscosity and vasoconstriction from the freezing temperatures. At some point, total cessation of blood flow may occur. Consequently, the tissues become deprived of needed oxygen (**hypoxia**), and metabolic **acidosis** shortly ensues. In combination with the crystalline formation occurring within the superficial tissues, damage to the cells and microscopic blood vessels results in tissue death.

Frostbite injuries may also be associated with hypothermia, which has priority in treatment. The treatment plans discussed earlier for hypothermia victims should be followed. In a situation of frostnip, the way to warm the affected part is to cover it (place a glove on an exposed hand, or place the hand under the armpit or in the groin area). Sensation and full function should be restored fairly quickly. It should be understood that repeated frostnip predisposes the individual to subsequent frostbite injury (Riddell 1984).

Rapid rewarming is important for frostbite injuries. This best way to accomplish this is to place the exposed area in warm water. Under no circumstances should the affected area be beaten, rubbed, or overheated. This includes rubbing the area with snow or ice or heating with excessively hot water (>111 °F [44 °C]); the affected area may suffer additional burn injury from **cold anesthesia** (Flora 1985). In addition, a freeze–thaw–freeze sequence could have disastrous results for future prognosis and should always be avoided. It may be better to keep the injured part frozen than to thaw it and take the chance that it may become frozen again while the victim is transported to a hospital. This is a real concern with injuries occurring in remote mountainous areas.

Immersion Foot

Immersion foot or **trench foot** is a cold injury associated with feet that are cold or wet for a prolonged period of time. It frequently occurs when the feet are in water or snow and may take from several hours to repeated exposures over several days to manifest. This is a nonfreezing injury that may cause lasting damage to the muscles and nerves. Although this type of injury can occur in the hands, it is much easier to check the hands and take appropriate measures to keep them warm and dry. Thus, it is more common to see this condition in the feet.

The exposure time needed to develop trench foot has been estimated to range from >12 h to 3 to 4 days in cold, wet environments (Imray and Castellani 2012). Trench foot generally develops when wet socks and shoes are worn continuously over many days. It progresses through several stages. Initially, the extremity is cold, discolored, and numb (Hamlet 1988). After several days at this stage, a tingling pain may occur that is associated with swelling, blister formation, desquamation, ulceration, and **gangrene**. This second stage may last between 2 to 6 weeks and may be followed by a final stage, which could last several months or a lifetime, involving a heightened sensitivity to the cold (**Raynaud's disease**) accompanied by severe pain. Amputation is a possibility in severe cases.

The primary preventive measure is wearing heavy socks in well-fitting boots or shoes. Be prepared with a pair of dry socks to replace socks that become wet or damp. Keep the feet out of water or snow, and make sure to keep the feet clean, dry, and warm. Numbness and tingling are signs of trench foot and should serve as a red flag to warm the feet immediately.

Chilblain

Chilblain, another nonfreezing injury, appears to affect women more than men. It is an inflammatory condition that develops as a result of cold exposure (Ward, Milledge, and West 1995). This injury appears to be more frequent in cold, humid conditions than in cold, dry conditions. Most lesions occur on the back of the extremities, between the joints, after an extended period of cold and dampness. Vasodilation and subcutaneous **edema** and swelling are characteristic of chilblain, and itching is the primary symptom during initial injury (Hamlet 1988). In chronic lesions, vasodilation may no longer be present, but the swelling remains and the itching is replaced by pain. **Pernio** is a more severe form of chilblain and is associated with a greater burning and pain sensation in the affected area. As the duration of the cold exposure lengthens and the severity of the injury increases, the individual is at a greater risk of trench foot.

Preventive Strategies for Cold-Related Injury

If there is no choice but to be exposed to a cold environment to work or compete, several strategies can be used to protect against hypothermia. The choice of clothing becomes a vital part of preparing for cold weather exposure. The goal is to wear clothing that will provide sufficient insulation to prevent hypothermia. Insulation is determined by how much air is effectively trapped by clothing. Using multiple layers of clothing and staying dry are key to maintaining body temperature. The use of layers of clothing allows air to be trapped and to act as insulation. As the person becomes active or ambient temperature increases, the layers can be removed. However, dressing for cold weather is more complicated than simply wearing thicker clothing. A concern with multiple layers of clothing is the potential for increasing moisture at the innermost layer due to sweating. To minimize this effect, the first layer of insulation (clothing) should be one that allows water vapor to be transported to outer layers for evaporation. If it does not, the clothing on the skin will become wet, removing the benefit of the insulation and increasing conductive heat loss.

Another concern for many athletes preparing to play in the cold is the bulkiness of thicker clothing and potential impact it may have on performance. However, participants in a competitive game in cold weather requires less clothing than do people exposed to the cold but not active. The increase in metabolic heat production during exercise in the cold appears to be effective in minimizing the risk of hypothermia, especially when wind is not present (Makinen et al. 2001). However, as exercise intensity is elevated, the benefit of increasing metabolic heat production though exercise is realized even in the wind (Brajkovic and Ducharme 2006; Gavhed et al. 2003). Preparing to play in the cold should be a well-thought-out process that includes appropriate clothing layers, sufficient warm-up to elevate core temperature, and changes of clothing to prevent moisture buildup in the inner layers. In addition, the use of hats, mittens, gloves, and facial covering plays an important role in reducing the risk of cold-related injuries. Table 22.2 provides some minimal

Table 22.2 Recommendations for Exercise in a Cold Environment

- Do not wear tight clothing that may restrict circulation.
- Cover exposed areas of the body with gloves (mittens, preferably), scarves, wool hat, high-top shoes or boots.
- Try to avoid wetting clothing, shoes, and socks and be prepared to change wet clothing with dry socks, shoes, and undergarments.
- Avoid exercising in the snow or in wet areas.
- Avoid water-based skin lotions and use oil-based ones instead such as Vaseline or ChapStick.
- Make sure to cover legs and genitals when dressing for outdoor training by wearing sweat pants, long underwear, Lycra tights, Gore-Tex pants, or a combination of these garments.
- Avoid wearing excessive clothing. A large sweat buildup decreases the effectiveness of the insulatory ability of the clothing. As exercise concludes, make sure to quickly change into dry clothes.
- Use proper judgment in deciding to exercise outdoors. Perhaps the workout can occur under safer conditions inside.
- Remember that individuals with a previous cold injury may be more susceptible to subsequent injuries.
- Dehydration causes added physiological strain and limits your ability to generate metabolic heat.
- Be sure to maintain proper water balance.

Adapted from Armstrong 2000.

recommendations for preparing to exercise in a cold environment.

Summary

When exercise is performed in a cold environment, the primary response of the body is to maintain thermal homeostasis by defending core temperature. Athletes can usually prevent unnecessary and potentially dangerous exposure to the cold through behavioral modifications (e.g., wearing appropriate clothing). Sometimes, however, by choice or because of circumstances beyond the athletes' control, they find themselves in an environmental condition that places them at an increased risk of cold injury. Cold acclimatization does not appear to be easily achieved, and knowledge concerning cold acclimatization is not yet complete. With the technological advances being made in the sport clothing industry, many athletes can protect themselves from the environment without sacrificing their mobility.

REVIEW QUESTIONS

1. Explain the importance of the shivering response in increasing metabolic rate during cold exposure.
2. Can chronic exposure to a hypothermic environment result in acclimatization to cold exposure? Explain why or why not.
3. Describe the effect of playing in a cold environment on athletic performance.
4. Name five recommendations for avoiding cold-related injuries.
5. Describe the difference between subacute hypothermia and chronic hypothermia.

chapter 23

Altitude

CHAPTER OBJECTIVES

After reading this chapter you should be able to do the following:

- Describe the hypobaric environment.
- Understand the physiological strain associated with altitude.
- Explain the effects of altitude on athletic performance.
- Understand the effects of acclimatization to altitude and learn how to potentially use this knowledge to improve athletic performance.
- Describe the symptoms of hypoxia and understand the etiology of acute mountain sickness.

The acute response to a **hypobaric** environment can cause significant physiological and psychological changes that negatively affect athletic performance. However, prolonged exposure to a hypobaric environment results in **acclimatization**. The physiological adaptations that stimulate the acclimatization process not only help athletes perform at altitude, but may also have an **ergogenic effect** when they return to sea level. This chapter discusses the effect that a hypobaric environment has on physiological processes and the subsequent effect of these changes on performance. It also discusses the potential advantages of acclimatization for subsequent performances at sea level.

The Hypobaric Environment

As an individual ascends above sea level, **barometric pressure** is reduced relative to the magnitude of the elevation. Because the weight of the upper atmosphere compresses the air of the lower atmosphere, barometric pressure decreases rapidly as one ascends from sea level. Changes in pressure are also influenced by changes in ambient **temperature**. The relationships between pressure, temperature, and **volume** follow some of the basic laws of physics. For a better understanding of the hypobaric environment, a brief review of some of these basic laws should be helpful.

- **Boyle's law** refers to the relationship between pressure and volume. It states that the pressure of a given gas at a constant temperature is inversely proportional to its volume.
- **Charles' law** concerns the relationship between volume and temperature. It states that at constant pressure, the volume of a gas is proportional to its absolute temperature.
- **Dalton's law** states that each gas in a mixture exerts a pressure according to its own concentration, independently of the other gases present.

The pressure of each gas is referred to as its **partial pressure**. Gas molecules, because of their random motion, tend to distribute themselves uniformly in a given space until the partial pressure is the same everywhere. These laws governing the relationship between pressure and volume formed the basis for the development of the **standard atmosphere model** by the National Oceanic and Atmospheric Administration (1976) (figure 23.1).

During ascent, the atmospheric pressure continuously decreases; however, the composition of the air remains the same as it was at sea level (20.93% oxygen, 0.03% carbon dioxide, and 79.04% nitrogen). The partial pressure of each gas is reduced in direct proportion to the increase in altitude (table 23.1). The reduced partial pressure of oxygen (**PO_2**) results in a reduced **pressure gradient**, which impedes oxygen diffusion from the blood to the tissues.

In addition to reduced oxygen availability at altitude, other environmental issues appear to pose significant hurdles for performance at high elevations. The individual exercising at altitude may be at risk of cold injuries because of the low air temperatures common at high elevations. Ambient temperature drops at a rate of 1.8 °F (1 °C) for every 490 ft (150 m) of ascent. For example, the average temperature at the summit of Mount Everest is about −40 °F (−40 °C) in the winter. As a result, most climbs are performed during the summer months when temperature at the summit has been recorded at 16 °F (−9 °C) (Ward, Milledge, and West 1995). However, given the effects of windchill, it may feel even colder at these high altitudes. Wind velocities in excess of 93 mph (150 km/h) have been reported on Himalayan peaks (Ward, Milledge, and West 1995), suggesting that windchill may have even more significance on the cold stress associated with altitude.

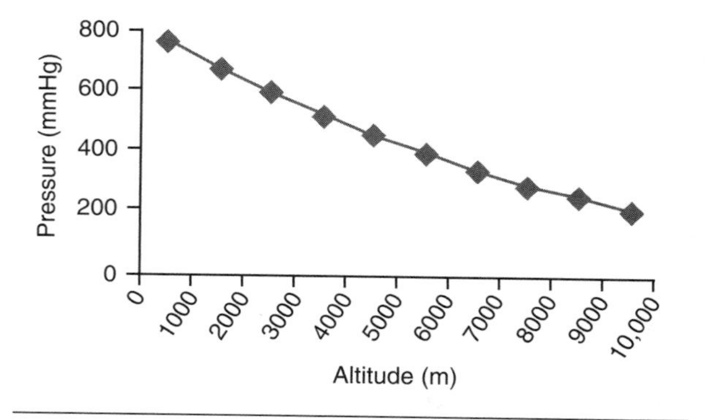

Figure 23.1 Standard atmosphere model.
Data from National Oceanic and Atmospheric Administration 1976.

As ambient temperature is reduced during elevation, the amount of water vapor per unit volume of gas is also reduced. At high altitudes, even if the air is fully saturated with water, the actual amount of water vapor is very small. For example, at 68 °F (20 °C), the water vapor pressure is 17 mmHg, but at −4 °F (−20 °C) it is only 1 mmHg. The extremely low humidity at altitude results in a large evaporative heat loss due to **ventilation** of the dry inspired air. The risk of dehydration at altitude is great even at rest, but during exercise, when the ventilation rate is further elevated, the risk becomes even more significant. A fluid loss from the lungs equivalent to 200 ml per hour was reported during moderate exercise at elevations of 18,000 ft (5500 m) (Pugh 1964). Acclimatization seems unable to reduce this chronic volume depletion, even in individuals who have been living at altitude for a prolonged period of time. Blume and colleagues (1984), during an ascent to Mount Everest, reported significantly elevated serum osmolality in subjects residing at 20,700 ft (6300 m) compared with sea level, even when fluids

Table 23.1 Changes in Barometric Pressure (P_B) and Partial Pressure of Oxygen (PO_2) at Varying Altitudes

Altitude (m)	P_B (mmHg)	PO_2 (mmHg)
0	760	159
1000	674	141
2000	596	125
3000	526	110
4000	463	97
5000	405	85
6000	354	74
7000	308	65
8000	267	56
9000	231	48

were readily available and no significant changes in exercise or diet were apparent. This study highlighted the risk of dehydration during exercise at altitude, emphasizing the need for high fluid intake during such environmental extremes even when the thirst drive is absent.

Physiological Response to Altitude

The physiological stress associated with altitude is manifested primarily in its effect on oxygen availability to the tissues. When oxygen levels are reduced in arterial blood, inspired gases, or tissues, a situation of **hypoxia** is present. Acute hypoxia, which occurs when an individual is initially exposed to altitude, results in changes in various physiological systems in the body. An overview of these changes is depicted in figure 23.2. Although there is considerable variability, most changes observed at moderate altitudes involve primarily the central nervous system. At 4900 ft (1500 m), night vision becomes impaired. At about 6600 ft (2000 m), resting heart rate becomes elevated, and it continues to elevate as ascent continues. During ascents above sea level to approximately 9800 ft (3000 m), a subject at rest does not experience any noticeable symptoms. However, performance of novel tasks (e.g., reaction time in complex choice reaction tests) may be affected (Ernsting and Sharp 1978).

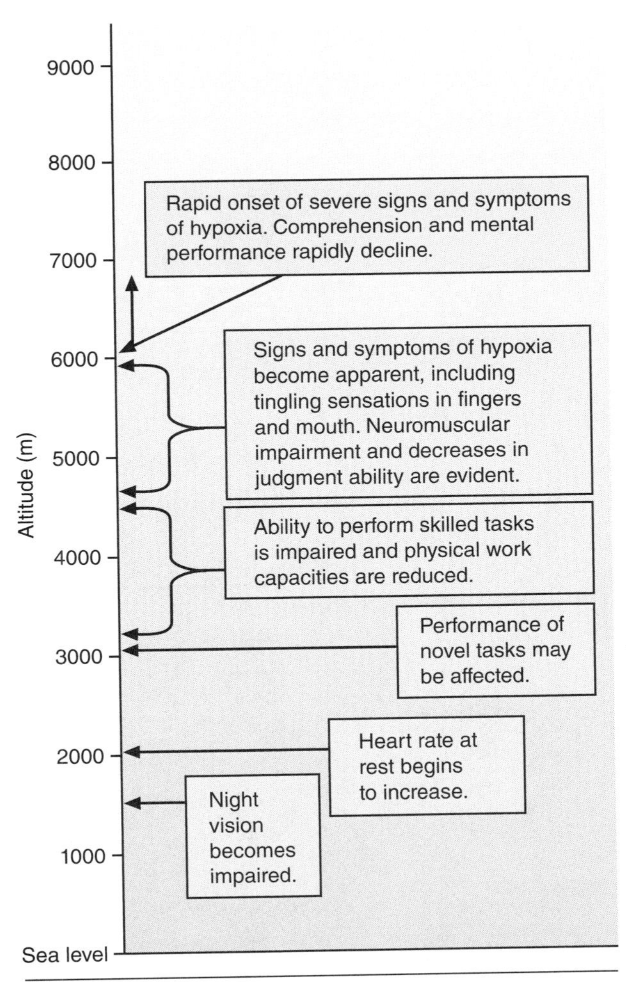

Figure 23.2 Effect of sudden exposure to various altitudes.

At elevations between 9800 and 14,800 ft (3000 and 4500 m), symptoms of hypoxia may not be apparent (see table 23.2 for a list of symptoms of hypoxia), but the ability to perform skilled tasks is impaired. In addition, physical work capacity is significantly diminished. If ambient temperature becomes colder than usual, symptoms of hypoxia may become apparent (Ernsting and Sharp 1978).

Between 14,800 and 19,700 ft (4500 and 6000 m), the signs and symptoms of hypoxia are present even at rest. Tingling sensations in the fingers and mouth may be noticed. In addition, higher mental processes and neuromuscular control are affected, with loss of critical judgment (Ernsting and Sharp 1978). Many of these symptoms resulting in deterioration of performance go unnoticed by the individual, which can present a very dangerous situation. A marked change in the emotional state is frequently seen, which may manifest itself as a release of the person's basic personality traits, ranging from euphoria to moroseness. Physiological strain is apparent and physical exertion may increase the severity of these symptoms, often resulting in a loss of consciousness.

At elevations above 19,700 ft (6000 m), subjects at rest rapidly experience severe signs and symptoms of hypoxia. Both comprehension and mental performance decline rapidly, and unconsciousness can occur with little to no warning (Ernsting and Sharp 1978). The time of **useful consciousness** (defined as the time between a reduction of the oxygen tension of inspired gas and the instant at which an impairment of performance occurs) is approximately 3.50 ± 1.36 min at 25,000 ft (7600 m) (Ernsting and Sharp 1978). The degree of performance impairment may vary from an inability to perform psychomotor tasks to failure to respond to simple commands.

Effect of Altitude on Respiratory Response

As one ascends above sea level, the partial pressure of oxygen (PO_2) becomes reduced. To compensate for the reduced PO_2 at altitude, the breathing rate is increased. However, as breathing rate is increased

Table 23.2 Symptoms of Hypoxia

Light	Medium	Severe
Euphoria	Anxiety	Dizziness
Loss of orientation	Desire to vomit	Delirium
Nausea	Chest pains	Coma
Headache	Apnea	Vomiting
Slightly increased blood pressure	Blood pressure is elevated	Respiration depressed and Cheyne-Stokes respiration possible; respiration may also cease
Increased, sometimes irregular heartbeat	Heart rate is reduced and irregular	Abrupt drop in breathing rate
Increased breathing rate	Muscles may spasm or become rigid	Blood pressure is weak and vanishing
Lack of muscular coordination	Skin becomes deeply cyanotic with possible heavy sweating	Heart rate is low
Skin may be lightly cyanotic	Pupils alternate in wideness	Muscles are relaxed, near paralysis
Pupils are irregular		Skin is gray and clammy
		Pupils are extremely dilated and rigid

(**hyperventilation**), the partial pressure of carbon dioxide (PCO_2) in the alveoli is reduced. As a result, the stimulus to maintain a high rate of ventilation may be removed because PCO_2 is an integral part of the driving force behind hyperventilation. The low PCO_2 also causes an elevation in blood pH, resulting in a condition known as **respiratory alkalosis**. To compensate for the increase in blood pH, the kidneys excrete more bicarbonate ions (HCO_3^-) to return the HCO_3^-/PCO_2 back to normal. The increased bicarbonate excretion from the kidneys appears to reduce the **buffering capacity** of the body. The importance of this will become apparent later.

The relationship between ventilation and PO_2 is hyperbolic, whereas the relationship between ventilation and **arterial saturation (SaO_2)** is linear (figure 23.3). Although variable from person to person, the increase in ventilation as one ascends does not appear to occur until inspired PO_2 reaches approximately 100 mmHg (corresponding to an alveolar PO_2 of 50 mmHg). This is equivalent to an elevation of about 9800 ft (3000 m) (Ward, Milledge, and West 1995). At this altitude, the saturation of oxygen to **hemoglobin** falls from 98% (normal levels seen at sea level) to less than 92%. The point at which **arterial PO_2 (PAO_2)** results in a marked increase in ventilation corresponds to the PO_2 at which the oxygen dissociation curve begins to steepen.

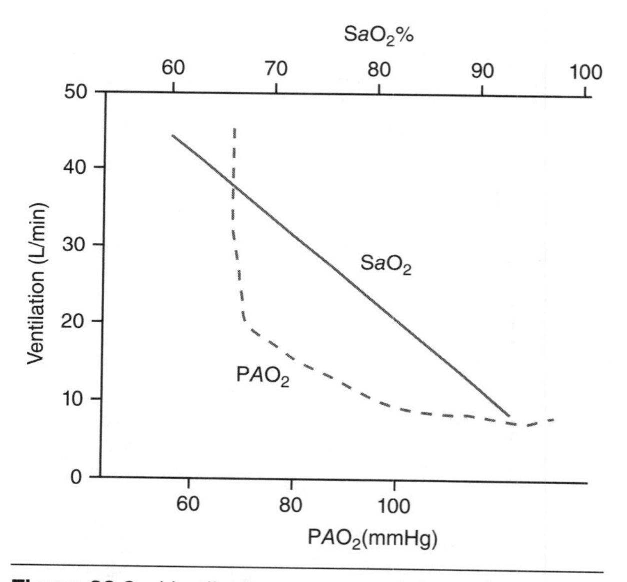

Figure 23.3 Ventilatory response to the partial pressure of alveolar oxygen (PAO_2) and arterial oxygen saturation (SaO_2).

High altitude medicine and physiology, M.P. Ward, J.S. Milledge, and J.B. West, Copyright 1995 Chapman & Hall Medical Publishers. Adapted by permission of Taylor & Francis Books UK.

The decreases in SaO_2 may not be the most significant factor that limits performance at altitude. The pressure gradient between arterial PO_2 and tissue

PO_2 is approximately 64 mmHg at sea level (the difference between an arterial PO_2 of 104 mmHg and a tissue PO_2 of 40 mmHg). This creates a pressure gradient, which causes oxygen to diffuse into the tissues. However, a reduction in PAO_2 at altitude results in a decrease in the pressure gradient, which decreases the diffusion capability of oxygen from the vasculature into the tissue. For instance, at an elevation of 8200 ft (2500 m), the PAO_2 drops to about 60 mmHg while the tissue PO_2 remains at 40 mmHg, thus creating a pressure gradient of only 20 mmHg. This 70% reduction in the pressure gradient causes a significant reduction in the speed at which oxygen moves between the capillaries and tissues. The reduction in the pressure gradient between the vasculature and the tissues has a much greater significance for performance decrements associated with altitude than the relatively small changes (~6-7%) in arterial saturation.

Effect of Altitude on Cardiovascular Response

Acute hypoxia results in an increase in **cardiac output** both at rest and during exercise. Obviously, this is a compensation for the reduced oxygen availability caused by the lower PO_2. The primary mechanism resulting in the increased cardiac output appears to be an increase in **heart rate**. Heart rate has been shown to increase 40% to 50% at rest without any change in **stroke volume** (Vogel and Harris 1967). Even during exercise, the increase in cardiac output appears to be primarily the result of an increase in heart rate. In contrast to what is normally seen during exercise at sea level, stroke volume decreases during exercise at altitude (Vogel and Harris 1967). The decrease in stroke volume is apparently caused by the reduction in **plasma volume** that has been observed within a short time after arrival at altitude (Singh, Rawal, and Tyagi 1990; Wolfel et al. 1991).

During initial exposure to altitude (observed at elevations between 9800 and 19,700 ft [3000 and 6000 m]), decreases in plasma volume appear to be the result of both **diuresis** (excretion of unusually large volumes of urine) and **natriuresis** (Honig 1983). The diuresis may be explained by the large evaporative heat loss caused by ventilation of dry inspired air at altitude, which has already been discussed. The natriuresis (increased sodium excretion in urine) appears to be the result of a neural stimulation of the kidneys to decrease the reabsorption of sodium caused by the hypoxic stimulus (Honig 1983). Even during prolonged altitude exposure, plasma volume remains below normal levels, and studies of individuals who reside at altitude have reported lower plasma volumes in comparison with residents of sea-level communities (Ward, Milledge, and West 1995). Thus, acclimatization does not appear to have any significant effect on a return of blood volume to preexposure levels.

Effect of Altitude on Metabolic Response

During exercise at altitude, increases in lactic acid at every given work rate reflect the increase in **anaerobic** metabolism observed at altitude. Considering the limitations of oxygen availability at altitude, such a response is not unexpected. It is interesting that during exposure to altitude, maximal lactic acid concentrations are generally reduced (Green et al. 1989; Sutton et al. 1988). This is likely related to the inability of the body to achieve a maximal effort because of a reduction in buffering capacity and limitations on energy production through glycolysis (West, 1986). As mentioned previously, respiratory alkalosis occurring during hyperventilation at altitude causes an increase in bicarbonate ions excreted from the kidneys to compensate for the high blood pH. The mechanism that can explain the reduced **glycolytic** efficiency during exercise at altitude is not well understood.

Effect of Altitude on Athletic Performance

The physiological changes seen during acute altitude exposure have significant detrimental effects on athletic performance. However, as altitude exposure is prolonged, physiological adaptations occur that not only enhance exercise performance at altitude but may also provide significant performance benefits on return to sea level.

Endurance Performance

Significant effects on endurance performance in athletes have been reported at relatively low to moderate levels of altitude (2000-3900 ft [600-1200 m]) (Gore et al. 1996, 1997; Terrados, Mizuno, and Andersen 1985). However, the effect of altitude on endurance performance may be related to the training status of the individual. At elevations that cause a significant decline in the **$\dot{V}O_2max$** of highly trained athletes, no significant changes were seen in untrained control subjects. Although a well-trained endurance athlete may be more sensitive to changes in the environment than an untrained subject, subsequent studies showed a large degree of variability even among highly trained athletes. Chapman, Emery, and Stager (1999) demonstrated that at a moderate altitude (3300 ft [1000 m]), eight endurance athletes had significant declines in $\dot{V}O_2$max while six others maintained their $\dot{V}O_2$max levels. The difference in the response between these subjects at altitude appeared to be related to differences in their SaO_2 at sea level. The study showed a significant correlation between SaO_2 of $\dot{V}O_2$max at sea level and the decline in $\dot{V}O_2$max at altitude. Additional data from the same laboratory showed that elite distance runners with the highest SaO_2 during maximal exercise at sea level had the smallest slowing of 3000 m run times when performing at 6900 ft (2100 m) (Chapman and Levine 2000). It appears that the ability to maintain arterial oxygen saturation during heavy exercise may be more related to the ability to maintain endurance performance at altitude than training status per se (Chapman and Levine 2000).

As the ascent continues, the effect on endurance performance becomes more pronounced. At altitudes exceeding 5200 ft (1600 m) (approximately the elevation of Denver, CO), decreases in $\dot{V}O_2$max are commonly observed. As altitude continues to increase, a linear decline in $\dot{V}O_2$max is seen (figure 23.4). An approximately 11% decline in $\dot{V}O_2$max is seen for every 3300 ft (1000 m) increase in elevation (Buskirk et al. 1967). In men with a $\dot{V}O_2$max of 50 ml · kg^{-1} · min^{-1} at sea level, a drop to approximately 5 ml · kg^{-1} · min^{-1} would be expected near the summit of Mount Everest, leaving these men with little to no ability for physical work without supplemental oxygen. In fact, a $\dot{V}O_2$max value that low would be barely sufficient to sustain resting oxygen requirements.

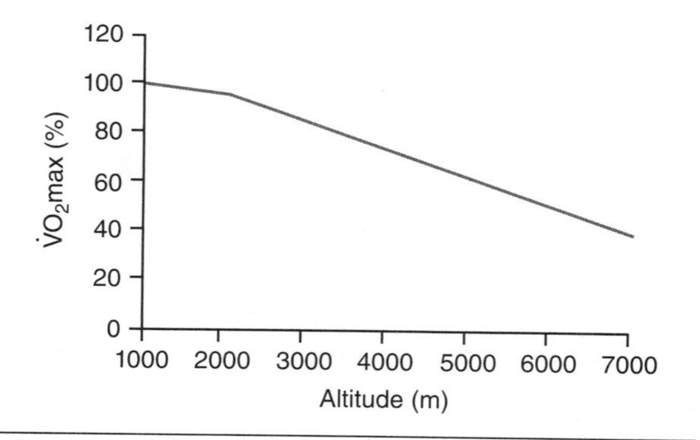

Figure 23.4 Changes in maximal aerobic power ($\dot{V}O_2$max) at differing altitudes.

Anaerobic Performance

Since the 1968 Olympics in Mexico City (elevation 7300 ft [2230 m]), sprinting, jumping, and throwing records have been set predominantly at altitude. The records have been set mostly in events that are largely driven by the adenosine triphosphate–phosphocreatine (ATP-PC) and glycolytic energy systems. The benefits seen in anaerobic exercise performance at altitude are likely related to the decrease in **drag** associated with the thinner air. Drag is the resistance that acts on a body in motion, either in air or in water. At sea level, drag is responsible for 3% to 9% of the energy cost of running (Pugh 1970) and for over 90% of the energy cost during cycling exercise performed at speeds >25 mph (40 km/h) (McCole et al. 1990). At altitude, the thinner air reduces the effects of drag, which results in faster times and a reduced energy cost. The performance benefits of reduced drag for the sprinter or cyclist is (unfortunately for the swimmer) not seen in water sports. The effects of hypobaria on the density of water are not significant enough to confer any performance benefit (Chapman and Levine 2000).

The effect of altitude on anaerobic sports that involve repeated high-intensity activity is not well understood. However, the decrease in buffering capacity of the muscle during acute exposure to at least a moderate altitude suggests that prolonged anaerobic activity may be affected. In addition, the lower lactic acid concentrations seen during maximal exercise at altitude may be suggestive of potential performance decrements. However, these potential negative effects on anaerobic performance have been discussed relative to acute altitude exposure. After prolonged exposure, the acclimatization experienced by the athlete is likely to result in improved performance.

Altitude Acclimatization

During prolonged altitude exposure, physiological adaptations that enhance the ability to perform at high elevations gradually occur. However, these adaptations never reach a point that fully compensates for hypobaric hypoxia. Although several physiological systems in the body are able to adapt to altitude, these systems have different time courses for adaptation.

Respiratory Changes After Prolonged Exposure to Altitude

During the first few days at altitude, changes in the respiratory response are seen. Initially, breathing rate increases and arterial PO_2 decreases. After a few days at altitude, arterial PO_2 begins to rise as PCO_2 values fall. However, the ventilatory rate continues to rise as a result of changes in the ventilatory response to CO_2 levels and in the sensitivity of the **carotid body**. The carotid body, situated above the bifurcation of the carotid artery, serves as a sensor of oxygen saturation in the blood. Its location is ideal for this role because it receives a large blood supply, allowing it to respond to oxygen saturation and not to oxygen content (Ward, Milledge, and West 1995). The increased sensitivity of the carotid body appears to have a biphasic response. The initial reaction may be a decrease in the hypoxic ventilatory response during the first 3 to 5 days of altitude exposure, but after these initial days of exposure, an increase in the ventilatory response is seen (Ward, Milledge, and West 1995).

In addition to changes in the ventilatory response, another component of respiratory adaptation to altitude may be **diffusion capacity**. During prolonged exposure (7-10 weeks) to altitude, diffusion capacities have been reported to increase 15% to 20% (West 1962). Although much of this improved diffusion capacity may be accounted for by increases in hemoglobin concentrations, other studies including highlanders (individuals permanently residing at altitude above 2600 m) and lowlanders (individuals residing at sea level) have also reported significant differences in diffusion capacities between these population groups (Dempsey et al. 1971). These differences may be related to the larger lung volumes developed through exposure to chronic hypoxia and the subsequent development of greater surface area for oxygen diffusion (Bartlett and Remmers 1971).

Diffusion ability appears to be affected during acute altitude exposure by a **ventilation–perfusion inequality** (Ward, Milledge, and West 1995). This is a limitation of the ability of oxygen to diffuse across from the alveoli to the pulmonary circulation and is reflected by a progressive decrease in arterial oxygen saturation at higher altitudes and at higher work levels. However, some evidence suggests that prolonged altitude exposure reduces the ventilation/perfusion inequality (Wagner, Saltzman, and West 1974; Wagner et al. 1987).

Cardiovascular and Hematological Changes After Prolonged Exposure to Altitude

After several weeks of altitude exposure, cardiac output remains similar to the values observed at sea level. However, similar to the situation during acute altitude exposure, increases in cardiac output after extended stays at elevation appear to be attributable primarily to increases in heart rate. Stroke volume continues to remain reduced (Reeves et al. 1987). A low stroke volume is also common among highlanders. When highlanders are compared to acclimatized lowlanders, similar stroke volumes between the two subject populations are commonly observed (Ward, Milledge, and West 1995). Thus, even during chronic altitude exposure, stroke volume does not improve, suggesting that other adaptations need to occur to compensate for the reduced plasma volume.

One of the best-known adaptations to prolonged altitude exposure is the increase in the number of red blood cells per unit volume of blood. Exposure to hypoxic conditions results in the release of the hormone **erythropoietin**. Erythropoietin is responsible for stimulating red blood cell (**erythrocyte**) production and is seen to increase within 2 h of exposure to altitude. It reaches a maximum rate of increase at about 24 to 48 h (Eckardt et al. 1989). After 3 weeks of exposure, erythropoietin concentrations appear to return to baseline levels, but not until they have contributed to an approximate 20% to 25% increase in packed cell volume (Milledge and Cotes 1985). Increases in red cell mass continue even after erythropoietin returns to normal levels (Milledge and Cotes 1985). The mechanism for stimulating erythropoietin concentrations appears to involve activation of the hypoxia-inducible factor 1α (HIF-1α) complex (Caro 2001; Samaja 2001). In the presence of oxygen, the half-life of HIF-1α is the shortest of any known protein (Jaakkola et al. 2001); but under hypoxic conditions, the HIF-1α complex is stable, stimulating erythropoietin synthesis. Given the large variability in erythropoietin concentrations as the result of hypoxic exposure, the idea that a genetic endowment could enable some to benefit more than others from hypoxia training has been considered. However, limited studies have been unable to identify a single marker linked to erythropoietin production or an association between the eight genes linked to erythropoietin regulation (Jedlickova et al. 2003).

As red cell volume increases, so does the hemoglobin concentration of the blood. This increase allows for a greater amount of oxygen to be carried per unit volume of blood. However, as red cell volume and hemoglobin concentration increase, the viscosity of the blood also increases, presenting an inherent danger associated with these physiological adaptations to altitude. The increase in hemoglobin concentration is proportional to the increase in elevation (Wilmore and Costill 1999). This is likely related to the greater diuresis and resulting loss in plasma volume associated with higher altitudes. The changes in the oxygen-carrying capacity of the blood after altitude exposure are shown in figure 23.5.

Cardiovascular and Hematological Changes After Short-Term Exposure to Altitude

Short-term exposure (24 h) to altitude results in a relatively quick increase in erythropoietin production that can be sustained during time of exposure. In a study by Ge and colleagues (2002), subjects spent 24 h in a decompression chamber at simulated altitudes of 1780 m (5840 ft), 2085 m (6841 ft), 2454 m (8051 ft), and 2800 m (9186 ft). Erythropoietin concentrations significantly increased at all four altitudes after 6 h exposure (figure 23.6) but plateaued at the first two altitudes. During exposure to both of the higher altitudes (above 2454 m), continued elevations in erythropoietin concentrations were seen. The investigators reported on large subject variability, with some subjects unable to increase erythropoietin concentrations even at the higher altitudes. At higher altitudes (4000-5500 m), increases in erythropoietin concentrations may be seen after 2 h of exposure (Knaupp et al. 1992). The physiological response to short-term hypoxic exposures suggests an interesting potential training strategy—intermittent hypoxic exposures to stimulate physiological adaptations for sea-level performance. Gore and colleagues (2006) examined 4 weeks of intermittent hypoxic exposure at a simulated altitude of 4000 to 5500 m for 3 h/day, 5 days a week. They reported a significant increase in erythropoietin concentrations, but no change in red cell volume or hemoglobin concentrations. The lack of change in erythropoiesis was attributed in part to the intermittent nature of the hypoxic exposure: 3 h per day may not be sufficient to sustain stimulus for erythropoiesis. It was suggested that upon return to sea level there would be a rapid degradation of HIF-1α and a decrease in red cell survival time. This would likely compromise the ability to increase red

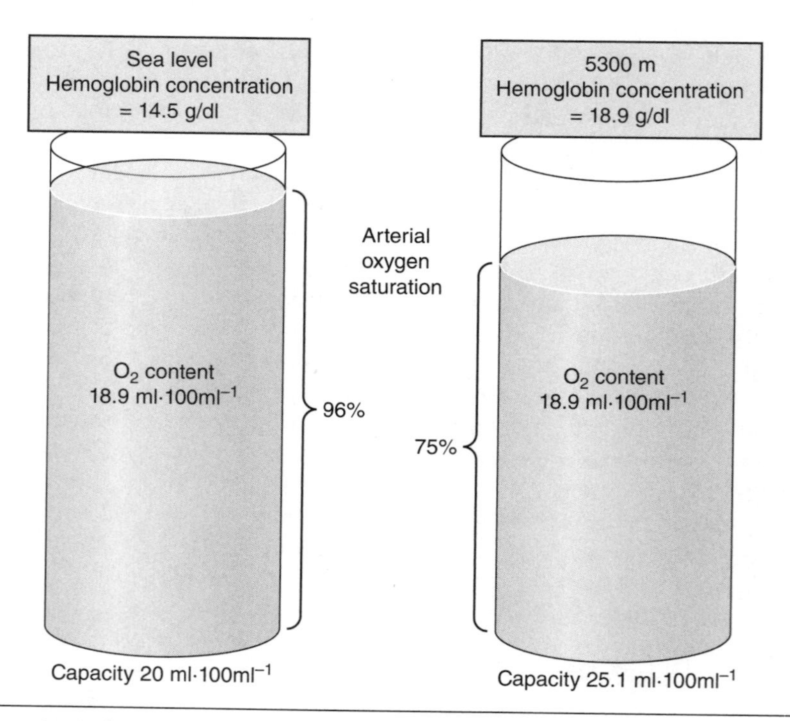

Figure 23.5 Oxygen content of arterial blood in acclimatized subjects.

High altitude medicine and physiology, M.P. Ward, J.S. Milledge, and J.B. West, Copyright 1995 Chapman & Hall Medical Publishers. Adapted by permission of Taylor & Francis Books UK.

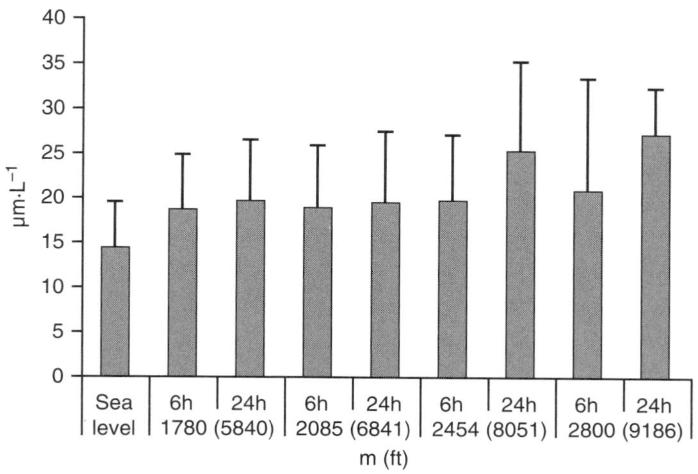

Figure 23.6 Erythropoietin changes from sea level and simulated altitude exposure.
Data from Ge et al. 2002.

cell volume and sustain the erythropoietic response. In addition, the 4-week intermittent hypoxic protocol using the same exposure times and frequencies was reported to have no effect on either swimming or running economy during submaximal exercise (Truijens et al. 2008).

Metabolic and Neuromuscular Changes After Prolonged Exposure to Altitude

Findings from early studies have led to the general conclusion that prolonged exposure to moderate altitudes (13,100-16,400 ft [4000-5000 m]) results in significant increases in **oxidative** enzymes with no change in glycolytic enzyme activity (Ward, Milledge, and West 1995). This is similar to what occurs during endurance training. When challenged by oxygen deficiency either during altitude exposure or through intense endurance training, the muscles compensate by increasing their ability for oxidative metabolism.

However, when individuals ascend to more extreme altitudes, the effect on both oxidative and glycolytic enzymes appears to be different. At elevations exceeding 19,700 ft (6000 m), several studies have shown that the oxidative enzymes decrease after prolonged exposure (Cerretelli 1987; Green et al. 1989; Howald et al. 1990). Green and colleagues (1989) reported decreases ranging from 21% to 53% in the oxidative enzymes (e.g., succinate dehydrogenase, citrate synthase, and hexokinase) during prolonged exposure. The results of these studies do not support the hypothesis that hypobaric hypoxia will lead to adaptations that maximize oxidative function. The effect of extreme altitude on the glycolytic enzymes is less clear. After expeditions to Lhotse Shar or a simulated expedition to Mount Everest, glycolytic enzymes were reported to decrease (Cerretelli 1987; Green et al. 1989). In contrast, studies by Howald and colleagues (1990) after expeditions to both Mount Everest and Lhotse Shar reported increases in the glycolytic enzymes, suggesting an aerobic to anaerobic shift in muscle energy metabolism. These contrasting results are difficult to explain and are likely to be examined further in future research.

Prolonged altitude exposure also appears to affect muscle structure and **capillary density**. After 40 days of progressive decompression simulating an ascent to the summit of Mount Everest, reductions in both type I (25%, $p < 0.05$) and type II (26%, $p > 0.05$) fiber areas were seen (MacDougall et al. 1991). These results confirmed earlier reports of muscle atrophy by other investigators examining ascents to altitudes exceeding 26,200 ft (8000 m) (Boutellier et al. 1983; Cerretelli et al. 1984). The effect of prolonged or chronic exposure to more moderate altitudes on muscle fiber size is less clear. After an 8-week expedition to the Himalayas (altitude >16,400 ft [5000 m]), significant reductions (−10%) in muscle cross-sectional area were observed (Hoppeler et al. 1990). However, in limited studies at lower elevations (~6600 ft [2000 m]), no significant differences in muscle fiber size have been seen (Saltin et al. 1995). In addition, endurance-trained athletes native to an altitude approximately 6600 ft (2000 m) above sea level did not differ from lowland endurance-trained athletes in muscle fiber size (Saltin et al. 1995). Several studies suggest that the magnitude of reduction in muscle cross-sectional area depends on the elevation (figure 23.7). However, the time frame for these adaptations to occur is less well understood.

The decrease in muscle fiber area during sojourns at moderate to high elevations may be an example of a physiological adaptation to provide the best opportunity to supply muscle with oxygen, thereby giving the body the maximum chance of survival. The decrease in fiber size occurs without any change in the number of capillaries within the muscle (Green et al. 1989; Hoppeler et al. 1990). As a result, the capillary density of the muscle is increased. This adaptation is similar to what is commonly seen after endurance training. The difference between the two is that increases in capillary density as a result of endurance training occur through a rise in the number of capillaries within the fiber, whereas increases in capillary density after altitude exposure are a result of muscle atrophy. Table 23.3 lists potential muscle structural and metabolic changes during prolonged altitude exposure.

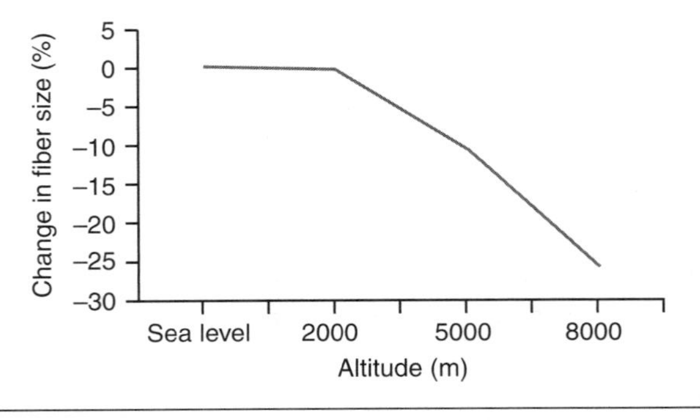

Figure 23.7 Effect of altitude on percent change in muscle fiber size.

Chronic Altitude Exposure and Benefits for Endurance Performance

The physiological adaptations associated with exposure to altitude suggest that there is an advantage for people who reside at altitude. Schmidt and colleagues (2002) compared trained and untrained lowlanders (individuals living at sea level) to trained and untrained highlanders (individuals who lived their entire lives at 2600 m). The trained participants were experienced cyclists. Table 23.4 illustrates the results of the investigation. Aerobic capacity levels were sig-

Table 23.3 Muscle Structural and Metabolic Changes During Prolonged Exposure to Altitude

Adaptation	Low to moderate altitude (1000-5000 m)	Moderate to high altitude (>5000 m)
Muscle fiber size	↔ or ↓	↓
Capillary number	↔	↔
Capillary density	↔ or ↑	↑
Oxidative enzymes	↑	↓
Glycolytic enzymes	↔	↑↓

Table 23.4 Hematology and Aerobic Capacity in Untrained and Trained Lowlanders and Highlanders

Subjects	$\dot{V}O_2$max (ml · min^{-1} · kg^{-1})	Total hemoglobin (g/kg)	Plasma volume (ml/kg)	Hematocrit (%)	Blood volume (ml/kg)
Untrained lowland	45.3 ± 3.2	11.0 ± 1.1	45.7 ± 5.3	45.8 ± 2.0	78.3 ± 7.9
Untrained altitude	39.6 ± 4.0	13.4 ± 0.9	48.1 ± 3.1	48.5 ± 1.7	88.2 ± 4.8
Lowland cyclists	68.2 ± 2.7	15.4 ± 0.9	60.8 ± 3.0	47.4 ± 1.9	107.0 ± 6.2
Altitude cyclists	69.9 ± 4.4	17.1 ± 1.4	64.9 ± 7.3	47.8 ± 1.5	116.5 ± 11.4

Adapted from Schmidt et al. 2002.

nificantly higher in the untrained lowlanders than the untrained highlanders but did not differ between the two groups of trained subjects. Total hemoglobin was influenced by training and by chronic exposure to altitude. Similar effects were noted in red cell volume. Plasma volume was not affected by altitude, but it was influenced by training. Thus blood volume was increased in the trained lowlanders and was highest in the trained highlanders. The authors concluded that endurance athletes who had lived at moderate altitude throughout their lives had experienced hematological changes that were influenced by both training and altitude exposure, providing a benefit to endurance performance.

Training at Altitude for Improved Performance at Altitude

The physiological adaptations that occur as part of the process of acclimatizing to altitude result in improved performance in exercise at altitude. These adaptations, which include increases in ventilation, hemoglobin concentration, capillary density, and tissue myoglobin concentration, would be most beneficial to endurance performance. However, the reduction in muscle fiber size and body mass would have a negative effect on absolute strength and power performance. Reductions in muscle cross-sectional area may be a concern only at elevations exceeding 16,400 ft (5000 m). Athletes preparing to compete at altitude may use a number of training strategies to enhance performance (see sidebar), but there are two primary options. One is to arrive at the event location within 24 h of the competition. Although this does not provide time for any acclimatization, it helps to prevent altitude sickness during competition. Symptoms of **acute mountain sickness** (discussed later) are not completely evident within the first 24 h of exposure to altitude. The second option is to arrive at least 2 weeks before the competition to begin acclimatization. If time permits, 4 to 6 weeks would allow a more complete acclimatization to occur.

In Practice

Performance Changes During a Weeklong Training Camp at Altitude for Alpine Skiers

Investigators from the University of Connecticut (Hydren et al. 2013) studied a team (four boys and seven girls) of junior alpine skiers (13.7 ± 0.5 years) who resided in the Northeast (New Hampshire, 160 m) during a weeklong training camp at altitude (Summit County, CO, 2829 m). Daily exposure during skiing also included greater elevations (<3800 m). During the training camp, the Lake Louise scoring system was used daily to test for acute mountain sickness (AMS). Subjects provide a score on specific signs and symptoms related to AMS (headache, gastrointestinal problems, fatigue or weakness, dizziness or light-headedness, sleep, and activity). The investigators evaluated each participant for changes in mental status, ataxia (lack of voluntary control of nervous system), and peripheral edema. In addition, body mass was recorded by participants; daily urine was analyzed for color; and measures of balance, quickness, and vertical jump were obtained on a daily basis. A battery of performance tests was performed at baseline (in New Hampshire) and on the last day of camp in Colorado; this included vertical jump, agility (T-test), flexibility (sit and reach), push-ups, sit-ups, and a multistage fitness test. In addition, a predictive test for $\dot{V}O_2max$ was performed. During the training camp, approximately 20% of the participants experienced signs and symptoms of AMS over the first 3 days. This is consistent with what has been reported in other studies. No performance decrements were noted in balance or quickness and reaction time; these performance measures actually improved toward the middle of the week (following 3 days of acclimatization). At the end of the training camp, participants demonstrated improvements in flexibility, vertical jump, quickness, speed, and local muscle endurance. Recommendations from the investigators emphasized a gradual increase in training volume during the first 3 days of a training camp at altitude to reduce the prevalence of AMS in young athletes.

Training Strategies for Competing at Altitude

- Arrive at competition site within 24 h of scheduled performance.
- If possible, arrive at competition site at least 2 weeks before performance for acclimatization. Ideally, for a more complete acclimatization, arrive 4 to 6 weeks before competition.
- Training should occur between 4900 ft and 9800 ft (1500 m and 3000 m). The lower altitude is considered the lowest elevation at which an acclimatization effect will occur, and the upper altitude is thought to be the highest level that is conducive to a productive training session.
- Training intensity should be reduced to 60% to 70% of sea-level intensity at first and gradually increased within 10 to 14 days.

Data from Wilmore and Costill 1999.

Training at Altitude for Improved Performance at Sea Level

The physiological adaptations seen during acclimatization at altitude have led many scientists to hypothesize that either living or training at altitude, or a combination of the two, will enhance endurance performance at sea level. Theoretically, the physiological adaptations to altitude caused by the hypobaric hypoxic conditions are similar to those caused by the endurance training stimulus. Several studies have examined strategies employed to elicit a training effect. With one strategy, "**living high/training low**," the athlete resides at a moderate altitude but trains at a lower altitude. Other strategies include "**living high/training high**," in which the athlete resides and trains at a moderate altitude, and "**living low/training high**," in which the athlete lives at sea level but trains at a moderate altitude.

Studies on the effects of "training high" have yielded mixed results (Chapman and Levine 2000). Many of these studies suffered from experimental design problems. In addition, athletes find it difficult to maintain the same exercise intensity at elevation as at sea level (Levine and Stray-Gundersen 1997). As a result, the quality of the workouts performed at elevation is reduced, so the training stimulus to enhance the physiological adaptations is inferior to that occurring at sea level. In a well-designed study, athletes living at a moderate altitude (8200 ft [2500 m]) but training at a low elevation (4100 ft [1250 m]) showed significantly greater performance improvements than athletes living high/training high or living low/training low (Levine and Stray-Gundersen 1997). In that study, 39 collegiate runners (27 men and 12 women) completed 6 weeks of preparation at sea level and were then randomly assigned to one of the three groups for 4 additional weeks of training. Both of the groups living at altitude significantly improved their $\dot{V}O_2max$ (5%), and these improvements were significantly correlated to increases in red cell volume (9%, $r = 0.37$, $p < 0.05$). In addition, race times for the 5K run were significantly reduced (13.4 ± 10 s), but only in the high/low group.

The same researchers then performed a retrospective examination of their studies by dividing the subjects into two groups: those they deemed responders to altitude training and those they deemed nonresponders (Chapman, Stray-Gundersen, and Levine 1998). Responders were those athletes who had improved their 5K run times by more than 14.1 s, whereas nonresponders had improved less than 14.1 s. Although both groups had significant increases in erythropoietin concentrations after 30 h at 8200 ft (2500 m), the responders had a significantly higher erythropoietin concentration than the nonresponders. After 14 days at altitude, erythropoietin levels had declined to baseline levels in nonresponders but were still significantly elevated in the responders. As a result, the responders had a 7.9% increase in red blood cell volume and a 6.5% increase in $\dot{V}O_2max$. Nonresponders did not have an increase in either red blood cell volume or $\dot{V}O_2max$. It is possible that erythropoietin needs to reach a threshold concentration before it is able to stimulate the production of additional red blood cells and confer an ergogenic effect for performance at sea level (Chapman and Levine 2000).

Newer data suggest that the benefits of a live high/train low routine may be dependent on baseline hemoglobin mass (Robach and Lundby 2012). However, one of the primary physiological adaptations with endurance training is elevations in hemoglobin mass; thus the value of this training strategy in elite-level endurance athletes may be questionable. Similarly, most studies on live low/train high strategies in trained individuals have indicated that hematocrit and hemoglobin concentrations remain largely unchanged from baseline levels (Hoppeler, Klossner, and Vogt 2008); but results suggest that this form of training may have more consistent benefits in the untrained individual. This may be an indication that acclimatization to altitude yields more consistent benefits for individuals who are fit but not highly trained than for the well-trained endurance athlete. Based on this consideration, the use of intermittent hypoxic exposure for tactical athletes (e.g., soldiers) has been recommended before deployment to mountainous regions (Muza 2007).

It has been suggested that the large variability in the physiological response to altitude training is related in part to genetic predisposition (Wilber, Stray-Gundersen, and Levine 2007). However, as discussed earlier, a specific gene linked to the variability of the erythropoietin response has yet to be identified. The benefits of training at altitude for performance at sea level are still being debated. Although there may be distinct advantages to altitude training strategies, several indications suggest that not all athletes will benefit from residing at altitude. More research appears to be needed to provide a better understanding of the factors that determine which athletes have the best opportunity to benefit from altitude training. The sidebar provides recommendations for training at altitude in preparation for sea-level competition.

Simulated Altitude and Ethical Issues

The use of simulated altitude devices and rooms has given many athletes who do not have access to hypobaric chambers or to residences at elevation an ability to incorporate the "live high/train low" strategy. Athletes can train during the day at sea level but spend their evenings in an altitude tent within their own bedroom or in an altitude-controlled room. Many Olympic training centers around the world have renovated their dormitories so that the oxygen concentration in a room can be controlled to simulate a particular altitude. This enables athletes to train at sea level but sleep and rest at designated altitudes. Theoretically, the benefits should be the same as those discussed earlier for athletes who descend for training and spend the evenings at higher elevations. This would eliminate the need for athletes to travel to remote locations in mountainous regions to train and would allow them to implement a live high/train low strategy. The benefits of simulated altitude rooms have been reported. Robertson and colleagues (2010) showed that simulated live high/train low exposures (2 × 3-week blocks of 14 h a day at ~3000 m) can induce reproducible improvements in aerobic capacity and hemoglobin concentrations in elite endurance athletes. However, reports have indicated that the use of hypoxic tents may disturb the sleep of some athletes, potentially negating the performance benefits as a consequence of poor rest (Pedlar et al. 2005).

Recommendations for Training at Altitude for Sea-Level Competitions

The erythropoietin response to altitude depends on elevation above sea level. The higher the altitude, the greater the erythropoietin response (Eckardt et al. 1989). Athletes who do not increase their erythropoietin concentrations at a particular altitude may need to ascend to a higher elevation; however, the higher the altitude, the greater the chance of experiencing acute mountain sickness (AMS). Residing at an altitude greater than 6600 ft (2000 m) appears necessary to elicit increases in red cell volume (Chapman and Levine 2000). Training should be performed at an altitude as close to sea level as possible, thereby minimizing the reductions in training velocity and oxygen availability. It appears that 4100 ft (1250 m) is low enough to provide a training stimulus sufficient to improve endurance performance while athletes are residing at higher altitudes (Levine and Stray-Gundersen 1997). Considering the biphasic time course of erythropoietin (initial increase within 24-48 h and a return to prealtitude levels within 3-4 weeks), it is thought that living at altitude for 3 to 4 weeks would be sufficient to maximize potential performance benefits (Chapman and Levine 2000).

Another concern for athletes using simulated altitude training is the potential for an increased propensity to illness (Gore, Clark, and Saunders 2007) as a result of a suppressed immune function associated with altitude and exercise (Mazzeo 2005). In addition, hypoxia is considered thrombogenic (i.e., has a tendency to form blood clots) (Lippi, Franchini, and Guidi 2007). Although most athletes who train at altitude would not be at risk (since the elevations rarely exceed 3000 m), those who reside at facilities simulating altitudes up to 5000 m may be at greater risk for developing thromboembolic complications (Lippi, Franchini, and Guidi 2007). In light of the potential medical issues associated with altitude training, and probably the greater concern that athletes are using simulated altitudes to replace blood doping and erythropoietin use (see chapter 20) and circumvent the World Anti-Doping Association's (WADA) prohibited substance list, the legality of the use of simulated altitude exposures has been debated (Wilber 2007). After much consideration, WADA decided to not prohibit the use of "artificially induced hypoxic conditions" by elite athletes. Although the WADA decision makers indicated that simulated altitude exposures may pose a health risk if not implemented under medical conditions, they did not believe it was appropriate at the time to place this method on a list of banned practices. Interestingly, the Italian Health ministry has banned the use of artificial hypoxic exposures due to an incident involving professional cyclists in 2005 (Wilber 2007). During the 2012 Olympics the use of simulated altitude training was not banned, and many athletes, including Michael Phelps, were very open about their use of altitude tents or altitude-controlled dorms.

Clinical Problems Associated With Acute Exposure to Altitude

On ascent to altitude, some people experience various symptoms that have been collectively referred to as AMS. Symptoms include headache, nausea, vomiting, fatigue or weakness, dizziness or lightheadedness, dyspnea (difficulty breathing), and insomnia (difficulty sleeping). Symptoms typically appear between 6 and 96 h after arrival at altitude. They start gradually and usually peak on the second or third day of exposure. Within 4 or 5 days, symptoms are usually gone and do not reoccur at the given altitude. AMS, if it does not progress, is not considered a life-threatening illness.

The incidence of AMS depends on several factors; the most important are rate of ascent and altitude reached. The lowest altitude at which symptoms might appear seems to be about 8200 ft (2500 m) (Ward, Milledge, and West 1995), a height that is experienced by some recreational skiers and hikers. The rate of incidence varies. If the stay at altitude is only 1 or 2 h, the symptoms of AMS are negligible (Ward, Milledge, and West 1995). Although fitness levels do not appear to have any significant relationship to AMS (Milledge et al. 1991), exercise may exacerbate the condition (Roach et al. 2000). In addition, it appears that certain individuals are more susceptible than others to symptoms of the illness (Forster 1984). That is, people who suffer from AMS during one exposure to altitude will likely experience similar symptoms during their next exposure. The consistency of the response of individuals ascending to altitude helps predict their future response to acute altitude exposure.

The mechanism underlying AMS appears to be related to disturbances in fluid or electrolyte homeostasis caused by the hypobaric hypoxic conditions (Hackett and Rennie 1979; Hackett et al. 1981). Hypoxia causes a decrease in PO_2 and an increase in PCO_2, resulting in vasodilation and subsequent water and sodium retention. Hypoxia also causes an increase in both cerebral and pulmonary blood flow. It may increase the microvascular permeability in these areas as well, resulting in a greater fluid leakage (Ward, Milledge, and West 1995). The increase of CO_2 in the tissues and the change in fluid volume, which cause a shift in fluid from intracellular to extracellular compartments, appear to be responsible for the associated symptoms. In addition, the increase in extracellular fluid may result in a severe progression of AMS to the more lethal **high-altitude cerebral edema (HACE)** or **high-altitude pulmonary edema (HAPE)**.

Symptoms of HACE include those associated with AMS; but the appearance of **ataxia**, irrationality, hallucinations, blurred vision, and clouding of consciousness may be a sign of the more lethal HACE. Symptoms of HAPE are again similar to those seen with AMS, but disruption of normal breathing as a result of pulmonary edema may be present, along with blueness of lips and fingernails, mental confusion, and loss of consciousness. The incidence of HAPE and HACE depends on the altitude, the rate of ascent, and the susceptibility of the individual. For example, the occurrence of HAPE increased from 2.5% to 15.5% with a rapid 18,000 ft (5500 m) ascent by airlift compared to a hike over 4 to 6 days (Bartsch 1999). In addition, the incidence of HAPE after a 22 h ascent to 15,100 ft (4600 m) was reported to be

10% in mountaineers without any previous history of HAPE; but it increased to 60% in mountaineers with previously documented HAPE (Bartsch 1999). The treatment plans for these severe forms of AMS are similar and include administration of supplemental oxygen and removal to lower altitude.

To minimize the risk of AMS, a slow rate of ascent is recommended. In addition, people should climb only 1000 ft (300 m) per day above elevations of 9800 ft (3000 m). If symptoms of AMS are present, the individual should remain at the same height until they subside. Drugs are also used prophylactically to prevent symptoms. Acetazolamide and dexamethasone have been used with success to reduce the incidence of AMS and to treat it when symptoms appear. Wagner and colleagues (2008), examining nearly 900 trekkers to Mount Whitney (4419 m; the highest point in the 48 contiguous U.S. states), suggested that age, hours spent above 3000 m in the 2 weeks before the ascent, and being female all reduced the risk of developing AMS. Taking analgesics (including aspirin, acetaminophen, and nonsteroidal anti-inflammatories) was associated with an increased risk of AMS; however, it was not known whether the analgesics were a cause of the AMS or whether they had been used to combat the symptoms.

Summary

Ascent to altitude causes a number of physiological changes that limit an athlete's ability to perform. These limitations have been seen primarily in the endurance athlete, whereas the anaerobic athlete may see performance improvements at altitude because of the reduced effects of drag and energy cost of performance. Prolonged altitude exposure (between 2 and 6 weeks) results in acclimatization, which enhances the athlete's ability to perform at altitude. The physiological adaptations associated with prolonged altitude exposure (e.g., increased ventilation, increased red blood cell volume, and increased hemoglobin content) have enticed athletes who compete at sea level to reside at altitude to benefit from the ergogenic effect. Brief altitude exposure does not appear to present any clinical problems. However, as duration of exposure is prolonged, symptoms may develop that can limit an individual's performance. These symptoms are indicative of an illness known as acute mountain sickness. AMS is not life threatening but may progress to a more dangerous clinical situation of high-altitude pulmonary edema or cerebral edema, which could indeed be life threatening if not treated.

In Practice

What Is the Lowest Effective Dose of Acetazolamide?

The effectiveness of acetazolamide for the prevention of AMS was examined in a meta-analysis (a method of research that focuses on combining and contrasting results from different studies) by Ritchie and colleagues (2012). The authors reviewed 36 studies, but for various methodological reasons they were able to include only 17 studies in the analysis. A total of 1765 participants who used either acetazolamide or a placebo were included. Each study in the analysis met the inclusion criteria as a double-blind, placebo-controlled trial. Three different doses of acetazolamide were used (250, 500, and 750 mg/day). Results indicated that acetazolamide was effective in reducing the incidence of AMS. There was no evidence of an asymmetry in the funnel plot that would have been suggestive of underlying bias; there was also no evidence of any difference in treatment effect by trial design. With regard to dosing, no differences were noted between any of the three dosages examined, suggesting that the lowest dose may be as effective as the highest. Appearance of adverse effects (reported to include numbness and tingling in the fingers and toes; taste alterations; and possible gastrointestinal disturbances such as nausea, vomiting, and diarrhea; polyuria; and occasional instances of drowsiness and confusion) was related to dose. In conclusion, the authors suggested that prophylactic use of acetazolamide in the prevention of AMS at a dose of 250 mg/day (divided) is similar in efficacy to the use of higher doses. Treatment appeared to be most beneficial in those individuals who had the highest risk of developing AMS.

REVIEW QUESTIONS

1. Explain Boyle's law and how it can potentially affect the exercise response as one ascends to 3000 m.
2. What are the physiological changes that occur as one ascends to altitude?
3. Compare endurance and strength–power performance changes at altitude.
4. Discuss and compare the strategies "live high, train low" and "live high, train high" with regard to physiological adaptation and performance improvement.
5. Explain and describe high-altitude pulmonary edema and high-altitude cerebral edema.
6. Provide recommendations to reduce the risk for acute mountain sickness.

part

V

MEDICAL AND HEALTH CONDITIONS

This section deals with the medical and health issues associated with sport and conditioning. An in-depth discussion on issues relating to overtraining includes specifics of methods used to monitor athlete performance. In addition, this section covers the etiology of both diabetes and exercise-induced bronchospasm, and discusses how athletes are able to manage these diseases and continue to play at high levels of performance. The final chapter in the section addresses issues relating to sudden death in sports.

chapter 24

Overtraining

CHAPTER OBJECTIVES

After reading this chapter you should be able to do the following:

- Describe the overtraining syndrome.
- Differentiate between overreaching, overcompensation, and overtraining.
- Understand the differences between endurance and strength–power athletes with regard to symptoms of overtraining.
- Understand various physiological and performance indicators of overtraining.
- Select a monitoring program that can be implemented to identify athletes at risk for overtraining.

The objective of all training programs is to optimally prepare the athlete or team for competition or, in the situation of a noncompetitive athlete, to achieve a specific training goal. The emphasis on either specific performance training (practices) or conditioning depends on the time of year and the type of athlete. An athlete who competes in a team sport (e.g., basketball, American football, or baseball) has a defined season, off-season, and preseason. Although sport-specific training and conditioning occur throughout each season, the emphasis on each variable may change depending on the time of the competitive year. During the off-season, greater emphasis is generally directed toward conditioning, whereas during the season the primary contact time between athletes and coaches is devoted to sport-specific improvement and game or competition preparedness.

The placement of emphasis—on an entire season of competition or on a specific meet—appears to be the primary difference between athletes who participate in team sports and athletes who compete in individual sports (e.g., track and field, swimming). Athletes involved in team sports are generally brought to peak condition immediately before the season, and this level of conditioning is maintained throughout the year by sport-specific practices and competitions. In contrast, the track and field athlete or swimmer may perform early-season competitions in less than peak physical condition or preparedness in order to peak for the more important competitions at the end of the season.

In both training scenarios, it is the goal of the coach to bring each athlete to peak condition at the appropriate time of the year. This is often accomplished through a periodized training program (see chapter 14) in which both training **intensity** and training **volume** are manipulated. During each phase of training, the athlete may experience a brief reduction in performance. As the athlete makes the necessary physiological **adaptations**, he or she experiences a **supercompensation** that results in improved performance (figure 24.1). If adequate **recovery** or rest does not occur, the ability of the body to adapt to the training stimulus is affected. The performance decrements that may be experienced during each new phase of training may become further exacerbated. In addition, as athletes get closer to their genetic or performance maximum, their potential for further adaptation and subsequent performance enhancements may be limited. Further increases in training volume or intensity, or maintenance of a high volume or intensity for a prolonged period of time, may also result in a decrease in performance. If these conditions persist, an **overtraining syndrome** may result. As can be seen in figure 24.1, there appears to be a fine line between peak performance and the potential for overtraining. It is a line that many coaches and athletes strive to reach but never to cross. One of the most difficult challenges that both coaches and athletes face is determining the training stimulus (both intensity and volume) that optimizes performance without causing any undesired training responses.

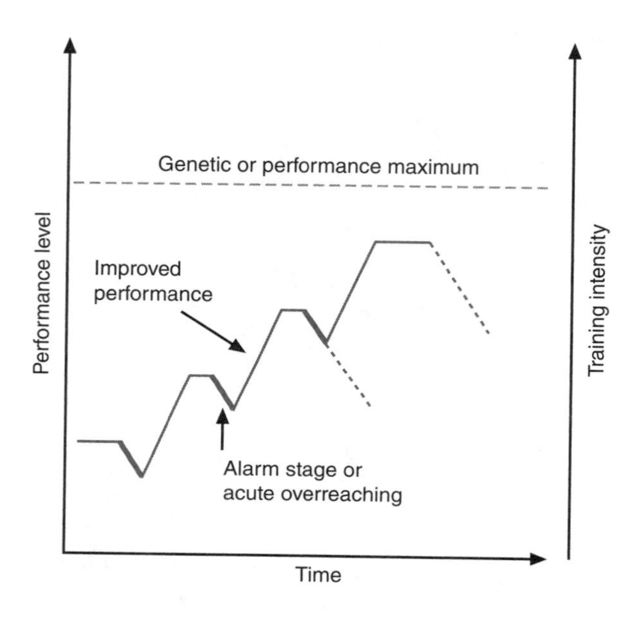

Figure 24.1 Hypothetical relationship between training adaptation, overtraining, and performance.

This chapter focuses on the different stages of overtraining and the factors that may contribute to these performance decrements. It also discusses methods of identifying, monitoring, and preventing overtraining that are specific to **endurance**, anaerobic, and **strength–power** sports.

Definitions of Overtraining

The overtraining syndrome may be considered a continuum of negative adaptations to training. Symptoms of overtraining appear when the training stimulus has reached the point at which training intensity or volume (or both) has become excessive, coupled with inadequate rest and recovery. Initial stages of overtraining are generally characterized by subjective feelings of **fatigue** and **staleness** that may or may not be accompanied by performance decrements. Farther along on the continuum, these subjective feelings of fatigue and staleness become associated with decreases in performance. When the training stimulus is excessive and recovery and adaptation do not occur within an anticipated time, the athlete is considered to be **overreaching**. With a decrease in the training stimulus and adequate rest, complete recovery usually occurs within 1 to 2 weeks (Kreider, Fry, and O'Toole 1998). The recovery may also coincide with an **overcompensation** and improved performance. Overreaching is a planned phase of many training programs.

When the imbalance between training and recovery continues for an indefinite period of time, the athlete progresses from a stage of overreaching and fatigue to the more serious problem of overtraining. Signs that are not acknowledged by an athlete or coach may be warnings of overtraining. Frequently a plateau or a decrement in performance is met with frustration on the part of the athlete or coach. However, the plateau or decline may be the initial symptom of overreaching. The coach or athlete may ignore the signs, thinking that what is needed is harder training to get past the plateau. Instead of a reduction in the training stimulus and resting, the training stimulus is increased. This results in a downward spiral of events that culminates in chronic fatigue and significant performance decrements associated with the overtraining syndrome. Recovery from overtraining can be a long process (possibly exceeding 6 months) (Kreider, Fry, and O'Toole 1998).

Contributing Factors

A number of factors contribute to an athlete's susceptibility to overtraining. These factors are related to training program issues, training environment, psychological issues (including stress from school for the student-athlete), nutritional concerns, and perhaps travel (jet lag). Although a single factor may be sufficient to cause overtraining, each additional factor increases the total stress experienced by the athlete.

Failure to allow for adequate recovery is thought to be the primary training-related factor that causes overtraining, or the progression from overreaching to the more serious overtraining syndrome (Fry, Morton, and Keast 1991). This is especially relevant when the training stimulus (i.e., volume or intensity) is increased. Whether intensity or volume imposes a greater stress and thereby makes the athlete more susceptible to overtraining is debatable. In addition, the impact that these training variables have on training stress may depend on the type of athlete (e.g., endurance or strength-power), as examined further in the next section of the chapter.

During periods of intense training, the athlete must consume adequate carbohydrates and calories to prevent the catabolization of muscle for amino acids as a fuel source for exercise. In addition, prolonged exercise under hypoglycemic conditions negatively affects performance (Kuipers and Keizer 1988). An athlete's appetite may be influenced by a host of factors such as the stress of training, personal problems, sleep, and environmental conditions (Berning 1998). Athletes exercising in a hot and humid environment may have elevated body temperatures throughout the day, resulting in a decrease in hunger (Berning 1998). Regardless of the mechanism causing reduced caloric intake, athletes increase their risk of overtraining if energy demands are not met.

Psychological factors may also play a role. Monotony of training or emotional demands related to family, school, or work may be a prelude to early stages of overreaching. Excessive expectations on the part of coaches, oneself, or the public may be another cause of emotional stress experienced by the athlete (Kuipers and Keizer 1988).

Comparison of Endurance and Strength–Power Athletes

For both endurance and strength–power athletes, achievement of training goals is accomplished through manipulation of training volume (increased distance and increased number of sets and repetitions, respectively) and training intensity (exercise at a higher percentage of maximal oxygen uptake and one-repetition maximum [1RM], respectively). However, the effect of each training variable on each type of athlete with respect to increasing susceptibility to overtraining may be different.

The results of studies designed to elicit overtraining in endurance athletes appear to be inconclusive. Several studies have suggested that increasing the volume of training over the short term does not result in any symptoms of overreaching or overtraining (Costill et al. 1988; Kirwin et al. 1988). In contrast, other researchers have reported that an acute increase in training volume may be a potent stimulator of overtraining (Hooper et al. 1993, 1995; Lehmann et al. 1992). Furthermore, an increase in training intensity in a short-term study looking to elicit overreaching resulted in improved performance (Lehmann et al. 1992). Although the endurance athlete manipulates both training volume and intensity, the literature suggests that an excessive increase in volume without an appropriate regeneration phase places the endurance athlete at greater risk for overreaching or overtraining. However, most studies have looked at overtraining over short durations and not over more prolonged periods of exercise, which are common to most endurance athletes experiencing overtraining.

In strength–power athletes, overtraining has been attributed to alterations in both training volume (Stone and Fry 1998) and intensity (Fry 1998). Apparently, changes in either variable without allowing for sufficient rest or recovery can cause overtraining in a relatively short period of time. Similar to the studies on endurance athletes, most overtraining studies of strength–power athletes have been of short duration and may not truly represent the influences responsible for overtraining over a prolonged season of practices. In addition, most of the research on overtraining and the strength–power athlete has dealt with weightlifters. Studies on athletes participating in anaerobic sports (e.g., football or basketball) that combine several methods of training (e.g., strength, power, speed, and endurance) are sorely lacking. This is especially unfortunate considering the huge numbers of athletes participating in these sports.

Differences between endurance and strength–power athletes may become more apparent when one examines symptoms of overtraining that appear to be specific to each. These differences are reviewed in the section on recognition of overtraining.

Susceptibility to Overtraining

It is difficult to determine which athletes may be more susceptible to overtraining because all athletes at all levels of performance are at risk. However, it is the highly motivated athlete who appears to be the most susceptible (Fry, Morton, and Keast 1991). The occurrence of fatigue and other stages of overtraining depends primarily on how the individual athlete responds to the specific training stimulus. This is unfortunate for athletes in a team sport in which the training program is developed for the team as a whole and not for the individual athletes. In this situation, although most of the team may be responding well to the practice regimen, a particular athlete may be having difficulty adapting to greater intensities or higher volumes of practices. Without appreciating the possibility that this athlete may be in a particular stage of overtraining, the coach may put undue pressure on the individual to work harder, which may cause further damage.

Track and field athletes, swimmers, and weightlifters may be in a better situation with regard to a coach's ability to recognize signs and symptoms of overtraining. Although it is not possible to predict who might be susceptible to overtraining, it is not difficult to monitor the training performance of these athletes. One can prevent overtraining by identifying potential markers and making appropriate adjustments to the training program. This depends primarily on the ability to objectively measure daily performance of the athletes. In contrast, athletes who participate in team sports may depend on more subjective means of evaluation to determine whether changes in performance are related to a physiological stress or the superior ability of the opponent. This difference is reflected by the relatively large volume of research on overtraining in endurance athletes (Flynn et al. 1994; Lehmann et al. 1995; Morgan et al. 1987), swimmers (Costill et al. 1988; Hooper et al. 1993; Raglin, Morgan, and O'Connor 1991), and even weightlifters (Fry, Kraemer, van Borselen, Lynch, Triplett, et al. 1994; Fry, Kraemer, van Borselen, Lynch, Marsit, et al. 1994) compared with the few studies on overtraining in team sports (e.g., basketball and football) (Hoffman, Epstein, et al. 1999; Verma, Mahindroo, and Kansal 1978).

Recognition of Overtraining

The primary indicator of either overreaching or overtraining is a decrement in performance. However, several reports have suggested that athletes whose performances have stagnated may also be overtrained (Hooper et al. 1995; Kuipers and Keizer 1988; Rowbottom, Keast, and Morton 1998). Much of the overtraining literature has been directed at identifying markers that would enable coaches to recognize overtraining at its earliest stages. It is thought that with early recognition, coaches could make the necessary training adjustments to prevent the more damaging and longer-lasting overtraining syndrome. A host of variables, categorized according to physiological, psychological, immunological, and biochemical manifestations, have been reported in the literature as symptoms associated with overtraining. The major biological symptoms of overtraining, as indicated by their prevalence in the literature, were first compiled by Fry, Morton, and Keast (1991) and are listed in table 24.1. One or possibly several of these symptoms may be seen in athletes experiencing performance decrements. Unfortunately, these symptoms have been recorded in the absence of a single objective measure identified as a consistent marker for overtraining. In addition, the appearance of these symptoms seems to confirm only that the athlete is in some stage of overtraining. It does not appear that the symptoms can indicate whether the athlete is on the threshold of overreaching or overtraining.

The following sections review some of the more prominent biological disturbances reported in athletes who are overtrained. Although some cause-and-effect relationships may exist between various biological disturbances, some of these imbalances may occur independently of one another.

Autonomic Nervous System Disturbances

Overtraining is thought to produce an autonomic nervous system imbalance, which results in either a **sympathetic** nervous system dominance or a **parasympathetic** nervous system dominance (Israel 1976). Sympathetic overtraining is associated with greater sympathetic activity in the resting state, whereas parasympathetic overtraining is associated with an inhibition of sympathetic activity and

Table 24.1 Major Symptoms of Overtraining as Indicated by Their Prevalence in the Literature

Physiological performance		
Decreased performance	Inability to meet previously attained performance standards or criteria	Prolonged recovery
Reduced tolerance of loading	Decreased muscular strength	Decreased maximum work capacity
Loss of coordination	Decreased efficiency or decreased amplitude of movement	Reappearance of mistakes already corrected
Reduced capacity to differentiate and correct technical faults	Increased difference between lying and standing heart rate	Abnormal T-wave pattern in ECG
Heart discomfort on slight exertion	Changes in blood pressure	Changes in heart rate at rest, exercise, and recovery
Increased frequency of respiration	Increase frequency of respiration	Decreased body fat

Increased oxygen consumption	Increased ventilation and heart rate at submaximal workloads	Shift of the lactate curve toward the *x*-axis
Decreased evening postworkout weight	Elevated basal metabolic rate	Chronic fatigue
Insomnia with and without night sweats	Feelings of thirstiness	Anorexia nervosa
Loss of appetite	Bulimia	Amenorrhea or oligomenorrhea
Headaches	Nausea	Increased aches and pains
Gastrointestinal disturbances	Muscle soreness or tenderness	Tendon complaints
Periosteal complaints	Muscle damage	Elevated C-reactive protein
Rhabdomyolysis		
Psychological or information processing		
Feelings of depression worsening	General apathy	Decreased self-esteem or feelings of self
Emotional instability	Difficulty in concentrating at work and training	Sensitivity to environmental and emotional stress
Fear of competition	Changes in personality	Decreased ability to narrow concentration
Increased internal and external distractibility	Decreased capacity to deal with large amounts of information	Giving up when the going gets tough
Immunological		
Increased susceptibility to and severity of illnesses, colds, and allergies	Flu-like illnesses	Unconfirmed glandular fever
Slow healing of minor scratches	Swelling of the lymph glands	One-day colds
Decreased functional activity of neutrophils	Decreased total lymphocyte counts	Reduced response to mitogens
Increased blood eosinophil count	Decreased proportion of null (non-T, non-B) lymphocytes	Bacterial infection
Reactivation of herpes viral infection	Significant variations in CD4:CD8 lymphocytes	
Biochemical		
Negative nitrogen balance	Hypothalamic dysfunction	Flat glucose tolerance curves
Depressed muscle glycogen concentration	Decreased bone mineral content	Delayed menarche
Decreased hemoglobin	Decreased serum iron	Decreased serum ferritin
Lowered total iron binding capacity	Mineral depletion (e.g., Zn, Co, Al, Mn, Se, Cu)	Increased urea concentrations
Elevated cortisol levels	Elevated ketosteroids in urine	Low free testosterone
Increased serum hormone binding globulin	More than 30% decrease in ratio of free testosterone to cortisol	Increased uric acid production

Data from Fry et al. 1991.

a greater parasympathetic activity both at rest and during exercise.

Sympathetic overtraining is associated with restlessness and irritability, sleep disturbances, weight loss, and an elevation in resting heart rate and blood pressure. A parasympathetic imbalance is associated with fatigue and depression; a reduction in resting heart rate and blood pressure; and a suppressed heart rate, glucose, and lactate response to exercise. Neuromuscular activity also appears to be impaired (Lehmann et al. 1998). The parasympathetic imbalance is thought to be more common in endurance sports, and the hyperexcitability associated with sympathetic overtraining is more associated with explosive, power sports (Lehmann et al. 1998). However, these two forms of overtraining may also represent a continuum of varying symptoms associated with different stages of overtraining (Flynn 1998; Kuipers and Keizer 1988). Although both forms of overtraining are associated with performance decrements, the parasympathetic form may be more difficult to distinguish because the symptoms are far less alarming and, in the initial stage, are similar to the positive adaptations associated with training (Fry, Morton, and Keast 1991).

Parasympathetic overtraining is thought to be a reflection of a more advanced form of overtraining and is closely associated with exhaustion of the **neuroendocrine system**. Sympathetic overtraining reflects a fatigued state seen during the initial stages of overtraining (Kuipers and Keizer 1988). In either type of overtraining, the change in the homeostatic balance is further reflected by changes in other physiological systems.

Neuroendocrine Disturbances

Any stress, including that accompanying exercise, causes marked changes in the neuroendocrine response. These changes can result from both acute and chronic training stresses. Acute changes can be caused by manipulation of an acute program variable (e.g., training intensity). As the body adapts to this new training stimulus, the hormonal response returns to baseline levels fairly quickly. During chronic training stresses, in which recovery is inadequate or the body is unable to adapt to the greater stress, disturbances to several hormonal axes (e.g., hypothalamic-pituitary-adrenal, hypothalamic-pituitary-gonadal) become apparent. Changes in the endocrine response to increases in training volume appear to be similar between endurance and resistance exercise. However, these similarities do not appear to exist when overtraining is seen following increases in training intensity (Fry 1998).

The hormones **testosterone** and **cortisol** are frequently monitored in connection with overtraining. Decreases in both total and free testosterone with an accompanying increase in cortisol concentrations are thought to reflect a greater catabolic state in athletes. The testosterone/cortisol ratio has been proposed as a potential monitor of training stress (Adlercreutz et al. 1986). If the ratio of free testosterone to cortisol declines more than 30% or is less than $0.35 \cdot 10^{-3}$, it is thought to represent an insufficient recovery from exercise and may be an indicator of overtraining. If such a decline occurs after 1 day of training, it is an indicator of insufficient recovery. If it occurs after 3 months of training, it may represent overtraining. A prolonged disruption in the anabolic to catabolic balance may result in decreases in body mass, primarily through a loss of lean body tissue. However, a number of researchers have been unable to find any change in the testosterone/cortisol ratio during overtraining or have not seen any correlation between changes in this ratio and performance (Fry et al. 1992; Fry, Kraemer, et al. 1993; Kirwin et al. 1988; Urhausen, Gabriel, and Kindermann 1995). Thus, changes in testosterone and cortisol concentrations may not be directly responsible for performance decrements and may be more reflective of changes in training stimuli.

Catecholamines appear to be very responsive to training stresses. The involvement of the catecholamines with a number of physiological systems suggests that they may be potent mediators of the overtraining syndrome. In overtrained endurance athletes, a reduction in the catecholamine response to exercise, paralleling impaired glycolytic energy mobilization, may be seen (Urhausen, Gabriel, and Kindermann 1995). In addition, overtrained endurance athletes may also have reduced nocturnal levels of both **epinephrine** and **norepinephrine** (Lehmann et al. 1998). This would be consistent with parasympathetic overtraining, common in this type of athlete. Furthermore, athletes suffering from parasympathetic overtraining are reported to show a 50% to 70% reduction in basal urinary catecholamine excretions (Lehmann et al. 1998). Decreases in catecholamine concentrations are negatively correlated to both fatigue (Lehmann et al. 1992) and latency of rapid eye movement (REM) sleep (Lehmann et al. 1998), suggesting an important role of catecholamine reduction in central fatigue.

Increases in catecholamine concentrations at rest (Hooper et al. 1995; Lehmann et al. 1992) and during exercise (Fry, Kraemer, van Borselen, Lynch, Triplett, et al. 1994; Lehmann et al. 1992) have also been reported in overtrained endurance and strength–power athletes. These increases in catecholamine concentrations may reflect a loss of sensitivity in the target organs for catecholamines, specifically norepinephrine. Decreases in **β-adrenergic receptors** have been reported after prolonged high-volume training (Jost, Weiss, and Weicker 1989). The elevated catecholamine concentrations, also seen in overtrained athletes performing high-intensity resistance training, likely also reflect a downregulation of the β-adrenergic receptors. Regardless of the exercise stimulus, a reduction in β-adrenergic receptors in muscle causes impaired muscle performance. This is highlighted by the significant relationship reported between immediate postexercise catecholamine concentrations and maximal strength performance in resistance-trained individuals (*r* ranging from 0.79 to 0.96). This relationship was not seen in overtrained subjects (Fry, Kraemer, van Borselen, Lynch, Triplett, et al. 1994).

Hypothalamic–pituitary dysfunction has been shown primarily in overtrained endurance athletes. In the limited amount of research available on overtrained resistance athletes, the hypothalamic–pituitary axis does not appear to be similarly affected (Fry 1998). Barron and colleagues (1985) were first to show a dysfunction in the hypothalamic–pituitary axis by reporting reduced adrenocorticotropic hormone (ACTH), growth hormone, and prolactin responses to insulin-induced hypoglycemia in overtrained marathon runners. Although those investigators reported that the dysfunction was seen primarily in the **hypothalamus**, they suggested that **pituitary** insensitivity was possible. Other studies have shown reduced pulsatile luteinizing hormone (LH) secretion (Hackney, Sinning, and Bruot 1990; MacConnie et al. 1986) and reduced β-endorphin, thyroid-stimulating hormone, and growth hormone concentrations (Fry, Morton, and Keast 1991; Keizer 1998; Urhausen, Gabriel, and Kindermann 1995) in overtrained athletes. The impaired pituitary response in overtrained athletes may cause reproductive abnormalities, such as menstrual cycle irregularities (e.g., amenorrhea) in females and reduced libido and sperm counts in males. Figure 24.2 reviews the hypothalamic–pituitary–adrenal–gonadal axes and overtraining and

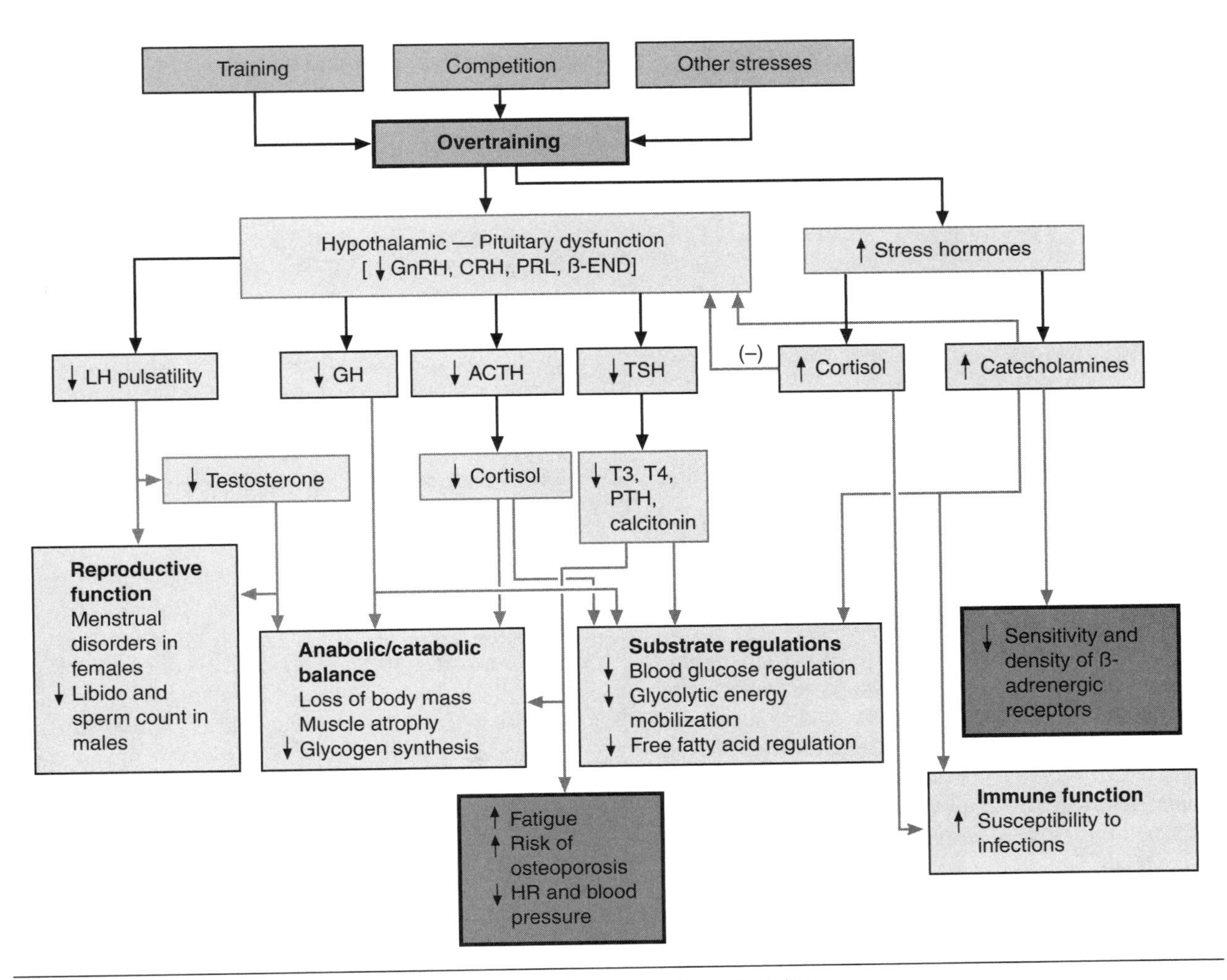

Figure 24.2 Hypothalamic–pituitary–adrenal–gonadal axes and overtraining.
Data from Fry et al. 1991; Urhausen et al. 1995.

also outlines some of the major symptoms associated with neuroendocrine dysfunction during overtraining.

Psychological Disturbances

Mood states have been reported to be sensitive to changes in training volume (Morgan et al. 1987; Raglin, Morgan, and Luchsinger 1990) and are thought to be potentially useful as a tool for monitoring training adaptations. The **Profile of Mood States (POMS)** is a self-report inventory frequently used to assess the mood states of athletes (Morgan et al. 1987). Athletes tend to score lower than the normal population on scales of tension, anxiety, anger, confusion, and fatigue and to score above the normal population in vigor. This mood pattern, called an "**iceberg profile**" because it has been suggested to resemble an iceberg, is typically observed at the onset of training. As a season progresses and training volume is increased, the mood profile of the athlete may become significantly altered, forming a pattern resembling that of the healthy nonathletic population. As training volume decreases during a taper, mood profiles return to preseason values. In contrast, stale or overtrained athletes do not respond to the taper and their mood state remains significantly altered, reflected by a flattened mood profile.

Overtrained athletes may also exhibit lower confidence in their ability to succeed. A study of overtrained resistance athletes showed that they did not exhibit any changes in their perception of the difficulty of the lifting task during 2 weeks of an overtraining protocol. However, within 8 days of the onset of the protocol, the athletes' confidence in their ability to successfully perform was significantly reduced (Fry, Fry, and Kraemer 1996). This study reflected the changes in self-efficacy that may be observed in overtrained athletes and may contribute to impaired performance.

Immunological Disturbances and Oxidative Stress

Overtrained endurance athletes appear to be at a greater risk of infectious illness, particularly upper respiratory tract infections (URTI), than their nonafflicted peers. Because most of the overtraining literature has focused on endurance athletes, it is unknown if overtrained strength–power athletes are at a similar risk. However, because immune suppression appears to be related to impaired neuroendocrine balance, it is likely that, on close examination, overtrained strength–power athletes would also prove to be more susceptible to infection.

Although changes in endocrine function (e.g., cortisol, catecholamines, and β-endorphins) are thought to be involved in the immune suppression reported during overtraining or during periods of intense training, the complexity of these changes suggests that more than one mechanism is responsible for alterations in immune function. As discussed in chapter 5, periods of intense training cause several changes in the immune system that may result in a greater susceptibility to illness in overtrained athletes. Reductions in resting leukocyte counts, lymphocyte counts, neutrophil activity, and immunoglobulin concentrations are likely some of the factors responsible for the increased risk of infectious disease in athletes who have overreached or overtrained. Recent work has suggested that neutrophil function can be used as a marker of overtraining in athletes (Yaegaki et al. 2008). Phagocytic activity (PA) and cytotoxic reactive oxygen species (ROS) were examined in both male and female judo athletes participating in a weeklong training camp. In general, PA and ROS compensate for each other to maintain neutrophil integrity; ROS often increase while PA decreases (Mochida et al. 2007). However, during intense and prolonged exercise (i.e., marathons), both PA and ROS may decrease (Pedersen, Rohde, and Bruunsgaard 1997). In the study on judo athletes, the weeklong training camp resulted in the typical response in males (increase in ROS and decrease in PA); however, in the female athletes, ROS production significantly decreased with no compensation from PA. The results indicated that the male athletes were able to maintain neutrophil function whereas the women were not, suggesting that the women were possibly overloaded during the training camp. The altered immune response may also cause athletes to be susceptible to URTI.

Several epidemiological studies have reported a higher incidence of URTI in athletes engaged in marathon events or very heavy endurance training (Heath et al. 1991; Nieman et al. 1990; Peters and Bateman 1983). Very few studies have directly examined the incidence of illness and its relationship to overtraining. Mackinnon and Hooper (1996) compared illness rates in athletes showing symptoms of overreaching (33% of athletes examined) to those who seemed to have adapted well to the intensified training. Greater incidences of URTI were observed in the athletes who did not show symptoms of overreaching (56%) compared with the athletes who were experiencing symptoms (12.5%). This study suggests that increases in URTI might depend more on changes in training intensity than on overreaching per se.

As already discussed, intense exercise typically results in an elevation of ROS indicative of oxida-

In Practice

How Important Are Rest and Sleep in Maximizing Recovery From Exercise?

One of the most overlooked issues relating to maximizing recovery from exercise or reducing the risk for developing the overtraining syndrome is adequate rest and sleep. A consensus statement by the European College of Sport Science and the American College of Sports Medicine recommended that at least one passive rest day be provided each week, primarily due to the relation seen between the absence of a recovery day and the onset of signs of overreaching and an inability to recover (Meeusen et al. 2013). The inclusion of a rest day can be prospectively incorporated into the athlete's weekly training program or provided when athlete assessment indicates that it is warranted. For some sports, such as American football, in which the weekly practice and competition schedule is relatively consistent, it becomes easier to incorporate predetermined off days to enhance recovery from competition. The day following a Sunday game in the National Football League, or a Saturday game in collegiate football, can be used to provide treatments for injury recovery or for film review, or just as a day off. In any case, a day off is often incorporated into the weekly schedule. For sports that do not have a set schedule, such as basketball, hockey or baseball, it becomes imperative that the coach identify rest periods within schedule simply to enhance recovery. The rest periods not only provide an opportunity to recover from the physiological stress associated with practice and competition; they also offer a distraction from the daily routine of training and potentially alleviate boredom that may be associated with the routine (Meeusen et al. 2013). In addition to appropriately placed rest days, adequate and healthful sleep is a critical part of fatigue management (Meeusen et al. 2013). It is believed that sleep is more important for its neurological benefit than for recovery of other biological functions (Horne and Pettitt 1984). Poor sleep patterns may impair cognitive function, especially the ability to concentrate. The recommended number of hours of sleep is highly individual and quite varied. Thus, the general recommendation for athletes has been to sleep the amount of time required to allow them to feel awake and alert during the day (Meeusen et al. 2013).

tive stress. Elevations in ROS are associated with an inflammatory response involved in the repair of muscle damage arising from the exercise stress (Evans and Cannon 1991). ROS likely stimulate neutrophil and macrophage increases within the muscle to participate in the repair and recovery of skeletal tissue. Although oxidative stress markers are not typically examined in connection with overtraining, impeded recovery is a response generally associated with overtraining; and measures of oxidative stress markers may provide additional insight into the overtraining syndrome. Margonis and colleagues (2007), in a 12-week resistance training study designed to elicit an overtraining stimulus, reported significant elevations in oxidative stress markers (isoprostane, TBARS [thiobarbituric acid reactive species], and protein carbonyls) in a dose-dependent proportion during the high-volume training period. In addition, glutathione status (both reduced and oxidized) was significantly lower during the overtraining period. Strength levels were significantly reduced as well. However, as training volume decreased, the ROS markers declined and strength levels returned.

Biochemical Disturbances

Elevations in creatine kinase and uric acid and decreases in **glycogen** levels and **lactate** concentrations have been reported after high-intensity training periods, and each has been suggested as a potential indicator of overtraining (Fry, Morton, and Keast 1991). However, many of these biochemical disturbances occur as part of the normal response to an acute exercise stress, and impairments may persist for several days postexercise. For example, muscle glycogen resynthesis may be impaired because of muscle damage, and full restoration of muscle glycogen after a marathon may take as long as a week (Sherman et al. 1983). Thus, caution is warranted when one is interpreting changes in biochemical function postexercise for evaluation of overtraining. A sufficient period of recovery should be allowed for proper assessment of the athlete.

Performance Indicators

Several investigations have looked at the use of performance measures as an indicator of overtraining (Fry, Lawrence, et al. 1993; Fry, Kraemer, van Borselen, Lynch, Marsit, et al. 1994; Hoffman and Kaminsky 2000; Lehmann et al. 1992). Performance measures are an attractive method of monitoring training stresses. A multitude of biological variables have been identified as potential markers of overtraining; however, none have been identified as consistent predictors. In addition, the cost associated with many of these measures makes these tests infeasible for many athletic teams.

Rowbottom, Keast, and Morton (1998) reviewed a number of studies employing laboratory performance measures such as maximal oxygen uptake, running speed at 4 mmol/L lactate, maximal-effort time trials, and treadmill runs to exhaustion. It is imperative to select a performance test with established validity and reliability. Overtrained athletes may have performance decrements as large as 29%; however, performance differences are likely to be more subtle during the initial stages of overreaching (Rowbottom, Keast, and Morton 1998). Endurance athletes have been the primary focus of these performance measures. It is more complicated to select an appropriate test to monitor overreaching in strength–power athletes participating in a team sport. Hoffman and Kaminsky (2000) reported on a testing battery they used to monitor national-level youth basketball players. A major concern of the coaching staff was the number of teams the athletes were playing for. In addition to the national team, all the players were competing on their respective high school or club team. In some cases, the athletes also played on their club's adult basketball team. Thus, a number of performance tests, representing the various components of fitness considered important for a basketball player (e.g., strength, speed, and agility), were used to monitor the athletes on a monthly basis. The results of these tests, including the athletes' training volume and subjectively rated training intensity, are shown in figure 24.3.

In this study, the 27 m sprint appeared to be the most sensitive test for highlighting players who were fatigued. When a "red flag" was seen (increase in sprint time greater than 0.15 s from the player's best time), a further analysis of the player's training log showed an increase in both training volume and training intensity for the 2 weeks preceding the testing period. Subsequently, the coaches of the player's club and high school teams were asked to reduce the player's practice volume (excuse the player from practices). During the next testing session (1 month later), the player's sprint time returned to normal (see table 24.2).

Methods Used to Monitor Athletic Performance

The key in preventing overtraining is to monitor the athlete or team continuously throughout the year. The major questions, though, are what methods can be used to perform these assessments and whether budgetary restraints, if they exist, limit the potential for successful monitoring. This section presents examples of successful athletic monitoring in situations of varying budget levels. However, as will become evident, the limiting factor is generally not financial but more likely time. The coach must make a concerted effort to provide the time needed to perform the monitoring program.

Developing an effective monitoring program does not require a large budget. What is needed is an assessment protocol that can be performed on a regular basis. Performance in games and practices tend to be difficult to quantify. Too many variables may affect game or practice performance (e.g., competition within practice or with the opponent during a game, different practice routines, inconsistent drills, lack of time within a practice session). What is needed is a time devoted solely to the assessments that will be used to monitor for overtraining. These need to be performed while the athlete is rested, for example before practice. Whatever assessment is chosen needs to be performed in the same way, at the same time, and by the same coaches, trainers, or assistants at each testing session to maintain test reliability (see chapter 16). Table 24.3 provides examples of simple, somewhat more complicated, and sophisticated assessment programs for overtraining in basketball. These are examples only and can easily be modified to fit specific teams or budgets.

Visual analog scales (VAS) are a very simple method of determining the athlete's subjective feelings of fatigue, energy, focus, and alertness. All that is needed is a line 15 cm in length with verbal anchors at each end representing extreme ratings (low and high). Statements to be completed are "My level of energy is . . .", "My level of focus is . . .", "My level of alertness is . . .", and "My level of fatigue is . . . ." The use of training logs is quite simple. As discussed in chapters 14 and 15, the in-season resistance training program is designed to maintain strength or perhaps

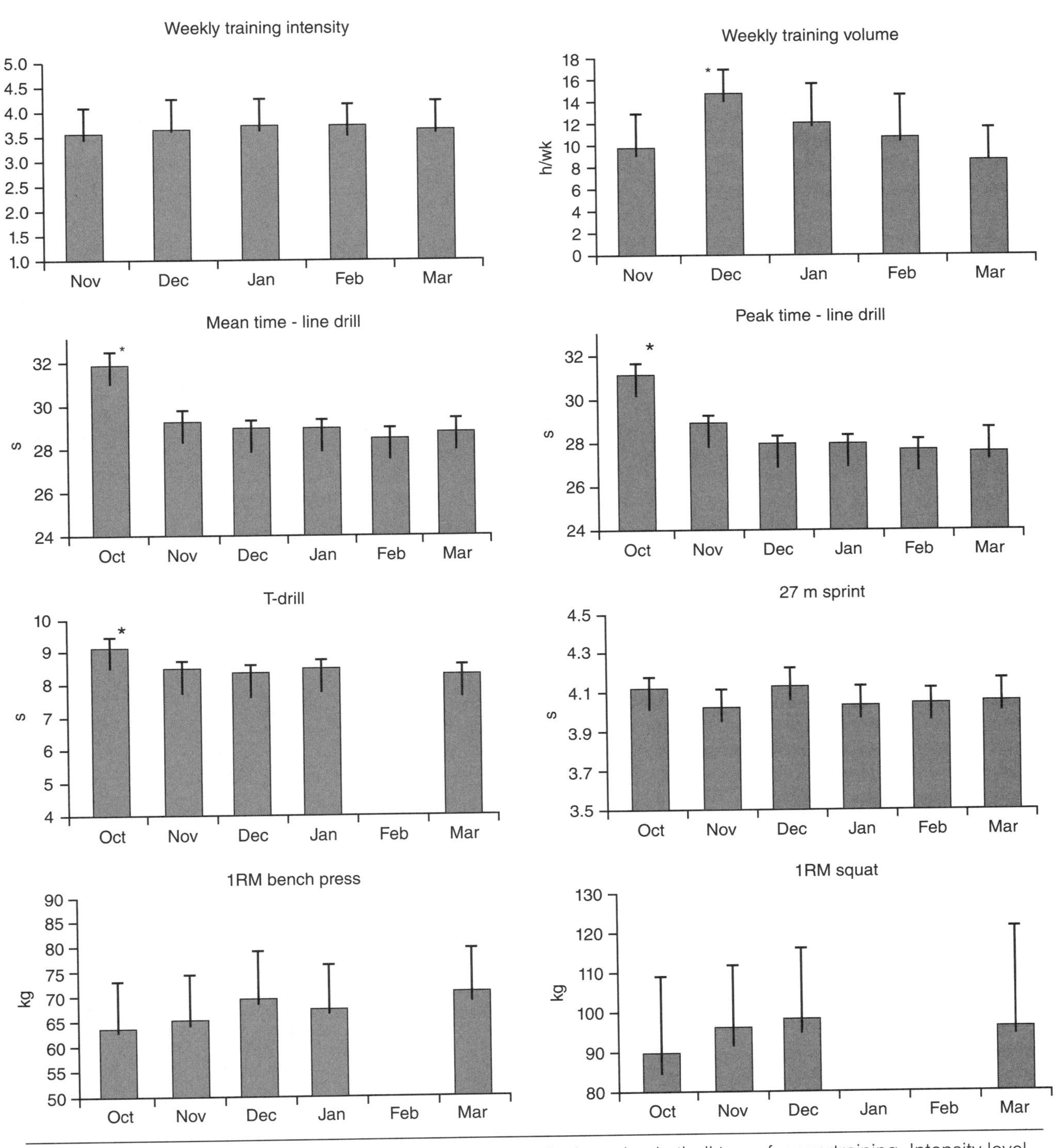

Figure 24.3 Example of athletic performance tests for monitoring a basketball team for overtraining. Intensity level based on subjective scale where 1 = very easy and 5 = very hard. * = Significantly different than all other time points.

Table 24.2 Example of "Red Flag" (in Bold) and Subsequent Performance Results When Training Volume Was Reduced

Month	27 m sprint (seconds)	Training volume (hours per week)	Training intensity
November	4.00	15.3	3.6
December	**4.17**	22.3	4.2
January	3.98	17.3	4.0

Intensity level was subjectively rated using a 5-point scale. The rating scale included the following: 1 = very easy, 2 = easy, 3 = average, 4 = hard, 5 = very hard. Results were averaged for the 2 weeks between national team practice sessions.

Table 24.3 Examples of Simple, More Complicated, and Sophisticated Assessment Programs to Monitor for Overtraining

Simple	More complicated	Sophisticated
Visual analog scale or ratings on a Likert scale for soreness, fatigue, energy, and focus.	Visual analog scale or ratings on a Likert scale for soreness, fatigue, energy, and focus.	Visual analog scale or ratings on a Likert scale for soreness, fatigue, energy, and focus.
Examine training logs for resistance training program and record changes in volume and intensity.	Examine training logs for resistance training program and record changes in volume and intensity. Add accelerometer to bar and record power performance. Determine the number of repetitions above 80% and 90% of the maximal power for each set. This provides a measure of the quality of each repetition.	Examine training logs for resistance training program and record changes in volume and intensity. Add accelerometer to bar and record power performance. Determine the number of repetitions above 80% and 90% of the maximal power for each set. This provides a measure of the quality of each repetition.
Perform three line drills with a 2 min rest between each sprint. Examine time per sprint and calculate fatigue rate (fastest time/slowest time).	Perform line drill; examine time per sprint and calculate fatigue rate (fastest time/slowest time). Add a heart rate monitor and determine heart rate recovery for 2 min between each sprint.	Perform line drill; examine time per sprint and calculate fatigue rate (fastest time/slowest time). Add a heart rate monitor and determine heart rate recovery for 2 min between each sprint. Test blood lactate concentrations by a finger or earlobe prick 2 min after sprint 3.
	Test countermovement vertical jump with the athlete placing hands on hips while an accelerometer is attached to waist. Power output is averaged over five consecutive jumps.	Test countermovement vertical jump with the athlete placing hands on hips while an accelerometer is attached to waist. Power output is averaged over five consecutive jumps.
		Take fasted blood draw or saliva for measures of testosterone and cortisol.
		Examine muscle quality (determining echo intensity) and architectural changes (muscle thickness, pennation angle, and cross-sectional area) via ultrasonography on both right and left legs.
		Examine visual and motor reaction time using commercial assessment devices.

increase it in novice athletes. Recommendations are for athletes to train twice per week using approximately 80% of their 1RM in the squat and bench press exercises. Athletes are generally required to use progressive overload. That is, as they are able to increase the number of repetitions, they should increase the weight; if they are unable to maintain the minimum number of repetitions, they may need to reduce the resistance. Since the lower body is stressed during the basketball season, the coach simply needs to monitor the intensity of training for the squat exercise; this may provide a good indication of lower body fatigue. Coaches who have access to an accelerometer (figure 24.4) can measure the power output per repetition to get a more sensitive indicator of changes in the quality of the workout. Even if training volume can be maintained, it is possible for the quality of repetitions to be reduced. This may occur before performance decrements are seen and may be an indicator of increased risk of fatigue or overreaching.

The line drill is often used as a conditioning drill during basketball practice (see chapter 16 for a complete description). Since it is familiar to most basketball players, the line drill can serve as a simple and relatively easy method of monitoring fatigue. The athlete performs three line drills with a 2 min rest between each one. The time of each sprint is recorded, and a fatigue rate can then be calculated. One can increase sensitivity by measuring heart rate deceleration (recovery heart rate) and, for even greater sophistication, measuring blood lactate concentrations following the final sprint. Changes in performance of the line drill could indicate fatigue. When combined with the VAS, the line drill may provide a good indicator of fatigue. Figure 24.5

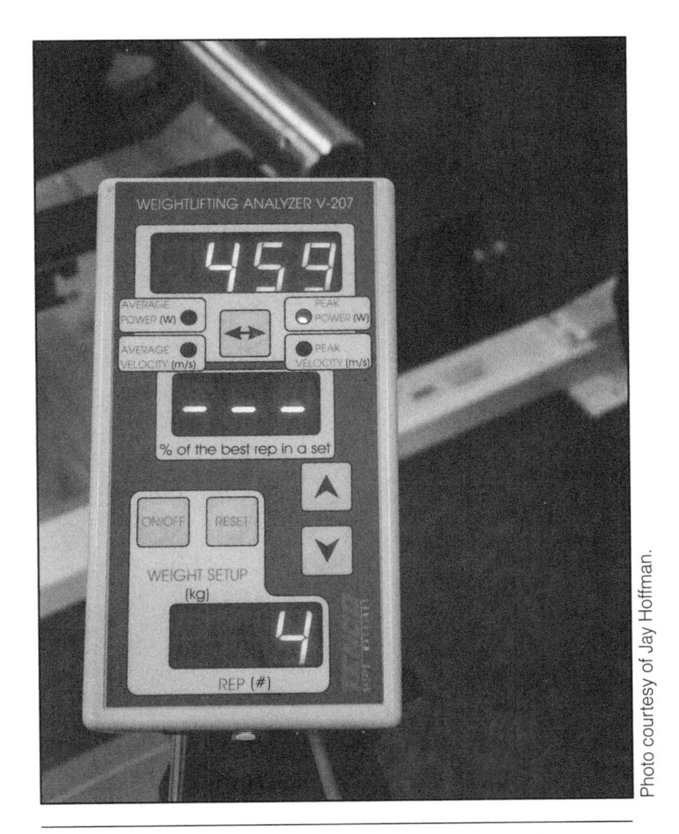

Photo courtesy of Jay Hoffman.

Figure 24.4 Accelerometer attached to a barbell measuring power output per repetition.

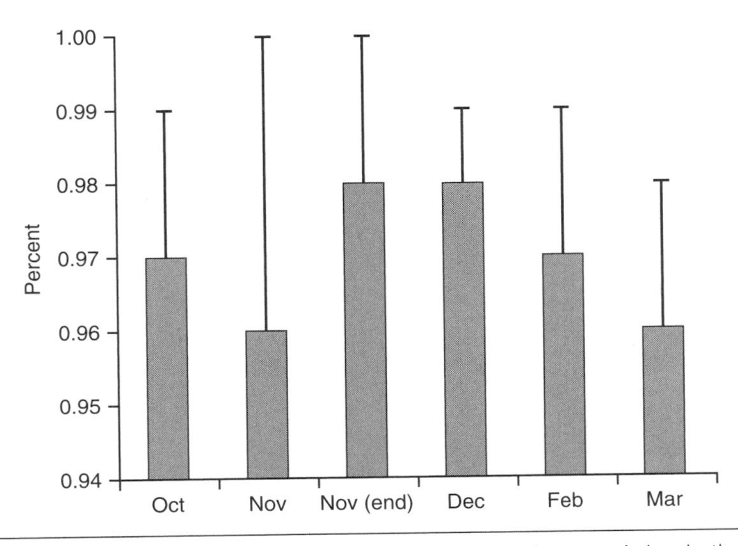

Figure 24.5 Changes in fatigue rate in the line drill in an NCAA Division I women's basketball team.

shows changes in the fatigue rate in an NCAA Division I women's basketball team over the course of a competitive season.

To quantify vertical jump power, players perform five consecutive countermovement jumps. During jumps, players have their hands on their waist at all times. They should be instructed to maximize the height of each jump while minimizing the contact time with the ground between jumps. Players wear a belt connected to an accelerometer. Power outputs for all five jumps are recorded and averaged. Figure 24.6*a* shows VAS for energy, focus, fatigue, and alertness in professional basketball players during a competitive season. Figure 24.6*b* shows vertical jump relative mean power output; and figure 24.6*c* shows squat power (averaged over eight repetitions). During February, a decrease in focus and an increase in fatigue were observed; with decrements in power outputs in both vertical jump and squat power performance (see arrows in figure 24.6, *a-c*). Given these indicators, the strength and conditioning staff told the head coach that the team was fatigued, suggesting that before this manifested as a full overtraining syndrome, reducing the training volume during practice would be prudent. The coach cooperated, backing off during practices, and the team was able to recover and reach the playoffs. This is a good example of how effective communication and belief in the system, as

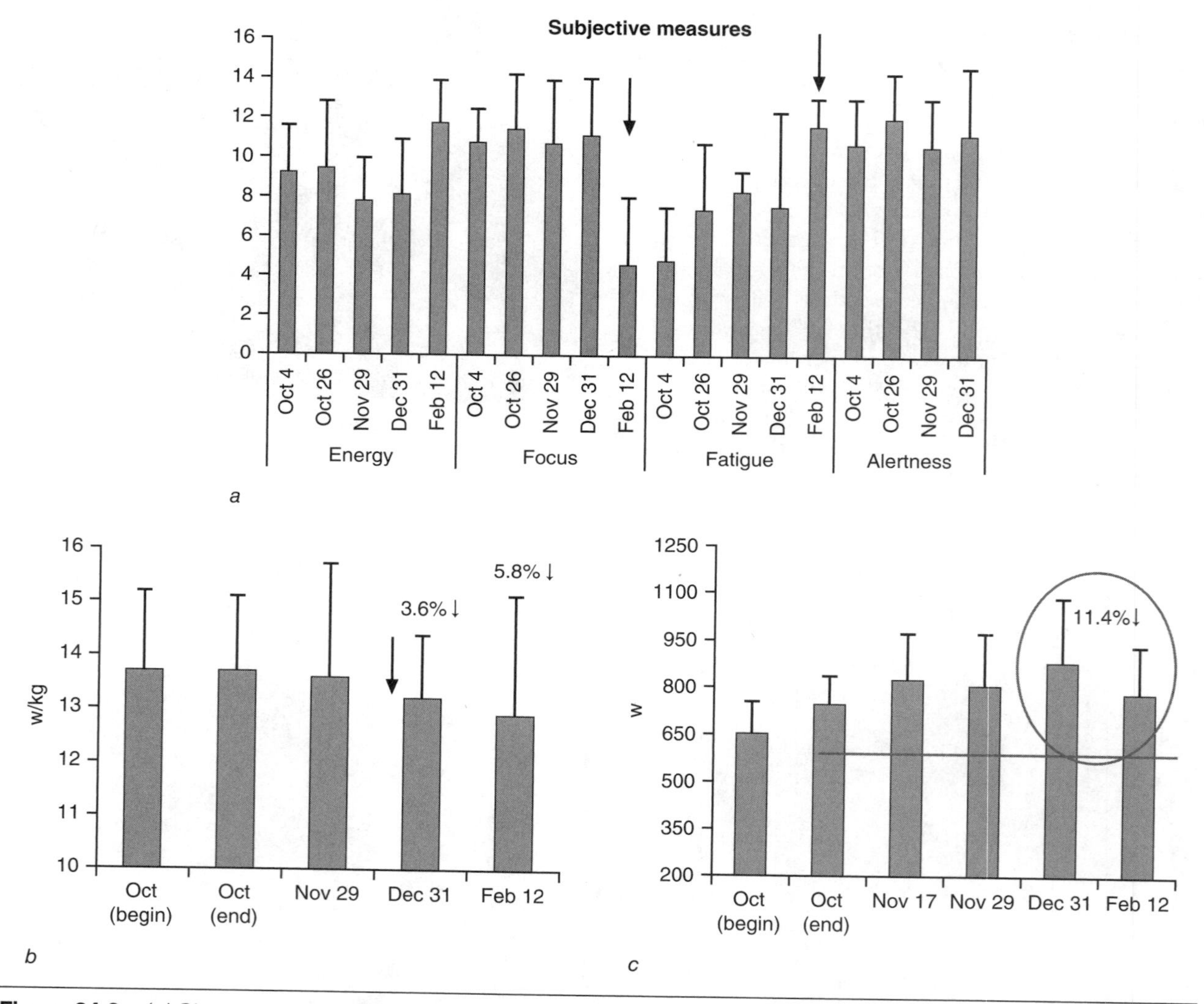

Figure 24.6 *(a)* Changes in visual analog scales during the course of a professional basketball season. *(b)* Changes in vertical jump mean power during a basketball season. *(c)* Changes in squat mean power during a basketball season.

In Practice

Considerations for Coaches and Health Care Providers Regarding the Overtraining Syndrome

The European College of Sport Science and the American College of Sports Medicine developed a consensus statement on the prevention, diagnosis, and treatment of the overtraining syndrome (Meeusen et al. 2013). One of the key issues discussed was the unavailability of a definitive diagnostic tool that could identify an athlete who is overtrained. Given this, it becomes very important that the coach, health care provider (i.e., athletic trainer, team physician), or both use performance measures and follow a number of recommendations to reduce the athlete's risk for overtraining. One of the committee's recommendations was accurately recording performance during training and competition. For this to be effective, though, baseline measures must be obtained. The coach then must be willing to adjust training intensity or volume to provide sufficient recovery. Other recommendations included avoiding monotony of training (perhaps by altering the exercise routine to maintain a fresh stimulus) and individualizing the intensity of training. It was also recommended that the training staff encourage optimal nutrition, hydration status, and sleep. Coaches also need to be aware that multiple stressors such as sleep deprivation, environmental change, and personal or family issues may also add to the stress of physical training. If overtraining syndrome is suspected, the most important treatment is reduced training or complete rest. Unfortunately, there are no definitive indicators of recovery, so resumption of training must be individualized. It becomes imperative to maintain a constant dialogue with athletes regarding their physical, mental, and emotional concerns. At times, the training staff may wish to include psychological questionnaires to assess the mental and emotional state of the athlete. Again, it is important to have baseline measures in order to place the specific feelings into the appropriate context. For the team physician, the consensus statement suggested that upper respiratory tract and other infections should result in giving serious consideration to reducing the training stimulus or eliminating it as long as the infection is present. In addition, the physician should always rule out an organic disease in cases of performance decrement, and viral infections should be investigated in situations in which the athlete experiences fatigue and is underperforming.

well as a testing program with a moderate level of sophistication, prevented a potentially serious overtraining issue from developing.

Even more sophisticated monitoring of athletes includes blood or saliva measures to examine anabolic and catabolic hormonal concentrations and determining the ratio between the two to ascertain recovery status. Saliva measures have been shown to be valid (Papacosta and Nassis 2011) and are preferable to blood draws for many athletes. Other sport scientists have recommended monitoring immune function or oxidative stress responses. Although interesting, the fact that the simpler monitoring methods provide quick answers and that it can be difficult for coaches and athletes to decipher the results of the more sophisticated measures may preclude acceptance of the latter. In addition to endocrine and biochemical markers are new noninvasive methods that can potentially provide very good information about changes in muscle quality, muscle architecture, and reaction times. Ultrasound measures are being used to assess changes in echo intensity; the echo intensity provides a measure of muscle quality. Presently this relatively new method is being examined for its sensitivity as a marker of the stresses of a season of competition. In addition, ultrasonography can provide measures of changes in muscle architecture (i.e., muscle thickness, pennation angle, and cross-sectional area) indicating potential atrophy or hypertrophy of skeletal tissue during the competitive year. Technological advances have also afforded an ability to examine visual and motor reaction response to visual stimuli (figure 24.7). It has recently been demonstrated that this is a sensitive measure of fatigue, even when gross motor skills are not affected (Hoffman et al. 2012).

Photo courtesy of Jay Hoffman.

Figure 24.7 Device for measurement of visual and motor reaction time.

Treatment of Overtraining

The most important aspect of treatment for overtraining is simply preventing it from occurring in the first place. Obviously, this would be the primary goal of the coach. To this end, the coach is likely to develop a training program based on the principles of periodization (see chapter 14) to minimize staleness, overreaching, or overtraining. However, regardless of the care taken in the design of the training program, individual athletes are unique in their responses to a training stimulus. This and other uncontrollable or unexpected influences (e.g., extreme environmental changes, exposure to altitude, and jet lag) may result in poor adaptation to training stresses, causing stagnant or decreased performance. Thus early detection becomes the key to treating overtraining.

During early stages of overtraining (i.e., overreaching), cessation of training for several days should be sufficient for complete recovery. After this rest period, the athlete can resume normal training. However, the cause of the staleness should be determined and appropriate adjustments should be made to the training program. During periods of competition, a decrease in the number or length of practice sessions should be considered. In addition, practice sessions during a busy competition schedule should focus primarily on technique and strategy rather than conditioning. However, a problem for the coach of a team sport is separating the individual from the collective team. Often the stress of practices and games is not equal for all team members (starters vs. nonstarters). Thus, reducing practice intensity and training volume may minimize the chance of staleness in the athletes playing most of the game, but it may also result in detraining the athletes who do not receive as much playing time. This highlights the importance for the coach of individualizing even practices in team sports to whatever extent is possible.

When an athlete is suffering from overtraining, training must be reduced drastically and competitions should be eliminated. Recovery for overtrained athletes is lengthy; it may exceed 6 months (Kreider, Fry, and O'Toole 1998). In such a situation, it is recommended that the athlete receive sufficient rest, sleep, and relaxation as well as proper nutrition (Kuipers and Keizer 1988). In addition, counseling may help athletes cope with the emotional conflicts and psychological demands they are facing.

Summary

Overtraining results when insufficient recovery accompanies an increase in the training stimulus (an increase in either volume or intensity). Inadequate regeneration occurs and performance can be affected. Overtraining is measured across a continuum of stages that increase in severity. During the initial stages, symptoms of fatigue may or may not be accompanied by decreased or stagnant performance. In later stages, several symptoms of overtraining become apparent, along with considerable performance decrements. The mechanisms that underlie overtraining may be different for endurance and strength–power athletes. During overreaching, several days of rest appear to be adequate for recovery and a return to full performance. In fact, many coaches design their training programs to include periods of overreaching in hopes of overcompensation during the recovery period. In the later stages of overtraining (overtraining syndrome), cessation of activity is required, and complete recovery may not be realized for more than 6 months.

REVIEW QUESTIONS

1. Describe the various stages of the overtraining syndrome.
2. Discuss the difference between sympathetic and parasympathetic dominance in overtraining.
3. Name several biochemical performance indicators used to monitor overtraining syndrome.
4. Provide an example of a monitoring system for overtraining in a strength–power athlete.
5. How would the monitoring system change for an endurance athlete?

chapter 25

Diabetes

CHAPTER OBJECTIVES

After reading this chapter you should be able to do the following:

- Define diabetes mellitus and describe the differences between type 1 and type 2 diabetes.
- Describe insulin resistance and the mechanisms associated with its etiology.
- Understand the treatment and management of diabetes.
- Explain the benefits of exercise training on insulin sensitivity.
- Discuss exercise prescription for the diabetic athlete.

Diabetes mellitus is the most common endocrine disorder, and it affects 25.8 million Americans or 8.3% of the American population (Centers for Disease Control and Prevention 2011). It is possible that an equal percentage of people have this disease but are unaware of it. Diabetes is a major health problem and one of the leading causes of death by disease. The direct relationship between diabetes and mortality is difficult to assess because of the host of vascular complications common to people with diabetes, which result in heart disease and stroke. In addition, diabetes is a leading cause of visual impairment and blindness and a significant contributor to kidney disease. The incidence and prevalence of diabetes increase with age, and it is more common in minority populations. Regardless of the devastating complications that may result from diabetes, most individuals can live full and productive lives if they are able to properly monitor and regulate their disease. With proper regulation, no limitations are imposed on people with diabetes. Many famous and successful athletes have competed at the highest levels while combating the disease. People with diabetes have played or are playing in the National Basketball Association (Chris Dudley of the New York Knicks), National Football League (Jay Leeuwenburg of the Cincinnati Bengals and Michael Sinclair of the Seattle Seahawks), National Hockey League (Bobby Clarke of the Philadelphia Flyers), and Major League Baseball (Jason Johnson of the Baltimore Orioles). In addition, Hall of Fame tennis player Bill Talbert lived with the disease for 70 years, and Olympic gold medalists Steve Redgrave (rowing) and Gary Hall Jr. (swimming) have also been diagnosed with diabetes. This chapter briefly reviews the physiology and pathophysiology of the two main types of diabetes and also discusses the role of exercise in its management.

Overview of Diabetes Mellitus

Diabetes mellitus is a disease associated with the inability of the body to regulate blood **glucose** levels and **carbohydrate** metabolism. This may be caused either by an inability of the **β-cells** within the islets of Langerhans of the **pancreas** to secrete a sufficient quantity of **insulin** (the hormone responsible for glucose transport and **glycogen synthesis**) or by an inability of hepatic or peripheral tissues to respond to adequate insulin concentrations. Thus, there are two basic forms of the disease: **type 1**, or insulin-dependent diabetes mellitus (**IDDM**), and **type 2**, or non-insulin-dependent diabetes mellitus (**NIDDM**).

Type 1 Diabetes

Type 1 diabetes is characterized by an inadequate secretion of insulin by the β-cells of the pancreas. It has a sudden and dramatic onset and is diagnosed primarily in children and adolescents. IDDM accounts for approximately 5% to 10% of diabetes cases (Ivy, Zderic, and Fogt 1999). People with type 1 diabetes are more likely to develop **ketosis** and require exogenous insulin for survival than are people without diabetes.

An insulin deficiency has drastic effects on carbohydrate, **lipid**, and protein metabolism. Reduced or inadequate insulin concentrations cause a decrease in glucose uptake by both skeletal muscle and the liver. As a result, a reduction in glycogen synthesis occurs, leading to an increase in both **glycogenolysis** and **gluconeogenesis**. This results in greater blood glucose concentrations (**hyperglycemia**). The high glucose concentration in the blood rises to a level that exceeds the reabsorption capacity of the kidneys, causing glucose to appear in the urine (**glucosuria**). The addition of glucose to the filtrate causes **osmotic diuresis** (pulling of water into the filtrate) and creates a greater need for urination. The increased frequency of urination results in a constant urge to drink. If fluid intake is inadequate, the diabetic person quickly becomes dehydrated, and if the dehydration is not addressed, it may eventually lead to circulatory failure. This cycle is depicted in figure 25.1.

As a consequence of the inability to secrete insulin, a decrease in **triglyceride** synthesis is seen,

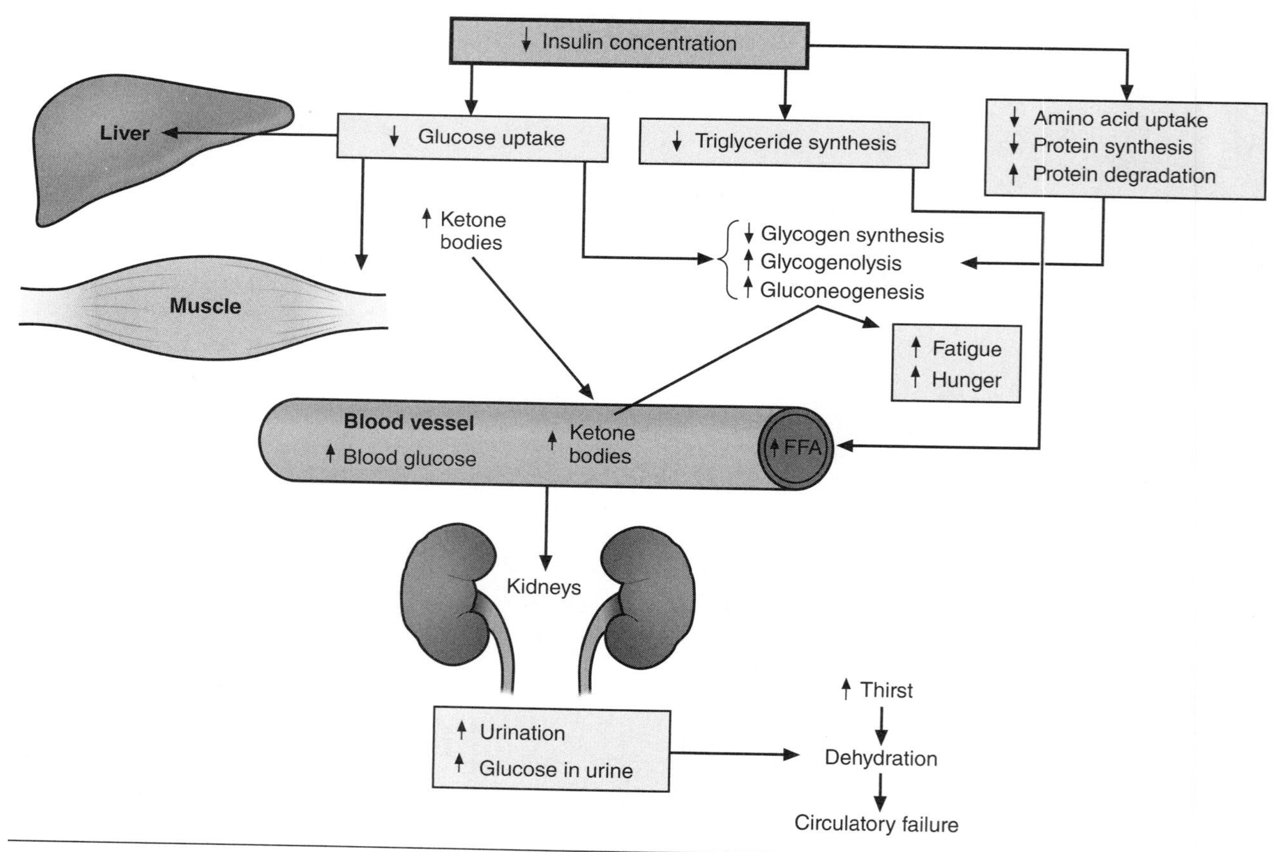

Figure 25.1 Acute effect of insulin deficiency.

which leads to an increase in circulating **free fatty acids**. This causes the liver to use a greater amount of free fatty acids as an energy substrate, leading to a greater production and release of **ketone bodies** into the circulation. Subsequently, people with type 1 diabetes are at a greater risk of ketosis and **metabolic acidosis**.

Type 2 Diabetes

Type 2 diabetes is associated with varying amounts of insulin production by the pancreas. Most often, insulin concentrations are normal, but at times they may exceed levels seen in the nondiabetic population. There is also a small subset of people with type 2 diabetes whose insulin levels are reduced because of a β-cell defect. However, the primary problem in NIDDM is a resistance to insulin action on the part of both the liver and skeletal muscle. NIDDM accounts for approximately 90% to 95% of the total diabetic population (Ivy, Zderic, and Fogt 1999). Type 2 diabetes is seen primarily in adults (occurrence after the age of 24 in 95% of the cases). The risk of NIDDM increases with advancing age; and, although the precise etiology of the disease remains unknown, its occurrence is thought to be related to obesity and a sedentary lifestyle (Ivy, Zderic, and Fogt 1999).

Insulin resistance is the inability to respond appropriately to insulin. This can occur via either reduced insulin responsiveness or reduced insulin sensitivity. Reduced responsiveness refers to a postreceptor defect that causes a reduced biological response to a maximally stimulating insulin concentration; reduced sensitivity refers to an **insulin receptor** defect that causes a reduced biological action at a given submaximal insulin concentration (Ivy, Zderic, and Fogt 1999). Obesity is a prime cause of insulin resistance. Both **hyperinsulinemia** and **downregulation** (reduction in number) of insulin receptors are associated with obesity. In most cases the individual is able to secrete a greater amount of insulin, and carbohydrate homeostasis can thus be maintained. However, in individuals who have a genetic predisposition for diabetes, the stress placed on the pancreas will eventually exhaust the reserve capacity of the β-cells, causing glucose intolerance (Hedge, Colby, and Goodman 1987).

Two mechanisms have been advanced by Ivy (1997) as potential causes of type 2 diabetes (figure 25.2). The first mechanism suggests that a sedentary lifestyle results in a positive calorie balance and a subsequent increase in fat storage and **adipocyte** hypertrophy. As the adipocytes enlarge, an insulin resistance develops as a result of the reduced

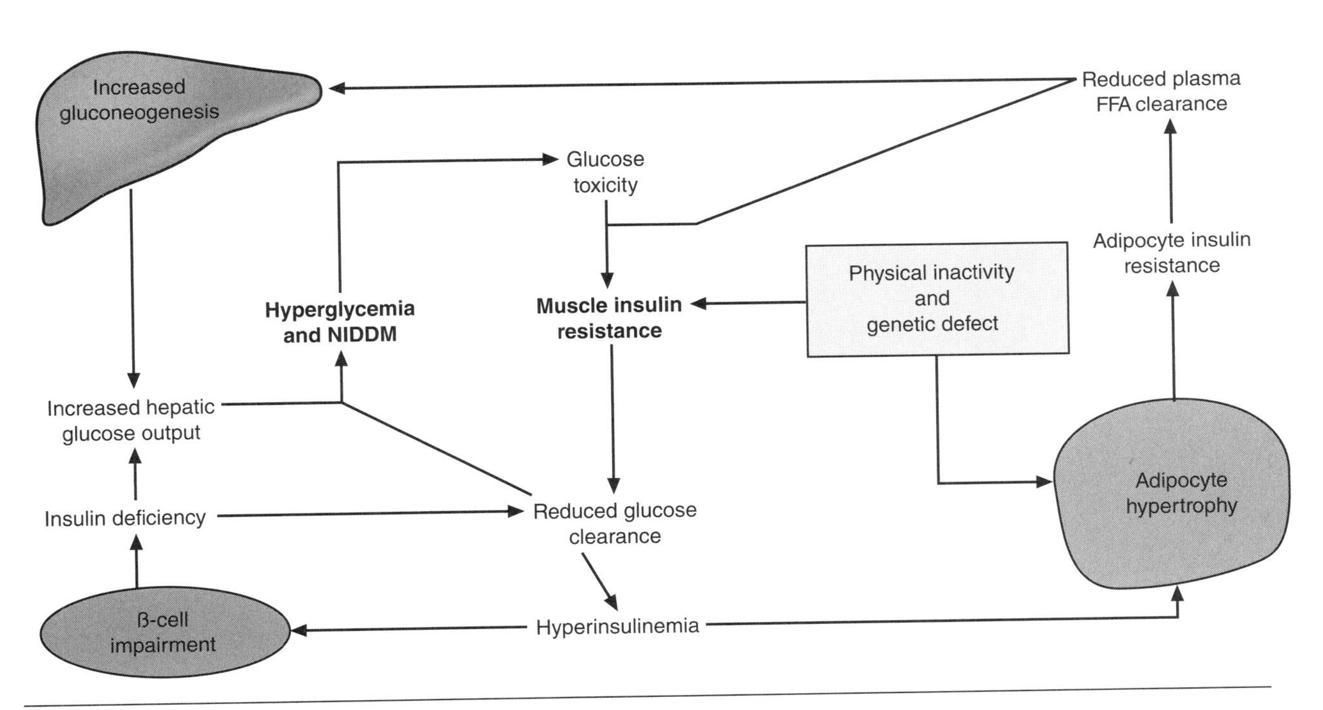

Figure 25.2 Two potential mechanisms for onset of NIDDM.

Adapted, by permission, from J.L. Ivy, T.W. Zderic, D.L. Fogt, 1999, "Prevention and treatment of non-insulin-dependent diabetes mellitus," *Exercise and Sport Science Reviews* 27:1-35.

insulin receptor density. In addition, research has demonstrated that insulin resistance may result from an inhibition or degradation of insulin receptor substrates involved in the signaling cascade that occurs consequent to insulin's binding to its receptor (Kalupahana, Moustaid-Moussa, and Claycombe 2012). Studies using transmission electron microscopy have indicated that intramyocellular lipid droplets are localized in fibers below the sarcolemma, in subsarcolemmal space, and between the myofibrils throughout the interior of the fiber (Nielsen et al. 2010). The concentration of the lipids in those areas was shown to be threefold higher in type 2 diabetics than in body mass index–matched control subjects and endurance athletes. Type 2 diabetics appear to be exposed to an increased content of lipids localized just beneath the sarcolemma, which likely affects regulation of skeletal muscle insulin sensitivity (Nielsen et al. 2010). In addition to the intramuscular effect, increases in blood lipids may also contribute to the etiology of type 2 diabetes. As free fatty acids (FFA) accumulate in the plasma, they begin to have several effects on blood glucose, including stimulating gluconeogenesis and hepatic glucose output and inhibiting insulin-stimulated muscle glucose clearance. FFA may also accumulate within muscle tissue, causing insulin resistance and a compensatory increase in β-cell insulin production. This cycle continues until the β-cells become impaired and insulin production is reduced. This in turn intensifies the insulin-resistant state and the reduced FFA clearance and also accelerates hepatic glucose output, resulting in type 2 diabetes.

The second possible mechanism suggests that a sedentary lifestyle exposes a genetic defect in skeletal muscle, which results in muscle insulin resistance. Similar to what occurs with the first mechanism, a hyperinsulinemia results in response to the elevated blood glucose. However, in this scenario, the hyperinsulinemia suppresses FFA oxidation and increases triglyceride storage and adipocyte hypertrophy. As a consequence, the adipocytes become insulin resistant, and an increase in FFA concentration is observed. Eventually, this cycle results in β-cell impairment and the development of type 2 diabetes.

A number of complications are associated with diabetes, especially NIDDM. These include an increased risk of coronary heart disease because of atherosclerotic changes to the coronary vasculature. The atherosclerotic changes common in people with diabetes are not limited to the coronary vasculature but are also seen in the cerebral and peripheral vascular beds. The high insulin concentrations associated with this disease cause an increase in lipid synthesis and deposition to arterial walls. In addition, microvascular lesions in the kidneys and retinas are commonly seen in people with diabetes, leading to a high prevalence of kidney disease, visual impairment, and blindness. Multiple neuropathies are also frequently found, resulting in dysfunction of peripheral nerves, the spinal cord, and the brain and causing disruption to most physiological systems in the body. The mechanisms directly responsible for these complications are not well understood but are largely believed to be the result of the inadequate insulin and the metabolic disturbances associated with diabetes. There is some debate about whether proper management of diabetes will reduce the incidence of these complications. Some theories suggest that the vascular lesions in diabetes are genetic in origin and that the occurrence of diabetes is associated with a genetic predisposition to the disease (Hedge, Colby, and Goodman 1987).

Treatment and Management of Diabetes

For individuals with type 1 diabetes, insulin is the only treatment available, and in some cases it may be the best therapeutic option for type 2 (NIDDM) diabetes. Insulin is available in many different formulations that vary in their onset of action, maximal activity, and duration of action. Depending on what chemical insulin is conjugated with, its action may become evident within 30 min or up to 24 h after administration. The large variability in the onset of insulin action is related to several factors, including site of injection, volume of insulin injected, and physical activity of the patient.

In NIDDM, the use of oral hypoglycemic agents may be the primary pharmacological means of control. The use of this medication is limited to individuals who are capable of secreting insulin. It is thought that these hypoglycemic agents enhance the binding of insulin to peripheral receptors, causing an increase in glucose utilization and a decrease in hepatic glucose production.

Exercise and Diabetes

The benefits of exercise as a treatment modality for diabetes have been known for centuries (Wallberg-Henriksson 1992). Even when insulin was not yet available, exercise was a generally prescribed course of treatment for people with the disease. The benefit of exercise has been seen to relate primarily to enhanced glucose uptake and insulin sensitivity, resulting in improved **glycemic control**. However, exercise may have even added importance for the person with type 2 diabetes in reducing the risk of coronary heart disease, which is common in the diabetic population.

Exercise and Carbohydrate Metabolism in Nondiabetic Individuals

Exercise places a large demand on carbohydrate metabolism. Although exercise and metabolism are covered in greater detail in chapter 3, it is important here to briefly review carbohydrate metabolism in the nondiabetic individual as a basis for appreciating the demands that exercise places on the diabetic person.

In people without diabetes, glucose uptake and utilization may increase more than 10-fold during exercise (Kanj, Schneider, and Ruderman 1988). However, the body is able to regulate blood glucose concentrations effectively so that only minimal changes are seen. During exercise, the increase in glucose demand is met by an accompanying increase in hepatic glucose production, which prevents a state of **hypoglycemia** (low blood sugar). Hepatic production of glucose is stimulated by an increase in the circulating concentrations of **catecholamines** and **glucagon** as well as a decrease in insulin concentration. Insulin and glucagon are both released from the pancreas, and they have opposite effects. As insulin concentrations in the circulation decrease, glycogen **phosphorylase** (the enzyme responsible for the breakdown of glycogen) activity in the liver increases, thereby enhancing glycogenolysis (the process of breaking down the glycogen molecule to its simplest form, glucose). At the same time, glucagon levels tend to elevate, which also has a glycogenolytic effect. When exercise ceases and the depleted glycogen stores need to be replaced, the reverse actions of insulin and glucagon on the liver take place. The interaction of these hormones on the liver is critical for regulating both glucose production and glycogen synthesis.

After exercise, a heightened sensitivity to insulin makes it easy to replenish depleted glucose stores at the next meal. The enhanced insulin sensitivity appears to persist for several hours after exercise and is also a hallmark of metabolic adaptation to endurance training (figure 25.3).

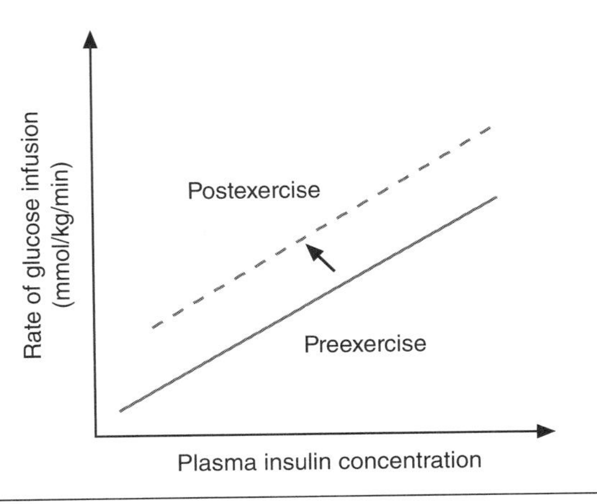

Figure 25.3 Enhanced insulin sensitivity after exercise.

Effect of Exercise on Type 1 (Insulin-Dependent) Diabetes

The primary concern for people with insulin-dependent diabetes is avoiding either hypo- or hyperglycemic conditions. The hypoglycemic state is the most frequent disturbance associated with exercise in this population. As previously mentioned, the person with type 1 diabetes lacks the ability to perform normal glucose regulation. If insulin levels are too high at the onset of exercise (possibly from too great an insulin injection dose or from an accelerated absorption of insulin from the injection site), insulin levels will not decrease in the normal physiological fashion during exercise. This prevents the liver from producing sufficient glucose to meet peripheral glucose demand, causing hypoglycemia. Exercise intensity and duration also appear to be determining factors in the magnitude of hypoglycemia (Wallberg-Henriksson 1992). As exercise intensity or duration increases, the risk of developing hypoglycemia increases as well. Hypoglycemia is not necessarily a problem only during exercise but may also occur 4 to 6 h after an exercise session (Campaign, Wallberg-Henriksson, and Gunnarsson 1987).

Hyperglycemia in people with type 1 diabetes is rarely seen during exercise but may occur if blood glucose levels are high when exercise begins (Wahren, Hagenfeldt, and Felig 1975). The lack of insulin impairs glucose transport into the exercising muscles, forcing a greater reliance on free fatty acids for fuel. An increase in ketone levels may result from a glucose–fatty acid cycle, which may be further accelerated by increases in the counterregulatory hormones (e.g., glucagon, catecholamines, and growth hormone). This in turn may exacerbate the hyperglycemic state and possibly lead to the development of ketosis. For this reason, people with diabetes must be under adequate control before beginning an exercise program. The American College of Sports Medicine and the American Diabetes Association (Colberg et al. 2010; Zinman et al. 2003) have suggested a number of precautions that a diabetic individual should follow before exercise to minimize the risk of unwanted reactions during exercise.

- Measure blood glucose before, during, and after exercise.
- Avoid exercise during periods of peak insulin activity.
- Unplanned exercise should be preceded by extra carbohydrates (e.g., 20-30 g per 30 min of exercise); insulin may have to be decreased after exercise.
- If exercise is planned, insulin dosages must be decreased before and after exercise according to the exercise intensity and duration as well as the personal experience of the patient; insulin dosage reductions may amount to 50% to 90% of daily insulin requirements.
- During exercise, easily absorbable carbohydrates may have to be consumed.
- After exercise, an extra carbohydrate-rich snack may be necessary.
- Be knowledgeable about the signs and symptoms of hypoglycemia.
- Exercise with a certified fitness trainer, especially at the beginning of a physical activity program.

Acute exercise in people with type 1 diabetes has been shown to reduce blood glucose concentrations (Wallberg-Henriksson 1992). The greater glucose uptake by the exercising muscles exerts an insulin-like effect. However, in long-term controlled studies, investigators have been unable to demonstrate that exercise training is capable of improving glycemic control in people with type 1 diabetes (Wallberg-Henriksson et al. 1982, 1986; Zinman, Zuniga-Guajardo, and Kelly 1984). Studies on children and adolescents have yielded conflicting results. An improved glycemic control has been reported in some studies (Campaign et al. 1984; Dahl-Jorgensen et al. 1980), whereas other studies have not shown any improvement (Hansen et al. 1989; Huttunen et al. 1989). These contrasting results may be related to the level of pretraining metabolic control (Wallberg-Henriksson 1992). Exercise programs for children with initially poor glycemic control appeared to be of benefit by bringing blood glucose levels under better control. However, in children who had good metabolic control before the onset of the exercise program, no further changes were seen.

Although exercise may not improve glycemic control, people with type 1 diabetes appear to benefit from long-term exercise by improving insulin sensitivity. Insulin sensitivity refers to a reduced concentration of insulin required to stimulate transport of a similar concentration of glucose into the muscle. Insulin sensitivity may increase 20% in the type 1 diabetic after 16 weeks of exercise (Wallberg-Henriksson 1992). Practically speaking, the person with type 1 diabetes will be able to reduce the dose concentration in the insulin injection.

People with type 1 diabetes appear able to achieve improvements in maximal aerobic capacity similar to those in persons without diabetes (Wallberg-Henriksson 1992). In addition, physiological adaptations associated with endurance training, such as increases in mitochondrial enzyme concentrations, are seen (Costill, Cleary, et al. 1979; Wallberg-Henriksson et al. 1984). However, several physiological adaptations associated with training may be compromised in the diabetic population with longstanding IDDM. For instance, the magnitude of the capillary density increase appears to be blunted in people with type 1 diabetes who have had the disease for more than 15 years in comparison with healthy individuals or even those who have had IDDM for less than 15 years (Wallberg-Henriksson et al. 1984). In addition, decreases in some glycolytic enzymes have been reported in people with type 1 diabetes (Wallberg-Henriksson 1992). The potential for physiological adaptation and improvement in exercise capacity may be impaired in persons with type 1 diabetes who have associated complications, such as autonomic neuropathies or nephropathy.

Effect of Exercise on Type 2 (Non-Insulin-Dependent) Diabetes

Several studies examining the effect of exercise training programs on people with type 2 diabetes have demonstrated consistent improvements in both glycemic control and insulin sensitivity (Dela et al. 1995; Holloszy et al. 1986; Reitman et al. 1984; Schneider et al. 1984). However, the magnitude of glycemic control, similar to what has been seen in

In Practice

Is There a Genetic Link to Type 2 Diabetes?

Much has been written about the effect of lifestyle on the onset of type 2 diabetes. If lifestyle was in fact the primary cause of type 2 diabetes, then we would have control over its onset. However, some evidence seems to suggest a genetic basis for the onset of type 2 diabetes that may predispose certain individuals to the disease. If that is the case, then regardless of the type of lifestyle (e.g., diet and exercise habits), the outcome for many may already be decided. Hagberg and colleagues (2012) have reviewed the relevant literature on this topic and suggest that at this time it may not be possible to draw any definitive conclusions. They noted that much of the research used very small sample sizes and lacked replication, and that only a small portion of potential glucose and insulin metabolism "candidate" genes have been studied. However, the authors indicated that there appear to be relatively strong and replicated data supporting a role for the peroxisome proliferator-activated receptor gamma (PPARγ) gene. The PPARγ gene has been implicated in mediating expression of fat-specific genes and in activating the program of adipocyte differentiation (Vidal-Puig et al. 1996). These genes have also been shown to control various signaling pathways that govern a network of metabolic genes regulating glucose and lipid metabolism, fatty acid transport, and inflammation (Dubuquoy et al. 2002). Hagberg and colleagues (2012) suggested that this gene or some of its variants are plausible candidates because of their role in several pathways involving the onset of type 2 diabetes. It is now the goal to conduct large-scale exercise training interventions on the PPARγ gene and to determine response mechanisms from exercise that may underlie the genotype-dependent training response.

those with type 1 diabetes, appeared to be related to the pretraining status (individual's level before the start of the exercise program). In subjects whose blood glucose levels were under proper control, the extent of improvement was minimal; but in subjects whose preexercise glycemic control was impaired, exercise appeared to provide substantial benefits for better control of blood glucose concentrations (Holloszy et al. 1986).

As already discussed, NIDDM is associated with obesity and a sedentary lifestyle. Individuals who are 20% to 30% overweight are at a higher risk of developing type 2 diabetes than are nonobese individuals (Wallberg-Henriksson 1992). Therefore, in addition to the benefits of exercise on glucose tolerance and insulin sensitivity, further benefits for the person with type 2 diabetes include reducing body fat, improving blood lipid profiles, and decreasing hypertension. These training adaptations reduce the inherently high risk of developing cardiovascular disease in this subject population as well as improving well-being and quality of life. In addition, the physiological adaptations resulting from chronic exercise training may also prevent or delay the cellular changes associated with the development of NIDDM (Ivy, Zderic, and Fogt 1999). The mechanisms that have been suggested to improve insulin action and glucose control during exercise training are depicted in figure 25.4.

A decrease in body fat, especially abdominal fat, has a significant role in reducing insulin resistance in obese individuals (Despres, Nadeau, and Bouchard 1988). Ivy, Zderic, and Fogt (1999) have suggested several potential explanations, although the exact mechanism is not known. Obese individuals with NIDDM are known to be resistant to insulin suppression of plasma FFA (Ivy, Zderic, and Fogt 1999). As FFA accumulate in the blood, an increase in both gluconeogenesis and hepatic glucose output is seen. The inability to properly regulate glucose transport into tissue is further diminished in these individuals, and this negative chain of events leads to the hyperglycemic state common in NIDDM. The size of the adipocytes is also related to the degree of insulin resistance. An inverse relationship has been demonstrated between adipocyte size and insulin receptor density (Craig et al. 1981). A downregulation in insulin receptors reduces the chances of insulin's interacting with its receptor and reduces insulin-stimulated glucose uptake. In addition, increases in plasma FFA may result in an increase in triglyceride

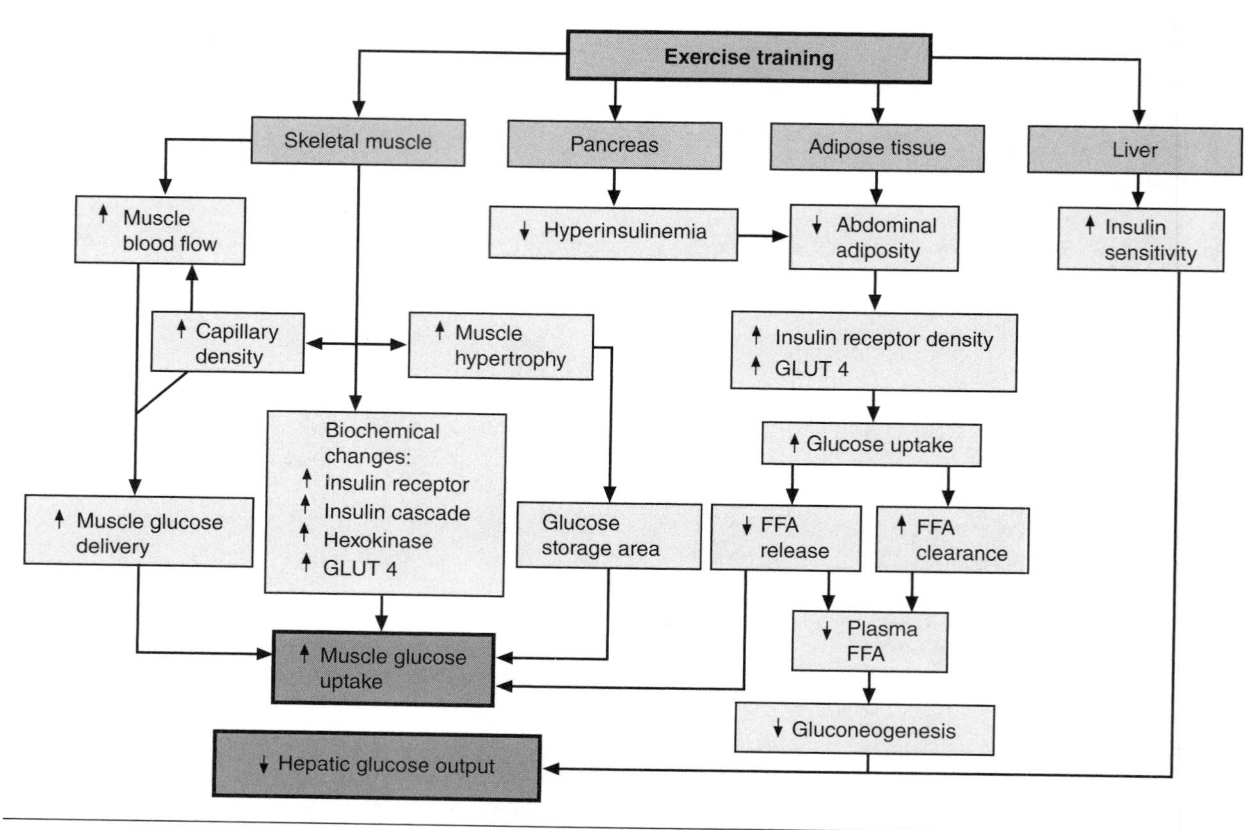

Figure 25.4 Effect of exercise on insulin action and glucose tolerance in NIDDM.

Adapted, by permission, from J.L. Ivy, T.W. Zderic, D.L. Fogt, 1999, "Prevention and treatment of non-insulin-dependent diabetes mellitus," *Exercise and Sport Science Reviews* 27:1-35.

formation and accumulation in muscle tissue. As muscle triglyceride concentration increases, an inverse relationship is seen with insulin-stimulated glucose uptake (Goodpaster et al. 1997; Phillips et al. 1996). This may result in peripheral insulin resistance and subsequent development of NIDDM (Ivy, Zderic, and Fogt 1999). Thus, the importance of reducing body fat through an exercise and diet program is easy to understand given the possible consequences when a sedentary lifestyle leads to obesity. Exercise training is beneficial for either preventing the disease or limiting its progression and improving its management.

Exercise training causes a number of skeletal muscle adaptations as reviewed in chapter 1. These physiological adaptations depend on the type of training program. Endurance training may result in an increase in skeletal muscle blood flow because of the increased capillary density associated with such training. Improved skeletal muscle blood flow would be beneficial for correcting any possible vascular deficiencies in NIDDM patients (Laakso et al. 1992). This may also be a potential mechanism for the improved insulin sensitivity and glucose uptake seen in people with type 2 diabetes after prolonged endurance training (Dela et al. 1995).

People with type 2 diabetes are also reported to have a reduced number of insulin receptors on the cell membrane (Caro et al. 1987; Olefsky 1976). This downregulation reduces the opportunity for interaction between circulating insulin and its receptor. Downregulation is also seen in obese individuals (Caro et al. 1987; Olefsky 1976) and may possibly be reversed during exercise training (Dohm, Sinha, and Caro 1987). However, research is still ongoing in this important area, and as yet, **upregulation** (increase in receptor number) has not been demonstrated in human NIDDM patients.

The insulin resistance in people with type 2 diabetes is not necessarily associated with a receptor–hormone interaction defect. For glucose to move across the cell membrane, a transporter protein is needed. This transporter protein (**GLUT 4**) is located intracellularly and is translocated to the cell membrane by insulin action. As the concentration of GLUT 4 increases, the rate of glucose transport across the cell membrane increases as well. A deficiency in GLUT 4 is not typically seen in those with type 2 diabetes (Lund et al. 1993). More likely, what is involved is a reduced ability of insulin to translocate GLUT 4 from its intracellular storage site to the cell membrane (Ivy, Zderic, and Fogt 1999). Since GLUT 4 has been shown to increase as a result of training (Houmard et al. 1991), the benefit for those with type 2 diabetes may have to do with increasing GLUT 4 concentrations within the cell. This possibly compensates for the defect in translocation by somehow positioning GLUT 4 closer to the cell membrane and enhancing glucose transport (Ivy, Zderic, and Fogt 1999).

Exercise Prescription for Athletes With Diabetes

Diabetes does not limit the ability to excel at any level of sport. As mentioned in the beginning of this chapter, a number of athletes have competed and are competing in the Olympics and in professional sport. Children, adolescents, and adults—all competitors regardless of age—need to be cognizant of their disease and take necessary precautions and measures to avoid hypoglycemia and ketoacidosis. The recommendations on exercise and diabetes presented earlier in the chapter are also relevant for the competitive athlete; however, the following are additional factors that need to be considered.

Timing of Insulin Injection and Insulin Absorption

- When regular insulin is injected before exercise, hypoglycemia will likely occur 2 to 3 h following the injection and between 40 and 90 min following a rapid-acting insulin analogue (Robertson et al. 2009).
- In choosing the injection site, athletes should avoid injecting into muscle that will be primarily active during the exercise period. The increased blood flow to the active muscle will result in a more rapid absorption and metabolic effect (Robertson et al. 2009). The abdomen may be preferable to a limb that will be used during exercise in general; the abdomen or arm may be a better choice than a leg for cycling. However, others suggest that changing the injection site may alter the time course of insulin absorption (Hornsby and Chetlin 2005). Thus, rotating the injection site around the region may be preferable. The athlete should probably choose the injection site that will generally be least affected by training or competition.
- Use of rapid-acting insulin analogues appears to prevent the increase in absorption rate associated with exercise and often seen with the use of regular insulin (Peter et al. 2005).
- High ambient temperature can increase insulin absorption and also place greater stress on the cardiovascular system, resulting in a greater reduction in blood glucose concentrations (Robertson et al. 2009).

Nuances of Competition

- When participating in all-day tournaments, the athlete should use a long-acting insulin the evening before (Robertson et al. 2009).
- Due to the effect that withholding insulin has on weight reduction, diabetic athletes who participate in sports in weight categories may withhold insulin until after their weigh-in. Metabolic control becomes problematic and the athlete is at serious risk for ketoacidosis (Hornsby and Chetlin 2005). This practice should be avoided.
- The athlete should make sure that a teammate knows something about diabetes and how to recognize and manage hypoglycemia.
- If participating at altitude, the diabetic athlete should recognize the potential risk that acetazolamide (used to prevent or treat altitude sickness) may contribute to risk of ketoacidosis (Moore et al. 2001).
- Competition or practice is constantly in flux. Changes in the environment, the intensity and duration of exercise, and emotional state all affect insulin absorption and glucose response. Given the broad range of potential responses, rapidly absorbable carbohydrates should be readily available for use.

The sidebar ahead featuring practical tips gives diabetic athletes tips for preparing to play and practice and for living. The information is adapted from Chris Dudley's website (www.chrisdudley.org). Chris played in the NBA for 16 years with diabetes. He developed a foundation to help raise money and provide education on playing with diabetes.

In Practice

Blood Glucose and Performance in Adolescent Athletes With Type 1 Diabetes

A study from Canada examined 27 boys and girls with type 1 diabetes who were participating in a diabetes sports camp (Kelly, Hamilton, and Riddell 2010). Their mean age was 11.4 ± 1.9 years, and the duration of diabetes ranged from 1 to 13 years. Children with type 1 diabetes experience fluctuating blood glucose levels ranging from hypoglycemia (low blood glucose levels) to hyperglycemia (high blood glucose levels). Exogenous insulin injections are not as efficient in maintaining balanced blood glucose concentrations as endogenous insulin secretions. As a result, the young athlete may experience broad changes in glucose concentrations that could drastically affect performance. The investigators examined changes in blood glucose concentrations and changes in sport skill performance and cognitive function (Stroop test). The Stroop test is a commonly used clinical tool, validated in children, that measures attention and speed and accuracy of cognitive functions affected by fatigue (Golden, Freshwater, and Golden 2003). Results of the study showed that compared to normal or high blood glucose concentrations, hypoglycemia has a significant detrimental effect on the ability of youth to perform basic sport skills. Reading skills and color recognition were lowest during periods of hypoglycemia compared to other blood glucose levels. However, these measures were not affected by hyperglycemia. In addition, nocturnal hypoglycemia does not influence subsequent daytime performance if blood glucose concentrations are normalized before exercise. In summary, the investigators concluded that maintaining normal blood glucose is imperative for maintaining athletic skill and cognitive function in youth with type 1 diabetes.

Practical Tips for the Diabetic Athlete Preparing to Compete

1. Test, don't guess: check your blood sugar (glucose) levels.
2. Be sure to have a plan: Know what you need to eat and drink before, during, and after practices and games. Know when to test and when to snack. Know the signs of hypoglycemia and correct if necessary. Routine is the key!
3. Check with your doctor before making any changes in your diabetes plan.
4. Keep the right tools handy. Always have with you the materials and tools necessary for checking and testing.
5. Don't load carbohydrates. Eat more calories during activity instead.
6. Pay attention to your body. Be aware of your own pattern of blood glucose response to exercise. Wear proper shoes and socks.
7. Let someone know. Let coaches, trainers, and teammates know how to recognize hypoglycemia and what to do.
8. Live your life. Be active and follow your dream; don't let diabetes stop you!

Adapted from Chris Dudley (www.chrisdudley.org).

Exercise Prescription for Noncompetitive, Recreational Athletes With Diabetes

There are no physical limitations on young, active people with type 1 diabetes who do not have any complications and who have good blood glucose control. As already mentioned, competitive athletes of all ages with type 1 diabetes participate in all sports at all levels, including elite and professional. The primary concern for these athletes is the proper adjustment of insulin dosage and diet to ensure safe participation and maximum performance. The concern for recreational athletes with diabetes, or for sedentary individuals with diabetes beginning an exercise program, is no different. It is also important that people with type 1 diabetes and their training partners or coaches are aware of management techniques and treatment of hypoglycemia.

For those with type 1 diabetes who are sedentary and have had IDDM for several years, as well as those with type 2 diabetes, it is highly recommended that a complete cardiovascular evaluation be performed before the person begins an exercise regimen. Many of these patients have asymptomatic coronary artery disease and need to be thoroughly evaluated before participating in any physical exercise program. In addition, selection of the mode of exercise depends on any complications. Certain exercises are contraindicated according to the medical limitation. For example, in diabetics with peripheral neuropathy, jogging may cause trauma to the lower extremities and would not be recommended for the exercise program.

Relatively few studies have examined the optimal intensity and duration of exercise for the diabetic patient. However, the American College of Sports Medicine in collaboration with the American Diabetes Association (Colberg et al. 2010) reported, based on several long-term training studies, that exercise at an intensity ranging from 50% to 80% of $\dot{V}O_2$ max, performed three or four times per week for 30 to 60 min per session, appears to have the greatest potential for eliciting the desired metabolic adaptations. Aerobic exercise (e.g., brisk walking, jogging, cycling, or swimming) appears to be the most desirable; however, resistance training has also been reported to improve glucose tolerance in diabetic patients (Miller, Sherman, and Ivy 1984). Nevertheless, for older people with diabetes and for those with other complications such as retinopathy, some concern has been raised about the advisability of performing high-intensity anaerobic exercise (Kanj, Schneider, and Ruderman 1988).

Summary

Much evidence demonstrates the importance of exercise as a treatment for both type 1 and type 2 diabetes. For people with type 1 diabetes, there do not appear to be any limitations to the exercise regimens they can participate in. The primary concern for these patients is adjusting their insulin dosage and diet so that they can exercise without increasing the risk of hypoglycemia.

Exercise is of vital importance in improving the metabolic dysfunction associated with the disease. However, exercise may play an even greater role in preventing the disease or in limiting its progression. Large muscle mass exercises such as brisk walking, jogging, swimming, or cycling appear most able to elicit the desired metabolic adaptations. Resistance training has also been shown to be of benefit as part of the exercise training regimen for people with type 2 diabetes.

REVIEW QUESTIONS

1. Detail the differences between type 1 and type 2 diabetes.
2. Describe the mechanisms suggested to contribute to non-insulin-dependent diabetes.
3. Explain the precautions that a diabetic individual should follow before exercise to minimize the risk of unwanted reactions during exercise.
4. What are some of the medical complications associated with non-insulin-dependent diabetes?
5. What is the importance of the transporter protein GLUT 4 in diabetes?

chapter

26

Exercise-Induced Bronchospasm

CHAPTER OBJECTIVES

After reading this chapter you should be able to do the following:

- Define exercise-induced bronchospasm (EIB).
- Describe the proposed mechanisms underlying EIB.
- Explain how EIB is diagnosed.
- Explain pharmacologic therapies used to treat EIB.
- Understand how athletes with EIB can best prepare for exercise and competition.

Exercise-induced bronchospasm (EIB) is a condition in which vigorous exercise stimulates an acute narrowing of the airway in individuals with heightened airway sensitivity. For many years, the bronchoconstriction provoked by exercise was termed exercise-induced **asthma**. Symptoms of EIB include cough, wheezing, and chest tightness during or after exercise (Storms 1999). EIB often occurs in otherwise healthy individuals who experience these symptoms only during exercise. However, up to 90% of people with asthma may experience EIB when exercising at a relatively high **intensity** (Godfrey 1988). The prevalence of EIB is reported to be two to three times greater than that of asthma and is thought to reflect the large subset of individuals whose airways are sensitive to an exercise stress (Smith and LaBotz 1998).

Asthma is one of the most common conditions seen in the United States. Recent reports indicate that 18.7 million people in the United States have asthma (Schiller et al. 2012), including 7 million children (Bloom, Cohen, and Freeman 2011). Athletes appear to be as susceptible to this disease as any other population group. However, asthma does not prevent athletes from achieving their maximum potential. During the 1984 Summer Olympics, 67 of the 597 American athletes (11.2%) were reported to have EIB (Voy 1986). Of these athletes, 41 won medals (15 gold, 20 silver, and 6 bronze) in 14 different sports. During the 1996 Summer Olympics, 20% of the participating American athletes were reported to have EIB (Weiler, Layton, and Hunt 1998). The American delegation has not been the only group with a high incidence of EIB. Australian teams reported that more than 9% of their athletes participating in the Olympic Games between 1976 and 1992 had asthma (Morton 1995). The occurrence of EIB in Winter Olympic athletes may be even greater. Wilber and colleagues (2000) reported that 23% of the athletes on the 1998 U.S. Winter Olympic team had EIB. This is consistent with other reports of a higher incidence of EIB among winter sport athletes (Heir and Oseid 1994; Larsson, Hemmingsson, and Boethius 1994). Some reports have suggested that the incidence or prevalence of EIB may have been overestimated or misdiagnosed (Dickinson et al. 2005), yet asthma is still the most common medical condition reported in Olympic athletes. More than 7% of athletes were confirmed to have asthma at the 2006, 2008, and 2010 Olympic Games (Morton and Fitch 2011).

In Practice

Exercise-Induced Bronchospasm: Why Is It So Common Among Olympic Athletes?

The relatively large number of Olympic athletes who report symptoms of EIB has created much interest in why these symptoms seem to be disproportionately associated with athletes at this level compared to other population groups. Moreira and colleagues (2011) suggested that genetic susceptibility, neurogenic-mediated immune inflammation, and epithelial sensitivity are all mechanisms that may contribute to the high prevalence of EIB in this elite athletic population. They postulated that the high-intensity training may make athletes at this level more susceptible to greater airway inflammation and a wear-and-tear effect that increases their risk. If athletes possess specific genes that have been implicated in EIB, this may further increase their susceptibility. Although some research is suggestive of an upregulation of a specific gene responsible for neurogenic inflammation (the PPT-1 gene) (Teixeira and De Lima 2003), the evidence is insufficient to convince scientists of a direct genotypic link between training stress, genetic predisposition, and EIB. Further research is needed to either help confirm or to refute this hypothesis.

This chapter focuses on the definition of EIB, the pathogenesis of its onset, and the ways in which athletes with this ailment can compete successfully. Pharmacologic intervention plays an important role in preventing or reducing the symptoms associated with EIB, and discussion also focuses on the various treatments for the condition.

What Is Exercise-Induced Bronchospasm?

Asthma is characterized by hypersensitive airways that overreact to air pollutants, **allergens** (e.g., pollen, dust, specific foods), psychological stress, and other triggers such as temperature changes or physical exertion. The response to these stimuli results in a reversible airway obstruction caused by **bronchoconstriction** or inflammation of the mucosal linings. As a result, the afflicted individual is unable to exhale completely, and both **residual volume** and **vital capacity** are reduced. The feeling that one might experience during a mild or even a moderate asthma attack at rest may be seen as nothing more than an inconvenience. However, if such an attack occurs during moderate physical activity, the individual may experience severe respiratory difficulty (Morton 1995). The majority of asthmatics experience bronchoconstriction that presents during exercise; others develop the symptoms only after exercise (Morton and Fitch 2011; Parsons et al. 2011). In the past, the during-exercise condition was generally referred to as exercise-induced asthma, and the after-exercise condition was referred to as exercise-induced bronchospasm. However, as discussed earlier, exercise-induced bronchospasm is the term most frequently used by experts today (Morton and Fitch 2011).

When physical activity is the trigger for inducing an asthmatic reaction, the condition is referred to as EIB. Exercise can be the primary and only stimulus that results in an asthmatic response. Often as intense exercise begins, the **peak expiratory flow rate (PEFR)**, **forced expiratory volume in the first second (FEV_1)**, and **forced expiratory volume percent (FEV%)** are elevated. This is common to all exercising individuals, asthmatic or not. However, in the individual with EIB, these values are reduced to below preexercise levels (10-15%) as the exercise bout continues (Fitch et al. 2008; Parsons et al. 2011). Specific values that are considered indicators of EIB are discussed further in the section on diagnosis of EIB.

The onset of EIB appears to occur only when a relatively high intensity of exercise is reached. Bronchospasms are most often reported when exercise intensity is greater than 65% of the individual's **maximal oxygen consumption** (Morton 1995). As exercise intensity elevates, an increase in **catecholamine** concentrations is seen (see chapter 2). The rise in catecholamine concentrations reflects the increase in **bronchodilation** in both asthmatic and nonasthmatic subjects at the onset of high-intensity exercise. The elevated catecholamine concentrations may serve a protective function for people with bronchoconstriction during exercise. This possibly explains the occurrence of EIB commonly seen after exercise when catecholamine concentrations return to baseline levels. It is important to note, however, that much of this information has come from traditional clinical testing for EIB, with most protocols calling for 6 to 8 min of strenuous exercise. Most athletic events have exercise durations that far exceed 6 to 8 min of intense activity and often result in the appearance of symptoms of EIB.

At cessation of exercise, FEV_1 and PEFR in the person with asthma drop by at least 15% of their preexercise values and reach their lowest points at about 3 to 10 min postexercise (Morton 1995). Recovery is gradual, with a return to baseline levels in about 60 min (figure 26.1). In some populations, primarily children, a second or late reaction may appear 3 to 4 h after exercise (Morton 1995).

When exercise produces an asthmatic reaction, there appears to be a period of time in which any further exercise results in a reduced bronchoconstrictive response in comparison with the initial episode of EIB. This period may last for up to several hours and is known as the **refractory period**. A possible mechanism responsible for the refractory period is an increase in **prostaglandins**, which results in an increase in bronchodilation (Manning, Watson, and O'Byrne 1993).

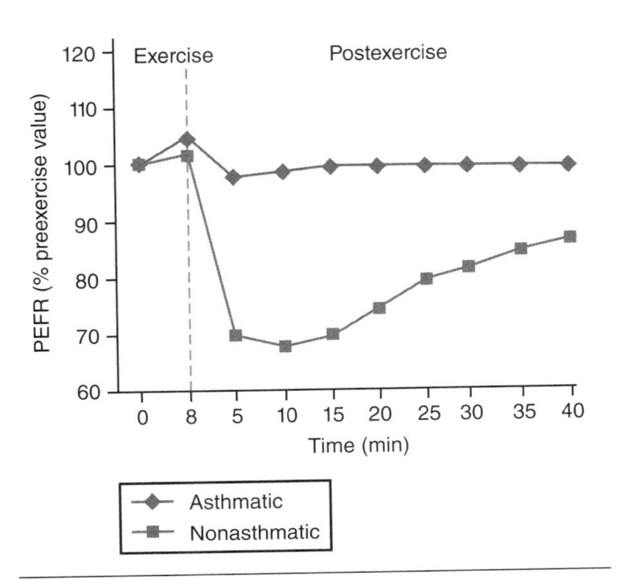

Figure 26.1 Changes in PEFR after exercise in asthmatic and nonasthmatic subjects.

Pathophysiology of EIB

The pathogenesis of EIB is not completely understood, although it is generally believed that EIB is closely associated with changes in both heat and water exchange in the **tracheobronchial tree** (McFadden and Gilbert 1994). **Hyperventilation** associated with intense exercise results in an inability to properly warm and humidify the inhaled air. As the air travels down the tracheobronchial tree, it remains cool and dry, which may cause an alteration in the airway fluid-layer homeostasis (Smith and LaBotz 1998). There is some debate about the relative contributions of bronchial heat and water loss in EIB, and this has resulted in two separate hypotheses of the pathophysiology of EIB.

The first of these hypotheses is the **hyperosmolarity theory**. During exercise, the ventilation rate is dramatically increased. This rapid breathing may result in an increased evaporation of mucosal surface water as the air travels from the upper to the lower airways. As the water loss increases, changes in intracellular osmolarity and temperature occur, although the mechanisms responsible for these changes are not completely understood. The hyperosmolarity of the airway surface water, or **airway drying**, appears to result in **mast cell** degranulation, **histamine** release, and airway smooth muscle constriction (McFadden and Gilbert 1994). Support for this theory has been generated primarily by studies showing bronchoconstriction after inhalation of hyperosmolar saline at rest (Storms 1999). However, no direct evidence demonstrates that airway drying develops, and some studies suggest that it may not (Gilbert, Fouke, and McFadden 1987, 1988).

The second hypothesis, referred to as the **thermal expenditure theory**, proposes that rapid rewarming of the airways after exercise results in bronchoconstriction (Storms 1999). This is believed to be caused by **hyperemia** (increased blood flow) in the vasculature of the airway, with a resulting edema, and does not involve constriction of the bronchial smooth muscle itself. During exercise, heat is transferred from the bronchiolar vasculature in an attempt to warm the inhaled air. After exercise, there is a rapid rewarming of the airways. As a result, the bronchiolar vessels become dilated (from the influx of blood for rewarming), causing the airways to narrow, and hence the bronchoconstriction. The plausibility of this hypothesis has gained support based on studies demonstrating that the capillary bed is more permeable in the asthmatic than in the nonasthmatic population (McFadden and Gilbert 1994). In addition, changes to bronchiolar blood vessels have been shown to influence the cooling and heating of the airways as well as affecting pulmonary function (Gilbert and McFadden 1992; McFadden and Gilbert 1994). As these vessels dilate, some fluid may leak into the tissue, leading to an inflammatory mediator release and resulting in bronchospasm (Storms 1999). Figure 26.2 depicts the two hypotheses that have been suggested as mechanisms leading to EIB.

Diagnosis of EIB

The diagnosis of EIB is frequently based on self-reported symptoms (chest tightness, **dyspnea** out of proportion to the exercise intensity, coughing, wheezing, and excess sputum) without pulmonary function testing. However, Rundell and colleagues

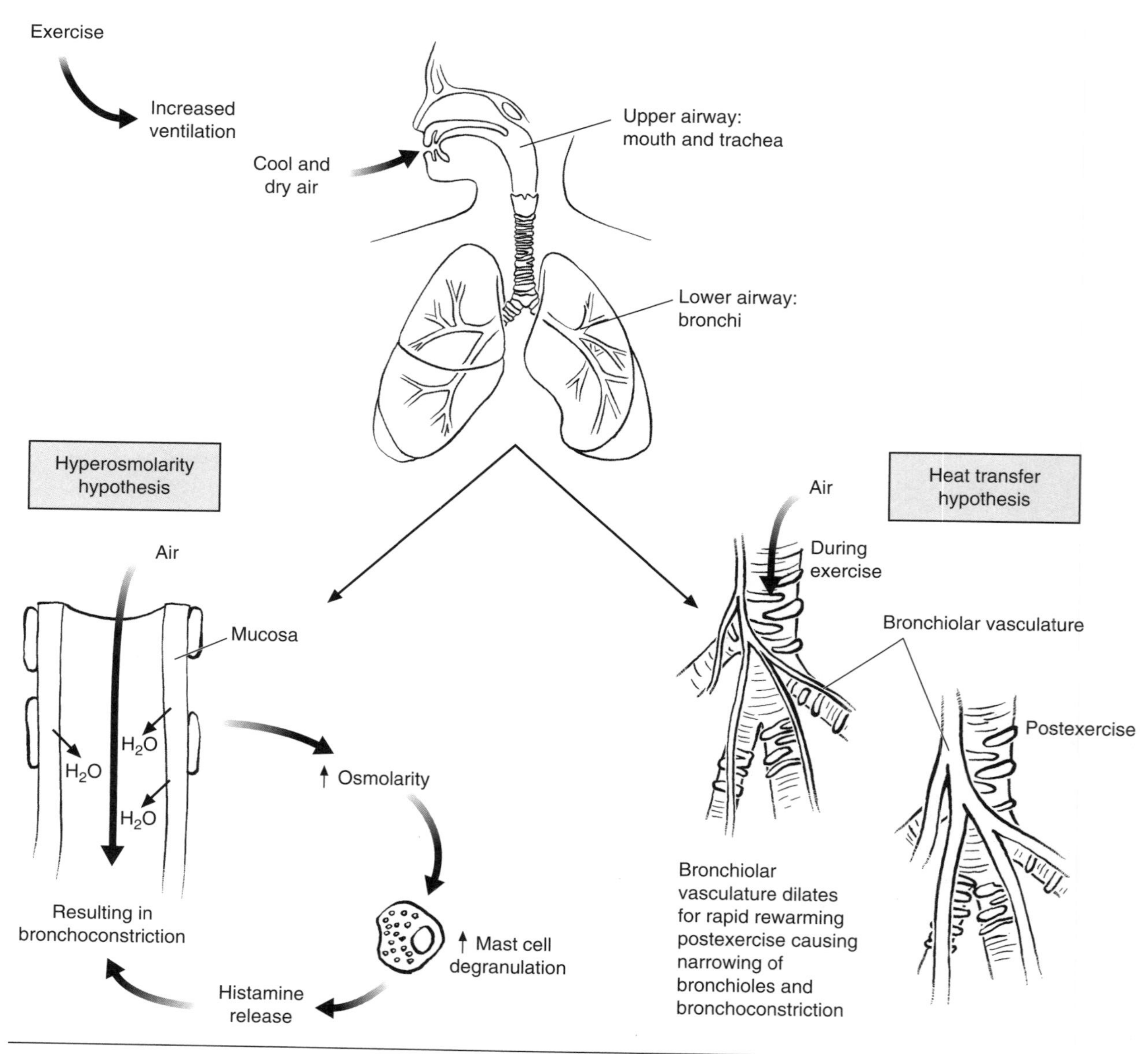

Figure 26.2 Suggested mechanisms for exercise-induced bronchospasm.

(2000) highlighted the inadequacy of relying only on symptoms of EIB for diagnosis. In their study, conducted on U.S. national-level cold weather athletes, the authors reported that 45% of the athletes who had normal pulmonary function exams reported symptoms of EIB, and only 61% of the athletes with positive pulmonary function tests reported these symptoms. The researchers concluded that the use of symptoms alone appears to be unreliable for the diagnosis of EIB.

A general guideline for assessment of EIB includes 6 to 8 min of exercise at an intensity approximately 85% of the subject's maximal predicted heart rate under standard laboratory conditions (e.g., normal room temperature). However, some athletes (especially cold weather athletes) with EIB do not experience any symptoms until exercise intensity reaches their race pace (90-100% maximal heart rate [HRmax]) and when the ambient temperature during exercise is cold (Rundell et al. 2000). Exercise protocols requiring the individual to exercise at a percentage of maximal oxygen consumption have also been recommended (McFadden and Gilbert 1994). When possible, the individual should be tested in the mode of activity and environment that induce the asthmatic symptoms. The diagnosis of EIB is confirmed with pulmonary function tests. A fall in PEFR or FEV_1 greater than 15% is indicative of airway flow obstruction (McFadden and Gilbert 1994; Fitch et al. 2008; Parsons et al. 2011), although some reports have suggested decrements as low as 10% to be of clinical significance (Tan and Spector 1998). Rundell and colleagues (2000) showed that field tests performed on cold weather athletes may be more sensitive than laboratory-based exercise challenges in the diagnosis of EIB. The field tests used were the actual competition or simulated competition. Exercise duration varied from approximately 1 min 20 s for speed skaters to over 1 h for cross-country skiers. Results of this study suggested not only that the field tests were more sensitive than laboratory measures of these athletes but also that accepted criteria for EIB diagnosis (discussed previously) may be too restrictive. The authors concluded that EIB may be diagnosed in cold weather athletes using a sport/environment field test based on postexercise decreases of −8.3% for forced vital capacity, −6.5% for FEV_1, and −12% for PEFR.

Dickinson and colleagues (2006) suggested that eucapnic voluntary hyperventilation (EVH) may be the most suitable method of diagnosing EIB in cold weather athletes. EVH is a laboratory assessment using an indirect airway challenge that controls for minute ventilation and environmental conditions. In this assessment, the athlete was required to hyperventilate for 6 min (30 × baseline FEV_1 breathing a gas mixture containing 5% carbon dioxide, 21% oxygen, and 74% nitrogen). The inspired air was at an ambient temperature of 19.1 °C with a relative humidity of >2%. The results were compared to those obtained with a laboratory exercise challenge (8 min treadmill run at 90% of heart rate maximum at 18 °C, 53% relative humidity) and a sport-specific exercise challenge (skating for 6 min at a pace of 250 m in 11 to 12 s at 8 °C, 35% relative humidity for speed skaters, and a 20 min race at 1 or 2 °C at a relative humidity of 34% for biathletes). Results indicated that the EVH was a more sensitive challenge for EIB, especially in asymptomatic athletes, than any of the other screening tests.

Factors Modifying the Asthmatic Response to Exercise

A number of variables influence the asthmatic response to a bout of exercise. As previously mentioned, if the goal is to induce an asthmatic response, it is most productive to use the mode of activity that the individual primarily performs. However, in general, running is the mode of exercise most likely to induce an asthmatic response, whereas swimming and walking are the activities least likely to induce EIB.

Both the duration and intensity of exercise are important variables with regard to triggering an asthmatic response. When exercise is of short duration (e.g., 8 min, as might be the case in an exercise challenge performed to induce EIB), bronchodilation is generally seen during the exercise period. However, bronchoconstriction often occurs during the recovery period. When exercise is of longer duration, bronchoconstriction may be evident after 15 min of activity and may remain constant until the end of the exercise session (Beck 1999). During intermittent types of exercise, similar to most sporting events, bronchodilation is frequently seen during the periods of high intensity, and bronchoconstriction is seen when the exercise intensity is reduced (Beck, Offord, and Scanlon 1994). Postexercise bronchoconstriction is related to the intensity of exercise. The maximum postexercise asthmatic response occurs when exercise intensity is 65% to 75% of maximal oxygen consumption or 75% to 85% of the predicted maximal heart rate (Morton 1995). The exercise intensity at which an asthmatic response occurs may be higher in elite-level athletes (Rundell et al. 2000).

The environment also has a significant influence on EIB. A cold, dry climate appears to be the most likely to induce an asthmatic response during exercise. This is reflected by the higher incidence of EIB reported in Olympic winter sport athletes than in Olympic summer sport athletes (Wilber et al. 2000). Other environmental factors that have been shown to influence the severity of EIB include the level of air pollution and allergens.

Treatment of EIB

The primary goal of treating individuals with EIB is prophylaxis (preventing an asthmatic attack), but therapy may also be based on symptoms. The treatment plans for individuals with EIB can be either pharmacologic or nonpharmacologic and are critical for helping both the recreational and competitive athlete perform. However, competitive athletes and their physicians need to be aware of the restrictions placed on certain medications used to treat asthma. Some of these therapies have proven ergogenic effects and are banned by major sport governing bodies.

Inhalation therapy is the most common form of treatment. **β-Agonists** are the first line of therapy in the prevention and treatment of EIB (Fitch et al. 2008; Smith and LaBotz 1998). These drugs bind to **β_2-adrenergic receptors**, causing bronchodilation. The binding properties of a β_2-agonist are similar to those of catecholamines. The major advantage of this particular pharmacologic therapy is the longer duration of effect of the β_2-agonists in comparison with the endogenous catecholamine response. Although side effects (tachycardia and muscle tremors) caused by cross-reactivity with α- and β_1-receptors were a concern when this therapy was introduced, most of the β_2-agonists used today have a high β_2-receptor selectivity. Albuterol, also referred to as salbutamol, is one of the most common β_2-agonists in current use. It acts over the short duration and is generally given prophylactically. Protection against EIB lasts for 1 to 2 h after administration (Smith and LaBotz 1998). If exercise is of longer duration, selection of a β_2-agonist with longer-lasting protection (e.g., bitolterol or salmeterol) could be considered (McFadden and Gilbert 1994; Smith and LaBotz 1998). β_2-Agonists may also be used during exercise when symptoms are slow to resolve.

An issue regarding regular treatment with β_2-agonists is the tolerance that may develop (Fitch et al. 2008). There is most likely a downregulation of the β_2-receptors on smooth muscle in the airway as well as on inflammatory cells such as mast cells. This downregulation appears to occur with both short- and long-acting β_2-agonists (Fitch et al. 2008). There appears to be no well-accepted treatment method to prevent the development of tolerance. The recommendation to minimize the risk of tolerance is to avoid the use of β_2-agonists on a daily basis, but this could be challenging for the athlete who trains daily. Use of other pharmacologic treatment options has been proposed; however, corticosteroids (Haney and Hancox 2006; Kalra et al. 1996) or higher doses of β_2-agonists (Haney and Hancox 2007) do not appear to be serviceable options. Fitch and colleagues (2008) have suggested that anti-inflammatory treatments may reduce the severity of EIB and perhaps reduce the need for β_2-agonists on a daily basis.

Other standard pharmacologic therapies used to combat EIB or asthma include the **khellin derivatives** (cromolyn sodium and nedocromil), **anticholinergics** (ipratropium bromide), and **glucocorticoids**. The khellin derivatives are often used as a second-line therapy for prophylaxis. Cromolyn inhibits mast cell degranulation, and nedocromil has anti-inflammatory properties. The relatively high degree of efficacy and minimal side effects of these medications may make them the first choice for prophylaxis in some situations (Smith and LaBotz 1998). The use of anticholinergic agents may be limited because of their slow onset of action in the treatment of EIB. Glucocorticoid treatment is common and effective for managing asthma, but its efficacy as a treatment specifically for EIB is relatively unknown. However, the use of inhaled glucocorticoids in combination with a β_2-agonist is thought to have possible benefits for the individual with severe EIB symptoms (Smith and LaBotz 1998).

A relatively newer method of treatment is the use of **leukotriene antagonists**. Leukotrienes are fatty compounds produced by the immune system that cause inflammation in asthma and bronchitis. Leukotriene antagonists, such as montelukast, has been shown to be effective in reducing EIB in both adults and children (Carlsen et al. 2008). Their use to treat EIB in competitive athletes has met with mixed results. Helenius and colleagues (2004) were unable to demonstrate efficacy for asthma-like symptoms in ice hockey players, whereas Rundell and colleagues (2005) demonstrated a protective effect in some but not all recreationally active individuals with EIB.

Bronchodilator agents effectively improve the ventilatory capacity of those with asthma. However, this improvement is generally reflected as a return to baseline; bronchodilators do not have the ability to increase ventilatory capacity beyond normal levels. β_2-Agonists do have a stimulatory effect, and it has been suggested that some (primarily clenbuterol) have anabolic effects. Because of the stimulatory effect associated with the use of β_2-agonists, the National Collegiate Athletic Association (NCAA) and the U.S. Olympic Committee (USOC) have banned the systemic use of these agents. Only with a therapeutic use exemption can albuterol, formoterol, terbutaline, and salmeterol be used in the inhaled form. However, khellin derivatives do not require written notification, and their use is permitted. Table 26.1 provides a list of permitted and prohibited treatment options for EIB according to the World Anti-Doping Association (WADA) and International Olympic Committee (IOC).

The issue of oral β-agonist therapy for treating the asthmatic athlete has not been without recent controversy. The biggest issues involve drugs used to treat an ailment, but that may also provide an ergogenic benefit. McKenzie and Fitch (2011) recently argued that supposed differences between permitted and banned β-agonists are not supported by sound scientific evidence. They suggest that recent changes to WADA's list of permitted and prohibited agents are difficult to justify given that pharmacologic differences between some of these drugs are nonexistent. Greater effort seems needed to determine appropriate treatment options for athletes—options that provide clinical relief but are not ergogenic—but more importantly to clearly differentiate prohibited and permitted medications based on scientific and clinical merit.

How to Exercise With Asthma (Nonpharmacologic Therapy)

As noted in the introduction to this chapter, EIB should not pose any limitations for athletes trying to reach their maximum performance potential. In addition, although pharmacologic therapy is a staple of many treatment plans, it is possible to significantly modify or prevent EIB using nonpharmacologic interventions (Smith and LaBotz 1998).

If an athlete is aware that exercise induces an asthmatic response, it is recommended that he or she precede all exercise sessions with a warm-up. The warm-up should consist of low-intensity activity for a period of time sufficient to raise body temperature (see chapter 7 for further details concerning the warm-up). Although the efficacy of the warm-up in producing a normal bronchial response to exercise in asthmatic people is not clear, it is thought to induce a refractory period such that bronchoconstriction during the exercise session is reduced. The refractory period may last between 1 and 4 h (Weiler, Layton, and Hunt 1998) and is specific to exercise only. The airways will still be sensitive to other stimuli that can easily cause bronchospasm.

Table 26.1 Permitted and Prohibited Treatments for Exercise-Induced Bronchospasm

Treatment	WADA	IOC
Antihistamines	Permitted	Permitted
Anti-leukotrienes	Permitted	Permitted
Oral corticosteroids	Prohibited in competition, require TUE approval	Prohibited in competition, require TUE approval
Topical corticosteroids	Require TUE approval	Require notification
Oral β-agonists: inhaled albuterol, formoterol, terbutaline, and salmeterol	Require an abbreviated TUE approval	Mandatory documentation of bronchial hyperresponsiveness, reversibility to inhaled bronchodilators, positive exercise test, eucapnic hyperventilation test, or cold air challenge
Immunotherapy	Permitted	Permitted
Inhaled or nasal ipratropium bromide	Permitted	Permitted
Disodium cromoglycate	Permitted	Permitted

TUE = Therapeutic use exemption.

Adapted from Carlsen et al. 2008.

It may also be possible to limit EIB by controlling the activity or the environment. Most recreational athletes can accomplish this easily by selecting an appropriate activity (e.g., swimming and walking are less likely to trigger an asthmatic response than running) or exercising indoors versus outdoors. It may be more difficult for competitive athletes to control these variables, as they are often not self-selected.

As previously mentioned, exercise in a cold, dry climate is the most potent stimulator of EIB. However, when people cannot exercise indoors, it is recommended that they warm the inspired air by wearing a scarf or a mask (Smith and LaBotz 1998). Among all activities, running sports have the greatest **asthmogenicity** (ability to cause an asthmatic response). Table 26.2 lists both high and low asthmogenic sports. If possible, it is best to select an activity with minimal asthmogenicity. In addition, low-intensity exercise performed intermittently may provide the greatest protection against EIB.

Table 26.2 High and Low Asthmogenic Activities

High asthmogenic activities	Low asthmogenic activities
Basketball	Baseball
Cross-country snow skiing	Boxing
Cycling	Football
Ice hockey	Golf
Ice skating	Gymnastics
Long-distance running	Karate
Rugby	Racket sports (tennis, racquetball)
Soccer	Sprinting
	Swimming and water sports
	Wrestling
	Weightlifting

Adapted from Storm 1999.

In Practice

The Effect of Interval Training in Children With Exercise-Induced Asthma Competing in Soccer

A study from Greece compared the effect of interval training and a standard soccer training program on respiratory function and endurance in youth soccer players (Sidiropoulou et al. 2007). It was hypothesized that an interval training program used as a warm-up for soccer would enhance the refractory period and would reduce triggering of the EIB symptoms. Twenty-nine boys (ages ranging from 10 to 14 years) with EIB were divided into two groups, an interval training group and the standard soccer training group. The standard soccer training consisted of continuous low-intensity warm-up activity (50-70% of maximal heart rate) that progressed to greater intensity (80-90% of maximal heart rate). This was the typical warm-up program used by these children during their soccer training. The interval training program (practice preparation) consisted of a 10 min warm-up using interval loading exercises of low intensity (50-60% maximal heart rate for 100 s) followed by higher-intensity training (80-90% of maximal heart rate for up to 20 s). The program was performed three times per week for 8 weeks. Results of the study indicated that the interval training program caused a significant improvement in forced expiratory volume in 1 second (FEV_1) and an improvement in endurance (assessed by a 6 min free running test). Further, the traditional low-intensity endurance group appeared to have impaired their respiratory function during the 8-week period, likely due to chronic inflammation. Thus, it appears that a controlled interval training program may be the best preparatory program for young soccer players.

Conditioning appears to be beneficial for people with asthma. Ram, Robinson, and Black (2000) have shown that aerobic conditioning programs improve aerobic capacity in this population. However, even more importantly for the athlete with EIB and asthma, a higher level of aerobic fitness may increase the tolerance and threshold levels so that a greater stimulus is necessary to elicit an asthmatic response (Morton 1995).

Between 2003 and 2008, a total of 868 therapeutic use exemptions were filed for β_2-agonists with the International Association of Athletics Federation. These numbers clearly demonstrate that asthma or EIB does not limit the potential of any athlete. A host of well-known athletes with asthma in all sports have achieved athletic greatness; examples include Jerome Bettis, who played for 13 years in the National Football League; Dennis Rodman, who played for 14 years in the National Basketball Association; and Jackie Joyner-Kersee, who medaled six times in the Olympics. However, living with asthma or EIB requires the individual to take some practical precautions to limit asthmatic attacks that can derail a competition. The sidebar provides practical tips for the asthmatic athlete preparing to play competitive sport.

Preparing to Play With EIB

1. Make sure that you use medications properly.
2. Identify and control asthmatic triggers. Possibly change practice site if good ventilation is not available or if allergens or irritants such as tobacco smoke and pollutants trigger asthma.
3. Use a peak flow meter to identify when peak flow drops; this is a sign of airway inflammation.
4. A rescue inhaler should be available during games and practices, and a certified athletic trainer or coach should have an extra rescue inhaler for administration during emergencies. In case of emergencies, a nebulizer should also be available.
5. A warm-up before the onset of practice or competition is important because it provides a refractory period, which can reduce the propensity to trigger EIB (McKenzie, McLuckie, and Stirling 1994).
6. Reduce the effects of cold air with the use of a mask during training.

Adapted from Fitch et al. 2008.

Summary

The prevalence of EIB is high, even in Olympic athletes, with a greater occurrence in winter sport athletes. EIB does not prevent an athlete from reaching peak athletic potential. The proper use of both pharmacologic and nonpharmacologic treatments minimizes discomfort and enhances performance during athletic competition. Competitive athletes should give consideration to the type of medication used to ensure that it falls within the confines established by the given athletic association's governing body.

Review Questions

1. Describe the refractory period following exercise in someone who suffers from EIB.
2. Explain how a diagnosis of EIB is confirmed.
3. Compare the hyperosmolarity theory to the thermal expenditure theory in the pathogenesis of EIB.
4. What are the possible pharmacologic treatments used to treat athletes with EIB?
5. Describe nonpharmacologic treatment plans for individuals with EIB.

chapter 27

Sudden Death in Sport

CHAPTER OBJECTIVES

After reading this chapter you should be able to do the following:

- Describe sickle cell anemia and its etiology.
- Understand what exertional rhabdomyolysis is and learn how appropriate exercise prescription can minimize its occurrence.
- Describe the pathophysiology of exertional heatstroke.
- Understand cardiovascular structural and electrical events that contribute to sudden cardiac arrest.
- Explain what traumatic brain injury is and discuss return to play following concussion.

During the years 2000 through 2010, 21 athletes participating in National Collegiate Athletic Association (NCAA) football off-season conditioning practices died from nontraumatic causes. Most of these deaths were exercise-related sudden death associated with sickle cell trait, exertional heatstroke, and cardiac events (Mueller and Colgate 2011). In addition, since 2000 there have been 91 cases of traumatic brain injury with incomplete recovery (Mueller and Cantu 2011). The majority of these incidents occurred at the high school level. With the tremendous changes in communication, many of these cases have been publicized on a national level, yielding a perception that there has been a sudden increase in the occurrence of sudden death among young athletes. Epidemiological assessments have determined that sudden deaths in young athletes have indeed significantly increased, at a rate of 6% per year, and that the proportion of sudden death events increased significantly from 1994 to 2006 compared to the 14-year period from 1980 to 1993 (Maron et al. 2009). The identification and tracking of these catastrophic events have resulted in the development of guidelines and action plans to try to minimize their occurrence. Ten conditions have been identified as potential causes of sudden death in sport (asthma, catastrophic brain injury, cervical spine injury, diabetes, exertional heatstroke, exertional hyponatremia, exertional sickling, head-down contact in football, lightning, and sudden cardiac arrest) (Casa, Guskiewicz, et al. 2012). This chapter focuses on sickle cell trait, exertional heatstroke, cardiac events, and traumatic brain injury. Some of the other conditions have been covered elsewhere in this textbook, though not necessarily in relation to sudden death in sport but more in connection with their etiology and description.

Sickle Cell Trait

Sickle cell disease is one of the most common of genetic diseases, affecting millions of people worldwide. It is primarily seen in individuals whose ancestors originated from sub-Saharan Africa, but it is also seen in smaller proportions in various other populations throughout the world. The basic problem in sickle cell disease lies within the hemoglobin molecule. Hemoglobin, existing in red blood cells, is responsible for carrying oxygen in the systemic circulation. In sickle cell disease, there is a mutation in the hemoglobin gene. When both alleles are affected, the individual is homozygous for the mutation that causes the disease. This gene mutation results in a decreased ability of the cell to carry oxygen, causing the cell to form a rigid, sickle shape. The reduced oxygen-carrying capacity results in an anemic response, hence the term sickle cell anemia. During transport in the bloodstream, if the red blood cells sickle they tend to occlude the vessel (figure 27.1), leading to sickle cell crisis, which is the hallmark of the disease. This is why sickle cell disease is generally considered incompatible with strenuous exercise (Harmon et al. 2012).

When only one of the alleles is affected by the gene mutation, the individual is said to have sickle cell trait. One gene is normal and the other gene is affected. Approximately 300 million people worldwide, and between 6% and 9% of African Americans residing in the United States, have sickle cell trait (O'Connor et al. 2012). Usually this is considered a benign situation, primarily due to the minimal complications and their mildness (O'Connor et al. 2012). However, it is becoming clear that sickle cell trait has the potential to be very serious, especially in the competitive athlete, and it has been implicated as a factor in exercise-related sudden death (Goldsmith et al. 2012; Harmon et al. 2012). Sickle cell trait was first reported as a cause of sudden death in 1970 during Army basic training (Jones, Binder, and Donowho 1970). Subsequent to that case were several other cases of sudden death in military personnel associated with sickle cell trait, all occurring during intense exercise (Anzalone et al. 2010). Retrospective examination indicated that African Americans with sickle cell trait had a 30-fold greater risk for exercise-related death than those without the trait (Goldsmith et al. 2012). Sickle cell trait is associated with a relative risk of death 37-fold higher in NCAA football athletes with the trait than in those without it, according to a database of 2 million athlete-years (Harmon et al. 2012).

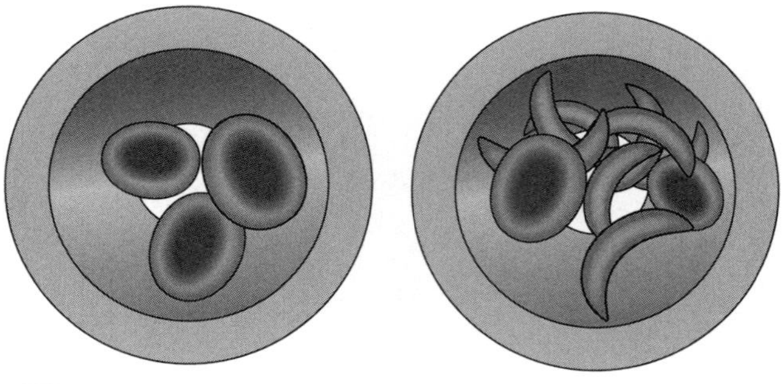

During repeated high-intensity exercise or at increased altitude, the red blood cells of individuals with sickle cell trait begin to change shape and take the rigid sickle form (Goldsmith et al. 2012; O'Connor et al. 2012). As the cells change shape they begin to get caught in the vessel and then slow down and aggregate. This results in vaso-occlusion within the muscle, which continues a cascade of events that may end in catastrophe. The hypoxic environment and lowering of pH (increased acidity) stimulate additional sickling. This cascade of events results in muscle cramps that are quite different from those experienced during heat cramping. The muscle cramps during sickling are more generalized to the muscle and are not pinpointed or excruciating (Casa, Guskiewicz, et al. 2012). Heat cramps cause athletes to stop in pain, whereas athletes whose cells are sickling may try to push through the pain until they collapse. The muscle of an athlete suffering from heat cramps is contracted and tight to the touch; in the case of an athlete experiencing a sickling episode, it looks and feels normal. The major problem with the athlete whose cells are sickling is the effort to continue through the pain and fatigue. If the athlete simply stops exercise and rests, the cells stop sickling and return to their normal shape. The cells become oxygenated, and the athlete soon recovers and feels good. The biggest problems arise when the athlete continues to struggle through the workout or is urged on by the coach to continue. These situations have often had lethal consequences.

Normal red blood cell flowing through blood vessel

Abnormal, sickled red blood cells getting stuck in blood vessel, preventing normal blood flow

Figure 27.1 Normal red blood cells and sickled red blood cells.

Exertional Rhabdomyolysis

The pathophysiology of sickle cell trait is not fully understood. The cardiovascular insult that occurs as red blood cells sickle during intense exercise or during travel to high altitudes leads to a clear hypoxic event occurring as the result of vascular occlusion. The subsequent impaired blood flow to muscle leads to ischemic rhabdomyolysis. Rhabdomyolysis causes hyperkalemia and acidosis, which have an adverse affect on cardiac function and lower the threshold for a lethal cardiac arrhythmia (Harris et al. 2012). Autopsy studies revealed that **exertional rhabdo-**

myolysis was present in 50% of the athletes who died from sickle cell trait (Harris et al. 2012); others have suggested that exertional rhabdomyolysis is increased about 200-fold in sickle cell trait (Anzalone et al. 2010).

Prevention, Recognition, and Treatment

There are no contraindications for participation in sport for the athlete with sickle cell trait (Casa et al. 2012). However, recognition of the trait and understanding and adhering to precautions can avoid catastrophic results. All trainers and coaches working with an athlete who has sickle cell trait need to know that the athlete has the genetic mutation. In addition, they need to be educated on the basics of the disease, its signs and symptoms, ways to minimize risk, and treatment. Interestingly, deaths caused by sickle cell trait have generally not occurred during actual competition. These events occur primarily during the initial days of strength and conditioning practice when the athletes are detrained. Often the exercise paradigm for the workout session is inappropriate. The duration and intensity of exercise, rest interval length, and environmental conditions all contribute to catastrophic events. The following are symptoms associated with sickle cell trait:

Muscle cramping
Pain
Swelling
Weakness
Tenderness
Inability to catch one's breath
Fatigue

Once any of these symptoms appears, activity needs to cease immediately to avoid a sickling episode. When athletes are allowed to set their own pace, they seem to avoid any occurrence of sickling (Casa, Guskiewicz, et al. 2012).

Strength and conditioning and sport coaches need to be cognizant that factors such as physical exertion accompanied by heat stress, dehydration, asthma, illness, and altitude predispose athletes with sickle cell trait to a crisis (Casa, Guskiewicz, et al. 2012). The National Athletic Trainers' Association has developed recommendations to minimize the risk for exertional sickling (Casa, Guskiewicz, et al. 2012). The work-to-rest ratio should be adjusted according to the environmental challenge. With greater heat strain, increased time for recovery between exercise bouts is required. Fluid should always be available, and frequent water breaks should be provided. Athletes with asthma should be sure that it is under control, and an athlete with sickle cell trait who is ill should be excused from the workout. Further, athletes with sickle cell trait should be closely monitored when exposed to a high-altitude environment. The training program should be modified and supplemental oxygen should be available for competition. The athlete should also be encouraged to report any signs and symptoms of leg or low back cramping, dyspnea (difficulty breathing), or fatigue.

Screening for sickle cell trait is a standard component of the preparticipation physical evaluation performed by the team physician before participation in organized team activities. The athletic trainer and coach should be familiar with the symptoms of sickle cell trait as listed previously and should be able to differentiate exertional sickling from other causes of collapse (Casa, Guskiewicz, et al. 2012). Upon any signs or symptoms of exertional sickling, the athlete should be removed from the activity. The athlete should be provided with high-flow oxygen at 15 L/min; vital signs should be monitored and the emergency action plan activated. Any sickling collapse should be treated as a medical emergency. The athletic trainer should also ensure that the athlete's treating physician is aware of the presence of sickle cell trait and prepared to treat the metabolic complications of exertional rhabdomyolysis (Casa, Guskiewicz, et al. 2012).

Exertional Heatstroke

As discussed in chapter 21, exertional heatstroke is on a continuum that includes heat exhaustion. Heatstroke is a condition that results in a systemic inflammatory response leading to multi-organ failure that may have a fatal outcome (Epstein and Roberts 2011). Therefore, early recognition and diagnosis become critical. Exertional heatstroke is defined as a core temperature above 104 °F (40.0 °C) with central nervous system dysfunction (Casa, Guskiewicz, et al. 2012). Central nervous system disturbances include disorientation, confusion, loss of balance, staggering, irrational or unusual behavior, apathy, aggressiveness, hysteria, delirium, loss of consciousness, or coma. Other signs and symptoms that may be present in exertional heatstroke are dehydration, hot and wet skin, hypotension, and hyperventilation. Most athletes diagnosed with heatstroke present with hot, sweaty skin, which contrasts with the dry skin often seen in classical heatstroke (i.e., nonexertional heat stroke) (Hubbard and Armstrong 1988). In addition, gastrointestinal disturbances including diarrhea, vomiting, or both are reported.

Exertional heatstroke is among the top three causes of death in athletes and may become the primary cause during the summer months. The actual number of exertional heatstroke incidents is likely not known, primarily because not all cases result in an athlete's death. Fatal outcomes are often reported, but nonfatal outcomes may not be publicized. Exertional heatstroke, though, is considered a rare event (Armstrong, Casa, Millard-Stafford, et al. 2007). Armstrong and colleagues (2007), in a pronouncement written for the American College of Sports Medicine on exertional heat illness during training and competition, reported that the incidence rate for fatal exertional heatstroke was estimated to be about 1 in 350,000 participants in sports such as American football, and less than 1 in 10,000 finishers in a marathon. However, the incidence rate in a marathon increases as the heat strain index rises. A popular road race held during the summer months under hot and humid conditions results in 10 to 20 cases of exertional heatstroke per 10,000 entrants (Armstrong et al. 2007). When the same race course is run during cooler conditions, no occurrence of exertional heatstroke is seen. This highlights the importance of choosing the appropriate time of the year to conduct a race.

Pathophysiology of Exertional Heatstroke

Environmental challenges and exercise lead to elevations in core temperature that cause blood flow to be diverted to the periphery. As perfusion to the gut, muscles, and other nonvital tissues decreases, the lack of sufficient blood flow to these areas leads to ischemia and potentially endotoxemia (O'Connor et al. 2010). A systemic inflammatory response may be initiated, which can lead to multi-organ failure. Exertional heatstroke results in greater morbidity and mortality than classical heatstroke and is more likely to be associated with rhabdomyolysis, renal failure, liver damage, hyperkalemia, hypercalcemia, and hypoglycemia (O'Connor et al. 2010).

Prevention, Recognition, and Treatment

The cause of exertional heatstroke appears to be multifactorial. Lopez and colleagues (2011) indicate that case studies on deaths attributed to exertional heatstroke usually point to more than one predisposing factor. These factors can be intrinsic, extrinsic, or both. Intrinsic factors that predispose individuals to exertional heatstroke include low level of fitness, underlying illness, and obesity. An examination of a series of case studies of individuals diagnosed with exertional heatstroke showed that an underlying illness was apparent in 18% of the 150 cases analyzed (Epstein et al. 1999). Athletes and coaches need to be educated on the risks associated with exercising in the presence of an underlying illness. Extrinsic factors associated with exertional heatstroke include training at the wrong time of day (i.e., during the hottest part of the day), insufficient rehydration regimen, inappropriate uniform ensembles for athletes during practice that increase uncompensable heat stress, improper work-to-rest cycles, and absence of appropriate medical personnel. Several factors could be considered both intrinsic and extrinsic—for instance, a mismatch between one's physical effort and one's physical fitness. This becomes a major issue when coaches err regarding the exercise prescription, demanding more than the athlete is physiologically capable of. The motivation of athletes to perform for the coach or pressure from both coaches and teammates may push athletes beyond what is safe. Other factors contributing to exertional heatstroke that are considered both intrinsic and extrinsic include sleep deprivation, improper heat acclimatization, and dehydration. Certain medications may also increase the risk for exertional heatstroke. The use of antihistamines, stimulants, anticholinergics, and antipsychotics is believed to affect thermoregulation, which can increase the risk for heat illness (Lopez et al. 2011). Although exertional heatstroke typically occurs in hot environments, it can also occur in cooler conditions, especially if several of the intrinsic or extrinsic factors are present (e.g., illness, sleep deprivation, dehydration) (O'Connor et al. 2010).

The two main criteria for diagnosis of exertional heatstroke are a core temperature above 104 °F (40.0 °C) and central nervous system dysfunction (see earlier discussion). Rectal temperature and gastrointestinal temperature measures are the only methods proven to be valid for accurate temperature measurement in patients with exertional heatstroke (Casa et al. 2012). One should not rely on an inferior measure in the absence of a valid device.

The treatment for exertional heatstroke centers primarily on lowering core temperature to under 102 °F (38.9 °C) as soon as possible (Armstrong et al. 2007; Casa et al. 2012). Cold-water immersion is considered the gold standard for treatment of the condition

(Armstrong et al. 2007; Casa et al. 2007). Adherence to the concept "cool first, transport second" is strongly encouraged. This will eliminate the delay in treatment associated with ambulance transport to a medical facility. The importance of rapid cooling supersedes the need to get the athlete to a facility. The critical window for enhancing survival is 30 to 60 min from the onset of exertional heatstroke (Casa et al. 2007). If possible, fluid can be provided intravenously to enhance rehydration. Once core temperature reaches the desired level (102 °F), the patient should be transported to the medical facility.

Cardiac Events

Sudden cardiac arrest from cardiovascular disease is the leading cause of death in athletes during sport participation (Drezner, Berger, and Campbell 2010; Harmon et al. 2011). Athletes usually display no symptoms before the episode, and relatively few athletes are identified as "at risk" before the event (Courson 2007). A sudden cardiac event resulting in death may have one of two different etiologies: a structural defect or an electrical defect. Here are some examples of both defects (Drezner, Berger, and Campbell 2010):

Structural or functional defects

- Hypertrophic cardiomyopathy
- Coronary artery anomalies
- Aortic rupture/Marfan syndrome
- Dilated cardiomyopathy
- Myocarditis
- Left ventricular outflow tract obstruction
- Mitral valve prolapse
- Coronary artery atherosclerotic disease
- Arrhythmogenic right ventricular cardiomyopathy
- Postoperative congenital heart disease

Electrical defects

- Long QT syndrome
- Wolf-Parkinson-White syndrome
- Brugada syndrome
- Catecholaminergic (polymorphic ventricular tachycardia)
- Short QT syndrome
- Complete heart block
- Other, including drugs and stimulants, commotio cordis, and primary pulmonary hypertension

The most common cause of sudden cardiac arrest in pediatric or young adult athletes is hypertrophic cardiomyopathy (Drezner, Berger, and Campbell 2010). **Hypertrophic cardiomyopathy** has accounted for 36% of the sudden cardiac deaths in athletic populations reported in the past 30+ years (Maron et al. 2009). Although some patients with hypertrophic cardiomyopathy present with symptoms such as chest pain, syncope, or exertional light-headedness, reports have indicated that 80% of athletes who die from this disease had no warning signs (Drezner, Berger, and Campbell 2010; Maron et al. 1996). Congenital coronary artery anomalies are the second leading cause of sudden cardiac death in athletes, accounting for 17% of the cases (Maron et al. 2009). Marfan syndrome is an autosomal dominant disorder of connective tissue. It is caused by an abnormality of fibrillin, a 350 kD glycoprotein, which is a structural component of microfibrils. The microfibrils provide a supporting scaffold for elastin throughout the body (Stuart and Williams 2007), including cardiac tissue. Virtually all adults with Marfan syndrome have an abnormal cardiovascular system. The most common cardiac abnormalities with Marfan syndrome are dilation of the aorta, aortic aneurysm, and mitral valve regurgitation (Pyeritz and McKusick 1979; Stuart and Williams 2007). The primary danger of an aortic aneurysm is that it can lead to aortic rupture, which accounts for about 3% of the sudden cardiac deaths in the United States (Maron et al. 2009).

Postmortem studies sometimes fail to identify a structural cause of death; it is increasingly recognized that in these cases an electrical event or ion channel disorder may be the cause of sudden cardiac arrest. Drezner and colleagues (2010) reported that 30% of sudden cardiac deaths in individuals 35 years or younger occurring in Australia, as well as 35% of the sudden cardiac deaths in recruits within the U.S. military, were related to electrical or ion channel disorders. Interestingly, only 3% of the sudden cardiac deaths were attributed to nonstructural causes in American athletes (Maron et al. 2009). Individuals with ion channel disorders may present with symptoms such as syncope or unexplained seizure activity (Drezner, Berger, and Campbell 2010). In addition to structural or electrical factors are several others that can cause a sudden cardiac event. The use of drugs, especially stimulants, can have a severe effect on cardiac function. In addition, a blow to the left precordium may induce a type of ventricular fibrillation referred to as commotio cordis (Maron et al. 2002).

Epidemiology of Sudden Cardiac Death in Athletes

In older individuals (>35 years), sudden cardiac death is predominantly caused by atherosclerotic coronary artery disease; in the younger athlete (<35 years), it is more likely related to one of the inherited diseases discussed previously. Over a 27-year period, a total of 1866 U.S. athletes who were participating in an organized team or individual sport involving regular competition against an opponent, and who had died suddenly or survived a cardiac arrest, were evaluated (Maron et al. 2009). This systematic evaluation from a large registry looked at cause of death, its frequency, and demographics and circumstances in order to provide a greater understanding of potential preventive strategies. Of the 1866 sudden death events, 1049 (56%) were linked to cardiovascular causes. The investigators were able to determine that 690 of these cases could be primarily attributed to known cardiovascular diseases. Figure 27.2 provides the percentile occurrences of these diseases. The type of cardiovascular disease reported was based on the diagnosis that occurred postmortem. The cardiovascular disease most commonly resulting in sudden death was hypertrophic cardiomyopathy. Hypertrophic cardiomyopathy is considered in an individual with apparent cardiac disease characterized by unexplained left ventricular hypertrophy, associated with nondilated ventricular chambers, in the absence of another cardiac or systemic disease. Clinically, hypertrophic cardiomyopathy is diagnosed by maximal left ventricular wall thickness ≥15 mm, with wall thicknesses between 13 to 14 mm considered borderline, particularly in the presence of other compelling information (e.g., family history of hypertrophic cardiomyopathy), based on echocardiography (Gersh et al. 2011). The range of left ventricular wall thickness was 15 mm to 40 mm (mean ± SD: 23 ± 5 mm). The heart weight averaged 521 ± 113 g. Coronary artery anomalies were the second most frequent reason for sudden cardiac death, while a variety of diseases contributed (≤6%) to the fatal cardiac event. Of the athletes examined, 75 had been diagnosed with cardiac disease while they were alive but continued to participate in organized sport (including six who were medically disqualified).

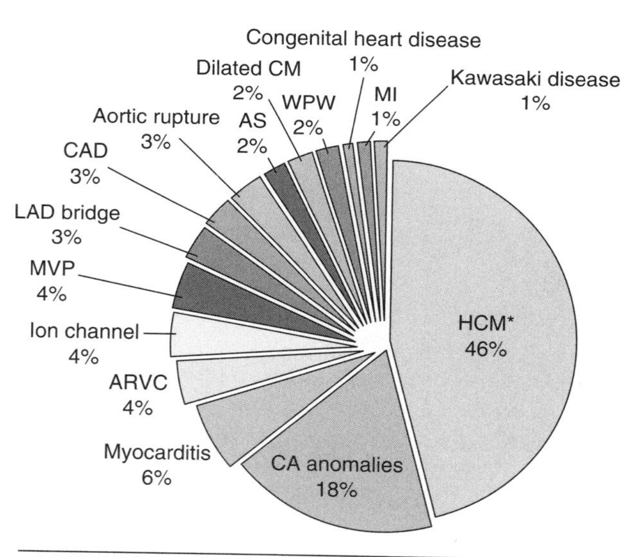

Figure 27.2 Causes of cardiovascular sudden deaths (% occurrence). HCM, hypertrophic cardiomyopathy; CA, coronary artery; ARVC, arrhythmogenic right ventricular cardiomyopathy; MVP, mitral valve prolapse; LAD, left anterior descending coronary disease; CAD, coronary artery disease; AS, aortic stenosis; CM, cardiomyopathy; WPW, Wolf-Parkinson-White; MI, myocardial infarction; * = Combined known HCM and possible (not definitive) HCM.

Adapted from Maron et al. 2009.

There appear to be differences in sudden death according to race. Deaths due to cardiac disease were more common in nonwhite than in white athletes, and the fraction of reported deaths attributed to hypertrophic cardiomyopathy and congenital coronary anomalies was higher in nonwhites (predominantly blacks) than in whites (20% vs. 10%, respectively) (Maron et al. 2009). The fraction of deaths attributed to ion channel disorders was greater in whites (2%) than in nonwhites (0.3%) (Maron et al. 2009). Sudden death from cardiovascular causes generally occurred during or just after physical exertion while the athlete was engaged in practice or competition (80%). The remaining athletes died during sleep or during other routine daily activities not associated with sport. In an examination of sudden deaths in NCAA athletes between 2004 and 2008, a total of 16% of the deaths recorded were cardiac (Harmon et al. 2011). This was the leading medical cause of death in NCAA athletes. Of these deaths, the vast majority were seen in males at the Division I level. Males had more than twice the risk of sudden cardiac death that females had (1:33,134 vs. 1:76,646 death rate per athlete years, respectively). In addition, the death rate among black athletes was nearly threefold greater than in white athletes (1:17,696 vs. 1:58,653 death rate per athlete years, respectively). Figure 27.3 highlights the incidence of sudden cardiac death in NCAA athletes by sport. Basketball has the highest incidence of all sports, with an overall incident rate of 1:11,394 athletes per year. Harmon and colleagues (2011) showed that the highest rate of occurrence was seen in male NCAA basketball players at the Division I level (1:3,126 athletes per year).

Prevention, Recognition, and Treatment

The primary method of screening for sudden cardiac death is the preparticipation physical examination that includes a standardized medical history and attention to episodes of exertional syncope or presyncope chest pain (Casa, Guskiewicz, et al. 2012). In addition, a personal or family history of sudden cardiac arrest or sudden death and exercise intolerance needs to be obtained. However, there is much debate in this area, as 80% of individuals who experienced sudden cardiac death were asymptomatic until the event (Casa, Guskiewicz, et al. 2012; Maron et al. 1996). Thus, screening by medical history and physical examination alone may have limited sensitivity for identifying at-risk athletes. The American Heart Association published a 12-point preparticipation cardiovascular screen (table 27.1) to be used for competitive athletes based on both medical history

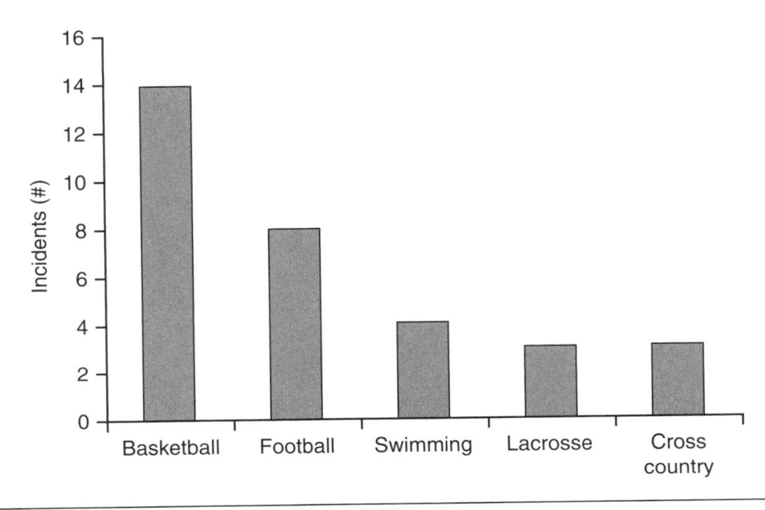

Figure 27.3 Incidence of sudden cardiac death in NCAA athletes by sport.

Adapted from Harmon et al. 2011.

Table 27.1 Twelve-Element American Heart Association Recommendations for Preparticipation Cardiovascular Screening of Competitive Athletes

Medical history		
Personal history	**Family history**	**Physical examination**
Exertional chest pain or discomfort	Premature death (sudden and unexpected or otherwise) before age 50 due to heart disease in one or more relatives	Heart murmur (in both the supine and standing position)
Unexplained syncope or near syncope judged not to be neurocardiogenic (vasovagal)	Disability from heart disease in a close relative under the age of 50	Femoral pulses to exclude aortic coarctation
Excessive exertional and unexplained dyspnea or fatigue associated with exercise	Specific knowledge of certain cardiac conditions in family members: hypertrophic or dilated cardiomyopathy, long QT syndrome or other ion channel disorders, Marfan syndrome, or clinically important arrhythmias	Physical stigmata of Marfan syndrome
Prior recognition of a heart murmur		Brachial artery blood pressure (seated)
Elevated systemic blood pressure		

Adapted from Maron et al. 2007.

and physical exam (Maron et al. 2007). Prevention of cardiac death is also based on preparedness; given that the vast majority of occurrences are in previously asymptomatic athletes, the availability of an automated defibrillator (AED) is essential. A goal of less than 3 to 5 min from time of collapse to delivery of the first shock is strongly recommended (Casa et al. 2012).

An athlete who suddenly collapses and is unresponsive should be assumed to be under cardiac arrest. CPR (cardiopulmonary resuscitation) should be provided and medical personnel should be summoned immediately. Airway, breathing, circulation, and heart rhythm (using the AED) should be assessed. A jerking or seizure-like activity is often present following collapse and should not be mistaken for a seizure only. It should be assumed until proven otherwise that the athlete is in cardiac arrest. Similarly, occasional or agonal gasping (labored breathing) should not be mistaken for normal breathing.

Traumatic Brain Injury

Traumatic brain injury occurs in a number of helmeted and nonhelmeted sports. It can be defined as a neurological dysfunction following head trauma (Sahler and Greenwald 2012). The vast majority of brain injuries in sport are cerebral concussions. These are often referred to as mild traumatic brain injuries (Casa et al. 2012; Herring et al. 2011; Mueller and Cantu 2011). According to the Centers for Disease Control and Prevention, 1.6 to 3.8 million sport-related concussions occur annually (Langlois, Rutland-Brown, and Wald 2006). Figure 27.4 provides the frequency of occurrence (in percentages) of concussions in NCAA sports. The sport with the highest rate of concussion is women's ice hockey (18.3% of all injuries), which also has the highest rate of injury per athlete exposure (0.91 per 1000 athlete exposures) (Daneshvar et al. 2011). However, the total athlete participation rate in both men's and women's ice hockey over 25 years has been relatively low (approximately 910,000 high school and college athletes), while the highest participation rate is seen in American football (approximately 37 million athletes participating over the same time span) (Daneshvar et al. 2011). Given the concussion frequency rate in American football of 6% (0.37 per 1000 athlete exposures), the highest absolute number of concussions is seen in American football.

Linear and rotational head accelerations resulting from either direct or inertial (i.e., whiplash) contact to the head have been suggested as the primary risk factors for concussion (Guskiewicz and Mihalik 2011). Acceleration of the head causes insult to brain tissue, potentially resulting in an injury. The magnitude and extent of the injury depend on a number of factors. The initial injury results from the direct impact and damage to neuronal and parenchyma tissue. A secondary or delayed phase of injury occurs as a result of the inflammatory response, edema, ischemia, and the effects of free radicals and ions that cause additional insult to the brain (Sahler and Greenwald 2012). The rash of media reports detailing problems of dementia and depression in former

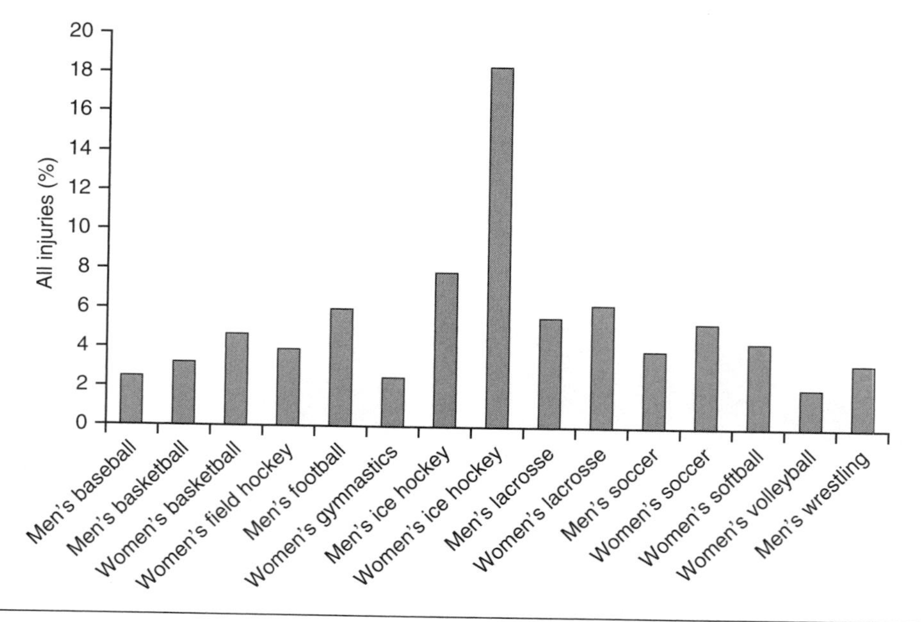

Figure 27.4 Frequency of concussion as a percentage of all injuries.

Adapted from Daneshvar et al. 2011.

professional football players, as well as the number of case studies of chronic traumatic encephalopathy in former athletes involved in contact sports, has resulted in a concerted effort to focus on the clinical and scientific study of concussion. To date, a precise threshold for causing concussive injury has yet to be determined. However, according to Guskiewicz and Mihalik (2011), research findings have shown that concussion results from lower impact magnitudes than previously thought; that linear acceleration may be just as important as angular acceleration in causing concussion; that athletes can sustain repeated head contacts during a season and not sustain a diagnosed concussion; and that clinicians should not use impact magnitude or location to predict clinical outcomes of symptom severity, neuropsychological function, and balance.

A concussive impact gives rise to a number of clinical signs and symptoms that range from subtle mood changes to a loss of consciousness (Sahler and Greenwald 2012). Signs of mild traumatic brain injury (i.e., concussion) include amnesia, behavior or personality changes, memory disturbance (i.e., confabulation), delayed verbal and motor responses, disequilibrium, loss of orientation, slurred or incoherent speech, or a vacant stare. Symptoms may include blurry or double vision, confusion, dizziness, excessive drowsiness, sleep difficulties, feeling as if the "head is in a fog" or haze, grogginess, headache, nausea, vomiting, inability to focus or concentrate, and an intolerance to light or noise (Kelly, Rosenberg, and Stevens 1997). In addition are systemic physiological effects from the impact that reduce cerebral blood flow and cerebrovascular reactivity (Len et al. 2011); these may have negative impacts on brain metabolism (Herring et al. 2011). The cerebrovascular abnormalities appear to persist even after neurocognitive function returns (Len et al. 2011). This warrants a concern to ensure that the athlete is indeed fully recovered from the concussion.

One of the greatest concerns about concussion is the neurological effects from repeated impacts. A major complication of a concussion is the **second-impact syndrome**, which can occur within a short time after a brain injury and often occurs in an athlete who has experienced a previous concussion (sometimes in the same game). The second impact causes a vascular engorgement leading to massive intracranial pressure and brain herniation resulting in severe brain damage or death (Herring et al. 2011). In addition to the second-impact syndrome, which has to do with multiple head injuries typically in the same event, another concern relates to repeated or multiple concussions over time. Repetitive concussions

In Practice

Depression and Neurocognitive Performance Following Concussion in Male and Female Athletes

There is a major push to investigate both short-term and long-term risks associated with concussion. Kontos and colleagues (2012) prospectively examined (over 2 years) the relationship between concussions occurring during sporting events and the short-term effect on depression and neurocognitive performance in high school and collegiate athletes. Baseline measures were obtained in all athletes. Once a concussion was diagnosed, the injured athlete was examined 2, 7, and 14 days postinjury. The dependent variables were a total depression score from the Beck Depression Inventory-II (a 21-question self-report that measures factors relating to depression) and computerized neurocognitive test scores for verbal memory, visual memory, reaction time, and processing speed. A total of 75 athletes from multiple sites received diagnoses of concussion and participated in the study. The results indicated that depression was elevated at 2, 7, and 14 days postinjury. In contrast to the investigators' a priori hypothesis that depressive symptoms would subside by 2 weeks postinjury, the findings suggested that this was not the case. It is important to note that there was no evidence of clinical depression, only that depression scores were elevated following injury. However, this indicates that athletes should be monitored following a concussion. In addition, neurocognitive results reflected significant decrements in visual memory and reaction time that lasted at least 14 days. The researchers concluded that both increases in depression and decreases in neurocognitive function are a concern in athletes who have had a concussion and that future studies should monitor these individuals for periods longer than 2 weeks. Furthermore, tracking of depression trajectories postinjury appears warranted.

In Practice

Repeated Concussions and Neurocognitive Function

A major concern among sports medicine clinicians is the long-term consequences of repeated concussion. Covassin and colleagues (2008) performed a multisite study from five universities in the northeastern United States. A total of 21 athletes with a history of two or more concussions were examined. Subjects who sustained a concussion were assessed on days 1 and 5 postinjury. Neurocognitive function was assessed using the Immediate Post-Concussion Assessment Cognitive Testing (ImPACT) computer program. The program consists of six tests used to measure attention, verbal recognition memory, visual working memory, visual processing speed, reaction time, numerical sequencing ability, and learning. A total of four scores were developed: verbal memory, visual memory, reaction time, and visual processing speed. At day 1 postinjury, 81% of the athletes with a history of concussion demonstrated a decline in reaction time, 57% in visual processing speed, 52% in verbal memory, and 48% in visual memory. At day 5, 57% of these athletes demonstrated at least one decline in both reaction time and verbal memory, a 48% decline in visual processing speed, and a 29% decline in visual memory. In a comparison group of concussed athletes with no prior history of concussion, the performance results were much lower. On day 1 they showed a 56% decline in visual processing speed, 50% in reaction time, 44% in visual memory, and 39% in verbal memory; but on day 5 they showed a 31% decline in processing speed, 22% in visual memory, 14% in reaction time, and 11% in verbal memory. These results suggest neurocognitive deficits in athletes with a history of concussion. The authors concluded that sports medicine professionals need to closely monitor athletes who have had a concussion before returning them to the playing field.

sustained over a professional football career have been associated with the rate of clinical depression and late-life cognitive impairment (Guskiewicz et al. 2005, 2007). Another concern, though, is the relation of repeated subconcussive impacts and concussion history to neurological function. Gysland and colleagues (2012) reported that college football players sustained 1177 ± 773 head impacts during a competitive football season, with an average of 12 ± 11 head impacts registering above a 90 *g* (gravitational force) "high-magnitude" impact threshold. The *g* forces were determined by linear accelerometers placed inside the football helmet. Only impacts registering above 10 *g* were reported. Defensive and offensive linemen registered the most head impacts. The majority of the impacts were found to not be clinically meaningful on tests used to identify neurological impairment. The repeated subconcussive impacts accumulated during a competitive season did not appear to have any link to any deficit on measures of neurological function. Football players appeared to be able to maintain neurological function despite the repeated head impacts. However, the authors suggested that additional research exploring years of playing exposure needs to be performed.

In addition to mild traumatic brain injury, since 1984 the National Center for Catastrophic Sports Injury has recorded 164 brain injuries with incomplete recovery (Mueller and Cantu 2011). The incidence rate for these injuries, recorded between 1984 and 2011 (28 years), was 0.36 per 100,000 participants for high school athletes and 0.52 per 100,000 participants for college football players. Interestingly, brain deaths have been decreasing over time, but brain injuries with disability have been increasing. It is possible that the inverse changes are related to better medical trauma capability at event sites, but also that rule changes and greater emphasis on improving coaching techniques have reduced the incidence of catastrophic brain injury.

Prevention, Recognition, Treatment, and Return to Play

The prevention of catastrophic brain injury starts with the use of certified protective equipment that meets the required standards for each sport. Educational outreach programs should alert coaches, players, and parents to the signs and symptoms associated with brain injuries and the potential dangers of ignoring these indicators. Emphasis needs to be placed on having all involved parties understand the risk of an athlete's returning to play while still symptomatic. It should be acknowledged that helmets do not prevent concussion but do reduce the incidence of skull fracture and major head trauma (Herring et al. 2011).

To provide a baseline of measures, it is recommended that all athletes be assessed during the preseason with a battery of objective cognitive function and balance tests to prevent premature clearance of the athlete to return to play (Casa, Guskiewicz, et al. 2012). Considering that nearly 50% to 75% of athletes do not report their concussions (McCrea et al. 2004), the use of objective measures becomes very important. At time of injury, the athlete should be thoroughly evaluated by medical personnel. If a concussion is suspected or confirmed, the athlete should be monitored for postconcussion signs and symptoms to determine if there is any deterioration of function. If symptoms continue or worsen, the athlete should be transported to a medical facility. If the brain injury is severe and involves brain or brainstem impairment, members of the sports medicine team must be prepared to perform manual ventilations through either endotracheal intubation or bag-valve-mouth resuscitation (Casa, Guskiewicz, et al. 2012). To reduce intracranial pressure, the sports medicine team may elevate the head to at least 30° and ensure that the head and neck are in the midline position to optimize venous outflow from the brain (Casa et al. 2012). The use of intravenous diuretics can also reduce intracranial pressure (Guha 2004).

Returning the athlete to full participation requires a gradual series of steps. With each step, the medical staff needs to determine that the activity did not produce any symptoms. Only if the athlete is able to perform the given step without symptoms can he or she proceed to the next step. Table 27.2 provides a graduated return-to-play sample protocol recommended by the National Athletic Trainers' Association (Casa, Guskiewicz, et al. 2012). Athletes should perform no more than two steps in a single day. This gives the medical personnel time to determine whether any symptoms appeared following the activity. Full participation should occur only after the athlete has remained symptom free for 24 h following step 4 of the program.

Summary

The tremendous benefits that come with sport participation greatly outweigh the risks of participation. Catastrophic injuries are the nightmare of all parents, coaches, and athletes; and minimizing their occurrence requires constant vigilance. Some traumatic events occur despite the best of intentions, and often they occur in the absence of any previous signs or symptoms. However, other traumatic injuries are completely avoidable; minimizing risk for these requires only the education of the coaches, use of appropriate preseason screening programs, and appropriate progression of training. Recent interassociation task forces have published a number of recommendations, including the hiring of accredited strength and conditioning coaches and the use of appropriate reporting structures, to ensure that training program progression is appropriate and minimizes the risk for traumatic injury in the competitive athletic population (Casa, Anderson et al. 2012).

Table 27.2 Graduated Return-to-Play Sample Program

Step	Activity
1	20 min stationary bike at 10 to 14 mph (16 to 23 km/h)
2	Interval bike: 30 s sprint at 18 to 20 mph (29 to 32 km/h), 30 s recovery × 10 repetitions Body weight circuit squats, push-ups, sit-ups, 20 s × 3 repetitions
3	60 yd (55 m) shuttle run × 10 repetitions with 40 s rest Plyometric workout: 10 yd (9 m) bounding, 10 medicine ball throws, 10 vertical jumps × 3 repetitions Noncontact sport-specific drills × 15 min
4	Limited, controlled return to practice with monitoring for symptoms
5	Full sport participation in practice

Adapted from Casa et al. 2012.

REVIEW QUESTIONS

1. What is sickle cell anemia, and how can it be avoided?
2. What role does the coach have in reducing the risk of sudden death?
3. Describe the second impact syndrome.
4. Describe both the intrinsic and extrinsic factors associated with exertional heatstroke.
5. Discuss methods that can be employed to reduce the risk for exertional heatstroke.

References

Aagaard, P., J.L. Andersen, M. Bennekou, B. Larsson, J.L. Olesen, R. Crameri, S.P. Magnusson, and M. Kjaer. 2011. Effects of resistance training on endurance capacity and muscle fiber composition in young top-level cyclists. *Scandinavian Journal of Medicine and Sports Science* 21:e298-e307.

Abe, H. 2000. Role of histidine-related compounds as intracellular proton buffering constituents in vertebrate muscle. *Biochemistry (Moscow)* 65:891-900.

Abe, T., K. Kumagai, and W.F. Brechue. 2000. Fascicle length of leg muscles is greater in sprinters than distance runners. *Medicine and Science in Sports and Exercise* 32:1125-1129.

Acheson, K.J., B. Zahorska-Markiewicz, P.H. Pittet, K. Anantharaman, and E. Jequier. 1980. Caffeine and coffee: Their influence on metabolic rate and substrate utilization in normal weight and obese individuals. *American Journal of Clinical Nutrition* 33:989-997.

Achten, J., M. Gleeson, and A.E. Jeukendrup. 2002. Determination of the exercise intensity that elicits maximal fat oxidation. *Medicine and Science in Sports and Exercise* 34:92-97.

Adamovich, D.R. 1984. *The heart: Fundamentals of electrocardiography, exercise physiology and exercise stress testing*. Freeport, NY: Sports Medicine Books.

Adams, D., J.P. O'Shea, K.L. O'Shea, and M. Climstein. 1992. The effect of six weeks of squat, plyometric, and squat-plyometric training on power production. *Journal of Applied Sport Science Research* 6:36-41.

Adams, G.R., B.M. Hather, K.M. Baldwin, and G.A. Dudley. 1993. Skeletal muscle myosin heavy chain composition and resistance training *Journal of Applied Physiology* 74:911-915.

Adams, W.C., R.H. Fox, A.J. Fry, and I.C. MacDonald. 1975. Thermoregulation during marathon running in cold, moderate and hot environments. *Journal of Applied Physiology* 38:1030-1037.

Adlercreutz, H., M. Harkonen, K. Kuoppasalmi, I. Huhtaniemi, H. Tikanen, K. Remes, A. Dessypris, and J. Karvonen. 1986. Effect of training on plasma anabolic and catabolic steroid hormones and their response during physical exercise. *International Journal of Sports Medicine* 7(suppl):227-228.

Aguilera, R., C.K. Hatton, and D.H. Catlin. 2002. Detection of epitestosterone by isotope ratio mass spectrometry. *Clinical Chemistry* 48:629-636.

Ahima, R.S. 2000. Leptin and the neuroendocrinology of fasting. *Frontiers of Hormone Research* 26:42-56.

Ahlborg, G., P. Felig, L. Hagenfeldt, R. Hendler, and J. Wahren. 1974. Substrate turnover during prolonged exercise in man: Splanchnic and leg metabolism of glucose, free fatty acids, and amino acids. *Journal of Clinical Investigation* 53:1080-1090.

Ainegren, M., P. Carlsson, M. Tinnsten, and M.S. Laaksonen. 2013. Skiing economy and efficiency in recreational and elite cross-country skiers. *Journal of Strength and Conditioning Research* 27:1239-1252.

Akil, H., S.J. Watson, E. Young, M.E. Lewis, H. Khachaturian, and J.M. Walker. 1984. Endogenous opioids: Biology and function. *Annual Review of Neuroscience* 7:223-255.

Alberici, J.C., P.A. Farrell, P.M. Kris-Etherton, and C.A. Shively. 1993. Effects of pre-exercise candy bar ingestion on glycemic response, substrate utilization, and performance. *International Journal of Sport Nutrition* 3:323-333.

Alen, M., and K. Hakkinen. 1985. Physical health and fitness of an elite bodybuilder during 1 year of self-administration of testosterone and anabolic steroids: A case study. *International Journal of Sports Medicine* 6:24-29.

Alen, M., K. Hakkinen, and P.V. Komi. 1984. Changes in neuromuscular performance and muscle fibre characteristics of elite power athletes self-administering androgenic and anabolic steroids. *Acta Physiologica Scandinavica* 122:535-544.

Alen, M., A. Pakarinen, and K. Hakkinen. 1993. Effects of prolonged training on serum thyrotropin and thyroid hormones in elite strength athletes. *Journal of Sports Sciences* 11:493-497.

Alen, M., M. Reinila, and R. Vihko. 1985. Response of serum hormones to androgen administration in power athletes. *Medicine and Science in Sports and Exercise* 17:354-359.

Alexander, M.J.L. 1989. The relationship between muscle strength and sprint kinematics in elite sprinters. *Canadian Journal of Sport Science* 14:148-157.

Alford, C., H. Cox, and R. Westcott. 2001. The effects of Red Bull energy drink on human performance and mood. *Amino Acids* 21:139-150.

Almond, C.S., A.Y. Shin, E.B. Fortescue, R.C. Mannix, D. Wypij, B.A. Binstadt, C.N. Duncan, D.P. Olson, A.E. Salerno, J.W. Newburger, and D.S. Greenes. 2005. Hyponatremia among runners in the Boston Marathon. *New England Journal of Medicine* 14:1550-1556.

Alter, M. 1996. *Science of flexibility and stretching*. Champaign, IL: Human Kinetics.

Alway, S.E., W.H. Grumbt, W.J. Gonyea, and J. Stray-Gunderson. 1989. Contrasts in muscle and myofibers of elite male and female body builders. *Journal of Applied Physiology* 67:24-31.

Alway, S.E., W.H. Grumbt, J. Stray-Gunderson, and W.J. Gonyea. 1992. Effects of resistance training on elbow flexors of highly competitive bodybuilders. *Journal of Applied Physiology* 72:1512-1521.

Alway, S.E., J.D. MacDougall, and D.G. Sale. 1989. Contractile adaptations in human triceps surae after isometric exercise. *Journal of Applied Physiology* 66:2725-2732.

Alway, S.E., P.K. Winchester, M.E. Davis, and W.J. Gonyea. 1989. Regionalized adaptations and fiber proliferation in stretch-induced muscle enlargement. *Journal of Applied Physiology* 66:771-781.

Amenta, F., and S.K. Tayebati. 2008. Pathway of acetylcholine synthesis, transport and release as targets for treatment of adult-onset cognitive dysfunction. *Current Medicinal Chemistry* 15:488-498.

American College of Sports Medicine (ACSM). 2000. The physiological and health effects of oral creatine supplementation. Consensus statement. *Medicine and Science in Sports and Exercise* 32:706-717.

American College of Sports Medicine (ACSM). 2013. *Guidelines for exercise testing and prescription,* 9th ed., ed. L.S. Pescatello. Philadelphia: Lippincott Williams & Wilkins.

Andersen, T., and J. Fogh. 2001. Weight loss and delayed gastric emptying following a South American herbal preparation in overweight patients. *Journal of Human Nutrition and Dietetics* 14:243-250.

Anderson, D.E. 2007. Reliability of air displacement plethysmography. *Journal of Strength and Conditioning Research* 21:169-171.

Anderson, L.L., G. Tufekovic, M.K. Zebis, R.M. Crameri, G. Verlaan, M. Kjaer, C. Suetta, P. Magnusson, and P. Aagaard. 2005. The effect of resistance training combined with timed ingestion of protein on muscle fiber size and muscle strength. *Metabolism* 54:151-156.

Anderson, M.A., J.B. Gieck, D. Perrin, A. Weltman, R. Rutt, and C. Denegar. 1991. The relationships among isometric, isotonic and isokinetic quadriceps and hamstring force and three components of athletic performance. *Journal of Orthopaedic and Sports Physical Therapy* 14:114-120.

Anderson, T., and J.T. Kearney. 1982. Effects of three resistance training programs on muscular strength and absolute and relative endurance. *Research Quarterly for Exercise and Sport* 2:27-30.

Anderson, W.A., M.A. Albrecht, and D.B. McKeag. 1993. *Second replication of a national study of the substance use/abuse habits of college student athletes. Report to NCAA*. Mission, KS: National Collegiate Athletic Association.

Antal, L., and C. Good. 1980. Effects of oxprenolol on pistol shooting under stress. *Practitioner* 224:755-760.

Antonio, J., and W.J. Gonyea. 1993. Skeletal muscle fiber hyperplasia. *Medicine and Science in Sports and Exercise* 25:1333-1345.

Antonio, J., and W.J. Gonyea. 1994. Muscle fiber splitting in stretch-enlarged avian muscle. *Medicine and Science in Sports and Exercise* 26:973-977.

Anzalone, M.L., V.S. Green, M. Buja, L.A. Sanchez, R.I. Harrykissoon, and E.R. Eichner. 2010. Sickle cell trait and fatal rhabdomyolysis in football training: A case study. *Medicine and Science in Sports and Exercise* 42:3-7.

Aoki, M.S., A.L.R.A. Almeida, F. Navarro, L.F.B.P. Costa-Rosa, and R.F.P. Bacurau. 2004. Carnitine supplementation fails to maximize fat mass loss induced by endurance training in rats. *Annals of Nutrition and Metabolism* 48:90-94.

Apel, J.M., R.M. Lacey, and R.T. Kell. 2011. A comparison of traditional and weekly undulating periodized strength training programs with total volume and intensity equated. *Journal of Strength and Conditioning Research* 25:694-703.

Appell, H.J., S. Forsberg, and W. Hollmann. 1988. Satellite cell activation in human skeletal muscle after training: Evidence for muscle fiber neoformation. *International Journal of Sports Medicine* 9:297-299.

Appleby, B., R.U. Newton, and P. Cormie. 2012. Changes in strength over a 2-year period in professional rugby union players. *Journal of Strength and Conditioning Research* 26:2538-2546.

Arenas, J., J.R. Ricoy, A.R. Encinas, P. Pola, S. D'Iddio, M. Zeviani, S. Didonato, and M. Corsi. 1991. Carnitine in muscle, serum, and urine of nonprofessional athletes: Effects of physical exercise, training, and L-carnitine administration. *Muscle and Nerve* 14:598-604.

Arlettaz, A., B. Le Panse, H. Portier, A.M. Lecoq, R. Thomasson, J. De Ceaurriz, and K. Collomp. 2009. Salbutamol intake and substrate oxidation during submaximal exercise. *European Journal of Applied Physiology* 105:207-213.

Armstrong, L.E. 1988. Research update: Fluid replacement and athlete hydration. *National Strength and Conditioning Journal* 10:69-71.

Armstrong, L.E. 2000. *Performing in extreme environments*. Champaign, IL: Human Kinetics.

Armstrong, L.E., D.J. Casa, C.M. Maresh, and M.S. Ganio. 2007. Caffeine, fluid-electrolyte balance, temperature regulation, and exercise-heat tolerance. *Exercise and Sport Sciences Reviews* 35:135-140.

Armstrong, L.E., D.J. Casa, M. Millard-Stafford, D.S. Moran, S.W. Pyne, and W.O. Roberts. 2007. Exertional heat illness during training and competition. *Medicine and Science in Sports and Exercise* 39:556-572.

Armstrong, L.E., D.J. Casa, M.W. Roti, E.C. Lee, S.A. Craig, J.W. Sutherland, K.A. Fiala, and C.M. Maresh. 2008. Influence of betaine consumption on strenuous running and sprinting in a hot environment. *Journal of Strength and Conditioning Research* 22:851-860.

Armstrong, L.E., D.L. Costill, and W.J. Fink. 1985. Influence of diuretic-induced dehydration on competitive running performance. *Medicine and Science in Sports and Exercise* 17:456-461.

Armstrong, L.E., A.E. Crago, R. Adams, W.O. Roberts, and C.M. Maresh. 1996. Whole-body cooling of hyperthermic runners: Comparison of two field therapies. *American Journal of Emergency Medicine* 14:355-358.

Armstrong, L.E., J.P. De Luca, and R.W. Hubbard. 1990. Time course of recovery and heat acclimation ability of prior exertional heatstroke patients. *Medicine and Science in Sports and Exercise* 22:36-48.

Armstrong, L.E., and J.E. Dziados. 1986. Effects of heat exposure on the exercising adult. In *Sports physical therapy*, ed. D.B. Bernhardt, 197-214. New York: Churchill Livingstone.

Armstrong, L.E., J.A. Herrera Soto, F.T. Hacker, D.J. Casa, S.A. Kavouras, and C.M. Maresh. 1998. Urinary indices during dehydration, exercise, and rehydration. *International Journal of Sport Nutrition* 8:345-355.

Armstrong, L.E., R.W. Hubbard, B.H. Jones, and J.T. Daniels. 1986. Preparing Alberto Salazar for the heat of the 1984 Olympic marathon. *Physician and Sportsmedicine* 14:73-81.

Armstrong, L.E., R.W. Hubbard, W.J. Kraemer, J.P. De Luca, and E. Christensen. 1987. Signs and symptoms of heat exhaustion during strenuous exercise. *Annals of Sports Medicine* 3:182-189.

Armstrong, L.E., E.C. Johnson, D.J. Casa, M.S. Ganio, B.P. McDermott, L.M. Yamamoto, R.M. Lopez, and H. Emmanuel. 2010. The American football uniform: Uncompensable heat stress and hyperthermic exhaustion. *Journal of Athletic Training* 45:117-127.

Armstrong, L.E., and C.M. Maresh. 1991. The induction and decay of heat acclimatization in trained athletes. *Sports Medicine* 12:302-312.

Armstrong, L.E., and C.M. Maresh. 1993. The exertional heat illnesses: A risk of athletic participation. *Medicine, Exercise, Nutrition and Health* 2:125-134.

Armstrong, L.E., and C.M. Maresh. 1996. Fluid replacement during exercise and recovery from exercise. In *Body fluid balance: Exercise and sport,* ed. E.R. Buskirk and S.M. Puhl, 259-282. New York: CRC Press.

Armstrong, L.E., C.M. Maresh, M. Whittlesey, M.F. Bergeron, C. Gabaree, and J.R. Hoffman. 1994. Longitudinal exercise-heat tolerance and running economy of collegiate distance runners. *Journal of Strength and Conditioning Research* 8:192-197.

Armstrong, L.E., and K.B. Pandolf. 1988. Physical training, cardiorespiratory physical fitness and exercise-heat tolerance. In *Human performance physiology and environmental medicine at terrestrial extremes*, ed. K.B. Pandolf, M.N. Sawka, and R.R. Gonzalez, 199-226. Indianapolis: Benchmark Press.

Artioli, G.G., B. Gualano, A. Smith, J.R. Stout, and A.H. Lancha Jr. 2010. The role of β-alanine supplementation on muscle carnosine and exercise performance. *Medicine and Science in Sports and Exercise* 42:1162-1173.

Arvat, E., L. Di Vito, F. Broglio, M. Papotti, G. Mucciolo, C. Dieguez, F.F. Casanueva, R. Deghenghi, F. Camanni, and E. Ghigo. 2000. Preliminary evidence that ghrelin, the natural GH secretagogue (GHS)-receptor ligand, strongly stimulates GH release in humans. *Journal of Endocrinological Investigation* 23:493-480.

Asmussen, E., F. Bonde-Peterson, and K. Jorgensen. 1976. Mechano-elastic properties of human muscles at different temperatures. *Acta Physiologica Scandinavica* 96:86-93.

Astorino, T.A., B.J. Martin, L. Schachsiek, K. Wong, and K. Ng. 2011. Minimal effect of acute caffeine ingestion on intense resistance training performance. *Journal of Strength and Conditioning Research* 25:1752-1758.

Astorino, T.A., and D.W. Roberson. 2010. Efficacy of acute caffeine ingestion for short-term high intensity exercise: A systematic review. *Journal of Strength and Conditioning Research* 24:257-265.

Astorino, T.A., M.N. Terzi, D.W. Roberson, and T.R. Burnett. 2010. Effect of two doses of caffeine on muscular function during isokinetic exercise. *Medicine and Science in Sports and Exercise* 42:2205-2210.

Åstrand, P.O. 1965. *Work tests with the bicycle ergometer.* Varberg, Sweden: AB Cykelfabriken Monark.

Atha, J. 1981. Strengthening muscle. *Exercise and Sport Sciences Reviews* 9:1-73.

Avois, L., N. Robinson, C. Saudan, N. Baume, P. Mangin, and M. Saugy. 2006. Central nervous system stimulants and sport practice. *British Journal of Sports Medicine* 40:16-20.

Ayalon, A., O. Inbar, and O. Bar-Or. 1974. Relationships among measurements of explosive strength and anaerobic power. In *International series on sport sciences, Vol. 1, Biomechanics IV*, ed. R.C. Nelson and C.A. Morehouse, 527-532. Baltimore: University Park Press.

Ayers, J.W.T., Y. Komesu, T. Romani, and R. Ansbacher. 1985. Anthropometric, hormonal, and psychological correlates of semen quality in endurance trained male athletes. *Fertility and Sterility* 43:917-921.

Bacurau, R.F.P., F. Navarro, R.A. Bassit, M.O. Meneguello, R.V.T. Santos, A.L.R. Almeida, and L.F.B.P. Costa Rosa. 2003. Does exercise training interfere with the effects of L-carnitine supplementation? *Nutrition* 19:337-341.

Baechle, T.R., R.W. Earle, and D. Wathen. 2008. Resistance training. In *Essentials of strength training and conditioning*, 3rd ed., ed. T.R. Baechle and R.W. Earle, 381-412. Champaign, IL: Human Kinetics.

Baguet, A., H. Reyngoudt, A. Pottier, I. Everaert, S. Callens, E. Achten, and W. Derave. 2009. Carnosine loading and washout in human skeletal muscles. *Journal of Applied Physiology* 106:837-842.

Bahrke, M.S., and C.E. Yesalis. 1994. Weight training. A potential confounding factor in examining the psychological and behavioural effects of anabolic-androgenic steroids. *Sports Medicine* 18:309-318.

Baker, D. 2001. The effects of an in-season of concurrent training on the maintenance of maximal strength and power in professional and collegiate-aged rugby league football players. *Journal of Strength and Conditioning Research* 15:172-177.

Baker, D., and S. Nance. 1999. The relation between running speed and measures of strength and power in professional rugby players. *Journal of Strength and Conditioning Research* 13:230-235.

Baker, D., S. Nance, and M. Moore. 2001a. The load that maximizes the average mechanical power output during explosive bench press throws in highly trained athletes. *Journal of Strength and Conditioning Research* 15:20-24.

Baker, D., S. Nance, and M. Moore. 2001b. The load that maximizes the average mechanical power output during jump squats power trained athletes. *Journal of Strength and Conditioning Research* 15:92-97.

Baker, D., G. Wilson, and R. Carlyon. 1994. Periodization: The effect on strength of manipulating volume and intensity. *Journal of Strength and Conditioning Research* 8:235-242.

Baldwin, K.M., P.J. Campbell, and D.A. Cooke. 1977. Glycogen, lactate, and alanine changes in muscle fiber types during graded exercise. *Journal of Applied Physiology* 43:151-157.

Baldwin, K.M., and F. Haddad. 2001. Effects of different activity and inactivity paradigms on myosin heavy chain gene expression in striated muscle. *Journal of Applied Physiology* 90:345-357.

Ballantyne, C.S., S.M. Phillips, J.R. MacDonald, M.A. Tarnopolsky, and J.D. MacDougall. 2000. The acute effects of androstenedione supplementation in healthy young men. *Canadian Journal of Applied Physiology* 25:68-78.

Ballor, D.L., and R.E. Keesey. 1991. A meta-analysis of the factors affecting exercise-induced changes in body mass, fat mass and fat free mass in males and females. *International Journal of Obesity and Related Metabolic Disorders* 15:717-726.

Ballor, D.L., and E.T. Poehlman. 1992. Resting metabolic rate and coronary-heart-disease risk factors in aerobically and resistance-trained women. *American Journal of Clinical Nutrition* 56:968-974.

Balsam, A., and L.E. Leppo. 1975. Effect of physical training on the metabolism of thyroid hormones in man. *Journal of Applied Physiology* 38:212-215.

Balsom, P.D., B. Ekblom, K. Soderlund, B. Sjoden, and E. Hultman. 1993. Creatine supplementation and dynamic high-intensity intermittent exercise. *Scandinavian Journal of Medicine and Science in Sports* 3:143-149.

Balsom, P.D., K. Soderlund, and B. Ekblom. 1994. Creatine in humans with special reference to creatine supplementation. *Sports Medicine* 3:143-149.

Barbero-Alvarez, J.C., A. Coutts, J. Granda, V. Barbero-Alvarez, and C. Castagna. 2010. The validity and reliability of a global positioning satellite system device to assess speed and repeated sprint ability (RSA) in athletes. *Journal of Science and Medicine in Sports* 13:232-235.

Barnett, C., D.L. Costill, M.D. Vukovich, K.J. Cole, B.H. Goodpaster, S.W. Trappe, and W.J. Fink. 1994. Effect of L-carnitine supplementation on muscle and blood carnitine content and lactate accumulation during high intensity sprint cycling. *International Journal of Sport Nutrition* 4:280-288.

Bar-Or, O. 1987. The Wingate anaerobic test: An update on methodology, reliability and validity. *Sports Medicine* 4:381-394.

Bar-Or, O., R. Dotan, O. Inbar, A. Rotstein, J. Karlsson, and P. Tesch. 1980. Anaerobic capacity and muscle fiber type distribution in man. *International Journal of Sports Medicine* 1:89-92.

Barr, S.I. 1999. Effects of dehydration on exercise performance. *Canadian Journal of Applied Physiology* 24:164-172.

Barron, J.L., T.D. Noakes, W. Levy, C. Smith, and R.P. Millar. 1985. Hypothalamic dysfunction in overtrained athletes. *Journal of Clinical Endocrinology and Metabolism* 60:803-806.

Bartlett, D., and J.E. Remmers. 1971. Effects of high altitude exposure on the lungs of young rats. *Respiratory Physiology* 13:116-125.

Bartsch, P. 1999. High altitude pulmonary edema. *Medicine and Science in Sports and Exercise* 31:S23-S27.

Barwell, C.J., A.N. Basma, M.A. Lafi, and L.D. Leake. 1989. Deamination of hordenine by monoamine oxidase and its action on vasa deferentia of the rat. *Journal of Pharmacy and Pharmacology* 41:421-423.

Basaria, S., J.T. Wahlstrom, and A.S. Dobs. 2001. Anabolic-androgenic steroid therapy in the treatment of chronic diseases. *Journal of Clinical Endocrinology and Metabolism* 86:5108-5117.

Bassett, D.R., J. Flohr, W.J. Ducy, E.T. Howley, and R.L. Pein. 1991. Metabolic responses to drafting during front crawl swimming. *Medicine and Science in Sports and Exercise* 23:744-747.

Bassett, D.R., and E.T. Howley. 1997. Maximal oxygen uptake: "Classical" versus "contemporary" viewpoints. *Medicine and Science in Sports and Exercise* 29:591-603.

Bassett, D.R., and E.T. Howley. 2000. Limiting factors for maximum oxygen uptake and determinants of endurance performance. *Medicine and Science in Sports and Exercise* 32:70-84.

Bauer, K., and M. Schulz. 1994. Biosynthesis of carnosine and related peptides by skeletal muscle cells in primary culture. *European Journal of Biochemistry* 219:43-47.

Baum, M., H. Liesen, and J. Enneper. 1994. Leucocytes, lymphocytes, activation parameters and cell adhesion molecules in middle-distance runners under different training conditions. *International Journal of Sports Medicine* 15:S122-S126.

Baumann, C.K., and M. Castiglione-Gertsch. 2007. Estrogen receptor modulators and down regulators: Optimal use in postmenopausal women with breast cancer. *Drugs* 67:2335-2353.

Baumann, G., J.G. MacCart, and K. Amburn. 1983. The molecular nature of circulating growth hormone in normal and acromegalic man: Evidence for a principal and minor monomeric forms. *Journal of Clinical Endocrinology and Metabolism* 56:946-952.

Bazzucchi, I., F. Felici, M. Montini, F. Figura, and M. Sacchetti. 2011. Caffeine improves neuromuscular function during maximal dynamic exercise. *Muscle and Nerve* 43:839-844.

Beck, K.C. 1999. Control of airway function during and after exercise in asthmatics. *Medicine and Science in Sports and Exercise* 31:S4-S11.

Beck, K.C., K.P. Offord, and P.D. Scanlon. 1994. Bronchoconstriction occurring during exercise in asthmatic subjects. *American Journal of Respiratory and Critical Care Medicine* 149:352-357.

Becker, G.F., R.C. Macedo, S. Cunha Gdos, J.B. Martins, O. Laitano, and A. Reischak-Oliveira. 2012. Combined effects of aerobic exercise and high-carbohydrate meal on plasma acylated ghrelin and levels of hunger. *Applied Physiology, Nutrition and Metabolism* 37:184-192.

Beelen, A., and A.J. Sargeant. 1991. Effect of lowered muscle temperature on the physiological response to exercise in men. *European Journal of Applied Physiology* 63:387-392.

Behm, D.G., A. Bambury, F. Cahill, and K. Power. 2004. Effect of acute static stretching on force, balance, reaction time, and movement time. *Medicine and Science in Sports and Exercise* 36:1397-1402.

Bell, A., K.D. Dorsch, D.R. McCreary, and R. Hovey. 2004. A look at nutritional supplement use in adolescents. *Journal of Adolescent Health* 34:508-516.

Bell, D.G., and I. Jacobs. 1999. Combined caffeine and ephedrine ingestion improves run times of Canadian forces warrior test. *Aviation, Space, and Environmental Medicine* 70:325-329.

Bell, D.G., I. Jacobs, and K. Ellerington. 2001. Effect of caffeine and ephedrine ingestion on anaerobic exercise performance. *Medicine and Science in Sports and Exercise* 33:1399-1403.

Bell, D.G., I. Jacobs, T.M. McLellan, and J. Zamecnik. 2000. Reducing the dose of combined caffeine and ephedrine preserves the ergogenic effect. *Aviation, Space, and Environmental Medicine* 71:415-519.

Bell, D.G., I. Jacobs, and J. Zamecnik. 1998. Effects of caffeine, ephedrine and their combination on time to exhaustion during high-intensity exercise. *European Journal of Applied Physiology and Occupational Physiology* 77:427-433.

Bell, G.J., S.R. Petersen, J. Wessel, K. Bagnall, and H.A. Quinney. 1991. Physiological adaptations to concurrent endurance training and low velocity resistance training. *International Journal of Sports Medicine* 12:384-390.

Bell, G., D. Syrotuik, T.P. Martin, R. Burnham, and H.A. Quinney. 2000. Effect of concurrent strength and endurance training on skeletal muscle properties and hormone concentrations in humans. *European Journal of Applied Physiology* 81:418-427.

Bell, G., D. Syrotuik, T. Socha, I. Maclean, and H.A. Quinney. 1997. Effect of strength training and concurrent strength and endurance training on strength, testosterone and cortisol. *Journal of Strength and Conditioning Research* 11:57-64.

Bellar, D., G.H. Kamimori, and E.L. Glickman. 2011. The effects of low-dose caffeine on perceived pain during a grip strength to exhaustion task. *Journal of Strength and Conditioning Research* 25:1225-1228.

Bemben, M.G., D.A. Bemben, D.D. Loftiss, and A.W. Khehans. 2001. Creatine supplementation during resistance training in college football athletes. *Medicine and Science in Sports and Exercise* 33:1667-1673.

Benardot, D. 2000. *Nutrition for serious athletes*. Champaign, IL: Human Kinetics.

Bendich, A. 1989. Carotenoids and the immune response. *Journal of Nutrition* 119:112-115.

Benoni, G., P. Bellavite, A. Adami, S. Chirumbolo, G. Lippi, G. Brocco, G.M. Guilini, and L. Cuzzolin. 1995. Changes in several neutrophil functions in basketball players before, during and after the sports season. *International Journal of Sports Medicine* 16:34-37.

Bentley, D.J., G.P. Millet, V.E. Vleck, and L.R. McNaughton. 2002. Specific aspects of contemporary triathlon. Implications for physiological analysis and performance. *Sports Medicine* 32:345-359.

Berger, A.J., and K. Alford. 2009. Cardiac arrest in a young man following excess consumption of caffeinated "energy drinks." *Medical Journal of Australia* 190:41-43.

Berger, R.A. 1962. Effect of varied weight training programs on strength. *Research Quarterly for Exercise and Sport* 33:168-181.

Berger, R.A. 1963. Comparative effects of three weight training programs. *Research Quarterly for Exercise and Sport* 34:396-398.

Bergeron, M.F. 1996. Heat cramps during tennis: A case report. *International Journal of Sport Nutrition* 6:62-68.

Bergh, U., and B. Ekblom. 1979. Influence of muscle temperature on maximal muscle strength and power output in human muscle. *Acta Physiologica Scandinavica* 107:332-337.

Bergh, U., A. Thorstensson, B. Sjodin, B. Hulten, K. Piehl, and J. Karlsson. 1978. Maximal oxygen uptake and muscle fiber types in trained and untrained humans. *Medicine and Science in Sports* 10:151-154.

Bergman, B.C., G.E. Butterfield, E.E. Wolfel, G.D. Lopashulk, G.A. Casazza, M.A. Horning, and G.A. Brooks. 1999. Muscle net glucose uptake and glucose kinetics after endurance training in men. *American Journal of Physiology* 277:E81-E92.

Bergstrom, J., L. Hermansen, E. Hultman, and B. Saltin. 1967. Diet, muscle glycogen and physical performance. *Acta Physiologica Scandinavica* 71:140-150.

Bergstrom, M., and E. Hultman. 1988. Energy cost and fatigue during intermittent electrical stimulation of human skeletal muscle. *Journal of Applied Physiology* 65:1500-1505.

Berk, L.S., D.C. Nieman, W.S. Youngberg, K. Arabatzis, M. Simpson-Westerberg, J.W. Lee, S.A. Tan, and W.C. Eby. 1990. The effect of long endurance running on natural killer cells in marathoners. *Medicine and Science in Sports and Exercise* 22:207-212.

Berning, J.M., K.J. Adams, and B.A. Stamford. 2004. Anabolic steroid usage in athletics: Facts, fiction, and public relations. *Journal of Strength and Conditioning Research* 18:908-917.

Berning, J.R. 1998. Energy intake, diet, and muscle wasting. In *Overtraining in sport,* ed. R.B. Kreider, A.C. Fry, and M.L. O'Toole, 275-288. Champaign, IL: Human Kinetics.

Bessman, S.P., and F. Savabi. 1990. The role of the phosphocreatine energy shuttle in exercise and muscle hypertrophy. In *Biochemistry of exercise VII,* ed. A.W. Taylor, P.D. Gollnick, H.J. Green, C.D. Ianuzzo, E.G. Noble, G. Metivier, and J.R. Sutton, 167-177. Champaign, IL: Human Kinetics.

Bhasin, S., O.M. Calof, T.W. Storer, M.L. Lee, N.A. Mazer, R. Jasuja, V.M. Montori, W. Gao, and J.T. Dalton. 2006a. Drug insight: Testosterone and selective androgen receptor modulators as anabolic therapies for chronic illness and aging. *Nature Clinical Practice: Endocrinology and Metabolism* 2:146-159.

Bhasin, S., O.M. Calof, T.W. Storer, M.L. Lee, N.A. Mazer, R. Jasuja, V.M. Montori, W. Gao, and J.T. Dalton. 2006b. Drug insights: Anabolic applications of testosterone and selective androgen receptor modulators in aging and chronic illness. *Nature Clinical Practice: Endocrinology and Metabolism* 2:133-140.

Bhasin, S., G.R. Cunningham, F.J. Hayes, A.M. Matsumoto, P.J. Snyder, R.S. Swerdloff, and V.M. Montori. 2006. Testosterone therapy in adult men with androgen deficiency syndromes: An Endocrine Society clinical practice guideline. *Journal of Clinical Endocrinology and Metabolism* 91:1995-2010.

Bhasin, S., T.W. Storer, N. Berman, C. Callegari, B. Clevenger, J. Phillips, T.J. Bunnell, R. Tricker, A. Shirazi, and R. Casaburi. 1996. The effects of supraphysiologic doses of testosterone on muscle size and strength in normal men. *New England Journal of Medicine* 335:1-7.

Bhasin, S., T.W. Storer, N. Berman, K.E. Yarasheski, B. Clevenger, J. Phillips, W.P. Lee, T.J. Bunnell, and R. Casaburi. 1997. Testosterone replacement increases fat-free mass and muscle size in hypogonadal men. *Journal of Clinical Endocrinology and Metabolism* 82:407-413.

Bhasin, S., T.W. Storer, M. Javanbakht, N. Berman, K.E. Yarasheski, J. Phillips, M. Dike, I. Sinha-Hikim, R. Shen, R.D. Hays, and G. Beall. 2000. Testosterone replacement and resistance exercise in HIV-infected men with weight loss and low testosterone levels. *Journal of the American Medical Association* 283:763-770.

Bhasin, S., L. Woodhouse, R. Casaburi, A.B. Singh, D. Bhasin, N. Berman, X. Chen, K.E. Yarasheski, L. Magliano, C. Dzekov, J. Dzekov, R. Bross, J. Phillips, I. Sinha-Hikim, R. Shen, and T.W. Storer. 2001. Testosterone dose-response relationships in healthy young men. *American Journal of Physiology: Endocrinology and Metabolism* 281:E1172-E1181.

Binkhorst, R.A., L. Hoofd, and A.C.A. Vissers. 1977. Temperature and force-velocity relationship of human muscles. *Journal of Applied Physiology: Respiratory, Environmental and Exercise Physiology* 42:471-475.

Biolo, G., R.Y.D. Fleming, and R.R. Wolfe. 1995. Physiologic hyperinsulinemia stimulates protein synthesis and enhances transport of selected amino acids in human skeletal muscle. *Journal of Clinical Investigation* 95:811-819.

Biolo, G., S.P. Maggi, B.D. Williams, K.D. Tipton, and R.R. Wolfe. 1995. Increased rates of muscle protein turnover and amino acid transport after resistance exercise in humans. *American Journal of Physiology and Endocrinology* 268:E514-E520.

Biolo, G., K.D. Tipton, S. Klein, and R.R. Wolfe. 1997. An abundant supply of amino acids enhances the metabolic effect of exercise on muscle protein. *American Journal of Physiology and Endocrinology* 273:E122-E129.

Birzniece, V., A.E. Nelson, and K.K. Ho. 2011. Growth hormone and physical performance. *Trends in Endocrinology and Metabolism* 22:171-178.

Bishop, D., D.G. Jenkins, L.T. Mackinnon, M. McEniery, and M.F. Carey. 1999. The effects of strength training on endurance performance and muscle characteristics. *Medicine and Science in Sports and Exercise* 31:886-891.

Bjorneboe, A., G.A. Bjorneboe, and C.A. Drevon. 1990. Absorption, transport and distribution of vitamin E. *Journal of Nutrition* 120:233-242.

Bjorntorp, P. 1991. Importance of fat as a support nutrient for energy: Metabolism of athletes. *Journal of Sports Sciences* 9:71-76.

Black, W., and E. Roundy. 1994. Comparisons of size, strength, speed and power in NCAA division I-A football players. *Journal of Strength and Conditioning Research* 8:80-85.

Blanco, C.E., P. Popper, and P. Micevych. 1997. Anabolic-androgenic steroid induced alterations in choline acyltransferase messenger RNA levels of spinal cord motoneurons in the male rat. *Neuroscience* 78:873-882.

Blannin, A.K., L.J. Chatwin, R. Cave, and M. Gleeson. 1996. Effects of submaximal cycling and long-term endurance training on neutrophil phagocytic activity in middle aged men. *British Journal of Sports Medicine* 30:125-129.

Bleisch, W.V., A.L. Harrelson, and V.N. Luine. 1982. Testosterone increases acetylcholine receptor number in the "levator ani" muscle of the rat. *Journal of Neurobiology* 13:153-161.

Blokland, A., W. Honig, F. Browns, and J. Jolles. 1999. Cognition-enhancing properties of subchronic phosphatidylserine (ps) treatment in middle-aged rats: Comparison of bovine cortex ps with eggs ps and soybean ps. *Nutrition* 15:778-783.

Blomqvist, C.G., and B. Saltin. 1983. Cardiovascular adaptations to physical training. *Annual Review of Physiology* 45:169-189.

Bloom, B., R.A. Cohen, and G. Freeman. 2011. Summary health statistics for U.S. children: National Health Interview Survey, 2010. *Vital Health Statistics* 10(250):1-80.

Bloom, S., R. Johnson, D. Park, M. Rennie, and W. Sulaiman. 1976. Differences in the metabolic and hormonal response to exercise between racing cyclists and untrained individuals. *Journal of Physiology* 258:1-18.

Blume, F.D., S.J. Boyer, L.E. Braverman, A. Cohen, J. Dirkse, and J.P. Mordes. 1984. Impaired osmoregulation at high altitude: Studies on Mt. Everest. *Journal of the American Medical Association* 252:524-526.

Bobbert, M.F., K.G.M. Gerritsen, M.C.A. Litjens, and A.J. Van Soest. 1996. Why is countermovement jump height greater than squat jump height? *Medicine and Science in Sports and Exercise* 28:1402-1412.

Bodine, S.C. 2006. mTOR signaling and the molecular adaptation to resistance exercise. *Medicine and Science in Sports and Exercise* 38:1950-1957.

Bogdanis, G.C., M.E. Nevill, L.H. Boobis, and H.K.A. Lakomy. 1996. Contribution of phosphocreatine and aerobic metabolism to energy supply during repeated sprint exercise. *Journal of Applied Physiology* 80:876-884.

Boirie, Y., M. Dangin, P. Gachon, M.P. Vasson, J.L. Maubois, and B. Beaufrere. 1997. Slow and fast dietary proteins differently modulate postprandial protein accretion. *Proceedings of the National Academy of Sciences U S A* 94:14930-14935.

Bompa, T.O. 1999. *Periodization: Theory and methodology of training.* Champaign, IL: Human Kinetics.

Boning, D., R. Beneke, and N. Maassen. 2005. Lactic acid still remains the real cause of exercise-induced metabolic acidosis. *American Journal of Physiology: Regulatory, Integrative and Comparative Physiology* 289:R902-R903.

Bonow, R.O. 1994. Left ventricular response to exercise. In *Cardiovascular response to exercise,* ed. G.F. Fletcher, 31-48. Mount Kisco, NY: Futura.

Boobis, L.H. 1987. Metabolic aspects of fatigue during sprinting. In *Exercise: Benefits, limitations and adaptations,* ed. D.R. Macleod, R. Maughan, M. Nimmo, T. Reilly, and C. Williams, 116-143. London: E&FN Spon.

Boone, J.B., C.P. Lambert, M.G. Flynn, T.J. Michaud, A. Rodriguez-Zayas, and F.F. Andres. 1990. Resistance exercise effects on plasma cortisol, testosterone and creatine kinase activity in anabolic-androgenic steroid users. *International Journal of Sports Medicine* 11:293-297.

Boots, A.W., G.R. Haenen, and A. Bast. 2008. Health effects of quercetin: From antioxidant to nutraceutical. *European Journal of Pharmacology* 585:325-337.

Boozer, C.N., P.A. Daly, P. Homel, J.L. Solomon, D. Blanchard, J.A. Nasser, R. Strauss, and T. Meredith. 2002. Herbal ephedra/caffeine for weight loss: A 6-month randomized safety and efficacy trial. *International Journal of Obesity and Related Metabolic Disorders* 26:593-604.

Borg, G.A.V. 1982. Psychophysical bases of perceived exertion. *Medicine and Science in Sports and Exercise* 14:377-381.

Boros-Hatfaludy, S., G. Fekete, and P. Apor. 1986. Metabolic enzyme activity in muscle biopsy samples in different athletes. *European Journal of Applied Physiology and Occupational Physiology* 55:334-338.

Borsheim, E., M.G. Cree, K.D. Tipton, T.A. Elliot, A. Aarsland, and R.R. Wolfe. 2004. Effect of carbohydrate intake on net muscle protein synthesis during recovery from resistance exercise. *Journal of Applied Physiology* 96:674-678.

Borsheim, E., K.D. Tipton, S.E. Wolf, and R.R. Wolfe. 2002. Essential amino acids and muscle protein recovery from resistance exercise. *American Journal of Physiology and Endocrinology* 283:E648-E657.

Bosco, C., and P.V. Komi. 1979. Mechanical characteristics and fiber composition of human leg extensor muscles. *European Journal of Applied Physiology* 24:21-32.

Bosco, C., P. Mognoni, and P. Luhtanen. 1983. Relationship between isokinetic performance and ballistic movement. *European Journal of Applied Physiology* 51:357-364.

Bosco, C., J.T. Viitalsalo, P.V. Komi, and P. Luhtanen. 1982. Combined effect of elastic energy and myoelectric potentiation during stretch-shortening cycle exercise. *Acta Physiologica Scandinavica* 114:557-565.

Bosco, C., and C. Vittori. 1986. Biomechanical characteristics of sprint running during maximal and supramaximal speed. *New Studies in Athletics* 1:39-45.

Bosquet, L., J. Montpetit, D. Arvisais, and I. Mujika. 2007. Effects of tapering on performance: A meta-analysis. *Medicine and Science in Sports and Exercise* 39:1358-1365.

Bottecchia, D., D. Bordin, and R. Martino. 1987. Effect of different kinds of physical exercise on the plasmatic testosterone level of normal adult males. *Journal of Sports Medicine and Physical Fitness* 27:1-5.

Boutellier, U., H. Howald, P.E. di Prampero, D. Giezendanner, and P. Cerretelli. 1983. Human muscle adaptations to chronic hypoxia. *Progress in Clinical Biology Research* 136:273-285.

Boyce, R.W., G.R. Jones, K.E. Schendt, C.L. Lloyd, and E.L. Boone. 2009. Longitudinal changes in strength of police officers with gender comparisons. *Journal of Strength and Conditioning Research* 23:2411-2418.

Brajkovic, D., and M.B. Ducharme. 2003. Finger dexterity, skin temperature, and blood flow during auxiliary heating in the cold. *Journal of Applied Physiology* 95:758-770.

Brajkovic, D., and M.B. Ducharme. 2006. Facial cold-induced vasodilatation and skin temperature during exposure to cold wind. *European Journal of Applied Physiology and Occupational Physiology* 96:711-721.

Brandenberger, G., V. Candas, M. Follenius, J.P. Libert, and J.M. Kahn. 1986. Vascular fluid shifts and endocrine responses to exercise in the heat. *European Journal of Applied Physiology* 55:123-129.

Brandsch, C., and K. Eder. 2002. Effect of L-carnitine on weight loss and body composition of rats fed a hypocaloric diet. *Annals of Nutrition and Metabolism* 46:205-210.

Brener, W., T.R. Hendrix, and P.R. McHugh. 1983. Regulation of the gastric emptying of glucose. *Gastroenterology* 85:76-82.

Brenner, M., J. Walberg-Rankin, and D. Sebolt. 2000. The effect of creatine supplementation during resistance training in women. *Journal of Strength and Conditioning Research* 14:207-213.

Briefel, R.R., and C.L. Johnson. 2004. Secular trends in dietary intake in the United States. *Annual Review of Nutrition* 24:401-431.

Brisswalter, J., C. Hausswirth, D. Smith, F. Vercruyssen, and J.M. Vallier. 2000. Energetically optimal cadence vs freely chosen cadence during cycling: Effect of exercise duration. *International Journal of Sports Medicine* 21:60-64.

Broeder, C.E. 2003. Oral andro-related prohormone supplementation: Do the potential risks outweigh the benefits? *Canadian Journal of Applied Physiology* 28:102-116.

Broeder, C.E., J. Quindry, K. Brittingham, L. Panton, J. Thomson, S. Appakondu, K. Breuel, R. Byrd, J. Douglas, C. Earnest, C. Mitchell, M. Olson, T. Roy, and C. Yarlagadda. 2000. The Andro Project: Physiological and hormonal influences of androstenedione supplementation in men 35 to 65 years old participating in a high-intensity resistance training program. *Archives of Internal Medicine* 160:3093-3104.

Brooks, G.A. 1986. The lactate shuttle during exercise and recovery. *Medicine and Science in Sports and Exercise* 18:360-368.

Brooks, G.A. 1991. Current concepts in lactate exchange. *Medicine and Science in Sports and Exercise* 23:895-906.

Brooks, G.A. 2009. Cell-cell and intracellular lactate shuttles. *Journal of Physiology* 587:5591-5600.

Brooks, G.A., H. Dubouchaud, M. Brown, J.P. Sicurello, and C.E. Butz. 1999. Role of mitochondrial lactic dehydrogenase and lactate oxidation in the "intra-cellular lactate shuttle." *Proceedings of the National Academy of Sciences U S A* 96:1129-1134.

Brooks, S., J. Burrin, M.E. Cheetham, G.M. Hall, T. Yeo, and C. Williams. 1988. The responses of the catecholamines and b-endorphin to brief maximal exercise in man. *European Journal of Applied Physiology* 57:230-234.

Brooks, S., M.E. Nevill, L. Meleagros, H.K.A. Lakomy, G.M. Hall, S.R. Bloom, and C. Williams. 1990. The hormonal responses to repetitive brief maximal exercise in humans. *European Journal of Applied Physiology* 60:144-148.

Brown, C.M., J.C. McGrath, J.M. Midgley, A.G. Muir, J.W. O'Brien, C.M. Thonoor, C.M. Williams, and V.G. Wilson. 1988. Activities of octopamine and synephrine stereoisomers on alpha-adrenoceptors. *British Journal of Pharmacology* 93:417-429.

Brown, G.A., E.R. Martini, B.S. Roberts, M.D. Vukovich, and D.S. King. 2002. Acute hormonal response to sublingual androstenediol intake in young men. *Journal of Applied Physiology* 92:142-146.

Brown, G.A., and D. McKenzie. 2006. Acute resistance exercise does not change the hormonal response to sublingual and androstenediol intake. *European Journal of Applied Physiology* 97:404-412.

Brown, G.A., M. Vukovich, and D.S. King. 2006. Testosterone prohormone supplements. *Medicine and Science in Sports and Exercise* 38:1451-1461.

Brown, G.A., M.D. Vukovich, E.R. Martini, M.L. Kohut, W.D. Franke, D.A. Jackson, and D.S. King. 2000. Endocrine responses to chronic androstenedione intake in 30- to 56-year-old men. *Journal of Clinical Endocrinology and Metabolism* 85:4074-4080.

Brown, G.A., M.D. Vukovich, T.A. Reifenrath, N.L. Uhl, K.A. Parsons, R.L. Sharp, and D.S. King. 2000. Effects of anabolic precursors on serum testosterone concentrations and adaptations to resistance training in young men. *International Journal of Sport Nutrition and Exercise Metabolism* 10:340-359.

Brown, G.A., M.D. Vukovich, R.L. Sharp, T.A. Reifenrath, K.A. Parsons, and D.S. King. 1999. Effect of oral DHEA on serum testosterone and adaptations to resistance training in young men. *Journal of Applied Physiology* 87:2274-2283.

Brown, L.E., and A.V. Khamoui. 2012. Agility training. In *NSCA's guide to program design,* ed. J.R. Hoffman, 143-164. Champaign, IL: Human Kinetics.

Brown, M.E., J.L. Mayhew, and L.W. Boleach. 1986. Effect of plyometric training on vertical jump performance in high school basketball players. *Journal Sports Medicine and Physical Fitness* 26:1-4.

Bruce, R.A., F. Kusumi, and D. Hosmer. 1973. Maximal oxygen uptake and nomographic assessment of functional aerobic impairment in cardiovascular disease. *American Heart Journal* 85:546-562.

Brzycki, M. 1993. Strength testing: Predicting a one-rep max from reps to fatigue. *Journal of Health, Physical Education, Recreation and Dance* 64:88-90.

Buchman, A.L., M. Sohel, M. Brown, D.J. Jenden, C. Ahn, M. Roch, and T.L. Brawley. 2001. Verbal and visual memory improve after choline supplementation in long-term total parenteral nutrition: A pilot study. *Journal of Parenteral and Enteral Nutrition* 25:30-35.

Buford, T.W., S.J. Rossi, D.B. Smith, and A.J. Warren. 2007. A comparison of periodization models during nine weeks with equated volume and intensity for strength. *Journal of Strength and Conditioning Research* 21:1245-1250.

Buick, F.J., N. Gledhill, A.B. Froese, L. Spriet, and E.C. Meyers. 1980. Effect of induced erythrocythemia on aerobic work capacity. *Journal of Applied Physiology* 48:636-642.

Bunt, J.C., R.A. Boileau, J.M. Bahr, and R.A. Nelson. 1986. Sex and training differences in human growth hormone during prolonged exercise. *Journal of Applied Physiology* 61:1796-1801.

Buono, M.J., J.E. Yeager, and J.A. Hodgdon. 1986. Plasma adrenocorticotropin and cortisol responses to brief high-intensity exercise in humans. *Journal of Applied Physiology* 61:1337-1339.

Burd, N.A., R.J. Andrews, D.W. West, J.P. Little, A.J. Cochran, A.J. Hector, J.G. Cashaback, M.J. Gibala, J.R. Potvin, S.K. Baker, and S.M. Phillips. 2012. Muscle time under tension during resistance exercise stimulates differential protein sub-fractional synthetic responses in men. *Journal of Physiology* 590:351-362.

Buresh, R., K. Berg, and J. French. 2009. The effect of resistive exercise rest interval on hormonal response, strength, and hypertrophy with training. *Journal of Strength and Conditioning Research* 23:62-71.

Burke, D.G., S. Silver, L.E. Holt, T. Smith-Palmer, C.J. Culligan, and P.D. Chilibeck. 2000. The effect of continuous low dose creatine supplementation on force, power, and total work. *International Journal of Sport Nutrition and Exercise Metabolism* 10:235-244.

Burke, L.M., G.R. Collier, and M. Hargreaves. 1993. Muscle glycogen storage after prolonged exercise: Effect of the glycemic index of carbohydrate feeding. *Journal of Applied Physiology* 75:1019-1023.

Burkett, L.N. 1970. Causative factors in hamstring strains. *Medicine and Science in Sports and Exercise* 2:39-42.

Buskirk, E.R., P.F. Iampietro, and D.E. Bass. 1958. Work performance after dehydration: Effects of physical conditioning and heat acclimatization. *Journal of Applied Physiology* 12:189-194.

Buskirk, E.R., J. Kollias, R.F. Akers, E.K. Prokop, and E.P. Reategui. 1967. Maximal performance at altitude and return from altitude in conditioned runners. *Journal of Applied Physiology* 23:259-266.

Butki, B.D., D.L. Rudolph, and H. Jacobsen. 2001. Self-efficacy, state anxiety, and cortisol responses to treadmill running. *Perceptual and Motor Skills* 92:1129-1138.

Cabasso, A. 1994. Peliosis hepatis in a young adult bodybuilder. *Medicine and Science in Sports and Exercise* 26:2-4.

Cadore, E.L., M. Izquierdo, M. Conceição, R. Radaelli, R.S. Pinto, B.M. Baroni, M.A. Vaz, C.L. Alberton, S.S. Pinto, G. Cunha, M. Bottaro, and L.F. Kruel. 2012. Echo intensity is associated with skeletal muscle power and cardiovascular performance in elderly men. *Experimental Gerontology* 47:473-478.

Caizzo, V.J., J.J. Perrine, and V.R. Edgerton. 1981. Training-induced alterations of the in vivo force-velocity relationship of human muscle. *Journal of Applied Physiology: Respiratory, Environmental and Exercise Physiology* 51:750-754.

Caldwell, J.E., E. Ahonen, and U. Nousiainen. 1984. Differential effects of sauna-, diuretic-, and exercise-induced hypohydration. *Journal of Applied Physiology: Respiratory, Environmental and Exercise Physiology* 57:1018-1023.

Calle, M.C., and M.L. Fernandez. 2010. Effects of resistance training on the inflammatory response. *Nutrition Research and Practice* 4:259-269.

Calof, O., A.B. Singh, M.L. Lee, R.J. Urban, A.M. Kenny, J.L. Tenover, and S. Bhasin. 2005. Adverse events associated with testosterone supplementation of older men. *Journal of Gerontology and Medical Science* 60:1451-1457.

Campaign, B.N., T.B. Gilliam, M.L. Spencer, R.M. Lampman, and M.A. Schork. 1984. Effects of a physical activity program on metabolic control and cardiovascular fitness in children with insulin dependent diabetes mellitus. *Diabetes Care* 7:57-62.

Campaign, B.N., H. Wallberg-Henriksson, and R. Gunnarsson. 1987. Glucose and insulin responses in relation to insulin dose and caloric intake 12 h after acute physical exercise in men with IDDM. *Diabetes Care* 10:716-721.

Campbell, B.I., M. Kilpatrick, C. Wilborn, P. La Bounty, B. Parker, B. Gomez, A. Elkins, S. Williams, and M.G. dos Santos. 2010. A commercially available energy drink does not improve peak power production on multiple 20-second Wingate tests. *Journal of the International Society of Sports Nutrition* 7(S1):10.

Campos, G.E.R., T.J. Luecke, H.K. Wendeln, K. Toma, F.C. Hagerman, T.F. Murray, K.E. Ragg, N.R. Ratamess, W.J. Kraemer, and R.S. Staron. 2002. Muscular adaptations in response to three different resistance-training regimens: Specificity of repetition maximum training zones. *European Journal of Applied Physiology* 88:50-60.

Canal, N., M. Franceschi, M. Alberoni, C. Castiglioni, P. De Moliner, and A. Longoni. 1991. Effect of L-alpha-glyceryl-phoshorylcholine on amnesia caused by scopolamine. *International Journal of Clinical Pharmacology, Therapy and Toxicology* 29:103-107.

Candow, D.G., P.D. Chilibeck, M. Facci, S. Abeysekara, and G.A. Zello. 2006. Protein supplementation before and after resistance training in older men. *European Journal of Applied Physiology* 97:548-556.

Cannell, J.J., B.W. Hollis, M. Zasloff, and R.P. Heaney. 2008. Diagnosis and treatment of vitamin D deficiency. *Expert Opinion on Pharmacotherapy* 9:107-118.

Cannon, J.G., S.N. Meydani, R.A. Fielding, M.A. Fiatarone, M. Meydani, N. Farhangmeh, S.F. Orencole, J.B. Blumberg, and W.J. Evans. 1991. Acute phase response in exercise II: Association with vitamin E, cytokines and muscle proteolysis. *American Journal of Physiology* 260:R1235-R1240.

Cannon, J.G., S.F. Orencole, R.A. Fielding, M. Meydani, S.N. Meydani, M.A. Fiatarone, J.B. Blumberg, and W.J. Evans. 1990. Acute phase response in exercise: Interaction of age and vitamin E on neutrophils and muscle enzyme release. *American Journal of Physiology* 259:R1214-R1219.

Cappon, J., J.A. Brasel, S. Mohan, and D.M. Cooper. 1994. Effect of brief exercise on circulating insulin-like growth factor I. *Journal of Applied Physiology* 76:2490-2496.

Cardinale, M., and M.H. Stone. 2006. Is testosterone influencing explosive performance? *Journal of Strength and Conditioning Research* 20:103-107.

Carlsen, K.H., S.D. Anderson, L. Bjermer, S. Bonini, V. Brusasco, W. Canonica, J. Cummiskey, L. Delgado, S.R. Del Giacco, F. Drobnic, T. Haahtela, K. Larsson, P. Palange, T. Popov, and P. van Cauwenberge. 2008. Treatment of exercise-induced asthma, respiratory and allergic disorders in sport and the relationship to doping: Part II of the report from the Joint Task Force of European Respiratory Society (ERS) and European Academy of Allergy and Clinical Immunology (EAACI) in cooperation with GA2LEN. *Allergy* 63:492-505.

Caro, J. 2001. Hypoxia regulation of gene transcription. *High Altitude Medicine and Biology* 2:145-154.

Caro, J.F., M.K. Sinha, S.M. Raju, O. Ittoop, W.J. Pories, E.G. Flickinger, D. Meelheim, and D. Dohm. 1987. Insulin receptor kinase in human skeletal muscle from obese subjects with and without noninsulin dependent diabetes. *Journal of Clinical Investigation* 79:1330-1337.

Carpene, C., J. Galitzky, E. Fontana, C. Atgie, M. Lafontan, and M. Berlan. 1999. Selective activation of beta3-adrenoreceptors by octopamine: Comparative studies in mammalian fat cells. *Naunyn Schmiedebergs Archives of Pharmacology* 359:310-321.

Carr, G. 1999. *Fundamentals of track and field*. Champaign, IL: Human Kinetics.

Carroll, J.F., V.A. Convertino, C.E. Wood, J.E. Graves, D.T. Lowenthal, and M.L. Pollack. 1995. Effect of training on blood volume and plasma hormone concentrations in the elderly. *Medicine and Science in Sports and Exercise* 27:79-84.

Carter, S.L., C.D. Rennie, S.J. Hamilton, and M.A. Tarnopolsky. 2001. Changes in skeletal muscle in males and females following endurance training. *Canadian Journal of Physiology and Pharmacology* 79:386-392.

Casa, D.J., S.A. Anderson, L. Baker, S. Bennett, M.F. Bergeron, D. Connolly, R. Courson, J.A. Drezner, E.R. Eichner, B. Epley, S. Fleck, R. Franks, K.M. Guskiewicz, K.G. Harmon, J.R. Hoffman, J.C. Holschen, J. Jost, A. Kinniburgh, D. Klossner, R.M. Lopez, G. Martin, B.P. McDermott, J.P. Mihalik, T. Myslinski, K. Pagnotta, S. Podda, G. Rodgers, A. Russell, L. Sales, D. Sandler, R.L. Stearns, C. Stiggins, and C. Thompson. 2012. The interassociation task force for preventing sudden death in collegiate conditioning sessions: Best practice recommendations. *Journal of Athletic Training* 47:477-480.

Casa, D.J., S.M. Becker, M.S. Ganio, C.M. Brown, S.W. Yeargin, M.W. Roti, J. Siegler, J.A. Blowers, N.R. Glaviano, R.A. Huggins, L.E. Armstrong, and C.M. Maresh. 2007. Validity of devices that assess body temperature during outdoor exercise in the heat. *Journal of Athletic Training* 42:333-342.

Casa, D.J., K.M. Guskiewicz, S.A. Anderson, R.W. Courson, J.F. Heck, C.C. Jimenez, B.P. McDermott, M.G. Miller, R.L. Stearns, E.E. Swartz, and K.M. Walsh. 2012. National athletic trainers' association position statement: Preventing sudden death in sports. *Journal of Athletic Training* 47:96-118.

Casa, D.J., C.M. Maresh, L.E. Armstrong, S.A. Kavouras, J.A. Herrera-Soto, F.T. Hacker, N.R. Keith, and T.A. Elliott. 2000. Intravenous versus oral rehydration during a brief period: Responses to subsequent exercise in the heat. *Medicine and Science in Sports and Exercise* 32:124-133.

Casa, D.J., B.P. McDermott, E.C. Lee, S.W. Yeargin, L.E. Armstrong, and C.M. Maresh. 2007. Cold water immersion: The gold standard for exertional heatstroke treatment. *Exercise and Sport Sciences Reviews* 35:141-149.

Casabona, A., M.C. Polizzi, and V. Perciavalle. 1990. Differences in H-reflex between athletes trained for explosive contractions and non-trained subjects. *European Journal of Applied Physiology and Occupational Physiology* 61:26-32.

Casaburi, R., S. Bhasin, L. Cosentino, J. Porszasz, A. Somfay, M.I. Lewis, M. Fournier, and T.W. Storer. 2004. Effects of testosterone and resistance training in men with chronic obstructive pulmonary disease. *American Journal of Respiratory and Critical Care Medicine* 170:870-878.

Casal, D.C., and A.S. Leon. 1985. Failure of caffeine to affect substrate utilization during prolonged running. *Medicine and Science in Sports and Exercise* 17:174-179.

Casavant, M.J., K. Blake, J. Griffith, A. Yates, and L.M. Copley. 2007. Consequences of use of anabolic androgenic steroids. *Pediatric Clinics of North America* 54:677-690.

Caselli, S., R. Di Pietro, F.M. Di Paolo, C. Pisicchio, B. di Giacinto, E. Guerra, F. Culasso, and A. Pelliccia. 2011. Left ventricular systolic performance is improved in elite athletes. *European Journal of Echocardiography* 12:514-519.

Castell, L.M., J.R. Poortmans, R. Leclercq, M. Brasseur, J. Duchateau, and E.A. Newsholme. 1997. Some aspects of the acute phase response after a marathon race, and the effects of glutamine supplementation. *International Journal of Sports Medicine* 75:47-53.

Castellani, J.W., C.M. Maresh, L.E. Armstrong, R.W. Kenefick, D. Riebe, M.E. Echegaray, D.J. Casa, and V.D. Castracane. 1997. Intravenous versus oral rehydration: Effects of subsequent exercise in the heat. *Journal of Applied Physiology* 82:799-806.

Castellani, J.W., A.J. Young, J.E. Kain, A. Rouse, and M. Sawka. 1999. Thermoregulation during cold exposure: Effects of prior exercise. *Journal of Applied Physiology* 87:247-252.

Cavanagh, P.R., and K.R. Williams. 1982. The effect of stride length variation on oxygen uptake during distance running. *Medicine and Science in Sports and Exercise* 14:30-35.

Centers for Disease Control and Prevention. 1983. Current trends hypothermia: United States. *Morbidity and Mortality Weekly Report* 32:46-48.

Centers for Disease Control and Prevention. 2011. *National diabetes fact sheet: National estimates and general information on diabetes and prediabetes in the United States, 2011*. Atlanta: U.S. Department of Health and Human Services, Centers for Disease Control and Prevention.

Cerretelli, P. 1987. Extreme hypoxia in air breathers. In *Comparative physiology of environmental adaptations*, ed. P. Dejours. Basel, Switzerland: Karger.

Cerretelli, P., C. Marconi, O. Deriaz, and D. Giezendanner. 1984. After-effects of chronic hypoxia on cardiac output and muscle blood flow at rest and exercise. *European Journal of Applied Physiology* 53:92-96.

Chandler, R.M., H.K. Byrne, J.G. Patterson, and J.L. Ivy. 1994. Dietary supplements affect the anabolic hormones after weight-training exercise. *Journal of Applied Physiology* 76:839-845.

Chang, F.E., W.G. Dodds, M. Sullivan, M.H. Kim, and W.B. Malarkey. 1986. The acute effects of exercise on prolactin and growth hormone secretion: Comparison between sedentary women and women runners with normal and abnormal menstrual cycles. *Journal of Clinical Endocrinology and Metabolism* 62:551-556.

Channell, B.T., and J.P. Barfield. 2008. Effect of Olympic and traditional resistance training on vertical jump improvement in high school boys. *Journal of Strength and Conditioning Research* 22:1522-1527.

Chanutin, A. 1926. The fate of creatine when administered to man. *Journal of Biology and Chemistry* 67:29-41.

Chapman, R.F., M. Emery, and J.M. Stager. 1999. Degree of arterial desaturation in normoxia influences the decline in $\dot{V}O_2$max in mild hypoxia. *Medicine and Science in Sports and Exercise* 31:658-663.

Chapman, R.F., and B.D. Levine. 2000. The effects of hypo- and hyperbaria on performance. In *Exercise and sport science*, ed. W.E. Garrett and D.T. Kirkendall, 447-458. Philadelphia: Lippincott Williams & Wilkins.

Chapman, R.F., J. Stray-Gundersen, and B.D. Levine. 1998. Individual variation in response to altitude training. *Journal of Applied Physiology* 85:1448-1456.

Charlton, G.A., and M.H. Crawford. 1997. Physiological consequences of training. *Cardiology Clinics* 15:345-354.

Chatard, J.C., D. Chollet, and G. Millet. 1998. Performance and drag during drafting swimming in highly trained triathletes. *Medicine and Science in Sports and Exercise* 30:1276-1280.

Chattoraj, S.C., and N.B. Watts. 1987. Endocrinology. In *Fundamentals of clinical chemistry*, ed. N.W. Tietz, 175-180. Philadelphia: Saunders.

Cheetham, M.E., L.H. Boobis, S. Brooks, and C. Williams. 1986. Human muscle metabolism during sprint running. *Journal of Applied Physiology* 61:54-60.

Chesley, A., G.J. Heigenhauser, and L.L. Spriet. 1996. Regulation of muscle glycogen phosphorylase activity following short-term endurance training. *American Journal of Physiology* 270:E328-E335.

Chester, N., T. Reilly, and D.R. Mottram. 2003. Physiological, subjective and performance effects of pseudoephedrine and phenylpropanolamine during endurance running exercise. *International Journal of Sports Medicine* 24:3-8.

Cheung, S.S. 2010. *Advanced environmental exercise physiology*, 69-88. Champaign, IL: Human Kinetics.

Cheung, S.S., and T.M. McLellan. 1998. Heat acclimation, aerobic fitness, and hydration effects on tolerance during uncompensable heat stress. *Journal of Applied Physiology* 84:1731-1739.

Cheung, S.S., D.L. Montie, M.D. White, and D. Behm. 2003. Changes in manual dexterity following short-term hand and forearm immersion in 10 degrees C water. *Aviation, Space, and Environmental Medicine* 74:990-993.

Chu, D. 1992. *Jumping into plyometrics*. Champaign, IL: Human Kinetics.

Chu, K.S., T.J. Doherty, G. Parise, J.S. Milheiro, and M.A. Tarnopolsky. 2002. A moderate dose of pseudoephedrine does not alter muscle contraction strength or anaerobic power. *Clinical Journal of Sport Medicine* 12:387-390.

Chun, O.K., S.J. Chung, and W.O. Song. 2007. Estimated dietary flavonoid intake and major food sources of U.S. adults. *Journal of Nutrition* 137:1244-1252.

Claremont, A.D., F. Nagle, W.D. Reddan, and G.A. Brooks. 1975. Comparison of metabolic, temperature, heart rate and ventilatory responses to exercise at extreme ambient temperatures (0° and 35°C). *Medicine and Science in Sports and Exercise* 7:150-154.

Clarke, N., and U.M. Kabadi. 2004. Optimizing treatment of hypothyroidism. *Treatments in Endocrinology* 3:217-221.

Clarkson, P.M. 1991. Minerals, exercise performance and supplementation in athletes. *Journal of Sports Sciences* 9:91-116.

Clarkson, P.M. 1997. Eccentric exercise and muscle damage. *International Journal of Sports Medicine* 18(suppl):S314-S316.

Clauson, K.A., K.M. Shields, C.E. McQueen, and N. Persad. 2003. Safety issues associated with commercially available energy drinks. *Journal of the American Pharmaceutical Association* 48:e55-e63.

Coast, J.R., P.S. Clifford, T.W. Henrich, J. Stray-Gundersen, and R.L. Johnson. 1990. Maximal inspiratory pressure following maximal exercise in trained and untrained subjects. *Medicine and Science in Sports and Exercise* 22:811-815.

Coffey, V.G., B. Jemiolo, J. Edge, A.P. Garnham, S.W. Trappe, and J.A. Hawley. 2009. Effect of consecutive repeated sprint and resistance exercise bouts on acute adaptive responses in human skeletal muscle. *American Journal of Physiology: Regulatory, Integrative and Comparative Physiology* 297:R1441-R1451.

Coffey, V.G., Z. Zhong, A. Shield, B.J. Canny, A.V. Chibalin, J.R. Zierath, and J.A. Hawley. 2006. Early signaling responses to divergent exercise stimuli in skeletal muscle from well-trained humans. *FASEB Journal* 20:190-192.

Coggan, A.R., and E.F. Coyle. 1989. Metabolism and performance following carbohydrate ingestion late in exercise. *Medicine and Science in Sports and Exercise* 21:59-65.

Coggan, A.R., and E.F. Coyle. 1991. Carbohydrate ingestion during prolonged exercise: Effects on metabolism and performance. *Exercise and Sport Sciences Reviews* 19:1-40.

Cohen, J., R. Collins, J. Darkes, and D. Gwartney. 2007. A league of their own: Demographics, motivations and patterns of use of 1,955 male adult non-medical anabolic steroid users in the United States. *Journal of the International Society of Sports Nutrition* 11:4-12.

Cohen, P.A. 2012. Assessing supplement safety—the FDA's controversial proposal. *New England Journal of Medicine* 366:389-391.

Colberg, S.R., A.L. Albright, B.J. Blissmer, B. Braun, L. Chasan-Taber, B. Fernhall, J.G. Regensteiner, R.R. Rubin, and R.J. Sigal. 2010. Exercise and type 2 diabetes: American College of Sports Medicine and the American Diabetes Association: Joint position statement. *Medicine and Science in Sports and Exercise* 42:2282-2303.

Collins, M.A., and T.K. Snow. 1993. Are adaptations to combined endurance and strength training affected by the sequence of training? *Journal of Sports Sciences* 11:485-491.

Conley, D.L., and G. Krahenbuhl. 1980. Running economy and distance running performance of highly trained athletes. *Medicine and Science in Sports and Exercise* 12:357-360.

Consitt, L.A., J.L. Copeland, and M.S. Tremblay. 2001. Hormone responses to resistance vs. endurance exercise in premenopausal females. *Canadian Journal of Applied Physiology* 26:574-587.

Constantini, N.W., R. Arieli, G. Chodick, and G. Dubnov-Raz. 2010. High prevalence of vitamin D insufficiency in athletes and dancers. *Clinical Journal of Sport Medicine* 20:368-371.

Convertino, V.A. 1991. Blood volume: Its adaptation to endurance training. *Medicine and Science in Sports and Exercise* 23:1338-1348.

Convertino, V.A., L.E. Armstrong, E.F. Coyle, G.W. Mack, M.N. Sawka, L.C. Senay, and W.M. Sherman. 1996. American College of Sports Medicine position stand: Exercise and fluid replacement. *Medicine and Science in Sports and Exercise* 28:i-vii.

Convertino, V.A., L.C. Keil, E.M. Bernauer, and J.E. Greenleaf. 1981. Plasma volume, osmolality, vasopressin, and renin activity during graded exercise in man. *Journal of Applied Physiology: Respiratory, Environmental and Exercise Physiology* 50:123-128.

Convertino, V.A., L.C. Keil, and J.E. Greenleaf. 1983. Plasma volume, renin and vasopressin responses to graded exercise after training. *Journal of Applied Physiology: Respiratory, Environmental and Exercise Physiology* 54:508-514.

Cook, E.E., V.L. Gray, E. Savinar-Nogue, and J. Medeiros. 1987. Shoulder antagonistic strength ratios: A comparison between college-level baseball pitchers and nonpitchers. *Journal of Orthopaedic and Sports Physical Therapy* 8:451-460.

Cooke, W.H., P.W. Grandjean, and W.S. Barnes. 1995. Effect of oral creatine supplementation on power output and fatigue during bicycle ergometry. *Journal of Applied Physiology* 78:670-673.

Cormie, P., G.O. McCaulley, and J.M. McBride. 2007. Power versus strength-power jump squat training: Influence on the load-power relationship. *Medicine and Science in Sports and Exercise* 39:996-1003.

Cormie, P., G.O. McCaulley, N.T. Triplett, and J.M. McBride. 2007. Optimal loading for maximal power output during lower-body resistance exercises. *Medicine and Science in Sports and Exercise* 39:340-349.

Cormie, P., M.R. McGuigan, and R.U. Newton. 2010. Adaptations in athletic performance after ballistic power versus strength training. *Medicine and Science in Sports and Exercise* 42:1582-1598.

Cormie, P., M.R. McGuigan, and R.U. Newton. 2011. Developing maximal neuromuscular power. Part 1—biological basis of maximal power production. *Sports Medicine* 41:17-38.

Cornelissen, V.A., and R.H. Fagard. 2005. Effects of endurance training on blood pressure, blood pressure-regulating mechanisms, and cardiovascular risk factors. *Hypertension* 46:667-675.

Cornelissen, V.A., R.H. Fagard, E. Coeckelberghs, and L. Vanhees. 2011. Impact of resistance training on blood pressure and other cardiovascular risk factors: A meta-analysis of randomized, controlled trials. *Hypertension* 58:950-958.

Cornwell, A., A.G. Nelson, G.D. Heise, and B. Sidaway. 2001. Acute effects of passive muscle stretching on vertical jump performance. *Journal of Human Movement Studies* 40:307-324.

Cortiella, J., D.E. Mathews, R.A. Hoerr, D.M. Bier, and V.R. Vernon. 1988. Leucine kinetics at graded intakes in young men: Quantitative fate of dietary leucine. *American Journal of Clinical Nutrition* 48:998-1009.

Costill, D.L. 1988. Carbohydrates for exercise: Dietary demands for optimal performance. *International Journal of Sports Medicine* 9:1-18.

Costill, D.L., P. Cleary, W.J. Fink, C. Foster, J.L. Ivy, and F. Witzmann. 1979. Training adaptations in skeletal muscle of juvenile diabetics. *Diabetes* 28:818-822.

Costill, D.L., R. Cote, and W.J. Fink. 1976. Muscle water and electrolytes following varied levels of dehydration in man. *Journal of Applied Physiology* 40:6-11.

Costill, D.L., R. Cote, W.J. Fink, and P. Van Handel. 1981. Muscle water and electrolyte distribution during prolonged exercise. *International Journal of Sports Medicine* 2:130-134.

Costill, D.L., E. Coyle, G. Dalsky, W. Evans, W.J. Fink, and D. Hoopes. 1977. Effects of elevated plasma FFA and insulin on muscle glycogen usage during exercise. *Journal of Applied Physiology* 43:695-699.

Costill, D.L., E.F. Coyle, W.J. Fink, G.R. Lesmes, and F.A. Witzmann. 1979. Adaptations in skeletal muscle following strength training. *Journal of Applied Physiology* 46:96-99.

Costill, D.L., G.P. Dalsky, and W.J. Fink. 1978. Effects of caffeine ingestion on metabolism and exercise performance. *Medicine and Science in Sports and Exercise* 10:155-158.

Costill, D.L., W.J. Fink, and M.L. Pollock. 1976. Muscle fiber composition and enzyme activities of elite distance runners. *Medicine and Science in Sports and Exercise* 8:96-100.

Costill, D.L., M.G. Flynn, J.P. Kirwin, J.A. Houmard, J.B. Mitchell, R. Thomas, and S.H. Park. 1988. Effects of repeated days of intensified training on muscle glycogen and swimming performance. *Medicine and Science in Sports and Exercise* 20:249-254.

Costill, D.L., and E.L. Fox. 1969. Energetics of marathon running. *Medicine and Science in Sports and Exercise* 1:81-86.

Costill, D.L., and B. Saltin. 1974. Factors limiting gastric emptying during rest and exercise. *Journal of Applied Physiology* 37:679-683.

Costill, D.L., R. Thomas, R.A. Robergs, D.D. Pascoe, C.P. Lampert, S.I. Barr, and W.J. Fink. 1991. Adaptations to swimming training: Influence of training volume. *Medicine and Science in Sports and Exercise* 23:371-377.

Costill, D.L., F. Verstappen, H. Kuipers, E. Jansson, and W. Fink. 1984. Acid-base balance during repeated bouts of exercise: Influence of HCO_3^-. *International Journal of Sports Medicine* 5:228-231.

Costrini, A. 1990. Emergency treatment of exertional heatstroke and comparison of whole body cooling techniques. *Medicine and Science in Sports and Exercise* 22:15-18.

Cottrell, G.T., J.R. Coast, and R.A. Herb. 2002. Effect of recovery interval on multiple-bout sprint cycling performance after acute creatine supplementation. *Journal of Strength and Conditioning Research* 16:109-116.

Courson, R. 2007. Preventing sudden death on the athletic field: The emergency action plan. *Current Sports Medicine Report* 6:93-100.

Coutts, A.J., and R. Duffield. 2010. Validity and reliability of GPS devices for measuring movement demands of team sports. *Journal of Science and Medicine in Sport* 13:133-135.

Coutts, A.J., J. Quinn, J. Hocking, C. Castagna, and E. Rampinini. 2010. Match running performance in elite Australian rules football. *Journal of Science and Medicine in Sport* 13:543-548.

Covassin, T., D. Stearne, and R. Elbin III. 2008. Concussion history and postconcussion neurocognitive performance and symptoms in collegiate athletes. *Journal of Athletic Training* 43:119-124.

Cowan, D.A., A.T. Kicman, C.J. Walker, and M.J. Wheeler. 1991. Effect of administration of human chorionic gonadotropin on criteria used to assess testosterone administration in athletes. *Journal of Endocrinology* 131:147-154.

Cox, A.J., D.B. Pyne, M. Gleeson, and R. Callister. 2009. Relationship between C-reactive protein concentration and cytokine response to exercise in healthy and illness-prone runners. *European Journal of Applied Physiology* 107:611-614.

Cox, A.J., D.B. Pyne, P.U. Saunders, R. Callister, and M. Gleeson. 2007. Cytokine response to treadmill running in healthy and illness-prone athletes. *Medicine and Science in Sports and Exercise* 39:1918-1926.

Cox, G.R., B. Desbrow, P.G. Montgomery, M.E. Anderson, C.R. Bruce, T.A. Macrides, D.T. Martin, A. Moquin, A. Roberts, J.A. Hawley, and L.M. Burke. 2002. Effect of different protocols of caffeine intake on metabolism and endurance performance. *Journal of Applied Physiology* 93:990-999.

Cox, G., and D.G. Jenkins. 1994. The physiological and ventilatory responses to repeated 60s sprints following sodium citrate ingestion. *Journal of Sports Science* 12:469-475.

Cox, G., I. Mujika, D. Tumilty, and L. Burke. 2002. Acute creatine supplementation and performance during a field test simulating match play in elite female soccer players. *International Journal of Sport Nutrition and Exercise Metabolism* 12:33-46.

Coyle, E.F., and J. Gonzalez-Alonso. 2001. Cardiovascular drift during prolonged exercise: New perspectives. *Exercise and Sport Sciences Reviews* 29:88-92.

Coyle, E.F., M.K. Hemmert, and A.R. Coggan. 1986. Effects of detraining on cardiovascular response to exercise: Role of blood volume. *Journal of Applied Physiology* 60:95-99.

Coyle, E.F., W.H. Martin III, S.A. Bloomfield, O.H. Lowry, and J.O. Holloszy. 1985. Effects of detraining on responses to submaximal exercise. *Journal of Applied Physiology* 59:853-859.

Coyle, E.F., W.H. Martin III, D.R. Sinacore, M.J. Joyner, J.M. Hagberg, and J.O. Holloszy. 1984. Time course for loss of adaptation after stopping prolonged intense endurance training. *Journal of Applied Physiology* 57:1857-1864.

Craig, B.W., G.T. Hammons, S.M. Garthwite, L. Jarett, and J.O. Holloszy. 1981. Adaptations of fat cells to exercise: Response of glucose uptake and oxidation to insulin. *Journal of Applied Physiology* 51:1500-1506.

Craig, B.W., and H.Y. Kang. 1994. Growth hormone release following single versus multiple sets of back squats: Total work versus power. *Journal of Strength and Conditioning Research* 8:270-275.

Craig, F.N., and E.G. Cummings. 1966. Dehydration and muscular work. *Journal of Applied Physiology* 21:670-674.

Craig, S.A.S. 2004. Betaine in human nutrition. *American Journal of Clinical Nutrition* 80:539-549.

Cramer, J.T., and J.W. Coburn. 2004. Fitness testing protocols and norms. In *NSCA's essentials of personal training,* ed. R.W. Earle and T.R. Baechle, 217-264. Champaign, IL: Human Kinetics.

Cribb, P.J., and A. Hayes. 2006. Effects of supplement timing and resistance exercise on skeletal muscle hypertrophy. *Medicine and Science in Sports and Exercise* 38:1918-1925.

Crowley, R., and L.H. Fitzgerald. 2006. The impact of cGMP compliance on consumer confidence in dietary supplement products. *Toxicology* 221:9-16.

Cunniffe, B., W. Proctor, J.S. Baker, and B. Davies. 2009. An evaluation of the physiological demands of elite rugby union using global positioning system tracking software. *Journal of Strength and Conditioning Research* 23:1195-1203.

Cunningham, D.A., and J.A. Faulkner. 1969. The effect of training on aerobic and anaerobic metabolism during a short exhaustive run. *Medicine and Science in Sports and Exercise* 1:65-69.

Cunningham, D.A., D.H. Paterson, C.J. Blimkie, and A.P. Donner. 1984. Development of cardiorespiratory function in circumpubertal boys: A longitudinal study. *Journal of Applied Physiology* 56:302-307.

Currell, K., and A.E. Jeukendrup. 2008. Superior endurance performance with ingestion of multiple transportable carbohydrates. *Medicine and Science in Sports and Exercise* 40:275-281.

Daanen, H.A.M. 2003. Finger cold-induced vasodilation. *European Journal of Applied Physiology and Occupational Physiology* 89:411-426.

Dacaranhe, C.D., and J. Terao. 2001. A unique antioxidant activity of phosphatidylserine on iron-induced lipid peroxidation of phospholipid bilayers. *Lipids* 36:1105-1110.

Dahl-Jorgensen, K., H.D. Meen, K.F. Hanssen, and O. Asgenaes. 1980. The effect of exercise on diabetic control and hemoglobin A_1 (HbA_1) in children. *Acta Paediatrica Scandinavica* 283(suppl):53-56.

Dalbo, V.J., M.D. Roberts, J.R. Stout, and C.M. Kerksick. 2008. Acute effects of ingesting a commercial thermogenic drink on changes in energy expenditure and markers of lipolysis. *Journal of the International Society of Sports Nutrition* 5:6.

Dalrymple, K.J., S.E. Davis, G.B. Dwyer, and G.L. Moir. 2010. Effect of static and dynamic stretching on vertical jump performance in collegiate women volleyball players. *Journal of Strength and Conditioning Research* 24:149-155.

Daly, R.C., T.P. Su, P.J. Schmidt, M. Pagliaro, D. Pickar, and D.R. Rubinow. 2003. Neuroendocrine and behavioral effects of high-dose anabolic steroid administration in male normal volunteers. *Psychoneuroendocrinology* 28:317-331.

Daneshvar, D.H., C.J. Nowinski, A. McKee, and R.C. Cantu. 2011. The epidemiology of sport-related concussion. *Clinics in Sports Medicine* 30:1-17.

Dangin, M., Y. Boirie, C. Guillet, and B. Beaufrere. 2002. Influence of the protein digestion rate on protein turnover in young and elderly subjects. *Journal of Nutrition* 132:3228S-3233S.

Daniels, J.T. 1985. A physiologist's view of running economy. *Medicine and Science in Sports and Exercise* 17:332-338.

Danielsson, U. 1996. Windchill and the risk of tissue freezing. *Journal of Applied Physiology* 81:2666-2673.

Danzl, D.F., R.S. Pozos, and M.P. Hamlet. 1995. Accidental hypothermia. In *Wilderness medicine: Management of wilderness and environmental emergencies,* ed. P.S. Auerbach, 51-103. St. Louis: Mosby.

Darden, E. 1983. The facts about anabolic steroids. *Athletic Journal* (March):100-101.

Dash, A.K., and A. Sawhney. 2002. A simple LC method with UV detection for the analysis of creatine and creatinine and its application to several creatine formulations. *Journal of Pharmacology and Biomedical Analysis* 29:939-945.

Davies, C.T.M., I.K. Mecrow, and M.J. White. 1982. Contractile properties of the human triceps surae with some observations on the effects of temperature and exercise. *European Journal of Applied Physiology* 49:255-269.

Davies, C.T.M., and K. Young. 1983a. Effect of temperature on contractile properties and muscle power of triceps surae in humans. *Journal of Applied Physiology: Respiratory, Environmental and Exercise Physiology* 55:191-195.

Davies, C.T.M., and K. Young. 1983b. Effects of training at 30 and 100% maximal isometric force on the contractile properties of the triceps surae of man. *Journal of Physiology* 336:22-23.

Davis, J.A., S. Dorado, K.A. Keays, K.A. Reigel, K.S. Valencia, and P.H. Pham. 2007. Reliability and validity of the lung volume measurement made by the BOD POD body composition system. *Clinical Physiology and Functional Imaging* 27:42-46.

Davis, J.A., M.H. Frank, B.J. Whipp, and K. Wasserman. 1979. Anaerobic threshold alterations caused by endurance training in middle-aged men. *Journal of Applied Physiology* 46:1039-1046.

Davis, J.M., E.A. Murphy, M.D. Carmichael, and B. Davis. 2009. Quercetin increases brain and muscle mitochondrial biogenesis and exercise tolerance. American Journal of Physiology Regulatory Integrative Comparative Physiology 296:R1071-1077.

Dawson, B., M. Cutler, A. Moody, S. Lawrence, C. Goodman, and N. Randall. 1995. Effects of oral creatine loading on single and repeated maximal short sprints. *Australian Journal of Science and Medicine in Sport* 27:56-61.

Dayton, W.R., and M.R. Hathaway. 1989. Autocrine, paracrine, and endocrine regulation of myogenesis. In *Animal growth regulation,* ed. D.R. Campion, G.J. Hausman, and R.J. Martin, 69-90. New York: Lenum Press.

De Hon, O., and B. Coumans. 2007. The continuing story of nutritional supplements and doping infractions. *British Journal of Sports Medicine* 41:800-805.

Dela, F., J.J. Larson, K.J. Mikines, and H. Galbo. 1995. Normal effect of insulin to stimulate leg blood flow in NIDDM. *Diabetes* 44:221-226.

Del Balso, C., and E. Cafarelli. 2007. Adaptations in the activation of human skeletal muscle induced by short-term isometric resistance training. *Journal of Applied Physiology* 103:402-411.

Del Coso, J., E. Estevez, and R. Mora-Rodriguez. 2009. Caffeine during exercise in the heat: Thermoregulation and fluid-electrolyte balance. *Medicine and Science in Sports and Exercise* 41:164-173.

Deldicque, L., J. Decombaz, H. Zbinden Foncea, J. Vuichoud, J.R. Poortmans, and M. Francaux. 2008. Kinetics of creatine ingested as a food ingredient. *European Journal of Applied Physiology* 102:133-143.

Delecluse, C. 1997. Influence of strength training on sprint running performance: Current findings and implications for training. *Sports Medicine* 24:147-156.

Delecluse, C., H.V. Coppenolle, E. Willems, M.V. Leemputte, R. Diels, and M. Goris. 1995. Influence of high-resistance and high-velocity training on sprint performance. *Medicine and Science in Sports and Exercise* 27:1203-1209.

Delecluse, C., M. Roelants, R. Diels, E. Koninckx, and S. Vershcueren. 2005. Effects of whole body vibration training on muscle strength and sprint performance in sprint-training athletes. *International Journal of Sports Medicine* 26:662-668.

del Favero, S., H. Roschel, G. Artioli, C. Ugrinowitsch, V. Tricoli, A. Costa, R. Barroso, A.L. Negrelli, M.C. Otaduy, C. da Costa Leite, A.H. Lancha-Junior, and B. Gualano. 2012. Creatine but not betaine supplementation increases muscle phosphorylcreatine content and strength performance. *Amino Acids* 42:2299-2305.

del Favero, S., H. Roschel, M.Y. Solis, A. Hayashi, G. Artioli, M.C. Otaduy, F.B. Benatti, R.C. Harris, J.A. Wise, C.C. Leite, R.M. Pereira, A.L. de Sa-Pinto, A.H. Lancha-Junior, and B. Gualano. 2012. Beta-alanine (Carnosyn™) supplementation in elderly subjects (60–80 years): Effects on muscle carnosine content and physical capacity. *Amino Acids* 43:49-56.

Delgado-Reyes, C.V., M.A. Wallig, and T.A. Garrow. 2001. Immunohistochemical detection of betaine-homocysteine S-methyltransferase in human, pig, and rat liver and kidney. *Archives of Biochemistry and Biophysics* 393:184-186.

Deligiannis, A., E. Zahopoulou, and K. Mandroukas. 1988. Echocardiographic study of cardiac dimensions and function in weight lifters and body builders. *Journal of Sports Cardiology* 5:24-32.

Dempsey, J.A., W.G. Reddan, M.L. Birnbaum, H.V. Forster, J.S. Thoden, R.F. Grover, and J. Rankin. 1971. Effects of acute though life-long hypoxic exposure on exercise pulmonary gas exchange. *Respiration Physiology* 13:62-89.

Dengel, D.R., P.G. Weyand, D.M. Black, and K.J. Cureton. 1992. Effect of varying levels of hypohydration on responses during submaximal cycling. *Medicine and Science in Sports and Exercise* 24:1096-1101.

De Palo, E.F., G. Antonelli, R. Gatti, S. Chiappin, P. Spinella, and E. Cappellin. 2008. Effects of two different types of exercise on GH/IGF axis in athletes. Is the free/total IGF-I ratio a new investigative approach? *Clinica Chimica Acta* 387:71-74.

Derave, W., M.S. Ozdemir, R.C. Harris, A. Pottier, H. Reyngoudt, K. Koppo, J.A. Wise, and E. Achten. 2007. Beta-alanine supplementation augments muscle carnosine content and attenuates fatigue during repeated isokinetic contraction bouts in trained sprinters. *Journal of Applied Physiology* 103:1736-1743.

Derman, R.J. 1995. Effects of sex steroids on women's health: Implications for practitioners. *American Journal of Medicine* 98(suppl):137S-143S.

Deschenes, M.R., J.A. Giles, R.W. McCoy, J.S. Volek, A.L. Gomez, and W.J. Kraemer. 2002. Neural factors account for strength decrements observed after short-term muscle unloading. *American Journal of Physiology: Regulatory, Integrative and Comparative Physiology* 282:R578-R583.

Deschenes, M.R., W.J. Kraemer, J.F. Crivello, C.M. Maresh, L.E. Armstrong, and J. Covault. 1995. The effects of different treadmill running programs on the muscle morphology of adult rats. *International Journal of Sports Medicine* 16:273-277.

Deschenes, M.R., W.J. Kraemer, C.M. Maresh, and J.F. Crivello. 1991. Exercise-induced hormonal changes and their effects upon skeletal tissue. *Sports Medicine* 12:80-93.

De Souza, E.O., V. Tricoli, C. Bueno Junior, M.G. Pereira, P.C. Brum, E.M. Oliveira, H. Roschel, M.S. Aoki, and C. Urginowitsch. 2013. The acute effects of strength, endurance and concurrent exercises on the Akt/mTOR/p70s6k1 and AMPK signaling pathway responses in rat skeletal muscle. *Brazilian Journal of Medical and Biological Research* 46:343-347.

DeSouza, M.J., C.M. Maresh, M.S. Maguire, W.J. Kraemer, G. Flora-Ginter, and K.L. Goetz. 1989. Menstrual status and plasma vasopressin, renin activity, and aldosterone exercise responses. *Journal of Applied Physiology* 67:736-743.

Despres, J.P., A. Nadeau, and C. Bouchard. 1988. Physical training and changes in regional adipose tissue distribution. *Acta Medica Scandinavica* 723(suppl):205-212.

Dessypris, A., K. Kuoppasalmi, and H. Adlercreutz. 1976. Plasma cortisol, testosterone, androstenedione and luteinizing hormone (LH) in a noncompetitive marathon run. *Journal of Steroid Biochemistry* 7:33-37.

Detopoulou, P., D.B. Panagiotakos, S. Antonopoulou, C. Pitsavos, and C. Stefanadis. 2008. Dietary choline and betaine intakes in relation to concentrations of inflammatory markers in healthy adults: The ATTICA study. *American Journal of Clinical Nutrition* 87:424-430.

Deuster, P.A., A. Singh, R. Coll, D.E. Hyde, and W.J. Becker. 2002. Choline ingestion does not modify physical or cognitive performance. *Military Medicine* 167:1020-1025.

Devlin, J.T., J. Calles-Escandon, and E.S. Horton. 1986. Effects of preexercise snack feeding on endurance cycle exercise. *Journal of Applied Physiology* 60:980-985.

Deyssig, R., H. Frisch, W.F. Blum, and T. Waldhor. 1993. Effect of growth hormone treatment on hormonal parameters, body composition and strength in athletes. *Acta Endocrinology (Copenhagen)* 128:313-318.

Dhar, R., C.W. Stout, M.S. Link, M.K. Homoud, J. Weinstock, and N.A. Estes. 2005. Cardiovascular toxicities of performance-enhancing substances in sports. *Mayo Clinic Proceedings* 80:1307-1315.

Dickerman, R.D., R.M. Pertusi, N.Y. Zachariah, D.R. Dufour, and W.J. McConathy. 1999. Anabolic steroid-induced hepatotoxicity: Is it overstated? *Clinical Journal of Sport Medicine* 9:34-39.

Dickerman, R.D., F. Schaller, I. Prather, and W.J. McConathy. 1995. Sudden cardiac death in a 20-year-old bodybuilder using anabolic steroids. *Cardiology* 86:172-173.

Dickinson, J.W., G.P. Whyte, A.K. McConnell, and M.G. Harries. 2005. Impact of changes in the IOC-MC asthma criteria: A British perspective. *Thorax* 60:629-632.

Dickinson, J.W., G.P. Whyte, A.K. McConnell, and M.G. Harries. 2006. Screening elite winter athletes for exercise induced asthma: A comparison of three challenge methods. *British Journal of Sports Medicine* 40:179-183.

Diepvens, K., K.R. Westerterp, and M.S. Westerterp-Plantenga. 2007. Obesity and thermogenesis related to the consumption of caffeine, ephedrine, capsaicin, and green tea. *American Journal of Physiology: Regulatory, Integrative and Comparative Physiology* 292:R77-R85.

Dimick, D.F., M. Heron, E.E. Baulieu, and M.F. Jayle. 1961. A comparative study of the metabolic fate of testosterone, 17 alpha-methyl-testosterone. 19-nor-testosterone. 17 alpha-methyl-19-nor-testosterone and 17 alpha-methylestr-5(10)-ene-17 beta-ol-3-one in normal males. *Clinica Chimica Acta* 6:63-71.

Dimri, G.P., M.S. Malhotra, J. Sen Gupta, T.S. Kumar, and B.S. Arora. 1980. Alterations in aerobic-anaerobic proportions of metabolism during work in heat. *European Journal of Applied Physiology* 45:43-50.

Ding, E.L., Y. Song, V.S. Malik, and S. Liu. 2006. Sex differences of endogenous sex hormones and risk of type 2 diabetes: A systematic review and meta-analysis. *Journal of the American Medical Association* 295:1288-1299.

Dintiman, G., and B. Ward. 2003. *Sports speed,* 3rd ed. Champaign, IL: Human Kinetics.

Di Paolo, M., M. Agozzino, C. Toni, A.B. Luciani, L. Molendini, M. Scaglione, F. Inzani, M. Pasotti, F. Buzzi, and E. Arbustini. 2007. Sudden anabolic steroid abuse-related death in athletes. *International Journal of Cardiology* 114:114-117.

Di Pasquali, M. 1992. Clenbuterol: A new anabolic drug. *Drugs in Sports* 1:8-11.

DiPerri, R., G. Coppola, L.A. Ambrosio, A. Grasso, F.M. Puca, and M. Rizzo. 1991. A multicentre trial to evaluate the efficacy and tolerability of alpha-glycerylphosphorylcholine versus cystosine diphosphocholine in patients with vascular dementia. *Journal of International Medical Research* 19:330-341.

Dodge, T.L., and J.J. Jaccard. 2006. The effect of high school sports participation on the use of performance-enhancing substances in young adulthood. *Journal of Adolescent Health* 39:367-372.

Doherty, M., and P.M. Smith. 2005. Effects of caffeine ingestion on rating of perceived exertion during and after exercise: A meta-analysis. *Scandinavian Journal of Medicine and Science in Sports* 15:69-78.

Doherty, T.J. 2003. Invited review: Aging and sarcopenia. *Journal of Applied Physiology* 95:1717-1727.

Dohm, G.L., M.K. Sinha, and J.F. Caro. 1987. Insulin receptor binding and protein kinase activity in muscle of trained rats. *American Journal of Physiology* 252:E170-E175.

Dolezal, B.A., and J.A. Potteiger. 1998. Concurrent resistance and endurance training influence basal metabolic rate in nondieting individuals. *Journal of Applied Physiology* 85:695-700.

Donevan, R.H., and G.M. Andrew. 1987. Plasma b-endorphin immunoreactivity during graded cycle ergometry. *Medicine and Science in Sports and Exercise* 19:229-233.

Dotan, R., and O. Bar-Or. 1980. Climatic heat stress and performance in the Wingate anaerobic test. *European Journal of Applied Physiology* 44:237-243.

Dotan, R., A. Rotstein, R. Dlin, O. Inbar, H. Kofman, and Y. Kaplansky. 1983. Relationships of marathon running to physiological, anthropometric and training indices. *European Journal of Applied Physiology* 51:281-293.

Dragoo, K.R., W.M. Silvers, K. Johnson, and E. Gonzalez. 2011. Effects of a caffeine-containing transdermal energy patch on aerobic and anaerobic exercise performance. *International Journal of Exercise Science* 4:141.

Drake, D.R., L. Herwaldt, N.W. Hines, K.C. Kregel, and D. Thoman. 2011. Report of the Special Presidential Committee to Investigate the January 2011 Hospitalization of University of Iowa Football Players. University of Iowa.

Dreyer, H.C., S. Fujita, J.G. Cadenas, D.L. Chinkes, E. Volpi, and B.B. Rasmussen. 2006. Resistance exercise increases AMPK activity and reduced 4E-BP1 phosphorylation and protein synthesis in human skeletal muscle. *Journal of Physiology* 576:613-624.

Drezner, J., S. Berger, and R. Campbell. 2010. Current controversies in the cardiovascular screening of athletes. *Current Sports Medicine Report* 9:86-92.

Drinkwater, B.L., and S.M. Horvath. 1972. Detraining effects in young women. *Medicine and Science in Sports and Exercise* 4:91-95.

Dubuquoy, L., S. Dharancy, S. Nutten, S. Pettersson, J. Auwerx, and P. Desreumaux. 2002. Role of peroxisome proliferator-activated receptor gamma and retinoid X receptor heterodimer in hepatogastroenterological diseases. *Lancet* 360:1410-1418.

Dudley, G.A., W.M. Abraham, and R.L. Terjung. 1982. Influence of exercise intensity and duration on biochemical adaptations in skeletal muscle. *Journal of Applied Physiology* 53:844-850.

Dudley, G.A., and R. Djamil. 1985. Incompatibility of endurance and strength training modes of exercise. *Journal of Applied Physiology* 59:1446-1451.

Dufaux, B., U. Order, and H. Liesen. 1991. Effect of a short maximal physical exercise on coagulation, fibrinolysis, and complement system. *International Journal of Sports Medicine* 12:S38-S42.

Duffield, R., M. Reid, J. Baker, and W. Spratford. 2010. Accuracy and reliability of GPS devices for measurement of movement patterns in confined spaces for court-based sports. *Journal of Science and Medicine in Sport* 13:523-525.

Dugan, E.L., T.L. Doyle, B. Humphries, C.J. Hasson, and R.U. Newton. 2004. Determining the optimal load for jump squats: A review of methods and calculations. *Journal of Strength and Conditioning Research* 18:668-674.

Dulloo, A.G., C.A. Geisler, T. Horton, A. Collins, and D.S. Miller. 1989. Normal caffeine consumption: Influence on thermogenesis and daily energy expenditure in lean and postobese human volunteers. *American Journal of Clinical Nutrition* 49:44-50.

Duncan, M.J., and S.W. Oxford. 2011. The effect of caffeine ingestion on mood state and bench press performance to failure. *Journal of Strength and Conditioning Research* 25:178-185.

Dunnett, M., and R.C. Harris. 1999. Influence of oral beta-alanine and L-histidine supplementation on the carnosine content of the gluteus medius. *Equine Veterinary Journal* 30(suppl):499-504.

DuRant, R.H., L.G. Escobedo, and G.W. Heath. 1995. Anabolic-steroid use, strength training, and multiple drug use among adolescents in the United States. *Pediatrics* 96:23-28.

DuRant, R.H., A.B. Middleman, A.H. Faulkner, S.J. Emans, and E.R. Woods. 1997. Adolescent anabolic-androgenic steroid use, multiple drug use, and high school sports participation. *Pediatric Exercise Science* 9:150-158.

Durnin, J.V.G.A., and J. Womersley. 1974. Body fat assessment from total body density and its estimation from skinfold thickness: Measurements on 481 men and women aged 16-72 years. *British Journal of Nutrition* 32:77-97.

du Vigneaud, V., S. Simonds, J.P. Chandler, and M. Cohn. 1946. A further investigation of the role of betaine in transmethylation reactions in vivo. *Journal of Biological Chemistry* 165:639-648.

Duvnjak-Zaknich, D.M., B.T. Dawson, K.E. Wallman, and G. Henry. 2011. Effect of caffeine on reactive time when fresh and fatigued. *Medicine and Science in Sports and Exercise* 43:1523-1530.

Earnest, C.P., M.A. Olson, C.E. Broeder, K.F. Breuel, and S.G. Beckham. 2000. In vivo 4-androstene-3, 17-dione and 4-androstene-3β,17β-diol supplementation in young men. *European Journal of Applied Physiology* 81:229-232.

Ebbeling, C.B., A. Ward, E.M. Puleo, J. Widrick, and J.M. Rippe. 1991. Development of a single-stage submaximal treadmill walking test. *Medicine and Science in Sports and Exercise* 23:966-973.

Ebben, W.P. 2008. The optimal downhill slope for acute overspeed running. *International Journal of Sports Physiology and Performance* 3:88-93.

Eckardt, K., U. Boutellier, A. Kurtz, M. Schopen, E.A. Koller, and C. Bauer. 1989. Rate of erythropoietin formation in humans in response to acute hypobaric hypoxia. *Journal of Applied Physiology* 66:1785-1788.

Eckerson, J.M., J.R. Stout, G.A. Moore, N.J. Stone, K. Nishimura, and K. Tamura. 2004. Effect of two and five days of creatine loading on anaerobic working capacity in women. *Journal of Strength and Conditioning Research* 18:168-172.

Eckert, H.M. 1968. Angular velocity and range of motion in the vertical and standing broad jump. *Research Quarterly for Exercise and Sport* 39:937-942.

Edge, J., D. Bishop, and C. Goodman. 2006. The effects of training intensity on muscle buffer capacity in females. *European Journal of Applied Physiology* 96:97-105.

Edge, J., S. Hill-Haas, C. Goodman, and D. Bishop. 2006. Effects of resistance training on H+ regulation, buffer capacity, and repeated sprints. *Medicine and Science in Sports and Exercise* 38:2004-2011.

Ehrnborg, C., K.H.W. Lange, R. Dall, J.S. Christiansen, P.A. Lundberg, R.C. Baxter, M.A. Boroujerdi, B.A. Bengtsson, M.L. Healey, C. Pentecost, and S. Longobardi. 2003. The growth hormone/insulin-like growth factor-I axis hormones and bone markers in elite athletes in response to a maximum exercise test. *Journal of Clinical Endocrinology and Metabolism* 88:394-401.

Ekblom, B. 1996. Effects of creatine supplementation on performance. *American Journal of Sports Medicine* 24:538-539.

Ekblom, B., and B. Berglund. 1991. Effect of erythropoietin administration on maximal aerobic power. *Scandinavian Journal of Medicine and Science in Sports* 1:88-93.

Eklund, M., E. Bauer, J. Wamatu, and R. Mosenthin. 2005. Potential nutritional and physiological functions of betaine in livestock. *Nutrition Research Reviews* 18:31-48.

Eldridge, F.L. 1994. Central integration of mechanisms in exercise hyperpnea. *Medicine and Science in Sports and Exercise* 26:319-327.

Eliakim, A., D. Nemet, and D.M. Cooper. 2005. Exercise, training, and the GH-IGF-I axis. In *The endocrine system in sports and exercise,* ed. W.J. Kraemer and A.D. Rogol. Malden, MA: Blackwell.

Ellenbecker, T.S. 1991. A total arm strength isokinetic profile of highly skilled tennis players. *Isokinetic Exercise Science* 1:9-21.

Elliot, T.A., M.G. Cree, A.P. Sanford, R.R. Wolfe, and K.D. Tipton. 2006. Milk ingestion stimulates net muscle protein synthesis following resistance exercise. *Medicine and Science in Sports and Exercise* 38:667-674.

Elliott, B.C., D.J. Wilson, and G.K. Kerr. 1989. A biomechanical analysis of the sticking region in the bench press. *Medicine and Science in Sports and Exercise* 21:450-462.

Enander, A. 1987. Effects of moderate cold on performance of psychomotor and cognitive tasks. *Ergonomics* 30:1431-1445.

Enoka, R.M. 1994. *Neuromechanical basis of kinesiology,* 2nd ed. Champaign, IL: Human Kinetics.

Epley, B. 1985. *Poundage chart.* Lincoln, NE: Boyd Epley Workout.

Epstein, Y. 1990. Heat intolerance: Predisposing factor or residual injury? *Medicine and Science in Sports and Exercise* 22:29-35.

Epstein, Y., D.S. Moran, Y. Shapiro, E. Sohar, and J. Shemer. 1999. Exertional heat stroke: A case series. *Medicine and Science in Sports and Exercise* 31:224-228.

Epstein, Y., and W.O. Roberts. 2011. The pathophysiology of heat stroke: An integrative view of the final common pathway. *Scandinavian Journal of Medicine and Science in Sports* 21:742-748.

Ernsting, J., and G.R. Sharp. 1978. Prevention of hypoxia at altitudes below 40,000 feet. In *Aviation medicine, physiology and human factors,* ed. J. Ernsting, 84-127. London: Tri-Med Books.

Erotokritou-Mulligan, I., E.E. Bassett, C. Bartlett, D. Cowan, C. McHugh, R. Seah, B. Curtis, V. Wells, K. Harrison, P.H. Sönksen, and R.I. Holt. 2008. The effect of sports injury on insulin-like growth factor-I and type 3 procollagen: Implications for detection of growth hormone abuse in athletes. *Journal of Clinical Endocrinology and Metabolism* 93:2760-2763.

Escobar-Morreale, H.F., J.I. Botella-Carretero, F. Escobar del Rey, and G. Morreale de Escobar. 2005. Review: Treatment of hypothyroidism with combinations of levothyroxine plus liothyronine. *Journal of Clinical Endocrinology and Metabolism* 90:4946-4954.

Esmarck, B., J.L. Andersen, S. Olsen, E.A. Richter, M. Mizuno, and M. Kjaer. 2001. Timing of postexercise protein intake is important for muscle hypertrophy with resistance training in elderly humans. *Journal of Physiology* 535:301-311.

Esperson, G.T., A. Elbaek, E. Ernst, E. Toft, S. Kaalund, C. Jersild, and N. Grunnet. 1990. Effect of physical exercise on cytokines and lymphocyte transformation and antibody formation. *Acta Pathologica, Microbiologica et Immunologica Scandinavica* 98:395-400.

Esperson, G.T., E. Toft, E. Ernst, S. Kaalund, and N. Grunnet. 1991. Changes of polymorphonuclear granulocyte migration and lymphocyte subpopulations in human peripheral blood. *Scandinavian Journal of Medicine and Science in Sports* 1:158-162.

Essig, D., D.L. Costill, and P.J. Van Handel. 1980. Effects of caffeine ingestion on utilization of muscle glycogen and lipid during leg ergometer cycling. *International Journal of Sports Medicine* 1:86-90.

Esteve-Lanao, J., C. Foster, S. Seiler, and A. Lucia. 2007. Impact of training intensity and distribution on performance in endurance athletes. *Journal of Strength and Conditioning Research* 21:943-949.

Ettema, G.J.C., A.J. Van Soest, and P.A. Huijing. 1990. The role of series elastic structures in prestretch-induced work enhancement during isotonic and isokinetic contractions. *Journal of Experimental Biology* 154:121-136.

Evans, N.A. 1997. Local complications of self administered anabolic steroid injections. *British Journal of Sports Medicine* 31:349-350.

Evans, W.J., and J.G. Cannon. 1991. The metabolic effects of exercise-induced muscle damage. *Exercise and Sport Sciences Reviews* 19:99-126.

Faerevik, H., and R.E. Reinertsen. 2012. Initial heat stress on subsequent responses to cold water immersion while wearing protective clothing. *Aviation, Space, and Environmental Medicine* 83:746-750.

Fagard, R.H. 2001. Exercise characteristics and the blood pressure response to dynamic physical training. *Medicine and Science in Sports and Exercise* 33(6)(suppl):S484-S492.

Fahey, T.D., and C.H. Brown. 1973. The effects of an anabolic steroid on the strength, body composition, and endurance of college males when accompanied by a weight training program. *Medicine and Science in Sports and Exercise* 5:272-276.

Fahey, T.D., R. Rolph, P. Moungmee, J. Nadel, and S. Martara. 1976. Serum testosterone, body composition, and strength of young adults. *Medicine and Science in Sports and Exercise* 8:31-34.

Faigenbaum, A.D. 2012. Dynamic warm-up. In *NSCA's guide to program design,* ed. J.R. Hoffman, 51-70. Champaign, IL: Human Kinetics.

Faigenbaum, A.D., A.C. Farrell, T. Radler, D. Zybojovsky, D. Chu, N.A. Ratamess, J. Kang, and J.R. Hoffman. 2009. "Plyo Play": A novel program of short bouts of moderate and high intensity exercise improves physical fitness in elementary school children. *Physical Educator* (Winter):37-44.

Faigenbaum, A.D., J. McFarland, F. Keiper, W. Tevlin, J. Kang, N. Ratamess, and J.R. Hoffman. 2007. Effects of a short-term plyometric and resistance training program on fitness performance in boys age 12–15 years. *Journal of Sports Science and Medicine* 6:519-525.

Faigenbaum, A.D., J.E. McFarland, N.A. Kelly, N.A. Ratamess, J. Kang, and J.R. Hoffman. 2010. Influence of recovery time on warm-up effects in adolescent athletes. *Pediatric Exercise Science* 22:266-277.

Faigenbaum, A.D., L.D. Zaichkowsky, D.E. Gardner, and L.J. Micheli. 1998. Anabolic steroid use by male and female middle school students. *Pediatrics* 101:6-14.

Falk, B., S. Radom-Isaac, J.R. Hoffman, Y. Wang, Y. Yarom, A. Magazanik, and Y. Weinstein. 1998. The effect of heat exposure on performance of and recovery from high-intensity, intermittent exercise. *International Journal of Sports Medicine* 19:1-6.

Falk, B., Y. Weinstein, R. Dotan, D.R. Abramson, D. Mann-Segal, and J.R. Hoffman. 1996. A treadmill test of sprint running. *Scandinavian Journal of Medicine and Science in Sports* 6:259-264.

Farrel, M., and J.G. Richards. 1986. Analysis of the reliability and validity of the kinetic communicator exercise device. *Medicine and Science in Sports and Exercise* 18:44-49.

Farrell, P.A., T.L. Garthwaite, and A.B. Gustafson. 1983. Plasma adrenocorticotropin and cortisol responses to submaximal and exhaustive exercise. *Journal of Applied Physiology: Respiratory, Environmental and Exercise Physiology* 55:1441-1444.

Farrell, P.A., M. Kjaer, F.W. Bach, and H. Galbo. 1987. Beta-endorphin and adrenocorticotropin response to supramaximal treadmill exercise in trained and untrained males. *Acta Physiologica Scandinavica* 130:619-625.

Farrell, P.A., J.H. Wilmore, E.F. Coyle, J.E. Billing, and D.L. Costill. 1979. Plasma lactate accumulation and distance running performance. *Medicine and Science in Sports and Exercise* 11:338-344.

Faulkner, J.A., D.R. Claflin, and K.K. McCully. 1986. Power output of fast and slow fibers from human skeletal muscles. In *Human muscle power,* ed. N.L. Jones, N. McCartney, and A.J. McComas, 81-94. Champaign, IL: Human Kinetics.

Febbraio, M.A., T.R. Flanagan, R.J. Snow, S. Zhao, and M.F. Carey. 1995. Effect of creatine supplementation on intramuscular TCr, metabolism and performance during intermittent, supramaximal exercise in humans. *Acta Physiologica Scandinavica* 155:387-395.

Fehr, H.G., H. Lotzerich, and H. Michna. 1988. The influence of physical exercise on peritoneal macrophage functions: Histochemical and phagocytic studies. *International Journal of Sports Medicine* 9:77-81.

Fehr, H.G., H. Lotzerich, and H. Michna. 1989. Human macrophage function and physical exercise: Phagocytic and histochemical studies. *European Journal of Applied Physiology* 58:613-617.

Felig, P., and J. Wahren. 1971. Amino acid metabolism in exercising man. *Journal of Clinical Investigation* 50:2703-2714.

Felsing, N.E., J.A. Brasel, and D.M. Cooper. 1992. Effect of low and high intensity exercise on circulating growth hormone in men. *Journal of Clinical Endocrinology and Metabolism* 75:157-162.

Fern, E.B., R.N. Bielinski, and Y. Schutz. 1991. Effects of exaggerated amino acid and protein supply in man. *Experientia* 47:168-172.

Fernandez-Fernandez, J., R. Zimek, T. Wiewelhove, and A. Ferrauti. 2012. High intensity interval training vs. repeated-sprint training in tennis. *Journal of Strength and Conditioning Research* 26:53-62.

Ferretti, G. 1992. Cold and muscle performance. *International Journal of Sports Medicine* 13(suppl):S185-S187.

Ferry, A., F. Picard, A. Duvallet, B. Weill, and M. Rieu. 1990. Changes in blood leukocyte populations induced by acute maximal and chronic submaximal exercise. *European Journal of Applied Physiology* 59:435-442.

Few, J.D. 1974. Effect of exercise on the secretion and metabolism of cortisol in man. *Journal of Endocrinology* 62:341-353.

Fielding, R.A., T.J. Manfredi, W. Ding, M.A. Fiatarone, W.J. Evans, and J.G. Cannon. 1993. Acute phase response in exercise III: Neutrophil and IL-1b accumulation in skeletal muscle. *American Journal of Physiology: Regulatory, Integrative and Comparative Physiology* 34:R166-R172.

Fimland, M.S., J. Helgerud, H.R. Knutsen, G. Leivseth, and J. Hoff. 2010. No effect of prior caffeine ingestion on neuromuscular fatiguing contractions. *European Journal of Applied Physiology* 108:123-130.

Fineschi, V., I. Riezzo, F. Centini, E. Silingardi, M. Licata, G. Beduschi, and S.B. Karch. 2007. Sudden cardiac death during anabolic steroid abuse: Morphologic and toxicologic findings in two fatal cases of bodybuilders. *International Journal of Legal Medicine* 121:48-53.

Fisher, A.G., and C.R. Jensen. 1990. *Scientific basis of athletic conditioning*. Malvern, PA: Lea & Febiger.

Fitch, K.D., M. Sue-Chu, S.D. Anderson, L.P. Boulet, R.J. Hancox, D.C. McKenzie, V. Backer, K.W. Rundell, J.M. Alonso, P. Kippelen, J.M. Cummiskey, A. Garnier, and A. Ljungqvist. 2008. Asthma and the elite athlete: Summary of the International Olympic Committee consensus conference, Lausanne, Switzerland, January 22-24, 2008. *Journal of Allergy and Clinical Immunology* 122:254-260.

Fitts, R.H. 1992. Substrate supply and energy metabolism during brief high intensity exercise: Importance in limiting performance. In *Energy metabolism in exercise and sport,* ed. D.R. Lamb and C.V. Gisolfi, 53-105. Carmel, IN: Brown & Benchmark.

Fitzsimmons, S., A. Tucker, and D. Martins. 2011. Seventy-five percent of national football league teams use pregame hyperhydration with intravenous fluid. *Clinical Journal of Sport Medicine* 21:192-199.

Fleck, S.J. 1988. Cardiovascular adaptations to resistance training. *Medicine and Science in Sports and Exercise* 20:S146-S151.

Fleck, S.J. 1999. Periodized strength training: A critical review. *Journal of Strength and Conditioning Research* 13:82-89.

Fleck, S.J., and J.E. Falkel. 1986. Value of resistance training for the reduction of sports injuries. *Sports Medicine* 3:61-68.

Fleck, S.J., C. Henke, and W. Wilson. 1989. Cardiac MRI of elite junior Olympic weight lifters. *International Journal of Sports Medicine* 10:329-333.

Fleck, S.J., and W.J. Kraemer. 2004. *Designing Resistance Training Programs, Third Edition*. Champaign, IL: Human Kinetics, 210–224.

Fleck, S.J., C. Mattie, and H.C. Martensen III. 2006. Effect of resistance and aerobic training on regional body composition in previously recreationally trained middle-aged women. *Applied Physiology, Nutrition, and Metabolism* 31:261-270.

Fleck, S.J., and R.C. Schutt. 1985. Types of strength training. *Clinics in Sports Medicine* 4:159-168.

Fletcher, I.M., and B. Jones. 2004. The effect of different warm-up stretch protocols on 20 meter sprint performance in trained rugby union players. *Journal of Strength and Conditioning Research* 18:885-888.

Flora, G. 1985. Secondary treatment of frostbite. In *High altitude deterioration,* ed. J.P. Rivolier, P. Cerretelli, J. Foray, and P. Segantini, 159-169. Basel, Switzerland: Karger.

Florini, J. 1985. Hormonal control of muscle growth. *Journal of Animal Science* 61(suppl):21-37.

Florkowski, C.M., G.R. Collier, P.Z. Zimmet, J.H. Livesey, E.A. Espiner, and R.A. Donald. 1996. Low-dose growth hormone replacement lowers plasma leptin and fat stores without affecting body mass index in adults with growth hormone deficiency. *Clinical Endocrinology* 45:769-773.

Flynn, M.G. 1998. Future research needs and discussion. In *Overtraining in sport,* ed. R.B. Kreider, A.C. Fry, and M.L. O'Toole, 373-384. Champaign, IL: Human Kinetics.

Flynn, M.G., F.X. Pizza, J.B. Boone, F.F. Andres, T.A. Michaud, and J.R. Rodriguez-Zayas. 1994. Indices of training stress during competitive running and swimming seasons. *International Journal of Sports Medicine* 15:21-26.

Foray, J. 1992. Mountain frostbite. *International Journal of Sports Medicine* 13(suppl):S193-S196.

Forbes, G.B. 1985. The effect of anabolic steroids on lean body mass: The dose response curve. *Metabolism* 34:571-573.

Forbes, S.C., D.G. Candow, J.P. Little, C. Magnus, and P.D. Chillibeck. 2007. Effect of Red Bull energy drink on repeated Wingate cycle performance and bench press muscle endurance. *International Journal of Sport Nutrition and Exercise Metabolism* 17:433-444.

Ford Jr., J.F., J.R. Puckett, J.P. Drummond, K. Sawyer, K. Gantt, and C. Fussell. 1983. Effects of three combinations of plyometric and weight training programs on selected physical fitness test items. *Perceptual and Motor Skills* 56:59-61.

Ford, L.E. 1976. Heart size. *Circulatory Research* 39:299-303.

Fornetti, W.C., J.M. Pivarnik, J.M. Foley, and J.J. Fliechtner. 1999. Reliability and validity of body composition measures in female athletes. *Journal of Applied Physiology* 87:1114-1122.

Forster, P. 1984. Reproducibility of individual response to exposure to high altitude. *British Medical Journal* 289:1269.

Fortney, S.M., C.B. Wenger, J.R. Bove, and E.R. Nadel. 1984. Effect of hyperosmolality on control of blood flow and sweating. *Journal of Applied Physiology* 57:1688-1695.

Foster, C., D.L. Costill, and W.J. Fink. 1979. Effects of preexercise feedings on endurance performance. *Medicine and Science in Sports and Exercise* 11:1-5.

Foster, C., D.L. Costill, and W.J. Fink. 1980. Gastric emptying characteristics of glucose and glucose polymers. *Research Quarterly for Exercise and Sport* 51:299-305.

Foster, C., L.L. Hector, R. Welsh, M. Schrager, M.A. Green, and A.C. Snyder. 1995. Effects of specific versus cross-training on running performance. *European Journal of Applied Physiology* 70:367-372.

Foster, N.K., J.B. Martyn, R.E. Rangno, J.C. Hogg, and R.L. Pardy. 1986. Leukocytosis of exercise: Role of cardiac output and catecholamines. *Journal of Applied Physiology* 61:2218-2223.

Fowler Jr., W.M., G.W. Gardner, and G.H. Egstrom. 1965. Effect of an anabolic steroid on physical performance in young men. *Journal of Applied Physiology* 20:1038-1040.

Fox, E.L., R.L. Bartels, C.E. Billings, D.K. Mathews, R. Bason, and W.M. Webb. 1973. Intensity and distance of interval training programs and changes in aerobic power. *Medicine and Science in Sports and Exercise* 5:18-22.

Fox III, S.M., J.P. Naughton, and W.L. Haskell. 1971. Physical activity and the prevention of coronary heart disease. *Annals of Clinical Research* 3:404-432.

Fragala, M.S., W.J. Kraemer, C.R. Denegar, C.M. Maresh, A.M. Mastro, and J.S. Volek. 2011. Neuroendocrine-immune interactions and responses to exercise. *Sports Medicine* 41:621-639.

Fraioli, F., C. Moretti, D. Paolucci, E. Alicicco, F. Crescenzi, and G. Fortunio. 1980. Physical exercise stimulates marked concomitant release of b-endorphin and ACTH in peripheral blood in man. *Experientia* 36:987-989.

Francesconi, R.P., M.N. Sawka, K.B. Pandolf, R.W. Hubbard, A.J. Young, and S. Muza. 1985. Plasma hormonal responses at graded hypohydration levels during exercise-heat stress. *Journal of Applied Physiology* 59:1855-1860.

Freund, B.J., E.M. Shizuru, G.M. Hashiro, and J.R. Claybaugh. 1991. Hormonal, electrolyte and renal responses to exercise are intensity dependent. *Journal of Applied Physiology* 70:900-906.

Friedl, K.E., J.R. Dettori, C.J. Hannan, T.H. Patience, and S.R. Plymate. 1991. Comparison of the effects of high dose testosterone and 19-nortestosterone to a replacement dose of testosterone on strength and body composition in normal men. *Journal of Steroid Biochemistry and Molecular Biology* 40:607-612.

Froberg, S.O. 1971. Effect of training and of acute exercise in trained rats. *Metabolism* 20:1044-1051.

Froiland, K., W. Koszewski, J. Hingst, and L. Kopecky. 2004. Nutritional supplement use among college athletes and the source of information. *International Journal of Sport Nutrition and Exercise Metabolism* 14:104-120.

Fry, A.C. 1998. The role of training intensity in resistance exercise overtraining and overreaching. In *Overtraining in sport,* ed. R.B. Kreider, A.C. Fry, and M.L. O'Toole, 107-130. Champaign, IL: Human Kinetics.

Fry, A.C., C.A. Allemeier, and R.S. Staron. 1994. Correlation between percentage fiber type area and myosin heavy chain content in human skeletal muscle. *European Journal of Applied Physiology* 68:246-251.

Fry, A.C., and W.J. Kraemer. 1991. Physical performance characteristics of American collegiate football players. *Journal of Applied Sport Science Research* 5:126-138.

Fry, A.C., W.J. Kraemer, and L.T. Ramsey. 1998. Pituitary-adrenal-gonadal responses to high intensity resistance exercise overtraining. *Journal of Applied Physiology* 85:2352-2359.

Fry, A.C., W.J. Kraemer, M.H. Stone, J.T. Kearney, S.J. Fleck, and C.A. Weseman. 1993. Endocrine and performance responses to high volume training and amino acid supplementation in elite junior weightlifters. *International Journal of Sport Nutrition* 3:306-322.

Fry, A.C., W.J. Kraemer, F. van Borselen, J.M. Lynch, J.L. Marsit, E.P. Roy, N.T. Triplett, and H.G. Knuttgen. 1994. Performance decrements with high-intensity resistance exercise overtraining. *Medicine and Science in Sports and Exercise* 26:1165-1173.

Fry, A.C., W.J. Kraemer, F. van Borselen, J.M. Lynch, N.T. Triplett, L.P. Koziris, and S.J. Fleck. 1994. Catecholamine responses to short-term high-intensity resistance exercise overtraining. *Journal of Applied Physiology* 77:941-946.

Fry, A.C., W.J. Kraemer, C.A. Weseman, B.P. Conroy, S.E. Gordon, J.R. Hoffman, and C.M. Maresh. 1991. The effects of an off-season strength and conditioning program on starters and non-starters in women's intercollegiate volleyball. *Journal of Applied Sport Science Research* 5:174-181.

Fry, A.C., and D.R. Powell. 1987. Hamstring/quadricep parity with three different weight training methods. *Journal of Sports Medicine and Physical Fitness* 27:362-367.

Fry, A.C., B.K. Schilling, R.S. Staron, F.C. Hagerman, R.S. Hikida, and J.T. Thrush. 2003. Muscle fiber characteristics and performance correlates of male Olympic-style weightlifters. *Journal of Strength and Conditioning Research* 17:746-754.

Fry, M.D., A.C. Fry, and W.J. Kraemer. 1996. Self-efficacy responses to short-term high-intensity resistance exercise overtraining. International Conference on Overtraining and Overreaching in Sport: Physiological, Psychological, and Biomedical Considerations. Memphis, TN.

Fry, R.W., S.R. Lawrence, A.R. Morton, A.B. Schreiner, T.D. Polglaze, and D. Keast. 1993. Monitoring training stress in endurance sports using biological parameters. *Clinical Journal of Sport Medicine* 3:6-13.

Fry, R.W., A.R. Morton, P. Garcia-Webb, G.P.M. Crawford, and D. Keast. 1992. Biological responses to overload training in endurance sports. *European Journal of Applied Physiology* 64:335-344.

Fry, R.W., A.R. Morton, and D. Keast. 1991. Overtraining in athletes. *Sports Medicine* 12:32-65.

Fugh-Berman, A., and A. Myers. 2004. Citrus aurantium, an ingredient of dietary supplements marketed for weight loss: Current status of clinical and basic research. *Experimental Biology and Medicine* 229:698-704.

Fukuda, D.H., A.E. Smith, K.L. Kendall, and J.R. Stout. 2010. The possible combinatory effects of acute consumption of caffeine, creatine, and amino acids on the improvement of anaerobic running performance in humans. *Nutrition Research* 30:607-614.

Fukumoto, Y., T. Ikezoe, Y. Yamada, R. Tsukagoshi, M. Nakamura, N. Mori, M. Kimura, and N. Ichihashi. 2012. Skeletal muscle quality assessed from echo intensity is associated with muscle strength of middle-aged and elderly persons. *European Journal of Applied Physiology* 112:1519-1525.

Furst, P. 2001. New developments in glutamine delivery. *Journal of Nutrition* 131(suppl):2562-2568.

Gabbett, T.J. 2010. GPS analysis of elite women's field hockey training and competition. *Journal of Strength and Conditioning Research* 24:1321-1324.

Gabriel, H., H.J. Miller, A. Urhausen, and W. Kindermann. 1994. Suppressed PMA-induced oxidative burst and unimpaired phagocytosis of circulating granulocytes one week after a long endurance exercise. *International Journal of Sports Medicine* 15:441-445.

Gabriel, H., L. Schwarz, P. Bonn, and W. Kindermann. 1992. Differential mobilization of leukocyte and lymphocyte subpopulations into the circulation during endurance exercise. *European Journal of Applied Physiology* 65:529-534.

Gabriel, H., A. Urhausen, and W. Kindermann. 1992. Mobilization of circulating leukocyte and lymphocyte subpopulations during and after short, anaerobic exercise. *European Journal of Applied Physiology* 65:164-170.

Gaitanos, G.C., M.E. Nevill, S. Brooks, and C. Williams. 1991. Repeated bouts of sprint running after induced alkalosis. *Journal of Sports Science* 9:355-370.

Gaitanos, G.C., C. Williams, L. Boobis, and S. Brooks. 1993. Human muscle metabolism during intermittent maximal exercise. *Journal of Applied Physiology* 75:712-719.

Galbo, H. 1981. Endocrinology and metabolism in exercise. *International Journal of Sports Medicine* 2:2203-2211.

Galbo, H. 1985. The hormonal response to exercise. *Proceedings of the Nutrition Society* 44:257-266.

Galbo, H., L. Hammer, I.B. Peterson, N.J. Christensen, and N. Bic. 1977. Thyroid and testicular hormone responses to gradual and prolonged exercise in man. *Journal of Applied Physiology* 36:101-106.

Galitzky, J., M. Taouis, M. Berlan, D. Riviere, M. Garrigues, and M. Lafontan. 1988. Alpha 2-antagonist compounds and lipid mobilization: Evidence for a lipid mobilizing effect of oral yohimbine in healthy male volunteers. *European Journal of Clinical Investigation* 18:587-594.

Gallagher, P.M., J.A. Carrithers, M.P. Godard, K.E. Schulze, and S. Trappe. 2000a. b-Hydroxy-b-methylbutyrate ingestion, part I: Effects on strength and fat free mass. *Medicine and Science in Sports and Exercise* 32:2109-2115.

Gallagher, P.M., J.A. Carrithers, M.P. Godard, K.E. Schulze, and S. Trappe. 2000b. b-Hydroxy-b-methylbutyrate ingestion, part II: Effects on hematology, hepatic and renal function. *Medicine and Science in Sports and Exercise* 32:2116-2119.

Galloway, S.D.R., and R.J. Maughan. 1997. Effects of ambient temperature on the capacity to perform prolonged cycle exercise in man. *Medicine and Science in Sports and Exercise* 29:1240-1249.

Galpin, A.J., Y. Li, C.A. Lohnes, and B.K. Schilling. 2008. A 4-week choice foot speed and choice reaction training program improves agility in previously non-agility trained, but active men and women. *Journal of Strength and Conditioning Research* 22:1901-1907.

Ganguly, S., S. Jayappa, and A.K. Dash. 2003. Evaluation of the stability of creatine in solution prepared from effervescent creatine formulations. *AAPS PharmSciTech* 4(2):119-128.

Garagioloa, U., M. Buzzetti, E. Cardella, F. Confaloneieri, E. Giani, V. Polini, P. Ferrante, R. Mancuso, M. Montanari, E. Grossi, and A. Pecori. 1995. Immunological patterns during regular intensive training in athletes: Quantification and evaluation of a preventive pharmacological approach. *Journal of International Medical Research* 23:85-95.

Garavik, N., E. Strahm, M. Garle, J. Lundmark, and L. Stahle. 2011. Long-term perturbation of endocrine parameters and cholesterol metabolism after discontinued abuse of anabolic-androgenic steroids. *Journal of Steroid Biochemistry and Molecular Biology* 127:295-300.

Garcia-Pallares, J., L. Sanchez-Medina, C. Esteban Perez, M. Izquierdo-Gabarren, and M. Izquierdo. 2010. Physiological effects of tapering and detraining in world-class kayakers. *Medicine and Science in Sports and Exercise* 42:1209-1214.

Gareau, R., M. Audran, R.D. Baynes, C.H. Flowers, A. Duvallet, L. Senecal, and G.R. Brisson. 1996. Erythropoietin abuse in athletes. *Nature* 380:113.

Garhammer, J., and R. Gregor. 1992. Propulsion forces as a function of intensity for weightlifting and vertical jumping. *Journal of Applied Sport Science Research* 6:129-134.

Garstecki, M.A., R.W. Latin, and M.M. Cuppett. 2004. Comparison of selected physical fitness and performance variables between NCAA division I and II football players. *Journal of Strength and Conditioning Research* 18:292-297.

Gavhed, D., T. Mäkinen, I. Holmér, and H. Rintamäki. 2003. Face cooling by cold wind in walking subjects. *International Journal of Biometerology* 47:148-155.

Ge, R.L., S. Witkowski, Y. Zhang, C. Alfrey, M. Sivieri, T. Karlsen, G.K. Resaland, M. Harber, J. Stray-Gundersen, and B.D. Levine. 2002. Determinants of erythropoietin release in response to short-term hypobaric hypoxia. *Journal of Applied Physiology* 92:2361-2367.

Gergley, T.J., W.D. McArdle, P. DeJesus, M.M. Toner, S. Jacobowitz, and R.J. Spina. 1984. Specificity of arm training on aerobic power during swimming and running. *Medicine and Science in Sports and Exercise* 16:349-354.

Gersh, B.J., B.J. Maron, R.O. Bonow, J.A. Dearani, M.A. Fifer, M.S. Link, S.S. Naidu, R.A. Nishimura, S.R. Ommen, H. Rakowski, C.E. Seidman, J.A. Towbin, J.E. Udelson, and C.W. Yancy. 2011. ACCF/AHA guideline for the diagnosis and treatment of hypertrophic cardiomyopathy: A report of the American College of Cardiology Foundation/American Heart Association Task Force on Practice Guidelines. *Circulation* 124:e783-e831.

Gettman, L.R. 1993. Fitness testing. In *ACSM's resource manual for guidelines for exercise testing and prescription,* 2nd ed., ed. J.L. Durstine, A.C. King, P.L. Painter, J.L. Roitman, and L.D. Zwiren, 229-246. Philadelphia: Williams & Wilkins.

Gettman, L.R., and M.L. Pollock. 1981. Circuit weight training: Critical review of its physiological benefits. *Physician and Sportsmedicine* 9:45-57.

Geyer, H., M.K. Parr, U. Mareck, U. Reinhart, Y. Schrader, and W. Schanzer. 2004. Analysis of non-hormonal nutritional supplements for anabolic-androgenic steroids–results of an international study. *International Journal of Sports Medicine* 25:124-129.

Giamberardino, M.A., L. Dragani, R. Valente, F. Di Lias, R. Saggini, and L. Vecchiet. 1996. Effects of prolonged L-carnitine administration on delayed muscle pain and CK release after eccentric effort. *International Journal of Sports Medicine* 17:320-324.

Gilbert, I.A., J.M. Fouke, and E.R. McFadden Jr. 1987. Heat and water flux in the intrathoracic airways and exercise-induced asthma. *Journal of Applied Physiology* 63:1681-1691.

Gilbert, I.A., J.M. Fouke, and E.R. McFadden Jr. 1988. Intra-airway thermodynamics during exercise and hyperventilation in asthmatics. *Journal of Applied Physiology* 64:2167-2174.

Gilbert, I.A., and E.R. McFadden Jr. 1992. Airway cooling and rewarming: The second reaction sequence in exercise-induced asthma. *Journal of Clinical Investigation* 90:699-704.

Gill, N.D., A. Shield, A.J. Blazevich, S. Zhou, and R.P. Weatherby. 2000. Muscular and cardiorespiratory effects of pseudoephedrine in human athletes. *British Journal of Clinical Pharmacology* 50:205-213.

Gillam, G.M. 1981. Effects of frequency of weight training on muscular strength. *Journal of Sports Medicine* 21:432-436.

Gillies, E.M., C.T. Putman, and G.J. Bell. 2006. The effect of varying the time of concentric and eccentric muscle actions during resistance training on skeletal muscle adaptations in women. *European Journal of Applied Physiology* 97:443-453.

Gillies, H., W.E. Derman, T.D. Noakes, P. Smith, A. Evans, and G. Gabriels. 1996. Pseudoephedrine is without ergogenic effects during prolonged exercise. *Journal of Applied Physiology* 81:2611-2617.

Giorgi, A., R.P. Weatherby, and P.W. Murphy. 1999. Muscular strength, body composition and health responses to the use of testosterone enanthate: A double blind study. *Journal of Science Medicine and Sport* 2:341-355.

Girandola, R.N., and F.L. Katch. 1976. Effects of physical training on ventilatory equivalent and respiratory exchange ratio during weight supported, steady-state exercise. *European Journal of Applied Physiology and Occupational Physiology* 21:119-125.

Girouard, C.K., and B.F. Hurley. 1995. Does strength training inhibit gains in range of motion from flexibility training in older adults? *Medicine and Science in Sports and Exercise* 27:1444-1449.

Gisolfi, C.V. 1973. Work-heat tolerance derived from interval training. *Journal of Applied Physiology* 35:349-354.

Gisolfi, C.V., and J.S. Cohen. 1979. Relationships among training, heat acclimation, and heat tolerance in men and women: The controversy revisited. *Medicine and Science in Sports and Exercise* 11:56-59.

Gisolfi, C.V., and S. Robinson. 1969. Relations between physical training, acclimatization, and heat tolerance. *Journal of Applied Physiology* 26:530-534.

Givoni, B., and R.F. Goldman. 1972. Predicting rectal temperature response to work, environment, and clothing. *Journal of Applied Physiology* 21:812-822.

Gleeson, M., W.A. McDonald, A.W. Cripps, D.B. Pyne, R.L. Clancy, and P.A. Fricker. 1995. The effect on immunity of long-term intensive training in elite swimmers. *Clinical and Experimental Immunology* 102:210-216.

Gleim, G.W., P.A. Witman, and J.A. Nicholas. 1984. Indirect assessment of cardiovascular "demands" using telemetry on professional football players. *American Journal of Sports Medicine* 9:178-183.

Gliottoni, R.C., and R.W. Motl. 2008. Effect of caffeine on leg-muscle pain during intense cycling exercise: Possible role of anxiety sensitivity. *International Journal of Sport Nutrition and Exercise Metabolism* 18:103-115.

Gmunder, F.K., P.W. Joller, H.I. Joller-Jemelka, B. Bechler, M. Cogoli, W.H. Ziegler, J. Muller, R.E. Aeppli, and A. Cogoli. 1990. Effect of herbal yeast food supplements and long-distance running on immunological parameters. *British Journal of Sports Medicine* 24:103-112.

Godfrey, S. 1988. Exercise-induced asthma. In *Allergic diseases from infancy to adulthood,* 2nd ed., ed. W.C. Bierman and D.S. Pearlman, 597. Philadelphia: Saunders.

Godt, R.E., and B.D. Lindly. 1982. Influence of temperature upon contractile activation and isometric force production in mechanically skinned muscle fibers of the frog. *Journal of General Physiology* 80:279-297.

Goforth, H.W., A.N. Campbell, J.A. Hodgdon, and A.A. Sucec. 1982. Hematologic parameters of trained distance runners following induced erythrocythemia. *Medicine and Science in Sports and Exercise* 14:174.

Goldberg, L., D.L. Elliot, and K.S. Kuehl. 1994. A comparison of the cardiovascular effects of running and weight training. *Journal of Strength and Conditioning Research* 8:219-224.

Golden, C.J., S.M. Freshwater, and Z. Golden. 2003. *Stroop color and word test children's version for ages 5–14.* Wood Date, IL: Stoetling.

Golden, F.S.C., I.F.G. Hampton, G.R. Hervery, and A.V. Knibbs. 1979. Shivering intensity in humans during immersion in cold water. *Journal of Physiology* 277:48.

Goldfarb, A.H., B.D. Hatfield, D. Armstrong, and J. Potts. 1990. Plasma beta-endorphin concentration: Response to intensity and duration of exercise. *Medicine and Science in Sports and Exercise* 22:241-268.

Goldfinch, J., L. McNaughton, and P. Davies. 1988. Induced metabolic alkalosis and its effects on 400-m racing time. *European Journal of Applied Physiology* 57:45-48.

Golding, L.A., J.E. Freydinger, and S.S. Fishel. 1974. The effect of an androgenic-anabolic steroid and a protein supplement on size, strength, weight and body composition in athletes. *Physician and Sportsmedicine* 2:39-45.

Goldman, R.F. 2007. Biomedical effects of clothing on thermal comfort and strain. In *Handbook of clothing,* 2nd ed., ed. R.F. Goldman and B. Kampmann, 2.1-2.19. NATO Research Study Group 7 on Bio-Medical Research Aspects of Military Protective Clothing. www.environmental-ergonomics.org/Handbook%20on%20Clothing%20-%202nd%20Ed.pdf.

Goldsmith, J.C., V.L. Bonham, C.H. Joiner, G.J. Kato, A.S. Noonan, and M.H. Steinberg. 2012. Framing the research agenda for sickle cell trait: Building on the current understanding of clinical events and their potential implications. *Journal of Hematology* 87:340-346.

Goldspink, G., A. Schutt, P.T. Loughna, D.J. Wells, T. Jaenicke, and G.F. Gerlach. 1992. Gene expression in skeletal muscle in response to stretch and force generation. *American Journal of Physiology* 262:R356-R363.

Goldstein, E.R., P.L. Jacobs, M. Whitehurst, T. Penhollow, and J. Antonio. 2010. Caffeine enhances upper body strength in resistance-trained women. *Journal of the International Society of Sports Nutrition* 7:18.

Goldstein, E.R., T. Ziegenfuss, D. Kalman, R. Kreider, B. Campbell, C. Wilborn, L. Taylor, D. Willoughby, J. Stout, B.S. Graves, R. Wildman, J.L. Ivy, M. Spano, A.E. Smith, and J. Antonio. 2010. International Society of Sports Nutrition position stand: Caffeine and performance. *Journal of the International Society of Sports Nutrition* 7:5.

Gollnick, P.D., R.B. Armstrong, C.W. Saubert, K. Piehl, and B. Saltin. 1972. Enzyme activity and fiber composition in skeletal muscle of untrained and trained men. *Journal of Applied Physiology* 33:312-319.

Gollnick, P.D., D. Parsons, M. Riedy, and R.L. Moore. 1983. Fiber number and size in overloaded chicken anterior latissimus dorsi muscle. *Journal of Applied Physiology: Respiratory, Environmental and Exercise Physiology* 54:1292-1297.

Gollnick, P.D., and B. Saltin. 1982. Significance of skeletal muscle oxidative enzyme enhancement with endurance training. *Clinical Physiology* 2:1-12.

Gollnick, P.D., and B. Saltin. 1988. Fuel for muscular exercise: Role of fat. In *Exercise, nutrition, and energy metabolism,* ed. E.S. Horton and R.L. Terjung, 72-88. New York: Macmillan.

Gollnick, P.D., B.F. Timson, R.L. Moore, and M. Riedy. 1981. Muscular enlargement and number of fibers in skeletal muscle of rats. *Journal of Applied Physiology: Respiratory, Environmental and Exercise Physiology* 50:936-943.

Gonyea, W.J. 1980a. Muscle fiber splitting in trained and untrained animals. *Exercise and Sport Sciences Reviews* 8:19-39.

Gonyea, W.J. 1980b. Role of exercise in inducing increases in skeletal muscle fiber number. *Journal of Applied Physiology: Respiratory, Environmental and Exercise Physiology* 48:421-426.

Gonyea, W.J., D.G. Sale, F. Gonyea, and A. Mikesky. 1986. Exercise induced increases in muscle fiber number. *European Journal of Applied Physiology* 55:137-141.

Gonzalez, A.M., A.L. Walsh, N.A. Ratamess, J. Kang, and J.R. Hoffman. 2011. Effect of a pre-workout energy supplement on acute multi-joint resistance exercise. *Journal of Sport Science and Medicine* 10:261-166.

Goodman, H.M. 1988. *Basic medical endocrinology.* New York: Raven Press.

Goodman, H.M., and J.C.S. Fray. 1988. Regulation of sodium and water balance. In *Basic medical endocrinology,* ed. H.M. Goodman, 153-174. New York: Raven Press.

Goodpaster, B.H., F.L. Thaete, J.A. Simoneau, and D.E. Kelly. 1997. Subcutaneous abdominal fat and thigh muscle composition predict insulin sensitivity independently of visceral fat. *Diabetes* 46:1579-1785.

Gore, C.J., S.A. Clark, and P.U. Saunders. 2007. Nonhematological mechanisms of improved sea-level performance after hypoxic exposure. *Medicine and Science in Sports and Exercise* 39:1600-1609.

Gore, C.J., A.G. Hahn, G.C. Scroop, D.B. Watson, K.I. Norton, R.J. Wood, D.P. Campbell, and D.L. Emonson. 1996. Increased arterial desaturation in trained cyclists during maximal exercise at 580 m altitude. *Journal of Applied Physiology* 80:2204-2210.

Gore, C.J., S.C. Little, A.G. Hahn, G.C. Scroop, K.I. Norton, P.C. Bourdon, S.M. Woolford, J.D. Buckley, T. Stanef, D.P. Campbell, D.B. Watson, and D.L. Emonson. 1997. Reduced performance of male and female athletes at 580 m altitude. *European Journal of Applied Physiology* 75:136-143.

Gore, C.J., F.A. Rodriguez, M.J. Truijens, N.E. Townsend, J. Stray-Gundersen, and B.D. Levine. 2006. Increased serum erythropoietin but not red cell production after 4 wk of intermittent hypobaric hypoxia (4,000-5,500 m). *Journal of Applied Physiology* 101:1386-1393.

Gossell-Williams, M., O. Simon, L. Young, and M. West. 2006. Choline supplementation facilitates short-term memory consolidation into intermediate long-term memory of young Sprague-Dawley rats. *West Indian Medical Journal* 55(1):4-8.

Graham, M.R., J.S. Baker, P. Evans, A. Kicman, D. Cowan, D. Hullin, N. Thomas, B. Bavies. 2008. Physical effect of short-term recombinant human growth hormone administration in abstinent steroid dependency. *Hormone Research* 69:343-354.

Graham, T.E. 2001. Caffeine and exercise: Metabolism, endurance, and performance. *Sports Medicine* 31:785-807.

Graham, T.E., J.W. Helge, D.A. MacLean, B. Kiens, and E.A. Richter. 2000. Caffeine ingestion does not alter carbohydrate or fat metabolism in human skeletal muscle during exercise. *Journal of Physiology* 529:837-847.

Graham, T.E., E. Hibbert, and P. Sathasivam. 1998. Metabolic and exercise endurance effects of coffee and caffeine ingestion. *Journal of Applied Physiology* 85:883-889.

Graham, T.E., and L.L. Spriet. 1995. Metabolic, catecholamine and exercise performance responses to varying doses of caffeine. *Journal of Applied Physiology* 78:867-874.

Grandjean, A.C., K.J. Reimers, K.E. Bannick, and M.C. Haven. 2000. The effect of caffeinated, non-caffeinated, caloric and non-caloric beverages on hydration. *Journal of the American College of Nutrition* 19:591-600.

Gravelle, B.L., and D.L. Blessing. 2000. Physiological adaptation in women concurrently training for strength and endurance. *Journal of Strength and Conditioning Research* 14:5-13.

Gray, A.B., R.D. Telford, M. Collins, and M.J. Weidemann. 1993. The response of leukocyte subsets and plasma hormones to interval exercise. *Medicine and Science in Sports and Exercise* 25:1252-1258.

Gray, A.J., D. Jenkins, M.H. Andrews, D.R. Taafe, and M.L. Glover. 2010. Validity and reliability of GPS for measuring distance travelled in field-based team sports. *Journal of Sports Sciences* 28:1319-1325.

Green, G.A., F.D. Uryasz, T.A. Petr, and C.D. Bray. 2001. NCAA study of substance use and abuse habits of college student-athletes. *Clinical Journal of Sport Nutrition* 11:51-56.

Green, H.J., J.R. Sutton, G. Coates, M. Ali, and S. Jones. 1991. Response of red cells and plasma volume to prolonged training in humans. *Journal of Applied Physiology* 70:1810-1815.

Green, H.J., J. Sutton, P. Young, A. Cymerman, and C.S. Houston. 1989. Operation Everest II: Muscle energetics during maximal exhaustive exercise. *Journal of Applied Physiology* 66:142-150.

Green, R.L., S.S. Kaplan, B.S. Rabin, C.L. Stanitski, and U. Zdiarski. 1981. Immune function in marathon runners. *Annals of Allergy* 47:73-75.

Greenhaff, P.L. 1995. Creatine and its application as an ergogenic aid. *International Journal of Sport Nutrition* 5:S100-S110.

Greenhaff, P.L. 1997. The nutritional biochemistry of creatine. *Journal of Nutritional Biochemistry* 11:610-618.

Greenhaff, P.L., K. Bodin, K. Soderlund, and E. Hultman. 1994. Effect of oral creatine supplementation on skeletal muscle phosphocreatine resynthesis. *American Journal of Physiology* 266:E725-E730.

Greenhaff, P.L., M.E. Nevill, K. Soderlund, K. Bodin, L.H. Boobis, C. Williams, and E. Hultman. 1994. The metabolic responses of human type I and II muscle fibres during maximal treadmill sprinting. *Journal of Physiology* 478:149-155.

Greenway, F.L., L. De Jonge, D. Blanchard, M. Frisard, and S.R. Smith. 2004. Effect of a dietary herbal supplement containing caffeine and ephedra on weight, metabolic rate, and body composition. *Obesity Research* 12:1153-1157.

Greer, F., D. Friars, and T.E. Graham. 2000. Comparison of caffeine and theophylline ingestion: Exercise metabolism and endurance. *Journal of Applied Physiology* 89:1153-1157.

Griffin, L., P.E. Painter, A. Wadhwa, and W.W. Spirduso. 2009. Motor unit firing variablility and synchronization during short-term light loading in older adults. *Experimental Brain Research* 197:337-345.

Grimsby, J., M. Toth, K. Chen, T. Kumazawa, L. Klaidman, J.D. Adams, F. Karoum, J. Gal, and J.C. Shih. 1997. Increased stress response and β-phenylethylamine in MAOB-deficient mice. *Nature Genetics* 17:206-210.

Grinspoon, S., C. Corcoran, K. Parlman, M. Costello, D. Rosenthal, E. Anderson, T. Stanley, D. Schoenfeld, B. Burrows, D. Hayden, N. Basgoz, and A. Klibanski. 2000. Effects of testosterone and progressive resistance training in eugonadal men with AIDS wasting. A randomized, controlled trial. *Annals of Internal Medicine* 133:348-355.

Gruber, A.J., and H.G. Pope. 1998. Ephedrine abuse among 36 female weightlifters. *American Journal of Addiction* 7:256-261.

Gruber, A.J., and H.G. Pope. 2000. Psychiatric and medical effects of anabolic-androgenic steroid use in women. *Psychotherapy and Psychosomatics* 69:19-26.

Guelfi, K.J., C.E. Donges, and R. Duffield. 2013. Beneficial effects of 12 weeks of aerobic compared with resistance exercise training on perceived appetite in previously sedentary overweight and obese men. *Metabolism* 62:235-243.

Guggenheimer, J.D., D.C. Dickin, G.F. Reyes, and D.G. Dolny. 2009. The effects of specific preconditioning activities on acute sprint performance. *Journal of Strength and Conditioning Research* 23:1135-1139.

Guglielmini, C., A.R. Paolini, and F. Conconi. 1984. Variations of serum testosterone concentrations after physical exercise of different duration. *International Journal of Sports Medicine* 5:246-249.

Guha, A. 2004. Management of traumatic brain injury: Some current evidence and applications. *Postgraduate Medical Journal* 80:650-653.

Guskiewicz, K.M., S.W. Marshall, J. Bailes, M. McCrea, R.C. Cantu, C. Randolph, and B.D. Jordan. 2005. Association between recurrent and late-life cognitive impairment in retired professional football players. *Neurosurgery* 57:719-726.

Guskiewicz, K.M., S.W. Marshall, J. Bailes, M. McCrea, H.P. Harding Jr., A. Matthews, J.R. Mihalik, and R.C. Cantu. 2007. Recurrent concussion and risk of depression in retired football players. *Medicine and Science in Sports and Exercise* 39:903-909.

Guskiewicz, K.M., and J.P. Mihalik. 2011. Biomechanics of sport concussion: Quest for the elusive injury threshold. *Exercise and Sport Sciences Reviews* 39:4-11.

Gysland, S.M., J.P. Mihalik, J.K. Register-Mihalik, S.C. Trulock, E.W. Shields, and K.M. Guskiewicz. 2012. The relationship between subconcussive impacts and concussion history on clinical measures of neurologic function in college football players. *Annals of Biomedical Engineering* 40:14-22.

Hack, B., G. Strobel, M. Weiss, and H. Weicker. 1994. PMN cell counts and phagocytic activity of highly trained athletes depend on training period. *Journal of Applied Physiology* 77:1731-1735.

Hackett, P.H., and D. Rennie. 1979. Rales, peripheral edema, retinal hemorrhage and acute mountain sickness. *American Journal of Medicine* 67:214-218.

Hackett, P.H., D. Rennie, R.F. Glover, and J.T. Reeves. 1981. Acute mountain sickness and the edemas of high altitude: A common pathogenesis? *Respiratory Physiology* 46:383-390.

Hackman, R.M., P.J. Havel, H.J. Schwartz, J.C. Rutledge, M.R. Watnik, E.M. Noceti, S.J. Stohs, J.S. Stern, and C.L. Keen. 2006. Multinutrient supplement containing ephedra and caffeine causes weight loss and improves metabolic risk factors in obese women: A randomized controlled trial. *International Journal of Obesity (London)* 30:1545-1556.

Hackney, A.C., and J.D. Dobridge. 2009. Thyroid hormones and the interrelationship of cortisol and prolactin: Influence of prolonged, exhaustive exercise. *Endokrynologia Polska* 60:252-257.

Hackney, A.C., W.E. Sinning, and B.C. Bruot. 1988. Reproductive hormonal profiles of endurance-trained and untrained males. *Medicine and Science in Sport and Exercise* 20:60-65.

Hackney, A.C., W.E. Sinning, and B.C. Bruot. 1990. Hypothalamic-pituitary-testicular axis function in endurance-trained males. *International Journal of Sports Medicine* 11:298-303.

Haff, G.G., K.B. Kirksey, M.H. Stone, B.J. Warren, R.L. Johnson, M. Stone, H. O'Bryant, and C. Proulx. 2000. The effect of 6 weeks of creatine monohydrate supplementation on dynamic rate of force development. *Journal of Strength and Conditioning Research* 14:426-433.

Hagberg, J.M., N.T. Jenkins, and E. Spangenburg. 2012. Exercise training, genetics and type 2 diabetes-related phenotypes. *Acta Physiologica* 205:456-471.

Hagberg, J.M., R.C. Hickson, A.A. Ehsani, and J.O. Holloszy. 1980. Faster adjustment to and recovery from submaximal exercise in the trained state. *Journal of Applied Physiology* 48:218-224.

Hagerman, F.C. 1984. Applied physiology of rowing. *Sports Medicine* 1:303-326.

Hagobian, T.A., C.G. Sharoff, and B. Braun. 2008. Effects of short-term exercise and energy surplus on hormones related to regulation of energy balance. *Metabolism: Clinical and Experimental* 57:393-398.

Hakkinen, K., M. Alen, and P.V. Komi. 1985. Changes in isometric force- and relaxation-time, electromyographic and muscle fibre characteristics of human skeletal muscle during strength training and detraining. *Acta Physiologica Scandinavica* 125:573-585.

Hakkinen, K., M. Alen, W.J. Kraemer, E. Gorostiaga, M. Izquierdo, H. Rusko, J. Mikkola, A. Hakkinen, H. Valkeinen, E. Kaarakainen, S. Romu, V. Erola, J. Ahtiainen, and L. Paavolainen. 2003. Neuromuscular adaptations during concurrent strength and endurance training versus strength training. *European Journal of Applied Physiology* 89:42-52.

Hakkinen, K., and P.V. Komi. 1983. Electromyographic changes during strength training and detraining. *Medicine and Science in Sports and Exercise* 15:455-460.

Hakkinen, K., and P.V. Komi. 1985a. Changes in electrical and mechanical behavior of leg extensor muscles during heavy resistance strength training. *Scandinavian Journal of Sports Sciences* 7:55-64.

Hakkinen, K., and P.V. Komi. 1985b. The effect of explosive type strength training on electromyographic and force production characteristics of leg extensor muscles during concentric and various stretch-shortening cycle exercises. *Scandinavian Journal of Sports Sciences* 7:65-76.

Hakkinen, K., and P.V. Komi. 1986. Training induced changes in neuromuscular performance under voluntary and reflex conditions. *European Journal of Applied Physiology* 55:147-155.

Hakkinen, K., P.V. Komi, and M. Alen. 1985. Effect of explosive type strength training on isometric force- and relaxation-time, electromyographic and muscle fibre characteristics of leg extensor muscles. *Acta Physiologica Scandinavica* 125:587-600.

Hakkinen, K., P.V. Komi, M. Alen, and H. Kauhanen. 1987. EMG, muscle fibre and force production characteristics during a 1 year training period in highly competitive weightlifters. *European Journal of Applied Physiology and Occupational Physiology* 56:419-427.

Hakkinen, K., and A. Pakarinen. 1991. Serum hormones in male strength athletes during intensive short term strength training. *European Journal of Applied Physiology* 63:191-199.

Hakkinen, K., and A. Pakarinen. 1993. Acute hormonal responses to two different fatiguing heavy-resistance protocols in male athletes. *Journal of Applied Physiology* 74:882-887.

Hakkinen, K., A. Pakarinen, M. Alen, H. Kauhanen, and P.V. Komi. 1987. Relationships between training volume, physical performance capacity, and serum hormone concentration during prolonged training in elite weightlifters. *International Journal of Sports Medicine* 8(suppl):61-5.

Hakkinen, K., A. Pakarinen, M. Alen, H. Kauhanen, and P.V. Komi. 1988. Neuromuscular and hormonal adaptations in athletes to strength training in two years. *Journal of Applied Physiology* 65:2406-2412.

Hakkinen, K., A. Pakarinen, M. Alen, and P.V. Komi. 1985. Serum hormones during prolonged training of neuromuscular performance. *European Journal of Applied Physiology* 53:287-293.

Hakkinen, K., A. Pakarinen, P.V. Komi, T. Ryushi, and H. Kauhanen. 1989. Neuromuscular adaptations and hormone balance in strength athletes, physically active males and females during intensive strength training. In *Proceedings of XII International Congress of Biomechanics,* No. 8, ed. R.J. Gregor, R.F. Zernicke, and W.C. Whiting, 889-898. Champaign, IL: Human Kinetics.

Hakkinen, K., A. Pakarinen, H. Kyrolainen, S. Cheng, D.H. Kim, and P.V. Komi. 1990. Neuromuscular adaptations and serum hormones in females during prolonged power training. *International Journal of Sports Medicine* 11:91-98.

Halbert, J.A., C.A. Silagy, P. Finucane, R.T. Withers, P.A. Hamdorf, and G.R. Andrews. 1997. The effectiveness of exercise training in lowering blood pressure: A meta-analysis of randomized controlled trials of 4 weeks or longer. *Journal of Human Hypertension* 11:641-649.

Haller, C.A., N.L. Benowitz, and P. Jacob. 2005. Hemodynamic effects of ephedra-free weight-loss supplements in humans. *American Journal of Medicine* 118:998-1003.

Haller, C.A., P. Jacob, and N.L. Benowitz. 2004. Enhanced stimulant and metabolic effects of combined ephedrine and caffeine. *Clinical Pharmacology and Therapeutics* 75:259-273.

Halliday, T.M., N.J. Peterson, J.J. Thomas, K. Kleppinger, B.W. Hollis, and D.E. Larson-Meyer. 2011. Vitamin D status relative to diet, lifestyle, injury, and illness in college athletes. *Medicine Science in Sports and Exercise* 43:335-343.

Halseth, A., P.J. Flakoll, E.K. Reed, A.B. Messina, M.G. Krishna, D.B. Lacy, P.E. Williams, and D.H. Wasserman. 1997. Effect of physical activity and fasting on gut and liver proteolysis in the dog. *American Journal of Physiology* 273:E1073-E1082.

Hamlet, M.P. 1988. Human cold injuries. In *Human performance physiology and environmental medicine at terrestrial extremes*, ed. K.B. Pandolf, M.N. Sawka, and R.R. Gonzalez, 435-466. Indianapolis: Benchmark Press.

Handelsman, D.J. 2006. Clinical review: The rationale for banning human chorionic gonadotropin and estrogen blockers in sport. *Journal of Clinical Endocrinology and Metabolism* 91:1646-1653.

Haney, S., and R.J. Hancox. 2006. Recovery from bronchoconstriction and bronchodilator tolerance. *Clinical Reviews of Allergy and Immunology* 31:181-196.

Haney, S., and R.J. Hancox. 2007. Overcoming beta-agonist tolerance: High dose salbutamol and ipratropium bromide: Two randomized controlled trials. *Respiratory Research* 8:19.

Hansen, A.A.P. 1973. Serum growth hormone response to exercise in non-obese and obese normal subjects. *Scandinavian Journal of Clinical Investigation* 31:175-178.

Hansen, J.B., and D.K. Flaherty. 1981. Immunological responses to training in conditioned runners. *Clinical Science* 60:225-228.

Hansen, J.B., L. Wilsgard, and B. Osterud. 1991. Biphasic changes in leukocytes induced by strenuous exercise. *European Journal of Applied Physiology* 62:157-161.

Hansen, K.R., S.M. Krasnow, M.A. Nolan, G.S. Fraley, J.W. Baumgartner, D.K. Clifton, and R.A. Steiner. 2003. Activation of the sympathetic nervous system by galanin-like peptide--a possible link between leptin and metabolism. *Endocrinology* 144:4709-4717.

Hansen, L.P., B.B. Jacobsen, P.E.L. Kofeod, M.L. Larsen, T. Tougaard, and I. Johansen. 1989. Serum fructosamine and HbA1c in diabetic children before and after attending a winter camp. *Acta Paediatrica Scandinavica* 78:451-452.

Harber, M.P., P.M. Gallagher, A.R. Creer, K.M. Minchev, and S.W. Trappe. 2004. Single muscle fiber contractile properties during a competitive season in male runners. *American Journal of Physiology: Regulatory, Integrative and Comparative Physiology* 287:R1124-R1131.

Harman, E.A., J. Garhammer, and C. Pandorf. 2000. Administration, scoring, and interpretation of selected tests. In *Essentials of strength and conditioning*, ed. T. Baechle and R. Earle, 287-318. Champaign, IL: Human Kinetics.

Harman, E.A., M.T. Rosenstein, P.N. Frykman, R.M. Rosenstein, and W.J. Kraemer. 1991. Estimation of human power output from vertical jump. *Journal of Applied Sport Science Research* 5:116-120.

Harmer, A.R., P.A. Ruell, M.J. McKenna, D.J. Chisholm, S.K. Hunter, J.M. Thom, N.R. Morris, and J.R. Flack. 2006. Effects of sprint training on extrarenal potassium regulation with intense exercise in type I diabetes. *Journal of Applied Physiology* 100:26-34.

Harmon, K.G., I.M. Asif, D. Klossner, and J.A. Drezner. 2011. Incidence of sudden cardiac death in National Collegiate Athletic Association athletes. *Circulation* 123:1594-1600.

Harmon, K.G., J.A. Drezner, D. Klossner, and I.M. Asif. 2012. Sickle cell trait associated with a RR of death of 37 times in national collegiate athletic association football athletes: A database with 2 million athlete-years as the denominator. *British Journal of Sports Medicine* 46:325-330.

Harris, K.M., T.S. Haas, E.R. Eichner, and B.J. Maron. 2012. Sickle cell trait associated with sudden death in competitive athletes. *American Journal of Cardiology* 110:1185-1188.

Harris, R.C., R.H.T. Edwards, E. Hultman, L.O. Nordesjo, B. Nylind, and K. Sahlin. 1976. The time course of phosphorylcreatine resynthesis during recovery of the quadriceps muscle in man. *Pflugers Archives* 367:137-142.

Harris, R.C., I.P. Kendrick, C. Kim, H. Kim, V.H. Dang, T.Q. Lam, T.T. Bui, and J.A. Wise. 2007. Effect of physical training on the carnosine content of v. lateralis using a one-leg training model [abstract]. *Medicine and Science in Sports and Exercise* 39:S91.

Harris, R.C., D.J. Marlin, M. Dunnett, D.H. Snow, and E. Hultman. 1990. Muscle buffering capacity and dipeptide content in the thoroughbred horse, greyhound dog and man. *Comparative Biochemistry and Physiology A* 97:249-251.

Harris, R.C., K. Soderlund, and E. Hultman. 1992. Elevation of creatine in resting and exercised muscle of normal subjects by creatine supplementation. *Clinical Science* 83:367-374.

Harris, R.C., M.J. Tallon, M. Dunnett, L. Boobis, J. Coakley, H.J. Kim, J.L. Fallowfield, C.A. Hill, C. Sale, and J.A. Wise. 2006. The absorption of orally supplied β-alanine and its effect on muscle carnosine synthesis in human vastus lateralis. *Amino Acids* 30:279-289.

Hartgens, F., H. van Straaten, S. Fideldij, G. Rietjens, H.A. Keizer, and H. Kuipers. 2002. Misuse of androgenic-anabolic steroids and human deltoid muscle fibers: Differences between polydrug regimens and single drug administration. *European Journal of Applied Physiology* 86:233-239.

Hartley, L.H., J.W. Mason, R.P. Hogan, L.G. Jones, T.A. Kotchen, E.H. Mougey, F.E. Wherry, L.L. Pennington, and P.T. Ricketts. 1972. Multiple hormonal responses to prolonged exercise in relation to physical training. *Journal of Applied Physiology* 33:607-610.

Harwood, M., B. Danielewska-Nikiel, J.F. Borzelleca, G.W. Flamm, G.M. Williams, and T.C. Lines. 2007. A critical review of the data related to the safety of quercetin and lack of evidence of in vivo toxicity including lack of genotoxic/carcinogenic properties. *Food and Chemical Toxicology* 45:2179-2205.

Hather, B.M., P.A. Tesch, P. Buchanan, and G.A. Dudley. 1991. Influence of eccentric actions on skeletal muscle adaptations to resistance training. *Acta Physiologica Scandinavica* 143:177-185.

Hausmann, R., S. Hammer, and P. Betz. 1998. Performance enhancing drugs (doping agents) and sudden death--a case report and review of the literature. *International Journal of Legal Medicine* 111:261-264.

Hausswirth, C., A.X. Bigard, R. Lepers, M. Berthelot, and C.Y. Gunzennec. 1995. Sodium citrate ingestion and muscle performance in acute hypobaric hypoxia. *European Journal of Applied Physiology* 71:362-368.

Hausswirth, C., D. Lehenaff, P. Dreano, and K. Savonen. 1999. Effects of cycling alone or in a sheltered position on subsequent running performance during a triathlon. *Medicine and Science in Sports and Exercise* 31:599-604.

Hausswirth, C., J.M. Vallier, D. Lehenaff, J. Brisswalter, D. Smith, G. Millet, and P. Dreano. 2001. Effect of two drafting modalities in cycling on running performance. *Medicine and Science in Sports and Exercise* 33:485-492.

Havenith, G. 1999. Heat balance when wearing protective clothing. *Annals of Occupational Hygiene* 43:289-296.

Hawley, J.A., E.J. Schabort, T.D. Noakes, and S.C. Dennis. 1997. Carbohydrate-loading and exercise performance. An update. *Sports Medicine* 24:73-81.

Hayashi, A.A., and C.G. Proud. 2007. The rapid activation of protein synthesis by growth hormone requires signaling through mTOR. *American Journal of Physiology: Endocrinology and Metabolism* 292:E1647-E1655.

Hayward, J.S., and J.D. Eckerson. 1984. Physiological responses and survival time prediction for humans in ice-water. *Aviation, Space, and Environmental Medicine* 55:206-212.

Hayward, J.S., J.D. Eckerson, and M.L. Collis. 1975. Effect of behavioral variables on cooling rate of man in cold water. *Journal of Applied Physiology* 38:1073-1077.

Healy, M.L., J. Gibney, C. Pentecost, P. Croos, D.L. Russell-Jones, P.H. Sonksen, and A.M. Umpleby. 2006. Effects of high-dose growth hormone on glucose and glycerol metabolism at rest and during exercise in endurance-trained athletes. *Journal of Clinical Endocrinology and Metabolism* 91:320-327.

Heath, G.W., E.S. Ford, T.E. Craven, C.A. Macera, K.L. Jackson, and R.R. Pate. 1991. Exercise and the incidence of upper respiratory tract infections. *Medicine and Science in Sports and Exercise* 25:186-190.

Heath, G.W., C.A. Macera, and D.C. Nieman. 1992. Exercise and upper respiratory tract infections: Is there a relationship? *Sports Medicine* 14:353-365.

Heck, H., A. Mader, G. Hess, S. Mucke, R. Muller, and W. Hollmann. 1985. Justification of the 4 mmol/L lactate threshold. *International Journal of Sports Medicine* 6:117-130.

Hedge, G.A., H.D. Colby, and R.L. Goodman. 1987. *Clinical endocrine physiology*. Philadelphia: Saunders.

Heigenhauser, G.J.F., and N.L. Jones. 1991. Bicarbonate loading. In *Perspectives in exercise science and sports medicine, Vol. 4, Ergogenics,* ed. D.R. Lamb and M.H. Williams, 183-212. Carmel, IN: Benchmark Press.

Heikkinen, A., A. Alaranta, I. Helenius, and T. Vasankari. 2011. Dietary supplementation habits and perceptions of supplement use among elite Finnish athletes. *International Journal of Sport Nutrition and Exercise Metabolism* 21:271-279.

Heir, T., and S. Oseid. 1994. Self-reported asthma and exercise-induced asthma symptoms in high-level competitive cross-country skiers. *Scandinavian Journal of Medicine and Science in Sports* 4:128-133.

Heled, Y., Y. Epstein, and D. Moran. 2004. Heat strain attenuation while wearing NBC clothing: Dry-ice vest compared to water spray. *Aviation, Space, and Environmental Medicine* 75:391-396.

Helenius, I., A. Lumme, J. Ounap, Y. Obase, P. Rytila, S. Sarna, A. Alaranta, V. Remes, and T. Haahtela. 2004. No effect of montelukast on asthma-like symptoms in elite ice hockey players. *Allergy* 59:39-44.

Hendrix, C.R., T.J. Housh, M. Miekle, J.M. Zuniga, C.L. Camic, G.O. Johnson, R.J. Schmidt, and D.J. Housh. 2010. Acute effects of caffeine-containing supplement on bench press and leg extension strength and time to exhaustion during cycle ergometry. *Journal of Strength and Conditioning Research* 24:859-865.

Hennessy, L.C., and A.W.S. Watson. 1994. The interference effects of training for strength and endurance simultaneously. *Journal of Strength and Conditioning Research* 8:12-19.

Herbold, N.H., I. Vazquez, E. Goodman, and S.J. Emans. 2004. Vitamin, mineral, and other supplement use by adolescents. *Topics in Clinical Nutrition* 19:266-272.

Herbold, N.H., B.K. Visconti, S. Frates, and L. Bandini. 2004. Traditional and nontraditional supplement use by collegiate female varsity athletes. *International Journal of Sport Nutrition and Exercise Metabolism* 14:586-593.

Herman, K., C. Barton, P. Malliaras, and D. Morrissey. 2012. The effectiveness of neuromuscular warm-up strategies, that require no additional equipment, for preventing lower limb injuries during sports participation: A systemic review. *BioMed Central Medicine* 10:75.

Hermansen, L., and M. Wachtlova. 1971. Capillary density of skeletal muscle in well-trained and untrained men. *Journal of Applied Physiology* 30:860-863.

Herring, S.A., R.C. Cantu, K.M. Guskiewicz, M. Putukian, and W.B. Kibler. 2011. Concussion (mild traumatic brain injury) and the team physician: A consensus statement–2011 update. *Medicine and Science in Sports and Exercise* 43:2412-2422.

Hervey, G.R., A.V. Knibbs, L. Burkinshaw, D.B. Morgan, P.R.M. Jones, D.R. Chettle, and D. Vartsky. 1981. Effects of methandienone on the performance and body composition of men undergoing athletic training. *Clinical Science* 60:457-461.

Heus, R., H.A.M. Daanen, and G. Havenith. 1995. Physiological criteria for functioning of hands in the cold. *Applied Ergonomics* 26:5-13.

Heymsfield, S.B., C. Arteaga, C. McManus, J. Smith, and S. Moffitt. 1983. Measurement of muscle mass in humans: Validity of the 24-hour urinary creatinine method. *American Journal of Clinical Nutrition* 37:478-494.

Heyward, V.H. 1997. *Advanced fitness assessment and exercise prescription*. Champaign, IL: Human Kinetics.

Heyward, V.H., and L.M. Stolarczyk. 1996. *Applied body composition assessment*. Champaign, IL: Human Kinetics.

Hickner, R.C., C.A. Horswill, J. Welker, J.R. Scott, J.N. Roemmich, and D.L. Costill. 1991. Test development for study of physical performance in wrestlers following weight loss. *International Journal of Sports Medicine* 12:557-562.

Hickson, R.C. 1981. Skeletal muscle cytochrome c and myoglobin, endurance, and frequency of training. *Journal of Applied Physiology* 51:746-749.

Hickson, R.C., B.A. Dvorak, E.M. Gorostiaga, T.T. Kurowski, and C. Foster. 1988. Potential for strength and endurance training to amplify endurance performance. *Journal of Applied Physiology* 65:2285-2290.

Hickson, R.C., K. Hidaka, C. Foster, M.T. Falduto, and R.T. Chatterton Jr. 1994. Successive time course of strength development and steroid hormone responses to heavy resistance training. *Journal of Applied Physiology* 76:663-670.

Hickson, R.C., M.A. Rosenkoetter, and M.M. Brown. 1980. Strength training effects on aerobic power and short-term endurance. *Medicine and Science in Sports and Exercise* 12:336-339.

Higgins, J.P., T.D. Tuttle, and C.L. Higgins. 2010. Energy beverages: Content and safety. *Mayo Clinic Proceedings* 85:1033-1041.

Hill, C.A., R.C. Harris, H.J. Kim, B.D. Harris, C. Sale, L.H. Boobis, C.K. Kim, and J.A. Wise. 2007. Influence of β-alanine supplementation on skeletal muscle carnosine concentrations and high intensity cycling capacity. *Amino Acids* 32:225-233.

Hirsch, M.J., J.H. Growdon, and R.J. Wurtman. 1978. Relations between dietary choline or lecithin intake, serum choline levels, and various metabolic indices. *Metabolism* 27:953-960.

Hirvonen, J., A. Nummela, H. Rusko, S. Rehunen, and M. Harkonen. 1992. Fatigue and changes of ATP, creatine phosphate, and lactate during the 400-m sprint. *Canadian Journal of Sport Science* 17:141-144.

Hirvonen, J., S. Rehunen, H. Rusko, and M. Harkonen. 1987. Breakdown of high-energy phosphate compounds and lactate accumulation during short supramaximal exercise. *European Journal of Applied Physiology* 56:253-259.

Hislop, M.S., B.D. Ratanjee, S.G. Soule, and A.D. Marais. 1999. Effects of anabolic-androgenic steroid use or gonadal testosterone suppression on serum leptin concentration in men. *European Journal of Endocrinology* 141:40-46.

Ho, K.W., R.R. Roy, C.D. Tweedle, W.W. Heusner, W.D. Van Huss, and R.E. Carrow. 1980. Skeletal muscle fiber splitting with weight-lifting exercise. *American Journal of Anatomy* 116:57-65.

Hodges, K., S. Hancock, K. Currell, B. Hamilton, and A.E. Jeukendrup. 2006. Pseudoephedrine enhances performance in 1500-m runners. *Medicine and Science in Sports and Exercise* 38:329-333.

Hoeger, W.W.K., S.L. Barette, D.F. Hale, and D.R. Hopkins. 1987. Relationship between repetitions and selected percentages of one repetition maximum. *Journal of Applied Sports Science Research* 1:11-13.

Hoeger, W.W.K., and S.A. Hoeger. 2000. *Lifetime physical fitness and wellness*. Englewood, CO: Morton.

Hoeger, W.W.K., D.R. Hopkins, S.L. Barette, and D.F. Hale. 1990. Relationship between repetitions and selected percentages of one repetition maximum: A comparison between untrained and trained males and females. *Journal of Applied Sports Science Research* 4:47-54.

Hoeger, W.W.K., D.R. Hopkins, S. Button, and T.A. Palmer. 1990. Comparing the sit and reach with the modified sit and reach in measuring flexibility in adolescents. *Pediatric Exercise Science* 2:156-162.

Hoene, M., and C. Weigert. 2010. The stress response of the liver to physical exercise. *Exercise Immunology Reviews* 16:163-183.

Hoffman, J.R. 1997. The relationship between aerobic fitness and recovery from high-intensity exercise in infantry soldiers. *Military Medicine* 162:484-488.

Hoffman, J.R. 2003. Physiology of basketball. In *Handbook on basketball: Olympic handbook of sports medicine,* ed. D.B. McKeag, 1-11. Oxford: Blackwell.

Hoffman, J.R. 2006. *Norms for fitness, performance, and health,* 3-115. Champaign, IL: Human Kinetics.

Hoffman, J.R. 2007. Protein intake: Effect of timing. *Strength and Conditioning Journal* 29:26-34.

Hoffman, J.R. 2010a. Creatine and β-alanine supplementation in strength/power athletes. *Current Topics in Nutraceutical Research* 8:19-32.

Hoffman, J.R. 2010b. Caffeine and energy drinks. *Strength and Conditioning Journal* 32:15-20.

Hoffman, J.R., L.E. Brown, and A.E. Smith. 2012. Training program implementation. In *NSCA's guide to program design,* ed. J.R. Hoffman, 259-288. Champaign, IL: Human Kinetics.

Hoffman, J.R., J. Cooper, M. Wendell, J. Im, and J. Kang. 2004. Effects of β-hydroxy β-methylbutyrate on power performance and indices of muscle damage and stress during high intensity training. *Journal of Strength and Conditioning Research* 18:747-752.

Hoffman, J.R., J. Cooper, M. Wendell, and J. Kang. 2004. Comparison of Olympic versus traditional power lifting training programs in football players. *Journal of Strength and Conditioning Research* 18:129-135.

Hoffman, J.R., S. Epstein, M. Einbinder, and Y. Weinstein. 1999. The influence of aerobic capacity on anaerobic performance and recovery indices in basketball players. *Journal of Strength and Conditioning Research* 13:407-411.

Hoffman, J.R., S. Epstein, M. Einbinder, and Y. Weinstein. 2000. A comparison between the Wingate anaerobic power test to both vertical jump and line drill tests in basketball players. *Journal of Strength and Conditioning Research* 14:261-264.

Hoffman, J.R., S. Epstein, Y. Yarom, L. Zigel, and M. Einbinder. 1999. Hormonal and biochemical changes in elite basketball players during a 4-week training camp. *Journal of Strength and Conditioning Research* 13:280-285.

Hoffman, J.R., A.D. Faigenbaum, N.A. Ratamess, R. Ross, J. Kang, and G. Tenenbaum. 2008. Nutritional and anabolic steroid use in adolescents. *Medicine and Science in Sports and Exercise* 40:15-24.

Hoffman, J.R., and M.J. Falvo. 2004. Protein—which is best? *Journal of Sports Science and Medicine* 3:118-130.

Hoffman, J.R., A.C. Fry, R. Howard, C.M. Maresh, and W.J. Kraemer. 1991. Strength, speed and endurance changes during the course of a division I basketball season. *Journal of Applied Sport Science Research* 5:144-149.

Hoffman, J.R., and J. Graham. 2012. Speed training. In *NSCA's guide to program design,* ed. J.R. Hoffman, 165-184. Champaign, IL: Human Kinetics.

Hoffman, J.R., J. Im, K. Rundell, J. Kang, S. Nioka, B.A. Spiering, R. Kime, and B. Chance. 2003. Effect of muscle oxygenation during resistance exercise on anabolic hormone response. *Medicine and Science in Sports and Exercise* 35:1929-1934.

Hoffman, J.R., and M. Kaminsky. 2000. Use of performance testing for monitoring overtraining in elite youth basketball players. *Strength and Conditioning* 22:54-62.

Hoffman, J.R., and J. Kang. 2003. Strength changes during an inseason resistance training program for football. *Journal of Strength and Conditioning Research* 17:109-114.

Hoffman, J.R., J. Kang, N.A. Ratamess, M.W. Hoffman, C.P. Tranchina, and A.D. Faigenbaum. 2009. Examination of a high energy, pre-exercise supplement on exercise performance. *Journal of the International Society of Sports Nutrition* 6:2.

Hoffman, J.R., J. Kang, N.A. Ratamess, P.F. Jennings, G. Mangine, and A.D. Faigenbaum. 2006. Thermogenic effect from nutritionally enriched coffee consumption. *Journal of the International Society of Sports Nutrition* 3:35-41.

Hoffman, J.R., J. Kang, N.A. Ratamess, P.F. Jennings, G. Mangine, and A.D. Faigenbaum. 2007. Effect of nutritionally enriched coffee consumption on aerobic and anaerobic exercise performance. *Journal of Strength and Conditioning Research* 21:456-459.

Hoffman, J.R., J. Kang, N.A. Ratamess, S.L. Rashti, and A.D. Faigenbaum. 2008. Thermogenic effect of a high energy, pre-exercise supplement. *Kinesiology* 40(2):207-213.

Hoffman, J.R., and S. Klafeld. 1998. The effect of resistance training on injury rate and performance in a self-defense course for females. *Journal of Strength and Conditioning Research* 12:52-56.

Hoffman, J.R., W.J. Kraemer, S. Bhasin, T. Storer, N.A. Ratamess, G.G. Haff, D.S. Willoughby, and A.D. Rogol. 2009. Position stand on androgen and human growth hormone use. *Journal of Strength and Conditioning Research* 23:S1-S59.

Hoffman, J.R., W.J. Kraemer, A.C. Fry, M. Deschenes, and M. Kemp. 1990. The effect of self-selection for frequency of training in a winter conditioning program for football. *Journal of Applied Sport Science Research* 3:76-82.

Hoffman, J.R., and C.M. Maresh. 2000. Physiology of basketball. In *Exercise and sport science,* ed. W.E. Garrett and D.T. Kirkendall, 733-744. Philadelphia: Lippincott Williams & Wilkins.

Hoffman, J.R., and C.M. Maresh. 2011. Nutrition and hydration issues for combat sport athletes. *Strength and Conditioning Journal* 33:10-17.

Hoffman, J.R., C.M. Maresh, and L.E. Armstrong. 1992. Isokinetic and dynamic constant resistance strength testing: Implications for sport. *Physical Therapy Practice* 2:42-53.

Hoffman, J.R., C.M. Maresh, L.E. Armstrong, C.L. Gabaree, M.F. Bergeron, R.W. Kenefick, J.W. Castellani, L.E. Ahlquist, and A. Ward. 1994. The effects of hydration status on plasma testosterone, cortisol, and catecholamine concentrations before and during mild exercise at elevated temperature. *European Journal of Applied Physiology* 69:294-300.

Hoffman, J.R., C.M. Maresh, L.E. Armstrong, and W.J. Kraemer. 1991. Effects of off-season and in-season resistance training programs on a collegiate male basketball team. *Journal of Human Muscle Performance* 1:48-55.

Hoffman, J.R., and N.A. Ratamess. 2006. Medical issues associated with anabolic steroid use: Are they exaggerated? *Journal of Sports Science and Medicine* 5:182-193.

Hoffman, J.R., N.A. Ratamess, J.J. Cooper, J. Kang, A. Chilakos, and A.D. Faigenbaum. 2005. Comparison of loaded and unloaded squat jump training on strength/power performance in college football players. *Journal of Strength and Conditioning Research* 19:810-815.

Hoffman, J.R., N.A. Ratamess, A.D. Faigenbaum, G.T. Mangine, and J. Kang. 2007. Effects of maximal squat exercise testing on vertical jump performance in college football players. *Journal of Sports Science and Medicine* 6:149-150.

Hoffman, J.R., N.A. Ratamess, A.D. Faigenbaum, R. Ross, J. Kang, and J.R. Stout. 2008. Short duration β-alanine supplementation increases training volume and reduces subjective feelings of fatigue in college football players. *Nutrition Research* 28:31-35.

Hoffman, J.R., N.A. Ratamess, A. Gonzalez, N.A. Beller, M.W. Hoffman, M. Olson, M. Purpura, and R. Jager. 2010. The effects of acute and prolonged CRAM supplementation on reaction time and subjective measures of focus and alertness in healthy college students. *Journal of the International Society of Sports Nutrition* 7:39.

Hoffman, J.R., N.A. Ratamess, and J. Kang. 2011. Performance changes during a college playing career in NCAA division III football athletes. *Journal of Strength and Conditioning Research* 25:2351-2357.

Hoffman, J.R., N.A. Ratamess, J. Kang, M.J. Falvo, and A.D. Faigenbaum. 2006. Effect of protein intake on strength, body composition and endocrine changes in strength/power athletes. *Journal of the International Society of Sports Nutrition* 3:12-18.

Hoffman, J.R., N.A. Ratamess, J. Kang, M.J. Falvo, and A.D. Faigenbaum. 2007. Effects of protein supplementation on muscular performance and resting hormonal changes in college football players. *Journal of Sport Science and Medicine* 6:85-92.

Hoffman, J.R., N.A. Ratamess, J. Kang, A.M. Gonzalez, N.A. Beller, and S.A.S. Craig. 2011. Effect of 15 days of betaine ingestion on concentric and eccentric force outputs during isokinetic exercise. *Journal of Strength and Conditioning Research* 25:2235-2241.

Hoffman, J.R., N.A. Ratamess, J. Kang, G. Mangine, A.D. Faigenbaum, and J.R. Stout. 2006. Effect of creatine and β-alanine supplementation on performance and endocrine responses in strength/power athletes. *International Journal of Sport Nutrition and Exercise Metabolism* 16:430-446.

Hoffman, J.R., N.A. Ratamess, J. Kang, S.L. Rashti, and A.D. Faigenbaum. 2009. Effect of betaine supplementation on power performance and fatigue. *Journal of the International Society of Sports Nutrition* 6:7.

Hoffman, J.R., N.A. Ratamess, J. Kang, S.L. Rashti, N. Kelly, A.M. Gonzalez, M. Stec, S. Andersen, B.L. Bailey, L.M. Yamamoto, L.L. Hom, B.R. Kupchak, A.D. Faigenbaum, and C.M. Maresh. 2010. Examination of the efficacy of acute L-alanyl-L-glutamine during hydration stress in endurance exercise. *Journal of the International Society of Sports Nutrition* 7:8.

Hoffman, J.R., N.A. Ratamess, M. Klatt, A.D. Faigenbaum, R. Ross, N. Tranchina, R. McCurley, J. Kang, and W.J. Kraemer. 2009. Comparison between different resistance training programs in Division III American college football players. *Journal of Strength and Conditioning Research* 23:11-19.

Hoffman, J.R., N.A. Ratamess, R. Ross, J. Kang, J. Magrelli, K. Neese, A.D. Faigenbaum, and J.A. Wise. 2008. β-Alanine and the hormonal response to exercise. *International Journal of Sports Medicine* 29:952-958.

Hoffman, J.R., N.A. Ratamess, R. Ross, M. Shanklin, J. Kang, and A.D. Faigenbaum. 2008. Effect of a pre-exercise "high-energy" supplement drink on the acute hormonal response to resistance exercise. *Journal of Strength and Conditioning Research* 22:874-882.

Hoffman, J.R., N.A. Ratamess, C.P. Tranchina, S.L. Rashti, J. Kang, and A.D. Faigenbaum. 2009. Effect of protein supplement timing on strength, power and body compositional changes in resistance-trained men. *International Journal of Sport Nutrition and Exercise Metabolism* 19:172-185.

Hoffman, J.R., N.A. Ratamess, C.P. Tranchina, S.L. Rashti, J. Kang, and A.D. Faigenbaum. 2010. Effect of protein ingestion on recovery indices following a resistance training protocol in strength/power athletes. *Amino Acids* 38:771-778.

Hoffman, J.R., H. Stavsky, and B. Falk. 1995. The effect of water restriction on anaerobic power and vertical jumping height in basketball players. *International Journal of Sports Medicine* 16:214-218.

Hoffman, J.R., J.R. Stout, M. Falvo, J. Kang, and N. Ratamess. 2005. The effect of low-dose, short-duration creatine supplementation on anaerobic exercise performance. *Journal of Strength and Conditioning Research* 19:260-264.

Hoffman, J.R., G. Tenenbaum, C.M. Maresh, and W.J. Kraemer. 1996. Relationship between athletic performance tests and playing time in elite college basketball players. *Journal of Strength and Conditioning Research* 10:67-71.

Hoffman, J.R., J. Vazquez, N. Pichardo, and G. Tenenbaum. 2009. Anthropometric and performance comparisons in professional baseball players. *Journal of Strength and Conditioning Research* 23:2173-2178.

Hoffman, J.R., M. Wendell, J. Cooper, and J. Kang. 2003. Comparison between linear and nonlinear inseason resistance training programs in freshman football players. *Journal of Strength and Conditioning Research* 17:561-565.

Hoffman, J.R., D.R. Williams, N.S. Emerson, M.W. Hoffman, D.M. McVeigh, A.J. Wells, W.P. McCormack, G.T. Mangine, A.M. Gonzalez, and M.S. Fragala. 2012. L-alanyl-L-glutamine ingestion maintains performance during a competitive basketball game. *Journal of the International Society of Sports Nutrition* 9:4.

Holick, M.F. 2004. Sunlight and vitamin D for bone health and prevention of autoimmune diseases, cancers, and cardiovascular disease. *American Journal of Clinical Nutrition* 80:1678S-1688S

Holick, M.F. 2007. Vitamin D deficiency. *New England Journal of Medicine* 357:266-281.

Holloszy, J.O. 1988. Metabolic consequences of endurance exercise training. In *Exercise, nutrition, and energy metabolism,* ed. E.S. Horton and R.L. Terjung, 116-131. New York: Macmillan.

Holloszy, J.O., and F.W. Booth. 1976. Biochemical adaptations to endurance exercise in muscle. *Annual Review of Physiology* 38:273-291.

Holloszy, J.O., and E.F. Coyle. 1984. Adaptations of skeletal muscle to endurance exercise and their metabolic consequences. *Journal of Applied Physiology* 56:831-838.

Holloszy, J.O., L.B. Oscai, I.J. Don, and P.A. Mole. 1970. Mitochondrial citric acid cycle and related enzymes: Adapted response to exercise. *Biochemical and Biophysical Research Communications* 40:1368-1373.

Holloszy, J.O., J. Schultz, J. Kusnierkiewicz, J.M. Hagberg, and A.A. Eshani. 1986. Effects of exercise on glucose tolerance and insulin resistance: A brief review and some preliminary results. *Acta Medica Scandinavica* 711(suppl):55-65.

Holt, L.E., T.M. Travis, and T. Okita. 1970. Comparative study of three stretching techniques. *Perceptual and Motor Skills* 31:611-616.

Honig, A. 1983. Role of arterial chemoreceptors in the reflex control of renal function and body fluid volumes in acute arterial hypoxia. In *Physiology of the peripheral arterial chemoreceptors,* ed. H. Acher and R.G. O'Regan, 395-429. Amsterdam: Elsevier.

Hooper, S., L.T. Mackinnon, R.D. Gordon, and A.W. Bachmann. 1993. Hormonal responses of elite swimmers to overtraining. *Medicine and Science in Sports and Exercise* 25:741-747.

Hooper, S., L.T. Mackinnon, A. Howard, R.D. Gordon, and A.W. Bachmann. 1995. Markers for monitoring overtraining and recovery in elite swimmers. *Medicine and Science in Sports and Exercise* 27:106-112.

Hoppeler, H., E. Kleinert, C. Schlegel, H. Claassen, H. Howald, S.R. Kayar, and P. Cerretelli. 1990. Morphological adaptations of human skeletal muscle to chronic hypoxia. *International Journal of Sports Medicine* 11:S3-S9.

Hoppeler, H., S. Klossner, and M. Vogt. 2008. Training in hypoxia and its effects on skeletal muscle tissue. *Scandinavian Journal of Medicine and Science in Sports* 18(suppl):38-49.

Hori, N., R.U. Newton, K. Nosaka, and M.H. Stone. 2005. Weightlifting exercises enhance athletic performance that requires high-load speed strength. *Strength and Conditioning Journal* 27:50-55.

Horne, J.A., and A.N. Pettitt. 1984. Sleep deprivation and the physiological response to exercise under steady-state conditions in untrained subjects. *Sleep* 7:168-179.

Hornsby, W.G., and R.D. Chetlin. 2005. Management of competitive athletes with diabetes. *Diabetes Spectrum* 18:102-107.

Horowitz, J.F., and E.F. Coyle. 1993. Metabolic responses to pre-exercise meals containing various carbohydrates and fat. *American Journal of Clinical Nutrition* 58:235-241.

Horswill, C.A., D.L. Costill, W.J. Fink, M.G. Flynn, J.P. Kirwin, J.B. Mitchell, and J.A. Houmard. 1988. Influence of sodium bicarbonate on sprint performance: Relationship to dosage. *Medicine and Science in Sports and Exercise* 20:566-569.

Horswill, C.A., R.C. Hickner, J.R. Scott, D.L. Costill, and K. Gould. 1990. Weight loss, dietary carbohydrate modifications and high intensity physical performance. *Medicine and Science in Sports and Exercise* 22:470-476.

Hortobagyi, T., J.A. Houmard, J.R. Stevenson, D.D. Fraser, R.A. Johns, and R.G. Israel. 1993. The effects of detraining on power athletes. *Medicine and Science in Sports and Exercise* 25:929-935.

Horvath, S.M. 1981. Exercise in a cold environment. *Exercise and Sport Sciences Reviews* 9:221-263.

Houmard, J.A., D.L. Costill, J.A. Davis, J.B. Mitchell, D.D. Pascoe, and R.A. Robergs. 1990. The influence of exercise intensity on heat acclimation in trained subjects. *Medicine and Science in Sports and Exercise* 22:615-620.

Houmard, J.A., P.C. Egan, P.D. Neufer, J.E. Friedman, W.S. Wheeler, R.G. Israel, and G.L. Dohm. 1991. Elevated skeletal muscle glucose transporter levels in exercise-trained middle-aged men. *American Journal of Physiology* 261:E437-E443.

Houmard, J.A., and R.A. Jones. 1994. Effects of taper on swim performance. Practical applications. *Sports Medicine* 17:224-232.

Housh, T.J., G.O. Johnson, L. Marty, G. Eichen, C. Eishen, and D. Housh. 1988. Isokinetic leg flexion and extension strength of university football players. *Journal of Orthopaedic and Sports Physical Therapy* 9:365-369.

Houston, M.E., H. Bentzen, and H. Larsen. 1979. Interrelationships between skeletal muscle adaptations and performance as studied by detraining and retraining. *Acta Physiologica Scandinavica* 105:163-170.

Houston, M.E., D.A. Marin, H.J. Green, and J.A. Thomson. 1981. The effect of rapid weight loss on physiological function in wrestlers. *Physician and Sportsmedicine* 9:73-78.

Houston, M.E., D.M. Wilson, H.J. Green, J.A. Thomson, and D.A. Ranney. 1981. Physiological and muscle enzyme adaptations to two different intensities of swim training. *European Journal of Applied Physiology* 46:283-291.

Howald, H., H. Hoppeler, H. Claassen, O. Mathieu, and R. Staub. 1985. Influence of endurance training on the ultrastructural composition of the different muscle fiber types in humans. *Pflugers Archives* 403:369-376.

Howald, H., D. Pette, J.A. Simoneau, A. Uber, H. Hoppeler, and P. Cerretelli. 1990. Effect of chronic hypoxia on muscle enzyme activities. *International Journal of Sports Medicine* 11:S10-S14.

Howard, R.L., W.J. Kraemer, D.C. Stanley, L.E. Armstrong, and C.M. Maresh. 1994. The effects of cold immersion on muscle strength. *Journal of Strength and Conditioning Research* 8:129-133.

Hubbard, R.W. 1990. Heatstroke pathophysiology: The energy depletion model. *Medicine and Science in Sports and Exercise* 22:19-28.

Hubbard, R.W., and L.E. Armstrong. 1988. The heat illnesses: Biochemical, ultrastructural, and fluid-electrolyte considerations. In *Human performance physiology and environmental medicine at terrestrial extremes,* ed. K.B. Pandolf, M.N. Sawka, and R.R. Gonzalez, 305-360. Indianapolis: Benchmark Press.

Hubbard, R.W., and L.E. Armstrong. 1989. Hyperthermia: New thoughts on an old problem. *Physician and Sportsmedicine* 17:97-113.

Hubbard, R.W., B.L. Sandick, W.T. Matthews, R.P. Francesconi, J.B. Sampson, M.J. Durkot, O. Maller, and D.B. Engell. 1984. Voluntary dehydration and alliesthesia for water. *Journal of Applied Physiology: Respiratory, Environmental and Exercise Physiology* 57:868-875.

Hubbard, R.W., P.C. Szlyk, and L.E. Armstrong. 1990. Influence of thirst and fluid palatability on fluid ingestion during exercise. In *Perspectives in exercise science and sports medicine, Vol. 3, Fluid homeostasis during exercise,* ed. C.V. Gisolfi and D.R. Lamb, 39-96. Carmel, IN: Benchmark Press.

Hubinger, L.M., L.T. Mackinnon, L. Barber, J. McCosker, A. Howard, and F. Lepre. 1997. The acute effects of treadmill running on lipoprotein (a) levels in males and females. *Medicine and Science in Sports and Exercise* 29:436-442.

Hughes, J.R., A.H. Oliveto, A. Liguori, J. Carpenter, and T. Howard. 1998. Endorsement of DSM-IV dependence criteria among caffeine users. *Drug and Alcohol Dependency* 52:99-107.

Hughes, R.J., G.O. Johnson, T.J. Housh, J.P. Weir, and J.E. Kinder. 1996. The effect of submaximal treadmill running on serum testosterone levels. *Journal of Strength and Conditioning Research* 10:224-227.

Hulmi, J.J., V. Kovanen, H. Selanne, W.J. Kraemer, K. Häkkinen, and A.A. Mero. 2009. Acute and long-term effects of resistance exercise with or without protein ingestion on muscle hypertrophy and gene expression. *Amino Acids* 37:297-308.

Hulsmann, W.C., and M.L. Dubelaar. 1988. Aspects of fatty acid metabolism in vascular endothelial cells. *Biochimie* 70:681-868.

Hulsmann, W.C., and M.L. Dubelaar. 1992. Carnitine requirement of vascular endothelial and smooth muscle cells in imminent ischemia. *Molecular and Cellular Biochemistry* 116:125-129.

Hultman, E., G. Cederblad, and P. Harper. 1991. Carnitine administration as a tool to modify energy metabolism during exercise. *European Journal of Applied Physiology* 62:450.

Hultman, E., K. Soderlund, J.A. Timmons, G. Cederblad, and P.L. Greenhaff. 1996. Muscle creatine loading in man. *Journal of Applied Physiology* 81:232-237.

Hunter, G.R. 1985. Changes in body composition, body build and performance associated with different weight training frequencies in males and females. *National Strength and Conditioning Association Journal* 7:26-28.

Hunter, G., R. Demment, and D. Miller. 1987. Development of strength and maximum oxygen uptake during simultaneous training for strength and endurance. *Journal of Sports Medicine and Physical Fitness* 27:269-275.

Hunter, G.R., J. Hilyer, and M. Forster. 1993. Changes in fitness during 4 years of intercollegiate basketball. *Journal of Strength and Conditioning Research* 7:26-29.

Hunter, J.R., B.J. O'Brien, M.G. Mooney, J. Berry, W.B. Young, and N. Down. 2011. Repeated sprint training improves intermittent peak running speed in team-sport athletes. *Journal of Strength and Conditioning Research* 25:1318-1325.

Hurley, B.F., J.M. Hagberg, W.K. Allen, D.R. Seals, J.C. Young, R.W. Cuddihee, and J.O. Holloszy. 1984. Effect of training on blood lactate levels during submaximal exercise. *Journal of Applied Physiology* 56:1260-1264.

Huttunen, N.P., S.L. Lankela, M. Knip, P. Lautala, M.L. Kaar, K. Laasonen, and P. Puuka. 1989. Effect of once-a-week training program on physical fitness and metabolic control in children with IDDM. *Diabetes Care* 12:737-740.

Huttunen, P., J. Hirvonen, and V. Kinnula. 1981. The occurrence of brown adipose tissue in outdoor workers. *European Journal of Applied Physiology* 46:339-345.

Huxley, H. 1969. The mechanism of muscular contraction. *Science* 164:1356-1366.

Huynh, M.L., V.A. Fadok, and P.M. Henson. 2002. Phosphatidylserine-dependent ingestion of apoptotic cells promotes tgf-β1 secretion and the resolution of inflammation. *Journal of Clinical Investigation* 109:41-50.

Hydren, J.R., W.J. Kraemer, J.S. Volek, C. Dunn-Lewis, B.A. Comstock, T.K. Szivak, D.R. Hooper, C.R. Denegar, and C.M. Maresh. 2013. Performance changes during a weeklong high-altitude alpine ski-racing training camp in lowlander young athletes. *Journal of Strength and Conditioning Research* 27:924-937.

Hymer, W.C., R.E. Grindeland, B.C. Nindl, and W.J. Kraemer. 2005. Growth hormone variants and human exercise. In *The endocrine system in sports and exercise,* ed. W.J. Kraemer and A.D. Rogol. Malden, MA: Blackwell.

Iaia, F.M., and J. Bangsbo. 2010. Speed endurance training is a powerful stimulus for physiological adaptations and performance improvements of athletes. *Scandinavian Journal of Medicine and Science in Sports* 20:11-23.

Idstrom, J.P., C.B. Subramanian, B. Chance, T. Schersten, and A.C. Bylund-Fellenius. 1985. Oxygen dependence of energy metabolism in contracting and recovery rat skeletal muscle. *American Journal of Physiology* 248:H40-H48.

Imray, C.H.E., and J.W. Castellani. 2012. Nonfreezing cold-induced injuries. In *Wilderness medicine,* 6th ed., ed. P.S. Auerbach, 171-180. Philadelphia: Elsevier Mosby.

Ingham, S.A., H. Carter, G.P. Whyte, and J.H. Doust. 2008. Physiological and performance effects of low-versus mixed intensity rowing training. *Medicine and Science in Sports and Exercise* 40:579-584.

Ingjer, F. 1979a. Capillary supply and mitochondrial content of different skeletal muscle fiber types in untrained and endurance trained men: A histochemical and ultra structural study. *European Journal of Applied Physiology* 40:197-209.

Ingjer, F. 1979b. Effects of endurance training on muscle fiber ATPase activity, capillary supply and mitochondrial content in man. *Journal of Physiology* 294:419-432.

Irving, L.M., M. Wall, D. Neumark-Sztainer, and M. Story. 2002. Steroid use among adolescents: Findings from project EAT. *Journal of Adolescent Health* 30:243-252.

Israel, S. 1976. Problems of overtraining from an internal medical and performance physiological standpoint. *Medizin Sport* 16:1-12.

Issurin, V.B., D.G. Liebermann, and G. Tenenbaum. 1994. Effect of vibratory stimulation training on maximal force and flexibility. *Journal of Sports Science* 12:561-566.

Itoh, M., B. Grassi, C. Marconi, P. Cerretelli, H. Araki, and K. Nishi. 2002. VE responses to VCO2 during exercise is unaffected by exercise training and different exercise limbs. *Japan Journal of Physiology* 52:489-496.

Iverson, P.O., B.L. Arvesen, and H.B. Benestad. 1994. No mandatory role for the spleen in the exercise-induced leucocytosis in man. *Clinical Science* 86:505-510.

Ivy, J.L. 1997. Role of exercise training in the prevention and treatment of insulin resistance and non-insulin-dependent diabetes mellitus. *Sports Medicine* 24:3221-3236.

Ivy, J.L., D.L. Costill, W.J. Fink, and R.W. Lower. 1979. Influence of caffeine and carbohydrate feedings on endurance performance. *Medicine and Science in Sports and Exercise* 11:6-11.

Ivy, J.L., A.L. Katz, C.L. Cutler, W.M. Sherman, and E.F. Coyle. 1988. Muscle glycogen synthesis after exercise: Effect of time of carbohydrate ingestion. *Journal of Applied Physiology* 64:1480-1485.

Ivy, J.L., T.W. Zderic, and D.L. Fogt. 1999. Prevention and treatment of non-insulin-dependent diabetes mellitus. *Exercise and Sport Sciences Reviews* 27:1-35.

Izquierdo, M., K. Hakkinen, J. Ibanez, W.J. Kraemer, and E.M. Gorostiaga. 2005. Effects of combined resistance and cardiovascular training on strength, power, muscle cross-sectional area, and endurance markers in middle-aged men. *European Journal of Applied Physiology* 94:70-75.

Izquierdo, M., J. Ibañez, J.A.L. Calbet, I. Navarro-Amezqueta, M. Gonzalez-Izal, F. Idoate, K. Hakkinen, W.J. Kraemer, M. Palacios-Sarrasqueta, M. Almar, and E.M. Gorostiaga. 2009. Cytokine and hormone responses to resistance training. *European Journal of Applied Physiology* 107:397-409.

Izquierdo, M., J. Ibañez, J.J. González-Badillo, K. Häkkinen, N.A. Ratamess, W.J. Kraemer, D.N. French, J. Eslava, A. Altadill, X. Asiain, and E.M. Gorostiaga. 2006. Differential effects of strength training leading to failure versus not to failure on hormonal responses, strength, and muscle power gains. *Journal of Applied Physiology* 100:1647-1656.

Izquierdo, M., J. Ibanez, J.J. Gonzalez-Badillo, N.A. Ratamess, W.J. Kraemer, K. Hakkinen, H. Bonnabau, C. Granados, D.N. French, and E.M. Gorostiaga. 2007. Detraining and tapering effects on hormonal responses and strength performance. *Journal of Strength and Conditioning Research* 21:768-775.

Jaakkola, P., D.R. Mole, Y.M. Tian, M.I. Wilson, J. Gielbert, S.J. Gaskell, A.V. Kriegsheim, H.F. Hebestreit, M. Mukherji, C.J. Schofield, P.H. Maxwell, C.W. Pugh, and P.J. Ratcliffe. 2001. Targeting of HIF-alpha to the von Hippel-Lindau ubiquitylation complex by O2-regulated prolyl hydroxylation. *Science* 292:468-472.

Jackson, A.S., and M.L. Pollock. 1985. Practical assessment of body composition. *Physician and Sportsmedicine* 13:76-90.

Jacobs, I. 1980. The effects of thermal dehydration on performance on the Wingate anaerobic test. *International Journal of Sports Medicine* 1:21-24.

Jacobs, I., and D.G. Bell. 2004. Effects of acute modafinil ingestion on exercise time to exhaustion. *Medicine and Science in Sports and Exercise* 36:1078-1082.

Jacobs, I., M. Esbjornsson, C. Sylven, I. Holm, and E. Jansson. 1987. Sprint training effects on muscle myoglobin, enzymes, fiber types, and blood lactate. *Medicine and Science in Sports and Exercise* 19:368-374.

Jacobs, I., H. Pasternak, and D.G. Bell. 2003. Effects of ephedrine, caffeine, and their combination on muscular endurance. *Medicine and Science in Sports and Exercise* 35:987-994.

Jaffee, W.B., E. Trucco, S. Levy, and R.D. Weiss. 2007. Is this urine really negative? A systematic review of tampering methods in urine drug screening and testing. *Journal of Substance Abuse and Treatment* 33:33-42.

Jäger, R., R.C. Harris, M. Purpura, and M. Francaux. 2007. Comparison of new forms of creatine in raising plasma creatine levels. *Journal of the International Society of Sports Nutrition* 4:17.

Jakeman, P., and S. Maxwell. 1993. Effect of antioxidant vitamin supplementation on muscle function after eccentric exercise. *European Journal of Applied Physiology* 67:426-430.

Janeway, C.A., and P. Travers. 1996. *Immunobiology: The immune system in health and disease,* 2nd ed. London: Current Biology Ltd.

Jankowski, R.J., B.M. Deasy, and J. Huard. 2002. Muscle-derived stem cells. *Gene Therapy* 9:642-647.

Jansson, E., M. Esbjornsson, I. Holm, and I. Jacobs. 1990. Increases in the proportion of fast-twitch muscle fibres in sprint training in males. *Acta Physiologica Scandinavica* 140:359-363.

Jansson, E., B. Sjodin, and P. Tesch. 1978. Changes in muscle fibre type distribution in man after physical training. *Acta Physiologica Scandinavica* 104:235-237.

Jansson, E., C. Sylven, and B. Sjodin. 1983. Myoglobin content and training in humans. In *Biochemistry of exercise, Vol. 13,* ed. H.G. Knuttgen, J.A. Vogel, and J. Poortmans, 821-825. Champaign, IL: Human Kinetics.

Jasuja, R., P. Ramaraj, R.P. Mae, A.B. Singh, T.W. Storer, J. Artaza, A. Miller, R. Singh, W.E. Taylor, M.L. Lee, T. Davidson, I. Sinha-Hikim, N. Gonzalez-Cadavid, and S. Bhasin. 2005. Δ-4-androstene-3,17-dione binds androgen receptor, promotes myogenesis in vitro, and increases serum testosterone levels, fat free mass, and muscle strength in hypogonadal men. *Journal of Clinical Endocrinology and Metabolism* 90:855-863.

Jedlickova, K., D.W. Stockton, H. Chen, J. Stray-Gundersen, S. Witkowski, G. Ri-Li, J. Jelinek, B.D. Levine, and J.T. Prchal. 2003. Search for genetic determinants of individual variability of the erythropoietin response to high altitude. *Blood Cells, Molecules, and Diseases* 31:175-182.

Jenkins, D.J.A., T.M.S. Wolever, R.H. Taylor, H. Baker, H. Fielden, J.M. Baldwin, A.C. Bowling, H.C. Newman, A.L. Jenkins, and D.V. Goff. 1981. Glycemic index of foods: A physiological basis for carbohydrate exchange. *American Journal of Clinical Nutrition* 34:362-366.

Jenkinson, C., A. Petroczi, J. Barker, and C.P. Naughton. 2012. Dietary green and white teas suppress UDP-glucuronsyltransferase UGT2B17 mediated testosterone glucuronidation. *Steroids* 77:691-695.

Jeukendrup, A.E. 2003. Modulation of carbohydrate and fat utilization by diet, exercise and environment. *Biochemical Society Transactions* 31:1270-1273.

Jeukendrup, A.E., F. Brouns, A.J.M. Wagenmakers, and W.H.M. Saris. 1997. Carbohydrate-electrolyte feedings improve 1 h time trial cycling performance. *International Journal of Sports Medicine* 18:125-129.

Jezova, D., M. Vigas, P. Tatar, R. Kevtnansky, K. Nazar, H. Kaciuba-Uscilko, and S. Kozlowski. 1985. Plasma testosterone and catecholamine responses to physical exercise of different intensities in men. *European Journal of Applied Physiology* 54:62-66.

Jockenhovel, F., C. Bullmann, M. Schubert, E. Vogel, W. Reinhardt, D. Reinwein, D. Muller-Wieland, and W. Krone. 1999. Influence of various modes of androgen substitution on serum lipids and lipoproteins in hypogonadal men. *Metabolism* 48:590-596.

Johansen, K.L., K. Mulligan, and M. Schambelan. 1999. Anabolic effects of nandrolone decanoate in patients receiving dialysis: A randomized controlled trial. *Journal of the American Medical Association* 281:1275-1281.

Johns, R.J., and V. Wright. 1962. Relative importance of various tissues in joint stiffness. *Journal of Applied Physiology* 17:824-828.

Johnson, B.L., K.J.W. Adamczy, K.O. Tennoe, and S.B. Stromme. 1976. A comparison of concentric and eccentric muscle training. *Medicine and Science in Sports and Exercise* 8:35-38.

Jones, M.T., B.M. Parker, and N. Cortes. 2011. The effect of whole-body vibration training and conventional strength training on performance measures in female athletes. *Journal of Strength and Conditioning Research* 25:2434-2441.

Jones, S.R., R.A. Binder, and E.M. Donowho Jr. 1970. Sudden death in sickle cell trait. *New England Journal of Medicine* 282:323-325.

Jost, J., M. Weiss, and H. Weicker. 1989. Comparison of sympatho-adrenergic regulation at rest and of the adrenoreceptor system in swimmers, long-distance runners, weight lifters, wrestlers, and untrained men. *European Journal of Applied Physiology* 58:596-604.

Jowko, E., P. Ostaszewski, M. Jank, J. Sacharuk, A. Zieniewicz, J. Wilczak, and S. Nissen. 2001. Creatine and β-hydroxy-β-methylbutyrate (HMB) additively increase lean body mass and muscle strength during a weight-training program. *Nutrition* 17:558-566.

Jurimae, J., P. Hofmann, T. Jurimae, R. Palm, J. Maestu, P. Purge, K. Sudi, K. Rom, and S. von Duvillard. 2007. Plasma ghrelin responses to acute sculling exercises in elite male rowers. *European Journal of Applied Physiology* 99:467-474.

Jurimae, J., J. Maestu, and T. Jurimae. 2003. Leptin as a marker of training stress in highly trained male rowers? *European Journal of Applied Physiology* 90:533-538.

Kalra, S., V.A. Swystun, R. Bhagat, and D.W. Cockcroft. 1996. Inhaled corticosteroids do not prevent the development of tolerance to the bronchoprotective effect of salmeterol. *Chest* 109:953-956.

Kalupahana, N.S., N. Moustaid-Moussa, and K.J. Claycombe. 2012. Immunity as a link between obesity and insulin resistance. *Molecular Aspects of Medicine* 33:26-34.

Kaminski, M., and R. Boal. 1992. An effect of ascorbic acid on delayed-onset muscle soreness. *Pain* 50:317-321.

Kammer, L., Z. Ding, B. Wang, D. Hara, Y. Liao, and J.L. Ivy. 2009. Cereal and nonfat milk support muscle recovery following exercise. *Journal of the International Society of Sports Nutrition* 6:11.

Kanakis, C., and R.C. Hickson. 1980. Left ventricular responses to a program of lower-limb strength training. *Chest* 78:618-621.

Kanehisa, H., and M. Miyashita. 1983. Specificity of velocity in strength training. *European Journal of Applied Physiology* 52:104-106.

Kaneko, M., T. Fuchimoto, H. Toji, and K. Suei. 1983. Training effect of different loads on the force-velocity relationship and mechanical power output in human muscle. *Scandinavian Journal of Sports Sciences* 5:50-55.

Kang, J., J.R. Hoffman, J. Im, B.A. Spiering, N.A. Ratamess, K.W. Rundell, S. Niokoa, J. Cooper, and B. Chance. 2005. Evaluation of physiological responses during recovery following three resistance exercise programs. *Journal of Strength and Conditioning Research* 19:305-309.

Kang, J., J.R. Hoffman, H. Walker, and E.C. Chaloupka. 2001. Regulating intensity of exercise using ratings of perceived exertion during treadmill exercise [abstract]. *Medicine and Science in Sports and Exercise* 33:S84.

Kanj, H., S.H. Schneider, and N.B. Ruderman. 1988. Exercise and diabetes mellitus. In *Exercise, nutrition and energy metabolism,* ed. E.S. Horton and R.L. Terjung, 228-241. New York: Macmillan.

Kanter, M. 1995. Free radicals and exercise: Effects of nutritional antioxidant supplementation. *Exercise and Sport Sciences Reviews* 23:375-398.

Kanter, M.M., L.A. Nolte, and J.O. Holloszy. 1993. Effects of an antioxidant vitamin mixture on lipid peroxidation at rest and postexercise. *Journal of Applied Physiology* 74:965-969.

Karagiorgos, A., J.F. Garcia, and G.A. Brooks. 1979. Growth hormone response to continuous and intermittent exercise. *Medicine and Science in Sports and Exercise* 11:302-307.

Karila, T.A., J.E. Karjalainen, M.J. Mantysaari, M.T. Viitasalo, and T.A. Seppala. 2003. Anabolic androgenic steroids produce dose-dependant increase in left ventricular mass in power athletes, and this effect is potentiated by concomitant use of growth hormone. *International Journal of Sports Medicine* 24:337-343.

Karslon, J., B. Dumont, and B. Saltin. 1971. Muscle metabolites during submaximal and maximal exercise in man. *Scandinavian Journal of Clinical Laboratory Investigation* 26:385-394.

Karvonen, M.J., E. Kentala, and O. Mustala. 1957. The effects of training on heart rate: A longitudinal study. *Annales Medicinae Experimentalis et Biologiae Fenniae* 35:307-315.

Katan, M.B., and E. Schouten. 2005. Caffeine and arrhythmia. *American Journal of Clinical Nutrition* 81:539-540.

Katch, V., A. Weltman, R. Martin, and L. Gray. 1977. Optimal test characteristics for maximal anaerobic work on the bicycle ergometer. *Research Quarterly for Exercise and Sport* 48:319-327.

Katz, A., D.L. Costill, D.S. King, M. Hargreaves, and W.J. Fink. 1984. Maximal exercise tolerance after induced alkalosis. *International Journal of Sports Medicine* 5:107-110.

Katz, A., and K. Sahlin. 1988. Regulation of lactic acid production during exercise. *Journal of Applied Physiology* 65:509-518.

Kaufman, W.C., and D.J. Bothe. 1986. Wind chill reconsidered, Siple revisited. *Aviation, Space, and Environmental Medicine* 57:23-26.

Kawamori, N., A.J. Crum, P.A. Blumert, J.R. Kulik, J.T. Childers, J.A. Wood, M.H. Stone, and G.G. Haff. 2005. Influence of different relative intensities on power output during the hang power clean: Identification of the optimal load. *Journal of Strength and Conditioning Research* 19:698-708.

Kawamori, N., and G.G. Haff. 2004. The optimal training load for the development of muscular power. *Journal of Strength and Conditioning Research* 18:675-684.

Kawamori, N., K. Nosaka, and R.U. Newton. 2013. Relationships between ground reaction impulse and sprint acceleration performance in team sport athletes. *Journal of Strength and Conditioning Research* 27:568-573.

Keen, P., D.A. McCarthy, L. Passfield, H.A.A. Shaker, and A.J. Wade. 1995. Leucocyte and erythrocyte counts during a multi-stage cycling race ("The Milk Race"). *British Journal of Sports Medicine* 29:61-65.

Keizer, H.A. 1998. Neuroendocrine aspects of overtraining. In *Overtraining in sport,* ed. R.B. Kreider, A.C. Fry, and M.L. O'Toole, 145-168. Champaign, IL: Human Kinetics.

Kelly, D., J.K. Hamilton, and M.C. Riddell. 2010. Blood glucose levels and performance in a sports cAMP for adolescents with type 1 diabetes mellitus: A field study. *International Journal of Pediatrics* 2010:pii: 216167.

Kelly, J.P., J.H. Rosenberg, and J.C. Stevens. 1997. Practice parameter: The management of concussion in sports (summary statement). *Neurology* 48:581-585.

Kendrick, I.P., R.C. Harris, H.J. Kim, C.K. Kim, V.H. Dang, T.Q. Lam, T.T. Bui, M. Smith, and J.A. Wise. 2008. The effects of 10 weeks of resistance training combined with beta-alanine supplementation on whole body strength, force production, muscular endurance and body composition. *Amino Acids* 34:547-554.

Kendrick, I.P., H.J. Kim, R.C. Harris, V.H. Dang, T.Q. Lam, T.T. Bui, and J.A. Wise. 2009. The effect of 4 weeks beta-alanine supplementation and isokinetic training on carnosine concentrations in type I and II human skeletal muscle fibres. *European Journal of Applied Physiology* 106:131-138.

Kenefick, R.W., S.N. Cheuvront, J.W. Castellani, and C. O'Brien. 2008. Thermal stress. In *Fundamentals of aerospace medicine,* 4th ed., ed. J.R. Davis, R. Johnson, J. Stepanek, and J.A. Fogarty, 206-220. Philadelphia: Wolter Kluwer, Lippincott Williams & Wilkins.

Kenny, G.P., A.R. Schissler, J. Stapleton, M. Piamonte, K. Binder, A. Lynn, C.Q. Lan, and S.G. Hardcastle. 2011. Ice cooling vest on tolerance for exercise under uncompensable heat stress. *Journal of Occupational and Environmental Hygiene* 8:484-491.

Kenny, W.L., J.H. Wilmore, and D.L. Costill. 2012. *Physiology of sport and exercise,* 5th ed. Champaign, IL: Human Kinetics.

Keogh, J.W.L., A.L. Payne, B.B. Anderson, and P.J. Atkins. 2010. A brief description of the biomechanics and physiology of a strongman event: The tire flip. *Journal of Strength and Conditioning Research* 24:1223-1228.

Kerner, J., and C. Hoppel. 2000. Fatty acid import into mitochondria. *Biochimica Biophysica Acta* 1486:1-17.

Kicman, A.T., R.V. Brooks, and D.A. Cowan. 1991. Human chorionic gonadotrophin and sport. *British Journal of Sports* Medicine 25:73-80.

Kierzkowska, B., J. Stanczyk, and J.D. Kasprzak. 2005. Myocardial infarction in a 17-year-old body builder using clenbuterol. *Circulation Journal* 69:1144-1146.

Kilic, M., A.K. Baltaci, M. Gunay, H. Gökbel, N. Okudan, and I. Cicioglu. 2006. The effect of exhaustion exercise on thyroid hormones and testosterone levels of elite athletes receiving oral zinc. *Neuroendocrinology Letters* 27:247-252.

Kimball, S.R., and L.S. Jefferson. 2004. Regulation of global and specific mRNA translation by oral administration of branched-chain amino acids. *Biochemical and Biophysical Research Communications* 313:423-427.

Kindermann, S., A. Schnabel, W.M. Schmitt, G. Biro, J. Cassens, and F. Weber. 1982. Catecholamines, growth hormone, cortisol, insulin and sex hormones in anaerobic and aerobic exercise. *European Journal of Applied Physiology* 49:389-399.

King, D.S., R.L. Sharp, M.D. Vukovich, G.A. Brown, T.A. Reifenrath, N.L. Uhi, and K.A. Parsons. 1999. Effect of oral androstenedione on serum testosterone and adaptations to resistance training in young men. *Journal of the American Medical Association* 281:2020-2028.

Kingsley, M., L.P. Kilduff, J. McEneny, R. Dietzig, and D. Benton. 2006. Phosphatidylserine supplementation and recovery following downhill running. *Medicine and Science in Sports and Exercise* 38:1617-1625.

Kingsley, M., M. Miller, L.P. Kilduff, J. McEneny, and D. Benton. 2006. Effects of phosphatidylserine on exercise capacity during cycling in active males. *Medicine and Science in Sports and Exercise* 38:64-71.

Kingsley, M., D. Wadsworth, L.P. Kilduff, J. McEneny, and D. Benton. 2005. Effects of phosphatidylserine on oxidative stress following intermittent running. *Medicine and Science in Sports and Exercise* 37:1300-1306.

Kirby, R.L., F.C. Simms, V.J. Symington, and J.B. Garner. 1981. Flexibility and musculoskeletal symptomatology in female gymnasts and age-matched controls. *American Journal of Sports Medicine* 9:160-164.

Kirkendall, D.T. 2000. Physiology of soccer. In *Exercise and sport science,* ed. W.E. Garrett and D.T. Kirkendall, 875-884. Philadelphia: Lippincott Williams & Wilkins.

Kirksey, K.B., M.H. Stone, B.J. Warren, R.L. Johnson, M. Stone, G.G. Haff, F.E. Williams, and Proulx, C. 1999. The effects of 6 weeks of creatine monohydrate supplementation on performance measures and body composition in collegiate track and field athletes. *Journal of Strength and Conditioning Research* 13:148-156.

Kirsch, K.A., H. von Ameln, and H.J. Wicke. 1981. Fluid control mechanisms after exercise dehydration. *European Journal of Applied Physiology* 47:191-196.

Kirwin, J.P., D.L. Costill, M.G. Flynn, J.B. Mitchell, W.J. Fink, P.D. Neufer, and J.A. Houmard. 1988. Physiological responses to successive days of intense training in competitive swimmers. *Medicine and Science in Sports and Exercise* 20:255-259.

Kitaura, T., N. Tsunekawa, and W.J. Kraemer. 2002. Inhibited longitudinal growth of bones in young male rats by clenbuterol. *Medicine and Science in Sports and Exercise* 34:267-273.

Kjaer, M. 1989. Epinephrine and some other hormonal responses to exercise in man: With special reference to physical training. *International Journal of Sports Medicine* 10:2-15.

Kjaer, M., J. Bangsbo, G. Lortie, and H. Galbo. 1988. Hormonal response to exercise in humans: Influence of hypoxia and physical training. *American Journal of Physiology* 254:R197-R203.

Kjaer, M., and H. Galbo. 1988. Effect of physical training on the capacity to secrete epinephrine. *Journal of Applied Physiology* 64:11-16.

Klinzing, J., and E. Karpowicz. 1986. The effects of rapid weight loss and rehydration on wrestling performance test. *Journal of Sports Medicine and Physical Fitness* 26:149-156.

Knapik, J.J., C.L. Bauman, B.H. Jones, J.M. Harris, and L. Vaughan. 1991. Preseason strength and flexibility imbalances associated with athletic injuries in female collegiate athletes. *American Journal of Sports Medicine* 19:76-81.

Knapik, J.J., R.H. Mawdsley, and M.V. Ramos. 1983. Angular specificity and test mode specificity of isometric and isokinetic strength testing. *Journal of Orthopaedic and Sports Physical Therapy* 5:58-65.

Knapp, P.E., T.W. Storer, K.L. Herbst, A.B. Singh, C. Dzekov, J. Dzekov, M. LaValley, A. Zhang, J. Ulloor, and S. Bhasin. 2008. Effects of supraphysiological dose of testosterone on physical function, muscle performance, mood, and fatigue in men with HIV-associated weight loss. *American Journal of Physiology: Endocrinology and Metabolism* 294:E1135-E1143.

Knaupp, W., S. Khilnani, J. Sherwood, S. Scharf, and H. Steinberg. 1992. Erythropoietin response to acute normobaric hypoxia in humans. *Journal of Applied Physiology* 73:837-840.

Knitter, A.E., L. Panton, J.A. Rathmacher, A. Petersen, and R. Sharp. 2000. Effects of β-hydroxy-β-methylbutyrate on muscle damage after a prolonged run. *Journal of Applied Physiology* 89:1340-1344.

Knobil, E., and J. Hotchkiss. 1964. Growth hormone. *Annual Review of Physiology* 26:47-74.

Knuttgen, H.G., and W.J. Kraemer. 1987. Terminology and measurement in exercise performance. *Journal of Applied Sports Science Research* 1:1-10.

Koehler, K., M.K. Parr, H. Geyer, J. Mester, and W. Schänzer. 2009. Serum testosterone and urinary excretion of steroid hormone metabolites after administration of a high-dose zinc supplement. *European Journal of Clinical Nutrition* 63:65-70.

Kohler, J.M., S.P. Flanagan, and W.C. Whiting. 2010. Muscle activation patterns while lifting stable and unstable loads on stable and unstable surfaces. *Journal of Strength and Conditioning Research* 24:313-321.

Kohn, T.A., B. Essen-Gustavsson, and K.H. Myburgh. 2011. Specific muscle adaptations in type II fibers after high-intensity interval training of well-trained runners. *Scandinavian Journal of Medicine and Science in Sports* 21:765-772.

Koivisto, V., R. Hendler, E. Nadel, and P. Felig. 1982. Influence of physical training on the fuel-hormone response to prolonged low intensity exercise. *Metabolism* 31:192-197.

Koivisto, V., V. Soman, E. Nadel, W.V. Tamborlane, and P. Felig. 1980. Exercise and insulin: Insulin binding, insulin mobilization and counterregulatory hormone secretion. *Federation Proceedings* 39:1481-1486.

Kojima, M., H. Hosoda, Y. Date, M. Nakazoto, H. Matsuo, and K. Kangawa. 1999. Ghrelin is a growth-hormone releasing acylated peptide from stomach. *Nature* 402:656-660.

Kokkonen, J., and A.G. Nelson. 1996. Acute stretching exercises inhibit maximal force performance. *Medicine and Science in Sports and Exercise* 28:S1130.

Komi, P.V. 1986. Training of muscle strength and power: Interaction of neuromotoric, hypertrophic and mechanical factors. *International Journal of Sports Medicine* 7:10-15.

Komi, P.V. 2003. Stretch-shortening cycle. In *The encyclopedia of sports medicine: Strength and power in sport,* 2nd ed., ed. P.V. Komi, 184-202. Oxford: Blackwell Science.

Komi, P.V., J. Karlsson, P. Tesch, H. Souminen, and E. Hakkinen. 1982. Effects of heavy resistance and explosive-type strength training methods on mechanical, functional and metabolic aspects of performance. In *Exercise and sport biology,* ed. P.V. Komi, 99-102. *International Series on Sports Sciences, Vol. 12.* Champaign, IL: Human Kinetics.

Kontos, A.P., T. Covassin, R.J. Elbin, and T. Parker. 2012. Depression and neurocognitive performance after concussion among male and female high school and collegiate athletes. *Archives of Physical Medicine and Rehabilitation* 93:1751-1756.

Koopman, R., B. Pennings, A.H. Zorenc, and L.J. van Loon. 2007. Protein ingestion further augments S6K1 phosphorylation in skeletal muscle following resistance exercise in males. *Journal of Nutrition* 137:1880-1886.

Koopman, R., A.J.M. Wagenmakers, R.J.F. Manders, A.H.G. Zorenc, J.M.G. Senden, M. Gorselink, H.A. Keizer, and L.J.C. Van Loon. 2005. Combined ingestion of protein and free leucine with carbohydrate increases postexercise muscle protein synthesis in vivo in male subjects. *American Journal of Physiology: Endocrinology and Metabolism* 288:E645-E653.

Kosaka, A., H. Takahashi, Y. Yajima, M. Tanaka, K. Okamura, R. Mizumoto, and K. Katsuta. 1996. Hepatocellular carcinoma associated with anabolic steroid therapy: Report of a case and review of the Japanese literature. *Journal of Gastroenterology* 31:450-454.

Kouri, E.M., S.E. Lukas, H.G. Pope, and P.S. Oliva. 1995. Increased aggressive responding in male volunteers following the administration of gradually increasing doses of testosterone cypionate. *Drug and Alcohol Dependency* 40:73-79.

Kovacs, E.M.R., J.H.C.H. Stegen, and F. Brouns. 1998. Effect of caffeinated drinks on substrate metabolism, caffeine excretion, and performance. *Journal of Applied Physiology* 85:709-715.

Kozak-Collins, K., E. Burke, and R.B. Schoene. 1994. Sodium bicarbonate ingestion does not improve performance in women cyclists. *Medicine and Science in Sports and Exercise* 26:1510-1515.

Koziris, L.P., W.J. Kraemer, J.F. Patton, N.T. Triplett, A.C. Fry, S.E. Gordon, and H.G. Knuttgen. 1996. Relationship of aerobic power to anaerobic performance indices. *Journal of Strength and Conditioning Research* 10:35-39.

Kraemer, R.R., and V.D. Castracane. 2007. Exercise and humoral mediators of peripheral energy balance: Ghrelin and adiponectin. *Experimental Biology and Medicine* 232:184-194.

Kraemer, W.J. 1992a. Endocrine responses and adaptations to strength training. In *Strength and power in sport,* ed. P.V. Komi, 291-304. London: Blackwell Scientific.

Kraemer, W.J. 1992b. Hormonal mechanisms related to the expression of muscular strength and power. In *Strength and power in sport,* ed. P.V. Komi, 64-76. London: Blackwell Scientific.

Kraemer, W.J. 1994. Neuroendocrine responses to resistance exercise. In *Essentials of strength training and conditioning,* ed. T. Baechle, 86-107. Champaign, IL: Human Kinetics.

Kraemer, W.J. 1997. A series of studies: The physiological basis for strength training in American football. *Journal of Strength and Conditioning Research* 11:131-142.

Kraemer, W.J., B.A. Aguilera, M. Terada, R.U. Newton, J.M. Lynch, G. Rosendaal, J.M. McBride, S.E. Gordon, and K. Hakkinen. 1995. Responses of IGF-1 to endogenous increases in growth hormone after heavy-resistance exercise. *Journal of Applied Physiology* 79:1310-1315.

Kraemer, W.J., B.A. Comstock, J.E. Clark, and C. Dunn-Lewis. 2012. Athlete needs analysis. In *NSCA's guide to program design,* ed. J.R. Hoffman, 1-22. Champaign, IL: Human Kinetics.

Kraemer, W.J., M.R. Deschenes, and S.J. Fleck. 1988. Physiological adaptations to resistance exercise. Implications for athletic conditioning. *Sports Medicine* 6:246-256.

Kraemer, W.J., J.E. Dziados, S.E. Gordon, L.J. Marchitelli, A.C. Fry, and K.L. Reynolds. 1990. The effects of graded exercise on plasma proenkephalin peptide F and catecholamine responses at sea level. *European Journal of Applied Physiology* 61:214-217.

Kraemer, W.J., J.E. Dziados, L.J. Marchitelli, S.E. Gordon, E. Harman, R. Mello, S.J. Fleck, P. Frykman, and N.T. Triplett. 1993. Effects of different heavy-resistance exercise protocols on plasma b-endorphin concentrations. *Journal of Applied Physiology* 74:450-459.

Kraemer, W.J., and S.J. Fleck. 2007. *Optimizing strength training: Designing nonlinear periodization workouts.* Champaign, IL: Human Kinetics.

Kraemer, W.J., S.J. Fleck, J.E. Dziados, E. Harman, L.J. Marchitelli, S.E. Gordon, R. Mello, P. Frykman, L.P. Koziris, and N.T. Triplett. 1993. Changes in hormonal concentrations after different heavy-resistance exercise protocols in women. *Journal of Applied Physiology* 75:594-604.

Kraemer, W.J., A.C. Fry, B.J. Warren, M.H. Stone, S.J. Fleck, J.T. Kearney, B.P. Conroy, C.M. Maresh, C.A. Weseman, N.T. Triplett, and S.E. Gordon. 1992. Acute hormonal responses in elite junior weightlifters. *International Journal of Sports Medicine* 13:103-109.

Kraemer, W.J., S.E. Gordon, S.J. Fleck, L.J. Marchitelli, R. Mello, J.E. Dziados, K. Friedl, E. Harman, C. Maresh, and A.C. Fry. 1991. Endogenous anabolic hormonal and growth factor responses to heavy resistance exercise in male and females. *International Journal of Sports Medicine* 12:228-235.

Kraemer, W.J., and L.A. Gotshalk. 2000. Physiology of American football. In *Exercise and sport science,* ed. W.E. Garrett and D.T. Kirkendall, 795-813. Philadelphia: Lippincott Williams & Wilkins.

Kraemer, W.J., K. Hakkinen, R.U. Newton, B.C. Nindl, J.S. Volek, M. McCormick, L.A. Gotshalk, S.E. Gordon, S.J. Fleck, W.W. Campbell, M. Putukian, and W.J. Evans. 1999. Effects of heavy-resistance training on hormonal response patterns in younger vs. older men. *Journal of Applied Physiology* 87:982-992.

Kraemer, W.J., D.L. Hatfield, J.S. Volek, M.S. Fragala, J.L. Vingren, J.M. Anderson, B.A. Spiering, G.A. Thomas, J.Y. Ho, E.E. Quann, M. Izquierdo, K. Hakkinen, and C.M. Maresh. 2009. Effects of amino acids supplement on physiological adaptations to resistance training. *Medicine and Science in Sports and Exercise* 41:1111-1121.

Kraemer, W.J., L. Marchitelli, S. Gordon, E. Harmon, J. Dziados, R. Mello, P. Frykman, D. McCurry, and S. Fleck. 1990. Hormonal and growth factor responses to heavy resistance exercise protocols. *Journal of Applied Physiology* 69:1442-1450.

Kraemer, W.J., and R.U. Newton. 2000. Training for muscular power. *Physical Medicine and Rehabilitation Clinics of North America* 11:341-368.

Kraemer, W.J., B.J. Noble, M.J. Clark, and B.W. Culver. 1987. Physiological responses to heavy-resistance exercise with very short rest periods. *International Journal of Sports Medicine* 8:247-252.

Kraemer, W.J., B. Noble, B. Culver, and R.V. Lewis. 1985. Changes in plasma proenkephalin peptide F and catecholamine levels during graded exercise in men. *Proceedings of the National Academy of Sciences U S A* 82:6349-6351.

Kraemer, W.J., J.F. Patton, S.E. Gordon, E. Harman, M.R. Deschenes, K. Reynolds, R.U. Newton, N. Travis-Triplett, and J.E. Dziados. 1995. Compatibility of high-intensity strength and endurance training on hormonal and skeletal muscle adaptations. *Journal of Applied Physiology* 78:976-989.

Kraemer, W.J., J.F. Patton, H.G. Knuttgen, C.J. Hannon, T. Kettler, S.E. Gordon, J.E. Dziados, A.C. Fry, P.N. Frykman, and E.A. Harman. 1991. Effects of high-intensity cycle exercise on sympathoadrenal-medullary response patterns. *Journal of Applied Physiology* 70:8-14.

Kraemer, W.J., N.A. Ratamess, J.S. Volek, K. Hakkinen, M.R. Rubin, D.N. French, A.L. Gomez, M.R. McGuigan, T.P. Scheet, R.U. Newton, B.A. Spiering, M. Izquierdo, and F.S. Dioguardi. 2006. The effects of amino acid supplementation on hormonal responses to overreaching. *Metabolism* 55:282-291.

Kraemer, W.J., B.A. Spiering, J.S. Volek, N.A. Ratamess, M.J. Sharman, M.R. Rubin, D.N. French, R. Silvestre, D.L. Hatfield, J.L. Van Heest, J.L. Vingren, D.A. Judelson, M.R. Deschenes, and C.M. Maresh. 2006. Androgenic responses to resistance exercise: Effects of feeding and L-carnitine. *Medicine and Science in Sports and Exercise* 38:1288-1296.

Kraemer, W.J., R.S. Staron, F.C. Hagerman, R.S. Hikida, A.C. Fry, S.E. Gordon, B.C. Nindl, L.A. Gothshalk, J.S. Volek, J.O. Marx, R.U. Newton, and K. Häkkinen. 1998. The effects of short-term resistance training on endocrine function in men and women. *European Journal of Applied Physiology and Occupational Physiology* 78:69-76.

Kraemer, W.J., J.S. Volek, K.L. Clark, S.E. Gordon, S.M. Puhl, L.P. Koziris, J.M. McBride, N.T. Triplett-McBride, M. Putukian, R.U. Newton, K. Hakkinen, J.A. Bush, and W.J. Sebastianelli. 1999. Influence of exercise training on physiological and performance changes with weight loss in men. *Medicine and Science in Sports and Exercise* 31:1320-1329.

Kraemer, W.J., J.S. Volek, D.N. French, M.R. Rubin, M.J. Sharman, A.L. Gomez, N.A. Ratamess, R.U. Newton, B. Jemiolo, B.W. Craig, and K. Hakkinen. 2003. The effects of L-carnitine L-tartrate supplementation on hormonal responses to resistance exercise and recovery. *Journal of Strength and Conditioning Research* 17:455-462.

Kreider, R.B., M. Ferreira, M. Wilson, and A.L. Almada. 1999. Effects of calcium β-hydroxy-β-methylbutyrate (HMB) supplementation during resistance-training on markers of catabolism, body composition and strength. *International Journal of Sports Medicine* 20:503-509.

Kreider, R.B., M. Ferreira, M. Wilson, P. Grindstaff, S. Plisk, J. Reinardy, E. Cantler, and A.L. Almada. 1998. Effects of creatine supplementation on body composition, strength, and sprint performance. *Medicine and Science in Sports and Exercise* 30:73-82.

Kreider, R.B., A.C. Fry, and M.L. O'Toole. 1998. *Overtraining in sport.* Champaign, IL: Human Kinetics.

Kruse, P., J. Ladefoged, U. Nielsen, P. Paulev, and J.P. Sorensen. 1986. b-blockade used in precision sports: Effect on pistol shooting performance. *Journal of Applied Physiology* 61:417-420.

Krustrup, P., J.J. Nielsen, B.R. Krustrup, J.F. Christensen, H. Pedersen, M.B. Randers, P. Aagaard, A.M. Petersen, L. Nybo, and J. Bangsbo. 2009. Recreational soccer is an effective health-promoting activity for untrained men. *British Journal of Sports Medicine* 43:825-831.

Kudlac, J., D.L. Nichols, C.F. Sanborn, and N.M. DiMarco. 2004. Impact of detraining on bone loss in former collegiate female gymnasts. *Calcified Tissue International* 75:482-487.

Kuipers, H., and H.A. Keizer. 1988. Overtraining in elite athletes: Review and direction for the future. *Sports Medicine* 6:79-92.

Kuipers, H., F.M. Peeze Binkhorst, F. Hartgens, J.A.G. Wijnen, and H.A. Keizer. 1993. Muscle ultrastructure after strength training with placebo or anabolic steroid. *Canadian Journal of Applied Physiology* 18:189-196.

Kumagai, K., T. Abe, W.F. Bruechue, T. Ryushi, S. Takano, and M. Mizuno. 2000. Sprint performance is related to muscle fascicle length in male 100 m sprinters. *Journal of Applied Physiology* 88:811-816.

Kuoppasalmi, K., and H. Adlercreutz. 1985. Interaction between catabolic and anabolic steroid hormones in muscular exercise. In *Exercise endocrinology,* ed. K. Fotherby and S.B. Pal, 65-98. Berlin: Walter de Gruyter.

Kuoppasalmi, K., H. Naveri, M. Harkonen, and H. Adlercreutz. 1980. Plasma cortisol, testosterone and luteinizing hormone in running exercise of different intensities. *Scandinavian Journal of Clinical Laboratory Investigation* 40:403-409.

Kurz, M.J., K. Berg, R. Latin, and W. DeGraw. 2000. The relationship of training methods in NCAA division I cross-country runners and 10,000-meter performance. *Journal of Strength and Conditioning Research* 14:196-201.

Kyriazis, G.A., J.D. Caplan, J. Lowndes, R.L. Carpenter, K.E. Dennis, S.A. Sivio, and T.J. Angelopoulos. 2007. Moderate exercise-induced energy expenditure does not alter leptin levels in sedentary obese men. *Clinical Journal of Sport Medicine* 17:49-51.

Laakso, M., S.V. Edelman, G. Brechtel, and A.D. Baron. 1992. Impaired insulin-mediated skeletal muscle blood flow in patients with NIDDM. *Diabetes* 41:1076-1083.

Lachowetz, T., J. Evon, and J. Pastiglione. 1998. The effect of an upper body strength program on intercollegiate baseball throwing velocity. *Journal of Strength and Conditioning Research* 12:116-119.

Lactorraca, S., P. Piersanti, G. Tesco, S. Piacentini, L. Amaducci, and S. Sorbi. 1993. Effect of phosphatidylserine on free radical susceptibility in human diploid fibroblasts. *Journal of Neural Transmission: Parkinson's Disease and Dementia Section* 6:73-77.

Landers, J. 1985. Maximum based on reps. *National Strength and Conditioning Association Journal* 6:60-61.

Lange, K.H., B. Larsson, A. Flyvbjerg, R. Dali, M. Bennekou, M.H. Rasmussen, H. Orskov, and M. Kjaer. 2002. Acute growth hormone administration causes exaggerated increases in plasma lactate and glycerol during moderate to high intensity bicycling in trained young men. *Journal of Clinical Endocrinology and Metabolism* 87:4966-4975.

Langlois, J.A., W. Rutland-Brown, and M.M. Wald. 2006. The epidemiology and impact of traumatic brain injury: A brief overview. *Journal of Head Trauma and Rehabilitation* 21:375-378.

Larsson, L.P., P. Hemmingsson, and G. Boethius. 1994. Self-reported obstructive airway symptoms are common in young cross-country skiers. *Scandinavian Journal of Medicine and Science in Sports* 4:124-127.

Larsson, P. 2003. Global positioning system and sport specific testing. *Sports Medicine* 33:1093-1101.

Latin, R.W., K. Berg, and T. Baechle. 1994. Physical and performance characteristics of NCAA division I male basketball players. *Journal of Strength and Conditioning Research* 8:214-218.

Lau, W.Y., and K. Nosaka. 2011. Effect of vibration treatment on symptoms associated with eccentric exercise-induced muscle damage. *American Journal of Physical Medicine and Rehabilitation* 90:648-657.

Laubach, L.L. 1976. Comparative muscular strength of men and women: A review of the literature. *Aviation, Space, and Environmental Medicine* 47:534-542.

Laurent, D., K.E. Schneider, W.K. Prusaczyk, C. Franklin, S.M. Vogel, M. Krssk, K.F. Petersen, H.W. Goforth, and G.I. Shulman. 2000. Effects of caffeine on muscle glycogen utilization and the neuroendocrine axis during exercise. *Journal of Clinical Endocrinology and Metabolism* 85:2170-2175.

Leder, B.Z., C. Longcope, D.H. Catlin, B. Ahrens, D.A. Schoenfeld, and J.S. Finkelstein. 2000. Oral androstenedione administration and serum concentrations in young men. *Journal of the American Medical Association* 283:779-782.

Lee, C.L., J.C. Lin, and C.F. Cheng. 2011. Effect of caffeine ingestion after creatine supplementation on intermittent high-intensity sprint performance. *European Journal of Applied Physiology* 111:1669-1677.

Lee, E.C., C.M. Maresh, W.J. Kraemer, L.M. Yamamoto, D.L. Hatfield, B.L. Bailey, L.E. Armstrong, J.S. Volek, B.P. McDermott, and S.A. Craig. 2010. Ergogenic effects of betaine supplementation on strength and power performance. *Journal of the International Society of Sports Nutrition* 7:27.

Lehmann, M., C. Foster, N. Netzer, W. Lormes, J.M. Steinacker, Y. Liu, A. Opitz-Gress, and U. Gastmann. 1998. Physiological responses to short- and long-term overtraining in endurance athletes. In *Overtraining in sport,* ed. R.B. Kreider, A.C. Fry, and M.L. O'Toole, 19-46. Champaign, IL: Human Kinetics.

Lehmann, M., E. Jakob, U. Gastmann, J.M. Steinacker, N. Heinz, and F. Brouns. 1995. Unaccustomed high mileage compared to high intensity-related performance and neuromuscular responses in distance runners. *European Journal of Applied Physiology* 70:457-461.

Lehmann, M., H. Mann, U. Gastmann, J. Keul, D. Vetter, J.M. Steinacker, and D. Haussinger. 1996. Unaccustomed high-mileage vs intensity training-related changes in performance and serum amino acid levels. *International Journal of Sports Medicine* 17:187-192.

Lehmann, M., W. Schnee, R. Scheu, W. Stockhausen, and N. Bachl. 1992. Decreased nocturnal catecholamine excretion: Parameter for an overtraining syndrome in athletes? *International Journal of Sports Medicine* 13:236-242.

Lehmkuhl, M., M. Malone, B. Justice, G. Trone, E. Pistilli, D. Vinci, E. Haff, J.L. Kilgore, and G.G. Haff. 2003. The effects of 8-weeks of creatine monohydrate and glutamine supplementation on body composition and performance measures. *Journal of Strength and Conditioning Research* 17:425-438.

Leiper, J.B., and R.J. Maughan. 1988. Experimental models for the investigation of water and solute transport in man: Implications for oral rehydration solutions. *Drugs* 36(suppl):65-79.

Leithead, C.S., and E.R. Gunn. 1964. The aetiology of cane cutters cramps in British Guiana. In *Environmental physiology and psychology in arid conditions,* 13-17. Liege, Belgium: UNESCO.

Lemon, P.W.R. 1995. Do athletes need more dietary protein and amino acids? *International Journal of Sport Nutrition* 5:S39-S61.

Lemon, P.W.R., M.A. Tarnopolsky, J.D. McDougall, and S.A. Atkinson. 1992. Protein requirements and muscle mass/strength changes during intensive training in novice bodybuilders. *Journal of Applied Physiology* 73:767-775.

Len, T.K., J.P. Neary, G.J.G. Asmundson, D.G. Goodman, B. Bjornson, and Y.N. Bhambhani. 2011. Cerebrovascular reactivity impairment after sport-induced concussion. *Medicine and Science in Sports and Exercise* 43:2241-2248.

Lentini, A.C., R.S. McKelvie, N. McCartney, C.W. Tomlinson, and J.D. MacDougall. 1993. Assessment of left ventricular response of strength trained athletes during weightlifting exercise. *Journal of Applied Physiology* 75:2703-2710.

Levenhagen, D.K., J.D. Gresham, M.G. Carlson, D.J. Maron, M.J. Borel, and P.J. Flakoll. 2001. Postexercise nutrient intake timing in humans is critical to recovery of leg glucose and protein homeostasis. *American Journal of Physiology: Endocrinology and Metabolism* 280:E982-E993.

Leveritt, M., and P.J. Abernethy. 1999. Acute effects of high-intensity endurance exercise on subsequent resistance activity. *Journal of Strength and Conditioning Research* 13:47-51.

Levin, G.T., M.R. McGuigan, and P.B. Laursen. 2009. Effect of concurrent resistance and endurance training on physiologic and performance parameters of well-trained endurance cyclists. *Journal of Strength and Conditioning Research* 23:2280-2286.

Levine, B.D., and J. Stray-Gundersen. 1997. "Living high-training low": Effect of moderate-altitude acclimatization with low-altitude training on performance. *Journal of Applied Physiology* 83:102-112.

Lewicki, R., H. Tchorzewski, A. Denys, M. Kowalska, and A. Golinska. 1987. Effect of physical exercise on some parameters of immunity in conditioned sportsmen. *International Journal of Sports Medicine* 8:309-314.

Lewicki, R., H. Tchorzewski, E. Majewska, Z. Nowak, and Z. Baj. 1988. Effect of maximal physical exercise on T-lymphocyte subpopulations and on interleukin 1 IL-1 and interleukin 2 IL-2 production in vitro. *International Journal of Sports Medicine* 9:114-117.

Lewis, S.F., W.F. Taylor, R.M. Graham, W.A. Pettinger, J.E. Schutte, and C.G. Blomqvist. 1983. Cardiovascular responses to exercise as functions of absolute and relative work load. *Journal of Applied Physiology* 54:1314-1323.

Lewis, T. 1930. Observations upon the reactions of the vessels of the human skin to cold. *Heart* 15:177-208.

Li, C.H., and H. Papkoff. 1956. Preparation and properties of growth hormone from human and monkey pituitary glands. *Science* 124:1293-1294.

Li, Y., B. Xu, F. Liu, L. Tan, and J. Li. 2006. The effect of glutamine-supplemented total parenteral nutrition on nutrition and intestinal absorptive function in a rat model. *Pediatric Surgery International* 22:508-513.

Lieberman, H.R. 2003. Nutrition, brain function and cognitive performance. *Appetite* 40:245-254.

Liesen, H., B. Dufaux, and W. Hollmann. 1977. Modifications of serum glycoproteins on the days following a prolonged physical exercise and the influence of physical training. *European Journal of Applied Physiology* 37:243-254.

Lima, A.A., G.H. Carvalho, A.A. Figueiredo, A.R. Gifoni, A.M. Soares, E.A. Silva, and R.L. Guerrant. 2002. Effects of an alanyl-glutamine-based oral rehydration and nutrition therapy solution on electrolyte and water absorption in a rat model of secretory diarrhea induced by cholera toxin. *Nutrition* 18:458-462.

Linde, F. 1987. Running and upper respiratory tract infections. *Scandinavian Journal of Sports Sciences* 9:21-23.

Lindlinger, M.I., J.M. Kowalchuk, and G.J.F. Heigenhauser. 2005. Applying physiochemical principles to skeletal muscle acid-base status. *American Journal of Physiology: Regulatory, Integrative and Comparative Physiology* 289:R891-R894.

Linossier, M.T., D. Dormois, P. Bregere, A. Geyssant, and C. Denis. 1997. Effect of sodium citrate on performance and metabolism of human skeletal muscle during supramaximal cycling exercise. *European Journal of Applied Physiology* 76:48-54.

Linossier, M.T., D. Dormois, A. Geyssant, and C. Denis. 1997. Performance and fibre characteristics of human skeletal muscle during short sprint training and detraining on a cycle ergometer. *European Journal of Applied Physiology and Occupational Physiology* 75:491-498.

Linossier, M.T., D. Dormois, C. Perier, J. Frey, A. Geyssant, and C. Denis. 1997. Enzyme adaptations of human skeletal muscle during bicycle short-sprint training and detraining. *Acta Physiologica Scandinavica* 161:439-445.

Lippi, G., M. Franchini, and G.C. Guidi. 2007. Prohibition of artificial hypoxic environments in sports: Health risks rather than ethics. *Applied Physiology, Nutrition, and Metabolism* 32:1206-1207.

Little, T., and A.G. Williams. 2005. Specificity of acceleration, maximum speed, and agility in professional soccer players. *Journal of Strength and Conditioning Research* 19:76-78.

Liu, H., D.M. Bravata, I. Okin, A. Friedlander, V. Liu, B. Roberts, E. Bendavid, O. Saynina, S.R. Salpeter, A.M. Garber, and A.R. Hoffman. 2008. Systematic review: The effects of growth hormone on athletic performance. *Annals of Internal Medicine* 148:747-758.

Liversedge, L.A. 1956. Glycocyamine and betaine in motor-neuron disease. *Lancet* 2:1136-1138.

Llewellyn, W. 2007. *William Llewellyn's anabolics 2007,* 6th ed. Jupiter, FL: Body of Science.

Lloyd, F.H., P. Powell, and A.P. Murdoch. 1996. Anabolic steroid abuse by body builders and male subfertility. *British Medical Journal* 313:100-101.

Lockie, R.G., A.J. Murphy, and C.D. Spinks. 2003. Effects of resisted sled towing on sprint kinematics in field-sport athletes. *Journal of Strength and Conditioning Research* 17:760-767.

Loftin, M., M. Sothern, C. Koss, G. Tuuri, C. Van Vrancken, A. Kontos, and M. Bonisi. 1997. Energy expenditure and influence of physiologic factors during marathon running. *Journal of Strength and Conditioning Research* 21:1188-1191.

Lohman, T.G. 1981. Skinfolds and body density and their relation to body fatness: A review. *Human Biology* 53:181-225.

Lopez, R.M., D.J. Casa, B.P. McDermott, M.S. Ganio, L.E. Armstrong, and C.M. Maresh. 2009. Does creatine supplementation hinder exercise heat tolerance or hydration status? A systematic review with meta-analyses. *Journal of Athletic Training* 44:215-223.

Lopez, R.M., D.J. Casa, B.P. McDermott, R.L. Stearns, L.E. Armstrong, and C.M. Maresh. 2011. Exertional heat stroke in the athletic setting. A review of the literature. *Athletic Training and Sports Health Care* 3:189-200.

Lorino, A.J., L.K. Lloyd, S.H. Crixell, and J.L. Walker. 2006. The effects of caffeine on athletic agility. *Journal of Strength and Conditioning Research* 20:851-854.

Loughton, S.J., and R.O. Ruhling. 1977. Human strength and endurance responses to anabolic steroids and training. *Journal of Sports Medicine and Physical Fitness* 17:285-296.

Lucia, A., J. Hoyos, and J.L. Chicharro. 2001. Preferred pedalling cadence in professional cycling. *Medicine and Science in Sports and Exercise* 33:1361-1366.

Lugar, A., B. Watschinger, P. Duester, T. Svoboda, M. Clodi, and G.P. Chrousos. 1992. Plasma growth hormone and prolactin responses to graded levels of acute exercise and to a lactate infusion. *Neuroendocrinology* 56:112-117.

Luhtanen, P., and P.V. Komi. 1978. Mechanical factors influencing running speed. In *Biomechanics VI-B,* ed. E. Asmussen and K. Jorgensen, 25. Baltimore: University Park Press.

Luke, J.L., A. Farb, R. Virmani, and R.H. Sample. 1990. Sudden cardiac death during exercise in a weight lifter using anabolic androgenic steroids: Pathological and toxicological findings. *Journal of Forensic Science* 35:1441-1447.

Lun, V., K.A. Erdman, T.S. Fung, and R.A. Reimer. 2012. Dietary supplementation practices in Canadian high-performance athletes. *International Journal of Sport Nutrition and Exercise Metabolism* 22:31-37.

Lund, S., H. Vestergaad, P.H. Anderson, O. Schmitz, L.B.H. Gotzsche, and O. Pedersen. 1993. GLUT-4 content in plasma membrane of muscle from patients with non-insulin-dependent diabetes mellitus. *American Journal of Physiology* 265:E889-E897.

Lundby, C., J.J. Thomsen, R. Boushel, M. Koskolou, J. Warberg, J.A.L. Calbet, and P. Robach. 2007. Erythopoietin treatment elevates haemoglobin concentration by increasing red cell volume and depressing plasma volume. *Journal of Physiology* 578:309-314.

Lusiani, L., G. Ronsisvalle, A. Bonanome, A. Visona, V. Castellani, C. Macchia, and A. Pagnan. 1986. Echocardiographic evaluation of the dimensions and systolic properties of the left ventricle in freshman athletes during physical training. *European Heart Journal* 7:196-203.

Lynch, T.J., A.S. Ryan, J. Evans, L.I. Katzel, and A.P. Goldber. 2007. Older elite football players have reduced cardiac and osteoporosis risk factors. *Medicine and Science in Sports and Exercise* 39:1124-1130.

Lyons, G.E., A.M. Kelly, and N.A. Rubinstein. 1986. Testosterone induced changes in contractile protein isoforms in the sexually dimorphic temporalis muscle of the guinea pig. *Journal of Biology and Chemistry* 261:13278-13284.

Lyons, T.J., and J. French. 1991. Modafinil: The unique properties of a new stimulant. *Aviation, Space, and Environmental Medicine* 62:432-435.

Macchi, G. 1987. Long-term effects of competitive sport activity on the heart of professional ex-athletes. *Italian Journal of Cardiology* 17:505-510.

MacConnie, S.E., A. Barkin, R.M. Lampman, M.A. Schork, and I.Z. Beitins. 1986. Decreased hypothalamic gonadotropin-releasing hormone secretion in male marathon runners. *New England Journal of Medicine* 315:411-417.

MacDougall, J.D. 1992. Hypertrophy or hyperplasia. In *Strength and power in sport,* ed. P.V. Komi, 230-238. London: Blackwell Scientific.

MacDougall, J.D. 1994. Blood pressure responses to resistive, static, and dynamic exercise. In *Cardiovascular response to exercise,* ed. G.F. Fletcher, 155-174. Mount Kisco, NY: Futura.

MacDougall, J.D., H.J. Green, J.R. Sutton, G. Coates, A. Cymerman, P. Young, and C.S. Houston. 1991. Operation Everest II: Structural adaptations in skeletal muscle in response to extreme simulated altitude. *Acta Physiologica Scandinavica* 142:431-427.

MacDougall, J.D., R.S. McKelvie, D.E. Moroz, D.G. Sale, N. McCartney, and F. Buick. 1992. Factors affecting blood pressure response during heavy weightlifting and static contractions. *Journal of Applied Physiology* 73:1590-1597.

MacDougall, J.D., W.G. Reddan, C.R. Layton, and J.A. Dempsey. 1974. Effects of metabolic hyperthermia on performance during heavy prolonged exercise. *Journal of Applied Physiology* 36:538-544.

MacDougall, J.D., D.G. Sale, S.E. Alway, and J.R. Sutton. 1984. Muscle fiber number in biceps brachii in bodybuilders and control subjects *Journal of Applied Physiology: Respiratory, Environmental and Exercise Physiology* 57:1399-1403.

MacDougall, J.D., D.G. Sale, G.C.B. Elder, and J.R. Sutton. 1982. Muscle ultrastructural characteristics of elite powerlifters and bodybuilders. *European Journal of Applied Physiology* 48:117-126.

MacDougall, J.D., D.G. Sale, J.R. Moroz, G.C.B. Elder, J.R. Sutton, and H. Howard. 1979. Mitochondrial volume density in human skeletal muscle following heavy resistance exercise. *Medicine and Science in Sports and Exercise* 11:164-166.

MacDougall, J.D., D. Tuxen, D.G. Sale, J.R. Moroz, and J.R. Sutton. 1985. Arterial blood pressure response to heavy resistance exercise. *Journal of Applied Physiology* 58:785-790.

MacDougall, J.D., G.R. Ward, D.G. Sale, and J.R. Sutton. 1977. Biochemical adaptations of human skeletal muscle to heavy resistance training and immobilization. *Journal of Applied Physiology* 43:700-703.

Mackinnon, L.T. 1999. *Advances in exercise immunology.* Champaign, IL: Human Kinetics.

Mackinnon, L.T., T.W. Chick, A. van As, and T.B. Tomasi. 1988. Effects of prolonged intense exercise on natural killer cell number and function. In *Exercise physiology: Current selected research, Vol. 3,* ed. C.O. Dotson and J.H. Humphrey, 77-89. New York: AMS Press.

Mackinnon, L.T., T.W. Chick, A. van As, and T.B. Tomasi. 1989. Decreased secretory immunoglobulins following intense endurance exercise. *Sports Medicine, Training, and Rehabilitation* 1:209-218.

Mackinnon, L.T., E. Ginn, and G. Seymour. 1991. Effects of exercise during sports training and competition on salivary IgA levels. In *Behaviour and immunity,* ed. A.J. Husband, 169-177. Boca Raton, FL: CRC Press.

Mackinnon, L.T., E. Ginn, and G. Seymour. 1993. Decreased salivary immunoglobulin A secretion rate after intense interval exercise training in elite kayakers. *European Journal of Applied Physiology* 67:180-184.

Mackinnon, L.T., and S.L. Hooper. 1994. Mucosal (secretory) immune system responses to exercise of varying intensity and during overtraining. *International Journal of Sports Medicine* 15:S179-S183.

Mackinnon, L.T., and S.L. Hooper. 1996. Plasma glutamine concentration and upper respiratory tract infection during overtraining in elite swimmers. *Medicine and Science in Sports and Exercise* 28:285-290.

Mackinnon, L.T., S.L. Hooper, S. Jones, A.W. Bachmann, and R.D. Gordon. 1997. Hormonal, immunological and hematological responses to intensified training in elite swimmers. *Medicine and Science in Sports and Exercise* 29:1637-1645.

MacLaren, D.P.M., T. Reilly, I.T. Campbell, and C. Hopkin. 1999. Hormonal and metabolic responses to maintained hyperglycemia during prolonged exercise. *Journal of Applied Physiology* 87:124-131.

Macleod, H., J. Morris, A. Nevill, and C. Sunderland. 2009. The validity of a non-differential global positioning system for assessing player movement patterns in field hockey. *Journal of Sports Sciences* 27:121-128.

MacRae, J.C., P.A. Skene, A. Connell, V. Buchan, and G.E. Lobley. 1988. The action of the β_2-agonist clenbuterol on protein and energy metabolism in fattening wether lambs. *British Journal of Nutrition* 59:457-465.

Magazanik, A., Y. Weinstein, R.A. Dlin, M. Derin, and S. Schwartzman. 1988. Iron deficiency caused by 7 weeks of intensive physical exercise. *European Journal of Applied Physiology* 57:198-202.

Magel, J.R., G.F. Foglia, W.D. McArdle, B. Gutin, and G.S. Pechar. 1975. Specificity of swim training on maximal oxygen uptake. *Journal of Applied Physiology* 38:151-155.

Magnusson, S.P., E.B. Simonsen, P. Aagaard, and M. Kjaer. 1996. Biomechanical responses to repeated stretches in human hamstring muscle in vivo. *American Journal of Sports Medicine* 24:622-628.

Mahesh, V.B., and R.B. Greenblatt. 1962. The in vivo conversion of dehydroepiandrosterone and androstenedione to testosterone in the human. *Acta Endocrinologica* 41:400-406.

Mäkinen, T.T., D. Gavhed, I. Holmér, and H. Rintamäki. 2001. Effects of metabolic rate on thermal responses at different air velocities in -10°C. *Comparative Biochemistry and Physiology Part A* 128:759-768.

Malone, D.A. Jr., R.J. Dimeff, J.A. Lombardo, and R.H. Sample. 1995. Psychiatric effects and psychoactive substance use in anabolic-androgenic steroid users. *Clinical Journal of Sport Medicine* 5:25-31.

Maltin, C.A., M.I. Delday, J.S. Watson, S.D. Heys, I.M. Nevison, I.K. Ritchie, and P.H. Gibson. 1993. Clenbuterol, a beta-adrenoceptor agonist, increases relative muscle strength in orthopaedic patients. *Clinical Science* 84:651-654.

Manach, C., G. Williamson, C. Morand, A. Scalbert, and C. Remesy. 2005. Bioavailability and bioefficiency of polyphenols in humans. I. Review of 97 bioavailability studies. *American Journal of Clinical Nutrition* 81:230S-242S.

Mangine, G.T., J.R. Hoffman, M.S. Fragala, J. Vazquez, M.C. Krause, J. Gillett, and N. Pichardo. 2013. Effect of age on anthropometric and physical performance measures in professional baseball players. *Journal of Strength and Conditioning Research* 27:375-381.

Mangine, G.T., N.A. Ratamess, J.R. Hoffman, A.D. Faigenbaum, J. Kang, and A. Chilakos. 2008. The effects of combined ballistic and heavy resistance training on maximal lower- and upper-body strength in recreationally-trained men. *Journal of Strength and Conditioning Research* 22:132-139.

Mangine, R.E., F.R. Noyes, M.P. Mullen, and S.D. Baker. 1990. A physiological profile of the elite soccer athlete. *Journal of Orthopaedic and Sports Physical Therapy* 12:147-152.

Manning, P.J., R.M. Watson, and P.M. O'Byrne. 1993. Exercise-induced refractoriness in asthmatic subjects involving leukotriene and prostaglandin interdependent mechanisms. *American Review of Respiratory Disease* 148:950-954.

Mannion, A.F., P.M. Jakeman, and P.L. Willan. 1994. Effects of isokinetic training of the knee extensors on high-intensity exercise performance and skeletal muscle buffering. *European Journal of Applied Physiology and Occupational Physiology* 68:356-361.

Maresh, C.M., B.C. Wang, and K.L. Goetz. 1985. Plasma vasopressin, renin activity, and aldosterone responses to maximal exercise in active college females. *European Journal of Applied Physiology* 54:398-403.

Margaria, R., P. Aghemo, and E. Rovelli. 1966. Measurement of muscular power (anaerobic) in man. *Journal of Applied Physiology* 21:1662-1664.

Margonis, K., I.G. Fatouros, A.Z. Jamurtas, M.G. Nikolaidis, I. Douroudos, A. Chatzinikolaou, A. Mitrakou, G. Mastorakos, I. Papassotiriou, K. Taxildaris, and D. Kouretas. 2007. Oxidative stress biomarkers responses to physical overtraining: Implications for diagnosis. *Free Radical Biology and Medicine* 43:901-910.

Maron, B.J. 1986. Structural features of the athletic heart as defined by echocardiography. *Journal of the American College of Cardiology* 7:190-203.

Maron, B.J., J.J. Doerer, T.S. Haas, D.M. Tierney, and F.O. Mueller. 2009. Sudden deaths in young competitive athletes: Analysis of 1866 deaths in the United States, 1980-2006. *Circulation* 119:1085-1092.

Maron, B.J., T.E. Gohman, S.B. Kyle, N.A. Estes 3rd, and M.S. Link. 2002. Clinical profile and spectrum of commotio cordis. *Journal of the American Medical Association* 287:1142-1146.

Maron, B.J., J. Shirani, L.C. Poliac, R. Mathenge, W.C. Roberts, and F.O. Mueller. 1996. Sudden death in young competitive athletes. Clinical, demographic, and pathological profiles. *Journal of the American Medical Association* 276:199-204.

Maron, B.J., P.D. Thompson, M.J. Ackerman, G. Balady, D. Cohen, R. Dimeff, P.S. Douglas, D.W. Glover, A.M. Hutter Jr., M.D. Krauss, M.S. Maron, M.J. Mitten, W.O. Roberts, J.C. Puffer. 2007. Recommendations and considerations related to preparticipation screening for cardivascular abnormalities in competitive athletes: 2007 update: A scientific statement from the American Heart Association Council on Nutrition, Physical Activity, and Metabolism. Endorsed by the Americn College of Cardiology Foundation. *Circulation* 115:1643-1645.

Marroqui, L., T.M. Batista, A. Gonzalez, E. Viera, A. Rafacho, S.J. Colleta, S.R. Taboga, A.C. Boschero, A. Nadal, E.M. Carneiro, and I. Quesada. 2012. Functional and structural adaptations in the pancreatic α-cell and changes in glucagon signaling during protein malnutrition. *Endocrinology* 153:1663-1672.

Martel, G.F., M.L. Harmer, J.M. Logan, and C.B. Parker. 2005. Aquatic plyometric training increases vertical jump in female volleyball players. *Medicine and Science in Sports and Exercise* 37:1814-1819.

Martello, S., M. Felli, and M. Chiarotti. 2007. Survey of nutritional supplements for selected for illegal anabolic steroids and ephedrine using LC-MS/MS and GC-MS methods, respectively. *Food Additives and Contaminants* 24:258-265.

Martin, B., M. Heintzelman, and H.I. Chen. 1982. Exercise performance after ventilatory work. *Journal of Applied Physiology* 52:1581-1585.

Martin, D.E., and P.N. Coe. 1997. *Better training for distance runners*, 2nd ed. Champaign, IL: Human Kinetics.

Martin, W.H. III, E.F. Coyle, S.A. Bloomfield, and A.A. Ehsani. 1986. Effects of physical deconditioning after intense endurance training on left ventricular dimensions and stroke volume. *Journal of the American College of Cardiology* 7:982-989.

Masuda, K., K. Okazaki, S. Kuno, K. Asano, H. Shimojo, and S. Katsuta. 2001. Endurance training under 2500-m hypoxia does not increase myoglobin content in human skeletal muscle. *European Journal of Applied Physiology* 85:486-490.

Mathur, N., and B.K. Pedersen. 2008. Exercise as a mean to control low-grade systemic inflammation. *Mediators of Inflammation* doi: 10.1155/2008/109502.

Matthews, D.E., M.A. Marano, and R.G. Campbell. 1993a. Splanchnic bed utilization of leucine and phenylalanine in humans. *American Journal of Physiology: Endocrinology and Metabolism* 264:E109-E118.

Matthews, D.E., M.A. Marano, and R.G. Campbell. 1993b. Splanchnic bed utilization of glutamine and glutamic acid in humans. *American Journal of Physiology: Endocrinology and Metabolism* 264:E848-E854.

Matthews, M.M., and T.W. Traut. 1987. Regulation of N-carbamoyl-beta-alanine amidohydrolase, the terminal enzyme in pyrimidine catabolism, by ligand-induced change in polymerization. *Journal of Biological Chemistry* 262:7232-7237.

Maughan, R.J. 1991. Carbohydrate-electrolyte solutions during prolonged exercise. In *Perspectives in exercise science and sports medicine, Vol. 4, Ergogenics*, ed. D.R. Lamb and M.H. Williams, 35-86. Carmel, IN: Benchmark Press.

Maughan, R.J., C.E. Fenn, and J.B. Leiper. 1989. Effects of fluid, electrolyte and substrate ingestion on endurance capacity. *European Journal of Applied Physiology* 58:481-486.

Maughan, R.J., and T.D. Noakes. 1991. Fluid replacement and exercise stress. *Sports Medicine* 12:16-31.

Mayhew, J.L., T.E. Ball, and J.C. Bowen. 1992. Prediction of bench press lifting ability from submaximal repetitions before and after training. *Sports Medicine, Training and Rehabilitation* 3:195-201.

Mayhew, J.L., B. Levy, T. McCormick, and G. Evans. 1987. Strength norms for NCAA division II college football players. *National Strength and Conditioning Association Journal* 9:67-69.

Mayhew, J.L., J.S. Ware, M.G. Bemben, B. Wilt, T.E. Ward, B. Farris, J. Juraszek, and J.P. Slovak. 1999. The NFL-225 test as a measure of bench press strength in college football players. *Journal of Strength and Conditioning Research* 13:130-134.

Mazzeo, R.S. 2005. Altitude, exercise and immune function. *Exercise and Immunological Reviews* 11:6-16.

McArdle, W.D., R.M. Glaser, and J.R. Magel. 1971. Metabolic and cardiorespiratory response during free swimming and treadmill walking. *Journal of Applied Physiology* 30:733-738.

McArdle, W.D., F.I. Katch, and V.L. Katch. 2010. *Exercise physiology: Energy, nutrition, and human performance*, 7th ed. Philadelphia: Lippincott Williams & Wilkins.

McArdle, W.D., J.R. Magel, D.J. Delio, M. Toner, and J.M. Chase. 1978. Specificity of run training on V̇O2 max and heart rate changes during running and swimming. *Medicine and Science in Sports and Exercise* 10:16-20.

McBride, J.M., G.O. McCaulley, P. Cormie, J.L. Nuzzo, M.J. Cavill, and N.T. Triplett. 2009. Comparison of methods to quantify volume during resistance exercise. *Journal of Strength and Conditioning Research* 23:106-110.

McBride, J.M., J.P. Porcari, and M.D. Scheunke. 2004. Effect of vibration during fatiguing resistance exercise on subsequent muscle activity during maximal voluntary isometric contractions. *Journal of Strength and Conditioning Research* 18:777-781.

McCabe, S.E., K.J. Bower, B.T. West, T.F. Nelson, and H. Wechsler. 2007. Trends in non-medical use of anabolic steroids by U.S. college students: Results from four national surveys. *Drug and Alcohol Dependency* 90:243-251.

McCarthy, D.A., and M.M. Dale. 1988. The leucocytosis of exercise: A review and model. *Sports Medicine* 6:333-363.

McCarthy, J.P., J.C. Agre, B.K. Graf, M.A. Pozniak, and A.C. Vailas. 1995. Compatibility of adaptive responses with combining strength and endurance training. *Medicine and Science in Sports and Exercise* 27:429-436.

McCarthy, J.P., M.A. Poszniak, and J.C. Agre. 2002. Neuromuscular adaptations to concurrent strength and endurance training. *Medicine and Science in Sports and Exercise* 34:511-519.

McCartney, N. 1999. Acute responses to resistance training and safety. *Medicine and Science in Sports and Exercise* 31:31-37.

McCartney, N., R.S. McKelvie, J. Martin, D.G. Sale, and J.D. MacDougall. 1993. Weight-training-induced attenuation of the circulatory response of older males to weight lifting. *Journal of Applied Physiology* 74:1056-1060.

McCartney, N., L.L. Spriet, G.J.F. Heigenhauser, J.M. Kowalchuk, J.R. Sutton, and N.L. Jones. 1986. Muscle power and metabolism in maximal intermittent exercise. *Journal of Applied Physiology* 60:1164-1169.

McCole, S.D., K. Claney, J.C. Conte, R. Anderson, and J.M. Hagberg. 1990. Energy expenditures during bicycling. *Journal of Applied Physiology* 68:748-753.

McConnell, G.K., C.M. Burge, S.L. Skinner, and M. Hargreaves. 1997. Influence of ingested fluid volume on physiological responses during prolonged exercise. *Acta Physiologica Scandinavica* 160:149-156.

McCrea, M., T. Hammeke, G. Olsen, P. Leo, and K. Guskiewicz. 2004. Unreported concussion in high school football players: Implications for prevention. *Clinical Journal of Sport Medicine* 14:13-17.

McCue, B.F. 1953. Flexibility of college women. *Research Quarterly for Exercise and Sport* 24:316-324.

McCullough, E.A., and W.L. Kenney. 2003. Thermal insulation and evaporative resistance of football uniforms. 2003. *Medicine and Science in Sports and Exercise* 35:832-837.

McCurdy, K., G. Langford, J. Ernest, D. Jenkerson, and M. Doscher. 2009. Comparison of chain- and plate-loaded bench press training on strength, joint pain and muscle soreness in Division II baseball players. *Journal of Strength and Conditioning Research* 23:187-195.

McDowell, S.L., K. Chalos, T.J. Housh, G.D. Tharp, and G.O. Johnson. 1991. The effect of exercise intensity and duration on salivary immunoglobulin A. *European Journal of Applied Physiology* 63:108-111.

McDowell, S.L., R.A. Hughes, R.J. Hughes, D.J. Housh, T.J. Housh, and G.O. Johnson. 1992. The effect of exhaustive exercise on salivary immunoglobulin A. *Journal of Sports Medicine and Physical Fitness* 32:412-415.

McEvoy, K.P., and R.U. Newton. 1998. Baseball throwing speed and base running speed: The effects of ballistic resistance training. *Journal of Strength and Conditioning Research* 12:216-221.

McFadden Jr., E.R., and I.A. Gilbert. 1994. Exercise-induced asthma. *New England Journal of Medicine* 330:1362-1367.

McGee, D., T.C. Jessee, M.H. Stone, and D. Blessing. 1992. Leg and hip endurance adaptations to three weight-training programs. *Journal of Applied Sport Science Research* 6:92-95.

McInnes, S.E., J.S. Carlson, C.J. Jones, and M.J. McKenna. 1995. The physiological load imposed on basketball players during competition. *Journal of Sport Sciences* 13:387-397.

McKenna, M.J., G.J. Heigenhauser, R.S. McKelvie, G. Obminski, J.D. MacDougal, and N.L. Jones. 1997. Enhanced pulmonary and active skeletal muscle gas exchange during intense exercise after sprint training in men. *Journal of Physiology* 501:703-716.

McKenzie, D.C., and K.D. Fitch. 2011. The asthmatic athlete: Inhaled beta-2 agonists, sport performance, and doping. *Clinical Journal of Sport Medicine* 21:46-50.

McKenzie, D.C., S.L. McLuckie, and D.R. Stirling. 1994. The protective effects of continuous and interval exercise in athletes with exercise-induced asthma. *Medicine and Science in Sports and Exercise* 26:951-956.

McLellan, T.M., and S.S. Cheung. 2000. Impact of fluid replacement on heat storage while wearing protective clothing. *Ergonomics* 43:2020-2030.

McMaster, W.C., S.C. Long, and V.J. Caiozzo. 1991. Isokinetic torque imbalances in the rotator cuff of the elite water polo player. *American Journal of Sports Medicine* 19:72-75.

McMorris, T., R.C. Harris, J. Swain, J. Corbett, K. Collard, R.J. Dyson, L. Dye, C. Hodgon, and N. Draper. 2006. Effect of creatine supplementation and sleep deprivation with mild exercise on cognitive and psychomotor performance, mood state and plasma concentrations of catecholamines and cortisol. *Psychopharmacology* 185:93-103.

McMorris, T., G. Mielcarz, R.C. Harris, J.P. Swain, and A. Howard. 2007. Creatine supplementation and cognitive performance in elderly individuals. *Aging, Neuropsychology and Cognition* 14:517-528.

McMurray, R.G., and A.C. Hackney. 2005. Interactions of metabolic hormones, adipose tissue and exercise. *Sports Medicine* 35:393-412.

McNaughton, L., K. Backx, G. Palmer, and N. Strange. 1999. Effects of chronic bicarbonate ingestion on the performance of high-intensity work. *European Journal of Applied Physiology* 80:333-336.

Meckel, Y., D. Nemet, S. Bar-Sela, S. Radom-Aizik, D.M. Cooper, M. Sagiv, and A. Eliakim. 2011. Hormonal and inflammatory responses to different types of interval training. *Journal of Strength and Conditioning Research* 25:2161-2169.

Meeusen, R., M. Duclos, C. Foster, A.C. Fry, M. Gleeson, D. Nieman, J. Raglin, G. Rietjens, J. Steinacker, and A. Urhausen. 2013. Prevention, diagnosis, and treatment of the overtraining syndrome: Joint consensus statement of the European College of Sport Science and the American College of Sports Medicine. *Medicine and Science in Sports and Exercise* 45:186-205.

Melin, B., J.P. Eclache, G. Geelen, G. Annat, A.M. Allevard, E. Jarsaillon, A. Zebidi, J.J. Legros, and C. Gharib. 1980. Plasma AVP, neurophysin, renin activity, and aldosterone during submaximal exercise performed until exhaustion in trained and untrained men. *European Journal of Applied Physiology and Occupational Physiology* 44:141-151.

Menapace, F.J., W.J. Hammer, T.F. Ritzer, K.M. Kessler, H.F. Warner, J.F. Spann, and A.A. Bove. 1982. Left ventricular size in competitive weight lifters: An echocardiographic study. *Medicine and Science in Sports and Exercise* 14:72-75.

Mendel, C.M. 1989. The free hormone hypothesis: A physiologically based mathematical model. *Endocrine Reviews* 10:232-274.

Mendel, R.W., M. Blegen, C. Cheatham, J. Antonio, and T. Ziegenfuss. 2005. Effects of creatine on thermoregulatory responses while exercising in the heat. *Nutrition* 21:301-307.

Mero, A., P.V. Komi, and R.J. Gregor. 1992. Biomechanics of sprint running: A review. *Sports Medicine* 13:376-392.

Meydani, M. 1992. Protective role of dietary vitamin E on oxidative stress in aging. *Age* 15:89-93.

Meyer, C.R. 1967. Effect of two isometric routines on strength, size and endurance in exercise and non-exercise arms. *Research Quarterly for Exercise and Sport* 38:430-440.

Migeon, C.J. 1972. Adrenal androgens in man. *American Journal of Medicine* 53:606-626.

Mikines, K.J., B. Sonne, P.A. Farrell, B. Tronier, and H. Galbo. 1988. Effect of physical exercise on sensitivity and responsiveness to insulin in humans. *American Journal of Physiology: Endocrinology and Metabolism* 254:E248-E259.

Miles, M.P., W.J. Kraemer, D.S. Grove, S.K. Leach, K. Dohi, J.A. Bush, J.O. Marx, B.C. Nindl, J.S. Volek, and A.M. Mastro. 2002. Effects of resistance training on resting immune parameters in women. *European Journal of Applied Physiology* 87:506-508.

Milledge, J.S., J.M. Beeley, J. Broome, N. Luff, M. Pelling, and D. Smith. 1991. Acute mountain sickness susceptibility, fitness and hypoxic ventilatory response. *European Respiratory Journal* 4:1000-1003.

Milledge, J.S., and P.M. Cotes. 1985. Serum erythropoietin in humans at high altitude and its relation to plasma renin. *Journal of Applied Physiology* 59:360-364.

Miller, K.E., J.H. Hoffman, G.M. Barnes, D. Sabo, M.J. Melnick, and M.P. Farrell. 2005. Adolescent anabolic steroid use, gender, physical activity, and other problem behaviors. *Substance Use and Misuse* 40:1637-1657.

Miller, S.L., K.D. Tipton, D.L. Chinkes, S.E. Wolf, and R.R. Wolfe. 2003. Independent and combined effects of amino acids and glucose after resistance exercise. *Medicine and Science in Sports and Exercise* 35:449-455.

Miller, T.A., E.D. White, K.A. Kinley, J.J. Congleton, and M.J. Clark. 2002. The effects of training history, player position, and body composition on exercise performance in collegiate football players. *Journal of Strength and Conditioning Research* 16:44-49.

Miller, W.J., W.M. Sherman, and J.L. Ivy. 1984. Effect of strength training on glucose tolerance and post-glucose insulin response. *Medicine and Science in Sports and Exercise* 16:539-587.

Millet, G.P., B.A. Jaouen, F. Borrani, and R. Candau. 2002. Effects of concurrent endurance and strength training on running economy and VO_2 kinetics. *Medicine and Science in Sports and Exercise* 34:1351-1359.

Milner-Brown, H.S., R.B. Stein, and R. Yemm. 1975. Synchronization of human motor units: Possible roles of exercise and supraspinal reflexes. *Electroencephalography and Clinical Neurophysiology* 38:245-254.

Minkler, S., and P. Patterson. 1994. The validity of the modified sit-and-reach test in college-age students. *Research Quarterly for Exercise and Sport* 65:189-192.

Mochida, N., T. Umeda, Y. Yamamoto, M. Tanabe, A. Kojima, K. Sugawara, and S. Nakaji. 2007. The main neutrophil and neutrophil-related functions may compensate for each other following exercise–a finding from training in university judoists. *Luminescence* 22:20-28.

Mold, J., R. Gore, J. Lynch, and E. Schantz. 1955. Creatine ethyl ester. *Journal of the American Chemical Society* 77:178-180.

Moller, N., J.O.L. Jorgensen, K.G.M.M. Alberti, A. Flyvbjerg, and O. Schmitz. 1990. Short-term effects of growth hormone on fuel oxidation and regional substrate metabolism in normal man. *Journal of Clinical Endocrinology and Metabolism* 70:1179-1186.

Molnar, J., J. Marton, and B. Halasz. 1990. Central nervous system sites of action of an enkephalin analogue (d-met, pro)-enkephalin, and naloxone on the secretion of five anterior pituitary hormones of male rats. *Journal of Neuroendocrinology* 2:477-483.

Monroe, M.B., R.E. Van Pelt, B.C. Schiller, D.R. Seals, and P.P. Jones. 2000. Relation of leptin and insulin to adiposity-associated elevations in sympathetic activity with age in humans. *International Journal of Obesity and Related Metabolic Disorders* 24:1183-1187.

Montain, S.J., J.E. Laird, W.A. Latzka, and M.N. Sawka. 1997. Aldosterone and vasopressin responses in the heat: Hydration level and exercise intensity effects. *Medicine and Science in Sports and Exercise* 29:661-668.

Montain, S.J., M.N. Sawka, B.S. Cadarette, M.D. Quigley, and J.M. McKay. 1994. Physiological tolerance to uncompensable heat stress: Effects of exercise intensity, protective clothing, and climate. *Journal of Applied Physiology* 77:216-222.

Monteleone, P., L. Beinat, C. Tanzillo, M. Maj, and D. Kemall. 1990. Effects of phosphatidylserine on the neuroendocrine response to physical stress in humans. *Neuroendocrinology* 52:243-248.

Montgomery, J.C., and J.A. MacDonald. 1990. Effects of temperature on nervous system: Implications for behavioral performance. *American Journal of Physiology: Regulatory, Integrative and Comparative Physiology* 259:R191-R196.

Moon, Y.L., L. Wang, R. DiCenzo, and M.E. Morris. 2008. Quercetin pharmacokinetics in humans. *Biopharmaceutics and Drug Disposition* 29:205-217.

Moore, D.R., M.J. Robinson, J.L. Fry, J.E. Tang, E.I. Glover, S.B. Wilkinson, T. Prior, M.A. Tarnopolsky, and S.M. Phillips. 2009. Ingested protein dose response of muscle and albumin protein synthesis after resistance exercise in young men. *American Journal of Clinical Nutrition* 89:161-168.

Moore, K., N. Vizzard, C. Coleman, J. McMahon, R. Hayes, and C.J. Thompson. 2001. Extreme altitude mountaineering and type 1 diabetes: The Diabetes Federation of Ireland Kilimanjaro expedition. *Diabetes Medicine* 18:749-755.

Moore, M.A., and R.S. Hutton. 1980. Electromyographic investigation of muscle stretching techniques. *Medicine and Science in Sports and Exercise* 12:322-329.

Moran, D.S., and K.B. Pandolf. 1999. Wet bulb globe temperature (WBGT) – to what extent is GT essential? *Aviation, Space, and Environmental Medicine* 70:480-484.

Moran, D.S., K.B. Pandolf, Y. Heled, and R.R. Gonzalez. 2003. Combined environmental stress and physiological strain indices for physical training guidelines. *Journal of Basic Clinical Physiology and Pharmacology* 14:17-30.

Moran, D.S., A. Shitzer, and K.B. Pandolf. 1998. A physiological strain index to evaluate heat stress. *American Journal of Physiology* 275:R129-R134.

Moreira, A., L. Delgado, and K.H. Carlsen. 2011. Exercise-induced asthma: Why is it so frequent in Olympic athletes? *Expert Reviews in Respiratory Medicine* 5:1-3.

Morgan, W.P. 1985. Affective beneficence of vigorous physical activity. *Medicine and Science in Sports and Exercise* 17:94-100.

Morgan, W.P., D.R. Brown, J.S. Raglin, P.J. O'Connor, and K.A. Ellickson. 1987. Psychological monitoring of overtraining and staleness. *British Journal of Sports Medicine* 21:107-114.

Morganroth, J., B.J. Maron, W.L. Henry, and S.E. Epstein. 1975. Comparative left ventricular dimensions in trained athletes. *Annals of Internal Medicine* 82:521-524.

Moritani, M.T., and H.A. deVries. 1979. Neural factors vs. hypertrophy in time course of muscle strength gain. *American Journal of Physical Medicine and Rehabilitation* 58:115-130.

Morton, A.R. 1995. Asthma. In *Science and medicine in sport,* ed. J. Bloomfield, P. Fricker, and K.D. Fitch, 616-627. Victoria, Australia: Blackwell Science.

Morton, A.R., and K.D. Fitch. 2011. Australian Association for Exercise and Sport Science position statement on exercise and asthma. *Journal of Science and Medicine in Sport* 14:312-316.

Motl, R.W., P.J. O'Connor, and R.K. Dishman. 2003. Effect of caffeine on perceptions of leg muscle pain during moderate intensity cycling exercise. *Journal of Pain* 4:316-321.

Mounier, R., L. Lantier, J. Leclerc, A. Sotiropoulos, M. Foretz, and B. Viollet. 2011. Antagonistic control of muscle cell size by AMPK and mTORC1. *Cell Cycle* 10:2640-2546.

Mueller, F.O., and R.C. Cantu. 2011. Annual survey of catastrophic football injuries. National Center for Catastrophic Sport Injury Research. www.unc.edu/depts/nccsi. Accessed November 19, 2012.

Mueller, F.O., and B. Colgate. 2011. Annual survey of football injury research. National Center for Catastrophic Sport Injury Research. www.unc.edu/depts/nccsi. Accessed November 19, 2012.

Mujika, I., J.C. Chatard, L. Lacoste, F. Barale, and A. Geyssant. 1996. Creatine supplementation does not improve sprint performance in competitive swimmers. *Medicine and Science in Sports and Exercise* 28:1435-1441.

Mujika, I., and S. Padilla. 2001. Cardiorespiratory and metabolic characteristics of detraining in humans. *Medicine and Science in Sports and Exercise* 33:413-421.

Mujika, I., S. Padilla, J. Ibanez, M. Izquierdo, and E. Gorostiaga. 2000. Creatine supplementation and sprint performance in soccer players. *Medicine and Science in Sports and Exercise* 32:518-525.

Mulligan, S.E., S.J. Fleck, S.E. Gordon, L.P. Koziris, N.T. Triplett-McBride, and W.J. Kraemer. 1996. Influence of resistance exercise volume on serum growth hormone and cortisol concentrations in women. *Journal of Strength and Conditioning Research* 10:256-262.

Murray, R. 1987. The effects of consuming carbohydrate-electrolyte beverages on gastric emptying and fluid absorption during and following exercise. *Sports Medicine* 4:322-351.

Murray, R., W.P. Bartoli, D.E. Eddy, and M.K. Horn. 1997. Gastric emptying and plasma deuterium accumulation following ingestion of water and two carbohydrate-electrolyte beverages. *International Journal of Sport Nutrition* 7:144-153.

Muza, S.R. 2007. Military applications of hypoxic training for high-altitude operations. *Medicine and Science in Sports and Exercise* 39:1625-1631.

Nadel, E.R., S.M. Fortney, and C.B. Wenger. 1980. Effect of hydration on circulatory and thermal regulation. *Journal of Applied Physiology* 49:715-721.

Nadel, E.R., I. Holmer, U. Bergh, P.O. Åstrand, and J.A.J. Stolwijk. 1974. Energy exchanges of swimming men. *Journal of Applied Physiology* 36:465-471.

Nagajothi, N., A. Khraisat, J.L.E. Velazquez-Cecena, and R. Arora. 2008. Energy drink-related supraventricular tachycardia. *American Journal of Medicine* 121:e3-e4.

Nagashima, K., J. Wu, S.A. Kavouras, and G.W. Mack. 2001. Increased renal tubular sodium reabsorption during exercise-induced hypervolemia in humans. *Journal of Applied Physiology* 91:1229-1236.

Nakagawa, Y., and M. Hattori. 2002. Relationship between muscle buffering capacity and fiber type during anaerobic exercise in humans. *Journal of Physiological Anthropology* 21:129-131.

Nakazawa, K., Y. Kawakami, T. Fukunaga, H. Yano, and M. Miyashita. 1993. Differences in activation patterns in elbow flexor muscles during isometric, concentric and eccentric contractions. *European Journal of Applied Physiology* 66:214-220.

Nath, S.K., P. Dechelotte, D. Darmaun, M. Gotteland, M. Rongier, and J.F. Desjeux. 1992. (^{15}N) and (^{14}C) glutamine fluxes across rabbit ileum in experimental diarrhea. *American Journal of Physiology* 262:G312-G318.

National Collegiate Athletic Association (NCAA). 2001. *NCAA study of substance use habits of college student-athletes*. Indianapolis: National Collegiate Athletic Association. www.ncaa.org.

National Oceanic and Atmospheric Administration. 1976. *US standard atmosphere*. Washington, DC: NOAA.

Ndon, J.A., A.C. Snyder, C. Foster, and W.B. Wehrenberg. 1992. Effects of chronic intensive exercise training on the leukocyte response to acute exercise. *International Journal of Sports Medicine* 13:176-182.

Neal, C.M., A.M. Hunter, D. Stuart, and R. Galloway. 2011. A 6-month analysis of training-intensity distribution and physiological adaptation in Ironman triathletes. *Journal of Sport Sciences* 29:1515-1523.

Neary, J.P., D.C. McKenzie, and Y.N. Bhambhani. 2005. Muscle oxygenation trends after tapering in trained cyclists. *Dynamic Medicine* 4:4.

Nehlsen-Cannarella, S.L., D.C. Nieman, A.J. Balk-Lamberton, P.A. Markoff, D.B.W. Chritton, G. Gusewitch, and J.W. Lee. 1991. The effect of moderate exercise training on immune response. *Medicine and Science in Sports and Exercise* 23:64-70.

Nelson, A.G., J.D. Allen, A. Cornwell, and J. Kokkonen. 1998. Inhibition of maximal torque production by acute stretching is joint-angle specific. Paper presented at 26th annual meeting of the southeastern chapter of the ACSM, Destin, FL, January.

Nelson, A.G., and G.D. Heise. 1996. Acute stretching exercises and vertical jump stored elastic energy. *Medicine and Science in Sports and Exercise* 28:S156.

Nemet, D., P.H. Connolly, A.M. Pontello-Pescatello, C. Rose-Gottron, J.K. Larson, P. Galassetti, and D.M. Cooper. 2004. Negative energy balance plays a major role in the IGF-I response to exercise training. *Journal of Applied Physiology* 96:276-282.

Neptune, R.R., S.A. Kautz, and M.L. Hull. 1997. The effect of pedalling rate on coordination in cycling. *Journal of Biomechanics* 30:1051-1058.

Neubauer, O., D. Konig, and K.H. Wagner. 2008. Recovery after an ironman triathlon: Sustained inflammatory responses and muscular stress. *European Journal of Applied Physiology* 104:417-426.

Neufer, P.D., D.L. Costill, R.A. Fielding, M.G. Flynn, and J.P. Kirwin. 1987. Effect of reduced training on muscular strength and endurance in competitive swimmers. *Medicine and Science in Sports and Exercise* 19:486-490.

Neufer, P.D., D.L. Costill, W.J. Fink, J.P. Kirwin, R.A. Fielding, and M.G. Flynn. 1986. Effects of exercise and carbohydrate composition on gastric emptying. *Medicine and Science in Sports and Exercise* 18:658-662.

Neufer, P.D., M.N. Sawka, A.J. Young, M.D. Quigley, W.A. Latzka, and L. Levine. 1991. Hypohydration does not impair skeletal muscle glycogen resynthesis after exercise. *Journal of Applied Physiology* 70:1490-1494.

Neufer, P.D., A.J. Young, and M.N. Sawka. 1989. Gastric emptying during exercise: Effects of heat stress and hypohydration. *European Journal of Applied Physiology and Occupational Physiology* 58:557-560.

Newton, R.U., P. Cormie, and W.J. Kraemer. 2012. Power training. In *NSCA's guide to program design,* ed. J.R. Hoffman, 95-117. Champaign, IL: Human Kinetics.

Newton, R.U., and W.J. Kraemer. 1994. Developing explosive muscular power: Implications for a mixed methods training strategy. *Strength and Conditioning* 16:20-31.

Newton, R.U., W.J. Kraemer, and K. Hakkinen. 1999. Effects of ballistic training on preseason preparation of elite volleyball players. *Medicine and Science in Sports and Exercise* 31:323-330.

Newton, R.U., W.J. Kraemer, K. Häkkinen, B. Humphries, and A.J. Murphy. 1996. Kinematics, kinetics, and muscle activation during explosive upper body movements. *Journal of Applied Biomechanics* 12:31-43.

Newton, R.U., and K.P. McEvoy. 1994. Baseball throwing velocity: A comparison of medicine ball training and weight training. *Journal of Strength and Conditioning Research* 8:198-203.

Nielsen, H.B., N.H. Secher, N.J. Christensen, and B.K. Pedersen. 1996. Lymphocytes and NK cell activity during repeated bouts of maximal exercise. *American Journal of Physiology: Regulatory, Integrative and Comparative Physiology* 271:R222-R227.

Nielsen, J., M. Mogensen, B.F. Vind, K. Sahlin, K. Hojlund, H.D. Schroder, and N. Ortenblad. 2010. Increased subsarcolemmal lipids in type 2 diabetes: Effect of training on localization of lipids, mitochondria, and glycogen in sedentary human skeletal muscle. *American Journal of Physiology: Endocrinology and Metabolism* 298:E706-E713.

Nieman, D.C. 2000. Exercise, the immune system, and infectious disease. In *Exercise and sport science,* ed. W.J. Garrett Jr. and D.T. Kirkendall, 177-190. Philadelphia: Lippincott Williams & Wilkins.

Nieman, D.C., L.S. Berk, M. Simpson-Westerberg, K. Arabatzis, S. Youngberg, S.A. Tan, J.W. Lee, and W.C. Eby. 1989. Effects of long-endurance running on immune system parameters and lymphocyte function in experienced marathoners. *International Journal of Sports Medicine* 10:317-323.

Nieman, D.C., D.A. Hensen, G. Gusewitch, B.J. Warren, R.C. Dotson, and S.L. Nehlsen-Cannarella. 1993. Physical activity and immune function in elderly women. *Medicine and Science in Sports and Exercise* 25:823-831.

Nieman, D.C., D.A. Hensen, R. Johnson, L. Lebeck, J.M. Davis, and S.L. Nehlsen-Cannarella. 1992. Effects of brief, heavy exertion on circulating lymphocyte subpopulations and proliferative response. *Medicine and Science in Sports and Exercise* 24:1339-1345.

Nieman, D.C., D.A. Hensen, K.R. Maxwell, A.S. Williams, S.R. McAnulty, F. Jin, R.A. Shanely, and T.C. Lines. 2009. Effects of quercetin and EGCG on mitochondrial biogenesis and immunity. *Medicine and Science in Sports and Exercise* 41:1467-1475.

Nieman, D.C., D.A. Hensen, C.S. Sampson, J.L. Herring, J. Stulles, M. Conley, M.H. Stone, D.E. Butterworth, and J.M. Davis. 1995. The acute immune response to exhaustive resistance exercise. *International Journal of Sports Medicine* 16:322-328.

Nieman, D.C., L.M. Johanssen, and J.W. Lee. 1989. Infectious episodes in runners before and after a roadrace. *Journal of Sports Medicine and Physical Fitness* 29:289-296.

Nieman, D.C., L.M. Johanssen, J.W. Lee, and K. Arabatzis. 1990. Infectious episodes in runners before and after the Los Angeles Marathon. *Journal of Sports Medicine and Physical Fitness* 30:316-328.

Nieman, D.C., A.R. Miller, D.A. Hensen, B.J. Warren, G. Gusewitch, R.L. Johnson, D.E. Butterworth, J.L. Herring, and S.L. Nehlsen-Cannarella.

1994. Effect of high- versus moderate-intensity exercise on lymphocyte subpopulations and proliferative response. *International Journal of Sports Medicine* 15:199-206.

Nieman, D.C., A.R. Miller, D.A. Hensen, B.J. Warren, G. Gusewitch, R.L. Johnson, J.M. Davis, D.E. Butterworth, and S.L. Nehlsen-Cannarella. 1993. Effect of high- versus moderate intensity exercise on natural killer cell activity. *Medicine and Science in Sports and Exercise* 25:1126-1134.

Nieman, D.C., S. Simandle, D.A. Hensen, B.J. Warren, J. Suttles, J.M. Davis, K.S. Buckley, J.C. Ahle, D.E. Butterworth, O.R. Fagoaga, and S.L. Nehlsen-Cannarella. 1995. Lymphocyte proliferative response to 2.5 hours of running. *International Journal of Sports Medicine* 16:404-408.

Nieman, D.C., S.A. Tan, J.W. Lee, and L.S. Berk. 1989. Complement and immunoglobulin levels in athletes and sedentary controls. *International Journal of Sports Medicine* 10:124-128.

Nieman, D.C., A.S. Williams, R.A. Shanely, F. Jin, S.R. McAnulty, N.T. Triplett, M.D. Austin, and D.A. Hensen. 2010. Quercetin's influence on exercise performance and muscle mitochondrial biogenesis. *Medicine and Science in Sports and Exercise* 42:338-345.

Nindl, B.C. 2009. Insulin-like growth factor-I as a candidate metabolic biomarker: Military relevance and future directions for measurement. *Journal of Diabetes Science and Technology* 3:371-376.

Nindl, B.C., W.C. Hymer, D.R. Deaver, and W.J. Kraemer. 2001. Growth hormone pulsatility profile characteristics following acute heavy resistance exercise. *Journal of Applied Physiology* 91:163-172.

Nindl, B.C., W.J. Kraemer, P.J. Arciero, N. Samatallee, C.D. Leone, M.F. Mayo, and D.L. Hafeman. 2002. Leptin concentrations experience a delayed reduction after resistance exercise in men. *Medicine and Science in Sports and Exercise* 34:608-613.

Nindl, B.C., W.J. Kraemer, J.O. Marx, P.J. Arciero, K. Dohi, M.D. Kellogg, and G.A. Loomis. 2001. Overnight responses of the circulating IGF-I system after acute, heavy-resistance exercise. *Journal of Applied Physiology* 90:1319-1326.

Nindl, B.C., and J.R. Pierce. 2010. Insulin-like growth factor I as a biomarker of health, fitness, and training status. *Medicine and Science in Sports and Exercise* 42:39-49.

Nindl, B.C., M. Santtila, J. Vaara, K. Hakkinen, and H. Kyrolainen. 2011. Circulating IGF-I is associated with fitness and health outcomes in a population of 846 young healthy men. *Growth Hormone and IGF Research* 21:124-128.

Nindl, B.C., D.E. Scofield, C.A. Strohbach, A.J. Centi, R.K. Evans, R. Yanovich, and D.S. Moran. 2012. IGF-I, IGFBPs, and inflammatory cytokine responses during gender-integrated Israeli Army basic combat training. *Journal of Strength and Conditioning Research* 26(suppl)2:S73-S81.

Nissen, S.L., and N.N. Abumrad. 1997. Nutritional role of the leucine metabolite b-hydroxy b-methylbutyrate (HMB). *Nutritional Biochemistry* 8:300-311.

Nissen, S., R. Sharp, M. Ray, J.A. Rathmacher, D. Rice, J.C. Fuller, A.S. Connelly, and N. Abumrad. 1996. Effect of leucine metabolite b-hydroxy-b-methylbutyrate on muscle metabolism during resistance-exercise training. *Journal of Applied Physiology* 81:2095-2104.

Noakes, T.D. 1993. Fluid replacement during exercise. *Exercise and Sport Sciences Reviews* 21:297-330.

Noakes, T.D., R.J. Norman, R.H. Buck, J. Godlonton, K. Stevenson, and D. Pittaway. 1990. The incidence of hyponatremia during prolonged ultraendurance exercise. *Medicine and Science in Sports and Exercise* 22:165-170.

Nordstrom, A., T. Olsson, and P. Nordstrom. 2005. Bone gained from physical activity and lost through detraining: A longitudinal study in young males. *Osteoporosis International* 16:835-841.

Norkin, C.C., and D.J. White. 1995. *Measurement of joint motion: A guide to goniometry.* Philadelphia: Davis.

Norrelund, H., K.S. Nair, S. Nielsen, J. Frystyk, P. Ivarsen, J.O. Jorgensen, J.S. Christiansen, and N. Moller. 2003. The decisive role of free fatty acids for protein conservation during fasting in humans with and without growth hormone. *Journal of Clinical Endocrinology and Metabolism* 88:4371-4378.

Nosaka, K., and P.M. Clarkson. 1996. Changes in indicators of inflammation after eccentric exercise of the elbow flexors. *Medicine and Science in Sports and Exercise* 28:953-961.

Nose, H., T. Morimoto, and K. Ogura. 1983. Distribution of water losses among fluid compartments after dehydration in humans. *Japan Journal of Physiology* 33:1019-1029.

Nunan, D., G. Howatson, and K.A. van Someren. 2010. Exercise-induced muscle damage is not attenuated by β-hydroxy-β-methylbutyrate and α-ketoisocaproic acid supplementation. *Journal of Strength and Conditioning Research* 24:531-537.

Oberstar, J.Y., G.A. Bernstein, and P.D. Thuras. 2002. Caffeine use and dependence in adolescents: One year follow-up. *Journal of Child and Adolescent Psychopharmacology* 12:127-135.

O'Brien, C. 2005. Reproducibility of the cold-induced vasodilation response in the human finger. *Journal of Applied Physiology* 90:254-259.

O'Brien, C., J.W. Castellani, and M.N. Sawka. 2011. Thermal face protection delays finger cooling and improves thermal comfort during cold air exposure. *European Journal of Applied Physiology and Occupational Physiology* 111:3097-3105.

O'Bryant, H.S., R. Byrd, and M.H. Stone. 1988. Cycle ergometer performance and maximum leg and hip strength adaptations to two different methods of weight-training. *Journal of Applied Sport Science Research* 2:27-30.

O'Connor, D.M., and M.J. Crowe. 2003. Effects of beta-hydroxy-beta-methylbutyrate and creatine monohydrate supplementation on the aerobic and anaerobic capacity of highly trained athletes. *Journal of Sports Medicine and Physical Fitness* 43:64-68.

O'Connor, F.G., M.F. Bergeron, J. Cantrell, P. Connes, K.G. Harmon, E. Ivy, J. Kark, D. Klossner, P. Lisman, B.K. Meyers, K. O'Brien, K. Ohene-Frempong, A.A. Thompson, J. Whitehead, and P.A. Deuster. 2012. ACSM and CHAMP summit on sickle cell trait: Mitigating risks for warfighters and athletes. *Medicine and Science in Sports and Exercise* 44:2045-2056.

O'Connor, F.G., D.J. Casa, M.F. Bergeron, R. Carter, P. Deuster, Y. Heled, J. Kark, L. Leon, B. McDermott, K. O'Brien, W.O. Roberts, and M. Sawka. 2010. American College of Sports Medicine roundtable on exertional heat stroke – return to duty/return to play: Conference proceedings. *Current Sports Medicine Reports* 9:314-321.

O'Connor, P.J., and D.B. Cook. 1999. Exercise and pain: Neurobiology, measurement, and laboratory study of pain in relation to exercise in humans. *Exercise and Sport Sciences Reviews* 27:119-166.

Odland, L.M., J.D. MacDougall, M.A. Tarnopolsky, A. Elorriaga, and A. Borgmann. 1997. Effect of oral creatine supplementation on muscle [PCr] and short-term maximum power output. *Medicine and Science in Sports and Exercise* 29:216-219.

Olefsky, J.M. 1976. The insulin receptor: Its role in insulin resistance in obesity and diabetes. *Diabetes* 25:1154-1164.

Olsson, K.E., and B. Saltin. 1970. Variation in total body water with muscle glycogen changes in man. *Acta Physiologica Scandinavica* 80:11-18.

Olthof, M.R., T. van Vliet, E. Boelsma, and P. Verhoef. 2003. Low dose betaine supplementation leads to immediate and long term lowering of plasma homocysteine in healthy men and women. *Journal of Nutrition* 133:4135-4138.

Olthof, M.R., and P. Verhoef. 2005. Effects of betaine intake on plasma homocysteine concentrations and consequences for health. *Current Drugs and Metabolism* 6:15-22.

Omwancha, J., and T.R. Brown. 2006. Selective androgen receptor modulators: In pursuit of tissue-selective androgens. *Current Opinion in Investigational Drugs* 7:873-881.

Oopik, V., I. Saaremets, L. Medijainen, K. Karelson, T. Janson, and S. Timpmann. 2003. Effects of sodium citrate ingestion before exercise on endurance performance in well trained college runners *British Journal of Sports Medicine* 37:485-489.

Ortega, F.B., K. Silventoinen, P. Tynelius, and F. Rasmussen. 2012. Muscular strength in male adolescents and premature death: Cohort study of one million participants. *British Medical Journal* 345:e7279.

O'Shea, J.P. 1966. Effects of selected weight training programs on the development of strength and muscle hypertrophy. *Research Quarterly for Exercise and Sport* 37:95-102.

O'Shea, J.P. 1971. The effects of anabolic steroids on dynamic strength levels of weightlifters. *Nutritional Reports International* 4:363-370.

Osterud, B., J.O. Olsen, and L. Wilsgard. 1989. Effect of strenuous exercise on blood monocytes and their relation to coagulation. *Medicine and Science in Sports and Exercise* 21:374-378.

Ostlund, R.E. Jr., J.W. Yang, S. Klein, and R. Gingerich. 1996. Relation between plasma leptin concentration and body fat, gender, diet, age, and metabolic covariates. *Journal of Clinical Endocrinology and Metabolism* 81:3909-3913.

O'Sullivan, S.E., and C. Bell. 2000. The effects of exercise and training on human cardiovascular reflex control. *Journal of the Autonomic Nervous System* 81:16-24.

O'Toole, M.L., and P.S. Douglas. 1995. Applied physiology of a triathlon. *Sports Medicine* 19:251-267.

O'Toole, M.L., P.S. Douglas, and W.D.B. Hiller. 1989. Applied physiology of a triathlon. *Sports Medicine* 8:201-225.

Paddon-Jones, D., A. Keech, and D. Jenkins. 2001. Short-term beta-hydroxy-beta-methylbutyrate supplementation does not reduce symptoms of eccentric muscle damage. *International Journal of Sport Nutrition and Exercise Metabolism* 11:442-450.

Pakarinen, A., M. Alen, K. Hakkinen, and P. Komi. 1988. Serum thyroid hormones, thyrotropin and thyroxin binding globulin during prolonged strength training. *European Journal of Applied Physiology* 57:394-398.

Pandolf, K.B., R.L. Burse, and R.F. Goldman. 1977. Role of physical fitness in heat acclimatization, decay and reinduction. *Ergonomics* 20:399-408.

Panton, L.B., J.A. Rathmacher, S. Baier, and S. Nissen. 2000. Nutritional supplementation of the leucine metabolite beta-hydroxy-beta-methylbutyrate (HMB) during resistance training. *Nutrition* 16:734-739.

Papacosta, E., and G.P. Nassis. 2011. Saliva as a tool for monitoring steroids, peptide and immune markers in sport and exercise science. *Journal of Science and Medicine in Sport* 14:424-434.

Papaioannou, A., C.C. Kennedy, A. Cranney, G. Hawker, J.P. Brown, S.M. Kaiser, W.D. Leslie, C.J. O'Brien, A.M. Sawka, A. Khan, K. Siminoski, G. Tarulli, D. Webster, J. McGowan, and J.D. Adachi. 2009. Risk factors for low BMD in healthy men age 50 years or older: A systematic review. *Osteoporosis International* 20:507-518.

Park, S.H., J.N. Roemmick, and C.A. Horswill. 1990. A season of wrestling and weight loss by adolescent wrestlers: Effect on anaerobic arm power. *Journal of Applied Sport Science Research* 4:1-4.

Parkhouse, W.S., D.C. McKenzie, P.W. Hochachka, and W.K. Ovalle. 1985. Buffering capacity of deproteinized human vastus lateralis muscle. *Journal of Applied Physiology* 58:14-17.

Parkinson, A.B., and N.A. Evans. 2006. Anabolic androgenic steroids: A survey of 500 users. *Medicine and Science in Sports and Exercise* 38:644-651.

Parra, J., J.A. Cadefau, G. Rodas, N. Amigo, and R. Cusso. 2000. The distribution of rest periods affects performance and adaptations of energy metabolism induced by high-intensity training in human muscle. *Acta Physiologica Scandinavica* 169:157-165.

Parsons, J.P., T.J. Craig, S.W. Stoloff, M.L. Hayden, N.K. Ostrom, N.S. Eid, and G.L. Colice. 2011. Impact of exercise-related respiratory symptoms in adults with asthma: Exercise-induced bronchospasm landmark national survey. *Allergy and Asthma Proceedings* 32:431-437.

Parsons, K. 2003. *Human thermal environments,* 2nd ed. London: Taylor & Francis.

Parssinen, M., U. Kujala, E. Vartiainen, S. Sarna, and T. Seppaia. 2000. Increased premature mortality of competitive powerlifters suspected to have used anabolic agents. *International Journal of Sports Medicine* 21:225-227.

Paton, C., T. Lowe, and A. Irvine. 2010. Caffeinated chewing gum increases repeated sprint performance and augments increases in testosterone in competitive cyclists. *European Journal of Applied Physiology* 110:1243-1250.

Pattini, A., F. Schena, and G.C. Guidi. 1990. Serum ferritin and serum iron changes after cross-country and roller ski endurance races. *European Journal of Applied Physiology* 61:55-60.

Pavlatos, A.M., O. Fultz, M.J. Monberg, and A. Vootkur. 2001. Review of oxymetholone: A 17α-alkylated anabolic-androgenic steroid. *Clinical Therapy* 23:789-801.

Payne, J.R., P.J. Kotwinski, and H.E. Montgomery. 2004. Cardiac effects of anabolic steroids. *Heart* 90:473-475.

Pearson, A.C., M. Schiff, D. Mrosek, A.J. Labovitz, and G.A. Williams. 1986. Left ventricular diastolic function in weight lifters. *American Journal of Cardiology* 58:1254-1259.

Pearson, D.R., D.G. Hamby, W. Russel, and T. Harris. 1999. Long-term effects of creatine monohydrate on strength and power. *Journal of Strength and Conditioning Research* 13:187-192.

Pedegna, L.R., R.C. Elsner, D. Roberts, J. Lang, and V. Farewell. 1982. The relationship of upper extremity strength to throwing speed. *American Journal of Sports Medicine* 10:352-354.

Pedersen, B.K., T. Rohde, and H. Bruunsgaard. 1997. Exercise and cytokines. In *Exercise and immunology,* ed. B.K. Pedersen, 89-111. New York: Springer.

Pedersen, B.K., A. Steensberg, C. Fischer, C. Keller, P. Keller, P. Plomgaard, M. Febbraio, and B. Saltin. 2003. Searching for the exercise factor: Is IL-6 a candidate? *Journal of Muscle Research and Cellular Motility* 24:113-119.

Pedersen, B.K., and H. Ullam. 1994. NK cell response to physical activity: Possible mechanisms of action. *Medicine and Science in Sports and Exercise* 26:140-146.

Pedlar, C., G. Whyte, S. Emegbo, N. Stanley, I. Hindmarch, and R. Godfrey. 2005. Acute sleep responses in a normobaric hypoxic tent. *Medicine and Science in Sports and Exercise* 37:1075-1079.

Pelliccia, A., B.J. Maron, A. Spataro, M.A. Proschan, and P. Spirito. 1991. The upper limit of physiologic cardiac hypertrophy in highly trained elite athletes. *New England Journal of Medicine* 324:295-301.

Penafiel, R., C. Ruzafa, F. Monserrat, and A. Cremades. 2004. Gender-related differences in carnosine, anserine and lysine content of murine skeletal muscle. *Amino Acids* 26:53-58.

Pequignot, J.M., L. Peyrin, M.H. Mayet, and R. Flandrois. 1979. Metabolic adrenergic changes during submaximal exercise and in the recovery period in man. *Journal of Applied Physiology: Respiratory, Environmental and Exercise Physiology* 47:701-705.

Perrier, E.T., M.J. Pavol, and M.A. Hoffman. 2011. The acute effects of a warm-up including static or dynamic stretching on countermovement jump height, reaction time, and flexibility. *Journal of Strength and Conditioning Research* 25:1925-1931.

Perrine, J.J., and V.R. Edgerton. 1978. Muscle force-velocity and power-velocity relationships under isokinetic loading. *Medicine and Science in Sports and Exercise* 10:159-166.

Perry, P.J., B.C. Lund, M.J. Deninger, B.C. Kutscher, and J. Schneider. 2005. Anabolic steroid use in weightlifters and bodybuilders. An internet survey of drug utilization. *Clinical Journal of Sport Medicine* 15:326-330.

Pertusi, R., R.D. Dickerman, and W.J. McConathy. 2001. Evaluation of aminotransferase elevations in a bodybuilder using anabolic steroids: Hepatitis or rhabdomyolysis. *Journal of the American Osteopathic Association* 101:391-394.

Pescatello, L.S., B.A. Franklin, R. Fagard, W.B. Farquhar, G.A. Kelley, and C.A. Ray. 2004. American College of Sports Medicine position stand. Exercise and hypertension. *Medicine and Science in Sports and Exercise* 36:533-553.

Peter, R., S.D. Luzio, G. Dunseath, A. Miles, B. Hare, K. Backx, V. Pauvaday, and D.R. Owens. 2005. Effects of exercise on the absorption of insulin glargine in patients with type 1 diabetes. *Diabetes Care* 28:560-565.

Peters, E.M. 1990. Altitude fails to increase susceptibility of ultramarathon runners to postexercise upper respiratory tract infection. *South African Journal of Sports Medicine* 5:4-8.

Peters, E.M., and E.D. Bateman. 1983. Respiratory tract infections: An epidemiological survey. *South African Medical Journal* 64:582-584.

Peters, E.M., J.M. Goetzsche, B. Grobbelaar, and T.D. Noakes. 1993. Vitamin C supplementation reduces the incidence of postrace symptoms of upper-respiratory-tract infection in ultramarathon runners. *American Journal of Clinical Nutrition* 57:170-174.

Petersson, A., M. Garle, P. Holmgren, H. Druid, P. Krantz, and I. Thiblin. 2006. Toxicological findings and manner of death in autopsied users of anabolic androgenic steroids. *Drug and Alcohol Dependency* 81:241-249.

Petibois, C., and G. Deleris. 2003. Effects of short- and long-term detraining on metabolic response to endurance exercise. *International Journal of Sports Medicine* 24:320-325.

Petko, M., and G.R. Hunter. 1997. Four-year changes in strength, power and aerobic fitness in women college basketball players. *Strength and Conditioning* 19:46-49.

Petroczi, A., D.P. Naughton, G. Pearce, R. Bailey, A. Bloodworth, and M.J. McNamee. 2008. Nutritional supplement use by elite young UK athletes: Fallacies of advice regarding efficacy. *Journal of the International Society of Sports Nutrition* 5:22.

Pette, D., and R.S. Staron. 1990. Cellular and molecular diversities of mammalian skeletal muscle fibers. *Review of Physiology, Biochemistry and Pharmacology* 116:2-75.

Phillips, S.M., H.J. Green, M.J. MacDonald, and R.L. Hughson. 1995. Progressive effect of endurance training on VO_2 kinetics at the onset of submaximal exercise. *Journal of Applied Physiology* 79:1914-1920.

Phillips, S.M., X.X. Han, H.J. Green, and A. Bonen. 1996. Increments in skeletal muscle GLUT-1 and GLUT-4 after endurance training in humans. *American Journal of Physiology* 270:E456-E462.

Phillips, S.M., K.D. Tipton, A. Aarsland, S.E. Wolf, and R.R. Wolfe. 1997. Mixed muscle protein synthesis and breakdown after resistance exercise in humans. *American Journal of Physiology: Endocrinology and Metabolism* 273:E99-E107.

Piacentini, M.F., G. De Ioannon, S. Comotto, A. Spedicato, G. Vernillo, and A. La Torre. 2013. Concurrent strength and endurance training effects on running economy in master endurance runners. *Journal of Strength and Conditioning Research* 27:2295-2903.

Pierce, E.F., N.W. Eastman, R.W. McGowan, H. Tripathi, W.L. Dewey, and K.G. Olson. 1994. Resistance exercise decreases b-endorphin immunoreactivity. *British Journal of Sports Medicine* 28:164-166.

Pimental, N.A., H.M. Cosimini, M.N. Sawka, and C.B. Wenger. 1987. Effectiveness of an air-cooled vest using selected air temperature and humidity combinations. *Aviation, Space, and Environmental Medicine* 58:119-124.

Pineau, J.C., A.M. Guihard-Costa, and M. Bocquet. 2007. Validation of ultrasound techniques applied to body fat measurement. A comparison between ultrasound techniques, air displacement plethysmography and bioelectrical impedance vs. dual-energy X-ray absorptiometry. *Annals of Nutrition and Metabolism* 51:421-427.

Pipes, T.V. 1978. Variable resistance versus constant resistance strength training in adult males. *European Journal of Applied Physiology and Occupational Physiology* 39:27-35.

Piwonka, R.W., S. Robinson, V.L. Gay, and R.S. Manalis. 1965. Preacclimatization of men to heat by training. *Journal of Applied Physiology* 20:379-384.

Plinta, R., M. Olszanecka-Glinianowicz, A. Drosdzol-Cop, J. Chudek, and V. Skrzypulec-Plinta. 2012. The effect of three-month pre-season preparatory period and short-term exercise on plasma leptin, adiponectin, visfatin, and ghrelin levels in young female handball and basketball players. *Journal of Endocrinological Investigation* 35:595-601.

Plisk, S. 2008. Speed, agility, and speed endurance development. In *Essentials of strength training and conditioning,* 3rd ed., ed. T.R. Baechle, R.W. Earle, and NSCA, 471-491. Champaign, IL: Human Kinetics.

Plisk, S., and V. Gambetta. 1997. Tactical metabolic training: Part I. *Strength and Conditioning* 19:44-53.

Podolsky, A., K.R. Kaufman, T.D. Calahan, S.Y. Aleskinsky, and E.Y. Chao. 1990. The relationship of strength and jump height in figure skaters. *American Journal of Sports Medicine* 18:400-405.

Poliquin, C. 1988. Five ways to increase the effectiveness of your strength training program. *National Strength and Conditioning Association Journal* 10:34-39.

Pontifex, K.J., K.E. Wallman, B.T. Dawson, and C. Goodman. 2010. Effects of caffeine on repeated sprint ability, reactive agility time, sleep and next day performance. *Journal of Sports Medicine and Physical Fitness* 50:455-464.

Poortmans, J.R., H. Auquier, V. Renaut, A. Durussel, M. Saugy, and G.R. Brisson. 1997. Effects of short-term creatine supplementation on renal responses in man. *European Journal of Applied Physiology* 76:566-567.

Poortmans, J.R., and M. Francaux. 1999. Long-term oral creatine supplementation does not impair renal function in healthy athletes. *Medicine and Science in Sports and Exercise* 31:1108-1110.

Pope, H.G., G. Kanayama, M. Ionescu-Pioggia, and J.L. Hudson. 2004. Anabolic steroid users' attitudes towards physicians. *Addiction* 99:1189-1194.

Pope, H.G., and D.L. Katz. 1994. Psychiatric and medical effects of anabolic-androgenic steroid use. A controlled study of 160 athletes. *Archives of General Psychiatry* 51:375-382.

Pope, H.G. Jr., E.M. Kouri, and J.I. Hudson. 2000. Effects of supraphysiologic doses of testosterone on mood and aggression in normal men: A randomized controlled trial. *Archives of General Psychiatry* 57:133-140.

Porcerelli, J.H., and B.A. Sandler. 1998. Anabolic-androgenic steroid abuse and psychopathology. *Psychiatry Clinics of North America* 21:829-833.

Portington, K.J., D.D. Pascoe, M.J. Webster, L.H. Anderson, R.R. Rutland, and L.B. Gladden. 1998. Effect of induced alkalosis on exhaustive leg press performance. *Medicine and Science in Sports and Exercise* 30:523-528.

Posma, E., H. Moes, M.J. Heineman, and M.M. Faas. 2004. The effect of testosterone on cytokine production in the specific and non-specific immune response. *American Journal of Reproductive Immunology* 52:237-243.

Poston, J. 2005. Training for the half-marathon. *Strength and Conditioning Journal* 27:81-83.

Potteiger, J.A. 2000. Aerobic endurance exercise training. In *Essentials of strength training and conditioning,* 2nd ed., ed. T. Baechle and R. Earle, 493-508. Champaign, IL: Human Kinetics.

Potteiger, J., L. Judge, J. Cerny, and V. Potteiger. 1995. Effects of altering training volume and intensity on body mass, performance and hormonal concentrations in weight-event athletes. *Journal of Strength and Conditioning Research* 9:55-58.

Pottgiesser, T., Y.O. Schumacher, B. Wolfarth, A. Schmidt-Trucksass, and G. Bauer. 2012. Longitudinal observation of Epstein-Barr virus antibodies in athletes during a competitive season. *Journal of Medical Virology* 84:1415-1422.

Poulmedis, P., G. Rondoyannis, A. Mitsou, and E. Tsarouchas. 1988. The influence of isokinetic muscle torque exerted in various speeds of soccer ball velocity. *Journal of Orthopaedic and Sports Physical Therapy* 10:93-96.

Powrie, J.K., E.E. Bassett, T. Rosen, J.O. Jørgensen, R. Napoli, L. Sacca, J.S. Christiansen, B.A. Bengtsson, and P.H. Sönksen. 2007. Detection of growth hormone abuse in sport. *Growth Hormone and IGF Research* 17:220-226.

Prasad, A.S., C.S. Mantzoros, F.W. Beck, J.W. Hess, and G.J. Brewer. 1996. Zinc status and serum testosterone levels of healthy adults. *Nutrition* 12:344-348.

Prather, I.D., D.E. Brown, P. North, and J.R. Wilson. 1995. Clenbuterol: A substitute for anabolic steroids? *Medicine and Science in Sports and Exercise* 27:1118-1121.

Preatoni, E., A. Colombo, M. Verga, C. Galvani, M. Faina, R. Rodano, E. Preatoni, and M. Cardinale. 2012. The effects of whole-body vibration in isolation or combined with strength training in female athletes. *Journal of Strength and Conditioning Research* 26:2495-2506.

Preen, D.B. Dawson, C. Goodman, S. Lawrence, J. Beilby, and S. Ching. 2001. Effect of creatine loading on long-term sprint exercise performance and metabolism. *Medicine and Science in Sports and Exercise* 33:814-821.

Prentice, W.E. 1983. A comparison of static stretching and PNF stretching for improving hip joint flexibility. *Athletic Training* 18:56-59.

Pruden, E.L., O. Siggard-Anderson, and N.W. Tietz. 1987. Blood gases and pH. In *Fundamentals of clinical chemistry,* ed. N.W. Tietz, 624-644. Philadelphia: Saunders.

Pugh, L.G.C.E. 1964. Animals in high altitude: Man above 5000 meters—mountain exploration. In *Handbook of physiology: Adaptation to the environment,* 861-868. Washington, DC: American Physiological Society.

Pugh, L.G.C.E. 1966. Accidental hypothermia in walkers, climbers, and campers: Report to the Medical Commission on Accident Prevention. *British Medical Journal* 1:123-129.

Pugh, L.G.C.E. 1970. Oxygen uptake in track and treadmill running with observations on the effect of air resistance. *Journal of Physiology* 207:823-835.

Pye, M., A.C. Quinn, and S.M. Cobbe. 1994. QT interval dispersion: A non-invasive marker of susceptibility to arrhythmia in patients with sustained ventricular arrhythmias? *British Heart Journal* 71:511-514.

Pyeritz, R.E., and V.A. McKusick. 1979. The Marfan syndrome: Diagnosis and management. *New England Journal of Medicine* 300:772-777.

Pyne, D.B., I. Mujika, and T. Reilly. 2009. Peaking for optimal performance: Research limitations and future directions. *Journal of Sport Sciences* 27:195-202.

Pyne, D.B., P.U. Saunders, P.G. Montgomery, A.J. Hewitt, and K. Sheehan. 2008. Relationships between repeated sprint testing, speed, and endurance. *Journal of Strength and Conditioning Research* 22:1633-1637.

Radaelli, R., M. Bottaro, E.N. Wilhelm, D.R. Wagner, and R.S. Pinto. 2012. Time course of strength and echo intensity recovery after resistance exercise in women. *Journal of Strength and Conditioning Research* 26:2577-2584.

Radley, D., P.J. Gately, C.B. Cooke, S. Carroll, B. Oldroyd, and J.G. Truscott. 2003. Estimates of percentage body fat in young adolescents: A comparison of dual-energy X-ray absorptiometry and air displacement plethysmography. *European Journal of Clinical Nutrition* 57:1402-1410.

Raglin, J.S., W.P. Morgan, and A.E. Luchsinger. 1990. Mood state and self-motivation in successful and unsuccessful women rowers. *Medicine and Science in Sports and Exercise* 22:849-853.

Raglin, J.S., W.P. Morgan, and P.J. O'Connor. 1991. Changes in mood states during training in female and male college swimmers. *International Journal of Sports Medicine* 12:585-589.

Ram, F.S.F., S.M. Robinson, and P.N. Black. 2000. Effects of physical training in asthma: A systematic review. *British Journal of Sports Medicine* 34:162-167.

Ramis, J.M., B. Bibiloni, J. Moreiro, J.M. Garcia-Sanz, R. Salinas, A.M. Proenza, and I. Llado. 2005. Tissue leptin and plasma insulin are associated with lipoprotein lipase activity in severely obese patients. *Journal of Nutritional Biochemistry* 16:279-285.

Ramson, R., J. Jurimae, T. Jurimae, and J. Maestu. 2012. The effect of 4-week training period on plasma neuropeptide Y, leptin and ghrelin responses in male rowers. *European Journal of Physiology* 112:1873-1880.

Ransone, J., K. Neighbors, R. Lefavi, and J. Chromiak. 2003. The effect of β-hydroxy-β-methylbutyrate on muscular strength and body composition in collegiate football players. *Journal of Strength and Conditioning Research* 17:34-39.

Rasmussen, B.B., K.D. Tipton, S.L. Miller, S.E. Wolf, and R.R. Wolfe. 2000. An oral essential amino acid-carbohydrate supplement enhances muscle protein anabolism after resistance exercise. *Journal of Applied Physiology* 88:386-392.

Rasmussen, B.B., E. Volpi, D.C. Gore, and R.R. Wolfe. 2000. Androstenedione does not stimulate muscle protein anabolism in young healthy men. *Journal of Clinical Endocrinology and Metabolism* 85:55-59.

Rasmussen, C.H., R.O. Nielsen, M.S. Juul, and S. Rasmussen. 2013. Weekly running volume and risk of running-related injuries among marathon runners. *International Journal of Sports Physical Therapy* 8:111-120.

Ratamess, N.R. 2012. Resistance training. In *NSCA's guide to program design,* ed. J.R. Hoffman, 71-94. Champaign, IL: Human Kinetics.

Ratamess, N.R., C.M. Chiarello, A.J. Sacco, J.R. Hoffman, A.D. Faigenbaum, R.E. Ross, and J. Kang. 2012. The effects of rest interval length on acute bench press performance: The influence of gender and muscle strength. *Journal of Strength and Conditioning Research* 26:1817-1826.

Ratamess, N.A., A.D. Faigenbaum, G.T. Mangine, J.R. Hoffman, and J. Kang. 2007. Acute muscular strength assessment using free weight bars of different thickness. *Journal of Strength and Conditioning Research* 21:240-244.

Ratamess, N.A., M.J. Falvo, G.T. Mangine, J.R. Hoffman, A.D. Faigenbaum, and J. Kang. 2007. The effect of rest interval length on metabolic responses to the bench press exercise. *European Journal of Applied Physiology* 100:1-17.

Ratamess, N.A., J.R. Hoffman, R. Ross, M. Shanklin, A.D. Faigenbaum, and J. Kang. 2007. Effects of an amino acid/creatine/energy supplement on performance and the acute hormonal response to resistance exercise. *International Journal of Sport Nutrition and Exercise Metabolism* 17:608-623.

Ratamess, N.A., W.J. Kraemer, J.S. Volek, M.R. Rubin, A.L. Gomez, D.N. French, M.J. Sharman, M.R. McGuigan, T.P. Scheet, K. Hakkinen, R.U. Newton, and F.S. Dioguardi. 2003. The effects of amino acid supplementation on muscular performance during resistance training overreaching. *Journal of Strength and Conditioning Research* 17:250-258.

Read, M.T.F., and M.J. Bellamy. 1990. Comparison of hamstring/quadricep isokinetic strength ratios and power in tennis, squash, and track athletes. *British Journal of Sports Medicine* 24:178-182.

Reeds, P.J., S.M. Hay, P.M. Dorwood, and R.M. Palmer. 1986. Stimulation of growth by clenbuterol: Lack of effect on muscle protein biosynthesis. *British Journal of Nutrition* 56:249-258.

Reeves, J.T., B.M. Groves, J.R. Sutton, P.D. Wagner, A. Cymerman, M.K. Malconian, P.B. Rock, P.M. Young, and C.S. Houston. 1987. Operation Everest II: Preservation of cardiac function at extreme altitude. *Journal of Applied Physiology* 63:531-539.

Reeves, N.D., C.N. Maganaris, and M.V. Narici. 2004. Ultrasonographic assessment of human skeletal muscle size. *European Journal of Applied Physiology* 91:116-118.

Reissig, C.J., E.C. Strain, and R.R. Griffiths. 2009. Caffeinated energy drinks – a growing problem. *Drug and Alcohol Dependency* 99:1-10.

Reiter, E.O., and R.G. Rosenfeld. 2008. Normal and aberrant growth. In *Williams textbook of endocrinology,* 11th ed., ed. H.M. Kronenberg, S. Melmed, K.S. Polonsky, and P.R. Larsen, 849-968. New York: Elsevier.

Reitman, J.S., B. Vasquez, I. Klimes, and M. Naguelsparan. 1984. Improvement of glucose homeostasis after exercise training in non-insulin-dependent diabetes. *Diabetes Care* 7:431-441.

Remes, K., K. Kuoppasalmi, and H. Adlercreutz. 1980. Effect of physical exercise and sleep deprivation on plasma androgen levels. *International Journal of Sports Medicine* 6:131-135.

Repantis, D., P. Schlattmann, O. Laisney, and I. Heuser. 2010. Modafinil and methylphenidate for neuroenhancement in healthy individuals: A systematic review. *Pharmacological Research* 62:187-206.

Rhea, M.R., S.D. Ball, W.T. Phillips, and L.N. Burkett. 2002. A comparison of linear and daily undulating periodized programs with equated volume and intensity for strength. *Journal of Strength and Conditioning Research* 16:250-255.

Rhea, M.R., and J.G. Kenn. 2009. The effect of acute applications of whole-body vibration on the iTonic platform on subsequent lower-body power output during the back squat. *Journal of Strength and Conditioning Research*. 2358-2361.

Rhea, M.R., J.G. Kenn, and B.M. Dermody. 2009. Alterations in speed of squat movement and the use of accommodated resistance among college athletes training for power. *Journal of Strength and Conditioning Research* 23:2645-2650.

Ricci, G., D. Lajoie, R. Petitclerc, F. Peronnet, R.J. Ferguson, M. Fournier, and A.W. Taylor. 1982. Left ventricular size following endurance, sprint, and strength training. *Medicine and Science in Sports and Exercise* 14:344-347.

Richter, E.A., K.J. Mikines, H. Galbo, and B. Kiens. 1989. Effect of exercise on insulin action in human skeletal muscle. *Journal of Applied Physiology* 66:876-885.

Riddell, D.I. 1984. Is frostnip important? *Journal of the Royal Naval Medical Service* 70:140-142.

Riesenhuber, A., M. Boehm, M. Posch, and C. Aufricht. 2006. Diuretic potential of energy drinks. *Amino Acids* 31:81-83.

Rimmer, E., and G. Sleivert. 2000. Effects of a plyometric intervention program on sprint performance. *Journal of Strength and Conditioning Research* 14:295-301.

Ritchie, N.D., A.V. Baggott, and W.T.A. Todd. 2012. Acetazolamide for the prevention of acute mountain sickness – a systemic review and meta-analysis. *Journal of Travel Medicine* 19:298-307.

Roach, R.C., D. Maes, D. Sandoval, R.A. Robergs, M. Icenogle, H. Hinghofer-Szalkay, D. Lium, and J.A. Loeppky. 2000. Exercise exacerbates acute mountain sickness at simulated high altitude. *Journal of Applied Physiology* 88:581-585.

Robach, P., and C. Lundby. 2012. Is live high-train low altitude training relevant for elite athletes with already high total hemoglobin mass? *Scandinavian Journal of Medicine and Science in Sports* 22:303-305.

Robbins, D.W., T.L. Goodale, F.E. Kuzmits, and A.J. Adams. 2013. Changes in the athletic profile of elite college American football players. *Journal of Strength and Conditioning Research* 27:861-874.

Robergs, R.A., F. Ghiasvand, and D. Parker. 2004. Biochemistry of exercise-induced metabolic acidosis. *American Journal of Physiology: Regulatory, Integrative and Comparative Physiology* 287:R502-R516.

Robergs, R.A., and R. Landwehr. 2002. The surprising history of the "HRmax=220-age" equation. *Journal of Exercise Physiology* 5:1-10.

Roberts, M.D., V.J. Dalbo, S.E. Hassell, J.R. Stout, and C.M. Kerksick. 2008. Efficacy and safety of a popular thermogenic drink after 28 days of ingestion. *Journal of the International Society of Sports Nutrition* 5:19.

Roberts, M.D., L.W. Taylor, J.A. Wismann, C.D. Wilborn, R.B. Kreider, and D.S. Willoughby. 2007. Effects of ingesting JavaFit Energy Extreme functional coffee on aerobic and anaerobic fitness markers in recreationally-active coffee consumers. *Journal of the International Society of Sports Nutrition* 4:25.

Robertson, E.Y., P.U. Saunders, D.B. Pyne, R.J. Aughey, J.M. Anson, and C.J. Gore. 2010. Reproducibility of performance changes to simulated live high/train low altitude. *Medicine and Science in Sports and Exercise* 42:394-401.

Robertson, K., P. Adolfsson, M. Riddell, G. Scheiner, and R. Hanas. 2009. Exercise in children and adolescents with diabetes. *Pediatric Diabetes* 10(suppl):154-168.

Robertson, R.J., R. Gilcher, K.F. Metz, C.J. Casperson, T.G. Allison, R.A. Abbott, G.S. Skrinar, J.R. Krause, and P.A. Nixon. 1984. Hemoglobin concentration and aerobic work capacity in women following induced erythrocythemia. *Journal of Applied Physiology* 57:568-575.

Rodnick, K.J., W.L. Haskell, A.L. Swislocki, J.E. Foley, and G.M. Reaven. 1987. Improved insulin action in muscle, liver and adipose tissue in physically trained human subjects. *American Journal of Physiology* 253:E489-E495.

Rodriguez, N.R., N.M. DiMarco, and S. Langley. 2009. Nutrition and athletic performance. Position stand of the American College of Sports Medicine, American Dietetics Association and Dieticians of Canada. *Medicine and Science in Sports and Exercise* 41:709-731.

Rogol, A.D. 2010. Sex steroids, growth hormone, leptin and the pubertal growth spurt. *Endocrine Development* 17:77-85.

Rogol, A.D. 2011. Clinical and humanistic aspects of growth hormone deficiency and growth-related disorders. *American Journal of Managed Care* 17(suppl)18:eS4-eS10.

Roitt, I., J. Brostoff, and D. Male. 1993. *Immunology,* 3rd ed. St. Louis: Mosby.

Rome, L.C. 1990. Influence of temperature on muscle recruitment and muscle function in vivo. *American Journal of Physiology: Regulatory, Integrative and Comparative Physiology* 259:R210-R222.

Romijn, J.A., E.F. Coyle, L.S. Sidossis, A. Gastaldelli, J.F. Horowitz, E. Endert, and R.R. Wolfe. 1993. Regulation of endogenous fat and carbohydrate metabolism in relation to exercise intensity and duration. *American Journal of Physiology: Endocrinology and Metabolism* 28:E380-E391.

Rosenbaum, D., and E.M. Henning. 1997. Reaction time and force development after passive stretching and a 10-minute warm-up run. *Deutsch Zeitschrift Sportmedizin* 48:95-99.

Rosenbaum, M., A. Pietrobelli, J.R. Vasselli, S.B. Heymsfield, and R.L. Leibel. 2001. Sexual dimorphism in circulating leptin concentrations is not accounted for by differences in adipose tissue distribution. *International Journal of Obesity and Related Metabolic Disorders* 25:1365-1371.

Ross, A., M. Leveritt, and S. Riek. 2001. Neural influences on sprint running: Training adaptations and acute responses. *Sports Medicine* 31:409-425.

Ross, R.E., N.A. Ratamess, J.R. Hoffman, A.D. Faigenbaum, J. Kang, and A. Chilakos. 2009. The effects of treadmill sprint training and resistance training on maximal running velocity and power. *Journal of Strength and Conditioning Research* 23:385-394.

Rothstein, A., E.F. Adolph, and J.H. Wells. 1947. Voluntary dehydration. In *Physiology of man in the desert,* ed. E.F. Adolph, 254-720. New York: Interscience.

Rowbottom, D.G., D. Keast, and A.R. Morton. 1998. Monitoring and preventing overreaching and overtraining in endurance athletes. In *Overtraining in sport,* ed. R.B. Kreider, A.C. Fry, and M.L. O'Toole, 47-68. Champaign, IL: Human Kinetics.

Rowell, L.B. 1986. *Human circulation regulation during physical stress.* New York: Oxford University Press.

Rowland, T.W., and G.M. Green. 1988. Physiological responses to treadmill exercise in females: Adult-child differences. *Medicine and Science in Sports and Exercise* 20:474-478.

Roy, B.D., J.R. Fowles, R. Hill, and M.A. Tarnopolsky. 2000. Macronutrient intake and whole body protein metabolism following resistance exercise. *Medicine and Science in Sports and Exercise* 32:1412-1418.

Roy, B.D., M.A. Tarnopolsky, J.D. MacDougall, J. Fowles, and K.E. Yarasheski. 1997. Effect of glucose supplement timing on protein metabolism after resistance training. *Journal of Applied Physiology* 82:1882-1888.

Roy, G. 2013. Personal communication. February, 2013.

Rozenek, R., C.H. Rahe, H.H. Kohl, D.N. Marple, G.D. Wilson, and M.H. Stone. 1990. Physiological responses to resistance-exercise in athletes self-administering anabolic steroids. *Journal of Sports Medicine and Physical Fitness* 30:354-360.

Rubin, M.R., W.J. Kraemer, C.M. Maresh, J.S. Volek, N.A. Ratamess, J.L. Vanheest, R. Silvestre, D.N. French, M.J. Sharman, D.A. Judelson, A.L. Gómez, J.D. Vescovi, and W.C. Hymer. 2005. High-affinity growth hormone binding protein and acute heavy resistance exercise. *Medicine and Science in Sports and Exercise* 37:395-403.

Rubin, M.R., J.S. Volek, A.L. Gomez, N.A. Ratamess, D.N. French, M.J. Sharman, and W.J. Kraemer. 2001. Safety measures of L-carnitine L-tartrate supplementation in healthy men. *Journal of Strength and Conditioning Research* 15:486-490.

Rundell, K.W., B.A. Spiering, J.M. Baumann, and T.M. Evans. 2005. Effects of montelukast on airway narrowing from eucapnic voluntary hyperventilation and cold air exercise. *British Journal of Sports Medicine* 39:232-236.

Rundell, K.W., R.L. Wilber, L. Szmedra, D.M. Jenkinson, L.B. Mayers, and J. Im. 2000. Exercise-induced asthma screening of elite athletes: Field versus laboratory exercise challenge. *Medicine and Science in Sports and Exercise* 32:309-316.

Ryan, A.J., G.P. Lambert, X. Shi, R.T. Chang, R.W. Summers, and C.V. Gisolfi. 1998. Effect of hypohydration on gastric emptying and intestinal absorption during exercise. *Journal of Applied Physiology* 84:1581-1588.

Sabelli, H., P. Fink, J. Fawcett, and C. Tom. 1996. Sustained antidepressant effect of PEA replacement. *Journal of Neuropsychiatry and Clinical Neuroscience* 8:168-171.

Saez, J.M., and M.G. Forest. 1979. Kinetics of human chorionic gonadotropin-induced steroidogenic response of the human testis. I. Plasma testosterone: Implications for human chorionic gonadotropin stimulation test. *Journal of Clinical Endocrinology and Metabolism* 49:278-283.

Safdar, A., J.P. Little, A.J. Stokl, B.P. Hettinga, M. Akhtar, and M.A. Tarnopolsky. 2011. Exercise increases mitochondrial PGC-1alpha content and promotes nuclear-mitochondrial cross-talk to coordinate mitochondrial biogenesis. *Journal of Biological Chemistry* 286:4135-4140.

Sahler, C.S., and B.D. Greenwald. 2012. Traumatic brain injury in sports: A review [published online ahead of print July 9, 2012]. *Rehabilitation Research and Practice* doi: 10.1155/2012/659652. Accessed November 23, 2012.

Sahlin, K., R.C. Harris, B. Nylind, and E. Hultman. 1976. Lactate content and pH in muscle samples obtained after dynamic exercise. *Pflugers Archives* 367:143-149.

Sale, D.G. 1988. Neural adaptation to resistance training. *Medicine and Science in Sports and Exercise* 20:S1135-S1145.

Sale, D.G., J.D. MacDougall, and S. Garner. 1990. Comparison of two regimens of concurrent strength and endurance training. *Medicine and Science in Sports and Exercise* 22:348-356.

Sale, D.G., J.D. MacDougall, I. Jacobs, and S. Garner. 1990. Interaction between concurrent strength and endurance training. *Journal of Applied Physiology* 68:260-270.

Sale, D.G., D.E. Moroz, R.S. McKelvie, J.D. MacDougall, and N. McCartney. 1993. Comparison of blood pressure response to isokinetic and weight-lifting exercise. *European Journal of Applied Physiology* 67:115-120.

Sale, D.G., D.E. Moroz, R.S. McKelvie, J.D. MacDougall, and N. McCartney. 1994. Effect of training on the blood pressure response to weight lifting. *Canadian Journal of Applied Physiology* 19:60-74.

Saltin, B., and P.O. Åstrand. 1967. Maximal oxygen uptake in athletes. *Journal of Applied Physiology* 23:353-358.

Saltin, B., and P.O. Åstrand. 1993. Free fatty acids and exercise. *American Journal of Clinical Nutrition* 57:S752-S757.

Saltin, B., J. Henriksson, E. Nygaard, and P. Anderson. 1977. Fiber types and metabolic potentials of skeletal muscles in sedentary man and endurance runners. *Annals of the New York Academy of Sciences* 301:3-29.

Saltin, B., C.K. Kim, N. Terrados, H. Larsen, J. Svedenhag, and C.J. Rolf. 1995. Morphology, enzyme activities and buffer capacity in leg muscles of Kenyan and Scandinavian runners. *Scandinavian Journal of Medicine and Science in Sports* 5:222-230.

Saltin, B., and L.B. Rowell. 1980. Functional adaptations to physical activity and inactivity. *Federation Proceedings* 39:1506-1513.

Samaja, M. 2001. Hypoxia-dependent protein expression: Erythropoietin. *High Altitude Medicine and Biology* 2:155-163.

Sapir, D.G., O.E. Owen, T. Pozefsky, and M. Walser. 1974. Nitrogen sparing induced by a mixture of essential amino acids given chiefly as their keto analogs during prolonged starvation in obese subjects. *Journal of Clinical Investigation* 54:974-980.

Sarabon, N. 2012. Balance and stability training. In *NSCA's guide to program design,* ed. J.R. Hoffman, 185-212. Champaign, IL: Human Kinetics.

Sargeant, A. 1987. Effect of muscle temperature on leg extension force and short-term power output in humans. *European Journal of Applied Physiology* 56:693-698.

Sargeant, A.J., E. Hoinville, and A. Young. 1981. Maximum leg force and power output during short-term dynamic exercise. *Journal of Applied Physiology* 26:188-194.

Sato, A., A. Kamei, E. Kamihigashi, M. Dohi, Y. Komatsu, T. Akama, and T. Kawahara. 2012. Use of supplements by young elite Japanese athletes participating in the 2010 youth Olympic games in Singapore. *Clinical Journal of Sport Medicine* 22:418-423.

Saugy, M., N. Robinson, C. Saudan, N. Baume, L. Avois, and P. Mangin. 2008. Human growth hormone doping in sport. *British Journal of Sports Medicine* 40(suppl):i35-i39.

Sawka, M.N. 1992. Physiological consequences of hypohydration: Exercise performance and thermoregulation. *Medicine and Science in Sports and Exercise* 24:657-670.

Sawka, M.N., R.G. Knowlton, and J.B. Critz. 1979. Thermal and circulatory responses to repeated bouts of prolonged running. *Medicine and Science in Sports and Exercise* 11:177-180.

Sawka, M.N., and K.B. Pandolf. 1990. Effects of body water loss on physiological function and exercise performance. In *Fluid homeostasis during exercise: Perspectives in exercise science and sports medicine, Vol. 3,* ed. C.V. Gisolfi and D.R. Lamb, 1-38. Indianapolis: Benchmark Press.

Sawka, M.N., C.B. Wenger, A.J. Young, and K.B. Pandolf. 1993. Physiological responses to exercise in the heat. In *Nutritional needs in hot environments,* ed. B.M. Marriott, 55-74. Washington, DC: National Academy Press.

Sawka, M.A., and A.J. Young. 2000. Physical exercise in hot and cold climates. In *Exercise and sport science,* ed. W.E. Garrett and D.T. Kirkendall, 385-400. Philadelphia: Lippincott Williams & Wilkins.

Sawka, M.N., A.J. Young, B.S. Cadarette, L. Levine, and K.B. Pandolf. 1985. Influence of heat stress and acclimation on maximal power. *European Journal of Applied Physiology* 53:294-298.

Sawka, M.N., A.J. Young, R.P. Francesconi, S.R. Muza, and K.B. Pandolf. 1985. Thermoregulatory and blood responses during exercise at graded hypohydration levels. *Journal of Applied Physiology* 59:1394-1401.

Sawka, M.N., A.J. Young, S.R. Muza, R.R. Gonzalez, and K.B. Pandolf. 1987. Erythrocyte reinfusion and maximal aerobic power: An examination of modifying factors. *Journal of the American Medical Association* 257:1496-1499.

Schabort, E.J., A.N. Bosch, S.M. Weltan, and T.D. Noakes. 1999. The effect of a preexercise meal on time to fatigue during prolonged cycling exercise. *Medicine and Science in Sports and Exercise* 31:464-471.

Schaefer, M.E., J.A. Allert, H.R. Adams, and M.H. Laughlin. 1992. Adrenergic responsiveness and intrinsic sinoatrial automaticity of exercise-trained rats. *Medicine and Science in Sports and Exercise* 24:887-894.

Schiaffino, S., and C. Reggiani. 2011. Fiber types in mammalian skeletal muscles. *Physiological Reviews* 91:1447-1531.

Schiller, J.S., J.W. Lucas, B.W. Ward, and J.A. Peregoy. 2012. Summary health statistics for U.S. adults: National Health Interview Survey, 2010. *Vital Health Statistics* 10(252):1-207.

Schilling, B.K., M.H. Stone, A. Utter, J.T. Kearney, M. Johnson, R. Coglianese, L. Smith, H.S. O'Bryant, A.C. Fry, M. Starks, R. Keith, and M.E. Stone. 2001. Creatine supplementation and health variables: A retrospective study. *Medicine and Science in Sports and Exercise* 33:183-188.

Schmidt, A., C. Maier, G. Schaller, P. Nowotny, M. Bayerle-Eder, B. Buranyi, A. Luger, and M. Wolzt. 2004. Acute exercise has no effect on ghrelin plasma concentrations. *Hormone and Metabolic Research* 36:174-177.

Schmidt, W., K. Heinicke, J. Rojas, J. Manuel Gomez, M. Serrato, M. Mora, B. Wolfarth, A. Schmid, and J. Keul. 2002. Blood volume and hemoglobin mass in endurance athletes from moderate altitude. *Medicine and Science in Sports and Exercise* 34:1934-1940.

Schmidtbleicher, D., A. Gollhofer, and U. Frick. 1988. Effects of a stretch-shortening type training on the performance capability and innervation characteristics of leg extensor muscles. In *Biomechanics XI-A,* ed. G. de Groot, A. Hollander, P. Huijing, and G. Van Ingen Schenau, 185-189. Amsterdam: Free University Press.

Schneider, M.B., and H.J. Benjamin. 2011. Clinical report–sports drinks and energy drinks for children and adolescents: Are they appropriate? *Pediatrics* 127:1182-1189.

Schneider, S.H., L.F. Amorosa, A.K. Khachadurian, and N.B. Ruderman. 1984. Studies on the mechanism of improved glucose control during regular exercise in type 2 (non-insulin-dependent) diabetes. *Diabetologia* 26:355-360.

Schneiker, K.T., D. Bishop, B. Dawson, and L.P. Hackett. 2006. Effects of caffeine on prolonged intermittent-sprint ability in team-sport athletes. *Medicine and Science in Sports and Exercise* 38:578-585.

Schurmeyer, T., K. Jung, and E. Nieschlag. 1984. The effect of an 1100 km run on testicular, adrenal and thyroid hormones. *International Journal of Andrology* 7:276-282.

Schutz, Y., and R. Herren. 2000. Assessment of speed on human locomotion using a differential satellite global positioning system. *Medicine and Science in Sports and Exercise* 32:642-646.

Schwarz, L., and W. Kindermann. 1989. Beta-endorphin, catecholamines, and cortisol during exhaustive endurance exercise. *International Journal of Sports Medicine* 10:324-328.

Segal, S., R. Narayanan, and J.T. Dalton. 2006. Therapeutic potential of the SARMs: Revisiting the androgen receptor for drug discovery. *Expert Opinion on Investigative Drugs* 15:377-387.

Seminick, D. 1990. The T-test. *National Strength and Conditioning Association Journal* 12:36-37.

Seminick, D. 1994. Testing protocols and procedures. In *Essentials of strength training and conditioning,* ed. T. Baechle, 258-73. Champaign, IL: Human Kinetics.

Senay, L. 1979. Temperature regulation and hypohydration: A singular view. *Journal of Applied Physiology* 47:1-7.

Shapiro, L. 1997. The morphological consequences of systemic training. *Cardiology Clinics* 15:373-379.

Sharp, R.L., D.L. Costill, W.J. Fink, and D.S. King. 1986. Effects of eight weeks of bicycle ergometer sprint training on human muscle buffer capacity. *International Journal of Sports Medicine* 7:13-17.

Sharp, R.L., J.P. Troup, and D.L. Costill. 1982. Relationship between power and sprint freestyle swimming. *Medicine and Science in Sports and Exercise* 14:53-56.

Shave, R., G. Whyte, A. Siemann, and L. Doggart. 2001. The effects of sodium citrate ingestion on 3,000-meter time-trial performance. *Journal of Strength and Conditioning Research* 15:230-234.

Shaver, L.G. 1970. Maximum dynamic strength, relative dynamic endurance, and their relationship. *Research Quarterly for Exercise and Sport* 42:460-465.

Shek, P.N., B.H. Sabiston, A. Buguet, and M.W. Radomski. 1995. Strenuous exercise and immunological changes: A multiple-time-point analysis of leukocyte subsets, CD4/CD8 ratio, immunoglobulin production and NK cell response. *International Journal of Sports Medicine* 16:466-474.

Shekelle, P., M. Hardy, S. Morton, M. Maglione, M. Suttorp, E. Roth, and L. Jungvig. 2003. *Ephedra and ephedrine for weight loss and athletic performance enhancement: Clinical efficacy and side effects.* Evidence Report/Technology Assessment No. 76 (prepared by Southern California Evidence-based Practice Center, RAND, under Contract No. 290-97-0001, Task Order No. 9). AHRQ Publication No. 03-E022. Rockville, MD: Agency for Healthcare Research and Quality.

Shephard, R.J., T. Kavanaugh, D.J. Mertens, S. Qureshi, and M. Clark. 1995. Personal health benefits of masters athlete competition. *British Journal of Sports Medicine* 29:35-40.

Sherman, W.M., G. Brodowicz, D.A. Wright, W.K. Allen, J. Simonsen, and A. Dernback. 1989. Effect of 4 h pre-exercise carbohydrate feedings on cycling performance. *Medicine and Science in Sports and Exercise* 21:598-604.

Sherman, W., D.L. Costill, W.J. Fink, F. Hagerman, L. Armstrong, and T. Murray. 1983. Effect of a 42.2-km foot race and subsequent rest or exercise on muscle glycogen and enzymes. *Journal of Applied Physiology* 55:1219-1224.

Sherman, W., D.L. Costill, W.J. Fink, and J.M. Miller. 1981. Effect of exercise-diet manipulation on muscle glycogen and its subsequent utilization during performance. *International Journal of Sports Medicine* 2:114-118.

Shono, N., H. Urata, B. Saltin, M. Mizuno, T. Harada, M. Shindo, and H. Tanaka. 2002. Effects of low intensity aerobic training on skeletal muscle capillary and blood lipoprotein profiles. *Journal of Atherosclerosis and Thrombosis* 9:78-85.

Shvartz, E., Y. Shapiro, A. Magazanik, A. Meroz, H. Birnfeld, A. Mechtinger, and S. Shibolet. 1977. Heat acclimation, physical training, and responses to exercise in temperate and hot environments. *Journal of Applied Physiology: Respiratory, Environmental and Exercise Physiology* 43:678-683.

Sidiropoulou, M.P., E.G. Fotiadou, V.K. Tsimara, A.P. Zakas, and N.A. Zgelopoulou. 2007. The effect of interval training in children with exercise-induced asthma competing in soccer. *Journal of Strength and Conditioning Research* 21:446-450.

Silva, A.C., M.S. Santos-Neto, A.M. Soares, M.C. Fonteles, R.L. Guerrant, and A.A. Lima. 1998. Efficacy of a glutamine-based oral rehydration solution on the electrolyte and water absorption in a rabbit model of secretory diarrhea induced by cholera toxin. *Journal of Pediatric Gastroenterology and Nutrition* 26:513-519.

Simoneau, J.A., G. Lortie, M.R. Boulay, M. Marcotte, M.C. Thibault, and C. Bouchard. 1985. Human skeletal muscle fiber type alteration with high-intensity intermittent training. *European Journal of Applied Physiology* 54:240-253.

Simoneau, J.A., G. Lortie, M.R. Boulay, M. Marcotte, M.C. Thibault, and C. Bouchard. 1987. Effects of two high-intensity intermittent training programs interspaced by detraining on human skeletal muscle and performance. *European Journal of Applied Physiology* 56:516-521.

Singh, A., E. Moses, and P. Deuster. 1992. Chronic multivitamin-mineral supplementation does not enhance performance. *Medicine and Science in Sports and Exercise* 24:726-732.

Singh, M.V., S.B. Rawal, and A.K. Tyagi. 1990. Body fluid status on induction, reinduction and prolonged stay at high altitude in human volunteers. *International Journal of Biometeorology* 34:93-97.

Sinha-Hikim, I., J. Artaza, L. Woodhouse, N. Gonzalez-Cadavid, A.B. Singh, M.I. Lee, T.W. Storer, R. Casaburi, R. Shen, and S. Bhasin. 2002. Testosterone-induced increase in muscle size in healthy young men is associated with muscle fiber hypertrophy. *American Journal of Physiology: Endocrinology and Metabolism* 283:E154-E164.

Sjodin, A.M., A.H. Forslund, K.R. Westerterp, A.B. Andersson, J.M. Forslund, and L.M. Hambraeus. 1996. The influence of physical activity on BMR. *Medicine and Science in Sports and Exercise* 28:85-91.

Sjogaard, G. 1986. Water and electrolyte fluxes during exercise and their relation to muscle fatigue. *Acta Physiologica Scandinavica* 128(suppl):129-136.

Skaper, S.D., S. Das, and F.D. Marshall. 1973. Some properties of a homocarnosine-carnosine synthetase isolated from rat brain. *Journal of Neurochemistry* 21:1429-1445.

Slater, G., D. Jenkins, P. Logan, H. Lee, M. Vukovich, J.A. Rathmacher, and A.G. Hahn. 2001. Beta-hydroxy-beta-methylbutyrate (HMB) supplementation does not affect changes in strength or body composition during resistance training in trained men. *International Journal of Sport Nutrition and Exercise Metabolism* 11:384-396.

Sleeper, M.M., C.F. Kearns, and K.H. McKeever. 2002. Chronic clenbuterol administration negatively alters cardiac function. *Medicine and Science in Sports and Exercise* 34:643-650.

Smith, B.W., and M. LaBotz. 1998. Pharmacologic treatment of exercise-induced asthma. *Clinics in Sports Medicine* 17:343-363.

Smith, C.J., and G. Havenith. 2012. Body mapping of sweating patterns in athletes: A sex comparison. *Medicine and Science in Sports and Exercise* 44:2350-2361.

Smith, D.J., and D. Roberts. 1991. Aerobic, anaerobic and isokinetic measures of elite Canadian male and female speed skaters. *Journal of Applied Sport Science Research* 5:110-115.

Smith, J.A., R.D. Telford, M.S. Baker, A.J. Hapel, and M.J. Weidemann. 1992. Cytokine immunoreactivity in plasma does not change after moderate endurance-exercise. *Journal of Applied Physiology* 71:1396-1401.

Smith, J.A., R.D. Telford, I.B. Mason, and M.J. Weidemann. 1990. Exercise, training and neutrophil microbicidal activity. *International Journal of Sports Medicine* 11:179-187.

Smith, T.B., W.G. Hopkins, and N.A. Taylor. 1994. Respiratory responses of elite oarsmen, former oarsmen, and highly trained non-rowers during rowing, cycling and running. *European Journal of Applied Physiology and Occupational Physiology* 69:44-49.

Smolander, J., P. Kolari, O. Korhonen, and R. Ilmarinen. 1986. Aerobic and anaerobic responses to incremental exercise in a thermoneutral and a hot dry environment. *Acta Physiologica Scandinavica* 128:15-21.

Snow, R.J., M.J. McKenna, S.E. Selig, J. Kemp, C.G. Stathis, and S. Zhao. 1998. Effect of creatine supplementation on sprint exercise performance and muscle metabolism. *Journal of Applied Physiology* 84:1667-1673.

Snow, T.K., M. Millard-Stafford, and L.B. Rosskopf. 1998. Body composition profile of NFL football players. *Journal of Strength and Conditioning Research* 12:146-149.

Snyder, S. 2012. Running. In *Developing endurance,* ed. B. Reuter, 181-220. Champaign, IL: Human Kinetics.

Socas, L., M. Zumbardo, O. Perez-Luzardo, A. Ramos, C. Perez, J.R. Hernandez, and L.D. Boada. 2005. Hepatocellular adenomas associated with anabolic androgenic steroid abuse in bodybuilders: A report of two cases and a review of the literature. *British Journal of Sports Medicine* 39:e27.

Soe, K.L., M. Soe, and C. Gluud. 1992. Liver pathology associated with the use of anabolic-androgenic steroids. *Liver* 12:73-79.

Sole, C.C., and T.D. Noakes. 1989. Faster gastric emptying for glucose-polymer and fructose solutions than for glucose in humans. *European Journal of Applied Physiology* 58:605-612.

Sotiropoulos, A., M. Ohanna, C. Kedzia, R.K. Menon, J.J. Kopchick, P.A. Kelly, and M. Pende. 2006. Growth hormone promotes skeletal muscle cell fusion independent of insulin-like growth factor 1 up-regulation. *Proceedings of the National Academy of Sciences U S A* 103:7315-7320.

Spence, A.L., L.H. Naylor, H.C. Carter, C.L. Buck, L. Dembo, C.P. Murray, P. Watson, D. Oxborough, K.P. George, and D.J. Green. 2011. A prospective randomised longitudinal MRI study of left ventricular adaptation to endurance and resistance exercise training in humans. *Journal of Physiology* l589:5443-5452.

Spiering, B.A., W.J. Kraemer, J.L. Vingren, D.L. Hatfield, M.S. Fragala, J. Ho, C.M. Maresh, J.M. Anderson, and J.S. Volek. 2007. Responses of criterion variables to different supplemental doses of L-carnitine L-tartrate. *Journal of Strength and Conditioning Research* 21:259-264.

Spillane, M., R. Schoch, M. Cooke, T. Harvey, M. Greenwood, R. Kreider, and D.S. Willougby. 2009. The effects of creatine ethyl ester supplementation combined with heavy resistance training on body composition, muscle performance, and serum and muscle creatine levels. *Journal of the International Society of Sports Nutrition* 6:6.

Spirito, P., A. Pelliccia, M. Proschan, M. Granata, A. Spataro, P. Bellone, G. Caselli, A. Biffi, C. Vecchio, and B.J. Maron. 1994. Morphology of the "athlete's heart" assessed by echocardiography in 947 elite athletes representing 27 sports. *American Journal of Cardiology* 74:802-806.

Sporer, B.C., A.W. Sheel, and D.C. McKenzie. 2008. Dose response of inhaled salbutamol on exercise performance and urine concentrations. *Medicine and Science in Sports and Exercise* 40:149-157.

Sprenger, H., C. Jacobs, M. Nain, A.M. Gressner, H. Prinz, W. Wesemann, and D. Gemsa. 1992. Enhanced release of cytokines, interleukin-2 receptors, and neopterin after long-distance running. *Clinical Immunology and Immunopathology* 53:188-195.

Spriet, L.L. 1991. Blood doping and oxygen transport. In *Perspectives in exercise science and sports medicine, Vol. 4, Ergogenics,* ed. D.R. Lamb and M.H. Williams, 213-248. Carmel, IN: Benchmark Press.

Spriet, L.L. 1995. Caffeine and performance. *International Journal of Sport Nutrition* 5:S84-S99.

Spriet, L.L., D.A. MacLean, D.J. Dyck, E. Hultman, G. Cederblad, and T.E. Graham. 1992. Caffeine ingestion and muscle metabolism during prolonged exercise in humans. *American Journal of Physiology* 262:E891-E898.

Sproles, C.B., D.P. Smith, R.J. Byrd, and T.E. Allen. 1976. Circulatory responses to submaximal exercise after dehydration and rehydration. *Journal of Sports Medicine and Physical Fitness* 16:98-105.

Stamford, B.A., and T. Moffatt. 1974. Anabolic steroid: Effectiveness as an ergogenic aid to experienced weight trainers. *Journal of Sports Medicine and Physical Fitness* 14:191-197.

Stanley, D.C., W.J. Kraemer, R.L. Howard, L.E. Armstrong, and C.M. Maresh. 1994. The effects of hot water immersion on muscle strength. *Journal of Strength and Conditioning Research* 8:134-138.

Starks, M.A., S.L. Starks, M. Kingsley, M. Purpura, and R. Jäger. 2008. The effects of phosphotidylserine on endocrine response to moderate intensity exercise. *Journal of the International Society of Sports Nutrition* 5:11.

Staron, R.S., and R.S. Hikida. 1992. Histochemical, biochemical, and ultrastructural analyses of single human muscle fibers with special reference to C fiber protein population. *Journal of Histochemistry and Cytochemistry* 40:563-568.

Staron, R.S., and P. Johnson. 1993. Myosin polymorphism and differential expression in adult human skeletal muscle. *Comparative Biochemical Physiology* 106:B463-B475.

Staron, R.S., D.L. Karapondo, W.J. Kraemer, A.C. Fry, S.E. Gordon, J.E. Falkel, F.C. Hagerman, and R.S. Hikida. 1994. Skeletal muscle adaptations during early phase of heavy resistance training in men and women. *Journal of Applied Physiology* 76(3):1247-1155.

Staron, R.S., M.J. Leonardi, D.L. Karapondo, E.S. Malicky, J.E. Falkel, F.C. Hagerman, and R.S. Hikida. 1991. Strength and skeletal muscle adaptations in heavy resistance trained women after detraining and retraining. *Journal of Applied Physiology* 70(2):631-640.

Staron, R.S., E.S. Malicky, M.J. Leonardi, J.E. Falkel, F.C. Hagerman, and G.A. Dudley. 1989. Muscle hypertrophy and fast fiber type conversions in heavy resistance trained women. *European Journal of Applied Physiology* 60:71-79.

Stearns, R.L., H. Emmanuel, J.S. Volek, and D.J. Casa. 2010. Effects of ingesting protein in combination with carbohydrate during exercise on endurance performance: A systemic review with meta-analysis. *Journal of Strength and Conditioning Research* 24:2192-2202.

Steensberg, A., G. van Hall, T. Osada, M. Sacchetti, B. Saltin, and B. Klarlund Pedersen. 2000. Production of interleukin-6 in contracting human skeletal muscles can account for the exercise-induced increase in plasma interleukin-6. *Journal of Physiology* 529(pt 1):237-242.

Stein, R.B., T. Gordon, and J. Shriver. 1982. Temperature dependence of mammalian muscle contractions and ATPase activities. *Journal of Biophysiology* 40:97-107.

Steiner, A.Z., M. Terplan, and R.J. Paulson. 2005. Comparison of tamoxifen and clomiphene citrate for ovulation induction: A meta-analysis. *Human Reproduction* 20:1511-1515.

Stilger, V.G., and C.E. Yesalis. 1999. Anabolic-androgenic steroid use among high school football players. *Journal of Community Health* 24:131-145.

Stokes, K.A., D. Sykes, K.L. Gilbert, J.W. Chen, and J. Frystyk. 2010. Brief, high intensity exercise alters serum ghrelin and growth hormone concentrations but not IGF-I, IGF-II or IGF-I bioactivity. *Growth Hormone and IGF Research* 20:289-294.

Stolt, A., T. Karila, M. Viitasalo, M. Mantysaari, U.M. Kujala, and J. Karjalainen. 1999. QT interval and QT dispersion in endurance athletes and in power athletes using large doses of anabolic steroids. *American Journal of Cardiology* 84:364-366.

Stone, M.H., S.J. Fleck, N.T. Triplett, and W.J. Kraemer. 1991. Health- and performance-related potential of resistance training. *Sports Medicine* 11:210-231.

Stone, M.H., and A.C. Fry. 1998. Increased training volume in strength/power athletes. In *Overtraining in sport,* ed. R.B. Kreider, A.C. Fry, and M.L. O'Toole, 87-130. Champaign, IL: Human Kinetics.

Stone, M.H., R. Johnson, and D. Carter. 1979. A short term comparison of two different methods of resistive training on leg strength and power. *Athletic Training* 14:158-160.

Stone, M.H., J.K. Nelson, S. Nader, and D. Carter. 1983. Short-term weight training effects on resting and recovery heart rates. *Athletic Training* (Spring): 69-71.

Stone, M.H., H. O'Bryant, and J. Garhammer. 1981. A hypothetical model for strength training. *Journal of Sports Medicine* 21:342-351.

Stone, M.H., G.D. Wilson, D. Blessing, and R. Rozenek. 1983. Cardiovascular response to short-term Olympic style weight-training in young men. *Canadian Journal of Applied Sport Science* 8:134-139.

Storch, K.J., D.A. Wagner, and V.R. Young. 1991. Methionine kinetics in adult men: Effects of dietary betaine on L-[$^2H_3$-methyl-l-^{13}C] methionine. *American Journal of Clinical Nutrition* 54:386-394.

Storms, W.W. 1999. Exercise-induced asthma: Diagnosis and treatment for the recreational or elite athlete. *Medicine and Science in Sports and Exercise* 31:S33-S38.

Stout, J.R., J.T. Cramer, J. O'Kroy, M. Mielke, R. Zoeller, and D. Torok. 2006. Effects of β-alanine and creatine monohydrate supplementation on the physical working capacity at neuromuscular fatigue threshold. *Journal of Strength and Conditioning Research* 20:928-931.

Stout, J.R., J.T. Cramer, R.F. Zoeller, D. Torok, P. Costa, J.R. Hoffman, R.C. Harris, and J. O'Kroy. 2007. Effects of β-alanine supplementation on the onset of neuromuscular fatigue and ventilatory threshold in women. *Amino Acids* 32:381-386.

Stout, J.R., J. Eckerson, K. Ebersole, G. Moore, S. Perry, T. Housh, A. Bull, J. Cramer, and A. Batheja. 2000. Effect of creatine loading on neuromuscular fatigue threshold. *Journal of Applied Physiology* 88:109-112.

Stowers, T., J. McMillan, D. Scala, V. Davis, D. Wilson, and M. Stone. 1983. The short-term effects of three different strength-power training methods. *National Strength and Conditioning Association Journal* 5:24-27.

Strahan, A.R., T.D. Noakes, G. Kotzenberg, A.E. Nel, and F.C. de Beer. 1984. C reactive protein concentrations during long distance running. *British Medical Journal* 289:1249-1251.

Strauss, R.H., and C.E. Yesalis. 1991. Anabolic steroids in the athlete. *Annual Review of Medicine* 42:449-457.

Street, C., J. Antonio, and D. Cudlipp. 1996. Androgen use by athletes: A reevaluation of the health risks. *Canadian Journal of Applied Physiology* 21:421-440.

Stromme, S.B., H.D. Meen, and A. Aakvaag. 1974. Effects of an androgenic-anabolic steroid on strength development and plasma testosterone levels in normal males. *Medicine and Science in Sports and Exercise* 6:203-208.

Strydom, N.B., and C.G. Williams. 1969. Effect of physical conditioning on state of heat acclimatization of Bantu laborers. *Journal of Applied Physiology* 27:262-265.

Stuart, A.G., and A. Williams. 2007. Marfan's syndrome and the heart. *Archives of Disease in Childhood* 92:351-356.

Stuart, G.R., W.G. Hopkins, C. Cook, and S.P. Cairns. 2005. Multiple effects of caffeine on simulated high-intensity team sport performance. *Medicine and Science in Sports and Exercise* 37:1998-2005.

Stupka, N., S. Lowther, K. Chorneyko, J.M. Bourgeois, C. Hogben, and M.A. Tarnopolsky. 2000. Gender differences in muscle inflammation after eccentric exercise. *Journal of Applied Physiology* 89:2325-2332.

Sturmi, J.E., and D.J. Diorio. 1998. Anabolic agents. *Clinical Sports Medicine* 17:261-282.

Sue-Chu, M. 2012. Winter sports athletes: Long-term effects of cold air exposure. *British Journal of Sports Medicine* 46:397-401.

Sundaram, K., N. Kumar, C. Monder, and C.W. Bardin. 1995. Different patterns of metabolism determine the relative anabolic activity of 19-norandrogens. *Journal of Steroid Biochemistry and Molecular Biology* 53:253-257.

Sutton, J.R., M.J. Coleman, J. Casey, and L. Lazarus. 1973. Androgen responses during physical exercise. *British Medical Journal* 1:520-522.

Sutton, J., and L. Lazarus. 1976. Growth hormone in exercise: Comparison of physiological and pharmacological stimuli. *Journal of Applied Physiology* 41:523-527.

Sutton, J.R., J.T. Reeves, P.D. Wagner, B.M. Groves, A. Cymerman, M.K. Malconian, P.B. Rock, P.M. Young, S.D. Walter, and C.S. Houston. 1988. Operation Everest II: Oxygen transport during exercise at extreme simulated altitude. *Journal of Applied Physiology* 64:1309-1321.

Suzuki, K., S. Naganuma, M. Totsuka, K.J. Suzuki, M. Mochizuki, M. Shiraishi, S. Nakaji, and K. Sugawara. 1996. Effects of exhaustive endurance exercise and its one-week daily repetition on neutrophil count and functional status in untrained men. *International Journal of Sports Medicine* 17:205-212.

Suzuki, Y., O. Ito, H. Takahashi, and, K. Takamatsu. 2004. The effect of sprint training on skeletal muscle carnosine in humans. *International Journal of Sport Health Science* 2:105-110.

Suzuki, Y., I. Osamu, N. Mukai, H. Takahashi, and K. Takamatsu. 2002. High level of skeletal muscle carnosine contributes to the latter half of exercise performance during 30-s maximal cycle ergometer sprinting. *Japanese Journal of Physiology* 52:199-200.

Swain, R.A., D.M. Harsha, J. Baenziger, and R.M. Saywell. 1997. Do pseudoephedrine or phenylpropanolamine improve maximum oxygen uptake and time to exhaustion? *Clinical Journal of Sport Medicine* 7:168-173.

Swinton, P.A., A.D. Stewart, J.W. Keogh, I. Agouris, and R. Lloyd. 2011. Kinematic and kinetic analysis of maximal velocity deadlifts performed with and without the inclusion of chain resistance. *Journal of Strength and Conditioning Research* 25:163-174.

Syrotuik, D.G., and G.J. Bell. 2004. Acute creatine monohydrate supplementation: A descriptive physiological profile of responders vs. nonresponders. *Journal of Strength and Conditioning Research* 18:610-617.

Szlyk, P.C., I.V. Sils, R.P. Francesconi, and R.W. Hubbard. 1990. Patterns of human drinking: Effects of exercise, water temperature, and food consumption. *Aviation, Space, and Environmental Medicine* 61:43-48.

Szlyk, P.C., I.V. Sils, R.P. Francesconi, R.W. Hubbard, and L.E. Armstrong. 1989. Effects of water temperature and flavoring on voluntary dehydration in men. *Physiology and Behavior* 45:639-647.

Tabata, I., Y. Atomi, Y. Mutoh, and M. Miyahita. 1990. Effect of physical training on the responses of serum adrenocorticotropic hormone during prolonged exhausting exercise. *European Journal of Applied Physiology* 61:188-192.

Takaishi, T., Y. Yasuda, and T. Moritani. 1994. Neuromuscular fatigue during prolonged pedalling exercise at different pedalling rates. *European Journal of Applied Physiology* 69:154-158.

Takamata, A., G.W. Mack, C.M. Gillen, and E.R. Nadel. 1994. Sodium appetite, thirst, and body fluid regulation in humans during rehydration without sodium replacement. *American Journal of Physiology: Regulatory, Integrative and Comparative Physiology* 266:R1493-R1502.

Takeshima, N., M.E. Rogers, M.M. Islam, T. Yamauchi, E. Watanabe, and A. Okada. 2004. Effect of concurrent aerobic and resistance circuit exercise training on fitness in older adults. *European Journal of Applied Physiology* 93:173-182.

Tallon, M.J., R.C. Harris, L. Boobis, J. Fallowfield, and J.A. Wise. 2005. The carnosine content of vastus lateralis is elevated in resistance trained bodybuilders. *Journal of Strength and Conditioning Research* 19:725-729.

Tallon, M.J., R.C. Harris, N. Maffulli, and M.A. Tarnopolsky. 2007. Carnosine, taurine, and enzyme activities of human skeletal muscle fibres from elderly subjects with osteoarthritis and young moderately active subjects. *Biogerontology* 8:129-137.

Tan, R.A., and S.L. Spector. 1998. Exercise-induced asthma. *Sports Medicine* 25:1-6.

Tarnopolsky, M.A. 1994. Caffeine and endurance performance. *Sports Medicine* 18:109-125.

Tarnopolsky, M.A., S.A. Atkinson, J.D. MacDougall, A. Chesley, S. Phillips, and H.P. Schwarcz. 1992. Evaluation of protein requirements for trained strength athletes. *Journal of Applied Physiology* 73:1986-1995.

Taube, W., M. Gruber, and A. Gollhofer. 2008. Spinal and supraspinal adaptations associated with balance training and their functional relevance. *Acta Physiologica Scandinavica* 193:101-116.

Taylor, L.W., C.D. Wilborn, T. Harvey, J. Wismann, and D.S. Willoughby. 2007. Acute effects of ingesting Java Fit™ energy extreme functional coffee on resting energy expenditure and hemodynamic responses in male and female coffee drinkers. *Journal of the International Society of Sports Nutrition* 4:10.

Teixeira, R.M., and T.C.M. De Lima. 2003. Involvement of tachykinin NK 1 receptor in the behavioral and immunological responses to swimming stress in mice. *Neuropeptides* 37:307-315.

Telford, R., E. Catchpole, V. Deakin, A. Hahn, and A. Plank. 1992. The effect of 7-8 months of vitamin/mineral supplementation on athletic performance. *International Journal of Sport Nutrition* 2:135-153.

Tena-Sempere, M., P.R. Manna, F.P. Zhang, L. Pinilla, L.C. Gonzalez, C. Dieguez, I. Huhtaniemi, and E. Aguilar. 2001. Molecular mechanisms of leptin action in adult rat testis: Potential targets for leptin-induced inhibition of steroidogenesis and pattern of leptin receptor messenger ribonucleic acid expression. *Journal of Endocrinology* 170:413-423.

Terjung, R.L., P. Clarkson, E.R. Eichner, P.L. Greenhaff, P.J. Hespel, R.G. Israel, W.J. Kraemer, R.A. Meyer, L.L. Spriet, M.A. Tarnopolsky, A.J. Wagenmakers, and M.H. Williams. 2000. American College of Sports Medicine roundtable. The physiological and health effects of oral creatine supplementation. *Medicine and Science in Sports and Exercise* 32:706-717.

Terrados, N., M. Mizuno, and H. Andersen. 1985. Reduction in maximal oxygen uptake at low altitudes: Role of training status and lung function. *Clinical Physiology* 5:S75-S79.

Tesch, P.A. 1985. Exercise performance and b-blockade. *Sports Medicine* 2:389-412.

Tesch, P.A., and J. Karlsson. 1985. Muscle fiber types and size in trained and untrained muscles of elite athletes. *Journal of Applied Physiology* 59:1716-1720.

Tesch, P.A., P.V. Komi, and K. Hakkinen. 1987. Enzymatic adaptations consequent to long-term strength training. *International Journal of Sports Medicine* 8:66-69.

Tesch, P.A., and L. Larson. 1982. Muscle hypertrophy in bodybuilders. *European Journal of Applied Physiology* 49:301-306.

Tharp, G.D., and M.W. Barnes. 1990. Reduction of saliva immunoglobulin levels by swim training. *European Journal of Applied Physiology* 60:61-64.

Thoden, J.S., G.P. Kenny, F. Reardon, M. Jette, and S. Livingstone. 1994. Disturbance of thermal homeostasis during postexercise hyperthermia. *European Journal of Applied Physiology* 68:170-176.

Thomaes, T., M. Thomis, S. Onkelinx, W. Coudyzer, V. Cornelissen, and L. Vanhees. 2012. Reliability and validity of the ultrasound technique to measure the rectus femoris muscle diameter in older CAD-patients. *BMC Medical Imaging* 12:7.

Thomas, G.A., W.J. Kraemer, B.A. Comstock, C. Dunn-Lewis, J.S. Volek, C.R. Denegar, and C.M. Maresh. 2012. Effects of resistance exercise and obesity level on ghrelin and cortisol in men. *Metabolism: Clinical and Experimental* 61:860-868.

Thompson, J.L., M.M. Manore, and J.R. Thomas. 1996. Effects of diet and diet-plus-exercise programs on resting metabolic rate: A meta-analysis. *International Journal of Sport Nutrition* 6:41-61.

Thompson, R.L., and J.S. Hayward. 1996. Wet-cold exposure and hypothermia: Thermal and metabolic responses to prolonged exercise in rain. *Journal of Applied Physiology* 81:1128-1137.

Thomson, J.M., J.A. Stone, A.D. Ginsburg, and P. Hamilton. 1983. The effects of blood reinfusion during prolonged heavy exercise. *Canadian Journal of Applied Sport Science* 8:72-78.

Thorstensson, A. 1975. Enzyme activities and muscle strength after sprint training in man. *Acta Physiologica Scandinavica* 94:313-318.

Thorstensson, A., and J. Karlsson. 1976. Fatigability and fiber composition of human skeletal muscle. *Acta Physiologica Scandinavica* 98:318-322.

Tillin, N., and D. Bishop. 2009. Factors modulating post-activation potentiation and its effect on performance of subsequent explosive activities. *Sports Medicine* 39:147-166.

Tipton, K.D., T.A. Elliot, M.G. Cree, A.A. Aarsland, A.P. Sanford, and R.R. Wolfe. 2007. Stimulation of net muscle protein synthesis by whey protein ingestion before and after exercise. *American Journal of Physiology: Endocrinology and Metabolism* 292:E71-E76.

Tipton, K.D., T.A. Elliot, M.G. Cree, S.E. Wolf, A.P. Sanford, and R.R. Wolfe. 2004. Ingestion of casein and whey proteins result in muscle anabolism after resistance exercise. *Medicine and Science in Sports and Exercise* 36:2073-2081.

Tipton, K.D., A.A. Ferrando, S.M. Phillips, D. Doyle Jr., and R.R. Wolfe. 1999. Postexercise net protein synthesis in human muscle from orally administered amino acids. *American Journal of Physiology* 276:E628-E634.

Tipton, K.D., B.E. Gurkin, S. Matin, and R.R. Wolfe. 1999. Nonessential amino acids are not necessary to stimulate net muscle protein synthesis in healthy volunteers. *Journal of Nutrition and Biochemistry* 10:89-95.

Tipton, K.D., B.B. Rasmussen, S.L. Miller, S.E. Wolf, S.K. Owens-Stovall, B.E. Petrini, and R.R. Wolfe. 2001. Timing of amino acid-carbohydrate ingestion alters anabolic response of muscle to resistance exercise. *American Journal of Physiology: Endocrinology and Metabolism* 281:E197-E206.

Tiryaki, G.R., and H.A. Atterbom. 1995. The effects of sodium bicarbonate and sodium citrate on 600 m running time of trained females. *Journal of Sports Medicine and Physical Fitness* 35:194-198.

Tischler, M.E., M. Desautels, and A.L. Goldberg. 1982. Does leucine, leucyl-tRNA, or some metabolite of leucine regulate protein synthesis and degradation in skeletal and cardiac muscle? *Journal of Biology and Chemistry* 257:1613-1621.

Tomasi, T.B., F.B. Trudeau, D. Czerwinski, and S. Erredge. 1982. Immune parameters in athletes before and after strenuous exercise. *Journal of Clinical Immunology* 2:173-178.

Toner, M.M., E.L. Glickman, and W.D. McArdle. 1990. Cardiovascular adjustments to exercise distributed between the upper and lower body. *Medicine and Science in Sports and Exercise* 22:773-778.

Toner, M.M., and W.D. McArdle. 1988. Physiological adjustments of man to the cold. In *Human performance physiology and environmental medicine at terrestrial extremes,* ed. K.B. Pandolf, M.N. Sawka, and R.R. Gonzalez, 361-399. Indianapolis: Benchmark Press.

Toner, M.M., M.N. Sawka, M.E. Foley, and K.B. Pandolf. 1986. Effects of body mass and morphology on thermal responses in water. *Journal of Applied Physiology* 60:521-525.

Toussaint, H.M. 1990. Differences in propelling efficiency between competitive and triathlon swimmers. *Medicine and Science in Sports and Exercise* 22:409-415.

Townshend, A.D., C.J. Worringham, and I.B. Stewart. 2008. Assessment of speed and position during human locomotion using nondifferential GPS. *Medicine and Science in Sports and Exercise* 40:124-132.

Tran, Q.T., D. Docherty, and D. Behm. 2006. The effects of varying time under tension and volume load on acute neuromuscular responses. *European Journal of Applied Physiology* 98:402-410.

Trappe, S., M. Harber, A. Creer, P. Gallagher, D. Slivka, K. Minchev, and D. Whitsett. 2006. Single muscle fiber adaptations with marathon training. *Journal of Applied Physiology* 101:721-727.

Tremblay, A., J.A. Simoneau, and C. Bouchard. 1994. Impact of exercise intensity on body fatness and skeletal muscle metabolism. *Metabolism* 43:814-818.

Trice, I., and E.M. Haymes. 1995. Effects of caffeine ingestion on exercise-induced changes during high-intensity, intermittent exercise. *International Journal of Sport Nutrition* 5:37-44.

Tricker, R., R. Casaburi, T.W. Storer, B. Clevenger, N. Berman, A. Shirazi, and S. Bhasin. 1996. The effects of supraphysiological doses of testosterone on angry behavior in healthy eugonadal men--a clinical research center study. *Journal of Clinical Endocrinology and Metabolism* 81:3754-3758

Troup, J.P., J.M. Metzger, and R.H. Fitts. 1986. Effect of high-intensity exercise on functional capacity of limb skeletal muscle. *Journal of Applied Physiology* 60:1743-1751.

Truijens, M.J., F.A. Rodriguez, N.E. Townsend, J. Stray-Gundersen, C.J. Gore, and B.D. Levine. 2008. The effect of intermittent hypoxic exposure and sea level training on submaximal economy in well-trained swimmers and runners. *Journal of Applied Physiology* 104:328-337.

Tseng, Y.L., H.R. Han, F.H. Kuo, M.H. Shieh, and C.F. Chang. 2003. Ephedrines in over-the-counter cold medicines and urine specimens collected during sport competitions. *Journal of Analytical Toxicology* 27:359-365.

Tsubakihara, T., T. Umeda, I. Takahashi, M. Matsuzaka, K. Iwane, M. Tanaka, M. Matsuda, K. Oyamada, R. Aruga, and S. Nakaji. 2011. Effects of soccer matches on neutrophil and lymphocyte functions in female university soccer players. *Luminescence* 28:129-135.

Turcotte, L.P. 2000. Muscle fatty acid uptake during exercise: Possible mechanisms. *Exercise and Sport Sciences Reviews* 28:4-9.

Turcotte, L.P., J.R. Swenberger, M.Z. Tucker, and A.J. Yee. 1999. Training-induced elevation in $FABP_{pm}$ is associated with increased palmitate use in contractile muscle. *Journal of Applied Physiology* 87:285-293.

Turner, A.P., M.F. Sanderson, and L.A. Attwood. 2011. The acute effect of different frequencies of whole-body vibration on countermovement jump performance. *Journal of Strength and Conditioning Research* 25:1592-1597.

Tvede, N., M. Kappel, J. Halkjaer-Kristensen, H. Galbo, and B.K. Pedersen. 1993. The effect of light, moderate and severe bicycle exercise on lymphocyte subsets, natural and lymphokine activated killer cells, lymphocyte proliferative response and interleukin 2 production. *International Journal of Sports Medicine* 15:100-104.

Tvede, N., B.K. Pedersen, F.R. Hansen, T. Bendix, L.D. Christensen, H. Galbo, and J. Halkjaer-Kristensen. 1989. Effect of physical exercise on blood mononuclear cell subpopulations and in vitro proliferative responses. *Scandinavian Journal of Immunology* 29:383-389.

Ubertini, G., A. Grossi, D. Colabianchi, R. Fiori, C. Brufani, C. Bizzarri, G. Giannone, A.E. Rigamonti, A. Sartorio, E.E. Muller, and M. Cappa. 2008. Young elite athletes of different sport disciplines present with an increase in pulsatile secretion of growth hormone compared with non-elite athletes and sendentary subjects. *Journal of Endocrinological Investigation* 31:138-145.

Upton, D.E. 2011. The effect of assisted and resisted sprint training on acceleration and velocity in division IA female soccer athletes. *Journal of Strength and Conditioning Research* 25:2645-2652.

Urhausen, A., T. Albers, and W. Kindermann. 2004. Are the cardiac effects of anabolic steroid abuse in strength athletes reversible? *Heart* 90:496-501.

Urhausen, A., H. Gabriel, and W. Kindermann. 1995. Blood hormones as markers of training stress and overtraining. *Sports Medicine* 20:251-276.

Urhausen, A., and W. Kindermann. 1987. Behavior of testosterone, sex hormone binding globulin (SHBG), and cortisol before and after a triathlon competition. *International Journal of Sports Medicine* 8:305-308.

USA Track & Field. 2000. *USA Track & Field coaching manual*. Champaign, IL: Human Kinetics.

Vandenberg, P., D. Neumark-Sztainer, G. Cafri, and M. Wall. 2007. Steroid use among adolescents: Longitudinal findings from Project EAT. *Pediatrics* 119:476-486.

Vanderburgh, P.M., M. Kusano, M. Sharp, and B. Nindl. 1997. Gender differences in muscular strength: An allometric model approach. *Biomedical Scientific Instrumentation* 33:100-105.

Van der Merwe, P.J., F.R. Muller, and F.O. Muller. 1988. Caffeine in sport. Urinary excretion of caffeine in healthy volunteers after intake of common caffeine containing beverages. *South African Medical Journal* 74:163-164.

Van Gammeren, D., D. Falk, and J. Antonio. 2001. The effects of supplementation with 19-nor-4-androstene-3,17-dione and 19-nor-4-androstene-3,17-diol on body composition and athletic performance in previously weight-trained male athletes. *European Journal of Applied Physiology* 84:426-431.

Van Gammeren, D., D. Falk, and J. Antonio. 2002. Effects of norandrostenedione and norandrostenediol in resistance-trained men. *Nutrition* 18:734-737.

Van Handel, P.J., A. Katz, J.P. Troup, and P.W. Bradley. 1988. Aerobic economy and competitive swim performance of U.S. elite swimmers. In *Swimming science V*, ed. B.E. Ungerechts, K. Wilke, and K. Reischle, 219-227. Champaign, IL: Human Kinetics.

Van Helder, W., K. Casey, R. Goode, and W. Radomski. 1986. Growth hormone regulation in two types of aerobic exercise of equal oxygen uptake. *European Journal of Applied Physiology* 55:236-239.

Van Helder, W.P., R.C. Goode, and W. Radomski. 1984. Effect of anaerobic and aerobic exercise on circulating growth hormone in man. *European Journal of Applied Physiology* 52:255-257.

Van Helder, W.P., M.W. Radomski, and R.C. Goode. 1984. Growth hormone responses during intermittent weight lifting exercise in men. *European Journal of Applied Physiology* 53:31-34.

Van Loon, F.P., A.K. Banik, S.K. Nath, F.C. Patra, M.A. Wahed, D. Darmaun, J.F. Desjeux, and D. Mahalanabis. 1996. The effect of L-glutamine on salt and water absorption: A jejuna perfusion study in cholera in humans. *European Journal of Gastroenterology and Hepatology* 8:443-448.

Van Middelkoop, M., J. Kolkman, J. Van Ochten, S.M.A. Bierma-Zeinstra, and B.W. Koes. 2008. Risk factors for lower extremity injuries among male marathon runners. *Scandinavian Journal of Medicine and Science in Sports* 18:691-697.

Van Rosendal, S.P., M.A. Osborne, R.G. Fassett, and J.S. Coombes. 2010. Guidelines for glycerol use in hyperhydration and rehydration associated with exercise. *Sports Medicine* 40:113-129.

Van Rosendal, S.P., N.A. Strobel, M.A. Osborne, R.G. Fassett, and J.S. Coombes. 2012. Performance benefits of rehydration with intravenous fluid and oral glycerol. *Medicine and Science in Sports and Exercise* 44:1780-1790.

Van Someren, K., K. Fulcher, J. McCarthy, J. Moore, G. Horgan, and R. Langford. 1998. An investigation into the effects of sodium citrate ingestion on high-intensity exercise performance. *International Journal of Sport Nutrition* 8:357-363.

Van Thienen, R., K. Van Proeyen, B. Vanden Eynde, J. Puype, T. Lefere, and P. Hespel. 2009. Beta-alanine improves sprint performance in endurance cycling. *Medicine and Science in Sports and Exercise* 41:898-903.

Veicteinas, A., G. Ferretti, and D.W. Rennie. 1982. Superficial shell insulation in resting and exercising men in cold water. *Journal of Applied Physiology* 52:1557-1564.

Velloso, C.P., M. Aperghis, R. Godfrey, A.J. Blazevich, C. Bartlett, D. Cowan, R.I.G. Holt, P. Bouloux, S.D.R. Harridge, and G. Goldspink. 2013. The effects of two weeks of recombinant growth hormone administration on the response of IGF-1 and N-terminal pro-peptide of collagen type III (P-III-NP) during a single bout of high resistance exercise in resistance trained young men. *Growth Hormone and IGF Research* 23:76-80.

Verges, S., P. Flore, M.P. Blanchi, and B. Wuyam. 2004. A 10-year follow-up study on pulmonary function in symptomatic elite cross-country skiers—athletes and bronchial dysfunctions. *Scandinavian Journal of Medicine and Science in Sports* 14:381-387.

Verkhoshansky, Y.V. 1996. Principles for a rational organization of the training process aimed at speed development. *New Studies in Athletics* 11:155-160.

Verma, S.K., S.R. Mahindroo, and D.K. Kansal. 1978. Effect of four weeks of hard physical training on certain physiological and morphological parameters of basketball players. *Journal of Sports Medicine* 18:379-384.

Verthelyi, D. 2001. Sex hormones as immunomodulators in health and disease. *International Immunopharmacology* 1:983-993.

Vetter, R.E., and M.L. Symonds. 2010. Correlations between injury, training intensity, and physical and mental exhaustion among college athletes. *Journal of Strength and Conditioning Research* 24:587-596.

Vidal-Puig, A., M. Jimenez-Liñan, B.B. Lowell, A. Hamann, E. Hu, B. Spiegelman, J.S. Flier, and D.E. Moller. 1996. Regulation of PPAR gamma gene expression by nutrition and obesity in rodents. *Journal of Clinical Investigation* 97:2553-2561.

Vierck, J.L., D.L. Icenoggle, L. Bucci, and M.V. Dodson. 2003. The effects of ergogenic compounds on myogenic satellite cells. *Medicine and Science in Sports and Exercise* 35:769-776.

Viitasalo, J.T., H. Kyrolainen, C. Bosco, and M. Allen. 1987. Effects of rapid weight reduction on vertical jumping height. *International Journal of Sports Medicine* 8:281-285.

Vila-Ch⊠, C., D. Falla, and D. Farina. 2010. Motor unit behavior during submaximal contractions following six seeks of either endurance or strength training. *Journal of Applied Physiology* 109:1455-1466.

Visser, M., T. Fuerst, T. Lang, L. Salamone, and T. Harris. 1999. Validity of fan-beam dual-energy X-ray absorptiometry for measuring fat-free mass and leg muscle mass. *Journal of Applied Physiology* 87:1513-1520.

Vogel, J.A., and C.W. Harris. 1967. Cardiopulmonary responses of resting man during early exposure to high altitude. *Journal of Applied Physiology* 22:1124-1128.

Vogel, R.A., M.J. Webster, L.D. Erdmann, and R.D. Clark. 2000. Creatine supplementation: Effect on supramaximal exercise performance at two levels of acute hypohydration. *Journal of Strength and Conditioning Research* 14:214-219.

Volek, J.S., N.D. Duncan, S.A. Mazzetti, R.S. Staron, M. Putukian, A.L. Gomez, D.R. Pearson, W.J. Fink, and W.J. Kraemer. 1999. Performance and muscle fiber adaptations to creatine supplementation and heavy resistance training. *Medicine and Science in Sports and Exercise* 31:1147-1156.

Volek, J.S., K. Houseknecht, and W.J. Kraemer. 1997. Nutritional strategies to enhance performance of high-intensity exercise. *Strength and Conditioning* 19:11-17.

Volek, J.S., and W.J. Kraemer. 1996. Creatine supplementation: Its effect on human muscular performance and body composition. *Journal of Strength and Conditioning Research* 10:200-210.

Volek, J.S., W.J. Kraemer, M.R. Rubin, A.L. Gomez, N.A. Ratamess, and P. Gaynor. 2002. L-carnitine, L-tartrate supplementation favorably affects markers of recovery from exercise stress. *American Journal of Physiology: Endocrinology and Metabolism* 282:E474-E482.

Voy, R.O. 1986. The U.S. Olympic Committee experience with exercise-induced bronchospasm, 1984. *Medicine and Science in Sports and Exercise* 18:328-330.

Wade, C.H., and J.R. Claybaugh. 1980. Plasma renin activity, vasopressin concentration and urinary excretory responses to exercise in men. *Journal of Applied Physiology: Respiratory, Environmental and Exercise Physiology* 49:930-936.

Wagner, D.R., K. D'Zatko, K. Tatsugawa, K. Murray, D. Parker, T. Streeper, and K. Willard. 2008. Mt. Whitney: Determinants of summit success and acute mountain sickness. *Medicine and Science in Sports and Exercise* 40:1820-1827.

Wagner, P.D., H.A. Saltzman, and J.B. West. 1974. Measurement of continuous distributions of ventilation-perfusion ratios: Theory. *Journal of Applied Physiology* 36:588-599.

Wagner, P.D., J.R. Sutton, J.T. Reeves, A. Cymerman, B.M. Groves, and M.K. Malconian. 1987. Operation Everest II: Pulmonary gas exchange during a simulated ascent of Mt. Everest. *Journal of Applied Physiology* 63:2348-2359.

Wahren, J., L. Hagenfeldt, and P. Felig. 1975. Splanchnic and leg exchange of glucose, amino acids, and free fatty acids during exercise in diabetes mellitus. *Journal of Clinical Investigation* 55:1303-1314.

Walberg, J.L., M.K. Leidy, D.J. Sturgill, D.E. Hinkle, S.J. Ritchey, and D.R. Sebolt. 1988. Macronutrient content of a hypoenergy diet affects nitrogen retention and muscle function in weight lifters. *International Journal of Sports Medicine* 9:261-266.

Waldrop, T.G., F.L. Eldridge, G.A. Iwamoto, and J.H. Mitchell. 1996. Central neural control of respiration and circulation during exercise. In *Comprehensive Physiology*. American Physiological Society, 333-380.

Walklate, B.M., B.J. O'Brien, C.D. Paton, and W. Young. 2009. Supplementing regular training with short-duration sprint-agility training leads to substantial increase in repeated sprint-agility performance with national level badminton players. *Journal of Strength and Conditioning Research* 23:1477-1481.

Wallace, J.D., R.C. Cuneo, R. Baxter, H. Orskov, N. Keay, C. Pentecost, R. Dall, T. Rosén, J.O. Jørgensen, A. Cittadini, S. Longobardi, L. Sacca, J.S. Christiansen, B.A. Bengtsson, and P.H. Sönksen. 1999. Responses of the growth hormone (GH) and insulin-like growth factor axis to exercise, GH administration, and GH withdrawal in trained adult males: A potential test for GH abuse in sport. *Journal of Clinical Endocrinology and Metabolism* 84:3591-3601.

Wallace, M.B., J. Lim, A. Cutler, and L. Bucci. 1999. Effects of dehydroepiandrosterone vs androstenedione supplementation in men. *Medicine and Science in Sports and Exercise* 31:1788-1792.

Wallberg-Henriksson, H. 1992. Exercise and diabetes mellitus. *Exercise and Sport Sciences Reviews* 20:339-368.

Wallberg-Henriksson, H., R. Gunnarsson, J. Henriksson, R. Defronzo, P. Felig, J. Ostman, and J. Wahren. 1982. Increased peripheral insulin sensitivity and muscle mitochondrial enzymes but unchanged blood glucose control in type I diabetics after physical training. *Diabetes* 31:1044-1050.

Wallberg-Henriksson, H., R. Gunnarsson, J. Henriksson, J. Ostman, and J. Wahren. 1984. Influence of physical training on formation of muscle capillaries in type I diabetes. *Diabetes* 33:851-857.

Wallberg-Henriksson, H., R. Gunnarsson, R. Rossner, and J. Wahren. 1986. Long-term physical training in female type I (insulin-dependent) diabetic patients: Absence of significant effect on glycaemic control and lipoprotein levels. *Diabetologia* 29:53-57.

Wallin, D., B. Ekblom, R. Grahn, and T. Nordenborg. 1985. Improvement of muscle flexibility: A comparison between two techniques. *American Journal of Sports Medicine* 13:263-268.

Walsh, R.M., T.D. Noakes, J.A. Hawley, and S.C. Dennis. 1994. Impaired high-intensity cycling performance time at low levels of dehydration. *International Journal of Sports Medicine* 15:392-398.

Wang, C., G. Alexander, N. Berman, B. Salehian, T. Davidson, V. Mcdonald, B. Steiner, L. Hull, C. Callegari, and R.S. Swerdloff. 1996. Testosterone replacement therapy improves mood in hypogonadal men--a clinical research center study. *Journal of Clinical Endocrinology and Metabolism* 81:3578-3583.

Wang, L., H. Mascher, N. Psilander, E. Blomstrand, and K. Sahlin. 2011. Resistance exercise enhances the molecular signaling of mitochondrial biogenesis induced by endurance exercise in human skeletal muscle. *Journal of Applied Physiology* 111:1335-1344.

Warber, J.P., J.F. Patton, W.J. Tharion, S.H. Zeisel, R.P. Mello, C.P. Kemnitz, and H.R. Lieberman. 2000. The effects of choline supplementation on physical performance. *International Journal of Sport Nutrition and Exercise Metabolism* 10:170-181.

Ward, M.P., J.S. Milledge, and J.B. West. 1995. *High altitude medicine and physiology.* London: Chapman & Hall Medical.

Ward, P. 1973. The effect of an anabolic steroid on strength and lean body mass. *Medicine and Science in Sports and Exercise* 5:277-282.

Ware, J.S., C.T. Clemens, J.L. Mayhew, and T.J. Johnston. 1995. Muscular endurance repetitions to predict bench press and squat strength in college football players. *Journal of Strength and Conditioning Research* 9:99-103.

Warren, G.L., N.D. Park, R.D. Maresca, K.I. McKibans, and M.L. Millard-Stafford. 2010. Effect of caffeine ingestion on muscular strength and endurance: A meta-analysis. *Medicine and Science in Sports and Exercise* 42:1375-1387.

Wasserman, D.H., D.B. Lacey, D.R. Green, P.E. Williams, and A.D. Cherrington. 1987. Dynamics of hepatic lactate and glucose balances during prolonged exercise and recovery in the dog. *Journal of Applied Physiology* 63:2411-2417.

Wasserman, K., B.J. Whipp, and J.A. Davis. 1981. Respiratory physiology of exercise: Metabolism, gas exchange, and ventilatory control. *International Review of Physiology* 23:149-211.

Watson, G., D.J. Casa, K.A. Fiala, A. Hile, M.W. Roti, J.C. Healey, L.E. Armstrong, and C.M. Maresh. 2006. Creatine use and exercise heat tolerance in dehydrated men. *Journal of Athletic Training* 41:18-29.

Webster, M.J., M.N. Webster, R.E. Crawford, and L.B. Gladden. 1993. Effect of sodium bicarbonate ingestion on exhaustive resistance exercise performance. *Medicine and Science in Sports and Exercise* 25:960-965.

Webster, S., R. Rutt, and A. Weltman. 1990. Physiological effects of a weight loss regimen practiced by college wrestlers. *Medicine and Science in Sports and Exercise* 22:229-234.

Weight, L.M., D. Alexander, and P. Jacobs. 1991. Strenuous exercise: Analogous to the acute-phase response? *Clinical Science* 81:677-683.

Weiler, J., T. Layton, and M. Hunt. 1998. Asthma in United States Olympic athletes who participated in the 1996 Summer Games. *Journal of Allergy and Clinical Immunology* 102:722-726.

Weiss, A., S. Schiaffino, and L.A. Leinwand. 1999. Comparative sequence analysis of the complete human sarcomeric myosin heavy chain family: Implications for functional diversity. *Journal of Molecular Biology* 290:61-75.

Weltman, A., J.Y. Weltman, R. Schurrer, W.S. Evans, J.D. Veldhuis, and A.D. Rogol. 1992. Endurance training amplifies the pulsatile release of growth hormone: Effects of training intensity. *Journal of Applied Physiology* 72:2188-2196.

Weltman, A., J.Y. Weltman, C.J. Womack, S.E. Davis, J.L. Blumer, G.A. Gaesser, and M.L. Hartman. 1997. Exercise training decreases the growth hormone (GH) response to acute constant-load exercise. *Medicine and Science in Sports and Exercise* 29:669-676.

Wenger, C.B. 1988. Human heat acclimatization. In *Human performance physiology and environmental medicine at terrestrial extremes,* ed. K.B. Pandolf, M.N. Sawka, and R.R. Gonzalez, 153-198. Indianapolis: Benchmark Press.

Wesensten, N.J., G. Belenky, M.A. Kautz, D.R. Thorne, R.M. Reichardt, and T.J. Balkin. 2002. Maintaining alertness and performance during sleep deprivation: Modafinil versus caffeine. *Psychopharmacology* 159:238-247.

West, J.B. 1962. Diffusing capacity of the lung for carbon monoxide at high altitude. *Journal of Applied Physiology* 17:421-426.

West, J.B. 1986. Lactate during exercise at extreme altitude. *Federal Proceedings* 45:2953-2957.

West, D.W.D., N.A. Burd, A.W. Staples, and S.M. Phillips. 2010. Human exercise-mediated skeletal muscle hypertrophy is an intrinsic process. *International Journal of Biochemistry and Cell Biology* 42:1371-1375.

West, D.J., D.J. Cunningham, R.M. Bracken, H.R. Bevan, B.T. Crewther, C.J. Cook, and L.P. Kiduff. 2013. Effects of resisted sprint training on acceleration in professional rugby union players. *Journal of Strength and Conditioning Research* 27:1014-1018.

West, D.J., N.J. Owen, M.R. Jones, R.M. Bracken, C.J. Cook, D.J. Cunningham, D.A. Shearer, C.V. Finn, R.U. Newton, B.T. Crewther, and L.P. Kilduff. 2011. Relationships between force-time characteristics of the isometric midthigh pull and dynamic performance in professional rugby league players. *Journal of Strength and Conditioning Research* 25:3070-3075.

Wexler, J.A. and J.T. Sharretts. 2007. Thyroid and bone. *Endocrinology and Metabolism Clinics of North America* 36:673–705.

Wheeler, G.D., S.R. Wall, A.N. Belcastro, and D.C. Cumming. 1984. Reduced serum testosterone and prolactin levels in male distance runners. *Journal of the American Medical Association* 252:514-516.

Whipp, B.J. 1994. Peripheral chemoreceptor control of exercise hyperpnea in humans. *Medicine and Science in Sports and Exercise* 26:337-347.

Whitacre, C.C. 2001. Sex differences in autoimmune disease. *Nature Immunology* 2:777-780.

Whitsel, E.A., E.J. Boyko, A.M. Matsumoto, B.D. Anawalt, and D.S. Siscovick. 2001. Intramuscular testosterone esters and plasma lipids in hypogonadal men: A meta-analysis. *American Journal of Medicine* 111:261-269.

Whyte, L.J., J.M. Gill, and A.J. Cathcart. 2010. Effect of 2 weeks of sprint interval training on health-related outcomes in sedentary overweight/obese men. *Metabolism* 59:1421-1428.

Wibom, R., E. Hultman, M. Johansson, K. Matherei, D. Constantin-Teodosiu, and P.G. Schantz. 1992. Adaptation of mitochondrial ATP production in human skeletal muscle to endurance training and detraining. *Journal of Applied Physiology* 73:2004-2010.

Wickiewicz, T.L., R.R. Roy, P.L. Powell, J.J. Perrine, and B.R. Edgerton. 1984. Muscle architecture and force-velocity relationships in humans. *Journal of Applied Physiology: Respiratory, Environmental and Exercise Physiology* 57:435-443.

Wiersinga, W.M. 2001. Thyroid hormone replacement therapy. *Hormone Research* 56(suppl 1): 74-81.

Wiklund, U., M. Karlsson, M. Öström, and T. Messner. 2009. Influence of energy drinks and alcohol on post-exercise heart rate recovery and heart rate variability. *Clinical Physiology and Functional Imaging* 29:74-80.

Wilber, R.L. 2007. Application of altitude/hypoxic training by elite athletes. *Medicine and Science in Sports and Exercise* 39:1610-1624.

Wilber, R.L., K.W. Rundell, L. Szmedra, D.M. Jenkinson, J. Im, and S. Drake. 2000. Incidence of exercise-induced bronchospasm in Olympic winter sport athletes. *Medicine and Science in Sports and Exercise* 32:732-737.

Wilber, R.L., J. Stray-Gundersen, and B.D. Levine. 2007. Effect of hypoxic "dose" on physiological responses and sea-level performance. *Medicine and Science in Sports and Exercise* 39:1590-1599.

Wiles, J.D., S.R. Bird, J. Hopkins, and M. Riley. 1992. Effect of caffeinated coffee on running speed, respiratory factors, blood lactate and perceived exertion during 1500-m treadmill running. *British Journal of Sports Medicine* 26:116-120.

Wilkerson, J.E., S.M. Horvath, and B. Gutin. 1980. Plasma testosterone during treadmill exercise. *Journal of Applied Physiology: Respiratory, Environmental and Exercise Physiology* 49:249-253.

Wilkes, D., N. Gledhill, and R. Smyth. 1983. Effect of acute induced metabolic acidosis on 800-m racing time. *Medicine and Science in Sports and Exercise* 15:277-280.

Wilkinson, S.B., S.M. Phillips, P.J. Atherton, R. Patel, K.E. Yarasheski, M.A. Tarnopolsky, and M.J. Rennie. 2008. Differential effects of resistance and endurance exercise in the fed state on signalling molecule phosphorylation and protein synthesis in human muscle. *Journal of Physiology* 15:3701-3717.

Williams, B.D., R.R. Wolfe, D. Bracy, and D.H. Wasserman. 1996. Gut proteolysis provides essential amino acids during exercise. *American Journal of Physiology: Endocrinology and Metabolism* 270:E85-E90.

Williams, M.H. 1991. Alcohol, marijuana, and beta blockers. In *Perspectives in exercise science and sports medicine, Vol. 4, Ergogenics,* ed. D.R. Lamb and M.H. Williams, 331-372. Carmel, IN: Benchmark Press.

Williams, M.H., S. Wesseldine, T. Somma, and R. Schuster. 1981. The effect of induced erythrocythemia upon 5-mile treadmill run time. *Medicine and Science in Sports and Exercise* 13:169-175.

Willoughby, D.S. 1992. A comparison of three selected weight training programs on the upper and lower body strength of trained males. *Journal of Applied Research in Coaching Athletics* (March):124-146.

Willoughby, D.S. 1993. The effects of meso-cycle-length weight training programs involving periodization and partially equated volumes on upper and lower body strength. *Journal of Strength and Conditioning Research* 7:2-8.

Willoughby, D.S., and J. Rosene. 2001. Effects of oral creatine and resistance training on myosin heavy chain expression. *Medicine and Science in Sports and Exercise* 33:1674-1681.

Willougby, D.S., J.S. Stout, and C.D. Wilborn. 2007. Effects of resistance training and protein plus amino acid supplementation on muscle anabolism, mass and strength. *Amino Acids* 32:467-477.

Wilmore, J.H. 1974. Alterations in strength, body composition and anthropometric measurements consequent to a 10-week weight training program. *Medicine and Science in Sports and Exercise* 6:133-138.

Wilmore, J.H. and D. Costill. 1999. *Physiology of sport and exercise* (Champaign, IL: Human Kinetics).

Wilmore, J.H., P.R. Stanforth, J. Gagnon, A.S. Leon, D.C. Rao, J.S. Skinner, and C. Bouchard. 1996. Endurance exercise training has a minimal effect on resting heart rate: The HERITAGE study. *Medicine and Science in Sports and Exercise* 28:829-835.

Wilson, G.J., R.U. Newton, A.J. Murphy, and B.J. Humphries. 1993. The optimal training load for the development of dynamic athletic performance. *Medicine and Science in Sports and Exercise* 25:1279-1286.

Wilson, J.D., and D.W. Foster. 1992. *Williams textbook of endocrinology,* inside front cover. Philadelphia: Saunders.

Wilson, J.M., S.C. Grant, S.R. Lee, I.S. Masad, Y.M. Park, P.C. Henning, J.R. Stout, J.P. Loenneke, B.H. Arjmandi, L.B. Panton, and J.S. Kim. 2012. Beta-hydroxy-beta-methyl-butyrate blunts negative age related changes in body composition, functionality and myofiber dimensions in rats. *Journal of the International Society of Sports Nutrition* 9:18.

Wilson, J.M., J.S. Kim, S.R. Lee, J.A. Rathmacher, B. Dalmau, J.D. Kingsley, H. Koch, A.H. Manninen, R. Saadat, and L.B. Panton. 2009. Acute and timing effects of beta-hydroxy-beta-methylbutyrate (HMB) on indirect markers of skeletal muscle damage. *Nutrition and Metabolism* 6:6.

Winchester, J.B., J.M. McBride, M.A. Maher, R.P. Mikat, B.K. Allen, D.E. Kline, and M.R. McGuigan. 2008. Eight weeks of ballistic exercise improves power independently of changes in strength and muscle fiber type expression. *Journal of Strength and Conditioning Research* 22:1728-1734.

Winchester, J.B., A. Nelson, D. Landin, M. Young, and I.C. Schexnayder. 2008. Static stretching impairs sprint performance in collegiate track and field athletes. *Journal of Strength and Conditioning Research* 22:13-18.

Wise, C.K., C.A. Cooney, S.F. Ali, and L.A. Poirier. 1997. Measuring S-adenosylmethionine in whole blood, red blood cells and cultured cells using a fast preparation method and high-performance chromatography. *Journal of Chromatography B: Biomedical Sciences and Applications* 696:145-152.

Wolfel, E.E., B.M. Groves, G.A. Brooks, G.E. Butterfield, R.S. Mazzeo, L.G. Moore, J.R. Sutton, P.R. Bender, T.E. Dahms, R.E. McCullough, R.G. McCullough, S.-Y. Huang, S.-F. Sun, R.F. Glover, H.N. Hultgren, and J.T. Reeves. 1991. Oxygen transport during steady-state submaximal exercise in chronic hypoxia. *Journal of Applied Physiology* 70:1129-1136.

Wong, P., A. Chaouachi, K. Chamari, A. Dellal, and U. Wisloff. 2010. Effect of preseason concurrent muscular strength and high-intensity interval training in professional soccer players. *Journal of Strength and Conditioning Research* 24:653-660.

Woodcock, A.H. 1962. Moisture transfer in textile systems, part I. *Textile Research Journal* 32:628-633.

Woodhouse, L.J., S. Reisz-Porszasz, M. Javanbakht, T.W. Storer, M. Lee, H. Zerounian, and S. Bhasin. 2003. Development of models to predict anabolic response to testosterone administration in healthy young men. *American Journal of Physiology: Endocrinology and Metabolism* 284:E1009-E1017.

Worrell, T.W., D.H. Perrin, B.M. Gansneder, and J. Gieck. 1991. Comparison of isokinetic strength and flexibility measures between hamstring injured and noninjured athletes. *Journal of Orthopaedic and Sports Physical Therapy* 13:118-125.

Worrell, T.W., T.L. Smith, and J. Winegardner. 1994. Effect of hamstring stretching on hamstring muscle performance. *Journal of Orthopaedic and Sports Physical Therapy* 20:154-159.

Wright, G.A., P.W. Grandjean, and D.D. Pascoe. 2007. The effects of creatine loading on thermoregulation and intermittent sprint exercise performance in a hot humid environment. *Journal of Strength and Conditioning Research* 21:655-660.

Wright, V., and R.J. Johns. 1960. Physical factors concerned with the stiffness of normal and diseased joints. *Bulletin of Johns Hopkins Hospital* 106:215-231.

Wu Z., M. Bidlingmaier, R. Dall, and C.J. Strasburger. 1999. Detection of doping with human growth hormone. *Lancet* 353:895.

Wurtman, R.J., M.J. Hirsch, and J.H. Growdon. 1977. Lecithin consumption raises serum-free-choline levels. *Lancet* 2:68-69.

Yaegaki, M., T. Umeda, I. Takahashi, Y. Yamamoto, A. Kojima, M. Tanabe, K. Yamai, M. Matsuzaka, N. Sugawara, and S. Nakaji. 2008. Measuring neutrophil function might be a good predictive marker of overtraining in athletes. *Luminescence* 23:281-286.

Yaglou, C.P., and D. Minard. 1957. Control of heat casualties at military training centers. *Archives of Industrial Health* 16:302-305.

Yang, R.C., G.W. Mack, R.R. Wolfe, and E.R. Nadel. 1998. Albumin synthesis after intense intermittent exercise in human subjects. *Journal of Applied Physiology* 84:584-592.

Yeargin, S.W., D.J. Casa, D.A. Judelson, B.P. McDermott, M.S. Ganio, E.C. Lee, R.M. Lopez, R.L. Stearns, J.M. Anderson, L.E. Armstrong, W.J. Kraemer, and C.M. Maresh. 2010. Thermoregulatory responses and hydration practices in heat-acclimatized adolescents during preseason high school football. *Journal of Athletic Training* 45:136-146.

Yerg II, J.E., D.R. Seals, J.M. Hagberg, and J.O. Holloszy. 1985. Effect of endurance exercise training on ventilatory function in older individuals. *Journal of Applied Physiology* 58:791-794.

Yoo, J.H., B.O. Lin, M. Ha, S.W. Lee, S.J. Oh, Y.S. Lee, J.G. Kim. 2010. A meta-analysis of the effect of neuromuscular training on the prevention of the anterior cruciate ligament injury in female athletes. *Knee Surgery, Sports Traumatology, Arthroscopy* 18:824-830.

Young, A.J. 1990. Energy substrate utilization during exercise in extreme environments. *Exercise and Sport Sciences Reviews* 18:65-117.

Young, A.J., J.W. Castellani, C. O'Brien, R.L. Shippee, P. Tikuisis, L.G. Meyer, L.A. Blanchard, J.E. Kain, B.S. Cadarette, and M.N. Sawka. 1998. Exertional fatigue, sleep loss, and negative energy balance increase susceptibility to hypothermia. *Journal of Applied Physiology* 85:1210-1217.

Young, A.J., S.R. Muza, M.N. Sawka, R.R. Gonzalez, and K.B. Pandolf. 1986. Human thermoregulatory responses to cold air are altered by repeated cold water immersion. *Journal of Applied Physiology* 60:1542-1548.

Young, A.J., M.N. Sawka, L. Levine, B.S. Cadarette, and K.B. Pandolf. 1985. Skeletal muscle metabolism during exercise is influenced by heat acclimation. *Journal of Applied Physiology* 59:1929-1935.

Young, D.S. 1987. Implementation of SI units for clinical laboratory data. *Annals of Internal Medicine* 106:114-128.

Young, W.B. 1993. Training for speed/strength: Heavy versus light loads. *National Strength and Conditioning Journal* 15:34-42.

Young, W.B., and G.E. Bilby. 1993. The effect of voluntary effort to influence speed of contraction on strength, muscular power and hypertrophy development. *Journal of Strength and Conditioning Research* 7:172-178.

Young, W.B., M.H. McDowell, and B.J. Scarlett. 2001. Specificity of sprint and agility training methods. *Journal of Strength and Conditioning Research* 15:315-319.

Zahn, A., J.X. Li, Z.R. Xu, and R.Q. Zhao. 2006. Effects of methionine and betaine supplementation on growth performance, carcass composition and metabolism of lipids in male broilers. *British Poultry Science* 47:576-580.

Zeisel, S.H., M.H. Mar, J.C. Howe, and J.M. Holden. 2003. Concentrations of choline-containing compounds and betaine in common foods. *Journal of Nutrition* 133:1302-1307.

Zhao, J., W.A. Bauman, R. Huang, A.J. Caplan, and C. Cardozo. 2004. Oxandrolone blocks glucocorticoid signaling in an androgen receptor-dependent manner. *Steroids* 69:357-366.

Ziegenfuss, T.N., M. Rogers, L. Lowery, N. Mullins, R. Mendel, J. Antonio, and P. Lemon. 2002. Effect of creatine loading on anaerobic performance and skeletal muscle volume in NCAA division I athletes. *Nutrition* 18:397-402.

Zinman, B., N. Ruderman, B.N. Campaigne, J.T. Devlin, and S.H. Schneider. 2003. Physical activity/exercise and diabetes mellitus. *Diabetes Care* 26(suppl 1):S73-S77.

Zinman, B., S. Zuniga-Guajardo, and D. Kelly. 1984. Comparison of the acute and long-term effects of exercise on glucose control in type I diabetics. *Diabetes Care* 7:515-519.

Zmuda, J.M., J.A. Cauley, A. Kriska, N.W. Glynn, J.P. Gutai, and L.H. Kuller. 1997. Longitudinal relation between endogenous testosterone and cardiovascular disease risk factors in middle-aged men. A 13-year follow-up of former multiple risk factor intervention trial participants. *American Journal of Epidemiology* 146:609-617.

Index

Note: Page numbers followed by an italicized *f* or *t* refer to the figure or table on that page, respectively.

About the Author

© Human Kinetics

Jay Hoffman, PhD, is the director of the Institute of Exercise Physiology and Wellness and a professor of sport and exercise science at the University of Central Florida. He is also the chair of the education and human sciences department. Long recognized as an expert in the field of exercise physiology, Hoffman has more than 190 publications in refereed journals, book chapters, and books, and he has lectured at more than 370 national and international conferences and meetings. A former professional athlete, Hoffman has coached elite athletes and conducted research on them throughout his professional career. This combination of the practical and the theoretical provides him with a unique perspective on writing for both coaches and academic faculty. Hoffman was elected president of the National Strength and Conditioning Association (NSCA) in 2009. He was awarded the 2005 Outstanding Kinesiological Professional Award by the Neag School of Education at the University of Connecticut, the 2007 Outstanding Sport Scientist of the Year by the NSCA, and the 2000 Outstanding Junior Investigator Award by the NSCA. A fellow of both the American College of Sports Medicine and National Strength and Conditioning Association, Hoffman is the editor of *NSCA's Guide to Program Design* (Human Kinetics, 2012) and the author of *Norms for Fitness, Performance, and Health* (Human Kinetics, 2006). He earned his PhD in exercise science from the University of Connecticut.